NATO ASI Series

Advanced Science Institutes Series

A series presenting the results of activities sponsored by the NATO Science Committee, which aims at the dissemination of advanced scientific and technological knowledge, with a view to strengthening links between scientific communities.

The Series is published by an international board of publishers in conjunction with the NATO Scientific Affairs Division

A **Life Sciences** B **Physics**	Plenum Publishing Corporation London and New York
C **Mathematical and Physical Sciences** D **Behavioural and Social Sciences** E **Applied Sciences**	Kluwer Academic Publishers Dordrecht, Boston and London
F **Computer and Systems Sciences** G **Ecological Sciences** H **Cell Biology** I **Global Environmental Change**	Springer-Verlag Berlin Heidelberg New York London Paris Tokyo Hong Kong Barcelona Budapest

PARTNERSHIP SUB-SERIES

1. **Disarmament Technologies**	Kluwer Academic Publishers
2. **Environment**	Springer-Verlag/Kluwer Acad. Publishers
3. **High Technology**	Kluwer Academic Publishers
4. **Science and Technology Policy**	Kluwer Academic Publishers
5. **Computer Networking**	Kluwer Academic Publishers

The Partnership Sub-Series incorporates activities undertaken in collaboration with NATO's Cooperation Partners, the countries of the CIS and Central and Eastern Europe, in Priority Areas of concern to those countries.

NATO-PCO DATABASE

The electronic index to the NATO ASI Series provides full bibliographical references (with keywords and/or abstracts) to about 50000 contributions from international scientists published in all sections of the NATO ASI Series. Access to the NATO-PCO DATABASE compiled by the NATO Publication Coordination Office is possible in two ways:

- via online FILE 128 (NATO-PCO DATABASE) hosted by ESRIN, Via Galileo Galilei, I-00044 Frascati, Italy.

- via CD-ROM "NATO Science & Technology Disk" with user-friendly retrieval software in English, French and German (© WTV GmbH and DATAWARE Technologies Inc. 1992).

The CD-ROM can be ordered through any member of the Board of Publishers or through NATO-PCO, Overijse, Belgium.

Series I: Global Environmental Change, Vol. 49

Springer
Berlin
Heidelberg
New York
Barcelona
Budapest
Hong Kong
London
Milan
Paris
Santa Clara
Singapore
Tokyo

Third Millennium BC Climate Change and Old World Collapse

Edited by

H. Nüzhet Dalfes
Istanbul Technical University
Department of Geology
80626 Ayazağa, İstanbul, Turkey

George Kukla
Columbia University
Lamont Doherty Earth Observatory
Palisades, New York 10964, USA

Harvey Weiss
Yale University
Departments of Anthropology
and New Eastern Languages and Civilizations
New Haven, Connecticut 06520, USA

With 156 Figures

Springer
Published in cooperation with NATO Scientific Affairs Division

Proceedings of the NATO Advanced Research Workshop on Third Millennium BC Abrupt Climate Change and Old World Social Collapse, held at Kemer, Turkey, September 19–24, 1994

Cataloging-in-Publication Data applied for

Die Deutsche Bibliothek - CIP-Einheitsaufnahme

Third millennium BC climate change and old world collapse : [proceedings of the NATO Advanced Research Workshop on Third Millennium BC Abrupt Climate Change and Old World Social Collapse, held at Kemer, Turkey, September 19 - 24, 1994] / ed. by H. Nüzhet Dalfes... Publ. in cooperation with NATO Scientific Affairs Division. - Berlin ; Heidelberg ; New York ; Barcelona ; Budapest ; Hong Kong ; London ; Milan ; Paris ; Santa Clara ; Singapore ; Tokyo : Springer, 1996
(NATO ASI series : Ser. I, Global environmental change ; Vol. 49)

ISBN-13: 978-3-642-64476-4 e-ISBN-13: 978-3-642-60616-8

DOI: 10.1007/978-3-642-60616-8

NE: Dalfes, Hasan Nüzhet [Hrsg.]; Advanced Research Workshop on Third Millennium BC Abrupt Climate Change and Old World Social Collapse <1994, Kemer>; NATO: NATO ASI series / I

ISBN-13: 978-3-642-64476-4 Springer-Verlag Berlin Heidelberg New York

Softcover reprint of the hardcover 1st edition 1997

Typesetting: Camera ready by authors/editors
Printed on acid-free paper
SPIN: 10466193 31/3137 - 5 4 3 2 1 0

PREFACE

Third Millennium BC Climate Change and the Old World Social Collapse

It has been argued recently, based on the Greenland ice cores, that the climate of the last 10 thousand years has been uniquely benign, unparalleled in the previous interglacial. Mankind has to be particularly careful not to disturb the delicate balance maintaining the status quo of our planet.

Not everyone shares this view. Sahel tribes and other folks living in marginal environments do not consider the current climate to be stable and benign at all. The weather vagaries are pretty bad today and there is evidence that they were even worse before. A period of particularly unstable climate appears to have affected many parts of the world between about three and five thousand years ago. It was during that time, around 4 thousand years ago, that the advanced urban civilizations in Egypt, Mesopotamia and India mysteriously collapsed.

What happened? Did the Old World societies succumb to political disorder or other internal ills? Many archaeologists subscribe to such a view. Or did the lack of rains destroyed crops and spread hunger through the early towns? Was the drying of the ancient lands man made? Or, did the droughts in the far away steppes of central Asia and northern Africa drive the nomadic people into urban centers where they caused a break-down of social order? Or, perhaps, has the radioactive fallout from a crashed UFO in Colorado desert (fig.1) contaminated the atmosphere, spreading a world-wide havoc?

Puzzling discoveries recently made at the archeological excavation at Tell Leilan in Syria strongly support the suspected link of the collapse with climate. A symposium organized by Harvey Weiss at the annual meeting of the Society of American Archeologists in St. Louis, Missouri on April, 18, 1993 discussed the issue. The participants of the symposium thought that the problem should be studied further and suggested holding an international gathering of paleoclimatologists and archaeologists to review the relevant evidence.

Such meeting indeed took place between September 19 and 24, 1994 under the auspices of the NATO Scientific and Environmental Affairs Division, headed by Luis Veiga da Cunha. It was the Advanced Research Workshop 931587, held in Kemer, Turkey, and entitled "Third Millennium BC Climate Change and the Old World Social Collapse." It was directed by George Kukla of Columbia University in New York. Local arrangements were made by Nüzhet Dalfes of the Istanbul Technical University assisted by Lale Baştuğ of the Pyramid-Gala Agency.

Over forty researchers from around the world attended and discussed their findings. The meeting reviewed the information gathered by paleoclimatologists, geologists, archeologists, historians, and paleobotanists on the abrupt social changes in the Old World at about 4 thousand years ago and on their relation to the global climate shifts which took place between 5 and 3 thousand years before present. This volume represents the proceedings of the conference and provides insight into the available knowledge of this critical interval of the earth history. The dates labeled as "BP" are given as uncorrected radiocarbon years before AD 1950, those marked as "BP cal" are adjusted to tree-ring chronologies using mostly the Stuiver's and Becker's calibration program [Radiocarbon 35 (1993), 35-65]. Many other dates are given in calendar years AD or BC. Table 1 shows the relation of these different dating systems.

Individual manuscripts were reviewed and in most cases also linguistically adjusted. Rhodes Fairbridge of Columbia University deserves our special thanks for his effort in this field. Final editing was done by Nüzhet Dalfes assisted by Mete Tayanç and Mehmet Karaca in Istanbul and by Miriam Collwell at Columbia. We are grateful to Springer-Verlag for the meticulous printing of the book.

A careful reader will have to conclude that there are no clear answers yet on what exactly happened some 4 thousand years ago in the Old World. Although the paleoclimatic data suggest a major change of the precipitation patterns in many parts of the world at approximately the same time, the limited availability and accuracy of the dating is insufficient to convincingly show the degree to which the dry and cold climate oscillations in different parts of the globe were coeval or time transgressive, how many and how abrupt they were and how closely do they relate to the social collapse. The potential man-made impact on the environment was discussed by several participants but most of the paleoclimatic evidence was shown to be, in principle, of natural origin. Obtained evidence tends to support the close temporal link between the collapse of the Old World urban civilizations and the relatively short dry and cold climate oscillations, which occurred over many parts of the world around the turn of the third and fourth millennium BC. One out of several theories proposed at the conference was that the migrations of nomadic tribes from semi-deserts of central Asia and north Africa into urban areas intensified, as the grasslands got decimated by droughts. Nomadic infiltrations may have lead, at least in part, to the disintegration of social structure in urban communities.

Since the developments in the third millennium BC are seen as potential analogs of the impeding future natural climate shift and since the current large-scale population relocations make the past episode in some respect relevant from the sociological point of view, the conference called for the focused attention of research communities to this interval .

One conclusion is nevertheless inescapable: The Holocene climate was far from stable and benign. Bad things happened to climate and Man in the Holocene long before the advent of modern polluting, high-technology society. There is no guarantee that a similar natural shift with serious adverse impact on society can not happen again.

George Kukla
Lamont Doherty Earth Observatory of Columbia University
Palisades, New York 10964, USA

H. Nüzhet Dalfes
Department of Geology,
Istanbul Technical University
80626 Ayazağa, Istanbul, Turkey

Harvey Weiss
Yale University,
New Haven, Connecticut 06520, USA

Fig 1 : Front page of the tabloid widely available in American supermarkets. The journal is conspicuously dated three days after the end of the workshop. We categorically deny any relation of our conference with the report. The copy of the journal was on sale already on September 20, three days prior to the start of the meeting. (Repeated calls to the journal and to Pentagon failed to clarify whether the crash was dated in calibrated or uncalibrated radiocarbon years).

Date BC	Date (calibrated) BP	Date (uncalibrated) BP
500	**2450**	**2442**
600	2550	2527
700	2650	2471
800	2750	2638
900	2850	2772
1000	**2950**	**2840**
1100	3050	2926
1200	3150	2978
1300	3250	3050
1400	3350	3119
1500	**3450**	**3255**
1600	3550	3313
1700	3650	3450
1800	3750	3506
1900	3850	3581
2000	**3950**	**3620**
2100	4050	3697
2200	4150	3789
2300	4250	3859
2400	4350	3881
2500	**4450**	**4075**
2600	4550	4101
2700	4650	4165
2800	4750	4208
2900	4850	4308
3000	**4950**	**4368**
3100	5050	4485
3200	5150	4536
3300	5250	4490
3400	5350	4719
3500	**5450**	**4679**

Table 1: Comparison of the BP and BC dates in the interval 500 to 3500 BC. Linearly interpolated from Stuiver M. and Becker B., Radiocarbon 35 (1993) 35-65.

TABLE OF CONTENTS

Nile Floods and Political Disorder in Early Egypt

Fekri A. HASSAN

Institute of Archaeology
University College London, UK

Abstract

The collapse of centralized government about 2200 BC (3700 uncalibrated radiocarbon years bp) coincides with reduced Nile flood discharge, invasion of the Nile Valley by dune sand, and possibly degradation of the Delta floodplain as revealed by proxy data in Egypt. At the same time lower lake levels a reduction of arboreal vegetation and onset of cooler conditions in equatorial Africa strongly indicate drier conditions that severely diminished water supply to the Nile tributaries. A dramatic dry episode in western Tibet influenced by reduced monsoon activity at 3800 bp suggests that the reduction of Nile flood discharge was a response to a global event influencing the ITCZ belt.

Introduction

On the banks of the Nile resident fishermen and food-collectors, together with herders and farmers from the surrounding desert, in the wake of the major droughts of the seventh and sixth millennium before the present began to lay the foundation of Egyptian civilization (Hassan 1981) By 3000 BC, over two millennia of agrarian developments and political transformations from villages to provincial states prepared Egypt for a process of unification into a nation-state (Hassan 1988). Following a formative period from 3000 to 2700 BC, the grandeur of the new state became manifest in monumental pyramids towering over the farming land from hills west of the Nile near the nation's capital, Memphis. The stepped pyramid of Zoser (2680 BC) was soon followed during fourth dynasty by the great pyramids of Giza. The pyramids signaled great prosperity, and the emergence of an effective bureaucratic organization, great technical achievements in building, surveying, geometry, and astronomy as well as a royal ideology that linked the kings to the powers of the great gods of creation. The Egyptian vision of a cosmic order was well established by the time the [Pyramid] texts of the fifth dynasty were inscribed. However, prosperity and order were rudely shattered by 2200 BC as Egypt plunged into an age of disorder,

NATO ASI Series, Vol. I 49
Third Millennium BC Climate Change
and Old World Collapse
Edited by H. Nüzhet Dalfes, George Kukla,
and Harvey Weiss

civil upheaval, and regional conflicts. This period referred to as the First Intermediate or the First Intermediate has puzzled historians. In a daring contribution, Barbara Bell (1971) suggested that the collapse was primarily caused by an episode of low Nile floods.

The First Intermediate is dated to 2180-2134 BC or approximately 3700-3600 radiocarbon years bp.1 Over the short span of twenty years, the records report eighteen kings, and possibly one queen (Kemp 1983: 112). Thebes became prominent for the first time. It rivaled and went to war with the north. In addition to civil war, social revolts and rebellion were prevalent if we accept the literary testimony of the Admonitions of Ipuwer. It also appears that social mimesis, centered around a fascination with the power of the pharaoh, led to the spread of certain royal funerary symbols to "commoners."

The magnitude of the impact of low floods on Egypt are captured in the following excerpt from a number of such accounts (cited in Bell 1971):

All Upper Egypt was dying of hunger, to such an extent that everyone has come to eating his children... The entire country had become starved like a starved (?) grasshopper, with people going to the north and to the south (in search of grain).

This account highlights the fragility of Egypt's economic base. Nile floods discharge is highly variable, contrary to common belief (Butzer 1995). Episodic changes, as interpreted from historical records and fluctuations in Birket Qarun, Faiyum (Hassan 1981, 1986b), include intervals of 20-50, 120-180, 300-500, and 1000-1600 years. The magnitude of change can be drastic. For example, average Nile flood discharge at Aswan in 1879 was 129 billion cubic meters per day. By contrast, in 1913 the discharge fell to a minimum of 44 billion cubic meters per day (Hurst 1957:260-62). During this century, following the construction of the Aswan High Dam, a period of high Nile floods ensued. The highest level of the reservoir (Lake Nasser) rose from 127.6 m asl to 175.71 m asl (Shahin 1985:50). However, a series of low Nile flood levels led to a panic in 1987 as water level became so low that Egyptians feared lack of drinking water and electricity. Reservoir lake level feel to a record low of 156 meters, and the available stored water to less than 25 billion cubic meters (El-Gably 1988, p.35). The lowest flood discharge during this century was in

1 As a matter of convention I use 'bp' to refer to radiocarbon years before present using the conventional half-life. BC, and rarely BP, are used exclusively to refer to solar years before the Christian era. CBC refers to radiocarbon age estimates converted to equivalent solar years

1972/73 and 1982/84 (62 billion cubic meters). Egypt's official estimate of water needs are estimated at 63 billion cubic meters in the year 2000, and at 70/75 billion cubic meters by Waterbury (1979).

In this contribution, I will present additional data to buttress Bell's claim. The demise of the Old Kingdom might have been in part facilitated by the attenuation of the power of the Memphite rulers and their centralized government (Grimal 1994:139). My aim here is two fold; first to present evidence from a variety of proxy data for the incidence and severity of a climatic crisis ca. 2200 BC (3700 uncalibrated radiocarbon years); and second to assess the magnitude of the impact of variations of Nile floods on food yield, and the consequence of the failure of Nile floods on the social and political life during the 3rd millennium BC.

The low Nile flood events of the third millennium were not a unique occurrence. Egypt experienced severe droughts causing famines and civil disorder of devastating consequences during the period from the 10th to the 13th century (Hassan 1993). Such events were the result of abrupt climatic swings linked of global scale. By examining the climatic dynamics of Nile flood variability we can adduce more evidence to support inferences from proxy data in Egypt. In general, as I will attempt to show in the next section, low Nile flood discharge in Egypt is accompanied with low lake levels in Ethiopia and Equatorial Africa, the sources of Nile floodwater. Dramatic reduction in low Nile flood discharge may also be matched by a reduction in sea level and sea regression.

Climatic Change and Nile Floods

In this section, interpretation and reconstruction of Nile floods during the First Intermediate is based on (1) analogy with the historical record of Nile flood maxima and minima since the 7th century AD, (2) geologic history of Birket Qarun, a Nile-fed Lake in the Faiyum Depression, (3) examination of mid-Holocene Nile sediments in the Nile Valley and the Delta, (4) evaluation of historical records of floods and their consequences, and (5) correlations with lake levels in East and Equatorial Africa.

To understand the nature of Nile floods in antiquity, it is useful to examine the annual records of Nile floods recorded by Arab scholars in Egypt since the 7th century AD. (Hassan 1981, Hassan and Stucki 1987). During the span of this record, episodes of high Nile flood discharge is correlated with greater rainfall in equatorial

---

before the Christian era. For the chronology of Egypt, I follow Baines and Malek.

East Africa, as well as warmer climate in Europe. Increasing rainfall in Equatorial Africa is related to the poleward advance of the Inter-Tropical Converge Zone (Flohn and Nicholsson 1980, Hassan 1981, Kutzbach and Guetter 1986, COHMAP 1988). During these episodes, the Sahara experiences conditions of greater rainfall. During the early Holocene, rain related to this poleward advance of the ITCZ reached as far north as northern Egypt. Neumann (1993: 163) infers a displacement of 550-600 km on the basis of the wood species recovered from archaeological sites. Today the shift between dry and wet years is estimated at 200-300 km (Said 1993:86)..

Rain related to the ITCZ supplies the tributaries of the Nile with water in Ethiopia and equatorial Africa. There is a general increase of rain from east to west. Areas close to the equator have two seasons, with maximum rainfall around April-May and October-November. Over most of the area the April-May maximum brings more rainfall because the ITCZ is very broad in this period and moves only slowly northwards. Farther away from the equator, rainfall prevails around July-September (Griffths 1972).

The Faiyum and the Nile Valley

In the Nile Valley, the earliest farming communities date to the early decades of the sixth millennium radiocarbon years bp, or approximately 4800 BC at Merimda Beni Salama (Hassan 1984). At 5000 BC, or slightly earlier in the Faiyum, the inhabitants were mostly engaged in fishing. There is a curious gap in the archaeological record between the earliest so called "Neolithic" sites in the Faiyum dated to 6350 radiocarbon years bp and the final Palaeolithic "Qarunian" sites dated to 8220-7140 bp (Hassan 1988). It thus appears that the earliest farming settlements in Upper Egypt, as well as other sites of the final Paleolithic fishing communities, were located close to the channel. These sites were later destroyed or buried during episodes of high flood stages.

Geological investigations of the lakeshore deposits of Birket Qarun in the Faiyum (Hassan 1986b, with references) provided a good sequence for variation of Nile flood discharge over the last 10,000 years. Birket Qarun was fed by Nile floods over a sill that separates the Faiyum depression from the river Nile. Reconstruction of the Faiyum lake levels on the basis of geological proxy data correlates satisfactorily with estimates of Nile annual flow based on a macromodel of the Holocene Inter-Tropical Convergence and jetstream positions and rainfall in east Africa (Bryson and Bryson

1995:21). They are also congruent with the outline of Holocene palaeoclimate in the Eastern Sahara (see below).

Lake levels during the early Holocene were high, but a major drop occurred ca. 7000 bp. By 6500 bp water level was rising. However, this was a short-lived event. By the beginning of the sixth millennium bp, the lake experienced another recessional event (Hassan 1986b: 491). At about 4800 bp, after the end of the Neolithic and before the Old Kingdom the lake level was at 15 m below sea level, a drop of more than thirty meters from the Neolithic lake stand. Hoffman et al. (1986) also state that there is a stratigraphic evidence for a decline in the height of average annual Nile floods ca. 3500 BC. This coincided with the transition from Nagada I (Amratian) to Nagada II (Gerzean) in the Predynastic of Upper Egypt, a transition that marks the emergence of towns, increasing cultural contacts, intensification of trade, and the emergence of petty states. In general, there seems to be a shift in the location of Nagada II (Gerzean) settlements closer to the active channels of the Nile in response to the decline in the height of flood levels.

Both empirical data and the model clearly reveal

Records of the height of Nile floods from Pharaonic Egypt, examined by Bell (1970) and Butzer (1984), are interpreted as evidence of low floods during the Early Dynastic period. If the peak shown in the Brysons' model at ca. 3000 BC, is correct, then the Terminal Predynastic might have coincided with exceptionally high Nile floods followed by a rapid decline during the First and Second Dynasty. It was perhaps such a rapid decline in flood discharge during the early phase of Egyptian civilization that promoted the utilization of Nile gauges and the keeping of Nile flood records. A brief interval of high flood discharge at ca. 3000 BC is in accord with an inferred brief wet period at Hierakonpolis after 3200 BC and before 2500 BC cal BC (Hoffman and Mills 1993)

It also seems that the low Nile floods of the First and Second Dynasty were followed by a continuation of the decline in Nile discharge. The re-dating of the Qasr el-Sagha Temple and associated sediments (Kozlowski and Ginter 1993) suggests that this indeed might have been the case. However, Nile floods at some time(s) after the Old Kingdom were higher than the levels during the construction of the pyramids of Giza (2575 BC or approximately 3980 radiocarbon years bp). This is indicated by the presence of what appears to be the harbor and the Valley Temple of Khufu (Cheops) below four meters of Nile silt. It would appear that these deposits belong to a Middle Kingdom high stand and/or another period of high discharge beginning after 3500 years ago. Recently, statues of sphinxes, presumably from the Late Period and later, were recovered below Nile floodplain deposits at Tell Batram, Giza.

Toward the end of the Fifth Dynasty, a graphic scene of the effects of famine appears in a relief from the causeway of the Pyramid of Unas (Bell 1971: 5). It may be noted, however, that linking the scene of emaciated people suffering from hunger to poor Nile floods is not secure since they may represented starving desert dwellers (Z. Hawass, *pers. comm.*).

During the First Intermediate (2180-2134 BC or approximately 3700-3600 radiocarbon years bp), Nile floods were low as indicated by a drop in the level of Birket Qarun, as well as the invasion of the Valley by dune sands in Memphis (David Jeffreys, *pers. comm.*). Sand dunes are also believed to have invaded the floodplain in Middle Egypt (Butzer 1959). It thus appears that the country was undergoing a dry episode with active movement wind blown sand. In addition, the floodplain was dry as a result of low flood levels. Lack of high floods would have deprived Birket Qarun of its major source of water. Deprived of water and subjected to severe evaporation under dry climate, the lake level would drop rapidly. Even if occasional high floods spilled over to the lake, high evaporation rates would still lead to a major drop in the level of the lake.

The recession of the lake exposed an extensive area for land reclamation and construction. It was apparently brief, since it was apparently followed by a re-rise of the lake ca. 3600-3500 bp as indicated by historical re-dating of the Qasr el-Sagha to the Middle Kingdom and a radiocarbon date of 3580 ± 60 radiocarbon years bp (Kozlowski and Ginter 1993: 331). This probably coincided with the slightly moist interval in the Sahara ca. 3500 radiocarbon years bp (Vernet 1993: 93).

The occurrence of a period of severe low Nile flood discharge is indicated by an increase in sand in the sediments from a core in Birket Qarun dated to 3600/3300 radiocarbon years before present (Jux and Steuber (1990: 266). This observation must be confirmed by future work given that two radiocarbon dates are in reverse temporal order to what might be expected from the stratigraphy.

The level of the ground of the temples of Kiman Faris (a heap of ruins that has now been overtopped by a housing project in the center of the city of Faiyum) and the colossi of Amenemhat III (1840-1792 BC) at Biyahmu suggests that the lake was below 20 m asl. The lake appears to have risen after the construction of the colossi of

Biyahmu as indicated by the presence of lakeshore deposits below a collapsed block from the colossi. Facies analysis may provide a basis for reconstructing the height of the lake, which must have stood at least at 22 m asl. At Qasr el-Sagha the deposits once attributed to the Old Kingdom (Caton-Thompson and Gardner 1939, 1929: 46) and now regarded as Middle Kingdom in age (Kozlowski and Ginter 1993) are at 4 and 13 m asl, and may thus belong to a stand in a rising Middle Kingdom lake.

The Biyahmu colossi are believed to be the "pyramids" witnessed by Herodotus, who saw them from a distance in the middle of the lake. It seems that the Greek historian Herodotus (484?-425? BC), who visited Egypt during the fifth century BC, saw the last high stand of Birket Qarun (Moeris Lake) before its eventual decline in early Ptolemaic times. The Ptolemies took advantage of the decline and reclaimed a substantial area of what used to be the bottom of Birket Qarun. Subsequently the lake began to shrink progressively until water could no longer flow over the Faiyum-Nile divide terminating the active surface flow of Nile water into the Faiyum depression. Butzer has already alluded to the impact of the onset of droughts shortly after 1260 BC on agriculture in Nubia, where fields were abandoned, wind blown sand accumulated, and salt encrusted the floodplain. The onset of this arid episode also coincides with a record of famines in Libya in the reign of Merenpetah (1224-1214 BC) which led to several attacks on Egypt. Earlier climatic droughts at the time of Amenemhat III (approximately 3400 radiocarbon years bp) led a party of the Medja desert dwellers to express their plight from droughts in the cry; "The desert is dying of hunger."

By 1245 AD (ca. 3000 radiocarbon years bp) the lake stood at 30 m below sea level. The low lake level at that time would signals a period of low Nile flood discharge which would caused severe agricultural problems.

Today, only the massive pedestals of the colossi remain standing. Senusert I or one of his successors instituted land reclamation and irrigation improvements which were referred to later by Herodotus and Strabo. Since then, the Faiyum has become not only one of the most fertile regions in Egypt but also a favored resort spot.

The Nile Delta

In the Nile Delta, a layer of peat or highly organic mud was deposited at about 7000 bp north of Tell el-Fara'in. By ca. 5900 bp the area of peat formation reached its maximum southern extent. Subsequently the deposition of peat ceased after ca. 4500 bp (Wunderlich 1989: 130). De Witt (1993:316) also concludes that reduced Nile flooding occurred in the first half of the 5th millennium bp. The northern Delta, invaded by a transgressive sea during the early Holocene, shows a landward advancing coastal deposits. The timing of various events during this transgressive interval is not yet clear, but the cessation of peat deposition after 4500 bp coincides both with the onset of the present conditions of hyperaridity in the desert and a drastic drop in Nile flood discharge. There are also indications that about 2500 BC, i.e. at the time of the 5th Dynasty, organic rich material was deposited at lower levels than some 100 years earlier (Andres and Wunderlich 1986).

Subsurface deposits also indicate that a deep channel was cut during the fourth or third millennium BC after 3990 BC and before ca. 3200 bp or ca. 1500 BC when it was filled by silt (Andres and Wunderlich (1986:162)). The incision of the channel is thus likely to have taken place either during the 4500 bp (3200 BC) or the First Intermediate (2200 BC). It is also possible that floodplain deposits dating to the period from 3200 to 2200 BC, about a meter of silt might have been destroyed by the degradation of the Delta during the period of degradation associated with low Nile flood discharge around 2200 BC.

The subsurface stratigraphy of the Delta near Minshat Abu Omar (Krzyzaniak 1993, 153) suggests that a Predynastic settlement was established on sand hillocks (*gezira*) about two meters above the floodplain. Accretion of the floodplain during the Predynastic and the Old Kingdom (as indicated by several radiocarbon dates) apparently occurred during an episode of declining Nile floods, as suggested by the lower position of younger material. Vertical and perhaps lateral erosion seems to have occurred at a time prior to the siltation of these channels (presumably during the New Kingdom) under a new Nile regime characterized by a suite of dark brown Nile mud and sand. The older deltaic deposits consisted, by contrast, of finer dark, bluish gray clay, silt, and silty clay dated to 2585-2405 cal BC . One phase of degradation of the Deltaic floodplain seems thus to have occurred *after* 2500 BC and *before the younger silt. Another* episode of degradation occurred *during* the deposition of the top silt. The two episodes are indicated by the erosion and lateral displacement of reworked artifacts and anthropogenic deposits of the remains of the Predynastic occupation in two discrete events. My *guess* is that the first episode corresponds to the late Old Kingdom-First Intermediate drop in the level of average floods ca. 2200 BC. The second to the degradation associated with falling Nile levels during the first century BC.

Andres and Wunderlich (1986) provide an estimate of 1.5 mm/ year for the dark gray sediments, which places the transition to the modern brown Nile silt in the Delta to ca. 500 BC (2500 years calendric BP). My own estimate for this part of the Delta is about 1.2 mm/yr (Hassan, n.d.). Using this rate, the topmost Nile silt unit would have accumulated before 500 BC. years, perhaps at about 650 BC (3000 bp) . However, we should be cautious since the thickness of the silt near the sand islands might be much thinner than that in the lowlands. Average thickness of top Nile silt in the Delta of 9.8 meters suggests that the initial deposition of modern silt began during the 8th millennium BC.

East and Equatorial African Lake Levels

In the high resolution pollen sequences provided by Bonnefille and collaborators from Burundi and Ethiopia, there is a conspicuous episode of aridity dated to 3650 radiocarbon years bp at Kashiru, Burundi, at 3660 bp at Nudurum Ndu2, Burundi, and 3750 at Kuruyange, Burundi (Bonnefille et al. 1992, Jolly and Bonnefille 1992) . The retreat of the trees after the development of a forest maximum around 4000 bp was, as Bonnefille, Riollet and Buchet (1991:321) put it "une récession brutale des arbres." A numerical analysis of the high resolution Holocene pollen record from Burundi affirms that "a sharp decrease of arboreal pollen percentages and a noticeable increase of *Celtis*, a semi-deciduous tree", which indicates the opening of montane forest due to a drier climate occurred at 3800 bp (Jolly, Bonnefille, and Roux 1994:357). The decrease in arboreal pollen was associated with an increase in non-arboreal pollen (Compositae, Verononieae, Umbelliferae, *Impatiens*). This dry episode is also recorded in Rawanda (Hamilton 1982) and in the highlands of Uganda (Hamilton 1982, Taylor 1993), at Lake Victoria (Kendall 1969) and in Ethiopia (Bonnefille and Hamilton 1986, Umer 1992). Farther to the west, Lake Bosumtwi in Ghana fell at 3750 bp in response to arid conditions (Talbot *et al.* 1984). Other low lake levels in Africa at that time are noted by Gasse and Van Campo (in press), Butzer (1980), Casanova and Hilaire-Marcel (1992). The dry episode was marked by a decrease in temperature (Bonnefille et al. 1990). According to Gasse and Van Campo (in press), maximum aridity is also evident in Lake Sumxi, Western Tibet, where maximum aridity for the whole Holocene period is recorded at ca. 3800 years bp by a minimum in *Artemisia* /Chenopodiaceae pollen ratio and in carbonate content, as well as a minimum in ^{18}O contents of the lake water.

These data clearly amplify and resonate with the data from the Faiyum and the Delta suggesting that within the 3800-3600 bp time interval severe aridity accompanied by cooler conditions prevailed over equatorial and eastern Africa. The Nile flood discharge was heavily reduced causing a drop in the level of Birket Qarun in the Faiyum and degradational geomorphic changes in the Delta.

In a high resolution pollen sequence covering the period from 2500 to about 800 BP, at Lake Langeno (Ethiopia), Mohammed and Bonnefille (1991:284) also provide evidence for a dry period with a drop in lake level at ca. 2000 bp (75 BC) . Shoreline vegetation, including Cyperaceae and halophyte plants, such as Chenopodiaceae, Amaranthaceae, as well as dry evergreen bushland with Olea, Euclea, and Dodonaea viscosa expanded. This corresponds with the second erosional degradation of the Delta inferred above. Bonnefille and Umer (1994) also note, from a high resolution study of a core from the Arsi Mountain in Arussi province of southeastern Ethiopia, that a cooling of about 2 C happened at about 560 ± 120 yr bp (1100 to 1400 AD). This event favorably compares with a period of major low Nile flood discharge as indicated from the historical record of Nile floods at Rhoda (Hassan 1981), which lasted from 1180 to 1350 BC. There is thus clearly a correlation between cooler temperature, drier conditions in equatorial and eastern Africa, and low Nile flood discharge.

The Economic Consequences of Low Nile Floods

Under natural basin irrigation, low Nile floods are likely to deprive higher parts of the floodplain from water, thus reducing the area available for cultivation by natural basin irrigation which was in use until the 19th century. According to the Al-Maqrizi (1441, pp. 185-186), which provides a sophisticated land-use classification, the best land is called *Baq*. It is neither the lowland (*Mustabhir*), which is flooded with no (natural) drainage, or the upland (*Sharaqi*), which floodwaters do not reach.

Very high floods lead to a lot of destruction (e.g., of dikes and houses), and floodwaters may not recede from a substantial area of the floodplain in time for planting. Low floods, on the other hand, deprive previously naturally irrigated areas from water and may radically reduce the potentially cultivable area. Maintenance of local irrigation canals and drains, as well as strengthening of dikes, was a major activity in ancient Egypt, but there is no evidence of centralized management of nation-wide irrigation projects. Episodes of political turmoil and unrest are likely to have had adverse effects on the economic prosperity of Egypt.

Augustus set his army to the task of repairing and deepening the irrigation canals with the result that, as Strabo (Book XVII, 788) informs us; whereas before the [Roman] conquest an abundant harvest required a rise of fourteen cubits in the level of the Nile and one of eight meant famine, under the Romans, a rise of twelve brought a bumper harvest, and there was no scarcity even when it was only eight (Kees 1961, p. 51). This may be contrasted with the evidence for the siltation of canals as a result of neglect under the Memluks as shown by pollen from the Faiyum Lake (Mehringer, Petersen, and Hassan 1979). Canal neglect in the 1700s is indicated by a record of pond weed (*Potamogeton*) pollen. *Potamogeton* is a rooted aquatic. Its abundance suggests a prolonged shallow water phase. A decline in *Potamogeton* in the 1800s coincides with renewal of canal maintenance initiated by Mohamed Ali, as well as higher Nile floods (op.cit., p.245).

The effect of low floods on economic conditions may be appreciated by analogy to the famines under the Fatimids and, more recently in 1783-1786, when three successive years of exceptionally low floods led to famine and soaring wheat prices. This was followed in 1789 by the plague (*Ta'oun Ismail Bey*), which lasted for five months. In 1791, and again in 1792, a slight drop below the long-term mean and only two cubits or about one meter from the optimal level led to a severe famine (Zakry 1926). People were forced to eat dead horses and donkeys, and even children. Another series of low floods in 1794, 1795, and 1796 led to a revolt. In 1877 a low flood, two meters below average, left 62% of Qena Province and 75% of Girga Province unirrigated (Willcocks and Craig 1913).

To fathom the impact of the reduction in Nile discharge, we may calculate the consequences for food yield and the number of people who could be sustained. If the parameters of all other variables remain constant, a 50% reduction in land area may mean a 50% reduction in population and food yield. But, in fact, people may consume less wheat, reduce their total food intake (depending on food stored from the previous year), or import food. Thus if we assume that, on average, people consumed 200 kg of grain per year, a reduction of annual consumption of wheat per person to 100 kg may mean that people can survive a year when floods are poor if they also have 20-50% from stored substitute or imported food. However, if low floods persist, there will be little in storage, and export areas may also be affected.

Substitute wild foods are also likely to suffer from over-utilization. The situation may be improved by improving agricultural technology, but if capital is low and morbidity (physical and psychological) is high, such an alternative may not be readily possible. In addition, proper innovations may not be right there. In addition, continuation of a poor diet is likely to lead to an increase in morbidity and mortality. Continuation of poor floods eventually leads to high levels of mortality and most likely an increase in abortions and child neglect which have both immediate and subsequent effect on the size of the labor force then and in the future. A reduction in labor force would mean that, even if in the interim floods were occasionally high, there might not be enough people to cultivate the land, as reported historically.

The dire consequences of poor Nile floods is clearly demonstrated by the events of famines, plague, and political disorder during the 10th-12th millennium AD. For example, famine in the wake of the low Nile flood in AD 967 reportedly led to the death of 600,000 persons around Fustat (a capital predating Cairo), which is as much one fourth of the whole population of Egypt. The record may thus refer to the population loss for all of Egypt. Abdel-Latif Al-Baghdadi (n.d.), an Iraqi scholar who was visiting Cairo at the time of the famines of AD 1220-1201, reported that as many as one hundred to five hundred persons died daily from hunger in Cairo (see also Raymond 1993). At a rate of 100 persons per day, Cairo would have lost half of its inhabitants in about six months. The population of Egypt was redued from 2.4 million at the beginning of the 10th century to 1.5 million at the end of that century (Hassan, 1993, Said 1993:165

Social and Political Consequences

Famines & Low Revenues

One of the important function of the government in Egypt was to collect revenues for the support of the court and state projects (Trigger et al. 1983:82). To the elite who depend on the revenues from the farmers, successive years of low floods are very critical. Even if the elite have amassed grain that can be redistributed in times of emergency, no amount of stored grain, given the limitation of storage as well as the rates of taxation, can withstand a long-term decline in yield as a result of a sustained trend of increasingly diminishing flood discharge (unless countermanded by irrigation works or a change in agrarian strategy). In fact the elite will, in a few years, begin to use force in collecting taxes, a condition not likely to be appreciated by the starving peasants.

The population of Ancient Egypt at the beginning of the Old Kingdom is likely to have been small. The potential of increasing yield per capita for adults by adding more children to the family would have encouraged a rapid increase in population. The farmers had no way of knowing that they were on a crash course with progressively dwindling Nile flood discharge. The crash point might have coincided with the First Intermediate.

Also, local administrators and governors who collect taxes and then pass them on to the royal house, a system prevalent in Ancient Egypt, are likely to consider delaying or actually transferring the taxes to the royal capital when yields are poor. Consequently, revenues to the king would plummet gravely impairing the ability of the monarch to pay for an army or to support loyalists. In addition, the monarch is not likely to be able to provide the funds for major agrarian projects to ameliorate the economic situation.

We should not also overlook the impact of famines on craft specialists and artisans who depend on barter for their livelihood, and are not food producers. Being at a distinct disadvantage, their numbers are likely to dwindle fast. Al-Baghdadi reported that during the 1200-1201 famines, of the nine hundred weavers in Cairo, no more than nineteen were left alive at the end.

Rivals to the Throne

It is at such times of economic distress that the power of the monarch and the sources of the legitimacy of his/her rule are questioned. Rivals from elite centers of power such as the military, the priesthood, or the provincial governors are likely to emerge. They are likely to point out the excesses of the monarch; overexpenditure, corruption, and deviation from "fundamental" religious beliefs. In addition, underground or subordinate religious beliefs may emerge to challenge the existing dominant theology.

During the First Intermediate, pyramid building on a massive scale may have been abandoned in response to declining Nile flood discharge. Massive pyramid complexes continued from the Third Dynasty to the Sixth Dynasty. The last monarch of the Old Kingdom, Pepy II, built the last of the great pyramid complexes, even though the cemetery of his courtiers was impoverished, perhaps reflecting the rough economic conditions at the time (Kemp 1983: 112). The rise of a sun cult in the Sixth

Dynasty and the takeover by a priestly family suggests a power struggle that may have been rooted in economic stress. Alternatively, the conflict might have weakened the monarchy. Eventually the resurgence of the Osirian theology represented a challenge to absolute monarchy as expressed in the solar royal cosmogony.

Political Geography and Centralization

The collapse of centralized government during the First Intermediate was brought about by a combination of factors. Undoubtedly low Nile floods, the culmination of a long-term trend that may have been already underway since the First Dynasty, was decidedly instrumental in weakening and eroding the economic and ideological power of the monarchy. Other participating factors included the weak bonds of unity under the Old Kingdom. Egypt consisted of an amalgam of mini-states (that were in turn composed of basic administrative units, called nomes in Greek times). Egypt was not totally fused into a single nation-state eliminating provincial loyalties and regional religious beliefs. During times of stress Egypt was likely to break along this provincial line of weakness. Geographically, Egypt poses an administrative challenge because of its linear aspect. The capital in Memphis or Thebes could not extend its power over all of Lower and Upper Egypt effectively. The range of power, given the transport capabilities then available, created a power range of ca. 400 km (Hassan 1993: 567).

Also, we should not overlook the impact of a rapid increase in the ratio of administrators, bureaucrats, functionaries, middle-men, and other non-food producers on the economic prosperity of the whole country, especially if there was a progressive, gradual reduction of agricultural yield. The situation would have been even worse because such non-food producers progressively strive for higher levels of consumption in the pursuit of the better life and the positive feedback between economic power and political power. Such developments during the Old Kingdom, coinciding with diminishing resources, would have been detrimental.

Refugees and Invaders

In Egypt, ephemeral lakes, playas, and mud pans show evidence for pronounced climatic variability throughout the Holocene. Detailed investigations of the deposits of these playas revealed that the trend toward contemporary hyperaridity was heralded during the seventh millennium and again during the sixth millennium bp. Following a slightly moist phase from 5900 to 4500 bp, the present hyperarid regime prevailed with short, occasional episodes of meager rain. During conditions of maximum rainfall, the northern Sudan (15°-20° latitude) received as much as c. 400 mm/year. Farther north in southern Egypt rainfall was not much greater than 100 mm, where today there is virtually no rain at all (Hassan 1986a, Neumann 1993, Ritchie and Haynes 1987).

The last major wet event in the Egyptian Sahara associated with the formation of freshwater ephemeral lakes is well recorded at El Bahr Playa in the Farafra Oasis, where lake mud deposited from water carried through surface runoff and local drainage is capped by clastic limestone debris deposited under increasingly cold and dry conditions (geoarchaeological research by the author in collaboration with B. Barich in Farafra Oasis is in progress) . This final event is dated to ca. 6000 bp (charcoal from a hearth below the contact is dated to 6160 ± 110 bp). It was during the 7th and 6th millennium bp episodes of severe droughts that dwellers of the desert drifted toward the Nile Valley, and elsewhere, in search of water, food, and fodder (Hassan 1986a, 1988, Holmes 1993).

Nomadic cattle keeping in the Gilf el-Kebir was possible until ca. 4300 bp, but not later. In the Sudan from 5200 bp onwards, the savanna vegetation retreated to the south until 3300 bp when the present situation was established (Neumann 1993: 166-167). In Dakhla oasis, the Bashendi artifact suite associated with nomads who depended on the seasonal ponding of the desert ephemeral lakes (playas), are dated to the millennium ending 5000 bp (McDonald 1993). The fifth millennium bp signals the end of any significant occupation of the Egyptian Sahara.

In the Eastern Desert, stratified rock-shelter deposits at Wadi Deir provide evidence for several occupations from about 3630 BC to 1920 BC. Lack of occupation after a phase from 2150 to 1920 BC (3650-3500 bp), coinciding with the First Intermediate, may be an indication of the onset of hyperaridity. Animal bones recovered from the site included those of Nubian wild ass, ibex, sheep/goat, suggesting a mixed hunting-herding strategy.

The Holocene record of the Central Sahara correlates with that of Egypt. Rising lake levels afterward coincided with a mid-Holocene moist phase in the Sahara ending around 4500 bp. The onset of aridity in the Sahara coincided again with a reduction in lake levels. Some lakes totally dried up. Vernet (1993) also notes that after 4000 bp a new pattern of adaptation emerged reflecting the changing climatic conditions. Vernet (1995) recognizes a period of severe aridity at 7000 bp, and again between 5000 and 4000 bp.

In general, by 4500 years bp, conditions of hypraridty prevailed in the Sahara. By 3300 bp, as Neumann (1993) shows the southward retreat of the savanna vegetation reached its present limit. The major episodes of low Nile flood discharge at about 7000, 4500, and 3800/3600 uncalibrated years bp correlate with hyperarid spells (see also below). Comparison of Nile flood discharge and Sahel rainfall over the last 100 years (Said 1993:123) also reveals a positive correlation as shown for the period since 1967, the early 1940s, and 1916 when reduced Nile flood discharge is matched with a drop in rainfall.

From these comparisons, the First Intermediate would have witnessed not only low Nile floods, but also a reduction of rainfall in the Eastern Desert, Nubia, and the Levant. Thus at that time, Egypt at a period o of economic, social, and political distress was probably faced by refuges seeking help, as in the biblical account. Many "Asiatics" and "Nubians" might have thus "emigrated" to Egypt at that time (see below about Asiatics in the Egyptian army and domestic staff during the Middle Kingdom). Under a worse scenario, with marauders of nomadic warriors pressed by drying pastures and vanishing game. A weakened state, perhaps "softened" by previous centuries of prosperity and embattled with local strife and sectarian conflicts, is often no match for desperate warriors who have very little to lose and everything to gain.

Events in Nubia at the time of the "First Intermediate" period include the appearance of the so called "C-Group," as a cultural phenomenon associated with "immigrants" from the Nubian Desert, who settled on the west bank of the Nile (Bietak 1986: 5). The initial movements of the ancestors of the people of Kerma and the C-Group was probably related to the progressive drying of the northern fringes of the Sahel beginning ca. 4500 and worsening, except perhaps for a brief interval afterward. The immigrants might have moved into Nubia through Wadi Hawa, but there is no direct evidence for this. Bietak (*op. cit.*, p. 6) postulates that the C-Group and a related cultural group, the Kerma Culture, separated from an original stock which might have been located in the El-Gash Delta or a similar zone elsewhere. He places the time of separation at ca. 2300-2200 BC (**3775-3700 bp**) the low Nile floods of the First Intermediate. The C-Group, contemporary with the 6th Dynasty, occupied Lower Nubia from Kubannyia (near Aswan) to the Batn el-Haggar (called *Wawat* in Middle Egyptian records). Kerma Culture was in Upper Nubia (*Kush*). Nubia included several chiefdoms before the rise of state societies (O'Connor 1991).

Alabaster vessels of the 6th Dynasty suggest trade relations were established between Egypt and Kerma by that time. A series of radiocarbon dates provide two average calibrated estimates of 2120 ± 30 and 2064 ± 50 BC, suggesting that the early establishment of Kerma may date back to the 6th or perhaps even the 5th Egyptian Dynasty. Unification of local groups and the emergence of a powerful state society at Kerma occurred during the First Intermediate (Bonnet 1986: 7). The people of Kerma benefited from trade relations with Egypt. It may be postulated that the people of Kerma might have profited from the adverse economic and political conditions in Egypt. Competition for status imports from Nubia among rival home chiefs might have expanded the Kerma market. In addition, caravans of several hundred donkeys, the precursors of camel caravans, might have brought back to Egypt not only incense, ivory, ebony oil, and leopard skins (Bietak 1986), but also food products to alleviate the food shortage.

Hypothetically, low Nile floods and droughts in Nubia and northern Sudan, as well as climatic difficulties in Southwest Asia would have occurred again during the Ramesside period as a resullt of 3000 yr bp dry interval (ca. 1300 BC) and during the 25th dynasty during the 2500 yr bp droughts (approximately 650 BC).

The Continuity of Egypt

The idea of monarchy had a strong appeal to power aspirants who eventually managed to re-establish it. The transition from anarchy or dissolution of political centralization might depend on an improvement in economic conditions as Nile floods became more benevolent. Alternatively or simultaneously, the restoration of unity might depend on a policy of moderate taxation, reduction of government expenditure, allocation of resources to increase yield (e.g., through irrigation projects), improvement in transportation, and trade. The kings of the Middle Kingdom took advantage of the land that was once covered by Birket Qarun in the Faiyum as the height of floods diminished.

It is thus not surprising that settlements, temples, pyramids, and palaces from this period are in the Faiyum. The legend of the wealth of "Qarun" (detailed in the Koran), a mythical ruler of the Faiyum, may go back to the prosperity that resulted from the reclamation of the Faiyum. It also seems that many Asiatics immigrated to Egypt during the droughts of the First Intermediate. It is also possible that they were employed as mercenaries by provincial rulers or competing kings. During the 13th Dynasty, a list enumerated 45 Asiatics among a total of a domestic staff of 79, and documents from Kahun refer to an "officer in charge of the Asiatic troops" (Kemp 1983: 155). The association between the biblical story of Joseph and the Faiyum may refer to the employment of Asiatics in the irrigation projects of the Middle Kingdom kings. A survey of Mid-Holocene environmental change in the southern Levant (Rosen 1989) reveals that hyperarid conditions had prevailed since ca. 4000 bp, contemporaneous with the end of the Old Kingdom in Egypt.

The survival of Egyptian civilization in view of economic calamities from floods may be attributed to the lack of a strong articulation between the centralized state or the state in general and food production: the peasants did not, in reality, need the state for meeting their own needs. In fact, the continuation of Egyptian civilization was predicated upon the high yields that ensured the production of an excess beyond the basic necessities of peasant life. The central government does not seem to have "micro-managed" the affairs of the provincial administrative districts. According to Kemp (1983: 113) a series of fourteen decrees issued by some of them, mostly appointed members of a prominent local family, to positions in the provincial and temple administration suggest that the Old Kingdom apparatus of government still functioned.

The continuity of Egypt as a high civilization, in the face of economic adversities, also resided in the firm belief in the monarchy. Regardless of the background of kings, the idea of monarchy within a basic [deep] cosmogonic ideology provided a strong attraction (for aspirants). The norm of social justice and benevolence, consolidated perhaps in the context of the woes of the First Intermediate, might have also lessened the potentially devastating impact of the power elite (e.g., as far as heavy taxation and coercion) and thus minimized social unrest, revolts, and anarchy. The Tale of the Eloquent Peasant clearly declares a bill of the rights of ordinary citizens, and the responsibility of the state toward securing such rights even against powerful government officials.

"*...because you are*
a father to the orphan,
a husband to the widow,
a brother to her who is divorced,
a garment to the motherless,
let me make your name in this land

in accord with every good law--
a leader devoid of rapacity,
a magnate devoid of baseness,
who destroys falsehood and fosters truth,
and who comes at the voice of the caller.
I speak that you may hear:
Do Justice,
O praised one whom the praised ones praise.
W. K. Simpson 1972, *The Literature of Ancient Egypt*, p. 35.

References

Al-Baghdadi (Muwafaq el-Din Abdel-Latif Al-Baghdadi). n.d. Al-Ifada wa-Alitibar fi Al-Imour Al-Mushahada wal-Hawadith Al-Mo'yana fi Ard Mis (in Arabic) Cairo.

Al-Maqrizi, Taqi eldine Ahmed Ibi Ali (1441). Al-Mawa'z wal -I'tibar bi-zikr al-Khitat wal-Athatr (in Arabic). Bulaq, Egypt.

Andres, W. and Wunderlich, J. 1986. Untersuchungen zur Paläogeographie des westlichen Niledeltas im Holozän. Marburger Geographische Schriften 100, Marburg/Lahn.

Baines, J. and Malek, J. 1984. Atlas of Ancient Egypt. Oxford, Phaidon.

Bietak, Manfred. 1986. The C-Group and the Pan-Grave Culture in Nubia. In Sixth International Conference for Nubian Studies, Klassisk Institutt, Universitet i Bergen, Volume I.

Bell, Barbara. 1970. The oldest records of the Nile floods. Geography Journal 136: 569-573.

Bell, Barbara. 1971. The Dark Ages in ancient history: I. The first Dark Age in Egypt. *American Journal of Archaeology* 75: 1-26.

Bonnefille, R. and Ummer Mohammed, 1994. Pollen-inferred climatic fluctuations in Ethiopia during the last 3000 years. *Palaeogeography, Palaeoclimatology, Palaeoecology* 109: 331-343.

Bonnefille, R. and A. Hamilton, 1986. Quaternary and Late Tertiary history of Ethiopian vegetation. Acta Univ. Ups., Symb. Bot. Ups., 26(2):48-63.

Bonnefillet, R., J.C. Roeland, and J.Guiot 1990. Temperature and rainfall estimates for the past 40,000 years in equatorial Africa. *Nature* 346: 347-349.

Bonnefille, R, G. Riollet and G. Buchet, 1991. Nouvelle séquence pollinique d'une tourbière de la crête Zaïre-Nil (Burundi). *Review of Palaeobotany and Palynology* 67:315-330.

Bonnet, Charles. 1986. Kerma, royaume africain de Haute Nubie. In Sixth International Conference for Nubian Studies, Klassisk Institutt, Universitet i Bergen, Volume I.

Bryson, Reid A, and R. U. Bryson. 1995. An Archeoclimatology Workbook: High Resolution, Site-specific Climate Modeling for Field Scientists. Center for Climatic Research, The Unversity of Wisconsin, Madison, 78 pp.

Butzer, Karl W. 1959. Contributions to the Pleistocene geology of the Nile Valley. Erdkunde 13: 46-67.

Butzer, K. W. 1976. Early Hydraulic Civilization in Egypt: A Study in Cultural Ecology. The University of Chicago Press, Chicago and London.

Butzer, K. W. 1980. The Holocene lake plain of North Rudolph, East Africa. *Phys. Geography* 1(1(: 42-58).

Butzer, K. W. 1984. Long-term Nile flood variation and political discontinuities in Pharaonic Egypt. In Clark, J. D. and Brandt, S. A. (eds.), From Hunters to Farmers, University of California Press, Berkeley, pp. 102-112.

Butzer, K. W. 1995. Environmental Change in the Near East and Human Impact on the Land. In Civilizations of the Ancient Near East, Jack M. Sasson, Editor in Chief, Vol. I, Part 2: 123-151, Scribner/s Sons, New York.

Caton-Thompson, Gertrude and Gardner, E. W. 1929. Recent work on the problem of Lake Moeris. *Geography Journal* 73: 20-60.

Casanova, J. and C. Hillaire-Marcel, 1992. Late Holocene hydrological history of Lake Tanganyika (Burundi). Bull. Soc. Géol. Fr. Sér 8, 6(4):647-656.

Caton-Thompson, Gertrude and Gardner, E. W. 1934. The Desert Faiyum. Royal Anthropological Society, London, 2 volumes.

COHOMAP Members, 1988. Climatic changes of the last 18,000 years: Observations and model simulations. *Science* 241, 1043-1052.

De Wit, Huib E. 1993. The evolution of the Eastern Nile Delta as a factor in the development of human culture. In Krzyzaniak, Lech, Kobusiewicz, Michal, and Alexander, John (eds.), Environmental Change and Human Culture in the Nile Basin and Northern Africa until the Second Millennium B.C., Poznan Archaeological Museum, Poznan, pp. 305-320.

Dittmann, Andreas. 1993. Environmental and climatic change in the northern part of the Eastern Desert during Middle Palaeolithic and Neolithic times. In Krzyzaniak, Lech, Kobusiewicz, Michal, and Alexander, John (eds.), Environmental Change and Human Culture in the Nile Basin and Northern Africa until the Second Millennium B.C., Poznan Archaeological Museum, Poznan, pp. 145-152.

El-Gably, Moustafa 1988. Egypt and the Nile. In Nilewater Crisis (in Arabic), Markaz al-Bihouth Al-Arabia, Cairo, pp. 30-39.

Flohn, H. and S. Nicholson, 1980. Climatic fluctuations in the arid belt of the 'Old World' since the last glacial maximum: possible causes and future implications. In Palaeoecology of Africa and the Surrounding Islands, E. F. van Zinderen & S. R. Bakker, Balkema, 12:3-21.

Gasse, F. and Van Campo In press. Abrupt post-glacial climate events in West Asia and North Africa monsoon domains. Earth and Planetary Science Reviews.

Grimal, Nicolas 1994. A History of Ancient Egypt. Blackwell, Oxford.

Griffiths, J. E. (ed.) 1972. Climates of Africa. World Survey of Climatology, Vol. 10, H. E. Landesberg, Editor in Chief, Eslevier, Amsterdam

Grove, A. T. 1993. Africa's climate in the Holocene. In Shaw, Thurston, Sinclair, Paul, Andah, Bassey, and Okpoko, Alex (eds.), The Archaeology of Africa: Food, Metals and Towns, Routledge, London and New York, pp. 32-42.

Hamilton, A.C. 1982. Environmental History of East Africa: A Study of the Quaternary. Academic Press, New York.

Hassan, Fekri. 1981. Historical Nile floods and their implications for climatic change. *Science* 212: 1142-1145.

Hassan, Fekri A. 1984. Radiocarbon chronology of Neolithic and Predynastic sites in Upper Egypt and the Delta. *African Archaeological Review* 3:95-116.

Hassan, Fekri. 1986a. Desert environment and origins of agriculture in Egypt. *Norwegian Archaeological Review* 19: 63-76.

Hassan, Fekri. 1986b. Holocene lakes and Prehistoric settlements of the Western Faiyum. *Journal of Archaeological Science* 13: 483-501.

Hassan, Fekri. 1988. The Predynastic of Egypt. Journal of World Prehistory 2 (2) 135-185.

Hassan, Fekri. 1993. Population ecology and civilization in Ancient Egypt. In Crumley, Carole L. (ed.), Historical Ecology: Cultural Knowledge and Changing Landscapes, School of American Research, Santa Fe, New Mexico, pp. 155-181.

Hassan, Fekri A. n.d. The dynamics of a riverine civilization an Egyptian perspective. Manuscript in preparation.

Hassan, Fekri and Stucki. B. 1987. Nile floods and climatic change. In Rampino, M. R., Sanders, J. E., Newman, W. S., and Konigsson, L. K. (eds.), Climate, History, Periodicity and Predictability, Van Nostrand Reinhold, New York, pp. 37-46.

Holmes, D. 1993. The rise of the Nile Delta. *Nature* 363:402-403.

Hoffman, Michael A. and Mills, James O. 1993. Problems of assessing environmental impact on the Predynastic settlements at Hierakonpolis. In Krzyzaniak, Lech, Kobusiewicz, Michal, and Alexander, John (eds.), Environmental Change and Human Culture in the Nile Basin and Northern Africa until the Second Millennium B.C., Poznan Archaeological Museum, Poznan, pp. 359-370.

Hoffmann, Michael A., Hamroush, H. A., Allen, R. 1986. A model of urban development for the Hierakonpolis region from Predynastic through Old Kingdom times. *Journal of American Research Center* 23: 175-187.

Hurst, H.E. 1957. The Nile. Constable, London 1957 revised edition.

Jolly, D. and R. Bonnefille, 1992. Histoire et dynamique du marécage tropical de Ndurumu (Burundi), données polliniques. *Review of Palaeobotany and Palynology* 75 (1992): 133-151.

Jolly, D., R. Bonnefille, and M. Roux, 1994. Numerical interpretation of high resolution Holocene pollen record from Burundi. *Palaeogeography, Palaeoclimatology, Palaeoecology*. 109:357-370.

Jux, U and T. Steuber 1990. Der Karun-See im ägyptischen Fayum, seine Sedimente im Spiegel der Geschichte. Naturwissenschaften 77: 262-270.

Kees, Hermann, 1961. Ancient Egypt, A cultural Geography. University of Chicago Press, Chicao.

Kemp, Barry J. 1983. Old Kingdom, Middle Kingdom and Second Intermediate Period c. 2686-1552 BC. In Trigger, B. G., Kemp, B. J., O'Connor, D., and Lloyd, A. B., Ancient Egypt: A Social History, Cambridge University Press, Cambridge, pp. 71-182.

Kemp, B. 1989. Ancient Egypt. Anatomy of a Civilization. London: Routledge.

Kendall, R.L. 1969. An ecological history of the Lake Victoria Basin. Ecol. Monograph 39:121-176.

Kozlowski, Janusz K. and Ginter, Boleslaw. 1993. Holocene changes in the Faiyum: Lake Moeris and the evolution of climate in Northeastern Africa. In Krzyzaniak, Lech, Kobusiewicz, Michal, and Alexander, John (eds.), Environmental Change and Human Culture in the Nile Basin and Northern Africa until the Second Millennium B.C., Poznan Archaeological Museum, Poznan, pp. 327-336.

Kutzbach, J. E. and Guetter, P. J. 1986. The influence of changing orbital parameters and surface boundary conditions on climate simulations for the past 18,000 years. *Journal of Atmospheric Science* 43 (16):1726-1759.

McDonald, Mary M. A.. 1993. Cultural adaptations in Dakhleh Oasis, Egypt, in the Early to Mid-Holocene. In Krzyzaniak, Lech, Kobusiewicz, Michal, and Alexander, John (eds.), Environmental Change and Human Culture in the Nile Basin and Northern Africa until the Second Millennium B.C., Poznan Archaeological Museum, Poznan, pp. 199-209.

Mehringer, Jr. P.J., Petersen, K.L., and Hassan, F. A. 1979. A pollen record from Birket Qarun and the recent history of the Fayum, Egypt. *Quaternary Research* 11, 238-256.

Mohammed, Umer. and R. Bonnefille, 1991. The recent history of vegetation and climate around Lake Langeno (Ethiopia). The Palaeoecology of Africa, Vol. 22:275-286.

Neumann, Katharina. 1993. Holocene vegetation of the Eastern Sahara: Charcoal from prehistoric sites. In Krzyzaniak, Lech, Kobusiewicz, Michal, and Alexander, John (eds.), Environmental Change and Human Culture in the Nile Basin and Northern Africa until the Second Millennium B.C., Poznan Archaeological Museum, Poznan, pp. 153-169.

O'Connor, David. 1991. Early states along the Nubian Nile. In Davies, W. V. (ed.), Egypt and Africa: Nubia from Prehistory to Islam, British Museum Press, pp. 145-165.

Raymond, André 1993. Le Caire. Librairie Arthème Fayarad, Paris.

Ritchie, J. C. and Haynes, C. V. 1987. Holocene vegetation zonation in the Eastern Sahara. *Nature* 330: 645-647.

Said, Rushdi 1993. The River Nile, Geology, Hydrology, and Utilization . Pergampn Press, Oxford.

Shahin, M. 1985. Hydrology of The Nile Basin. Elsevier, Amsterdam.

Simpson, William Kelly (ed.). 1972. The Literature of Ancient Egypt. Yale University Press, New Haven and London.

Talbot, M. R., Livingston, D. A., Palmer, P. G., Maley, J., Melack, J. M., Dilibrias, G. and Gulliksen, S. 1984. Preliminary results from sediment cores from Lake Bosumtwi, Ghana. Palaeoecology of Africa 16: 173-192.

Taylor, D. M. 1993. Environmental change in montane southwest Uganda: a pollen record for the Holocene from Ahakagyezi Swamp. *The Holocene* 3 (4) 324-332.

Umer Mohammed. 1992. Paléoenvioronments et palécoloclimatologie des derniers millénaires en Ethiopie. Contribution palynologique Univ. Aix-Marseille 3. Thèse Doct. Univ. Sci. 219 p.

Trigger, B. G., B.J. Kemp, D. O'Connor, and A.B. Lloyd, 1983. Ancient Egypt: A Social History. Cambridge University Press, Cambridge.

Vernet, Robert. 1993. Le Néolithique récent dans le sud-ouest du Sahara. In Krzyzaniak, Lech, Kobusiewicz, Michal, and Alexander, John (eds.), Environmental Change and Human Culture in the Nile Basin and Northern Africa until the Second Millennium B.C., Poznan Archaeological Museum, Poznan, pp. 91-101.

Vernet, Robert 1995. Les Paleoenvioronments du Nord de l'Afrique depuis 600,000 Ans. 149-278. Dossiers et Recherches sur l'Afrique, No.3. Centre National de la Recherche Scientifique, Meudon-Cedex

Waterbury, John 1979. Hydropolitics of the Nile Valley. Syracuse University Press.

Willcocks, W. and Craig, J. I. 1913. Egyptian Irrigation, 3rd edition, Vols. 1 &2. E& F.N. Spon Ltd, London.

Wunderlich, Jürgen. 1989. Untersuchungen zur Entwicklung des westlichen Nildeltas im Holozän. Gedruckt bei Wenzel, Marburg.

Zakry, Antoune 1926. The Nile in the the times of the Pharaohs and the Arabs (in Arabic). Maktabet Al-Ma'arif, Cairo (reissued in 1995 by Maktabet Madbouli, Caior).

Environmental Change and Human Adaptational Failure at the End of the Early Bronze Age in the Southern Levant

Arlene M. ROSEN

Dept. of Archaeology
University of Calgary
2500 University Drive, NW
Calgary, Alberta T2N 1N4
Canada

Abstract

Failure to adapt to changing climatic conditions can be as much a social problem as a technological one. This is illustrated by the collapse of Early Bronze Age society in the southern Levant. The agricultural system which supported these socially stratified communities was adapted to floodwater farming of cereals and the production of cash crops including olives and grapes. The climatic change at the end of the third millennium BC resulted in a decrease in precipitation which led to lowered stream base levels, and the hydraulic regime changed from an aggrading system to one of stream incision. Floodwater farming was no longer possible, yields became lower and less predictable, leading to the collapse of the agricultural sector. Technological change to a system of canal irrigation from springs would have maintained higher yields, and such a system was indeed used by Middle Bronze Age inhabitants in the same marginal environment. Although canal irrigation was known, it was not implemented as an adaptive measure by the Early Bronze Age society. Some of the reasons for this failure to adapt might include slow response time in the perception of catastrophe, the ability of the elite managers to profit from short-term environmental stress, and the direction of energy toward increased religious activity rather than technological change.

Introduction

The aim of this volume is to assemble evidence for climatic change from the Near East and surrounding regions for the middle Holocene and relate this evidence to responses on the part of ancient human societies. The evidence for desiccation in the Levant at the end of the third millennium BC coincides with indications of drought in

NATO ASI Series, Vol. I 49
Third Millennium BC Climate Change
and Old World Collapse
Edited by H. Nüzhet Dalfes, George Kukla,
and Harvey Weiss

some of the other regions in the vicinity, notably Northern Mesopotamia and Egypt. This environmental change had an undeniable impact on the complex agricultural based societies of these regions. However, the human responses to these episodes of agricultural stress varied. In some instances the cultural and technological systems were preadapted to survive the impact of the droughts, in others adjustments were made which contributed to a quick recovery from the social stress and political changes caused by the desiccation, but in many cases societies failed to adapt and thus suffered massive population decline and collapse. These differences in past human responses are a most valuable source of information for those of us interested in learning how populations cope with drought situations today. In this paper I present a case study of the Early Bronze Age societies in the Southern Levant which illustrates failure to adapt to a changing environment. some of the geomorphological and isotopic evidence for mid-Holocene environmental change in the southern Levant and discuss the possible reasons for it's catastrophic impact on the human societies living in the region at the time.

Present Environmental Setting

The region today is a semi-arid zone with a Mediterranean climate characterized by warm, dry summers and cool, wet winters. Summer temperatures average around 24° C in the hilly regions to 25° C on the coastal plain. Winter temperatures average about 9° C in the hills to ca. 12° C on the coastal plain. The average yearly rainfall varies from up to 1100 mm in the Upper Galilee down to the 50 mm in some parts of the Negev Desert. The precipitation occurs almost exclusively between the months of October through April.

Topographically the region is quite diverse with the Judean Hills forming a mountainous ridge to the east. A foothill zone (Shephela) borders this ridge on the west and gradually converges with the flat coastal plain of the Mediterranean littoral. This topographical variance allows for a diverse mixture of agricultural and pastoral products today, and the same was true in the Early Bronze Age (Orni and Efrat 1971).

Mid-Holocene Paleoenvironments in the Southern Levant

There are several lines of evidence which attest to a moister climate at the beginning of the middle Holocene period than that of today. According to this information the moister climate in the middle Holocene shifted to the present-day semi-arid regime some time close to the end of the third millennium BC In order to illustrate this environmental change I will discuss proxy data which come from geomorphological sources and isotope studies of land snails. The geomorphological evidence attests to both environmental changes which occurred at the end of the third millennium BC as well as the nature of landscape changes at this time.

Geomorphological Evidence

A survey of abandoned stream terraces was conducted by Rosen (1986, 1991) in south central Israel at the boundary between the coastal plain and foothill zones. Several wadi systems were investigated including Nahal (Wadi) Lachish, Nahal Shiqma, and Nahal Adorayim (see Figure 1). These systems exhibited a suite of Late Pleistocene stream terraces dated by archaeological material and radiocarbon dates respectively to the Mousterian period, the Upper Paleolithic period (26,000 BP), and the Epipaleolithic period (13,000 BP). The episodes in which the terraces formed are interpreted as moist phases in which the streams exhibited an aggradational regime. They are characterized by units of well sorted gravels and fine-grained alluvium, in some cases bearing the signs of truncated paleosols. Each group of terraces is in turn followed by an extensive erosional phase.

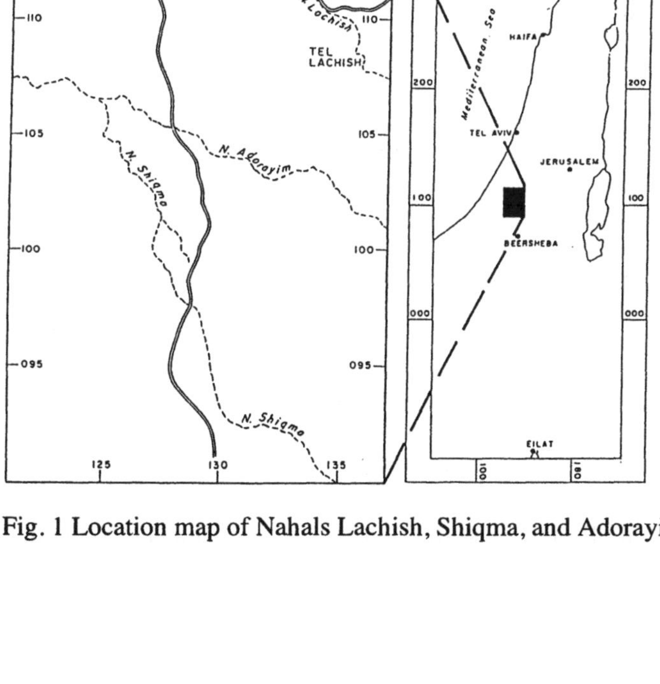

Fig. 1 Location map of Nahals Lachish, Shiqma, and Adorayim.

Two Holocene terraces are evident and are primarily dated by archaeological inclusions. These are the Chalcolithic/EB terrace of interest in this study and the Byzantine terrace (see Fig. 2). The Chalcolithic/EB terrace is a relatively low formation, generally no more than two meters high. In the vicinity of the aforementioned drainages it is characterized by two different lithologies. The earlier Chalcolithic deposits display very fine laminated silts which vary between horizontal and cross laminations. Another phase exhibited in this terrace contains channel gravels and massive silts. The later EB deposits are more consistent with the gravel and massive silts of the Chalcolithic deposit (Figure 3). This phase of alluviation created overbank flow and active floodplains in valleys which at the present time have incised channels with resticted flows. A good example of these landscape changes is the Nahal Lachish drainage in the vicinity of Tel Erani (Fig. 4) (Rosen 1991).

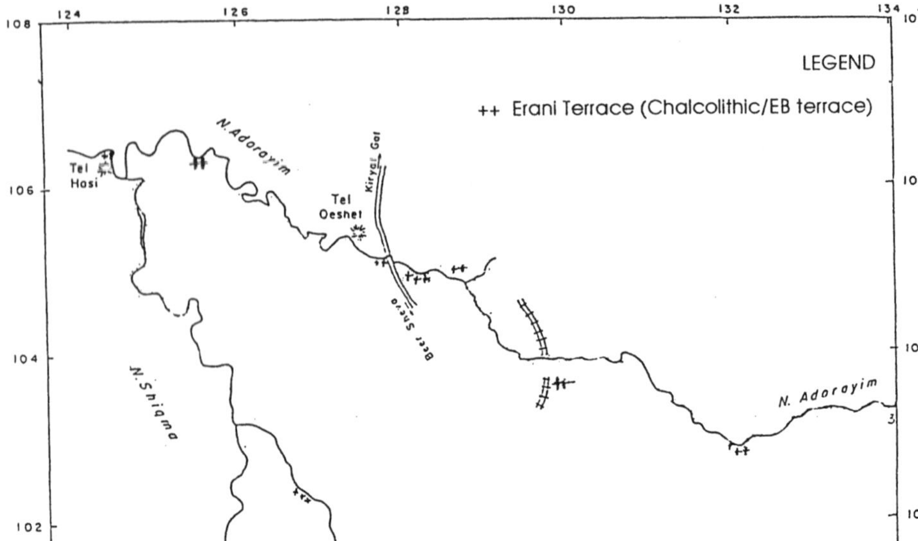

Fig. 2 Location of Chalcolithic/EB terrace systems in Nahals Adorayim and Shiqma.

The Mid-Holocene terraces represent fluctuations in the stream regimes over the two thousand year period of the Chalcolithic and Early Bronze Age. These fluctuations are attributed to the effects of changing rainfall patterns rather than the influence of human-induced landscape changes based on the character and lithology of the sediment units. Human-induced alluviation caused by the effects of deforestation and collapsed agricultural terraces takes on the character of poorly sorted deposits with a large component of angular gravels. This is inconsistent with the fine laminations and good sorting of the Mid-Holocene deposits. Similar deposits of Chalcolithic and EB age were described in the Negev Desert (Goldberg 1986, Goldberg and Rosen 1987, Goodfriend 1987). This is further evidence for climate induced stream changes, rather than human interference since agricultural activity along most of the Negev drainages was negligible at this time.

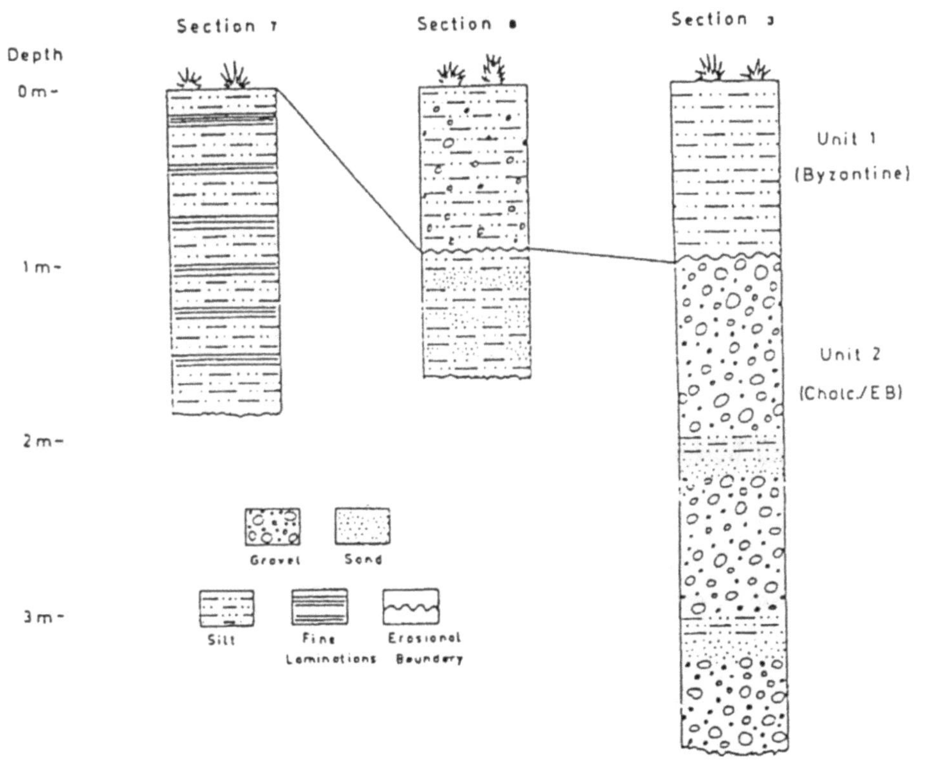

Fig. 3 Sections of Chalc./EB terraces from Nahal Lachish in the vicinity of Tel Erani.

Increased rainfall and higher sediment loads are most likely responsible for the existence of these mid-Holocene terraces. The fine laminations of the earlier terrace phases suggest well-sustained flows which resulted from rainfall episodes that were evenly distributed throughout the season or even throughout the year. This is much different from the present-day rainfall regime which consists of heavy downpours over a short period of time, inducing flash flooding. The alluvial phase represented by this terrace system ceased at the end of the third millennium BC and was replaced by a very long episode of stream deggradation.

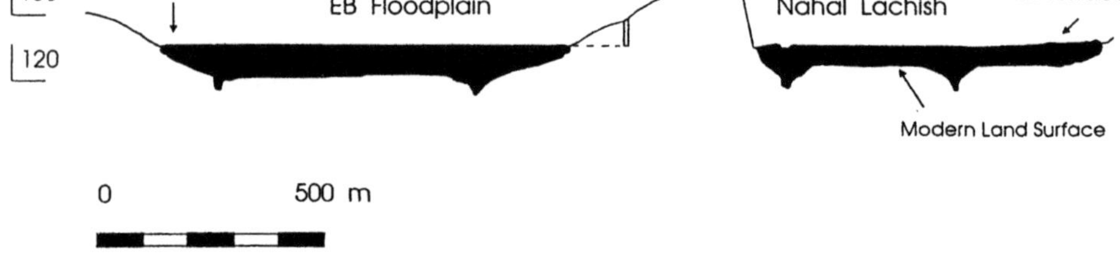

Fig. 4 Reconstruction of the Early Bronze Age floodplain near Tel Erani.

In recent geoarchaeological field work we identified an almost identical geomorphological occurrence at the Early Bronze Age site of Titrish near the town of Sanlıurfa in southeastern Turkey (Rosen and Goldberg, n.d.). Here, as in the southern Levant, we described an alluvial terrace system dating to the middle-late third millennium BC The height and lithology of this terrace suggested a higher stream base level and a more steady flow resulting from a moister rainfall regime. This terrace is contemporary with the florescence of Early Bronze Age occupation at the site of Titrish and vicinity. The drop in base level and subsequent stream incision and in one case channel abandonment, coincided with the demise of the urban occupation of the town, and general desertion of the locality.

The significance of these terraces goes beyond that of a form of proxy data for climatic change. The landscape alterations which took place both during the phase of alluviation and at its end were of greater import to the populations inhabiting the Southern Levant at the time, than the difference in rainfall quantities. During the phase of wadi alluviation the floodplains of drainages in the Southern Levant were actively aggrading and provided moist alluvial soils in which to cultivate cereals and produce high dependable yields. This was an essential component of Early Bronze Age agriculture because it allowed the production of surplus food supplies to feed the non-agricultural sector of the population and assured a supply of stored food to help bridge the normal occurrence of drought years (Rosen, n.d.).

Supporting Evidence

Some of the other evidence for climatic change in the Southern Levant at the end of the third millennium BC includes the analyses of $^{13}C/^{12}C$ ratios of organic matter in fossil land snails which indicates if the snails were subsisting on a vegetation community composed primarily of C_3 or C_4 plants. Goodfriend (1990) reported a more southerly distribution of C_3 plants for a period from ca. 6500-3000 BP. This indicates rainfall values for the Negev desert which are almost double those of the present for this mid-Holocene period.

In another study, Goodfriend (1991) demonstrated that mid-Holocene land snail shells contained depleted values of $\delta^{18}O$ compared to those of the Early Holocene and the present-day values. This is interpreted as indicating a shift in the atmospheric circulation pattern, and a resulting increase in storm frequencies originating from northeastern Africa. This mid-Holocene moist phase returns to the modern drier conditions between about 5000 and 3500 BP.

Lake cores from the Dead Sea also show a phase of increased runoff and higher lake levels in the Mid-Holocene (Neev and Emery 1967). This phase ends abruptly with a sharp increase in evaporites indicating a lower lake level and drier environment. Issar et al. (1992) combine ^{13}C and ^{18}O values from a Sea of Galilee core and Galilee speleothems to demonstrate that there were cool moist episodes from ca. 5700 - 5200 BP and again at ca. 4600 - 4200 BP (calibrated) corresponding respectively to the Chalcolithic period and Early Bronze Age. After 4200 warm/dry condition set in.

The above evidence strongly suggests changes in rainfall patterns which led to a moist episode in the middle Holocene that coincided with the rise of the first city-states in the Southern Levant. These conditions allowed the development of a diverse agricultural system including the production of cash crops and surplus subsistence products. This more benevolent climate switched to a drier regime, and the resulting landscape changes altered the agricultural system by removing the buffer against dry years that floodwater farming provided.

Early Bronze Age Adaptations

Climatic change can be a critical source of stress to a society, but we cannot assume that it will automatically and inevitably initiate the collapse of a civilization. Human societies can be considerably resilient in the face of external environmental stress. They can adapt to the changing environment through technological innovations which improve agricultural yields and increases the efficiency of storage, as well as adjust aspects of social and political organization to allow more efficient systems of food distribution. The more inflexible the society, the less likely that its adaptation will be successful. In the Near Eastern example, some city-states failed to adjust to the environmental disruption, and others continued to function after readjusting their technology and/or political organization. The point of interest in these events is the understanding of why some societies collapsed and others continued to function. It is therefore necessary to examine the mechanisms which contributed to the successful adjustment or the collapse of each particular civilization. In this study I examine the adaptive failure of the Early Bronze Age towns of the Southern Levant.

By the middle of the third millennium BC Early Bronze Age civilization in the southern Levant had developed into a secondary state society consisting of a tiered social hierarchy. The sites consisted of city-states in which the elite ruling classes obtained wealth and power from trade revenues gained from the production and export of agricultural products, primarily olive oil and wine (Stager 1985). The agrarian sector of subsistence farmers were integrated into the social network through redistribution of goods, taxation of agricultural produce, and a strong temple cult (Esse 1989, 1991, Joffe 1993; Kempinski 1978, 1989, de Miroschedji 1989).

Although there were several large population centers located in the hill country and the northern Negev desert, many of the important sites were located along the low foothills at the eastern edge of the coastal plain, (Gophna and Portugali 1988; Palumbo 1990: 44; Finkelstein and Gophna 1993). This locale allowed the settlements to exploit both the hill country for tree crops and the alluvial coastal plain for the production of cereals. The cash crops (olives and grapes) were presumably cultivated for the most part in the moister hill regions where the rocky soils are difficult to plow, whereas the subsistence crops (wheat and barley) were more easily produced on the lowland soils. The agricultural base of this society served not only to provide sustenance for the population, it also established a power base for the ruling classes.

Evidence from phytolith analysis strongly suggests that much of the lowland cultivation of cereals was conducted by floodwater farming the moist alluvial soils (Rosen and Weiner 1994). This type of farming was essential for acquiring high enough yields to support a large settled population in an environment which suffered from droughts on a regular basis (Rosen, n.d.). It also allowed for the storage and redistribution of surplus cereals, as indicated by the immense granaries at the site of Beit Yerah. Without this agricultural buffer, the society could not support its large population including non-producers who constituted the classes of administrators, craft specialists, and other laborers.

The climatic change which occurred in the late third millennium BC had a very marked influence on the hydrology of the Southern Levantine stream systems. Base levels dropped and the hydrological regime shifted from an aggradational to a degradational system. This downcutting left the previously well-watered floodplains dry and unirrigated. The Early Bronze Age subsistence farmers were therefore no longer able to exploit the lowland valleys with simple floodwater farming. Thus an important agricultural resource was eliminated from exploitation under the existing technological system.

Social Collapse in the Early Bronze Age

It is difficult to assign an exact date to the final occupation of each site, but the abandonment of EB III centers began in the early phases of the period with the desertion of major towns in the north of Israel. By the end of the third millennium, Early Bronze Age complex society had collapsed and most of the cities and towns in the southern Levant were abandoned (Seger 1989). Although the collapse of an agricultural system may appear inevitable given the above described landscape changes, the population did have the option of instituting technological changes which would have allowed them to recover from the initial impact of the environmental change. An example of one such innovation would be the institution of canal irrigation from upland springs. However, there appears to have been no attempt to transport water to the abandoned flood plains of the lowland valleys. The next large city-states to inhabit this region in the Middle Bronze Age did in fact employ canal irrigation (Mazar 1990).

A partial explanation for the failure of Early Bronze Age society to recover might lie in the fact that the elite managers were able to control the population through their function as food distributers. For them, the occasional drought strengthened their power by increasing the number of dependants and allowing them to acquire land and labor at low cost. This effect has been demonstrated in modern ethnographic situations (Rao 1974). This proved to be a successful adaptation to short-term drought, but may have slowed awareness of and response time to a more severe situation of long-term desiccation.

In a previous paper (Rosen, n.d.) I argued that the response of a given population to environmental change depends to a great extent on the belief systems, decision-making structure, and the way the environmental change is perceived by the society (see also Rappaport 1977). The response of a community to environmental stress is likely to be consistent with their framework for explaining natural phenomena. In the case of the Early Bronze Age peoples of the Southern Levant, technological innovation was perhaps not considered to be the most effective reaction to climatic change. With a fatalistic perspective on natural events in which the deities had totalitarian control over nature, the logical response would have been to direct the group energy to the propitiation of these deities. This would manifest itself physically in the construction of temples and cultic sanctuaries (see McGovern 1981 for an historical account of such reactions). Archaeologically, the last phases of the Early

Bronze Age are indeed typified by increased occurrences of temples and other cultic structures. There is no evidence of any attempt to dam streams or create canal systems.

The resulting social collapse was complete throughout the Southern Levant. The cities and town sites were abandoned and the diminished population retreated to the hinterlands where they settled in small village and pastoral camps primarily around perennial springs. The exception to this new settlement pattern were a few towns in Transjordan which were not abandoned at the end of the EB III. These towns were located on the banks of perennial streams (Prag 1974; Richard 1986) which presumably allowed the population to continue to farm the locality without significant disruption to yearly grain yields.

The adaptive failure on the part of the Southern Levantine population can be contrasted with the response of the Egyptian Old Kingdom society which was faced with drastically reduced Nile floods at this same time period (see Hassan, this volume). The low Nile floods initiated the First Intermediate Period, a time of political disruption and social chaos. This was a major factor in the collapse of the civilization, but the fall was not permanent. After a relatively short time political control was regained by new rulers who instituted social reform and technological innovations, a feat not achieved by the contemporary rulers of the Levantine states.

Conclusions

The collapse of Levantine Early Bronze Age complex society was an environmentally induced, but not environmentally determined event. When viewed diachronically we have a case study of an ancient economy based on the production of cash crops which supported a tiered social hierarchy. Their adaptation to a marginal farming environment successfully combined floodwater farming to buffer the periodic drought years, and a system of food distribution in times of famine. With the onset of extreme long-term desiccation, this system failed. Although relatively minor technological adjustments, such as the institution of canal irrigation, might have allowed the continuation of the Early Bronze Age economic and social organization, this option was not taken. Instead, there was a major depopulation and a re-adaptation in the way the population utilized the environment. The society transformed into a subsistence based economy of small village farmers and pastoralists.

A synchronic view of the effect of desiccation on other Near Eastern societies shows us that not all communities were similarly affected. Substantial city states continued to exist in areas of Transjordan with no apparent discontinuity. The Egyptian Old Kingdom civilization in the Nile Valley fell with the impact of low Nile floods, but rapidly recovered under the direction of the Middle Kingdom rulers. Instead of a monolithic pattern of civilizations collapsing in the wake of desiccation, there was a mosaic of societies that have heterogeneous responses and varying degrees of adaptational successes or failures. This variablility can perhaps guide us in understanding some of our own successes and failures in the face of modern climatic alterations.

Acknowledgments

I would like to thank Nüzhet Dalfes, George Kukla, and Harvey Weiss for organizing this most informative workshop. This paper benefited from discussions with Fekri Hassan, and the editorial comments of Steven Rosen. Yaniv Rosen assisted in the preparation of Fig. 3.

References

Esse DL (1989) Secondary state formation and collapse in Early Bronze Age Palestine. In *L'Urbanisation de la Palestine à l'Âge du Bronze Ancien*, edited by Pierre de Miroschedji, pp. 81-96. British Archaeological Reports, International Series, 527. BAR, Oxford.; Esse, Douglas L.

Esse DL (1991) *Subsistence, Trade, and Social Change in Early Bronze Age Palestine*. Studies in Ancient Oriental Civilization 50, Thomas A. Holland, general editor. University of Chicago, Chicago.

Finkelstein I. Gophna R (1993) Settlement, demographic, and economic patterns in the highlands of Palestine in the Chalcolithic and Early Bronze Periods and the beginning of urbanism. *Bulletin of the American Schools of Oriental Research* 289: 1-22.

Goldberg P (1986) Late Quaternary environmental history of the Southern Levant. *Geoarchaeology* 1: 225-244.

Goldberg P. Rosen AM (1987) Early Holocene paleoenvironments of Israel. In *Shiqmim I*, edited by TE Levy, pp. 23-33. BAR International, Oxford.

Goodfriend GA (1987) Chronostratigraphic studies of loessial sediments in the Negev Desert, using amino-acid epimerization analysis of land snails. *Quaternary Research* 28: 374-392.

Goodfriend GA (1990) Rainfall in the Negev Desert during the Middle Holocene, based on ^{13}C of organic matter in land snail shells. *Quaternary Research* 34: 186-197.

Goodfriend GA (1991) Holocene trends in ^{18}O in land snail shells from the Negev Desert and their implications for changes in rainfall source areas. *Quaternary Research* 35: 417-426.

Gophna R. Portugali J (1988) Settlement and demographic processes in Israel's coastal plain from the Chalcolithic to the Middle Bronze Age. *Bulletin of the American Schools of Oriental Research* 269: 11-28.

Issar AS. Govrin Y. Geyh MA. Wakshal E. Wolf M (1992) Climate changes during the Upper Holocene in Israel. *Israel Journal of Earth Sciences* 40: 219-223.

Joffe, AH (1993) *Settlement and Society in the Early Bronze Age I and II, Southern Levant*. Monographs in Mediterranean Archaeology 4, A. B. Knapp, general editor. Sheffield Academic Press, Sheffield.

Kempinski A (1978) *The Rise of an Urban Culture: The Urbanization of Palestine in the Early Bronze Age*. Israel Ethnographic Society, Jerusalem.

Kempinski A (1989) Urbanization and metallurgy in southern Canaan. In *L'Urbanisation de la Palestine à l'Âge du Bronze Ancien*, edited by P. de Miroschedji, pp. 163-168. BAR, International, Oxford.

McGovern TH (1981) The economies of extinction in Norse Greenland. In *Climate and History: Studies in Past Climates and their Impact on Man*, edited by TM. Wigley, MJ Ingram, and G Farmer, pp. 404-433. Cambridge University, Cambridge.

Mazar A (1990) *Archaeology of the Land of the Bible: 10,000-586 B.C.E.* Doubleday, New York.

de Miroschedji P (1989) Le processus d'urbanisation en Palestine au Bronze ancien: chronologie et rythmes. In *L'Urbanisation de la Palestine à l'Âge du Bronze Ancien*, edited by P. de Miroschedji, pp. 63-79. BAR, International, Oxford.

Neev D. Emery O (1967) *The Dead Sea: Depositional Processes and Environments of Evaporites*. Geological Survey of Israel, Bulletin 41. Geological Survey of Israel, Jerusalem, N.

Orni E. Efrat E (1971) *Geography of Israel*. Jewish Publication Society of America, Philadelphia.

Palumbo G (1990) *The Early Bronze Age IV in the Southern Levant: Settlement Patterns, Economy, and Material Culture of a 'Dark Age'*. Contributi e Materiali di Archeologia Orientale 3. Universita Degli Sudi Di Roma La Sapienza, Rome.

Prag K. (1974) The Intermediate Early Bronze-Middle Bronze Age: An interpretation of the evidence from Transjordan, Syria and Lebanon. *Levant* 6: 69-116, N.

Rao NVK (1974) Impact of drought on the social system of a Telengana village. *The Eastern Anthropologist* 27: 299-314.

Rappaport RA (1977) Maladaptation in social systems. In *The Evolution of Social Systems*, edited by J. Friedman, and M.J. Rowlands, pp. 49-71. Duckworth, London.

Richard S. (1986) Excavations at Khirbet Iskander, Jordan: A glimpse of settled life during the "Dark Age" in Palestinian archaeology. *Expedition* 28: 3-12.

Rosen AM (1986) *Quaternary Stratigraphy and Paleoenvironments of the Shephela, Israel*. Geological Survey of Israel, Reports 25/86. Geological Survey of Israel, Jerusalem.

Rosen AM (1991) Early Bronze Age Tel Erani: An Environmental Perspective. *Tel Aviv* 18: 192-204.

Rosen AM (n.d.) The social response to environmental change in Early Bronze Age Canaan. *Journal of Anthropological Archaeology*. In press.

Rosen AM. Goldberg P (n.d.) Paleoenvironmental investigations. In Algaze et al., Titris Hoyuk, a small EBA urban center in southeastern Anatolia: The 1994 season. *Anatolica*. In press.

Rosen AM. Weiner S (1994) Identifying ancient irrigation: a new method using opaline phytoliths from emmer wheat. *Journal of Archaeological Science* 21: 125-132.

Seger JD (1989) Some provisional correlations in EB III stratigraphy in southern Palestine. In *L'Urbanisation de la Palestine à l'Âge du Bronze Ancien*, edited by P. de Miroschedji, pp. 117-135. B.A.R., Oxford.

Stager LE (1985) The firstfruits of civilization. In *Palestine in the Bronze and Iron Ages: Papers in Honor of Olga Tufnell*, edited by J. Tubb, pp. 172-188. Institute of Archaeology, London.

Evidence for Mid-Holocene Environmental Change in the Western Khabur Drainage, Northeastern Syria

Frank HOLE

Department of Anthropology
Yale University
New Haven, CT, USA

Abstract

Vegetation on the semi-arid steppe of northern Syria responds quickly to fluctuations in patterns of precipitation and even small shifts may preclude subsistence agriculture. Archaeological survey in the western Khabur drainage has revealed periodic human settlement that responded to these fluctuations. During the late fifth-early fourth millennium B.C. the wadis held water perennially, indicating a more favorable water balance than today. Sites of this period were then buried beneath two meters of sands and gravels before the surface stabilized in the early third millennium. During this period, the numerous sites as large as 30 hectares indicate excellent conditions for agriculture. Shortly after 2500 B.C., all these sites were abandoned and the region was not again resettled until the first millennium. Subsequent occupations occured during the Roman/Byzantine and Islamic periods (after A. D. 622). The short periods of settlement were each separated by 1000 or more years of abandonment. After Medieval times, until recently, the region was occupied primarily by nomadic tribes of sheep herders. Changes in precipitation patterns probably account for some of these changes in the landscape and human settlement.

Introduction

Situated in the northeaast corner of Syria, the Khabur River drainage heads in the mountains of Turkey and ends at the Euphrates River near Deir ez Zor, crossing from north to south, fertile, well-watered agricultural land, and semi-arid and arid steppe (Fig. 1). Because of the combination of low precipitation and high interannual variability, traditional, subsistence farming was restricted to the northern portion of the basin or along the floodplain of the river. Only the Khabur River and its major tributary the Jaghjagh held water perennially and the numerous wadis were strictly

NATO ASI Series, Vol. I 49
Third Millennium BC Climate Change
and Old World Collapse
Edited by H. Nüzhet Dalfes, George Kukla,
and Harvey Weiss

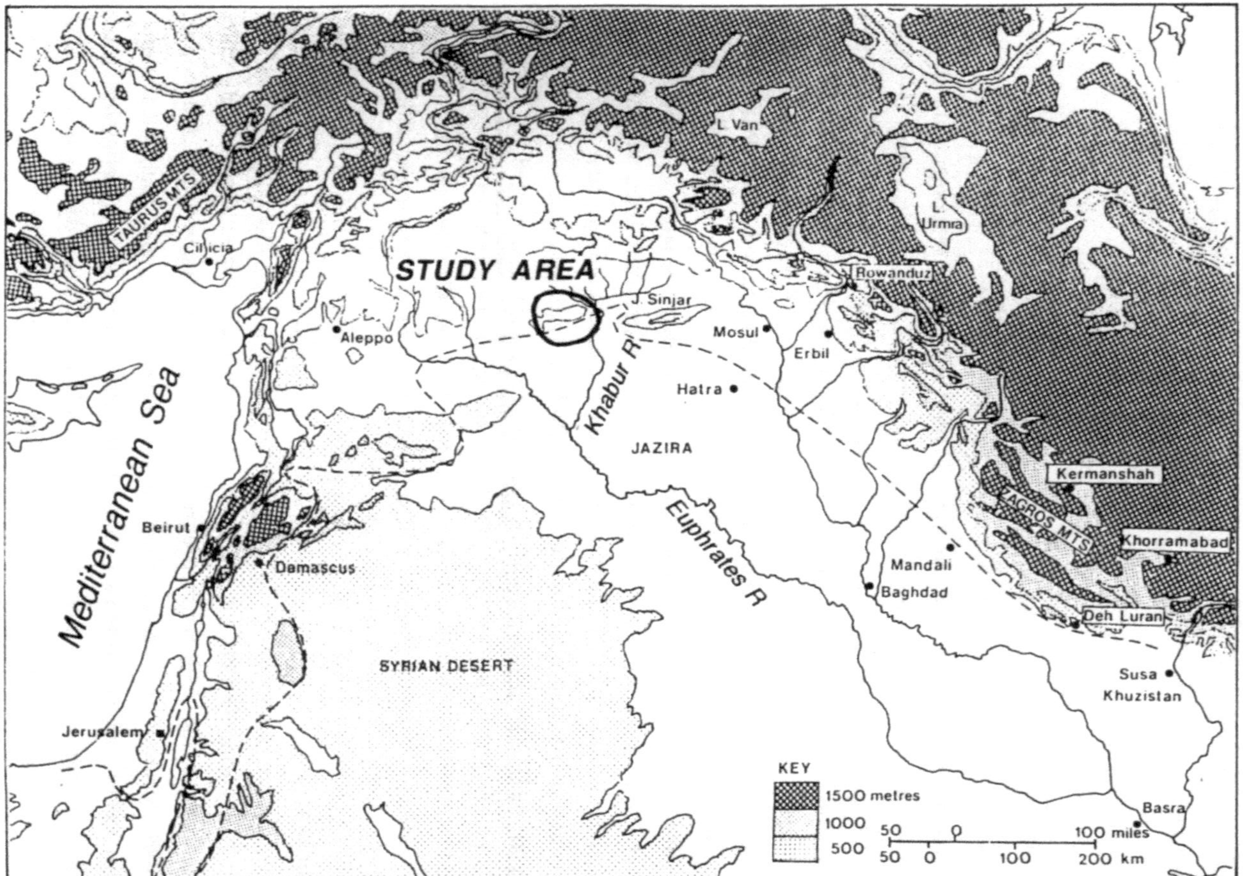

1. The study area of the Khabur drainage in northeastern Syria.

seasonal. Nomadic people occupying the semi-arid steppe depended on winter rains and wells for domestic needs and their livestock. Few people actually lived year-round in this region.

Syria, like the rest of the northern hemisphere, was affected by the early Holocene Climatic Optimum, which raised summer temperatures and affected the amount and distribution of rainfall. Although the timing and specific effects in the Khabur region remain to be determined, Courty sees evidence of summer precipitation engendered by increased insolation and shifts in the storm tracks. By 6000 BP, a Mediterranean climate, dominated by dry summers and cool to cold winters, was pervasive throughout the Near East. Apparently there were no other major global events but a small number of low amplitude perturbations such as the Little Ice Age that had purely local effects have been detected. As mid-to-late Holocene records are examined in more detail it becomes apparent that these small-scale changes were frequent, often abrupt, and some lasted up to 300 years. Examples of these are in records as disparite as snow accumulation in the GISP2 core (Meese, Gow et al. 1994) , methane in the GRIP ice core (Blunier, Chappellaz et al. 1995) rings in American bristlecone pine trees (Feng and Epstein 1994) , dust in Arabian Sea cores (Sirocko, Sarnthein et al. 1993), lake levels throughout the mid-latitudes (Harrison, Prentice et al. 1993; Magny 1993; Gasse and Van Campo 1994), and levels of the Dead Sea. These records all indicate significant periods of climate change in the Holocene, even though the changes are not entirely synchronous or in the same direction. In fact, one would not expect exact synchroneity or effect when the sites in question are under the influence of entirely different weather systems. Rather, local conditions determine the precise effects, but episodic and perhaps periodic fluctuations occurred throughout the Holocene. In view of the consistency of all these records in recording sharp changes, it is probable that some, if not all, affected settlement in the Khabur region.

Our archaeological surveys have discovered evidence that conditions were substantially better for agriculture and human settlement at several times in the past than they are now. These brief periods of settlement may coincide with the widespread pulsations of climate that are seen in the various records of the Holocene.

Briefly, we discovered two instances where remnants of early fourth millennium B.C. sites have been buried by silts, sands and gravels, and the sites themselves may have been effaced by violent flooding. A similar scenario has been proposed for the

Tell Leilan region, where Courty's micromorphological studies indicate a period of upland erosion, violent flooding and the destruction of sites (Courty lecture 1993). In her detailed report on the geomorphology of sites in the Wadi Awedj, north of the Khabur River, she denotes this period as Phase 3. She sees strong erosion and the incision of the wadis early in the Phase, with concomitant deposition of these sediments elsewhere in the drainage. By the end of Phase 3 she sees a stable landscape, no longer subject to erosion, and more favorable agricultural conditions than today (Courty 1994:45-47). With her limited sampling, she could date these events only within the range of cultural Late Ubaid to Late Uruk, roughly 5000-3000 B.C. The sites we report provide a *terminus ante quem* for the erosion and deposition event at just before 4000 B.C. These buried sites are evidence for a period of conditions considerably more favorable for settlement than those of the period immediately following, for which we have no evidence of settlement. In other words, it appears that as late as 4000 B.C. there was perennial water in wadis in the semi-arid steppe, but a sharp change in precipitation resulted in strong erosion that stabilized sometime before 3000 B.C. In as much as the regions examined by Courty and ourselves are in different parts of the Khabur drainage, the cycles of erosion and depositon may be somewhat out of phase, but in all likelihood were caused by the same climatic events.

A second, brief period of intensive settlement occurred during the early-to-mid-third millennium BC, after which this arid region was abandoned until the first millennium, and then settled only sporadically until modern times. The archaeological evidence suggests that the major periods of settlement were separated by 1000-1500 year intervals of abandonment. These data bear on interpretations of the growth and demise of northern cities such as Tell Leilan and Mozan, settlements along the Middle Khabur, and even those on the Euphrates.

The Archaeological Survey

These results derive from the Yale University Khabur Basin Project survey of a large region extending south from the Khabur River to the Jebel Abd al-Aziz and south of the Jebel to approximately the 36th parallel (Figs. 1,2). The survey reached as far west as a line south from Ras al 'Ain. This region had not been systematically surveyed previously, although Oppenheim reported some of the larger sites (Moortgat Correns 1972), Van Liere and Lauffrey examined historic settlements in

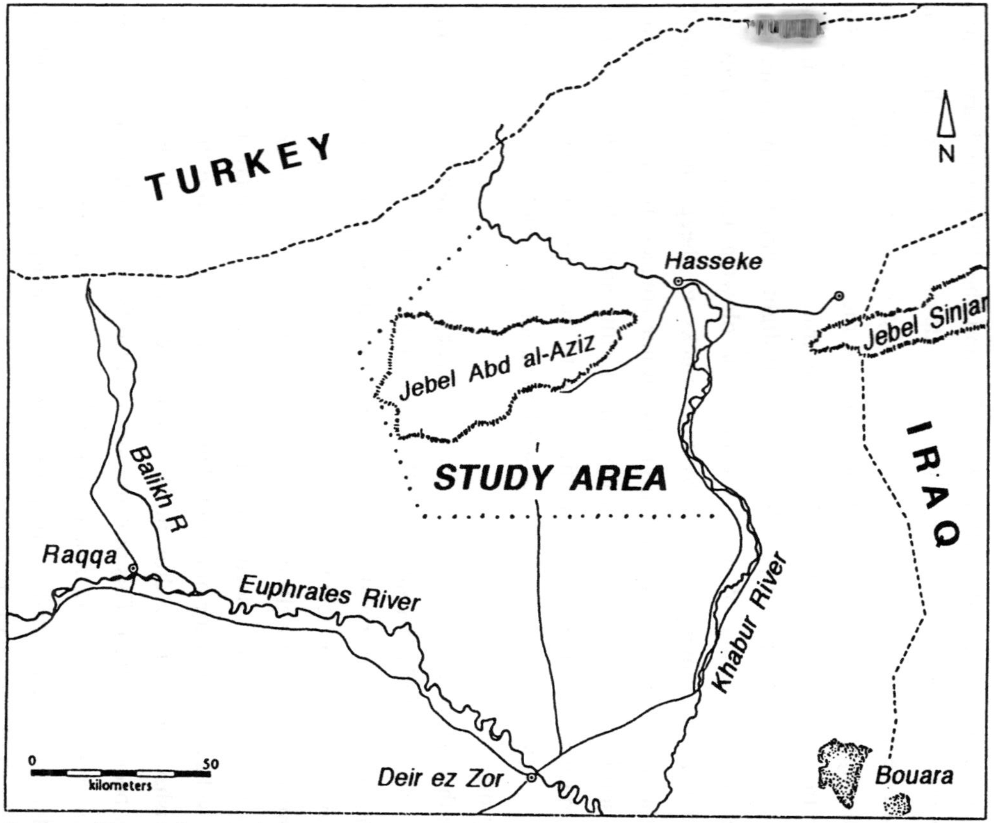

2. Location of the study area relative to the major river valleys and modern political boundaries. The survey was carried out in the semi-arid steppe surrounding the Jebel Abd al-Aziz.

relation to geography (Liere and Lauffray 1954-55), and the Tübingen Survey also visited some sites south of the Jebel. Until the past decade or two this region has been occupied nearly exclusively by migratory Bedouin of the Baggara tribe, one section of which primarily used the north side of the Jebel and the other the south side. For the most part this land was not cultivated. It lies in the transitional zone between dry farming and grazing land, and is especially sensitive to slight changes in precipitation. With the recent introduction of tractors capable of plowing tracts hundreds of hectares in extent, the moving of water by tank trucks to sheep and agricultural camps, and the use of diesel pumps to tap ground water sources, the region is now dotted with small hamlets and nearly all the land is cultivated. Dry wells scattered across the landscape attest to former practices and to the currently rapidly falling water table. Few flowing springs in this region yield sweet water and these are on the south side of the Jebel.

Although this entire area is under cultivation today, agriculture can be carried out on this scale only with full mechanization. While sufficient rain falls to produce crops in some years, interannual variability is excessive for subsistence farming (Figs. 3,4). The lack of available water precludes settlements unless they can get water by truck or diesel pumps. Abandoned wells exist in the vicinity of many of these ancient sites. Stone-lined and up to 9 meters deep, none of these produces water today. Indeed, it is notable that wells at some of the modern villages have failed and those that still exist are all brackish. As a result of a drop in the water table, some recently built villages have already been abandoned. This probably has resulted from excessive pumping of ground water, but the picture of abandoned villages reinforces the impression that this land does not yield readily to permanent settlement, even with modern technology. In short, this is a region that responds quickly and dramatically to even slight shifts in precipitation and ground water level.

The Jebel abd al-Aziz is the headwaters for all of the wadis that currently flow through this region. Today the mountain ridge is nearly devoid of trees, but one hundred years ago it was well-forested and in the past it was considered to be the primary source for terebinth, a product of the pistachio trees. In the absence of wood cutting and heavy grazing, botanists reckon that the trees would finger down the wadis out onto the plain, and the continuous ground cover under and among the open parkland tree cover would have controlled erosion. Today, only a few years into protecting of the land from grazing, a good ground cover has already developed.

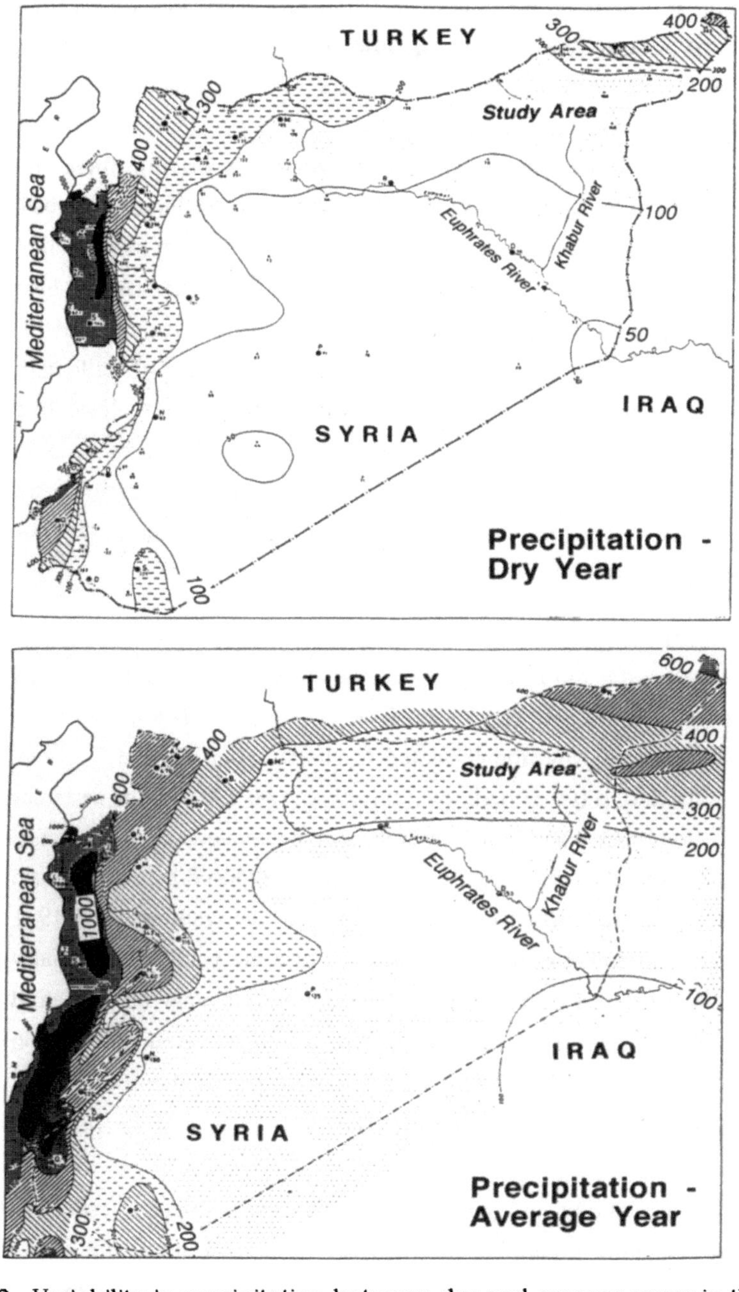

3. Variability in precipitation between dry and average years in the Khabur drainage. Redrawn from Kühne, ed., (1991: figs. 08,09).

Under such conditions of vegetation we may expect that the wadis will bring down less coarse sediment and perhaps run for longer periods following rains. Nevertheless, under today's climatic regime, it is scarcely conceivable that permanently flowing streams would extend out onto the plain.

The contrast with the northern part of the Khabur triangle is striking: there the plain is coursed by water channels that snake past hundreds of tells, large and small, that testify to settlement throughout history. In the region of our survey, tells are rare and pertain to only short periods. We believe that this episodic history can be best explained by invoking climatic/environmental change as a major determining factor. We contend that the evidence from geomorphology, fauna and patterns of settlement all imply striking environmental changes. Although one can advance alternative explanations for each observed effect, we maintain that the most reasonable interpretation is that there were significant episodic shifts in precipitation and intensity of storms against a background of prevailing aridity and a marked evapo-precipitation deficit similar to that of today.

Fifth-Fourth Millennium

During survey in 1988 and again in 1994 we discovered remnants of sites exposed in the bottom of a wadi that flows past Tell Khazneh (K157) (Fig. 5). The latest of the buried sites, K107, has ceramics provisionally dated to shortly after 4000 B.C. (Hole, Johnson et al. in press) , and K160 is perhaps century or two older, with ceramics probably descended from the post-Ubaid assemblages excavated at Kuran and Mashnaqa (Hole and Kouchoukos in press) . Both sites were exposed by winter flash floods. K107 was discovered and dug in 1988, but a search of the wadi again in 1991 and 1994 failed to reveal any traces of the site, indicating that the wadi is continually eroding its outer banks, alternately exposing or destroying remnants of sites, and possibly reburying others. K160, in the same wadi bed, was discovered and dug in 1994. In both instances the remnant deposits were exposed just at the level of the current wadi bottom and were overlain by more than two meters of gravels and sands.

Both sites consisted of pits dug into sterile silt and clay and in neither case was there evidence of any structure or cultural material above the level of the top of the pit. At K160 where part of the pit lay beneath the wadi bank it was possible to ascertain that the pit had not been dug through any of the overlying layers (Fig. 6).

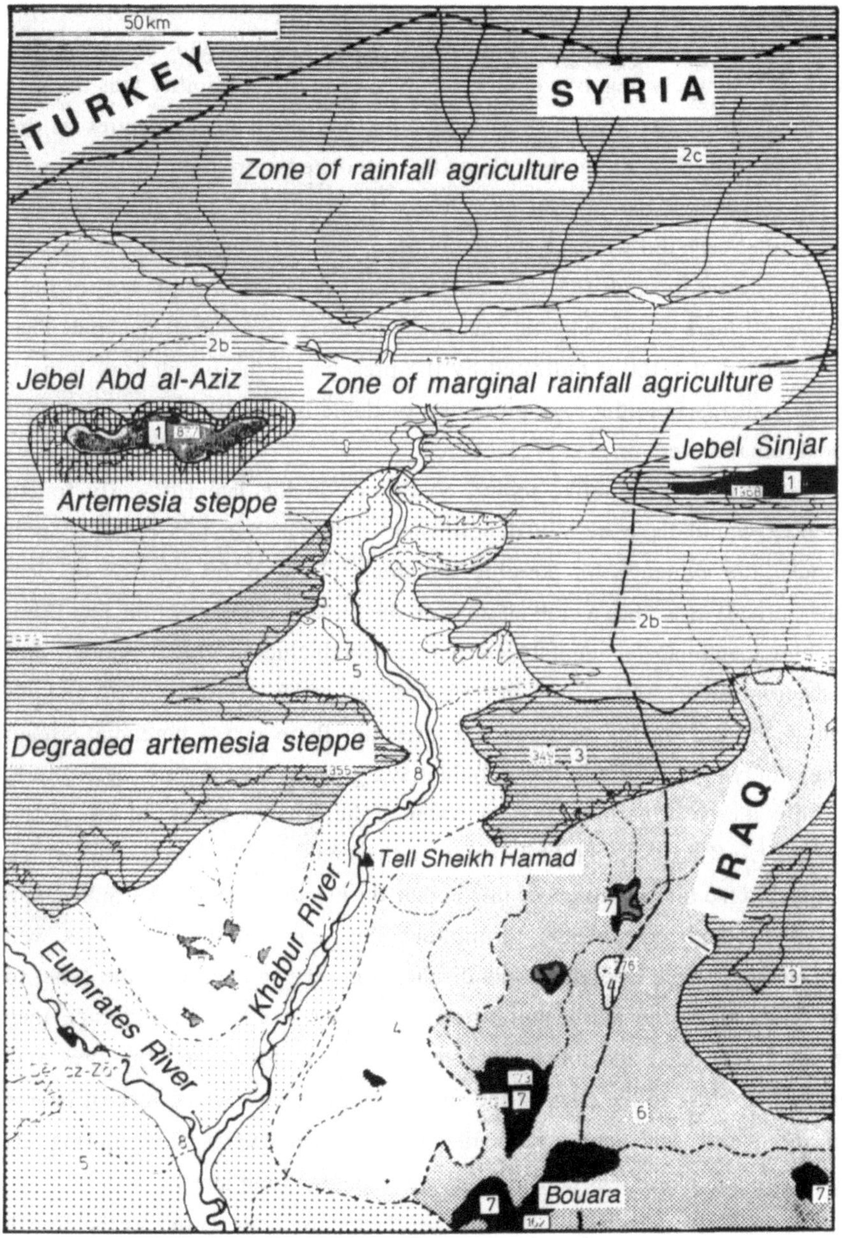

4. Vegetation types in the arid section of the Khabur drainage. Redrawn from Kühne, ed., (1991: fig. 49).

At K107, where the pit was entirely within the wadi bottom, we could not make this determination. This largely negative evidence leaves open the possibility that the pits had been partially truncated by subsequent flooding and that any structures at the sites had been destroyed by the violent flows that deposited more than two meters of sands and gravels above the pits. In each site we carried out a brief excavation, although in neither case did we excavate the entire contents of the pit. Part of the fill from K160 was sieved through 0.5 cm mesh, and the remainder was picked through by hand. Some 40 liters of the ashy fill was floated to recover charred material. That houses once existed at K160 is indicated by the presence in the pit fill of clay brick fragments. Only sherds and lithics are presently available from K107, although faunal and botanical remains were originally recovered.

These sites suggest that, unlike today, conditions in the late fifth-early fourth millennium B.C. would have permitted permanent settlement in this region. Apart from the bricks, the evidence for possibly sedentary life at K160 consists of clam shells (many burned) as well as a crab claw, from the midden contents of the pit. There are also plant remains (so far unanalyzed but clearly including cereal grains), and domestic sheep. Since clams, *Unio sp.,* as well as crabs, *Potamon sp.,* require oxygenated water and muddy bottoms, we infer that there was a perennial stream near the site, implying conditions that would have permitted sedentary agriculture. Today the only source of permanent river water is the Khabur River, 14 km to the north, while the Bir Khazne, several kilometers to the south, is only a tank into which water is pumped. One interpretation of this is that during the early fourth millennium rainfall was distributed more evenly throughout the year and the Jebel was well vegetated. Trees may have extended some distance onto the steppe, from the Jebel to the Khabur River along the permanent streams. In other words, the landscape may have been much less open and much more vegetated than today and the sites may have been situated close to the edge of the forest. As indicated by the lack of coarse gravels in the sterile soil into which the pit was cut, water flow during its aggradational pre-site phase was considerably more benign than later.

Apart from these sites, the region we surveyed lacks any sites dating to the fourth millennium despite the large number of sites in the third. Perhaps violent fluctuations in the local environment impacted some sites themselves, particularly when they were on river banks or in the path of massive alluviation. Our finding that more than two meters of sediments lie above the early fourth millennium sites north of the Jebel

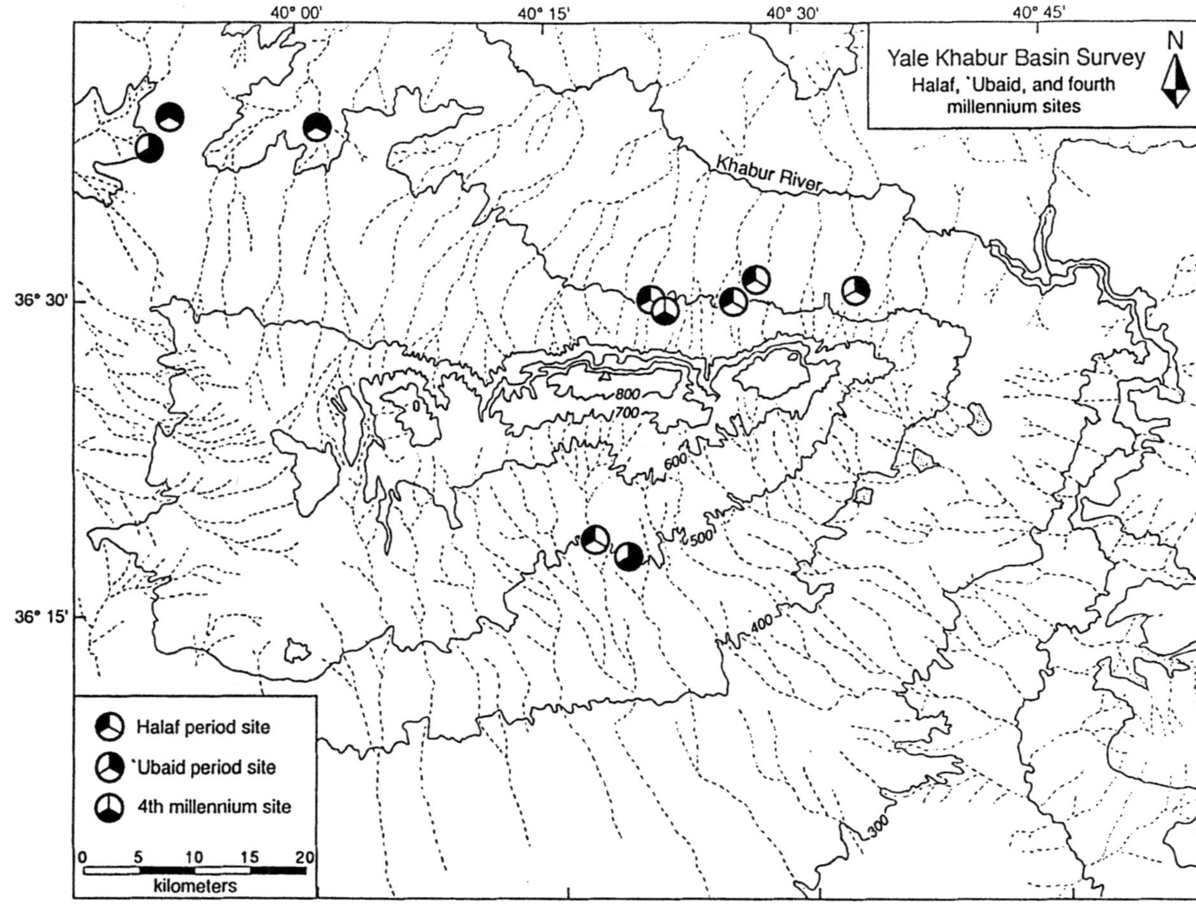

5. Fifth-fourth millennium BC archaeological sites in the survey area.

suggests that some sites may have been entirely buried or even destroyed by violent flooding during or after the fourth millennium, but this would apply only to sites situated close enough to the mountain to suffer the abrasion wrought by surges of sands and gravels. Moreover, this pertains to only part of the area we surveyed, but sites prior to the third millennium are conspicuously absent everywhere. At the moment, logic suggests that where sites are not visible today, they did not exist in the past or were small enough to be buried. If either is true, the situation is very different from what is characteristic of the wetter parts of the Khabur drainage where numerous tells showing repeated occupation are conspicuous topographic features. In our arid zone, as much as 1000 years may have elapsed before the region was resettled. The paucity of sites implies that much of the region may have been depopulated for nearly a millennium, and this applies to other parts of the Khabur drainage, for nowhere are fourth millennium sites numerous except along the major rivers.

We did find a few older sites, dating to the 6th-5th millennia B.C., denoted by Halafian and Ubaid sherds, mostly from the surfaces of third millennium sites, but these were so rare as to suggest that only small sites are present. Close to K107 we also found surface indications of an Halafian site (K109) that may lie on the same buried surface as K107. The sherds seem to have been brought to the surface when burials were placed in a very low mound which, at least at the surface, seems to be a natural feature. During survey we found only one actual mounded settlement (K203) dating from the Halaf-Ubaid-early fourth millennium, and this was in a region relatively unaffected by alluviation, in the northwest quadrant of our survey. Again, this reinforces the impression that this region was suitable for permanent settlement only rarely. Courty sees this as a cooler, wetter period with a slowly aggrading plain, punctuated by an intense period of aridification estimated to be 7800-7200 bp. We have little evidence for local conditions except for the presence of clam shells on the surface of K203.

Sometime after K160, but perhaps as early as during the occupation of K107, the climate must have become more severe subjecting the hillsides to violent flooding that brought down and deposited successive layers of sands, gravels and silts to a depth of two meters. In view of the sparse evidence for human settlement, there is no plausible way that this could have resulted from human activities; the alternative is that precipitation was reduced, or had became more seasonal, or more intense. Today

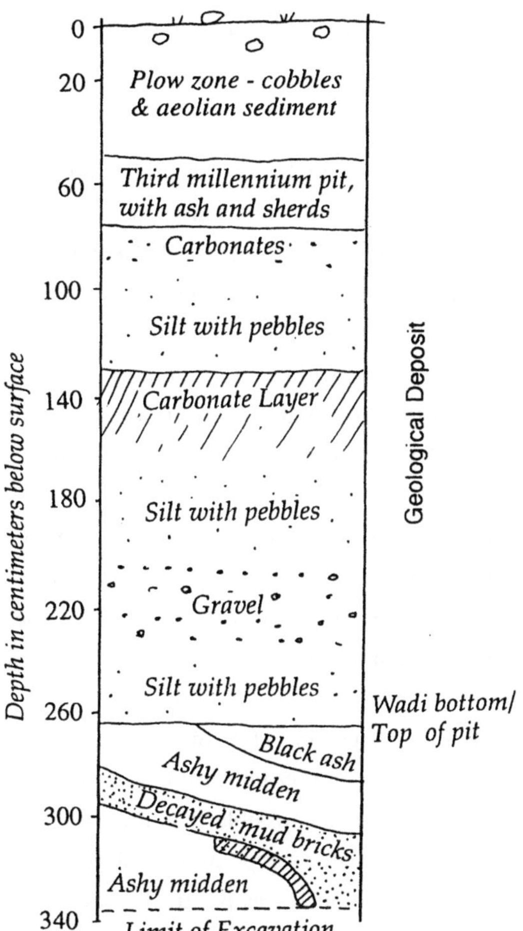

6. Section of pit and wadi at buried site K 160.

these deposits are capped by a mantle of aeolian silt and coarse rocks to a depth of about 50 cm, again suggesting mass movement of sediments, perhaps related to the periodic landslides that occur on the north side of the Jebel. The third millennium house walls were constructed of these coarse rocks and lie above the sands and gravels. In short, conditions had stabilized by the third millennium, after an hiatus of some 1000 years, and settlement resumed. Now for the first time there are many tell-like sites, most of which were occupied during the third millennium. A number of these date to the earlier half of the third millennium, most are mid-third millennium in age, and all had ceased to be occupied well before the abandonment of Tell Leilan and other sites around 2200 B.C.

Third Millennium

The number of third millennium sites that are situated in this arid, nearly featureless steppe on both the north and south sides of the Jebel is truly impressive. We recorded as many as 36 sites, ranging in size from a few houses to towns of 30 hectares. Present evidence is that some may date to the beginning of the millennium, but all date to the first half of the third millennium, and perhaps only half of the total lasted until 2500 B.C. Even today, with industrial scale agriculture and support systems, there are no settlements comparable to those of the third millennium. In fact, there are only two thirds as many Iron Age and Roman/Byzantine sites, and only half as many of the Islamic period (Figs. 8,9).

If this pattern of episodic settlement, with each successive period falling short of the previous is sustained through further survey and analysis, it suggests an alternative explanation for the rapid and unprecedented growth of sites that culminated in the urban centers of Leilan, Mozan and others situated in the wettest zones of the Khabur. It now seems possible that climatic conditions allowed for expansion of agriculture early in the third millennium, and the higher yields contributed to rapid population growth. Rather than a purely political explanation for the emergence of cities ca. 2600-2400 BC, it now seems possible that they grew in part because people moved from the desiccating southern steppe into the dependably wet north. There is no doubt that the system that developed was politically successful in the short-term, yet it apparently collapsed over a very brief period around 2200 BC Evidence reported from Leilan suggests a deterioration of climate marked by increased seasonality of discharge and reduced precipitation overall, beginning as early as 2500 B.C. and increasing in severity before the final abandonment of most of the site 300 years later (Weiss 1993; Weiss and Courty 1993). Our survey results do not contradict this inference and may provide strong support for it. Evidence now indicates that occupation of the sites in the arid zone ended around 2500 B.C., implying that conditions had already become untenable, a situation that may not have catastrophically affected the Leilan region for another

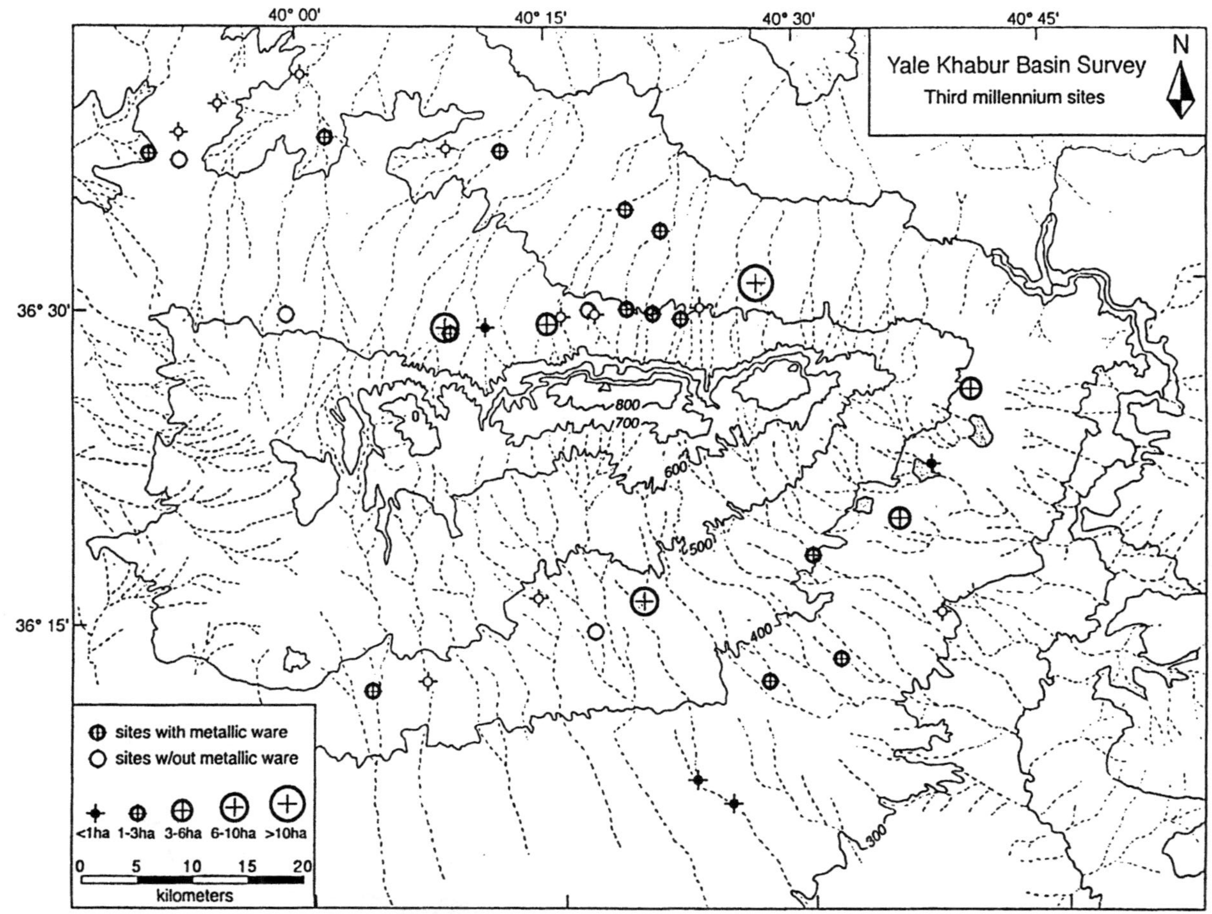

7. Third millennium BC archaeological sites in the survey area.

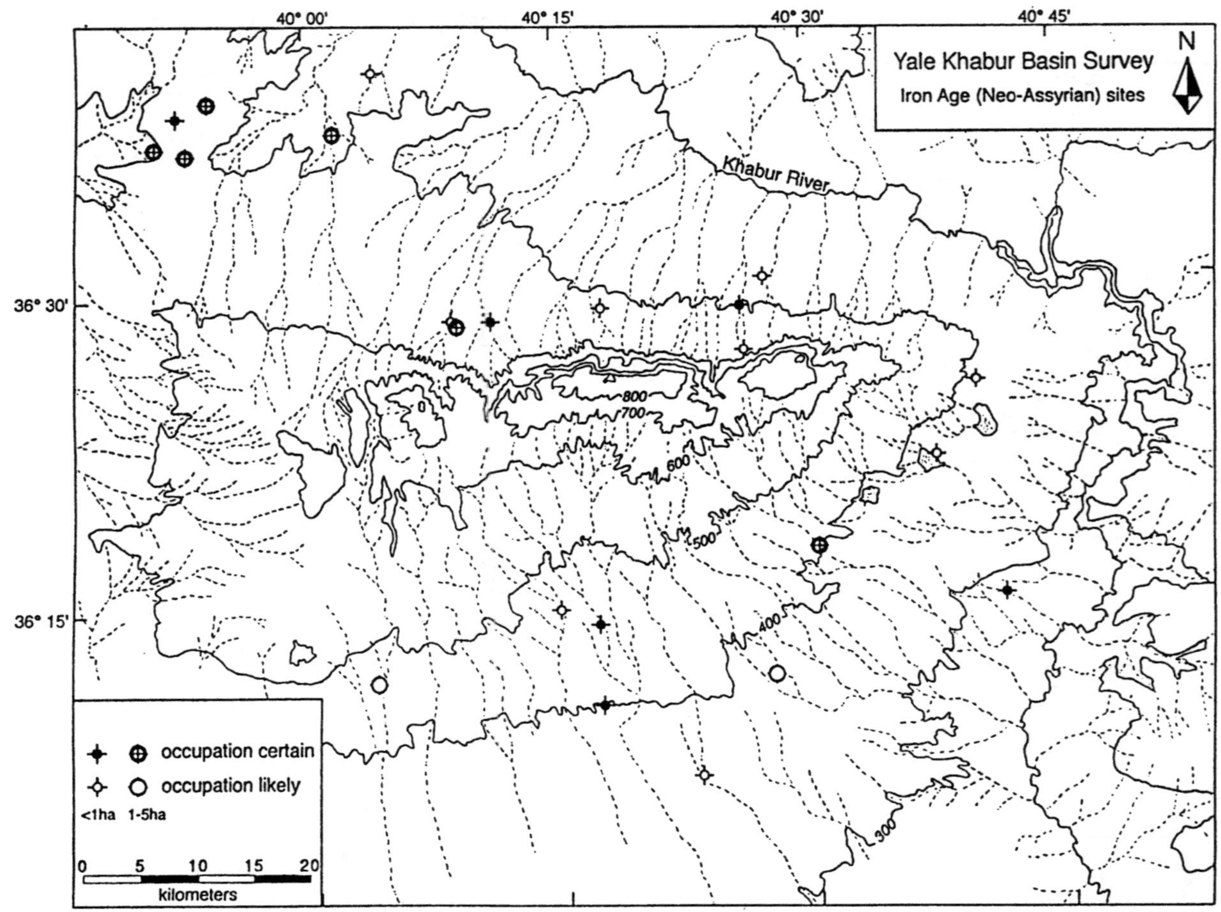

8. First millennium BC archaeological sites in the survey area.

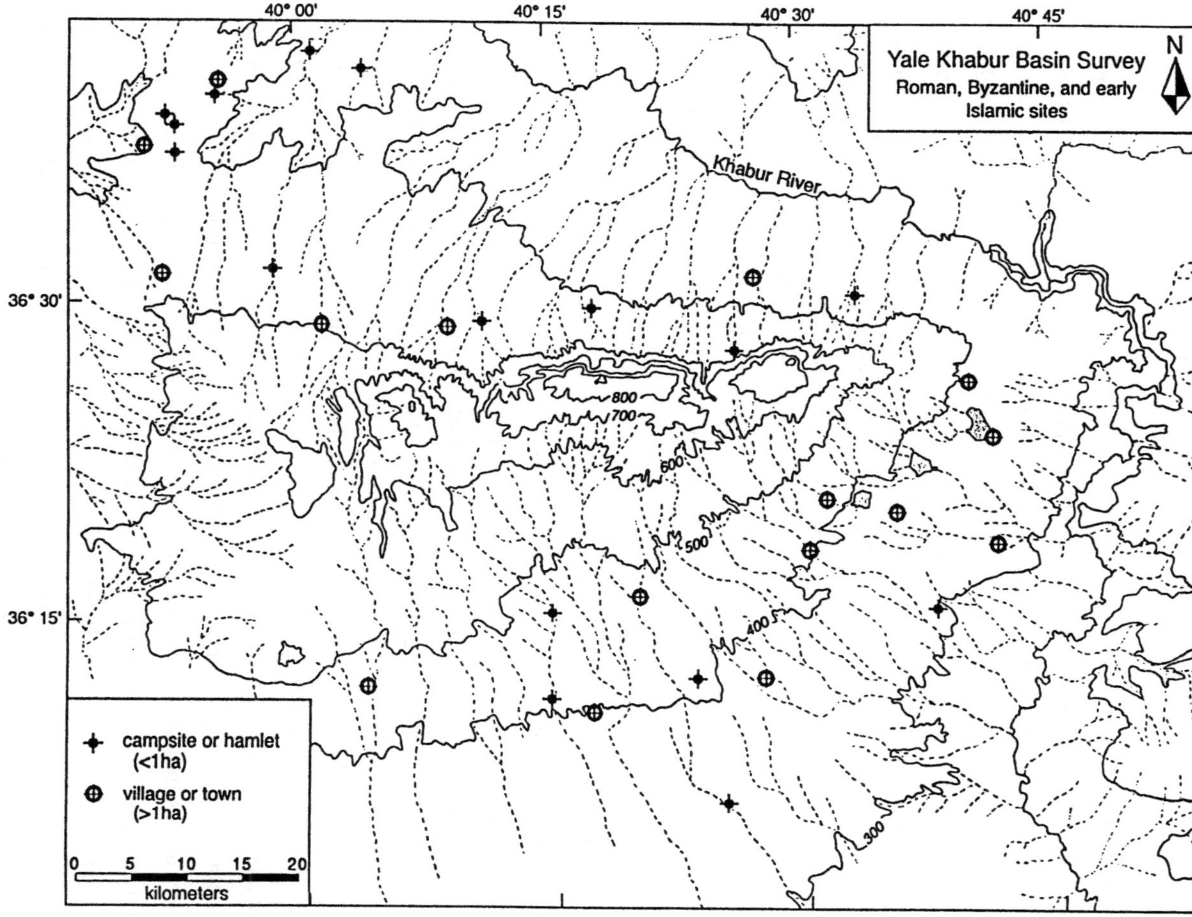

9. Roman/Byzantine archaeological sites in the survey area.

200 years and only after the added burden of an influx of refugees from the arid steppe.

Our archaeological survey provides no direct evidence for climatic or environmental conditions and one might propose a purely political or economic explanation for the observed changes in settlement, if it were not for the fact that under today's conditions the third millennium settlement we discovered could not occur. It is inescapable that this was one of only a few brief pulses of time when precipitation was adequate to support settled life in this region. Fortunately, it should be possible to establish these conditions in a relatively straight-forward manner through geology, micromorphology and geochemistry of the sites and their environs. These studies, coupled with judicious excavation of third millennium sites to recover plant and faunal remains, and carbon for dating, should make it possible to reconstruct with a high degree of accuracy the cycles of climate, vegetation, and human activities.

Wider Implications

Recently Weiss and his collaborators (Weiss, Courty et al. 1993) proposed that an intense desiccation triggered the demise of the Akkadian empire and a massive relocation of human populations into southern Mesopotamia. The occupational hiatus coincides "with an episode of tephra fall, followed by an abrupt arid phase marked by intensification of wind circulation that seems to match, in duration and amplitude, a major short-term climatic change. The reinstallation of climatic conditions similar to those before the tephra fall preceded reestablishment of Khabur Plains sedentary human occupation in the 19th century B. C." (Weiss, Courty et al. 1993:1001) . The apparently similar, synchronous social disruptions from Egypt to Greece, across the Near East and into the Indus Valley around 2200 B.C. lacked a unifying explanation until Courty discovered evidence of desiccation in the Khabur region and Weiss related this to the historical events. The cause(s) for this aridification remain unknown, and some scholars have dismissed the evidence or adopted a wait-and-see attitude (Cleuziou 1994; Glassner 1994). The physical evidence consists of layers of wind-blown sediment indicating volcanic activity, trapped in archaeological sites, and a near total abandonment of Akkadian cities and towns. In the Khabur, settlement and "normal" conditions resumed during the Leilan I period, about 1900 B.C. The tenability of the desiccation hypothesis rests on whether the entire region was abandoned simultaneously, and on additional evidence of desiccation or other environmental changes.

Our survey results contribute to this interpretation by stressing that climatic conditions during the third millennium when Leilan grew to city size were not like those of today; rather they were much better, at least until the mid-third millennium. Then they grew progressively worse until around 1900 B.C., in a resumption of what today are considered "normal" conditions. Our results suggest not only that there was an unusually arid period in the late third, but that the first half of the third was an unusually favorable time whose unsustainable prosperity that resulted in population increases, contributed to the later socio-political downfall.

In addition to these observations about the Leilan region, the survey may also contribute to an understanding of settlement on the middle Khabur River, where a dozen sites dating to the third millennium have been excavated. There has been debate concerning the role of these sites in a regional economy. Hole (Hole 1991a) has maintained that production was for local uses, while others have argued that these sites grew grain for export to one or another of the cities (Margueron 1991; Fortin 1993; Margueron 1993; Schwartz 1993) , implying that surplus grain was grown by means of irrigation and shipped to sites that were too large or in such unfavorable locales that they could not be self-sufficient. It now seems likely that conditions were unusually favorable throughout the region and that the land in the immediate vicinities of the large sites could easily have supported their populations. Indeed, if conditions were better, the apparently anomalous situation of Mari on the middle Euphrates, where it could not be self-sufficient today, can also be explained. If so, only after precipitation failed and Mari had already gained urban size, was it necessary to import its basic subsistence needs, as reported in 18th century texts (Margueron 1991; Margueron 1993).

Following this reasoning, another explanation for the storage facilities at the sites on the Middle Khabur can be entertained. Rather than shipping grain to southern Mesopotamia via Mari, as Fortin (Fortin and Cooper 1994:34) has argued, I propose that these sites grew grain for local consumption and fodder for livestock that were fed during the wetter months on the steppe beyond the range of agriculture. Indeed, I regard it as likely that a substantial part of the local populations may have resided in tents rather than in the specialized sites that served their common needs both on the river and in the steppe. It is notable that there are significant differences between the ceramics found on the north and south sides of the Jebel, suggesting that the people operated in different spheres of interaction. This would not be unexpected, for a similar situation existed in recent times, with the people on either side of the Jebel belonging to different sections of the Baggara tribe, while those settled on the river affiliated with still other tribes (Lewis 1987; Buccellati 1990). However, these issues cannot be settled until we know much more about the variability in local economies and can assess the actual extent of production and possible surpluses.

Other Environmental Studies

Apart from Courty's micromorphological analyses and trenches that reveal cycles of erosion and deposition, there have been only a few other substantial attempts to detect environmental changes: two based on geomorphological study of river sediments; the other a palynological analysis of sediments in the Khabur region.

If the land was much wetter during part of the third millennium, resulting in a larger flow into the Khabur River, this would support the finding of Michel Blackburn that the river was as much as 4 meters higher when the sites of 'Atij and Gudeda in the middle Khabur valley were occupied from 2700-2500 B.C. (Fortin 1990; Blackburn and Fortin 1994) . Under this scenario, Atij may have been an island, and shipping goods down the Khabur may have been feasible, although it remains to determine what these goods may have been. Higher river flows might also explain the elevation of the granaries and the protective walls surrounding some of them.

Another study of the river sediments was made in the Dur Katlimmu area, on the lower Khabur River. The difficulty of farming in this region is outlined forcefully by Ergenzinger et al. (1988) who point out that Dur Katlimmu is approximately at the line where the precipitation in 50% of the years falls below 200 mm (Fig.4) Because of interannual variability, even in the 250-300 mm zone, precipitation reaches 250 mm in only 66% of the years and "it remains fairly risky to grow crops." ((Ergenzinger, Frey et al. 1988:110) . Because this low precipitation may come at the wrong time of the year for agriculture, and the soils are gypsiferous, with little water retention capability, the region around Dur Katlimmu "is completely dependent on irrigation" (Ergenzinger, Frey et al. 1988:111) .

In his later detailed report on the geomorphology of the Khabur River at Tell Bderi and Dur Katlimmu, (Ergenzinger 1991:35) observes that 2 meters of silts overlie Early Holocene (meaning "before the Uruk period," i.e., late fourth millennium) sands and gravels, after which the river began to follow a meandering rather than a braided course. Further, he finds no evidence for substantial changes in discharge volume (as seen in flood deposits) since then. He relates the silt load to heavy erosion resulting from agriculture. However, the apparent timing of this deposit of sand and gravel corresponds to the burial of sites north of the Jebel; whereas the subsequent deposition of two meters of silts may relate to the rise in water flow attributed to the third millennium. Given the apparent importance of determining the dates of these events accurately, appropriate studies should be carried out expeditiously.

The other source of regional environmental information comes from pollen cores taken from a number of sites, between Lake Khatouniyeh in the east (300 mm precipitation) and the Bouara salt lake in the south (100 mm) (Fig.10). Unfortunately, the sites did not yield much suitable pollen and the best series came from the Bouara sediments, a locale well south of the modern dry farming region and outside the zone of settlement in all periods. Somewhat farther to the north in the region of 150 mm precipitation, in the drainages of several wadis that flow from the Iraqi Jebel Sinjar, there are no third millennium sites, and only a few dating to the Iron Age; however Bernbeck found one site, Wadi Agrubba, approximately the same age as our buried K160 (Bernbeck 1993) .

Despite its distance from known archaeological sites, the Bouara core may be more informative than it first appeared to be on the basis of only its pollen content. According to Gremmen and Bottema (Gremmen and Bottema 1991), five pollen zones could be distinguished in the 432 cm sediment core. One radiocarbon date, from the interface of zones 1 and 2, is $5730^{+}120$ bp (4545 B.C.), or in the range of Late Ubaid, approximately the same age as the Wadi Agrubba site. The interface between Zones 1 and 2 is denoted in the sediments by a 10 cm thick layer of peat. This is significant because it lies above some 80 cm of clay and is followed by another 115 cm of clay before another thin layer (15 cm) of peaty clay; followed in turn by clay and sandy clay to the top of the core. In other words, there are two intervals in this long core when there was perennial water and sufficient vegetation to create a peat layer.

At the same time as the peat layer (Zone 1-2) formed, a sudden increase occurred in graminae pollen, and *Quercus cerris*. Gremmen and Bottema (Gremmen and Bottema 1991:110) contend: "The high values for Graminae in the pollen assemblage zones 2 and 4 could be interpreted as a result of increased humidity or precipitation. An argument against such an increase in moisture is that at the same time

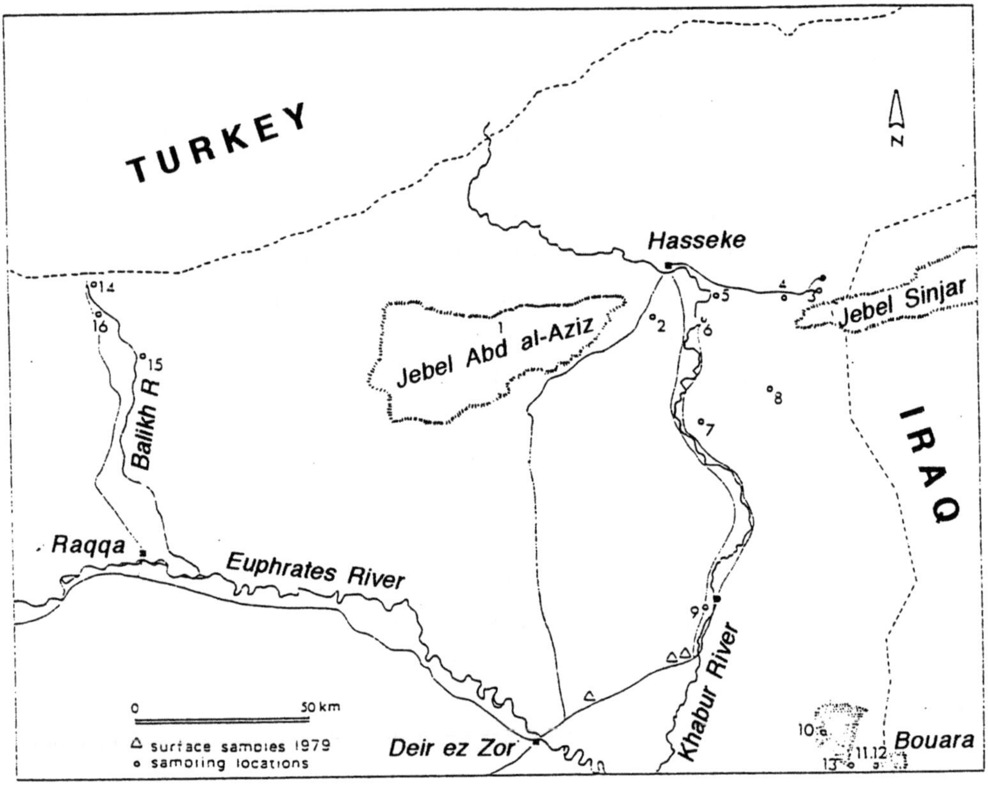

10. Location of pollen samples in the Khabur drainage and Syrian Jezireh. Figure redrawn from Kühne, ed., (1991: fig. 61).

Chenopodiaceae show very high values, which has to be considered as an indication for dry conditions. It is more plausible to consider the peaks in the curve of the Graminae in pollen assemblage zones 2 and 4 as changes in the local vegetation. A decrease of grazing in the Bouara area could explain the increase of the Graminae values."

In view of the apparently very sparse human occupation of this region, a decrease of grazing is unlikely to account for the values. I tentatively favor the climatic explanation which would also accord with the peat formation. Although there is no radiocarbon date for zone 4, I propose that it might pertain to the mid-third millennium. It would be helpful to obtain a date on this layer.

In summary, the evidence that can be drawn together about the local region, based on archaeological surveys and some limited palynological and geomorphological work, suggests several short-term changes in effective precipitation during the first half of the Holocene. A period late in the fifth millennium when conditions may have been much wetter was followed by a period of intense erosion in the fourth millennium, followed by a significantly wetter period in the first half of the third; perhaps a period of hyper desiccation late in the third; and finally the establishment of less variable, more "modern" conditions beginning in the early second millennium. An argument for unusually wet conditions in the early third millennium may be that settlement did not rebound in our area during the second, despite its presence elsewhere--in other words, following the desiccation, climate rebounded only part way to its optimum during the first half of the third. However, by the first millennium, technology for canal irrigation was well developed and apparently allowed settlement in the lower Khabur, but in the absence of flowing water in the area of our survey, such techniques would have been of no avail. In this arid zone settlement may have been related not just to technology, but to precipitation as well.

Environment, Settlement and Cyclical Changes

In the region of the buried sites today the land is relatively smoothly sloping toward the Khabur River, and gives the impression of being "flat" as one drives across it. The prevailing flatness is interrupted occasionally by low, rounded natural hills, many of which have third millennium and later occupation on them. The buried sites lie some 3-6 km out from the talus slopes of the Jebel Abd al-Aziz and presumably were buried during a prolonged episode of intensive slope wash that blanketed the upper plain with alluvium. The two sites we know about may be only a hint of what lies beneath this accumulated sediment and presumably more sites will occasionally be exposed as the wadis continue the headward and lateral erosion that has developed into a dendritic pattern proceeding upward from the Khabur River. An analogous situation of sites buried in an aggrading environment has been observed in the American Southwest where many sites, often undisturbed by this burial, lie beneath meters of sediment (Waters 1992:148-9) .

In the present drainage pattern the wadis stretch from the base of the Jebel to the Khabur River. They are deeply incised in the upper reaches where they cut through the coarse colluvium, but in the lower stretches they are often broadly meandering swales rather than entrenched streams. The gradient decreases markedly as the colluvium merges with alluvial fans, and the entrenched linear channels give way to a more dendritic wadi pattern with active headward erosion. Many large and small third millennium sites are situated on and beneath the upper portions of these fans which lie close to the 400 m contour in the region of the buried sites. The geomorphic situation is typical of arid regions and the main questions concern the environmental situation before, during, and after the deposition of sediments on top of the early fourth millennium sites.

There are a number of possibilities. As Waters (Waters 1992:158) remarks, "changes in a fluvial system are primarily triggered by changes in climate, tectonic activity, a rise or fall in the sea level, internal geomorphic adjustments, and human land use." Of these possible causes, we can probably rule out human land use, for there is no evidence that any settlements were very large at this time. Nor does a fall in sea level have bearing here for the base level of the Khabur River is established by a series of basalt dikes along its course. Tectonic activity remains a possibility for contributing to sediment flow, but this has not been investigated. Climate change seems the most likely, for the effects observed can all be attributed to changes in precipitation. As far as recent changes, such as a drop in groundwater level are concerned, the lowering of the Khabur River resulting from intensive use for irrigation and consumption in the villages and towns, must be considered, along with the pumping of water from deep wells.

The third millennium settlement pattern that we can presently observe conforms to that described for American Indian agriculturists in the Southwest. There, Indians chose the heads of alluvial fans at the terminus of ephemeral gullies on the bajadas (alluvial piedmonts) for their houses and farms because water coursing down the arroyos would distribute across the fans, making floodwater farming feasible. It is important to note that the Indians avoided locations where the gullies were incised because water could not be easily distributed to fields from these (Waters 1992:172-3). Had these techniques been used during the third millennium, it implies that the wadis were not incised, although obviously if there had been sufficient precipitation to support perennial streams, floodwater farming may not have been employed. In any event, despite the abundance of ruined stone-walled houses dating from the third and second millennia visible on the surface of sites, we saw no traces of check dams or other water diversion features.

More important, perhaps than a potential for floodwater farming, is the availability of perennial fresh water for human and animal consumption. As I have noted, such sources are rare today without the use of pumps. However, in the fourth millennium, when the presence of clams and crabs attest to a dependable water supply, it is possible that there were marshes or springs close to the sites, as well as permanently flowing streams which, at some point in their history, must have connected to the Khabur River.

What is difficult to understand, given present circumstances, is the character of the fluvial regime during the third and first millennia. Given the sizes of the communities, it is hard to believe that the residents depended only on wells for drinking water during the dry months. Therefore, we must entertain the possibility that during these times as well as in the fourth millennium, there was a perennial flow in the streams and that they may not have been so deeply incised as today. Clearly we need excavation at some of these sites to examine the geology and to recover biological and zoological remains that may tell specifically what grew in the vicinity, as well as what techniques were used to supply water, for all of these may have changed over time.

The environmental change that occurred in the early fourth millennium B.C. is superficially similar to what occurred in the American Southwest after the 1890s. There, a shift in rainfall patterns, coupled with intense grazing of the natural grasslands by domestic cattle and sheep that destroyed much of the vegetation cover, led to a lowering of the water table and a cycle of intense erosion. Today the landscape looks much like the arid region of the Khabur, with incised arroyos. However, before this change, the land was well vegetated, with perennial or seasonal streams lined with shade trees, and pools of water remained throughout the year. Apparently, beginning around 1890, winter rains declined and summer precipitation changed from gentle showers to violent downpours (Butzer 1976: 136-137). The combination of overgrazing and the shift in rainfall created conditions where erosion outstripped the regeneration of ground cover, resulting in a fall in the water table and gullying.

Following this model, one could postulate that rather than a marked shift in absolute precipitation, the principal changes in the Khabur region were cyclical variations in seasonality - from today's pattern when rainfall is confined to the winter months to a cycle when it was distributed over a longer period. The evidence appears now to suggest cycles on the order of 1000-1500 years. Such a series of climatic episodes - both better and worse for agriculture than today - accords with global and regional indications of sharp, but relatively low amplitude, Mid-to-Late Holocene climatic shifts. With a concerted multidisciplinary research project it should be possible to identify and date the cycles and their causes in this environmentally sensitive region. The outcome of these investigations would make a major contribution to understanding the environmental and cultural history of the Near East and perhaps wider regions still.

Acknowledgments

We are grateful to the Director General of Antiquities and Museums, Dr. S. Muhesen, the Director of Excavations, Dr. A. Bounni, Mr. J. S. Lazar, Director of Antiquities in Hasseke, and Mr. I. Murrad, our representative, for permission to carry out this research and for assistance in the field.

The project was made possible by Grant 5247-94, awarded by the National Geog. Soc., and a travel grant from the Williams Fund of Yale University. We gratefully acknowledge these sources of financial support. In the field, housing, vehicles and miscellaneous services were provided by G. and B. Taza, of Hasseke, all of which were essential to the success of the survey. Dr. R. Brakenridge, Dartmouth College, lent two GPS receivers to the Project, enabling us to make full use of the satellite imagery, previously processed in the Center for Earth Observation at Yale University.

The Project team consisted of Prof. F. Hole, Yale University, Dr. G. Johnson, Hunter College, and N. Kouchoukos, Department of Anthropology, Yale University. I. Murrad served as representative of the Department of Antiquities in Hasseke. Subsequent to the fieldwork, Dr. M. Zeder, of Smithsonian Institution, the Project archaeozoologist, identified faunal remains from K160 and secured its radiocarbon date. The processing of satellite imagery and preliminary assessment of the ceramics reported here was carried out by N. Kouchoukos.

References

Bernbeck, R. (1993) Steppe als Kulturlandschaft. Berliner Beiträge zum Vorderen Orient - Ausgrabungen, 1. Berlin, Dietrich Reimer Verlag.

Blackburn, M. and M. Fortin (1994) "Geomorphology of Tell 'Atij, northern Syria." Geoarchaeology **9**(1): 57-74.

Blunier, T., J. et al. (1995) "Variations in atmospheric methane concentration during the Holocene epoch." Science **Nature**(374):

Buccellati, G. (1990). 'Riverbank', 'high country', and 'pasture land': the growth of nomadism on the middle Euphrates and the Khabur. Tall al-Hamidiya 2, S. Eichler and M. Wäfler (ed.), Freiburg, Universitätsverlag.

Butzer, K. W. (1976) Geomorphology From the Earth. New York, Harper and Row.

Cleuziou, S. (1994). "La chute de l'empire d'Akkadé: Hommes et milieux au Moyen-Orient." Les Nouvelles de l'archéologie **56**: 45-48.

Courty, M.-A. (1994) "Le cadre paléogéographique des occupations humaines dans le bassin du Haut-Khabour, Syrie du Nord-Est. Premiers résultats." Paléorient **20**(1): 21-59.

Ergenzinger, P. (1991) Geomorphologische Untersuchungen im Unterlauf des Khabur. Die Rezente Umwelt von Tall Seh Hamad und Daten zur Umweltrekonstruktion der Assyrischen Stadt Dur-Katlimu, H. Kühne (ed.), 35-50. Berlin, Dietrich Reimer Verlag.

Ergenzinger, P. J., W. Frey, et al. (1988). The reconstruction of environment, irrigation and development of settlement on the Khabur in north-east Syria. Conceptual Issues in Environmental Archaeology, J. L. Bintliff, D. A. Davidson and E. G. Grant (ed.), 108-128. Edinburgh, Edinburgh University Press.

Feng, X. and S. Epstein (1994) "Climatic implications of a 8000-year hydrogen isotope time series from bristlecone pine trees." Science **265**: 1079-1081.

Fortin, M. (1990)."Rapport préliminaire sur le 3e campagne de fouilles à Tell 'Atij et la 2e à Tell Gudeda, sur le Khabour." Syria **67**(Automne): 535-577.

Fortin, M. (1993) Economie et société dans la moyenne vallée du Khabour au IIIème millénaire av. J. C. La Djezire et l'Euphrate Syriens de la ptotohistoire à la fin du second millenaire av. J. C.: tendances dans l'interpretation historique des donées nouvelles, Paris,

Fortin, M. & L. Cooper (1994) "Canadian excavations at Tell 'Atij „Syria 1992-1993." Bulletin, The Canadian Society for Mesopotamian Studies **27**(May): 33-50.

Gasse, F. and E. Van Campo (1994) "Abrupt post-glacial climate events in West Africa and North Africa monsoon domains." Earth and Planetary Science Letters **126**: 435-456.

Glassner, J.J. (1994) "La chute de l'empire d'Akkadé, les volcans d'Anatolie et la désertification de la vallée du Khabur." Les Nouvelles de 'lArchéologie **56**: 49-51.

Gremmen, W. H. E. and S. Bottema (1991). Palynological investigations in the Syrian Gazira. Die Rezente Umwelt von Tall Seh Hamad und Daten zur Umweltrekonstruktion der Assyrischen Stadt Dur-Katlimu, H. Kühne (ed.), 105-116. Berlin, Dietrich Reimer Verlag.

Harrison, S. P., I. C. Prentice, et al. (1993). "Climatic controls on Holocene lake-level changes in Europe." Climate Dynamics **8**: 189-200.

Hole, F. (1991a). "Middle Khabur settlement and agriculture in the Ninevite 5 Period." Bulletin of the Canadian Society for Mesopotamian Studies **21**: 1-15.

Hole, F., G. A. Johnson, et al. (in press). "Preliminary report on the joint American-Danish archaeobiological sampling of sites in the Khabur Basin (1990)." Annales Archéologiques Arabes Syrienne:

Hole, F. and N. Kouchoukos (in press). "Stratigraphic soundings at Tell Mashnaqa on the Khabur River 1991." Annales Archéologiques Arabes Syriennes:

Lewis, N. (1987). Nomads and Settlers in Syria and Jordan, 1800-1980. Cambridge, Cambridge University Press.

Liere, W. J. & J. Lauffray (1954-55). "Nouvelle prospection archéologique dans la Haute Jezireh Syrienne." Annales Archéologiques Arabes Syriennes **4-5**: 129-148.

Magny, M. (1993). "Solar influences on Holocene climatic changes illustrated by correlations between past lake-level fluctuations and the atmospheric 14C record." Quaternary Research **40**: 1-9.

Margueron, J.-C. (1991). "Mari, l'Euphrate, et le Khabur au milieu du IIIe millenaire." Bulletin of the Canadian Society for Mesopotamian Studies **21**: 79-100.

Margueron, J.-C. (1993). Mari et le Khabour:enjeux, possibilités et difficultés d'une approche archéologique. La Djezire et l'Euphrate Syriens de la ptotohistoire à la fin du second millenaire av. J. C.: tendances dans l'interpretation historique des donées nouvelles, Paris,

Meese, D. A., A. J. Gow, et al. (1994). "The accumulation record from the GISP2 core as an indicator of climate change throughout the Holocene." Science **266**: 1680-1682.

Moortgat Correns, U. (1972). Die Bildwerke vom Djebelet el Beda in ihrer räumlichen und zeitlichen Umwelt. Berlin, Dietrich Reimer Verlag.

Schwartz, G. M. (1993). "Rural archaeology in early urban Mesopotamia." National Geographic Research & Exploration **9**(1): 120-131.

Sirocko, F., M. Sarnthein, et al. (1993). "Century-scale events in monsoonal climate over the past 24,000 years." Nature **364**(322-324):

Waters, M. R. (1992). Principles of Geoarchaeology: A North American Perspective. Tucson, University of Arizona Press.

Weiss, H. (1993). Causality and chance: the origins and collapse of third millennium Mesopotamian civilization. La Djezire et l'Euphrate Syriens de la protohistoire à la fin du second millenaire av. J. C.: tendances dans l'interpretation historique des donées nouvelles, O. Rouault and m. Wäfler (ed.), Paris,

Weiss, H. and M.-A. Courty (1993). The genesis and collapse of the Akkadian Empire: The accidental refraction of historical law. Annual Meeting of the Society for American Archaeology, St. Louis,

Weiss, H., M. et al. (1993). "The genesis and collapse of third millennium north Mesopotamian civilization." Science **261**(995-1004):

Environmental Fluctuations, Agricultural Production and Collapse: A View from Bronze Age Upper Mesopotamia

Tony J. WILKINSON,

The Oriental Institute, University of Chicago

Abstract

The relationship between human settlements, their associated land use zones and changing climate, is not straightforward. Following a summary of the relationship between present day land-use and environment, this paper describes how land-use productivity (and therefore the population that potentially could be supported) can vary with crop management practices. In a climate characterized by high inter-annual variability, agricultural production will fluctuate thereby influencing potential carrying capacity of the land. Simple models that employ fluctuating levels of production indicate that population levels will also vary unless necessary steps are taken to store and import food, as well as to intensify production. As population increases and greater reliance is placed on such strategies the system becomes more stressed and shows a greater propensity to collapse. It is suggested that Bronze Age settlement and land-use systems of upper Mesopotamia were brittle systems and were therefore vulnerable to collapse. Under conditions of maximum production it is likely that even a short-term dry period *may* have resulted in collapse, but certainly a significant run of dry years would have resulted in considerable production deficits which in turn could precipitate collapse.

Introduction

Archaeological sites form a conspicuous and long-lived element in the landscape, and in marginal environments shifts or breaks in settlement may be related to adverse climatic change. Site distributions can therefore be employed as a proxy indicator of past climate, although it must be stressed that because they form part of a complex system of economic, social, political and environmental interactions such use is fraught with pit-falls. Nevertheless, in Upper Mesopotamia Neolithic (Hassuna) sites extending south of the present limit of dry farming may attest to a moister period a little before 9,000 years BP (Oates 1982). Alternatively, region wide hiatuses in the settlement record may suggest population collapse (Weiss et al. 1993). However,

NATO ASI Series, Vol. I 49
Third Millennium BC Climate Change
and Old World Collapse
Edited by H. Nüzhet Dalfes, George Kukla,
and Harvey Weiss

because settlements are supported by local agriculture and pasture, land use systems may form a more viable proxy indicator of climate change, but even then, owing to the numerous buffers and feedback mechanisms that are involved, such features should be used with caution. Changes in total population and its distribution may affect land-use strategies through time, and these in turn can influence the ability of those societies to deal with fluctuations in climate. A one-to-one relationship between settlement and climate should not therefore be assumed.

The objective of this paper is first to describe the relationship between land use and the present environment in Upper Mesopotamia, and to demonstrate how crop yield may depend upon a number of factors, in particular strategies of crop management. Next I will examine the response of land use to both demographic growth and to short-term climatic change. From this it is then possible to infer the potential responses of human communities to climatic fluctuations of longer return periods or of greater magnitude and to assess the likelihood of crop failure and collapse. A key factor to be examined is the tendency for societies in this agriculturally marginal environment to develop from a fairly resilient[1] state when population was low, towards highly productive but unstable economies which exhibit a greater propensity to suffer crop failure and perhaps collapse when population densities increase. This paper will not deal with cultural origins for state collapse, but rather will attempt to establish a general framework for assessing changes in ancient land use, settlement, and demography within the context of a climate of substantial inter-annual variability.

When examining the effects of recent climatic fluctuations on agriculture, earlier models focused upon the one-way impact of climate change on human activity. These have since been replaced by more sophisticated approaches that examine two-way interactions in which both population increase and land-use strategies are seen to influence food production so that a number of feedback relationships can develop (Parry 1990: 24-36). For example there is a continuing debate as to whether the recent Sahelian drought of 1968-84 was solely a climatic event or rather was a crisis that was pre-conditioned by economic and political factors over the previous decade and was triggered by a run of exceptionally dry years (Goudie 1986: 46; Mainguet

[1] Resilience being the ability of a biological system to maintain its structure and pattern of behaviour in the face of disturbance, contrasts with stability which is a state of equilibrium towards which a system will constantly return. Thus resilience emphasises a concern for long term survival even at the expense of numerical strength

1991: 6-16; Rapp 1987). Therefore increased vulnerability of land use systems was arguably a necessary pre-condition for the Sahelian famines, of the 1970s and 80s but was not a sufficient condition. Without the prolonged drought the disaster would not have occurred, but if the economic system had been less stressed it would have been more resilient and the effects of the drought might have been lessened (Parry 1986: 382). Therefore a key factor influencing the continuity of human settlement is their ability to survive exogenous shocks (Mokyr 1983, in the context of the Irish potato famine).

Here it is argued that in the rain-fed farming zone of the Near East, as urban centers grew in the 3rd millennium BC, changes in agricultural production may have increased the fragility of land-use systems which would have increased the likelihood of crop failure and the propensity to population collapse. By determining the potential likelihood of such intrinsic reasons for collapse it should then be possible to assess the magnitude of external events that might be necessary to precipitate collapses of still higher magnitude. Following a summary of the prevailing agronomic conditions in the rain-fed farming zone of Upper Mesopotamia, I will examine some of the scenarios that might result from population increase, land-use intensification and minor climate fluctuations. Finally selected scenarios will be tested against the field evidence as drawn from an archaeological survey within the region, and then assessed in terms of medium-term climatic fluctuations.

Regional Environment and Land-use

Upper Mesopotamia forms an arc of marginal dry land steppe extending from the Euphrates river in the west through to the mountains of Iran on the east (Fig. 1). Today most of Upper Mesopotamia falls within Syria, the remainder being within Turkey and Iraq. It forms an undulating steppe between 350 and 500 m above sea level and is developed upon Tertiary sedimentary rocks. These are punctuated by rare outcrops of volcanic hills, some of which have been active at various times within the Quaternary. Rainfall varies from 500-700 mm in the north to 150 mm or less within the desert, the limit of viable dry farming being in the region of 200-300 mm rainfall per annum depending on hydrology and other local conditions (Fig. 1). Inter-annual fluctuations of rainfall are considerable, varying from >60% in the desert to 20-30% in the moister areas of western Syria (Kerbe 1987, map 1). Because most precipitation

---

(for discussion, see Holling 1978; 1986: 296-97 and Adams 1978).

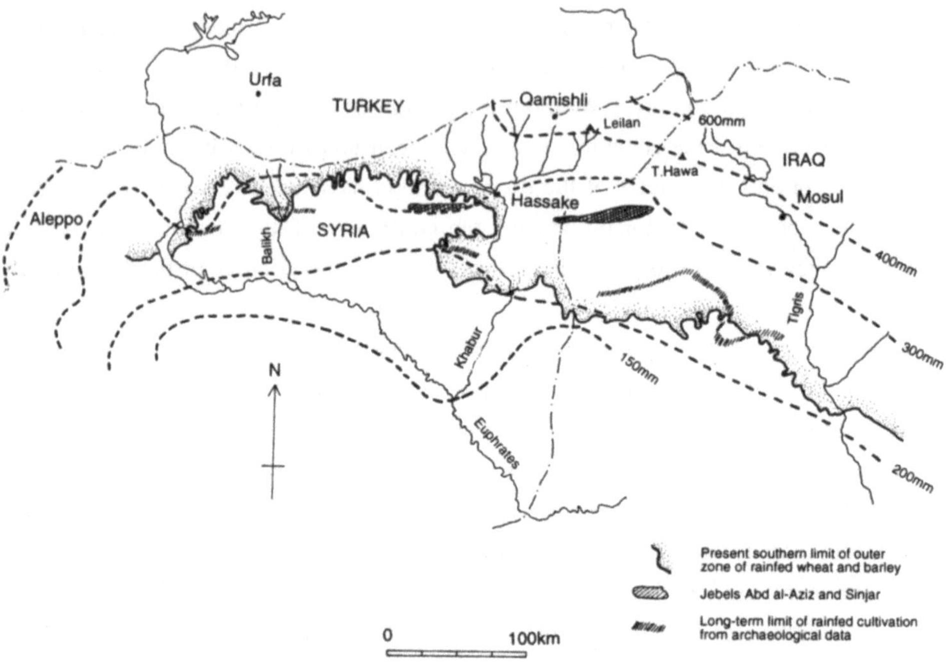

Fig.1 Upper Mesopotamia, present day limit of cultivation with respect to key rainfall isohyets (From TAVO (1985), Kerbe (1987) and various other sources).

comes from cool-season (October-April) westerly depressions, there is a high region-wide correlation of rainfall such that for example when Aleppo receives a high rainfall, distant places such as Qamishle also register high falls, and vice versa (Fig. 2: top). Such correlations do not necessarily extend into western Anatolia where changes can be in opposite directions (Weiss 1982; Roberts and Wright 1993). Within Upper Mesopotamia however, and under current conditions of general circulation, a drought in one area will probably be matched by drought over most of Upper Mesopotamia with the result that if inhabitants took to the road to move to lands of plenty, they would probably have had to flee a long way to find food (e.g., Jean 1947: 70, for an example from cuneiform texts).

Within the dry farming zone soils are mainly calcic xerosols[2] characterized by a horizon of calcium carbonate accumulation at depths of > 30 cm below the soil

[2]Using the terminology of the FAO soil classification system (UNESCO 1974); according to more traditional systems approximate equivalents are as follows: calcic xerosols = brown calcimorphic soils; yermosols =gray

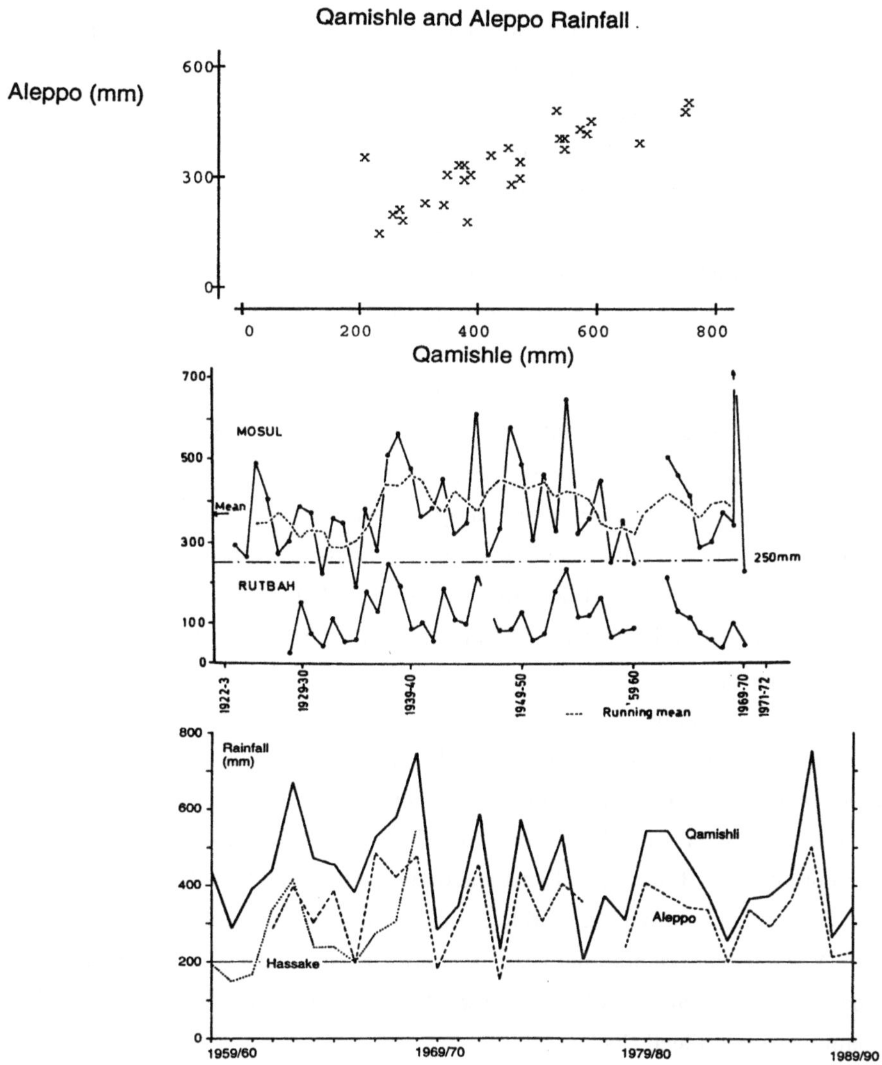

Fig.2 Top: Rainfall correlation between Qamishle and Aleppo. Center: Rainfall at Mosul and Rutbah (Iraq). Below Rainfall at Qamishle, Hasseke, and Aleppo (Syria).

surface (UNESCO 1974). To the south these grade into xerosols with gypsic subsurface B horizons, and ultimately into the gypsic yermosols of the true desert. Conversely to the north where rainfall is > 500 mm, calcic xerosols are replaced by

desert soils; chromic luvisols=terra rossa (Buringh 1979).

chromic luvisols in which the calcium carbonate B horizon is less well developed. In general these soils are fertile and, given sufficient rainfall, these only require applications of nitrogen and phosphorous to give sustained high crop yields (UNESCO 1977: 68). Although originally covered by grass steppe with scattered trees (*Pistachia, Crataegus, Amygdalus, Prunus, Pyrus* and *Rhamnus* spp; Guest 1966, Pabot 1955; Kaul and Thalen 1979: 245) which increase in amount to the north, most of the area is now an agricultural steppe devoid of trees as a result of long-term fuel cutting, burning and overgrazing. Such degradation has a long history extending back to at least 5000 BC (McCorriston 1993; Miller 1990).

Northern Syria and Iraq can be sub-divided into five agro-ecological zones as follows (Pabot 1956; Cocks et al. 1988; see also Fig. 3 based on Jones 1993, fig. 3):

Zone 1a: With rainfall > 600 mm, this area exhibits a wide range of crops. Within Turkey, zone 1a is characterized predominantly by wheat/barley and lentils grown in alternating years, with vineyards occupying marginal soils (Wilkinson 1990).

Zone 1b: Rainfall between 350-600 mm and not less than 300 mm in 2 of every 3 years. Main crops comprise wheat, chickpeas, lentils, fruit, vegetables and non-irrigated summer crops.

Zone 2: Mean rainfall of 250-300 mm, and above 250 mm in two of every three years. Crops are barley, wheat, some food legumes and summer crops; livestock, mainly sheep and goats are of increasing importance.

Zone 3: Mean annual rainfall greater than 250 mm and not less than 250 mm in one year out of two. Barley and livestock (mainly sheep and goats) are the main products, wheat is also grown but food legumes are of little importance.

Zone 4: With average rainfall between 200 and 250 mm, and not less than 200 mm during half the years, barley and sheep or goats remain the primary products, barley being mainly grown to feed flocks.

Zone 5: Even less rainfall than Zone 4 and extending into true desert, this area is not used for cultivation but is used for grazing, primarily by nomadic or semi-nomadic pastoralists.

Where mean annual rainfall exceeds about 325 mm wheat is the main cereal, although barley may be important on poorer soils (Jones 1993: 131). Further south barley predominates until the pastoral zone 5 takes over. The present day limit of dry farming is very ragged, and depends on a range of cultural and economic factors as well as on local soil and hydrological conditions (Fig. 1). For example, soils along

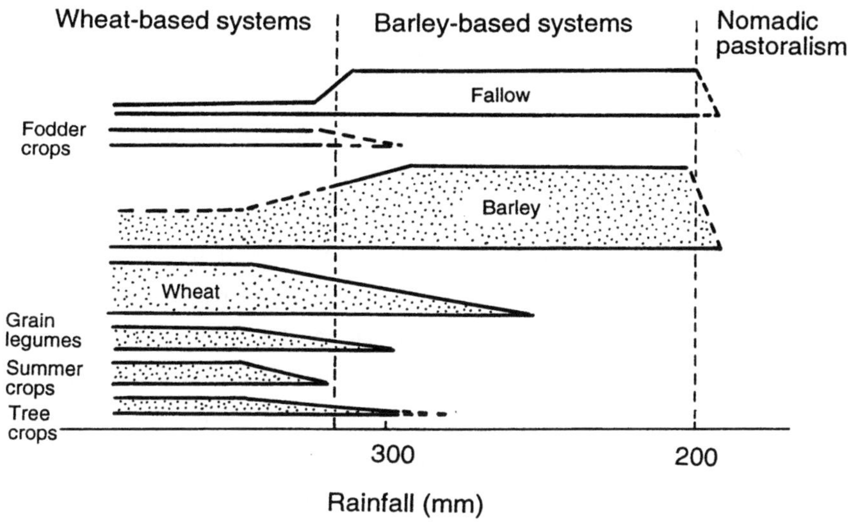

Fig.3 Agro-ecological zones in northern Syria (re-drawn from Jones, in Jones et al. 1993: 130).

wadis may remain moist for greater parts of the year with the result that lobes of dry-farming extend southward into the true desert. In drier areas most land is under barley, usually with an alternating year of fallow, but increasingly it is being grown as an annual crop. Because crop-production is not viable every year, grazing of animals increases towards the southern limit of Zone 4, and in some years barley is not even harvested but is directly grazed by the sheep and goats. This underscores the elasticity of such an economy, in which food production is not necessarily the primary goal but where there is a flexible arrangement such that standing barley is often regarded as an enhanced animal feed, rather than a failed food crop (Nordblom 1983). If necessary, and depending upon grain availability, animals can then be exchanged for grain produced in more productive areas. Therefore today at least, we see a range of agricultural practices within Upper Mesopotamia and a necessary degree of inter-dependence between adjacent zones. As a result of this flexible strategy of grazing standing barley in dry years, it should be appreciated that the perception of crop failure is much more blurred than it would be for a true commercial economy. Such factors of resilience and related feed-back mechanisms will be elaborated below.

Climate Fluctuations

Long-term climatic statistics are not available for Upper Mesopotamia, but a 45 year record from Mosul, two 30 year runs for Qamishle and Aleppo, and an even shorter run for Hasseke, provide a general picture of rainfall conditions for much of this century (Fig. 2). Mosul, with mean annual rainfall of 384 mm and a wide inter-annual fluctuation, rarely experienced rainfall less than 250 mm in any wet season[3]. Five year running means suggest two broad medium range variations (i.e. > 10 years duration), an earlier one between 1927 and 1937 of rather lower rainfall, followed by a moister (albeit wildly fluctuating) episode between the late 1930s and the early 1950s. Unfortunately disruptions in rainfall records around 1960 make interpretations beyond this point difficult. Whether such dry and moist periods represent significant climatic trends or not is debatable and rather may simply be an example of "persistence" in which runs of dry or wet years can appear grouped together over relatively short time periods (Gommes 1993: 72). Little is added by the records from Aleppo and Qamishle in Syria (mean annual rainfall of 334 and 444 mm respectively) except that there is a correlation between the rainfall records at both stations (correlation coefficient 0.834) and that dry years (such as occurred in 1969/70 and 1983/84) would have affected both stations, but Aleppo more seriously. If fallow was operating to successfully conserve soil moisture (see below), the earlier drought would have been less serious than that in 1983/84 because of the preceding wet year in 1968/69 (see below for a discussion of fallow as a moisture conservation measure). At Hasseke, the driest station of the four, a dry period from 1958/59 to 1960/61 would be potentially serious not because of its extremity but because moisture carry-over during the fallow year would be from dry year to dry year, with the result that the percentage carry-over would probably be insufficient to save crop failure (see below).

The foregoing statistics demonstrate that both Aleppo and Mosul are in marginal locations, but in neither case do the climatic records show runs of critically dry years. Nevertheless places with rainfall within the range 250-330 mm would experience more years with significantly lower rainfall as well as back-to-back pairs of dry years

---

[3] Rainfall is calculated in terms of hydrological years so that all rainfall falling within any given wet season is calculated together.

either of which could have produced major crop deficits[4] . In Syria, as rainfall decreases towards the limit of rain-fed cultivation, the percentage of years with no harvest increase as follows:

>350 mm p.a. 0%; 250-350 mm 1%; ca. 250 mm 36%; 200-250 mm 46%; <200 mm 64% (Oram and de Haans 1995: table 3.1).

The pollen record for Holocene climatic change in Upper Mesopotamia is meager but suggests that there has been little significant climate change over the past 6000 years (Bottema 1989, Gremmen and Bottema 1991). On the other hand, recent pedological studies suggest that the region has experienced broad phases of wetter and drier climate culminating in a progression towards greater aridity after 3800 BP (Courty 1994). Raised water tables within archaeological sites or increased river flow provide some support for moister conditions between 6000 and 3000 BC (uncalibrated dates; Rosen and Goldberg 1995; Wilkinson and Tucker 1995: 31 and 85), after which present or even drier conditions were attained. However, because variations in precipitation and human influences on vegetation can change runoff hydrology and groundwater conditions (Butzer 1995: 124), these interpretations should be regarded as preliminary . The region probably experienced a number of prolonged droughts per century which could have had a significant affect on agricultural production. Cuneiform texts, for example, provide a gloomy record of early Assyria, where drier conditions with droughts and famines at roughly 10 year intervals occurred in the 11th and 10th centuries BC. These were then followed by a moister spell during the early 1st millennium BC (Neumann and Parpola 1987). In comparison, a 1,000 year tree ring record from an area of comparable climate in Morocco, shows that clusters of dry years signifying major droughts (defined as at least 6 years of drought) occurred usually once or twice in 6 of the last 10 centuries (Stockton and Meko 1990: fig. 1.15[5]). If conditions comparable to those of the last millennium in Morocco prevailed in Upper Mesopotamia, it is therefore likely that between five and ten droughts of this order would have occurred during the millennium-long Early Bronze Age. Such episodes would only be compounded by outbreaks of infectious disease or plague that could episodically result in significant population decline (Issawi 1988: 96-97).

---

[4] Pairs of dry years occurred at Mosul in 1927/28-1928/29 and 1923/24- 24/25).

[5] see also Duvick and Blasing 1981 for evidence of major droughts in the American great plains approximately twice per century.

Land-use and climatic fluctuation

In the context of human perceptions and land use Parry has suggested three scales of climatic fluctuation: Climatic variability, that is the year-to-year differences in climatic variables to which economic and social systems generally adjust; medium-term climatic changes (of duration 10-200 years) to which society probably needs to adapt in order to avoid undesirable impact, and long-term climatic changes which occur on scales too large to be considered significant for the planning horizons of most societies (Parry 1986: 380). In addition of course, extreme events, either short-term or of longer duration, can have a major impact on crop production systems.

In terms of the biology of crop growth, water scarcity causes plants to close their stomata which results in reduced transpiration and intake of carbon dioxide. Consequently photosynthesis is minimal and crop growth is retarded (Parry and Carter 1988: 29). As a result, most case studies comparing crop yield and climate show a marked positive relationship with rainfall (ibid.). However such a straightforward correlation between climate and land-use should not be assumed. Rather, for cereal crops there are two main components that influence yield variability, namely technological change and climatic variability, to which must be added an often significant amount of random "noise" (McQuigg 1981: 122; Carter and Konijn 1988: 70). When assessing yields of US food grains McQuigg found that technology "explained" 70-90% of total variance, whereas weather fluctuations only 5-25% (McQuigg 1981: 126). In this case changing technology took the form of increased dressings of nitrogen between 1950 and 1980. Similarly, climate/land-use scenarios when projected into the 21st century AD suggest that the effect of technological change on most crop yields is likely to be several fold larger than that of climate change (NDA 1980: 51). Although such an extreme situation should not necessarily be applied to early societies, technological change cannot be ignored, and during times of food stress the application of manure or compost to fields may have raised crop yields sufficiently to override minor climate fluctuations.

Although technological factors can therefore complicate the relationship between climate and crop production, data from Syria indicate that it is possible to account for between 50 and 70% of the variability in wheat and barley yields by various estimates of seasonal rainfall (Harris 1991: 24; Van Oosterom et al. 1993). Using computer simulation, Harris has shown that simulated grain yield correlates well with rainfall, but rather less with length of growing season. Hence at Muslimiyah, a test station of ICARDA[6], rainfall accounted for 85% of variations in simulated crop yield (Harris 1991: 31). Similarly, in an area of southern Russia with a climate comparable to that of northern Syria there is a broad positive correlation between rainfall and actual crop yield, although as will become apparent below, land use practice can result in significant deviations from this trend (Parry and Carter 1988: fig.1.13c). Other climate factors that significantly influence crop yields are seasonal distribution of rainfall, length of growing season, and temperature. For example, frost can damage crops when they are actively growing, and extreme high temperatures in spring, especially when associated with high winds, can be equally damaging.

Similar correlations obtain in Tunisia, where in the arid zone (<350 mm per annum) there was a strong linear correlation of barley yield with rainfall from September to May (Ben Mechlia 1993: 122). However, in slightly wetter areas (rainfall > 450 mm), the scatter of points is greater and there is plateau at 15/quintals/ha (i.e. 1500 kg/ha), which Ben Mechlia regards as the probable maximum barley yield in Tunisia under present growing conditions. Further south where rainfall is < 250 mm, the use of runoff collection and other methods of soil moisture enhancement result in a lower correlation between winter rainfall and barley yield. Applying such principles to N Syria and Iraq, it is feasible that there might be less correlation between rainfall and crop yield in the moist steppe areas of zone 1, as well as in the drier areas (zones 4 and 5), while the intermediate two zones 2 and 3 would be expected to register the strongest correlations.

The relationship between land use and climate is illustrated by the extension of settlement and cultivation that occurred throughout the Jazira in the present century. Figure 4, reconstructed from various historical sources (Wilkinson and Tucker 1995: fig. 5; Thalen 1979) indicates that in 1938 the limit of cultivation in north west Iraq extended only a short distance south of the Jebel Sinjar. During the following 40 years a massive southward colonization occurred until by the 1970s settlement reached line D on Fig. 4, some 50-75 km south of its former limit. This colonization, which resulted in some 47, 200 people being established in 400 new villages between Hatra and Ba'aj (south of Jebel Sinjar), is ascribed to the sedentarization of Shammar bedu (Thalen 1979: 297), although it was almost certainly encouraged by

[6]International Center for Agricultural Research in the Dry Areas (Aleppo). I am especially grateful to the staff of ICARDA, particularly Tom Nordblum and Scott Christiansen for providing valuable reference material on dry farming systems.

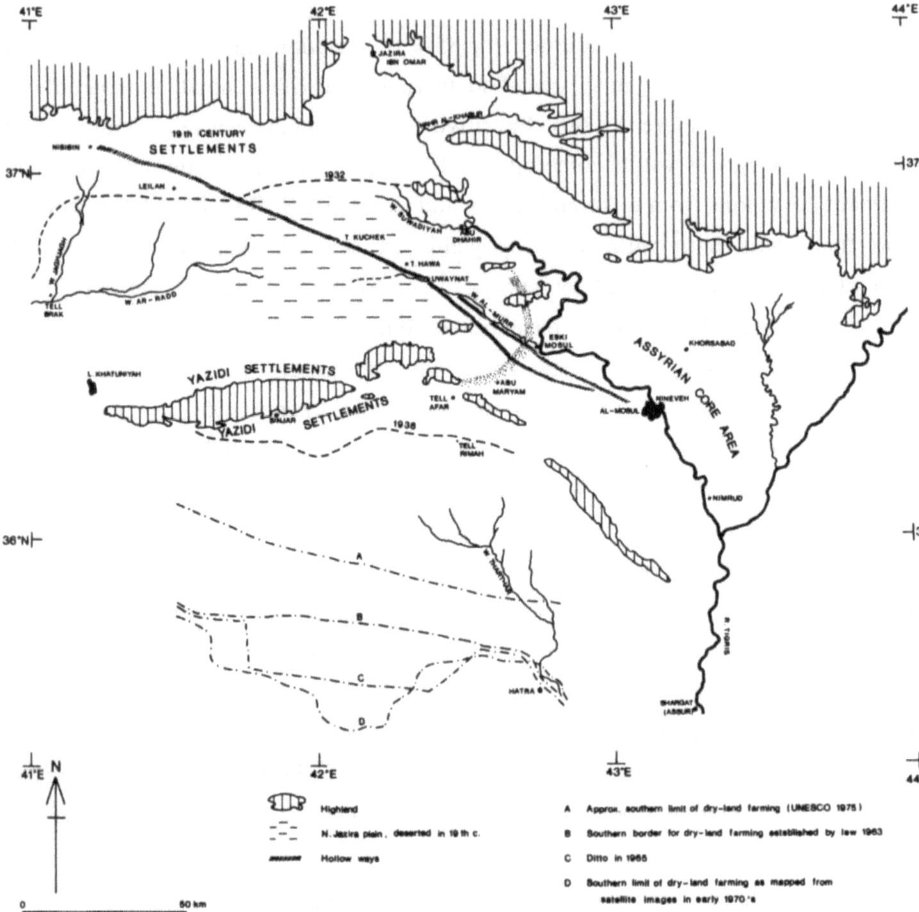

Fig.4 Historical reconstruction of northwest Iraq showing estimated limits of cultivation as follows (from N-S): 1932: southern limit, after Dillemann 1962. 1938: southern limit around Jebel Sinjar, sketched from Lloyd 1938. A-D southern limits showing the progressive southward movement as a result of colonization of steppe after World War II. "Yazidi settlements" and "19th century settlements" (near Nisibis) recorded by various 19th century travelers. Stippled arc near Eski Mosul indicates roughly the NW limit of cultivation as recorded by 19th century travelers. (From Wilkinson and Tucker 1995: fig.5).

the run of moist years that occurred in the 1940s and 50s (Fig. 2, Mosul curve). The unsustainability of such land-use practices is emphasized by attempts by various governments to restrain cultivation to within the moister northern areas of the steppe (lines A, B and C Fig. 4) and today a line established by the government of Iraq has managed to restrict cultivation within the northern band. In the Balikh valley of Syria satellite images and topographic maps indicate a similar southward shift in cultivation, in this case between 1945 and 1980 (Lewis 1988). This area, forming part of Rowton's zone of dimorphic chiefdoms (Rowton 1973), originally consisted of steppe grazing away from the valley and cultivated and often irrigated enclaves along the river. Thus pastoralism was balanced by cultivation each of which responded to different climatic parameters. The resilience of such communities is emphasized by the severe winter of 1911 during which, although some 80% of the sheep in the vilayet of Aleppo died from cold or hunger, human survival was possible because harvests were moderately good (Lewis 1988: 689). Gradually however this duel economy was disrupted by the sedentarization of nomadic groups and the movement of people out from the valleys, both of which resulted in the massive extension of cropland of the 1930s, 1940s and 1950s (Lewis 1988: 692). Although this extension of settlement is ascribed to conditions when a central government ensured secure, peaceful and economic conditions, again the existence of moister winters at this time must have enabled settlement to take place. The extension of settlement into more marginal lands has seen the replacement of a resilient traditional duel economy by cereal monoculture (with cash crops along the river). Southward colonization took place under both favorable political and economic conditions <u>and</u> improved rainfall. By replacing the resilient traditional system by a fragile rainfall-dependent monoculture, the system has become more vulnerable to crop failure. This has been made yet more likely because of soil degradation which decreases potential crop yields by removing soil and lessening its soil moisture-holding capacity. Finally the unrealistic extension of cultivation into more marginal lands again makes such systems extremely vulnerable to collapse.

Land use and risk management

In addition to climate, soil conditions and crop management practices can significantly influence yields and therefore carrying capacity. For example, deep soils provide significant reservoirs of soil moisture, and if root systems are long enough to penetrate to below 100 cm these lower horizons can supply crucial water at times of soil moisture deficit (Shaoxian 1993: 378). The above observation from China is supported by crop yield data from northern Iraq where on dry-farmed lands near Tell

'Afar, thinner soils on the rolling steppe provided significantly lower yields than the deeper soils of the plain (Ali 1955: 147).

A more widespread but controversial method of increasing crop production is the practice of fallowing which entails leaving some land uncultivated in alternate cropping years (in Syria 14-16 months). As a result, that portion of rain that falls on the land, rather than being consumed by crop growth, is held over for the next cropping year. The resultant bonus (normally in the range 10-25%) can be crucial by supplying supplementary soil moisture especially in dry years. However, because both nutrients and moisture can be limiting factors on crop growth it is not always clear which factor provides the bonus in crop growth (Loomis 1983: 359). Nevertheless, on the American Great Plains for example, fallow can have dramatic effects on yield:

"The general consensus about wheat fallow rotations in the American Great Plains are that fallow practices strongly reduce the risk of crop failure and the relative variability of years over years (USDA 1970).......Yields <400 kg/ha (failure) were obtained in 10 of the 27 years of continuous wheat, but never after fallow."

Loomis 1983: 360.

For example, in the North Platte region the coefficient of variation for continuous wheat was 74% but only 33% for wheat which followed fallow (ibid.). Of critical importance for marginal areas is that fallowing provides a crucial insurance, especially during dry years. For example in a steppe area of southern Russia (Stavropol immediately north of the Caucasus) a 20 year run of field trials demonstrated that when winter wheat was preceded by a rotated crop (i.e. no fallow and without fertilizer) yields were reduced by 31% in cool humid years, by 60% in years with recurrent drought and by 75% in years of severe drought (Nikonov et al: 1988: 598).

In Syria fallowing is a long-continued tradition in areas with <300 mm annual rainfall (Harris et al: 238), and in the Mediterranean region in general the practice certainly goes back to the Roman period (White 1970). In upper Mesopotamia fallowing was practiced in the Neo-Assyrian period (early 1st millennium BC; Fales 1990: 119) and it is also considered to have been a normal cropping practice as far back as the 15th century BC and probably earlier at Nuzi in NE Iraq (Zaccagnini 1975: 219-20). An explicit reference in a Neo-Assyrian cuneiform text declares:

..he shall have the use of the field for six years, three under cultivation and three in fallow"

Chicago Assyrian Dictionary: *karaphu*; see also C.W. Johns, Assyrian deeds and documents.

Although the efficacy of fallowing has been disputed by Loizides (1980), who concluded that fallow in Mediterranean climates does not result in any significant increases in amount of water available to next crop, more detailed studies at ICARDA demonstrated that storage of rainwater does take place under fallow in rainy years (Jabbour and Naji 1991: 163). Some 8% of the previous winter season's rain may be stored after a dry year, whereas maybe 37% is available after a wet year. Consequently if a dry year follows wet, the moisture surplus for the wet year helps provide a safety net for the dry year. On the other hand if there are two years of drought, moisture carry-over will be negligible and the drought effect can be severe. The difference between crop yields of biennially fallowed land and monoculture (fertilized or unfertilized) is indicated on Fig. 5 (from Cocks et al 1988 table 4, with additional rainfall data from Aleppo). Deeper soils (i.e. those > 90 cm) are more effective for fallowing than shallow soils owing to the reservoir effective of soil moisture stored at depths greater than 1 m (Harris et al. 1991: 239; based on Guler and Caraca 1988).

For fallowing to operate effectively, weeding or plowing is necessary to keep the fallow clean of weeds that can otherwise deplete soil moisture reserves. Because fallowed soils subjected to wetting and drying cycles can result in rapid nitrification and accelerated loss of soil organic matter[7] (Russell 1961: 304; Duff, Rasmussen and Smiley 1995: 108), the practice can result in soil deterioration, which in the long-term can counteract the advantages of soil moisture conservation and nutrient accumulation. Alternatively, fallowing can reduce the incidence of soil-borne diseases, so that in Syria, where continuous cereal growing is practiced, increases in infestation of crops can occur (Harris et al. 1991: 241).

As a complement to fallowing, the application of fertilizer, manure or compost can significantly enhance crop growth. Of these, organic amendments can produce more microbial biomass than inorganic fertilizers, thereby enhancing structural stability (Rassmussen and Collins 1991: 119). Both phosphorus and nitrogen stimulate substantial increases in dry matter and grain yield, and by increasing the growth of the plant canopy, shading can result and water use efficiency is increased (Cooper

[7] Alternatively, annual cropping results in the increased return of organic residues to the soil, thereby limiting the long-term decline in soil organic matter (Rasmussen and Collins 1991: 91-112).

1991: 139; Gregory 1991: 13). Furthermore, by increasing the development of roots, fertilizer applications may increase the ability of the plant to extract water from the soil in the spring and early summer growth periods (Cooper 1991: 139), or from soil moisture reserves at depth. Cropping trials in Syria indicate that positive responses to phosphate applications can occur in all parts of the dry-farming zone, even in dry areas (Cooper 1991, figs. 5 and 6), although where there is a previous history of extensive use of fertilizers, response to renewed phosphate applications may be negligible (Pala et al. 1992: 98). Nitrogen, on the other hand appears to be more effective in moister areas or in wetter years, as is the response to fertilizer applications in general (Pala et al. 1992: 97). Because of the absence of chemical fertilizers in antiquity, nitrogen and phosphate enhancement would have to come either from manure or from compost, ashes etc. applied as a soil dressing or by the grazing of animals . The latter, attested in the irrigated lands of southern Mesopotamia from Sumerian times (Civil 1994), may not however have been sufficient to replace lost nutrients in dry farming lands which did not benefit from rejuvenating accumulations of silt. Although in the past the effectiveness of manure applications to arid soils has been questioned owing to the dryness of the soil (Buringh 1960: 252-53), when applied as compost to moist soils such additions would have been effective (Diaz et al. 1994) . By adding badly needed humic material to the soil, manure or compost applications would also offset the loss of organic matter mineralized during fallowing, and in turn would increase the structural stability and moisture-holding capacity of the soil, thus further enhancing crop growth (Rassmussen and Collins 1991: 94).

An alternative way of raising production is by extending the area sown. In many dry-farming areas farmers can judge from the amount of autumn rain (i.e. October to December) the degree of risk entailed in planting. If the autumn is wet the following crop may be expected to be good and, if labor is available, larger areas can be sown. In Tunisia between 1979 and 1987 there was an increase in area sown to barley in wetter years and a decrease in drier years (Ben Mechlia 1993, fig. 3). Alternatively in northern Iraq autumn rainfall provides an indicator of the expected annual rainfall, with rainfall in September to December being approximately correlated with overall rainfall in that hydrological year (r=0.716 for a run of 45 years). As autumn rains usually provide 20 to 40% of total yearly rainfall, this is not sufficient to guarantee a successful crop or to act as a predictor of crop failure, but by judging the situation in

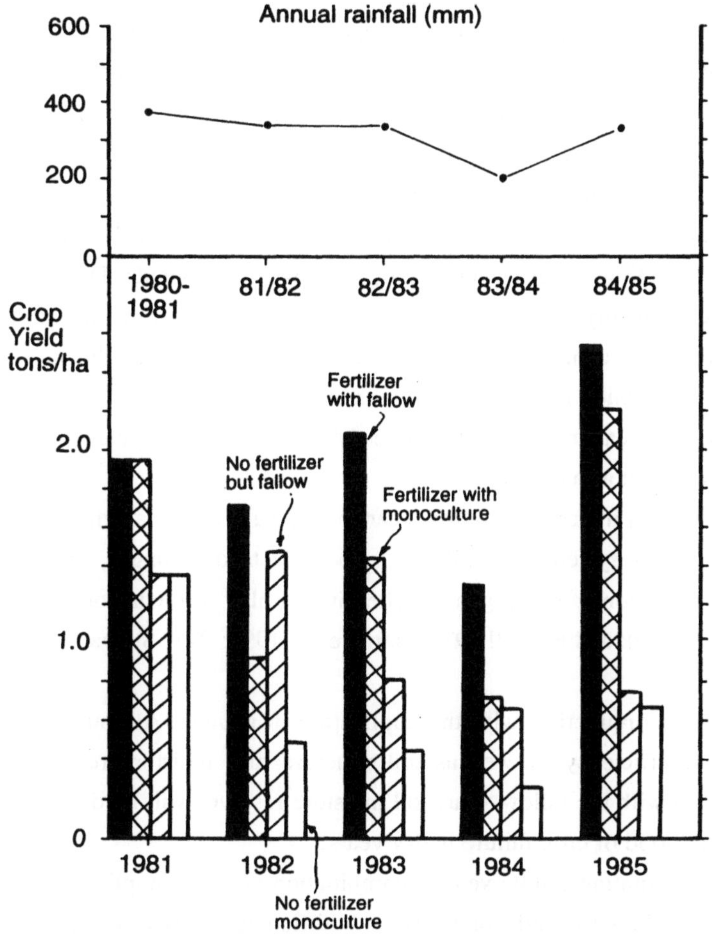

Fig. 5. Barley yield under different crop management systems in part of northern Syria compared with rainfall for the same years at Aleppo (top). (Based on data in Cocks et al. 1988 and various climatic data).

late February, [by which time some 65% of annual rainfall has been received (Thalen 1979: 53)], the likelihood of crop failure can be roughly gauged. As a result farmers can choose, if failure seems certain, to turn the crop over to grazing. Later in the season they can then recoup part of their loss by selling off spare animals and purchasing grain on the market.

In addition, a number of other crop management techniques can mitigate the effect of dry years, although in such a semi-arid environment, the versatility of crop management practices is limited, especially near the arid margins. Alternative practices include the variation of: total cropped area and the distribution of crops; the amount of intercropping; plant density; substitution of crops with low moisture demands; the date of planting, as well as the use of mulches to preserve soil moisture.

Of the above techniques, fallow and fertilization can mitigate some of the worst effects of drought and here I suggest an ordering of traditional cropping practices from those that are most liable to crop failure (A) to those that favor soil moisture conservation, nutrient recycling, water use efficiency and sustainable yields (D).

A) Cereals planted annually. In this practice soil nutrient reserves are not replenished and soil moisture cannot be carried over from one season to the next. This system, although *sometimes* providing more cereal production, simply because the land is cropped every year, is especially vulnerable following dry years when there is no soil moisture to act as a reserve.

B) Cereals planted with an intervening fallow year. As a result of the carrying over of soil moisture this ancient system can mitigate some effects of drought, but by encouraging the mineralization of soil organic matter, it can also result in long-term deterioration of the soil (Clark and Russell 1977; Harris et al. 1991: 241; Stewart et al. 1993: 71).

C) Annual cropping in conjunction with the application of manure. Although this may improve water use efficiency, as well as soil structure and moisture retention properties, by dispensing with the fallow year, soil moisture reserves will be depleted thus increasing the likelihood of crop failure in dry years.

D) Planting with an alternating fallow year in combination with the application of manure and or composts. Manure and composts, in addition to improving soil tilth and water retention properties, should replace or improve organic matter lost as a result of mineralization during the fallow year.

Of the above, strategy D may have been the most sustainable practice in antiquity. Cropping trials in the Stavropol area of southern Russia (with chemical rather than organic fertilizers) showed that clean fallow combined with fertilizer application resulted in significantly higher yields (5.3-6.1 tons/ha!), than either cereal and fallow (4.0-5.5 tons/ha) and wheat preceded by a rotated crop (2.5-3.6 tons/ha). In such areas of fertilized and fallowed plots, crop yield remained a significant 87% of maximum production in even the driest years (Nikonov 1988: 598).

Land use and soil degradation

Of the land use practices described above, fallowing, although beneficial in the short term, by encouraging depletion of organic matter and the associated mineralization of nitrogen, leads to situations that are unstable in the long term (Harris et al. 1991: 241; Clarke and Russell 1977: 287; Stewart et al. 1993: 71). Not only are nutrients lost, but also soil structure deteriorates and soils become more vulnerable to erosion by both wind and water. In turn, by removing the lighter soil fractions (silt/clay aggregates, organic matter), deflation can result in further deterioration of the soil. The existence of extensive areas of unvegetated (fallowed) land, either plowed to discourage weeds, or unplowed and crusted, can then encourage either wind or water erosion (Blaikie and Brookfield 1987: 131). Similarly, rapid declines in organic nitrogen and carbon can result if crop residues are not returned to the soil, a situation which is exacerbated by the increased soil erosion that can take place on clean fallow (Bolton 1991). Therefore in the past, if more crop residues (stubble etc.) were removed by grazing animals for stall feeding of animals, or as temper for mud brick, fuel etc., fertility would decline and erosion hazard would increase. This is a further negative feedback process that might result during Bronze Age urbanization and which would need to be addressed in order to avoid long-term declines in crop productivity[8].

Today the northern dry-farmed steppe within Syria is subject to moderate soil erosion from water, whereas areas further to the south are more prone to eolian degradation (UNEP 1992: 42-45). As a result of long-term often human-induced degradation, there is a gradual decrease of organic matter in topsoils such that marginal soils under native pasture exhibit most organic matter, more severely eroded marginal lands less, improved cultivation with farmyard manure less and cultivated land (un-manured) least (Cocks et al. 1988 table 5). Such is the pressure on land today that dust storms are frequent and the recent expansion of cultivation to the north of Jebel Bishri has resulted in eolian deposits forming over former grazing lands, as well as covering railway tracks in the Khabur valley (ibid.: 44-45). Today periodic

[8]Frequent plowing, by increasing the availability of nutrients, can however result in short-term increases in crop yield but at the expense of nutrient reserves. The complex inter-relationships between stubble and tillage management and soil moisture/nutrient reserves, are however beyond the scope of this paper. For discussion and references see various papers in Harris et al. 1991.

dust storms testify to the massive mobilization of topsoil that can occur from extensive areas of plowed steppe.

The increase in population that occurred through the Holocene in the marginal dry-farming zone of Upper Mesopotamia, by increasing cropping intensity as well as the area of cultivated land, encouraged soil degradation. This, together with the large-scale removal of vegetation by grazing animals and for fuel, must have contributed to the present heavily degraded landscape. Such processes of vegetation removal, by influencing surface albedo, can also result in complex feedback mechanisms which in turn can influence microclimate. For example, when shade-giving plants and trees are removed, local microclimates will become more extreme which can result in localized increases in aridity (Kaul and Thalen 1979: 258). Conversely, increased surface reflectivity can induce a complex range of responses that can increase surface warming which in turn may either encourage the development of localized convectional rain or conversely further drying (Balling 1988). Increased intensity of cultivation during the 3rd millennium BC may also have encouraged the development of periodic dust storms, which by increasing atmospheric dust may also have encouraged a number of feedback mechanisms which in turn may have influenced local climate (Bryson and Baerris 1967: 141). Unfortunately, because of uncertainties regarding the direction and significance of such mechanisms it is not yet clear how significant they are on a global scale (for further discussion see Williams and Balling 1996).

Demography, land use and collapse

Climate-induced population collapse can be analyzed by examining interactions between long-term population growth, a limited range of climatically-determined crop production values, and various simple crop production strategies which incorporate minor changes of technology. By holding the fluctuation of crop production values within a range currently expected for the region, we can examine how population might increase under normal conditions of random climatic fluctuation. This is equivalent to a constant climate model in which only random rainfall fluctuations influence crop yield. We can therefore determine under what conditions production shortfalls might be an inherent property of the system, and then whether climatic fluctuations of a given order would be necessary conditions to initiate crop failure and perhaps population collapse. These simple models only attempt to deal with the crop-production/climate system; other socio-economic factors potentially influencing collapse are not considered here (see for example Tainter 1988 and Yoffee 1988). In spite of these simplifications the model does serve to highlight a number of features of settlement system behavior and thus may enable researchers to assess what climate fluctuations might be sufficient to produce a collapse in population.

Because the use of population growth as an autonomous driving force is ultimately questionable (Cowgill 1974) it is employed here simply as a hypothetical background and initial condition. Growth rates, which are held constant during periods of population increase, are considered to fluctuate if crop production falls short of that required to support the population. As will become apparent, at each stage of development a number of outcomes are possible. Each outcome allows the system to remain stable, to collapse or for a new higher level of the system to operate. Although in essence this is a version of an earlier model (Smith and Young 1972), the original model assumed population growth as being the driving force behind all subsequent change. The proposed model instead, incorporates a range of crop production scenarios that allow for the intensification of production, as well as for enlarged zones of supply or improved technology. In its more developed stage the model also incorporates fluctuating levels of population and potentially allows for population growth to both lead to and result from changes in production (Netting 1993; Morrison 1994: 118).

Assumptions:

1) Early Bronze Age settlements are nucleated and each significant settlement is surrounded by a production zone of 5 km radius. This production territory, which is based on the assumption that territorial boundaries occur approximately half way between contemporaneous centers, coincides with a reasonable return walk from the main sites. Model parameters derive from archaeological surveys conducted in the region of Tell al-Hawa in northern Iraq where primary and secondary centers 9-12 km apart result in production areas of approximately this size. Territories of this scale are also supported by the fade-out point of ancient tracks radiating from the central site and from surface sherd scatters considered to be the result of manuring in antiquity (Wilkinson 1994).

2) Long-term population growth rate is 0.1% per year (equivalent to 10% per century). This is a commonly cited figure for the prehistoric period (Hassan 1981: 221) and does not conflict with demographic growth rates computed for the area of the

North Jazira Survey. Because constant population increase as a prime mover is an unrealistic assumption (Cowgill 1974: 507) this is only used for the heuristic Model I. A second model (Model II) is then generated on the assumption that when production falls below potential population (i.e. the growth curve), population will fall to the new, lower level, from which point population gradually recovers again at the assumed growth rate of 10% per century. Although still an over simplification, this scenario provides a more realistic picture than that of a constant growth rate.

3) In rain-fed Upper Mesopotamia cereal production varies from year-to-year and is primarily proportional to rainfall fluctuation. Here rather than model for rainfall fluctuation and then calculate the population that this can support, I have simply assumed a random fluctuation of production, initially as a uniform random distribution between a range of upper and lower values (model I) and second, as a random distribution normally distributed around a mean with specified standard deviation (model II). Carrying capacity is estimated directly as the number of people the production could support within the specified territory based on the assumption that annual consumption is approximately 250 kg of grain (or its wheat equivalent) per year. If another food product is eaten, partly as a substitute for wheat, wheat equivalent is counted as the amount of wheat used to exchange for that food (see Hillman, 1973). Fluctuations in potential carrying capacity are then applied to the same diagram as the demographic growth curve in order to depict their co-variation through elapsed time from a hypothetical origin (Figure 6). The scenario for Model II (Fig. 7), regime B assumes a crop yield range of 350-750 kg/ha minus wastage etc. of 25-30% . Net crop yields range therefore from 250-500 kg/ha with a mean yield of 375 kg/ha every 2nd year (for discussion see Russell 1988). This gives a mean carrying capacity for the 5 km radius catchment of 5850 with a standard deviation of 1464 people. This would be the standard production system of dry-farmed cereal cultivation under alternating years of fallow.

4) When production is considered to fluctuate at random between fixed upper and lower values, it is assumed that carrying capacity and population are determined by years of low yield. This concept of the "normal surplus" (Allen 1965, Halstead 1989) is based on the observation that traditional societies plant sufficient area to gain enough food in a year of low yield (i.e. not a year of mean yield which would result in periodic famines). This concept is also incorporated in the models that employ

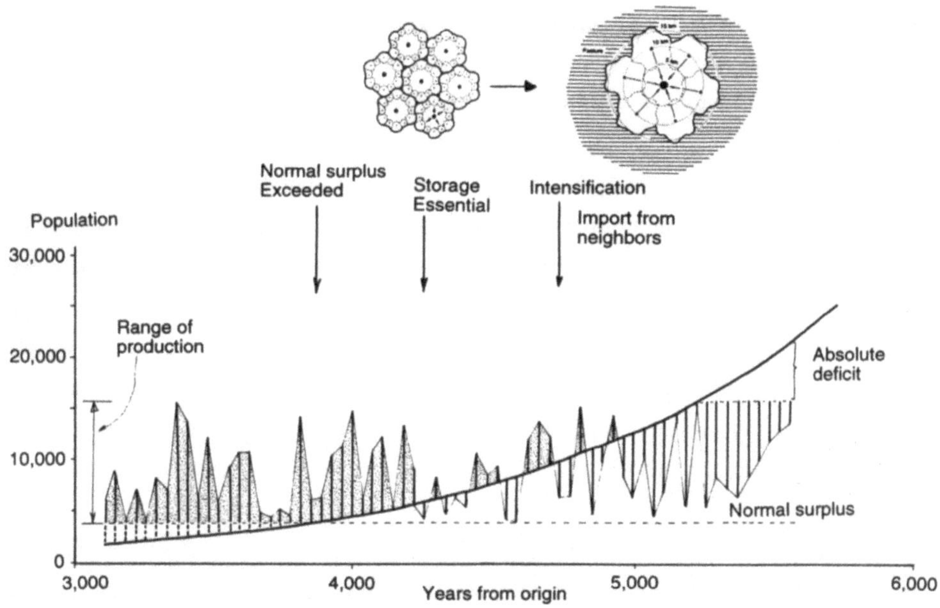

Fig. 6 Model I. Simplified model showing random fluctuations in gross carrying capacity within a 5 km catchment and strategies required to counteract production deficits as population increases through time. For details see text.

normally distributed random production figures, but in such cases it can be seen that the population growth curve, by declining when production falls below hypothetical population, automatically adjusts population to the appropriate low yield, with the result that there is a "normal surplus" in most years.

5) In model I, the basic cultivation strategy is that of biennial cropping without fertilization. A cereal yield of 250 kg/ha is assumed every 2nd year with intervening years of fallow. In good years a yield 1,000 kg/ha is assumed, again every 2nd year, which gives a catchment carrying capacity range from 3,900 to 15,600 people. Uniform random productions within this range were then generated for Model I (Figurė 5). Thresholds which indicate the approximate point where additional strategies (major storage, import from neighboring territories, and crop intensification) become necessary are indicated. In model II, rather than assuming that the same crop-production strategy prevailed through time, increased production figures have been assigned when crop production falls below population more than 2 years in a row.

Regime B is biennial fallow as given above in 3). Regime C assumes crop production of moderate intensity, namely manuring with biennial fallow. Cocks et al. (1988) in a study of dry-farming near Aleppo, Syria, give range of 1,300-2,500 kg/ha which is double the local yield for traditional unfertilized biennial fallow. The range of production for regime C is therefore now assumed to be double that of regime B, that is between 500 and 1,000 kg/ha, every second year which gives a mean carrying capacity of 11,700 people with a standard deviation of 2,925. The final most intensive scenario, regime D operates under conditions of maximum demand. High intensity production with manuring without fallow is assumed across the entire area. Cocks et al. (1988) give cereal productions for this practice that are *c*. x 1.5 the traditional unfertilized biennial fallow system. However, being practiced every year this gives the highest gross production. Here a yield range of 375 -750 kg/ha every year is assumed, which raises mean carrying capacity to 17,550 with a standard deviation of 4,388.

The above strategies therefore allow population to rise as a result of increased levels of production. The above device of introducing new production regimes, although seemingly arbitrary, is employed to mimic assumed behavior; that is when production potentially falls short, appropriate strategies are employed to raise production in the successive stages B, C and D (note that an initial stage A, relevant to early pre-urban stages of settlement has been omitted for reasons of space).

For simplicity the somewhat unrealistic assumption has been made that there is spatial uniformity of land use intensity. In other words there is, for example, no increase in fallowing away from the site (see Wilkinson 1994 for models that make allowance for such variations). Stages B, C and D should not be seen as discreet evolutionary steps but rather as a continuum that is progressively adopted through time. Therefore, for example, during stage B, some areas of any given territory will only be at stage A levels of production whereas others, (probably those nearest the central settlement) would be producing at stage C. Such factors would blur the stage changeover points on the models.

Model outcome

Figure 6 shows Model I with an arbitrary scale in years along the horizontal axis and a vertical axis expressed in terms of carrying capacity of the suggested 5 km radius catchment[9]. The time scale is in relation to a hypothetical origin, from which the curve was calculated at a growth rate of 0.1% per annum. Owing to the wide range of crop yields that would result from such a fluctuating pattern of rainfall, an equally large variation in population can be potentially supported from the given area; but because early communities probably operated on the principal of a normal surplus (Halstead 1989), most years of crop production would theoretically be in excess of needs. In reality such large surpluses would not result, either because the entire field area would not be sown, or if it was, some crop would be left perhaps to be grazed by animals. In any case limitations on available labor would tend to constrain the amount of land that could be planted or harvested (Wilkinson 1994: 495-96). Storage of any excess grain would tend to minimize waste as well as ensuring against crop failures in future.

For Model I, as population increases an initial threshold is reached where the "normal surplus" is exceeded. There then follow occasional "years" when randomized production falls below carrying capacity. This would represent no critical problem for the inhabitants because storage would enable such bad years to be ridden out with relatively little stress. From this point on however, storage would become crucial for survival.

With further population growth there is an increased probability that crop production within the site catchment would be less than potential population until a point would be reached when occasional years show two successive years of deficit. Although individual drought years would be survivable, either because of storage or because moisture carry-over in fallow would provide more stable crop yields, such back-to-back droughts (as defined by Parry and Carter 1988: 23) would not. At such times stored surpluses would be seriously strained and at this stage if not before, the inhabitants might adopt the expedient of either cropping every year, introducing irrigation (if feasible) or of introducing some form of fertilization (see below). The last-named practice, if combined with fallow would represent a more sustainable high yield production system (see above), but inevitably there may have been a temptation to cultivate annually. The introduction of fertility enhancement would raise system carrying capacity thus increasing the chance of survival for some time. Alternatively,

[9]Because insertion of data points at annual intervals would result in prohibitively large numbers of data points, for this purely illustrative model I have adopted an arbitrary spacing of carrying capacity values at 40 years apart. Actual thresholds and/or collapse points would therefore occur at rather different positions along the curve.

the inhabitants (or the elites) could have opted to import food from neighboring territories if this was possible. Such tactics may have been the norm for extended periods of time, but with increased pressure on resources such transactions would tend to become more directional and of greater long-term importance. If annual cropping was introduced in conjunction with (or without) fertilization, by dispensing with the fallow year the soil moisture surplus of the fallow would be lost. Consequently crop failure would be more likely (Loomis 1983: 360). At this stage it is evident that there are at least two outcomes for the system: a) survival at a new higher level of production, or b) collapse.

The alternative Model II introduces at key points increased cultivation intensity, which in turn results in increased production. These new levels are assigned at points where production falls below the carrying capacity of the site territory. As a result, population potential would rise further until again a stress point would be reached at which point collapse would be imminent or a new crop production strategy would be adopted. Because it is unrealistic to assume constant increase in population (i.e. the hypothetical smooth curve), secondary curves have been generated on the assumption that when the curve of randomly fluctuating production falls below the population curve, population declines to the new production level and then subsequently rises at 0.1% per annum until production again falls below the rising population curve. As a result of these episodic disparities between potential production and population the secondary curves (B, C and D Fig. 7) have a saw-toothed appearance. Rather than being viewed a mass die-offs of population, such mutual adjustments could represent changes in fertility, death rate or simply emigration out of the area. Furthermore, the illustrated declines would probably be smoothed by storage, the import of food, or by other measures to ensure continuity of food supply. The three curves consistently fall below the hypothetical curve of population increase and fluctuate around what must have been the long-term carrying capacity of the 5 km radius territory. In one fortuitous case the intermediate curve C rises above the higher production curve D, but eventually settles as expected between curves B and D. As an alternative to a strategy of intensification, food could have been imported from neighboring territories, but this would only have been possible if such neighbors had spare production, and if social and political relations between neighbors permitted it. The necessity of securing adequate supplies of food would therefore have encouraged cooperation

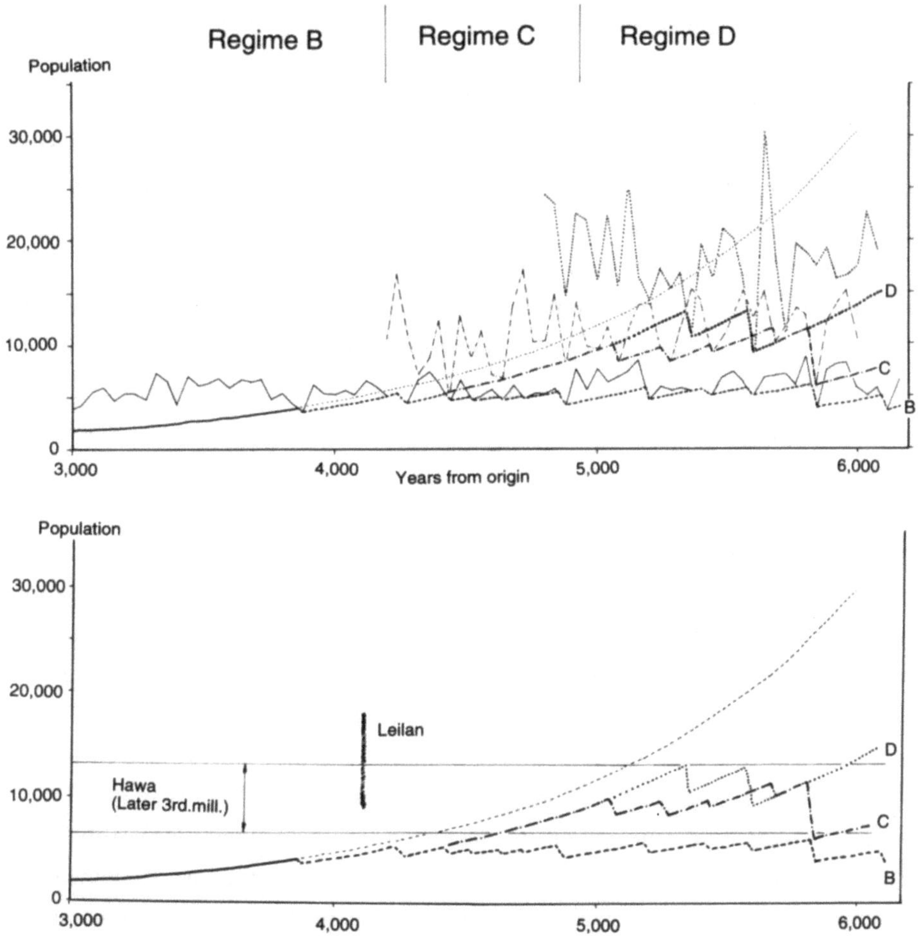

Fig. 7 Model II. As for Model I but incorporating occasional population declines resulting from interactions between fluctuating crop production population. The higher curves represent raised production levels and carrying capacities resulting from more intensive management practices (see text for details).

and trade between neighboring centers, with the result that perhaps a larger regional polity would result in which there would be greater interdependence between individual components of the system. This would represent the crossing of a fundamental threshold, and would potentially allow population of the center to rise still further, perhaps up to 10,000- 20,000 people, a figure which approximates to the maximum size of Early Bronze Age towns in Upper Mesopotamia (Wilkinson 1994). At this point each system might comprise for example a large central town of up to 100-120 ha area, several secondary towns of 10-20 ha each up to a maximum of six, together with a number of smaller places of lower rank.

Beyond a certain point, to raise production still further, the most intensive cropping system, annual cropping with manuring could have been adopted. By dispensing with the fallow year soil moisture levels would be depressed and nutrients would be extracted annually rather than every other year. The system, although more productive in the short term, would therefore have been more prone to ultimate decline. At such a point, if further import from neighbors was impossible, either because of political reasons or because they too were experiencing food stress, there would have been fewer alternative outcomes and at this stage famine and population collapse would be most likely.

The above scenarios only represent a first attempt at simulating potential population trajectories within bounded catchment areas. Because different land-use strategies would result in higher or lower values of crop production, and population estimates from the central sites are only approximate, the scenarios remain tentative. However, it is evident that these population trajectories fluctuate within the estimated population range of large Early Bronze Age centers (see Leilan and Hawa on Fig. 7) which suggests that each catchment was capable of supporting such settlements, with perhaps some importation of food from their neighbors. Changes in population may result either from immigration or emigration, or fluctuations of fertility/death rates. The peaks and troughs illustrated would probably either be smoothed by variations in storage, cropping intensity and the import of food from neighbors, or exaggerated by negative feedback effects. Furthermore, more advanced models could program in a wider range of population growth rates which would simulate more realistically real world conditions. Nevertheless, by showing how population might rise and fall in no predictable manner in response to random climate fluctuations, such models may eventually make it possible to gauge the impact of larger scale climatic events.

Land-use and Population in Upper Mesopotamia

To test the above models, data from the North Jazira Project (Iraq) are employed to provide estimates of land use and long-term population trends. It is assumed that sites supported a significant sedentary population and that the minimum amount of

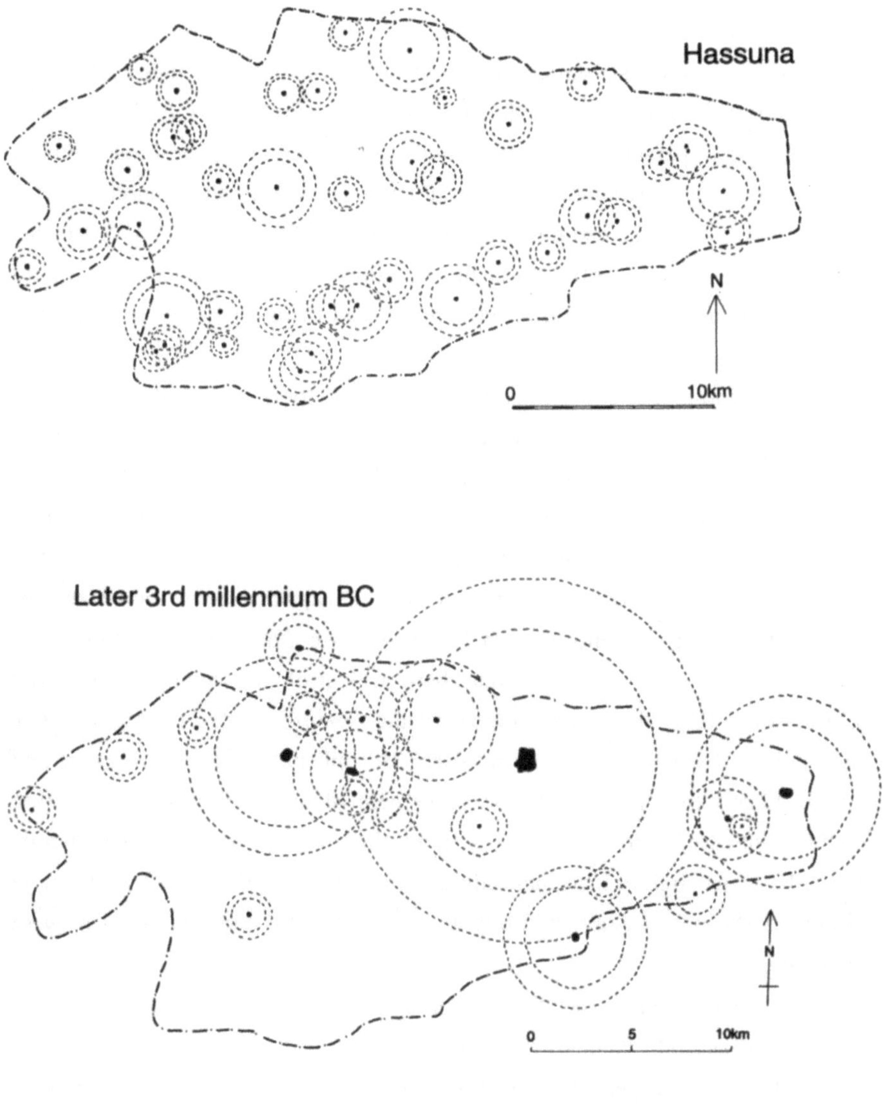

Fig. 8 Calculated sustaining (i.e. cultivated) areas for Top: Hassuna (6000-5000 BC uncalibrated) and Below: later 3rd millennium BC (ca. 2500-2000BC) for the area of the north Jazira survey around Tell al-Hawa in northern Iraq.

cultivated land necessary to support the inhabitants can be calculated according to the following assumptions: site population densities were probably between 100 and 200 persons per hectare (see Kramer 1980 and Sumner 1989 for discussions). However, based on estimates from architectural data from 3rd millennium BC Abu Salabikh in southern Iraq, Postgate (1994) provides an exceptionally high estimated range of 258 to 1205 persons per ha, which if it obtained in Upper Mesopotamia would lead to much high production deficits than indicated by these models. As a result collapse would be even more likely than predicted. Conversely low estimates of urban population density (60 persons/ha to allow for open space and public buildings) by Stein and Wattenmaker (1990) would lead to lower production deficits (see below).

Calculated sustaining areas, which were 18-36% of total surveyed area in the Hassuna period (Fig. 8a: 6000-5000 BC uncalibrated), had increased to between 56 and 127% of the area by the Uruk (roughly the 4th millennium BC). Although the figures remain approximately constant for the 2nd half of the third millennium, there was a shift from a much more evenly distributed population in the Uruk period to one that was concentrated in the area around Tell al-Hawa during the later 3rd millennium. With such a concentration in larger sites we find that sustaining areas merge and overlap to such a degree that there was probably an increase in the intensity of land use (Fig. 8b). Abandonment of the western part of the survey area, which occurred between the later 4th and early 3rd millennium BC (Wilkinson and Tucker 1995: 43-50) may result from the movement of population to the growing towns of the Ninevite 5 period (that is a trend that may have a social/economic cause). Less likely, the slightly marginal rolling terrain in the western area could have been abandoned during a sustained drought during the 4th millennium. This major settlement change might
correspond, to a "major social disruption" recorded in southern Mesopotamia towards the end of the Uruk period (Adams 1981; Postgate 1986: 96).

By the Middle Assyrian period (1300-1000 BC), following a phase of settlement growth during the early 2nd millennium (Old Assyrian period), settlement again declined so that somewhere between 29 and 57% of the survey area was cultivated. Again there is no direct evidence to indicate whether this decline was a result of socioeconomic or environmental causes.

Table 1

Period	Calculated aggregate cultivated area	% of total survey area
Hassuna (6000-5000BC[1])	84.5 - 169 sq. km	18-36%
Uruk/Late Chalcolithic (3700-3000BC)	268 - 604[2] sq km	56 - 127%
Later 3rd Millennium BC (2500-2000BC)	292 - 584 sq km	61 - 123%[3]
Middle Assyrian (1300-1000BC)	136 - 273 sq km	29 - 57%

1. Dates are BC uncalibrated.

2. The wide range is due to uncertainties in the upper size limit of occupation of Tell al-Hawa during the 4th millennium BC. The upper limit is less likely being based upon the presence of the outlying Uruk/Late Chalcolithic occupation in a single trench under Tell al-Hawa village (Area LP: see Ball et al. 1989: 31-33 and 39).

3) When calculated using the statistics of Stein and Wattenmaker (1990) this would give total sustaining areas of 304 sq km amounting to 64% of the survey area.

Long-term trajectories of population can be approximated at a coarse level of resolution from the aggregate settlement area estimated from archaeological surveys on the assumption that overall population is directly proportional to aggregate site area (Fig. 9). Although earlier settlements will be under-estimated as a result of their burial beneath later occupations, there is a general impression of a long-term increase in aggregate settlement area from the Hassuna (6000-5000 BC) to the Khabur period (early 2nd millennium BC). The disruption between the Uruk and Ninevite 5 is indicated by a fall in total settlement number and a slight decrease in aggregate settlement area; at the same time the abandonment of settlements increased and settlement continuity declined (Wilkinson and Tucker 1995: 78-80). Particularly noteworthy is that the upward trend in aggregate settlement area that peaked in the 3rd and early 2nd millennium did not continue after the mid-2nd millennium. A substantial decline in both settlement numbers and aggregate settlement area during

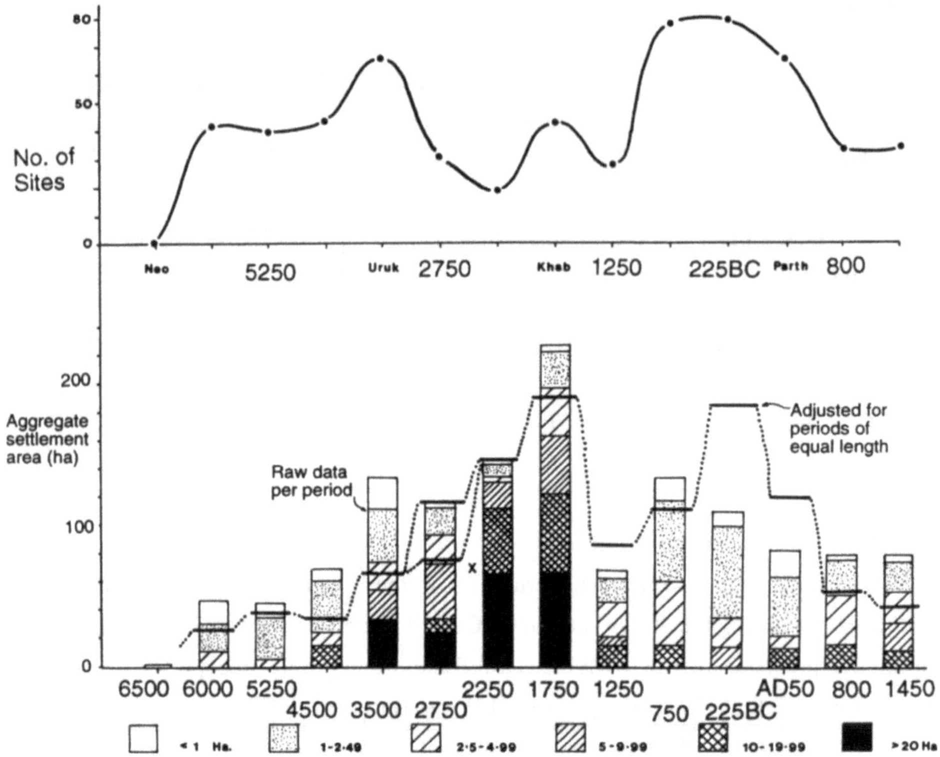

Fig. 9 Aggregate site area taken as proxy for regional population for part of northern Iraq. Symbols refer to the occupied area of sites; survey area 475 sq km in area of Tell al-Hawa (amended from Wilkinson and Tucker 1995: fig. 50).

the Middle Assyrian period resulted from the dwindling of existing settlements. This decline continued until there was a significant rejuvenation of settlement in the Late Assyrian or Hellenistic periods (1st millennium BC) depending upon whether the raw or the adjusted population proxy data are employed. There followed a steady decline in settlement until the plain was virtually deserted by the 18th/19th century AD (Wilkinson and Tucker 1995). Such a long-term decline suggests that after the 2nd millennium BC settlement, although viable, could not occur on the scale that existed during the 4th, 3rd and early 2nd millennium.

Although data from archaeological surveys is not yet sufficiently consistent to allow the results from the Iraqi north Jazira to be extended over the remainder of Upper Mesopotamia, the extent of Bronze Age settlement (i.e. ca. 3000-1300 BC) can be roughly estimated from the distribution of tells (a large number of which were occupied during this period), combined with the estimated area of contemporaneous cultivation. These estimates suggest that the area of Bronze Age occupation and cultivation differ little from that of present day limit of cultivation indicated on Figure 1. Nevertheless, if the trajectory of aggregate settlement area for this greater region corresponds even approximately to that illustrated for the Tell Hawa area (Fig. 9) it would seem that although the limit of cultivation may have changed relatively little since the Bronze Age, the actual population within the rain-fed zone may have declined significantly. If intensive cultivation was continued for sustained periods over such an area it is likely that soil deterioration would also have resulted. Periodic droughts or longer dry spells may have contributed a significant amount of dust to the atmosphere from the rain-fed zone, thereby exacerbating soil degradation and potentially influencing local climate (cf. Bryson and Baerris 1967).

Discussion

The above underscores the complexity of land use-climate relationships, from which it is possible to distill a number of themes. Settlement and land use systems can fluctuate between a resilient state during which it would be feasible to over-ride minor climatic events and "brittle" systems which would be prone to collapse. The latter state would have been especially prevalent during the early and middle Bronze Age phases of urbanization. Runs of wetter years may have allowed settlement to extend into drier marginal areas, if political and socioeconomic conditions allowed, and conversely during drier intervals settlement declines might have occurred. However, because of high inter-annual rainfall fluctuations, negative feedback effects can influence levels of production so that continued cultivation would encourage loss of nutrients, the fine earth fraction and organic matter. As a result, soil structure would deteriorate and soils would be vulnerable to deflation thereby resulting in further depletion of productive capacity (Mainguet 88-96). Other feedback effects include the possible entrainment of dust into the atmosphere which might have influenced regional or local climates (cf. Bryson and Baerris 1967). As population increased during the Holocene, urbanizing parts of the northern Jazira in Iraq and Syria must have experienced increased land use intensity with biennial cropping being replaced either by irrigation (if feasible), by localized annual cropping, and/or by cropping systems (e.g. manuring) that returned nutrients to the soil. If practiced, annual cropping by abolishing fallow, would have decreased soil moisture carry-over thereby destabilising yields. Such systems would periodically at least be more dependent upon neighboring territories for food, the import of which would incur greater socio-economic ties between communities and would result in increased integration between settlement systems. Parallel with this however, the import of food from neighbors would result in the export of nutrients from the soils of those neighbors, which if practiced over a long period would result in depletion of their soils.

Although population growth, urbanization and land use intensification can increase the likelihood of collapse, regional studies and archaeological excavations also suggest phases of significant population decline or shift that might result from a climatic change. Ultimately it is possible that a number of inter-related factors may have increased the instability of Holocene crop-production and settlement systems in the dry farming area of the Near East. First, an apparent trend towards lower mean annual rainfall, during or shortly after the 3rd millennium BC (Courty 1994, Frumkin et al. 1994, Geyh 1994) could have led to both an overall decline in crop production, as well as to increased inter-annual variability[10]. Second the domestication of plants and the associated localization of production in specific ecological areas, although increasing food yield, could also intensify the potential instability of the system, again increasing the likelihood of crop failure and collapse in production (Rhindos 1984: 273; see also Portugali 1994). Third, a range of socio-economic and political factors might have inhibited adaptation of Bronze Age communities to changing environmental conditions (Rosen 1995). The above factors combined with those already discussed must have significantly contributed to instabilities in crop production, instabilities which may have been tripped by minor events such as back-to-back droughts, or short-term runs of dry years. The ensuing production short-falls during such events would result in the perturbations evident on Figs. 7 and 9. Furthermore, the effect of a decline in rainfall (and associated catchment carrying capacity) equivalent to, for example from regime D to C (Fig. 7), may have resulted in a complex saw-tooth population trajectory with a lower mean population, rather than simple collapse. Given these complex inter-relationships, it is hardly surprising that

[10] According to Kerbe (1987, map 1) inter-annual variation in the Syrian desert (< 150mm rainfall) is in the range 50-60% compared with 30-40% in the moist steppe with rainfall > 400mm.

occasional population collapses occurred and that some were perhaps substantial and wide-ranging.

References

Abate, Tom 1994 Overview of collapse in Bioscience 44: 516-19. (QH1 A.15 but in Crerer in alphabetical order)

Adams R McC (1978) Strategies of maximisation, stability, and resilience in Mesopotamian society, settlement, and agriculture, Proceedings of the American Philosophical Society 122: 329–35

Adams R McC (1981) Heartland of Cities. University of Chicago Press, Chicago

Ali, Hassan Mohammed (1955) Land Reclamation and Settlement in Iraq Baghdad Printing Press, Iraq

Allen W (1965) The African husbandman, Edinburgh, Oliver and Boyd

Ben Mechlia N (1993) Variations in the crop area and yield of barley in relation to rainfall under a Mediterranean climate. in Jones M, Mathys G, and Rijks D (eds) The Agrometeorology of Rainfed Barley-based Farming Systems: 121-26. ICARDA, Aleppo, Syria

Blaikie P, Brookfield H (1987) Land Degradation and Society, London, Methuen.

Bolton FE (1991) Tillage and stubble management, in Harris et al (eds): 39-47

Bottema S (1989) Notes on the prehistoric environment of the Syrian Djezireh, in OMC Haex, Curvers, HH, and Akkermans PMMG (eds) To the Euphrates and Beyond. Archaeological Studies in Honour of Maurits N van Loon Brookfield, Rotterdam

Brichambaut G Perrin D and Wallin CC (1963) A Study of Agroclimatology in the Near East World Meteorological Organization Technical Note 56, Geneva

Bryson RA, and Baerris DA (1967) Possibilities of major climatic modification and their implications: North-west India, a case study, Bulletin of the American Meteorological Society 48, no. 3: 136-42

Buringh P (1960) Soils and Soil Conditions in Iraq, Republic of Iraq, Ministry of Agriculture, Baghdad

Butzer K (1995) Environmental change in the Near East and human impact on the land in J Sasson (ed) Civilizations of the Ancient Near East vol. 1: 123-51. New York, Scribners/Macmillan

Butzer, KW 1980. Civilizations: Organisms or systems? *American Scientist* 68: 517–23.

CAD The Assyrian Dictionary of the Oriental Institute of the University of Chicago

Civil M (1994) The Farmer's Instructions, Barcelona,

Clarke AL and Russell JS (1977) Crop sequential practices, in Russell JS and Greacen EL (eds) Crop Production in a Semi-arid Environment: 279-300, The Australian Society of Soil Science and Queensland University Press, Brisbane, Australia

Cocks PS, Thompson EF, Somel K and Abd el-Moneim A (1988) Degradation and Rehabilitation of Agricultural Land in North Syria. ICARDA, Aleppo, Syria

Cooper PJM (1991) Fertilizer use, crop growth, water use and water use efficiency in Mediterranean rainfed farming systems. in Harris HC et al. (eds): 135-52

Courty M-A 1994 Le cadre paléogéographique des occupations humaines dans le bassin du Haut-Khabur (Syrie du Nord-est). Premier resultats, Paléorient 20.1: 21-59

Cowgill G (1974) On causes and consequences of ancient and modern population changes. American Anthropologist 77: 505-25

Davies DH (1957) Observations on land use in Iraq, Economic Geography 33: 122-34

Diaz E, Roldan A, Lax A, and Albaladejo (1994) Formation of stable aggregates in degraded soil by amendment with urban refuse and peat. Geoderma 63: 277-88

Dillemann L (1962) Haute Mesopotamie Orientale et Pays Adjacents. Paris Institut Français D'Archeologie de Beyrouth 72

Duff B, Rasmussen PE, and Smiley RW (1995) Wheat/Fallow systems in semi-arid regions of the Pacific NW Ameirica. in Barnett V, Payne R, and Steiner R (eds) Agricultural Sustainability: Economic, Environmental and Statistical Considerations: 85-109. New York, John Wiley

Fales FM (1990) The rural landscape of the Neo-Assyrian empire: a survey. Bulletin of the State Archives of Assyria: 81-142, Helsinki

FAO/UNESCO (1977) Soil Map of the World, Volume VII with accompanying handbook, UNESCO, Paris.

Finkbeiner U and Rollig W (1986) Gamdet Nasr: Period or Regional Style? TAVO, L Reichart Wiesbaden

Gommes R (1993) Current climate and population, constraints on world agriculture, in HM Kaiser and TE Drennen (eds) Agricultural Dimensions of Global Climate Change: 67-86, Florida, St Lucia Press

Goudie A (1986) The Human Impact on the Natural Environment, Cambridge, Mass. MIT Press.

Gregory PJ (1991) Concepts of Water Use Efficiency. in Harris HC et al. (eds): 9-20

Gremmen WHE and S Bottema (1991) Palynolological investigations in the Syrian Gazira, in: Die Rezente Umwelt von Tall Seh Hamad und Daten zur Umweltrekonstruktion der Assyrischen Stadt Dur-Katlimmu, H Kuhne (ed),: 105–16. Berlin, Dietrich Reimer Verlag

Guest E (1966) Flora of Iraq, Republic of Iraq, Ministry of Agriculture, Baghdad.

Halstead P (1989) The economy has a normal surplus: Economic stability and social change among early farming communites of Thessaly, Greece, in Bad year economics, Paul Halstead and John O'Shea (eds): 68–80. Cambridge: Cambridge University Press

Harris HC (1991) Implications of climatic variability, in Harris HC et al (eds): 21-34

Harris HC, Cooper PJM, Pala M (1991) Soils and Crop Management for Improved Water Use Efficiency in Rainfed Areas. ICARDA, Aleppo, Syria

Harris HC, Osman AE, Cooper PJM and Jones MJ (1991) The management of crop rotations for greater water use efficiency, in Harris et al: 237-50

Hassan Fekri A (1981) Demographic archaeology, New York, Academic Press

Hillman G (1973) Agricultural productivity and past populations at Asvan, Anatolian Studies 23: 225-40

Holling CS (1986) The resilience of terrestrial ecosystems: local surprises and global change in Clark WC and Munn RE (eds) Sustainable Development of the Biosphere: 292-320, Cambridge University Press

Issawi C (1988) The Fertile Crescent 1800-1914 Oxford, Oxford University Press

Jabbour E, Naji M (1991) Soil and crop management for improved water use efficiency in rainfed areas of Syria, in Harris HC et al (eds): 159-67

Jean Charles-F, (1947) Lettres de Mari IV Transcrites and traduites, Revue d'Assyriologie 42: 53-78

Johns CW (1901) An Assyrian Domesday Book or Liber Censualis of the district around Harran in the seventh century BC, Liepzig

Jones MJ (1993) Barley-based farming systems of the Mediterranean in Jones MJ (ed) The Agrometeorology of Rainfed Barley-based Farming Systems: 129-44, ICARDA, Aleppo, Syria

Kaul RN and Thalen DCP (1979) South-west Asia in DW Goodall and RA Perry (eds) Aridland Ecosystems: Structure, Functioning and Management vol. 1, Cambridge University Press

Kerbe J (1987) Climat Hydrologie et Amenagements Hydro-agricoles de Syrie, Presses Universitaires de Bordeaux (2 vols)

Kramer C (1980) Estimating prehistoric populations, an ethnographic approach, in: L'Archeologie de l'Iraq, CNRS 580: 315-34, Paris

Lewis N (1988) The Balikh valley and its people, in M van Loon (ed): Hammam et-Turkman I, vol. II: 683-95, Nederlands Historisch-Archeologisch Instituut, Istanbul.

Lloyd S (1938) Some ancient sites in the Sinjar district, Iraq 5: 123-42

Loizides P (1979) Crop rotations under rainfed conditions in a Mediterranean climate in relation to soil moisture and fertilizer requirements, Proceedings of Regional Seminar on Rainfed Agriculture in the Near East and North Africa: 23-38, Amman, Jordan. FAO, Rome

Loomis RS (1983) Crop manipulations for efficient use of water: an overview, in Taylor HM, Jordan WR, Sinclair TR (eds) Limitations to Efficient Water Use in Crop Production: 345-74. American Society of Agronomy, Madison, Wisconsin.

Mainguet Monique (1991) Desertification: Natural Background & Human Mismanagement Berlin, Springer Verlag

McCorriston J (1992) The Halaf environment and human activities in the Khabur drainage, Syria Jornal of Field Archaeology 19: 315-33

McQuigg JD (1981) Climate variability and crop yield in high and low temperate regions, in: Bach W, Pankrath J and Schneider SH (eds) Food-climate Interactions: 121-37, Dordrecht, Netherlands, Reidel

Miller N (1990) Clearing land for farmland and fuel: Archaeobotanical studies of the ancient Near East, MASCA 1990: 71-78

Minnis, PE 1985. *Social adaptation to food stress: A prehistoric Southwestern example*. Chicago: University of Chicago Press.

Mokyr, J (1983) Why Ireland Starved: A Quantitatitative and Analytical History of the Irish Econmy, 1800-1850 London and Boston, Allen and Unwin

Morrison K (1994) The intensification of production: archaeological approaches, Journal of Archaeological Method and Theory 1, no. 2: 111-59

NDA (1980) Crop Yield and Climatic Change to the Year 2000, National Defence Agency, Washington DC

Neumann J and Parpola S (1987) Climatic change and the 11th-10th century eclipse of Assyria and Babylonia, Journal of Near Eastern Studies 46: 161-82

Nikonov AA, Petrova LN, Stolyaranon M, Lebedev V, Siptits SO, Milyutin NN, and Konijin NT (1988) The effect of climatic variations on agriculture in the semi-arid zone of the European USSR, in ML Parry, Carter TR and Konijin NT (eds) The Impact of Climatic Variations on Agriculture (vol.2): 581-627, Kluwer Academic Publications, Dordrecht and London

Nordblom TL (1983) Livestock-crop interactions: the decision to harvest or to graze mature grain crops, Discussion paper No. 10, Farming Systems Program, ICARDA, Aleppo, Syria

Oates J (1982) Archaeological evidence for settlement patterns in Mesopotamia and eastern Arabia in relation to possible environmental conditions, in JL Bintliff and W van Zeist, (eds) Palaeoclimates, Palaeoenvironments and Human Communities in the Eastern Mediterranean Region in Later Prehistory: 359-98. BAR International series 133, Oxford

Oosterom, EJ van, Ceccarelli S, and Peacock JM (1993) Yield response of barley to rainfall and temperature in Mediterranean environments, Journal of Agricultural Science 121: 307-13

Oram, PA and de Haan C (1995) Technologies for Rainfed Agriculture in Mediterranean Climates Washington DC, World Bank Technical Paper No. 300

Pabot H (1956) Rapport au Gouvernement de Syria sur l'ecologie et ses applications, Rome FAO

Pala M, Matar A, Mazid A, and el-Hajj K (1992) Wheat response to nitrogen and phosphorous fertilization under various environmental conditions of northern Syria, in Ryan J and Matar Abdallah (eds) Fertilizer Use Efficiency under Rain-fed Agriculture in West Asia and North Africa: 92-105. ICARDA, Aleppo, Syria

Parry ML (1986) Some implications of climatic change for human development, in Clark WC and Munn RE (eds) Sustainable Development of the Biosphere: 278-409, Cambridge University Press

Parry ML (1990) Climate Change and World Agriculture, London, Earthscan

Parry ML and Carter TR (1988) The assessment of the effects of climatic variations on agriculture: aims, methods and summary of results, in Parry et al (eds): 11-96

Parry ML, Carter TR and Konijin NT (eds) (1988) The Impact of Climatic Variations on Agriculture (vol.2): 581-627, Kluwer Academic Publications, Dordrecht and London

Portugali J (1994) Theoretical sepculations on the transition from nomadism to monarchy, in Finkelstein I and Na'aman N (eds) From Nomadism to Monarchy, Archaeological and Historical Aspects of Early Israel: 203-17, Washington DC, Biblical Archaeology Society

Postgate JN (1986) The transition from Uruk to Early Dynastic: continuities and discontinuities in the record of settlement, in U Finkbeiner and Rollig W, (eds) Gemdet Nasr: Period or Regional Style? 90-106, TAVO Wiesbaden.

Postgate N (1994) How Many Sumerians per Hectare? - Probing the anatomy of an early city, Cambridge Archaeological Journal 4.1: 47-65

Rapp A (1987) Reflections on desertification 1977-1987: Problems and Prospects, Desertification Control Bulletin 15: 27-33, UN Environmental Program, Nairobi, Kenya

Rasmussen PE, and Collins HP (1991) Long-tern impacts of tillage, fertilizer, and crop residue on soil organic matter in temperate semiarid regions. Advances in Agronomy 45: 93- 134

Rhindos D (1984) The Origins of Agriculture, an Evolutionary Perspective New York, Academic Press

Rosen A (1995) The social response to environmental change in early Bronze Age Canaan, Journal of Anthropological Archaeology 14: 26-44

Rosen AM and Goldberg P (1995) Paleoenvironmental investigations. in G.Algaze (ed) Titris Hoyuk, A small urban center in SE Anatolia. The 1994 season. Anatolica 21: 32-37.

Rowton MB (1973) Urban autonomy in a nomadic environment, Journal of Near Eastern Studies 32: 201-15

Russell EJ (1961) Soil Conditions and Plant Growth, London, Longman

Russell, KW 1988. *After Eden: The behavioural ecology of early food production in the Near East and North Africa.* British Archaeological Reports International Series 391.

Shaoxian G (1993) Climate and the northern China dry-land agriculture, in Cheng Chunsu (ed) Climate and Agriculture in China: 367-90, Beijing

Smith PEL and Cuyler Young T, (1972) The evolution of early agriculture and culture in greater Mesopotamia, in B Spooner (ed) Population Growth: Anthropological Implications, Cambridge, MIT press

Stein G and Wattenmaker P (1990) The 1987 Tell Leilan regional survey: Preliminary report MASCA Research Papers in Science and Archaeology, supplement to vol. 7: 5-18

Stewart BA, Jones OR, and Unger PW (1993) Moisture management in semiarid temperate regions, in Agicultural and environmental challenges: Proceedings of the 13th agricultural sector symposium, JP Srivastava and H Alderman (eds), 67-80 Washington DC, The World Bank

Stockton Charles W, and David M Meko (1990) Some aspects of the hydroclimatology of arid and semi-arid lands, in Human intervention in the climatolgy of arid lands, Donald R Haragan (ed): 1-26, Albuquerque: University of New Mexico Press

Sumner W (1989) Population and settlement area: an example from Iran, American Anthropologist 91: 631–41

Tainter JA (1988) The Collapse of Complex Societies, Cambridge University Press

TAVO (1985) Tubinger Atlas des Vorderen Orients (Land Utilization) Map AX no.4, Wiesbaden, Dr Ludwig Reichart Verlag

Thalen DCP (1979) Ecology and Utilization of Desert Shrub Rangelands in Iraq, Dr. W Junk BU, The Hague, Netherlands

UNEP World Atlas of Desertification, United Nations Environment Programme, London, Edward Arnold

van Liere WJ and Lauffray J (1954) Nouvelle prospection archeologique dans la Haute Jezireh Syrienne, Les Annales Archeologiques de Syrie 4/5: 129-48

Weiss B (1982) The decline of Late Bronze Age civilization as a possible response to climatic change, Climatic Change 4 (2): 173-98

Weiss H, M-A Courty, W Wetterstrom, F Guichard, L Senior, R Meadow, A Curnow (1993) The genesis and collapse of third millennium north Mesopotamian civilization, Science 261: 995–1004

White KD (1970) Fallowing, crop rotation and crop yields in Roman times, Agricultural History 44: 281-90

Wilkinson TJ (1982) The definition of ancient manured zones by means of extensive sherd sampling techniques, Journal of Field Archaeology 9: 323-33

Wilkinson TJ (1990) Town and Country in Southeastern Anatolia vol.1: Settlement and Land Use at Kurban Hoyuk and other Sites in the Lower Karababa Basin, Oriental Institute Publications 109, Chicago

Wilkinson TJ (1994) The structure and dynamics of dry-farming states in Upper Mesopotamia, Current Anthropology 35: 483-520

Wilkinson, TJ and Tucker D (1995) Settlement Development in the North Jazira, Iraq. A Study of the Archaeological Landscape British School of Archaeology in Iraq, Aris and Phillips, Warminster

Williams MAJ, and Balling RC (1996) Interactions of Desertification and Climate WMO/UNEP. London, Arnold.

Yoffee Norman (1988) The collapse of ancient Mesopotamian states and civilizations, in The collapse of ancient states and civilizations, N Yoffee and G. L. Cowgill (eds): 44–66, Tucson: University of Arizona Press

Zaccagnini C (1975) The yield of fields at Nuzi, Orient Antiquus 14: 181-225

The Scenario of Environmental Degradation in the Tell Leilan Region, Ne Syria, During the Late Third Millennium Abrupt Climate Change

Marie-Agnès COURTY* and Harvey WEISS**

*C.N.R.S., C.R.A. Lab. de Science des Sols et Hydrologie INA P-G78850 Grignon FRANCE

**Departments of Anthropology and Near Eastern Languages and Civilizations, Yale University, New Haven CT 06520-8236 USA.

Abstract

This paper refines the characterization of the 2200-1900 BC abrupt climate change identified within soil proxy data retrieved on the Habur Plains (N.E. Syria). We compare a selection of soil stratigraphic data from: (1) archaeological contexts in which abandonment sequences above the last Tell Leilan period IIb (2300-2200 B.C.) occupation floors provide a continuous record of the depositional dynamics 2200-1900 BC; (2) natural soil contexts in the environs of Tell Leilan that parallel the local site record with changes of soil-forming conditions and landform dynamics at a micro-regional level.

An interpretive model is elaborated to define the environmental variables involved in the dynamic behavior of regional soil systems. Soil attributes produced by redistribution of calcium carbonate, biological activity, movements of solid particles at the soil surface and through the soil, wetting-drying cycles, and wind activity, are all examined.

The soil record of the abrupt climate change reveals a weakening of pedological transformations, an increase of surface crusting and wind intensity, and an aerosol fallout rich in glass shards and calcitic spherules. These features characterize an aridification event unique in comparison to the other environmental changes recorded over the last 8000 years on the Habur Plains. These changes are both greater and different in nature than the effects of the droughts common to the region's semi-arid Mediterranean climate. The soil degradation differs from the short-term fragilization of soil equilibrium induced by the abandonment of cultivation and land management. The increase of atmospheric dust loading rapidly initiated the

NATO ASI Series, Vol. I 49
Third Millennium BC Climate Change
and Old World Collapse
Edited by H. Nüzhet Dalfes, George Kukla,
and Harvey Weiss

hostile environmental conditions that constrained and reduced late third millennium intensified agro-production in North Mesopotamia, and thereby forced the collapse of Akkadian agro-imperialism. General circulation model linkages suggest both the spatial extent and the regional variability of the 2200-1900 BC abrupt climate change.

Introduction

The occurrence of an abrupt climate change 2200-1900 B.C. has been identified by changes in the dynamic of soil landscapes of the Habur Plains (Weiss et al. 1993). The chronostratigraphic record at Tell Leilan and regional survey links the rapid establishment of drastic arid conditions with site and regional abandonment. The soil properties of the 300 year long occupational hiatus stratum suggest that the climatic disturbance persisted until a "normal" climatic pattern, broadly similar to the present, was re-established ca. 1900 B.C. when the Tell Leilan region was re-occupied.

The soil-sedimentary record of the hiatus stratum presents: (1) an intensification of wind turbulence and a marked increase in dust fall rich in glass shards, (2) a drastic reduction of normal winter rainfall and an increase in erratic, torrential heavy rainstorms. Both reflect a significant departure from normal climate boundary conditions, and are probably linked to weakened cyclogenesis in East Mediterranean regions, and possibly linked to a stronger subtropical jet and more intensive circulation of northerly winds. Variability of dust mineralogical composition and particle size distribution over the 300 year long period suggest signs of irregularity of dust flux, wind intensity and air humidity. An aridity peak, shortly after the begining of the climate change, ended with a gradual increase of humidity and decrease of wind strength (Weiss et al. 1993). The drastic decrease of soil water reserves, the increase of aeolian deflation, and the atmospheric dust loading severely constrained and reduced late third millennium agro-production, and thereby forced collapse of Akkadian imperialism that derived revenues from both rain- fed and irrigation agriculture Mesopotamian lands (Weiss et al. 1993; Weiss and Courty 1993; Weiss 1996b).

Assuming that anomalous climatic conditions could not have persisted for several hundred years solely on the Habur Plains, and following the regional linkages of Mediterranean westerlies and the Indian monsoon, the we hypothetized that the abrupt climate change identified in this region was a large-scale phenomenon. We suggested that synchronous socio-economic disturbances previously documented in ecologically varied contexts, Egypt, Palestine, the Indus and the Aegean regions, were manifestations of the same event (Weiss et al. 1993; Weiss and Courty 1993).

Our conclusions have been challenged by several archaeologists and epigraphers. Wilkinson (1994a) denied that high amount of volcanic ash in the hiatus soilhorizon was exceptional for the region we have studied and simply relegated the climate change we have identified to the rank of "a long run of dry years" common to regional weather variability (Wilkinson 1994b). Ignoring the inter-regional linkages of climate dynamics and the synchronous collapse of varied agro-systems in Egypt, Palestine, the Indus and the Aegean regions, Wilkinson (1994b) alternatively explained the late third millennium disruption of agrosystems as a phenomena restricted to the region of Tell Leilan, in response to the climate enhanced fragility of overexploited soil landscapes.

Glassner (1994) reordered the succession of environmental and societal events we identified and thereby was able to interpret late third millennium landscape degradation as an anthropogenic phenomenon. Cleuziou (1994) misunderstood our soil data to have been retrieved solely from archaeological contexts in which human activities might alter natural signals and thereby disputed the reliability, regional significance and climatic relevance of our data. Cleuziou (1994) also disputed the chronological record for synchronous collapse.

We remain convinced, however, that both site and regional soil data comprise a proxy record for interregional climate change and provide one high resolution signal to link environmental and societal phenomena that occurred synchronously across much of the Old World (Weiss and Courty 1994; Weiss 1996a). Here, we refine our previous conclusions with complementary information for the following:

(1) Dynamics of soil landscapes and sustainability of Tell Leilan period IIb (2300-2200 BC) agrosystems to examine the strengths and vulnerabilities of pre-collapse land management;

(2) Causes, mechanisms and timing of the phenomena that precipitated the collapse of late third millennium North Mesopotamian agrosystems in order to:

(i) further document the nature and the climate significance of the 2200 - 1900 B.C dust fall;

(ii) refine characterization of the 2200 B.C. glass shard rich layer so as to establish large scale correlations;

(iii) understand how hostile conditions were maintained before the return to a "normal" climate regime at ca. 1900 B.C.

Study Area

The Tell Leilan region is the eastern side of the Jaghjagh River basin that consists of a gently undulating flood plain (Fig. 1). The undulations relate to the folding of the Miocene substratum, caused by the gradual uplift of the Taurus foothills to the north and of the jebel Abd-el-Aziz to the south, during the Pliocene and Pleistocene (Ponemarev and Bivshev 1966). The core of the undulations is formed of strongly cemented calcretes that are complex alluvio-lacustrine deposits formed during the Plio-Pleistocene period. The calcretes are blanketed by a sedimentary unit of calcareous clay loam of irregular thickness, ranging between 1 to 6 meters, that has been deposited during the early Holocene by extensive mudflow and sheetwash processes (Courty 1994). Over the middle and late Holocene, the original alluvial mantle has suffered a series of incision/aggradation cycles in response to variations of the hydrological regime, a dense network of ephemeral streams predominantly controlled by rainfall fluctuations in the catchment basin (Courty 1994). This evolution has resulted in the formation of a regional landscape system comprised of two major geomorphic units: - the sigmoidal part of the undulations that have suffered continuous micro-relief readjustment caused by surficial runoff and aeolian addition, with concommittant soil development; - the flood plain of each ephemeral stream that consists of the complex imbrication of alluvial units of different ages as the result of repeated vagrancies of the watersheds in their weakly differentiated basin.

The modern landscape presents a gradual transition between the sigmoidal undulations and the lowlands due to slope wash processes that have smoothed the topography created during the last phase of alluvial incision. The meandering course of the ephemeral streams (wadis) and their slightly entrenched channels reflect the low water discharge from the upper catchment area.

Surface soils, that range from Typic Ustrochrepts to Aridic Calisutolls (Soil Survey Staff 1975), present weak vertical pedogenic differentiation, essentially marked by structural changes and variability of secondary carbonates with depth. Subtle differences of soil surface properties and the spatial soil variability in the present

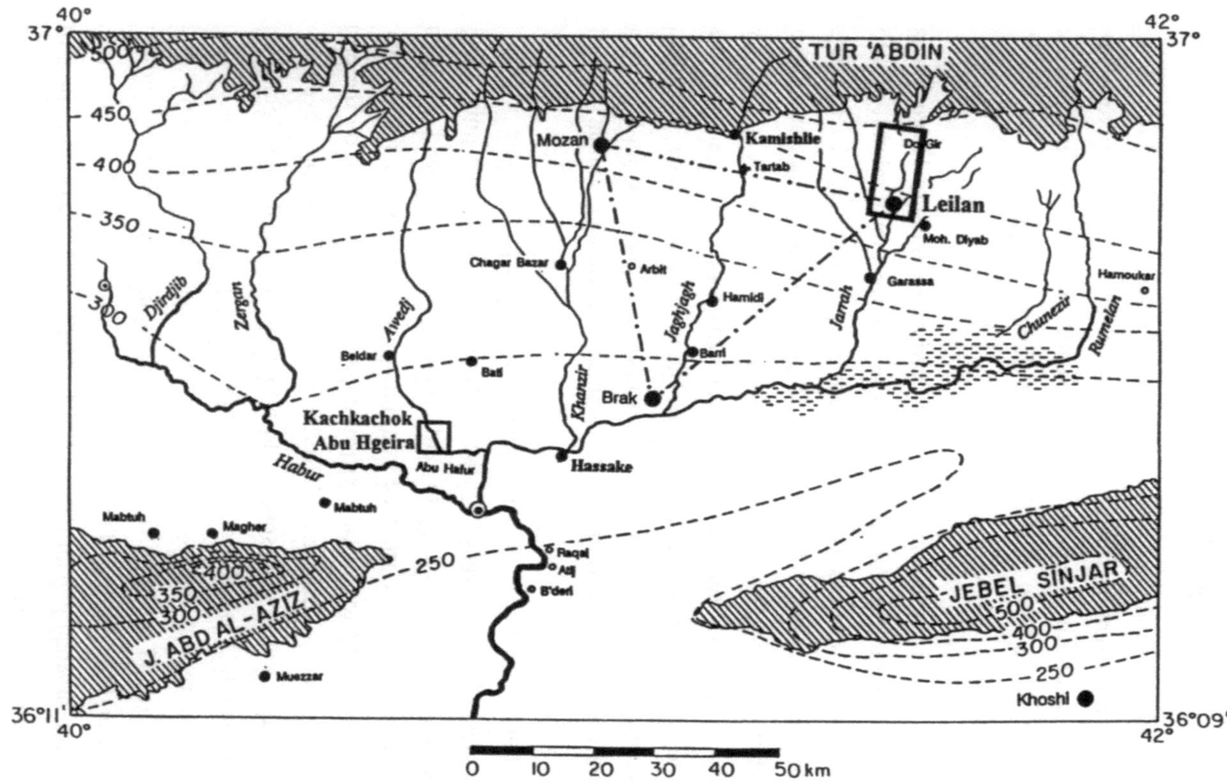

Figure 1. Upper Habur Plains, Northeast Syria, urban settlement system at 2600 to 2200 B.C. Diagonal lines are elevation >500 mabove sea level; dotted lines are modern rainfall isohyets (inmillimeters). The rectangle indicates location of the study areaaround Tell Leilan and the square the previously studied regionof Kashkashuk/Abu Hgeira 2 in which the glass shard rich layer ofTell Leilan was first identified (Weiss et al., 1993).

landscape are the result of a complex imbrication of geomorphic surfaces of different ages.

The site of Tell Leilan is located on the western edge of a sigmoidal landform along the flood plain of the wadi Jarrah. This position provides the opportunity to study how the different landform elements have reacted locally to geomorphic modifications induced by the 2200-1900 B.C. abrupt climate change.

Climate

The region of Tell Leilan is located in the area of maximum rainfall of Habur Plains due to the orographic effect of the Tur Abdin hills. The average annual rainfall of 440 mm rapidly decreases to the south to reach a minimum of approximately 220 mm/year in the more desertic Radd basin. Winter rainfall is predominantly associated with Mediterranean cyclone depressions, with regional distributions affected by orographic effects. The cold season is, however, frequently dry due to the establishment of high pressures controlled by the Eurasiatic anticyclone. Northern penetration by southern arid Mediterranean influences is responsible for the high interannual rainfall variability. Rains are often irregularly distributed, violent, and of short duration.

The reegional climate is linked to the wind circulation regime that is more effective during the dry season when influenced by the subtropical jet while the polar front and western circulation are pushed northward. Summer is characterized by high frequency, zonal, sandy winds that originate from intense warming of the soil surface. Ascending air movements first form local, vertical dustbowls, transformed by the end of the day into a continuous dust screen (Kerbe 1987: 173). These dry hazes can often last for a few days, cause atmospheric dust loading, and lead to a significant decrease of solar radiation. Violent wind storms have an average frequency of two times per year. These exceptional weather situations last for a few days and considerably reduce ground visibility. They are synchronous with extremely violent storms on the centre and the north of Turkey caused by convergence of the southern Saharan and the northern Anatolian air masses (Kerbe 1987: 180). Maximum effect of aeolian erosion is recorded in summer when the strongly desiccated bare soils are exposed to strong winds. Wind situations from October to February cool the region.

Methodology

The stratigraphy and properties of the soil-sedimentary layers have been studied from both archaeological/anthropogenic and natural soil contexts. First, excavations within the perimeter of Tell Leilan have provided diverse situations for analysis of the evolution of environmental conditions through occupational time scales. From eight loci studied (Fig. 2), sections have been selected here to extract an environmental signal of regional significance from anthropogenic deposits. Deposits vary with each context's occupational function and specific post-collapse history. The archaeological sequences studied are framed within time intervals defined by radiometric datings of materials retrieved from occupation floors (Weiss et al., 1993):

(i) period IIb (2300-2200 B.C.) occupation floors and construction materials;

(ii) Habur hiatus I (2200-1900 B.C.);

(iii) period I (1900-1700 B.C.) occupation floors.

The environmental transformations of anthropogenic debris and natural deposits are identified within soil/sedimentary attributes. In addition, the study of ancient soils buried below finely dated 3rd millennium BC archaeological strata has provided data to define (1) soil landscape conditions before Leilan period IIa (2400-2300 B.C.) and (2) the response of soil landscapes to the intensification of agro-production during the Leilan IIb Akkadian imperialization period.

Secondly, backhoe trenches to 6 meter depths at locations surrounding Tell Leilan have provided soil stratigraphic data from the dynamic components of the undulating landforms (summit, footslope and toeslope). Study site selection (Fig. 3) was aimed to reconstruct the pre-collapse, collapse and post-collapse palaeogeography at a micro-regional scale in order to identify environmental changes and the mechanisms that modified geomorphic and pedogenic dynamics. Basic principles of soil stratigraphy were followed to frame within a relative chronological sequence the successive episodes of deposition and concurrent periods of soil development. The uppermost part of the backhoe trenches has provided a range of surface soils needed to calibrate the dynamic of the soil system with present climate conditions.

Characters inherited from former pedogenic phases, often closely imbricated with functional characters, ones have been isolated through a comparison of buried and surface soils. This discriminative method is useful in the study region because the homogeneity of parent materials and the weak topographical variations minimize the effects of local factors on the spatial variability of soil properties. Because the entire

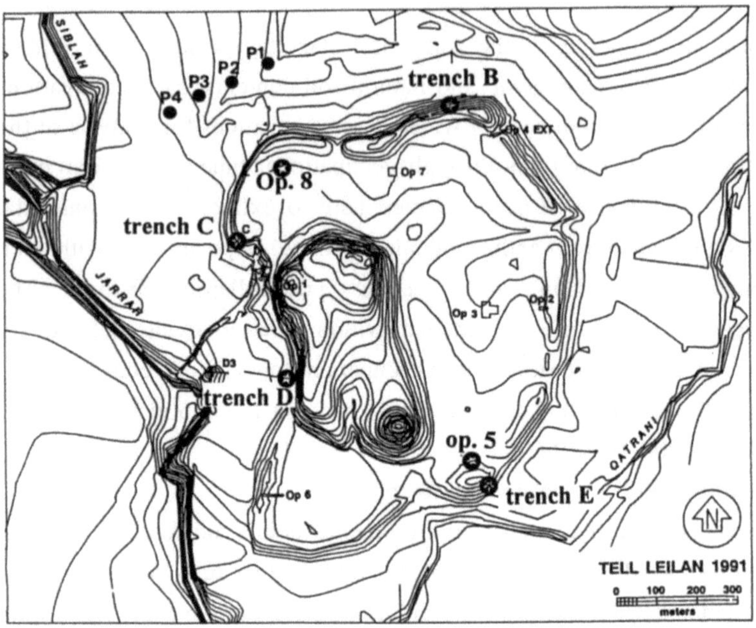

Figure 2. Topographic map of Tell Leilan, showing areas ofexcavations and location of the natural profiles discussed in the text.

Holocene soil system of the Habur Plains is weakly differentiated (Courty 1994), study of soil properties commonly used to both estimate the rate of soil profile development and investigate effects of climate changes on soils (Birkeland 1984; Mc Fadden and Knuepfer 1990; Birkeland et al. 1991; Wilkinson 1990) cannot be applied here. Therefore, we have adopted a methodology that first obtains maximum information from field observation. Field features have been further elucidated by a micro-level characterization of all soil/sedimentary attributes that establish linkages between soil properties, elementary soil-forming processes and environmental variables.

The study of thin sections under the petrographic microscope recognizes (1) the sequential order of soil/sedimentary attributes, (2) the dynamic processes that caused the formation of each attribute, (3) the environmental variables that influenced the occurrence of a given set of attributes, and (4) the signatures of anthropogenic transformation (Courty and Fedoroff 1985; Courty et al. 1989; Fedoroff et al. 1990; Macphail et al. 1990; Courty 1994; Courty et al. 1994a). As a result, the final

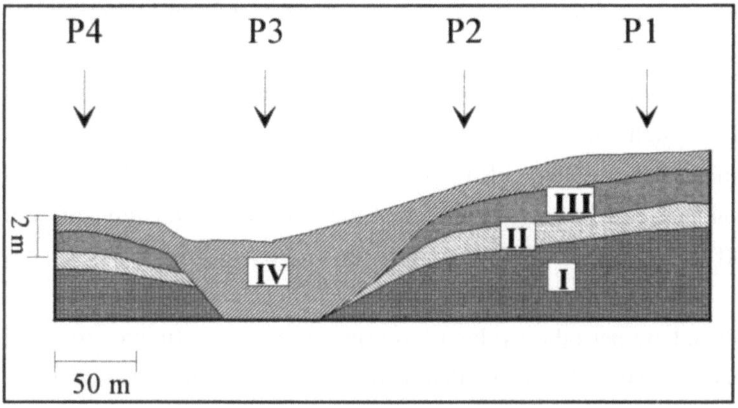

Fig. 3. Schematic cross section through the Jarrah flood plain asreconstructed from stratigraphic correlations between P1, P2, P3 andP4 soil profiles.
Soil unit I: Parent materials deposited before Leilan Period II fromthe combined effect of flooding and fine dust accession.
Soil unit II: Parent materials deposited during Leilan Period II fromfine dust accession and transformed by pedogenesis during this timeinterval.
Soil unit III: formed during the abrupt climate change from rapiddust accession and surficial runoff.
Soil unit IV: formed after the abrupt climate change from rapidflooding for profile 3, from dust accession for profile 1 and from thecombined effect of slope wash and dust accession for profiles 2 and 4.

interpretation documents the relative timing, magnitude, and causes of variations of geomorphic and pedogenic events involved in soil and landform evolution.

Soil systems of the study region have a dynamic behavior sensitive to environmental variables, especially to seasonal water movement (Courty1994). Therefore, the "factor function" approach (Retallack 1994) detects low amplitude effects of the water related variables on soil microfabrics. Five sets of soil features that are sensitive indicators of soil-water interactions, and therefore of climate variability, as illustrated by a wide range of previous studies, have been studied:

(1) features produced by the redistribution of calcium carbonate (dissolution and precipitation features) in response to percolation regime, seasonal gradient of evapotranspiration and degassing of CO2 (Courty 1994; Courty et al. 1987; Courty et al. 1994);

(2) features produced by the activity of the soil meso-fauna (biological features), such as earth-worms, that are closely linked to the seasonality of soil hydric regime (Courty and Fedoroff 1985; Courty 1990);

(3) features produced by the redistribution of solid particles at the soil surface and their translocation through the soil profile (textural features) that reflect the seasonality of rainfall, the velocity of raindrops (i.e. intensity of rainfall) and properties of the soil surface (covering potential of the vegetation and type of root network), (Fedoroff and Courty 1987; Fedoroff et Courty 1989; Bresson and Valentin 1994);

(4) features produced by periodic cycles of wetting and drying (micro-structural patterns) that reflect the seasonality of rainfall distribution and temperature gradient (Jim 1986);

(5) features produced by saltation transport of particles eroded from the soil surface by wind and re-incorporated into the subsurface horizons; their degree of preservation and abundance record the soil moisture content of subsurface horizons, wind intensity and the covering potential of vegetation (Courty 1994).

A total of 200 thin sections made from undisturbed samples collected in excavation and natural soil contexts have been studied following the terminology of Bullock et al. (1985) for soil features and Courty et al. (1989) for anthropogenic deposits.

Results And Discussion

Soil data obtained from excavation and natural stratigraphic contexts are integrated to present the evolution of environmental conditions during the following phases: (1) pre-collapse occupation, (2) collapse/abandonment (3) post-collapse occupation as provided by the chronological data of the archaeological sequence. The following discussion is arranged to highlight the exceptional character of the regional climate during the 2200-1900 B.C. interval in comparison to the "normal" preceding and subsequent situations.

1. Soil Landscape Dynamics Before The Abrupt Climate Change

Field examination of ancient soils buried below finely dated 3rd millennium B.C. archaeological strata and of soils exposed in deep backhoe trenches in the surrounding landscape emphasizes the striking uniformity of the Holocene soil covers developed in the region. This is expressed by the lack of visible lithological discontinuities, textural and mineralogical homogeneity of parent materials, and weak pedogenic differentiation. Consequently, the appearance in the field is likely to be interpreted as the characteristic of intact soil profiles that would have formed from materials accumulated during one major depositional event, and linearly transformed through time by chronic pedological processes. This would reflect the overall stability of climate throughout the Holocene and the inability of soil systems to manifest significant changes under minor fluctuations.

Comparison of the micro-fabrics of the early to middle 3rd millennium B.C. soils sealed at successive time intervals (from a few decades to a few 100 years) from Tell Leilan Trench B (Fig. 2) indicates, however, that the vertical differentiation did not result from a continuous pedological development. It is, in fact, the consequence of a step-wise deposition of sediments, synchronous with pedogenic transformations in response to modifications of the soil hydric regime. The intense bioturbation observed throughout all the studied profiles seems responsible for the lack of lithological discontinuities and detectable sedimentary features that define the mode and deposition rate of parent materials.

Parent materials of the soils that form the substratum of the 3rd millennium extension of Tell Leilan to the north and the 2 meter thick soil unit on the summit of the surrounding undulating landscape (unit II, Fig. 4 & 5, Table 1) consist of silt-sized calcite particles (average 68%) with a minor proportion of clay (22%) and reduced amount of quartzitic and calcareous fine sands (10%), which mineralogically and texturally resemble the fine alluvium accumulated in the active lowlands during recent flooding episodes. However, they display a more open micro-fabric that rules out deposition by massive flooding. The underlying soil unit (unit I) is formed from parent materials similar to those of soil unit II, but display a minor increase in fine clay with depth.

The similar chemical and morphological characteristics between soil unit I and II indicates that conditions of soil development remained broadly unchanged over half a century. The unit I soils are, however, characterized by a denser micro-fabric in comparison to unit II soils (Table 1) that is likely to reflect extensive flood deposition during early stages of landscape evolution. This conclusion is supported by the observation of fresh flooded deposits sealed below Leilan period IIId archaeological strata in a section studied in the Lower Town South at the east of the site (Op. 5, Fig. 2). In this location, burial occurred shortly after the accumulation of flooded deposits

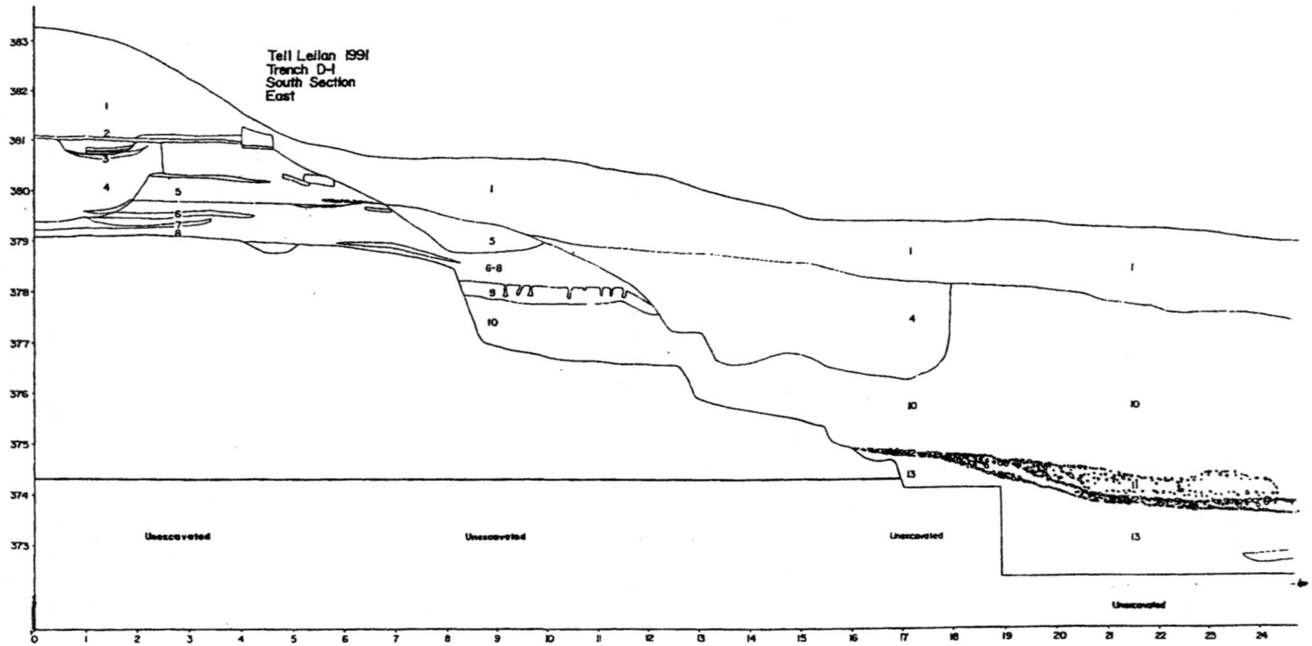

Figure 4. Schematic cross-section of trench D-I showing from the east to thewest: virgin soil (see Fig. 5, column 2) below the occupational hiatus(8 to 5) and period I archaeological deposits (4 to 1); embankmentswith a succession of human-made soils (10) and an upper one withplough marks (9) below the occupational hiatus (8 to 5); sandyalluvium and irregularly stratified, coarse textured alluviumdeposited during period II (13 and 12).

Table 1

Unit	Depth	Colour Texture	Microstructure	Pedogenic features
IIa1	0-25 cm	7.5YR 5/4 calcitic silt	Spongy to fine fissural; weakly packed, sub-rounded, centimetric aggregates.	Common, dense brown, rounded, millimetric, micritic nodules and rare, finely cracked, millimetric, microsparitic nodules. Abundant, millemetric, sub-rounded sherd, brick and herbivore dung fragments and millimetric charred plant remains. Rare gypsic infillings Abundant ferro-manganic thin hypocoatings and diffuse impregnations.
IIa2	25-40 cm	7.5YR 5/4 calcitic silt	Spongy to channel; densely packed, sub-rounded, melted, centimetric aggregates.	Common, pale grey, rounded, millimetric, microsparitic and micritic nodules. Common, millimetric, rounded, sherd, brick and herbivore dung fragments and millimetric charred plant remains. Abundant ferro-manganic thin hypocoatings and diffuse impregnations. Reworked biogenic aggregates. Fine mass made of weakly packed, sand-sized, rounded aggregates; channels filled of loosely packed, sand-sized, rounded aggregates.
IIa3	40-55 cm	7.5YR 5/4 calcitic silt	Fine fissural to massive.	Common, pale grey, rounded, finely cracked, centimetric, microsparitic nodules. Abundant charred plant remains. Common ferro-manganic diffuse impregnations. Fine mass made of densely packed, sand-sized, rounded aggregates.
IIb1	55-70 cm	7.5YR 6/4 calcitic silt	Channel to fine fissural; dense packing of sub-angular, centimetric aggregates.	Abundant, brownish grey, rounded, complex, centimetric, finely cracked, micritic/ microsparitic nodules. Abundant charred plant remains. Rare integrated slaking crusts. Common ferro-manganic diffuse impregnations. Fine mass made of melted,sand-sized, aggregates.
IIb2	70-110 cm	7.5YR 6/4 calcitic silt	Dense channel to spongy and fine fissural	Very abundant, brownish grey, rounded, complex, centimetric, finely cracked, micritic/ microsparitic nodules with pendent morphology. Massive fine mass.
I	110-150 cm	7.5YR 6/4 calcitic clay silt	Very fine fissural to massive.	Total excremental fabric Abundant, rounded, pale grey, finely cracked, microsparitic complex nodules with an internal micro-nodular fabric. Abundant diffuse micritic impregnations. Calcitic particles and features slightly corroded. Rare depleted patches integrated to the fine mass and along fissures. Massive fine mass. Abundant ferro-manganic impregnations.

Table 1. Main characteristics of the buried soil from trench D (seelocation in figure 4 and schematic stratigraphy in figure 5)

at ca. 2600 B.C., before they were transformed by pedological processes. In the north of the site, soils have remained exposed during approximatively 200 years before they were themselves sealed by Leilan period IIa (2400-2300 B.C.), then IIb (2300-2200 B.C.), archaeological strata.

In addition to the evidence provided by the characteristics of the micro-fabrics, the stratigraphic comparison of the deep trenches indicates that the 1 m thick soil unit formed during Leilan period IIId-IIb time interval (2600-2200 B.C.) cannot be of alluvial origin. The trench D section exposed at the west of the site has revealed that the 3rd millennium B.C. channel incised the formerly deposited flood plain (Fig. 4). Accumulation irregularly stratified coarse sands and lenses of rounded gravels indicates a torrential hyrological regime during Leilan period II that contrasts with the former period of regular, extensive floods. The altitude difference between bottom of the period II channel and surface of period II soils at the northern extension of the site precludes the possibility that the 60 cm to 1m thick soil unit II could be flooded deposits. Consequently, the deposits contemporaneous with Leilan period IIId-IIb (unit II, Fig. 5) is interpreted to derive from a steady-state accretion of silt-sized aeolian dust synchronous with soil development. This reflects the slow re-equilibration of soil-landforms subsequent to the previous period of extensive flooding that would be caused by the progressive drying of large marshlands originally present in the lower flood plain. Fine dust would have been easily picked up by aeolian deflation from the weakly structured soils of the lowlands that were no longer protected by a dense marshy vegetation.

The fact that the particle size distribution and mineralogical composition of the period IIId-IIb silty topsoil match those of local soils supports a short distance aeolian transport of the dust. The few angular coarse silt-sized to fine sand-sized mineral grains with well-oriented yellowish-brown clay coatings, that are absent in the original flood deposits, are the only evidence of a long distance transport, although they never represent a significant contribution to the dust accretion.

Difficulty in recognizing the original morphology of the aeolian particles is due to their continuous incorporation into the top soil layer by bioturbation. Absence of sand-sized rounded aggregates suggests that deflation of the soil surface by strong winds and saltation transport remained limited. It seems more likely that biogenic aggregates, abundantin the surface horizons, have been the source of dust eroded

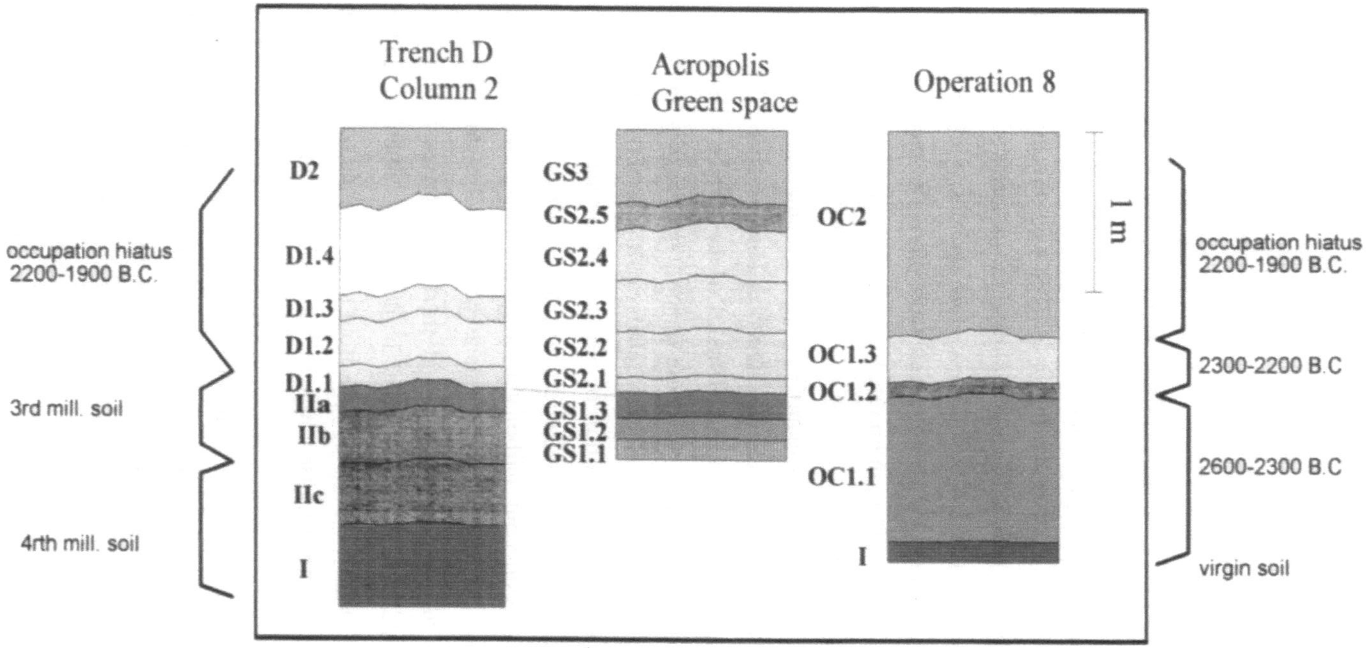

Figure 5. Schematic stratigraphical section of three of thearchaeological sequences studied at Tell Leilan.Trench D, column 2 (see Fig. 4): virgin soil formed before the abruptclimate change (soil units I to IIa) sealed by deposits accumulatedduring the abrupt climate change (D1.1: dust accumulated at thebegining of the abrupt climate change; D1.2 to D1.3: coarselystratified layers deposited during the maximum of aridity; D2:massive layer accumulated during the final phase of the abruptclimate change).
Acropolis Green Space: GS 1 layers correspond to the succession ofhuman-made soils formed during Period II; G.S 2.1: dustaccumulated at the begining of the abrupt climate change; GS 2.2. to2.5: finely stratified layers deposited during the maximum of aridity;GS3: massive layer accumulated during the final phase of the abruptclimate change).

from the lowlands in summer when the natural vegetation was reduced or when soils were no longer protected by crops.

The synchronous increase of dust accumulation with intensification ofagricultural exploitation during period II, as indicated by archaeological data (Weiss et al., 1993) raises the question of a possible anthropogenic enhancement of a geomorphic trend, originally initiated by a change of the hydrological conditions. Comparison of unit II soil layers formed during period IIa indicates a regular rate of dust accumulation (approximatively estimated at 0.2 cm/year) that did not increase during Leilan period IIb when agricultural intensification was at a maximum. The well developed biogenic microstructure of the period IIb subsurface soil horizons indicates that moisture conditions favorable to the proliferation of soil mesofauna and micro-organisms have been maintained during the period of agricultural intensification. The increase of dust accumulation rate cannot be, therefore, interpreted as a sign of anthropogenic land-degradation. In agreement with a recent review of effects of airborne accessions of dust on soils (Simonson, 1995), it seems more likely that accumulation of dust on summit and footslope positions had beneficial effects on land use potential during period II. It regenerated the amount of nutrient elements of the surface horizons, and created a soft top layer more accessible to the dense fine root system of graminae and more easily workable by a traditional technology relative to the underlying, strongly coherent, soil units. The continuous addition of non-altered silt-sized particles has, furthermore, increased the availability of soil water for plant growth as shown by the morphological change of pedogenic carbonates from abundant, well expressed nodules to rare ones. The addition of fresh calcite particles also contributed to decrease bulk density of biogenic aggregates, and therefore increased their water storage capacity, while reducing risks of topsoil hardening by stuctural collapse caused by rapid wetting. Moreover, removal of silt from the lowlands could not have induced a significant loss of soils because only a limited part of the flood deposits were redistributed.

In addition to the identification of the mode of deposition of period II soil units, study of their microfabrics reveals absence of significant effects of surficial hydric erosion during this time period, as shown by the rare occurrence of slaking crusts in the soils studied from summit positions (e.g. profile 1, Fig. 6 & Table 2 and the buried soil of trench D, Fig.4 and Table 1). This reflects the rare occurrence of heavy rainstorms in early autumn when soils were bare or poorly protected by the plant

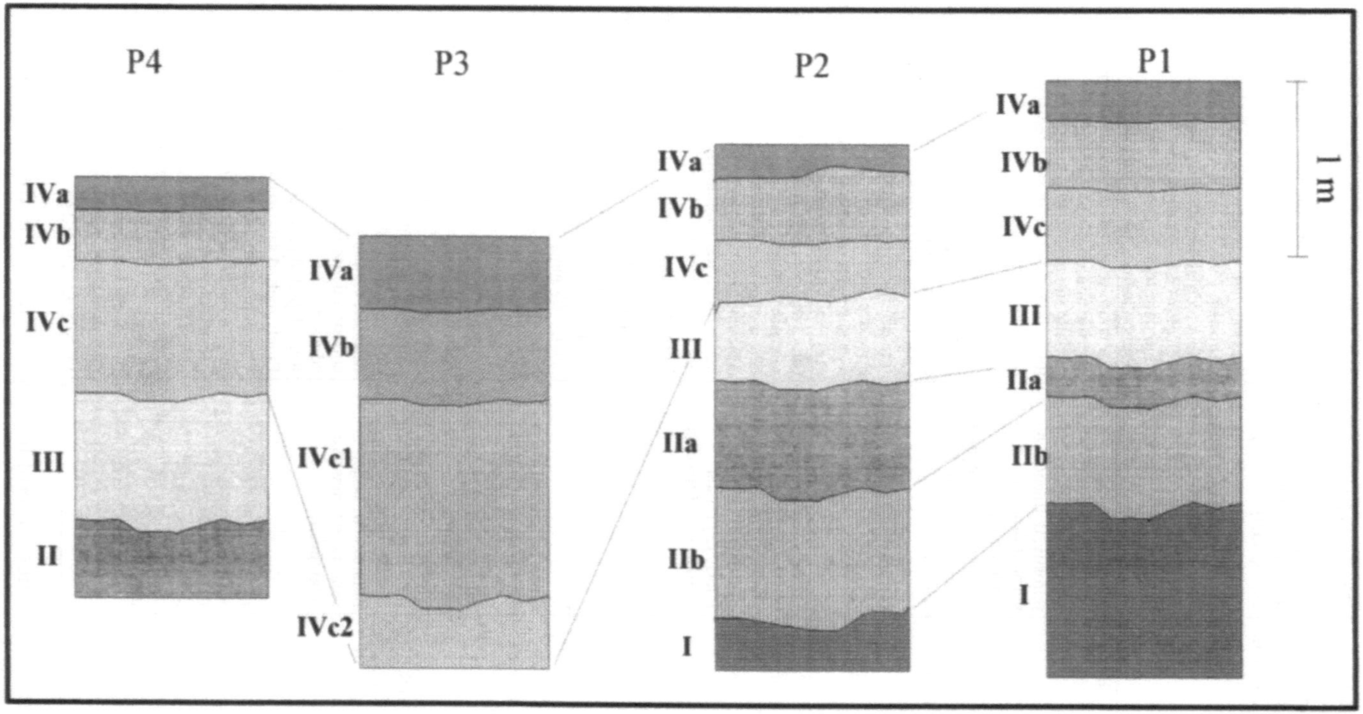

Figure 6. Schematic stratigraphical sections of the four natural soilprofiles studied at the west of the site (see location on figure 2;characteristics and age of the different soil units are described in the text).

Unit	Depth	Colour Texture	Microstructure	Pedogenic features
IVa	0-20 cm	7.5YR 5/4 Calcareous sandy loam	Loosely packed millimetric aggregates at the top merging to open packing of centimetric, dense aggregattes with abundant small channels, and delineated by fine fissures.	Millimetric, abundant, randomly distributed, well rounded micritic nodules; rare, dense micritic impregnations. Fragmented, coarse textured slaking crusts; sandy infillings A few reddish brown soil fragments Melted pseudo-sands Abundant root fragments Abundant fine charred fragments
IVb	20-58 cm	7.5YR 6/4 Calcareous silt	2 partial fabrics: (1) centimetric, massive aggregates with a channel porosity, delineated by vertical fissures; (2) loosely packed millimetric aggregates.	Common, centimetric, subangular, grey, dense micritic and microsparitic nodules; common micritic impregnations; rare calcitic hypocoatings. Rare, integrated small depleted zones Melted pseudo-sands
IVc	58-101 cm	7.5YR 6/4 Calcareous silty	Open channel to fine fissural. Loosely packed . centimetric aggregates.	Abundant, well rounded, centimetric, finely cracked, pale grey, micritic to microsparitic nodules; abundant, diffuse micritic impregnations; rare calcitic hypocoatings. Rare depleted zones Various types of melted pseudo-sands: calcitic grey, yellowish brown, orange.
III	101-154 cm	7.5YR 6/3 Calcareous silt	Densely packed sand-sized aggregates to massive; fine fissural and rare closed channels. Poorly expressed, subhorizontal fissures.	Rare, well rounded, millimetric, brownish grey, micritic nodules. Ghost of gypsic infillings. Abundant millimetric, rounded sherd, brick and herbivore dung fragments; abundant fine charred organic particles. Very abundant, rounded, partially melted pseudo-sands. Abundant, silt-sized, reddish brown clay coated minerals and weathered glass shards. Common silty clay slaking crusts. Sharp boundary with the underlying unit.
IIa	154-177 cm	7.5YR 6/4 Calcareous silt	Open packing of randomly distributed centimetric massive subrounded aggregates and millimetric subangular aggregates.	Heterogenoeus mixing of dense micritic grey silt, open calcitic silt and yellowish brown clayey silt. Abundant, well rounded, centimetric dense, complex, micritic nodules. Abundant, centimetric to millemetric subangular, sherd and brick fragments.
IIb	177-237 cm	7.5YR 6/4 Calcareous clayey silt	Fine fissural and rare closed channels. Subangular, large centimetric, massive aggregates.	Common, centimetric, finely cracked pure microsparitic to dense micritic nodules; common microsparitic infillings; common dense micritic impregnations with diffuse edges. Rare depleted patches along fissures. Finely melted sand-sized aggregates.
I	237-337cm	7.5YR 6/6 Calcareous silty clay	Fine fissural with a vertical pattern. Dense packing of subangular, large centimetric, very massive aggregates.	Common,millimetric, rounded, dense, microsparitic and micritic nodules; very abundant, greyish brown, dense micritic impregnations; abundant pure microsparitic infillings. Common integrated depleted patches along fissures. Integrated silty clay intercalations. Common ferro-manganic impregnations increasing with depth.

Table 2. Main characteristics of profile 1 (see location in figure 2 and schematic stratigraphy in figure 6)

cover. Evidence for moderate intensity rainfall is also demonstrated by the regular thickness of the top soil layer formed during period II along the slopes, and lack of signs indicating significant movement of soil materials by overland flow in the profiles studied from footslope position (Table 3). The slight structural change observed in footslope soils in comparison to summit soils reflects a moderate trend to the coalescence of biogenic aggregates in response to lateral circulation through flow, which, however, did not induce an important drainage decrease. In addition, the absence of sand-sized aeolian aggregates in soil unit II indicates that violent deflation and saltation transport, which may occur during long droughts, were not frequent during Leilan Period II.

In conclusion, study of the spatial variability of the top layers formed during period IIId-II indicates maintenance of highly resilient soils, mainly due to the good structural stability of the different landform units. No evidence of a human-induced degradation of soil properties has been recognized and addition of organic mineral constituents, possibly used as manure to increase soil fertility, has only been observed in the period II soils within the perimeter of the site, and always of an amount not exceeding a few percent. This suggests that the process of agricultural intensification during period II was not been limited by a reduction of soil fertility. The significant reduction of the hydrological regime, therefore, appears more likely to have been the major constraint upon natural resource exploitation. The artificial embankment of the natural channel at the western foot of Tell Leilan indicates that the water course has been modified locally, possibly to increase flow capacity or to protect the lowlands from channel vagrancy.

Comparison of soils formed during period IIId-IIb with soils that developed during the earlier episode of extensive flooding (units I, Table 1) indicates that the change of hydrological regime coincided with a moderate reduction of water reserves. The geomorphic evolution of Tell Leilan region during period IIId-IIb thus reflects a progressive decrease of rainfall, in agrement with previous observations in other areas of the Habur Plains (Courty 1994). This can be related to the global aridification trend initiated at ca. 5000 BP (Kutzbach 1981; Berger 1984; Street-Perrot and Perrot 1990; Cheddadi et al. 1991; Prentice and Sarnthein 1993; Courty et al. 1994; Gasse and Van Campo 1994; Petit-Maire et al. 1994;). The continuous accretion of aeolian dust appears to have had a positive feed-back effect on soil structure by increasing wettability and water availability for plant growth, although water reserves were

Unit	P4: west footslope	P3: toeslope	P2: east footslope
IVa	Sandy silt; fine vughy; densely packed micro-aggregates; rare, millimetric, well rounded, micritic and microsparitic nodules.	Sandy silt; fine vughy; densely packed micro-aggregates; rare, millimetric, well rounded, micritic nodules.	Densely packed micro-aggregates.
IVb	Sandy silt; subhorizontal fine fissural; densely packed micro-aggregates; common, millimetric, well rounded, micritic and microsparitic nodules; integrated silty clay slaking crusts.	Sandy silt; subhorizontal fine fissural, massive aggregates; rare micritic hypocoatings; integrated silty clay slaking crusts.	Dense, massive, centimetric aggregates.
IVc	Sandy silt; massive, channel; common, millimetric, well rounded, micritic and microsparitic nodules.	Sandy silt; subvertical fine fissural; very dense, massive, aggregates; rare, small, micritic nodules; integrated silty clay slaking crusts.	Finely cracked, well delineated, microsparitic infillings and nodules.
IVd	Not present	Sandy silt; subvertical fine fissural; very dense, massive, aggregates; diffuse, micritic impregnations, rare, micritic nodules.	Not present
III	Fine vughy to fine fissural; densely packed, sand-sized micro-aggregates; rare micritic infillings of nodular morphology; integrated silty clay slaking crusts.	Not observed	(a) Open fissural and polyconcave vughs. Mixing of (1) centimetric, isotropic, pale yellow, massive aggregates; (2) organic rich, yellowish brown, rounded aggregates; (3) pale grey aggregates made of densely packed calcareous pseudo-sands. Disaggregated microsparitic nodules. (b) Subhorizontal fine fissures. Very developed, poorly sorted, pale grey yellow, infillings, coatings and intercalations. Concentration and juxtaposition of various types calcitic nodules.
IIa	Not clearly defined	Not observed	Not clearly defined
IIb	Sandy silt; massive and fine fissural; abundant, diffuse semi-continuous micritic impregnations; very abundant dark brown ferro-manganic impregnations.	Not observed	Massive, densely packed, aggregated fine mass. Dense, continuous, micritic impregnations. Rare greyish brown micritic nodules.
I	Sandy silt; channel and sub-vertical fissure; dense packing of fine aggregates; integrated silty textural features; diffuse, large, yellowish brown, ferro-manganic impregnations; no calcitic features.	Not observed	Rare complex micritic nodules; strongly expressed, dense brown, micritic impregnation of nodular morphology; abundant depleted zones (associated with biogenic calcite).

Table 3. Main characteristics of profiles 2, 3 and 4 described bycomparison with the ones of profile 1 (see location in figure 2 andschematic stratigraphy in figure 6).

lower in comparison to the previous period. The rainfall reduction during Leilan period IIId-IIb did not significantly alter the natural equilibrium of the soil landscapes. Soils with high agronomic potential and minimal degradation risk remained available during the period of agricultural intensification.

2. Soil Landscape Dynamics During The Abrupt Climate Change

Originally recorded from Tell Leilan archaeological sequences (Weiss etal. 1993), the effects of the 2200-1900 B.C. abrupt climate change are also clearly observed in all the Tell Leilan region natural sections ranging from period IIb to the present time. They are marked by a sharp stratigraphic boundary and significant modification of texture and micro-fabrics in comparison to earlier soil units (Tables 1 & 2). Spatial diversity of of the soil features reflects modification of the mode of deposition of parent materials, their nature, and the conditions of soil development. The chronological correlation between natural and archaeological sequences is based on the abundance of similar silt-sized glass shards, associated with high amounts of fine charcoal, silt-sized phytoliths and unusual calcitic spherules found in both contexts. Glass shards have never been detected, even as traces, in the parent materials deposited during earlier periods. Glass shards are also absent from the surface soils that have developed from recent alluvium, but present as traces in the ones that formed on soil materials reworked during the abrupt climate change. This indicates that the Tell Leilan region recorded over the Holocene a single event of glass shard fallout, similar to the soil record studied in other areas of the Habur Plains (Courty 1994). As commonly used in regions affected by tephra deposition (Busacca et al. 1992), the sudden increase of glass shards appears to be a reliable stratigraphic marker to compare the soil record of environmental conditions in the natural profiles with those of Tell Leilan archaeological sequences 2200-1900 B.C.

Data from the two different contexts provide complementary evidence 1 to understand environmental dynamics during the collapse period. The record of the abrupt climate change from natural profiles, first, indicates the geomorphic modifications that affected soil lanscapes of the Tell Leilan region at the begining of the event by identification of the processes which caused a stratigraphic discontinuity between period IIb soil surface horizons and overlying strata. The morphology of the soil layer formed during the collapse period, then, defines the general trend during this time interval. Structural modifications induced by subsequent bioturbation have, however, significantly blurred out the soil-sedimentary signal of the evolution recorded during the collapse period. A high resolution signal is, therefore, provided by archaeological sequences because re-occupation ofthe site during period I rapidly sealed the layer formed during the collapse period.

Data From Natural Profiles

In the section studied from the summit position of the undulation (profile 1, Fig. 6 & Table 2), the abrupt stratigraphic boundary between units II and III is clearly an erosional contact. The erosion is, first, indicated by the occurrence of a top soil layer thinner than those of the contemporaneous buried soils within the perimeter of the site that were sealed by Leilan period IIb occupation layers, or by soil materials rapidly accumulated during the natural collapse of nearby structures. Signs of erosional disturbance are also expressed by the heterogeneity of the IIa soil unit that results from mixing by colluviation of surface horizons.

This differs from the homogeneity expected in an intact plough layer produced by 100 years of intense cultivation. Characteristics of the overlying soil unit III indicate a change of deposition and transformation conditions by soil-forming processes. This allows us to interpret the stratigraphic boundary to be both a lithogenic and a pedogenic limit. Abundance of sand-sized, rounded, particles reveals intensification of the local surface soil horizons by aeolian deflation and predominance of saltation processes by strong winds. As mentioned above, such aerologic situations implies that the soil surface was no longer protected by plant cover and that its soil moisture content was insufficient to maintain cohesion between large soil aggregates.

The sand-sized soil particles can be interpreted to be derived from biogenic aggregates detached by strong winds from the surface horizons formed during Leilan period IIb, and to be fragile because of the nearly complete destruction of the root network. The preservation of pseudo-sand morphology within a 50 cm thick unit indicates that seasonal effects of deflation and saltation have not been counteracted by the wetting that should have occurred if the seasonal rainfall pattern would have remained unchanged in spite of the increased wind velocity.

The common occurrence of surface crusting evidence throughout soil unit III, together with weak effects of bioturbation and rare signs of carbonate redistribution, provide complementary data to support the interpretation that the change of wind pattern was accompanied by a significant modification of rainfall regime. Important suurface crusting can be interpreted to result from effects of heavy rainstroms inducing surficial runfoff on bare soils because water percolation was reduced by the dryness of the soil surface. Random distribution of the slaking crusts indicates that this process has remained active all along accretion of soil unit III by coarse dust, and has not been counteracted by the activity of the soil fauna which should have occurred if conditions of more regular rainfall would have been restored. None of the calcitic features appear to have been formed in situ; their morphology and distribution is typical of a sedimentary origin by surface runfoff that has reworked previously formed calcic horizons exposed at the ground surface by erosion. Size and morphology of anthropogenic residues attest to a similar mode of deposition, most probably from short distance saltation transport of materials eroded from the site. It is, however, possible that herbivore dung fragments would represent signs of extensive grazing related to pastoral activies which could have taken place in abandoned cultivated lands. Abundance of the fine charred organic particles can also be explained by an aeolian and hydric incorporation of materials eroded from the site, although they may be evidence of aridity-induced wild fires. Both glass shards and clay coated minerals derive from long distance aeolian transport, as shown by their absence in the local geological context.

Identification of an erosional discontinuity in the summit position is supported by the lateral change of soil unit III along the slope. Characteristics of this layer in the P2 soil profile indicates the combined effect of aeolian sedimentation similar to Pl soil profile and transportation of soil materials eroded from summit position by overland flow (Fig. 6 & Table 3). Abundance of poorly sorted textural features reveals the effect of high intensity drop impact on poorly vegetated soils that have broken up soil aggregates. Continuation of the soil unit III on the west footslope of the wadi suggess that its formation represents a regional landscape feature (Fig. 6). Spatial variability of the morphological characteristics of the soil unit III along the toposequence indicates that most of the soil materials have a similar origin and mode of deposition (Table 3). Absence of anthropogenic constituents in the footslope profiles indicates that they are not evidence of manuring, but more likely derived from short distance transport of materials eroded from the site. Soil unit III could not be traced in the toeslope P3 soil profile because alluvial processes subsequent to its deposition have considerably reworked the lower flood plain (Fig. 6).

Data From Archaeological Sequences

In all the archaeological sequences studied, the 2200-1900 B.C. time interval is represented by a well defined stratigraphic unit which display similar characteristics for origin of the soil materials, mode of deposition, and conditions of transformation, although some spatial variability derives from local topography and archaeological structures (Fig. 4). The most informative record is provided by the Acropolis Northwest cultic quarter wall sequence (Weiss, Courty, and Zeder in press) because proximity of the city wall prevented down slope erosion. The original rather flat topography of this area and absence of mud brick constructions offered conditions for trapping local and allocthonous sediments with limited materials from the collapse of mudbrick walls. The sequence studied can be subdivied into three soil complexes (Fig. 4).

SOIL COMPLEX 1. The lower complex consists of yellowish brown, calcareous, sandy loam, decimetic layers with common calcareous and quartzitic, rounded gravels and abundant human-derived residues (mudbricks, sherds, herbivore dung, bones, phytoliths, charcoal). The lowermost unit (G.S. 1.1, 6 cm observed) displays a dense channel microstructure of biological origin in which intense earthworm turbation is responsible for the fine homogenization of two types of soil materials:

(1) dark, yellowish brown, sandy loam derived from mud bricks;

(2) pale grey, calcitic silt with fine laminations; diffuse calcitic impregnations are common.

The bottom part of G.S. 1.2 consists of a 5 cm thick massive, yellowish brown, sandy loam unit, with abundant very fine charred fragments and lenses of well-rounded fine sand-sized particles. The core ofG.S. 1.2 (12 cm) merges to a unit with a spongy fabric, with more abundant calcitic silt, a lower amount of mud brick derived materials, and common, finely laminated dusty clay coatings. G.S. 1.3 layer (15 cm) consists of an alternation of (1) dense spongy, yellowish brown, sandy loam unit, finely homogenised by biological activity, with abundant, dusty, clay coatings and (2) horizontal lenses made of densely packed,fine sand-sized particles with abundant charred remains, rounded, millimetric sherds and gravels which are interlayered with silty clay, laminated slaking crusts. The toppermost part of G.S. 1.3 displays abundant, large sized, charred root remains, well preserved in biogenic channels.

All these characteristics help recognize a human-made deposit which is derived both from slow disaggregation of mud bricks and from regular dumping of organic-mineral residues. The biological mixing, textural features and development of secondary carbonates document maintenace of high moisture content soil accumulation. The considerable development of water-induced features (calcitic silt, slaking crusts, clay coatings) suggests that watering not only relates to effects of rains, but also to local irrigation with water of high suspended charge.

The situation between a public quarter formed of a cultic platform surrounded by a complex of well designed rooms and the Acropolis wall suggests that this area was a carefully maintained open "Green Space," in which plants with a surficial root system were growing. Occurrence of charred root remains in the toppermost part of G.S.1 indicates that in situ burning of the plants occurred during a very last phase of maintenance of this area.

In addition to the identification of horizontally stratified period IIa and IIb sherds within the complex I soil units, stratigraphic correlation with the surrounding occupation layers and archaeological structures attests that maintenance of the Green Space is contemporaneous with Leilan period IIa (G.S. 1.1), and then period IIb cultural phases (G.S. 1.2 and 1.3).

SOIL COMPLEX 2. The very abrupt contact between the burnt topsoil layer of the soil complex 1 and the overlying soil complex 2, and significant changes of type and micro-fabrics of soil complex 2 reveals an abrupt change of the mode of deposition and conditions of soil development. Soil complex 2 consists of 1 meter thick stratified sequence made of centimetric layers that display vertical evolution of their sedimentary facies (Table 4): type I for the lowermost layer (G.S. 2.1), type II for the overlying ones (G.S. 2.2 to 2.4) and type III for the uppermost ones (G.S. 2.5 to 3).

The upper layers have been considerably disturbed by deep and large extra- and intra-mural modern garbage pits dug by three generations of Tell Leilan villagers. Only a few traces of Leilan period I occupation have been observed across a 600 square meter excavation area. Soil complex 2 is stratigraphically correlated with the occupational hiatus 2200-1900 BC identified in previously excavated areas (Weiss et al., 1993).

The facies of the basal layer of soil complex 2 indicates a massive deposition of dust that consists of a fine mixture of materials from two different sources: (1) abundant silt-sized calcite particles and a minor amount of coarser grains that are clearly of local origin as shown by their mineralogical and morphological resemblance to the pre-existing surface soils. Absence of signs of surficial runoff and water-related deposition rules out the possibility for the dust to derive from the rainfall-induced decay of the surrounding mud brick constructions which may have rapidly collapsed

Type	Facies description	Constituents and microscopic features
I	Pale brown (10 YR 6/3) silty facies: (a) massive to spongy, undifferentiated; 1 to 2 cm thick. (b) spongy to fissural, coarse layered centimetric to 1/2 centimetric, graded-bedded massive layers; (c) fissural, finely laminated, graded-bedded, millimetric layers; thickness varying between 2 to 6 cm.	dominantly made of (I) very fine, (ii) fine silt-sized and (ii) rare-fine sand-sized particles and coarser particles; 64% angular calcite particles (5-15μm). 20% angular silt-sized phytoliths, glass shards, mineral grains coated with yellowish brown clay, charred organic particles, occasional diatoms and rare calcitic micro-fossils (25-30 μm). 10% well rounded, sand-sized, yellowish brown soil fragments, grey carbonate grains. 5% rounded coarse fragments (500 μm-1 mm) of bricks, sherds, large connected silica remains (thached roofs); 1% Infillings of gypsic lenticular crystals, a few sand-sized calcitic spherules.
II	Yellowish brown (10YR 6/3) layered sandy silt facies: (a) regular alternation of millimetric layers of (i) vughy, loosely packed, rounded, coarse sands and gravels, with common silty clay coatings, (ii) vughy to sub-horizontal fine fissural, graded-bedded, medium to fine sand-sized grains; (ii) fine, fissural, laminated fine silt. (b) irregular alternation of the same kind of layers; common occurrence of disorganized, centlmetric fragments of finely fissured, silty layers embedded in developed calcitic, silty clay coatings. (c) vughy, finely fissural fabric with a platy microstructure; irregular imbrication of very fine to fine, silt-sized, millimetric layers interstratified with loose sands.	Gravels; rounde fragments of burnt bricks, baked bricks, md floors, sherds, micritic carbonate nodules, Sands and coarse silt: 60% grey micritic concretions, yellowish brown, silty clay soil aggregates, fragments of sparitic/microsparitic nodules; 20% well sorted, sub-rounded, charred organic fragments and phytoliths; 10% quartz, felspars, gypsum crystals coated with yellowish brown clay; 10% flaky, slightly weathered, glass shards, Fine and very fine silt-sized fraction similar to the facies I. The terminal laminae of silty clay coatings often consists of dark, rounded, aggregated organic particles (5-10 μm).
III	Complex aggregated facies: (a) heterogeneous vughy to vesicular fabric; loosely packed millimetric to sand-sized, sub-rounded aggregates with common, large fragments of grey, calcitic, laminated, silty clay slaking crusts and laminated, silty coatings at the underside of large vughs; mixed with irregular lenses and concentrations of resistant, millimetric, rounded aggreagates (b) heterogeneously loose packing of finely fragmented, centimetric, sub-angular aggregates with common channels filled with loose sands. (c) dense channel to spongy fabric of partially melted, centimetric, sub-rounded aggregateswith common gypsic infillings, interstratified with irregularly layered lenses made of silt-sized particles and well sorted, subrounded aggregates (500 μm - 1 mm); abundant integrated, calcitic, silty clay slaking crust fragments; common silty clay, laminated coatings, (d) Vughy, massive fabric with large fissures made of partially melted centimetric to millimetric aggregates; integrated fragments of calcitic, silty clay slaking crusts; common channles infilled with earthworm excrements.	Variable proportion of contruction materials depending upon the nature of surrounding structures (burnt bricks, sherds, large charred plant remains and burnt fragments of thached roofs). Coarser grains generally made of the most resistant constituants (carbonate nodules, burnt bricks, sherds, bones) Millimetric to sand-sized grains dominantly derived form the fragmentation of baked bricks and mud-floors; also commonly charred plant remains. Silt-sized fraction similar to the one of facies I

Table 4. Classification of the sedimentary facies characterizing theLeilan hiatus (2200-1900 yr B.C.) in archaeological contexts.

after their abandonment. (The calcite silts are more likely to be fine aerosols brought by heavy dust storms which reveal the establishment of an original wind pattern from the very begining of the abrupt climate change. Such a wind regime has never observed in the previous soil record.) (2) The non-calcareous component, predominantly glass shards with allocthonous clay-coated mineral grains and unusual calcitic spherules, clearly wind-transported from unknown sources.

Although phytoliths and charred remains were abundant in the human residues from the surrounding archaeological deposits, similarity of their size with the local and allocthonous dust, their rounded shape, their good sorting, and their fine imbrication with the other mineral constituents suggest that they may also be of aeolian origin. This suggests the occurrence of wild fires of cultivated fields and steppic vegetation synchronous with the increase of the atmospheric dust loading.

The very sharp contact of the uppermost unit of soil complex 1, which displays evidence for an in situ combustion of the very last plant cover, with the overlying dust layer suggests that the combustion and the first dust accumulation occurred at the same time. The massive character of the first dust layer indicates one episode of accumulation not interrupted by effects of rainfall. This can be explained by two different scenarios: 1. either a single episode of dust accessions that would have taken place during the wet season when the atmospheric dust load was washed by rain (2.5 cm in one seasonal event), followed by rapid desiccation and surficial reworking during the dry season; 2. an episode of a few years dominated by rainwash of aerosolic dust not interrupted by drying. Weak stratification of the lowermost massive dust layer would support the second scenario. Clearly, we require precise annual dating of the first dust layer to further test this hypothesis.

The fine silt-sized texture of the lowermost dust layer indicates strong detachment of elementary particles, originally eroded as sand-sized and coarser soil aggregates from the surrounding surface horizons. This effect is tentatively explained as a rise in kinetic energy of raindrops due to prevalent horizontal winds, as suggested by a recent experiment that demonstrates the enhancement of precipitation intensity by wind speed (Pedersen and Hasholt, 1995). This sedimentary facies differs from the contemporaneous one observed at Abu Hgeira 2 (Weiss et al., 1993; Courty, 1994). As previously described, the rounded pseudo-sands of Abu Hgeira 2 together with the excellent preservation of sand-sized tephra shards are more likely to have been formed by vertical turbulence with moderate energy input from wind and raindrops.

Presence of silty coatings around the sands, however, confirms that tephra fallout also first occurred in conditions of sufficient soil moisture content for producing links between calcitic silt by cementing agents. It is only during the following stage of dust accession that aridity and wind speed increased as shown by the textural trend towards non cemented, looselypacked, silt to fine sand-sized aeolian deposit. This regional difference suggests that initiation of the abrupt climate change was marked by a complex wind pattern at the regional level explained perhaps as the influence of the Tur Abdin on the wind regime.

The silt-sized texture of the glass shards confirms the difference previously noticed between the dust falolout at Tell Leilan and the contemporary tephra deposit identified in the Abu Hgeira 2 sequence (Weiss et al., 1993). The glass shards of the first dust layer of the Green Space soil complex 2 appear to have been strongly fragilised. This is shown by their total fragmentation into particles no larger than 30 fm under moderate mechanical treatment aimed to study particle size composition of the dust, as checked by observation of the sorted particles under the binocular. The fine fragmentation of the glass shards may be the consequence of aeolian transport with high energy turbulences. It is also possible that the accumulation which affected the Tell Leilan region would be derived from a different component of the plume, richer in fine dust and in high temperature gas in comparison to the one identified in the region of Abu Hgeira 2. This would explain preservation of finely fragmented glass shards that should have been easily dispersed when transported over long distances and locally reworked by wind. The abundance of individual flakey glass and mineral particles in the Leilan deposit that were not identified in the Abu Hgeira 2 samples supports the hypothesis that regional variability in nature of the dust fallout may relate to different composition and cooling processes of the plume depending upon local turbulence.

Occurrence of similar glass shards together with silt-sized, well sorted phytoliths, charred organic particles and calcite dust confirm previous observations from other locations that indicated continuation of aerosolic fallout while the construction materials were going through a gradual collapse (Weiss et al., 1993). Because ofthe absence of construction materials in the Green Space, the finely stratified sequence of soil unit 2 provides a better record to understand the evolution of the dust accessions and their transformation conditions. The regular stratification (facies of type IIa, Table 4) indicates a cyclical pattern of deposition that may possibly relate to a seasonal rhythmicity. The silt-sized layers would have been possibly deposited at the maximum of the rainy season when the rainfall could efficiently wash the atmosphere and when wetting conditions were sufficient for fragmentation of the fine aggregates.

The occurrence of a terminal laminae made solely of very fine silt-sized charred particles in the clay coatings present in the silty layers suggests wild fire contributions to the atmospheric aerosols. The sand-sized layers would have been formed at the end of the rainy season, when episodes of wetting would favor formation of puddles in small depressions and would alternate with rapid drying. The coarser layers would have been deposited at the very beginning of the rainy season when occassional heavy spells can only induce partial fragmentation of the dry soil materials and transportation by surficial runoff of the sand-sized aggregates formed during the previous dry season by desiccation of the finer deposits.

Abundance of archaeological debris in the coarser layers also indicates transport of materials eroded from the surrounding contruction, the disaggregation of which was facilitated by intense desiccation during the previous dry season. The Acropolis wall explains preferential trapping of materials both from local sources by surficial runoff and from allochthonous sources by dust accessions. The preservation of the fine stratification indicates that water effects were only surficial and were never sufficient to wetten the subsurface layers. This limited wetting is possibly linked to the presence of hydrophobic substances, such as waxes, known to be produced by high intensity firing and to increase water repellency that results in stronger surficial erosion and reduces water percolation (Scott and Van Wyk 1990).

The occurrence throughout soil unit 2 of sub-horizontal fine fissures at the contact between layers of different texture and within each layer reflects intense drying after each depositional episode. Development of this microstructre may have also been enhanced by moderate surface freezing.

Absence of biological reworking and lack of calcitic features produced by vertical solute movements provide complementary evidence for the maintenance of extremely dry conditions during the accumulation of soil unit 2. Occasional disturbance of the finely stratified deposits by torrential flow is shown by the repeated occurrence of disorganized heterogeneous lenses. Variability through time of dust accession and hydric effects is also expressed by the irregular thickness of the fine textured and coarser layers. As a consequence, a precise counting of the bi-phased or three phased deposit that would possibly represent an annual cycle is nearly impossible. By taking an average thickness of 1 cm a year, accumulation of the soil unit 2 can be roughly estimated to represent a minimum duration of 100 years.

SOIL UNIT 3. An increase of soil moisture during deposition of soil unit 3 (facies type IIIc to IIId; Table 4) is indicated by the trend towards a fine channel microstructure, then a vughy and massive one, although moderate bioturbation and preservation of the fine aggregation show that wetting remained moderate while aeolian dust continued to accumulate. The effects of longer duration wetting marked by structural collapse and earthworm turbation is only observed at the uppermost part of soil unit 3. The exact thickness of the last deposit cannot be evaluated because of modern pit disturbance.

Deposits similar to the Acropolis sequence have been observed in Trench B and Trench D (Fig. 5) where they are sandwiched between the period IIb virgin soil and Leilan period I occupation materials. The fine stratification that is typical of soil unit 2 is, however, never so clearly expressed because the accumulation took place near mudbrick structures where irregular collapse has interferred with the cyclical dust accessions. It is, however, possible to distinguish a first episode of accumulation that took place under extremely arid conditions with short wetting followed by rapid drying (vesicular microstructure of facies type IIIa and loose packing of facies type IIIb; Table 4), then a final episode of deposition marked by an increasing soil moisture content (facies type IIId; Table 4). This succession reinforces the exceptional character of environmental conditions that prevailed 2200-1900 BC. The facies described strongly differ from other abandonment deposits of the same region, formed from the collapse of construction materials and occupation debris during previous periods, such as the ones observed in the Uruk period site of Nasran I or at Leilan within period IIId layers. Although the latter display a similar aggregated fabric typical of the gradual fragmentation of human-derived materials, they always show a densely coalescent assemblage, an intense biological reworking, diffuse calcitic features and partial melting of aggregates, all features which indicate the effects of wetting and water percolation during the disaggregation process.

Climatic Implications

We originally suggested that the soil record of the 2200-1900 B.C. event, as deciphered from both Abu Hgeira 2 and Tell Leilan data, could be interpreted as the consequence of increased atmospheric instability caused by air warming, in addition to attenuated seasonality and extended aridification (Weiss et al., 1993). The warming effect was, in fact, deduced from present conditions when drought periods coincide with a warmer dry season, and more active wind turbulence with common sand storms. The new data provided here suggest that analysis of modern situations might not offer satisfactory analogs for reconstructing the scenario of the abrupt climate change.

Two unique qualities of the abrupt climate change have been isolated: (1) its coincidence with a glass shard rich dust fallout, and possibly wildfires, that still remain to be explained; (2). aerosol loading and wind intensification during the rainy season, features unknown within the instrumental record (Kerbe 1987). Three hundred years marked by high aerosol loading, air mass instability, short rainy season and strong soil moisture deficit, all describe a shift from normal climate boundary conditions that cannot be assigned to a long run of dry years as Wilkinson (1994b) has suggested. Scenarios able to explain the rapid initiation of the climate shift, its causative linkage with the original dust fallout and possibly wild fires, disorganisation of the seasonality and of the interannual variability and more common drought still remain to be produced. These disturbances, however, document a substantial weakening of the Mediterranean cyclogenesis itself possibly induced by a modification of the land/sea temperature gradient in the Mediterranean basin (Rind 1993).

The soil evidence suggests that the driving mechanism of the climate disturbance was the persistence of high dust atmospheric loading known to induce a decrease of solar radiation and atmospheric warming (Tuco and Golitsyn 1988; Charlson et al. 1992; Penner et al. 1992), the long term consequences of which on air mass dynamics have not been simulated. By using short term climate simulations, we can only formulate a tentative scenario that roughly links the various phenomena identified but is unable to describe the interaction through time between atmopsheric dynamics and evolution at the ground level (Fig. 7).

In addition to the unsolved question regarding initiation and progression of the event, its ending also remains obscure. The progressive increase of soil moisture content, synchronous with a decrease of aerosol fallout suggests the gradual restoration of a "normal" trend that would have allowed a slight increase in rainfall efficiency to stimulate regeneration of the plant cover, thus reducing dust atmospheric loading.

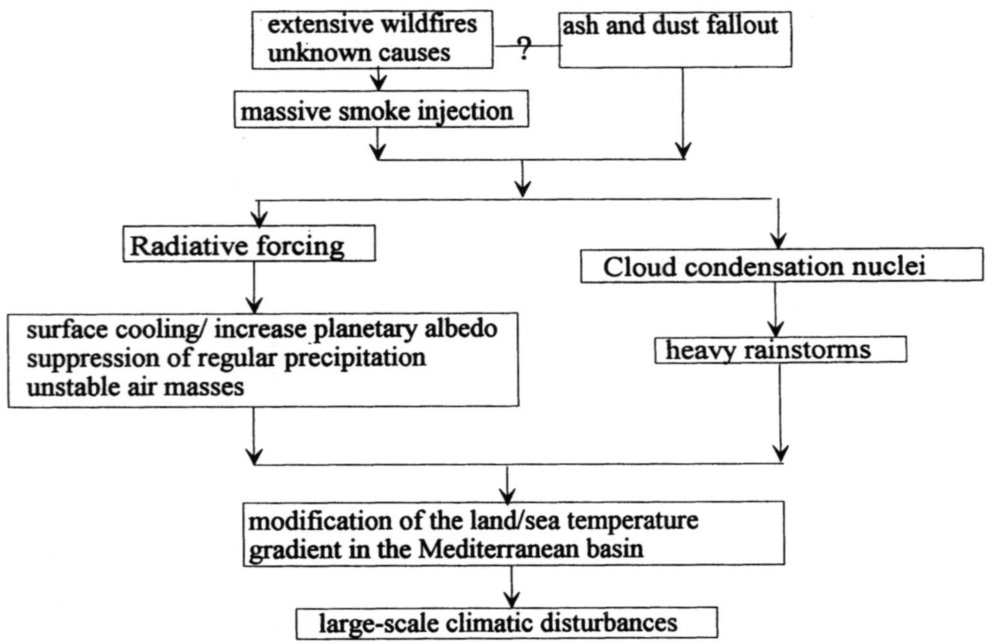

Figure 7. Tentative model describing the regional climatic effects ofthe synchroneous events recorded during the abrupt climate change.

Although the data presented here have been obtained from specific locations of a restricted area, the strong linkages with the climatic behavior of nearby regions suggests that the abrupt climate change recognised in Northern Mesopotamia represents a major shift from normal climate boundary conditions. This conclusion is now supported by abundant data reviewed and reported in this volume.

Land Degradation and Societal Dynamics

Data reported here reinforce our previous hypotheses of a direct linkage between desertion of the Habur Plains between ca. 2200 and 1900 B.C. and collapse of the Akkadian Empire, historically documented population movements among both nomadic and sedentary peoples, and the brief history of the Ur III dynasty, the replacement polity in southern Mesopotamia (Weiss et al., 1993; Weiss and Courty 1993).

Habur Plains populations abruptly faced a new environmental regime marked by a persistently cloudy weather in winter caused by high aerosol loading, low soil moisture efficiency due to high wind speed, deterioration of air quality because of the dusty atmosphere, and possibly wild fires. Lowered agro-production if not crop failure ensued (Fig. 8) and would have reduced the flow of imperial taxes that sustained Akkadian imperialization (Weiss 1996b). Channel flow might have remained sufficient at the begining of the event to provide a water surplus only before the regional decrease of rainfall affected the upper catchment area in the Tur Abdin.

The Leilan abandonment suggests that soil moisture reduction was not redressed with increased water flow, although the Akkadian rulers of the pre-collapse city had sufficient knowledge of hydraulic technologies to properly manage the channel system and control the seasonal irregularity of the fluvial discharge (Weiss et al., 1993). City and regional abandonment, therefore, are difficult to explain as a simple consequence of a limited technological capacity or ossified class structures, as claimed for contemporary Palestine (Rosen 1995).

The remnant urban populations of northern Mesopotamia, for instance those documented epigraphically at Brak and Nineveh, remain to be examined (Weiss and Courty 1993). Documentation for the recent claim that Brak was not partially abandoned is not available, while the robust evidence for its significant post-Akkadian abandonment, and the abandonment of villages in its environs, continues to mount (Mallowan 1947: 22; Matthews, Matthews and McDonald 1994;

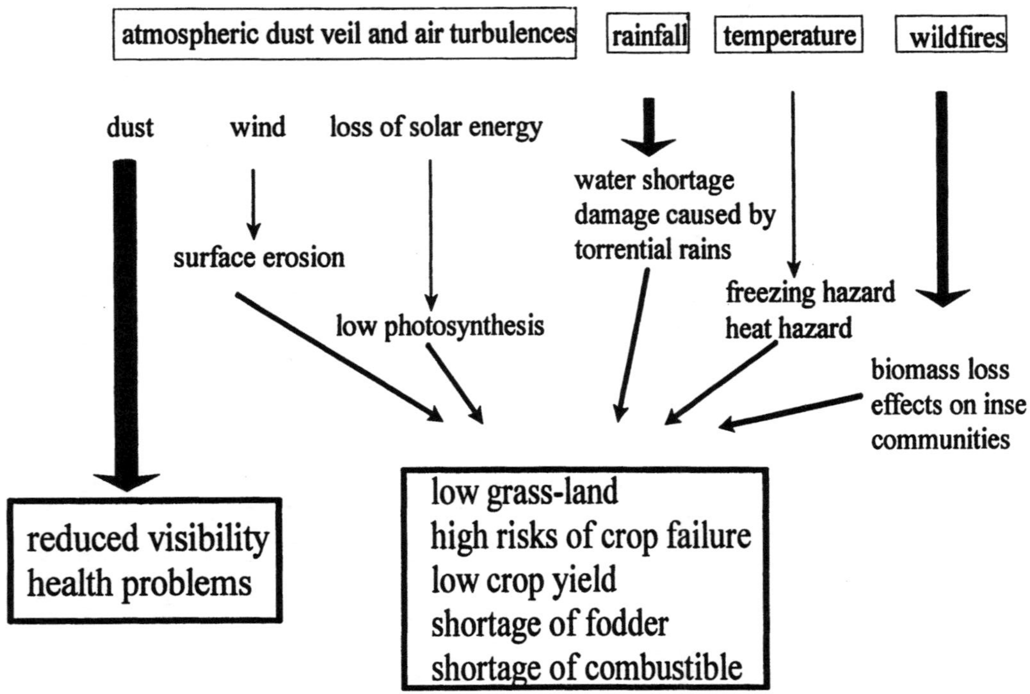

Figure 8. Tentative model explaining effects of the climatic deterioration and synchroneous phenomena on the disastrous declineof farming potential during the abrupt climate change.

Warburton and Eidem in press). Similar urban remnants, lacking rural support, are also known from EB IV Palestine (e.g., Khirbet Iskander).

Collapse and emigration, both adaptive responses, fill the contemporary archaeological record across western Asia (Weiss 1996a). Populations moved into the riverine environments of Ur III southern Mesopotamia where the effects of persistently cloudy weather, increased wind turbulence and and soil moisture deficits were tempered by the sustainability of wet lands, marsh, and extended irrigation agriculture for agro-pastoral use. Discriminating the direct effects of the environmental change from secondary disturbances of the regional politico-economic entities is difficult in the region of Tell Leilan because of its drastic devastation, but could be more easily achieved in possibly less degraded areas.

3. Soil Landscape Dynamics Subsequent To The Abrupt Climate Change

The stratigraphic discontinuity in natural soil profiles between the soil unit formed during the abrupt climate change (unit III, Table 2) and its successors (unit IV, Table 2) suggests rapid restoration of a "normal" climate, possibly within a few decades. Establishment of more regular rainfall, the decrease of wind effects and the stabilisation of the soil surface by a vegetation cover are marked in the lower layer of soil unit IV (Table 2, profile 1, unit IVc) by the biogenic fabric, occurrence of calcitic features, melting of the aeolian saltated sand-sized grains, and lack of surface crusting.

The soil landscapes of the region may have absorbed the effects of the disturbance as soon as hydric and wind stresses came to an end. The considerable increase of aerosol dust in soils during the abrupt climate change appears to have enhanced soil quality during the post-collapse period through input of nutrients favorable to plant growth, especially those from weathered glass shards, and a low density surface layer favorable to the regeneration of annual and then perennial plant species.

The thick accumulation of fine textured flood deposits in the toeslope position indicates the readjustment of landforms in response to the hydrographic disorganisation and slope disequilibrium generated by the abrupt climate change (Profile 3 in Fig. 6, Table 3). The lack of subsequent entrenchment of the main channel to the west of Tell Leilan and the fine texture of the flooded deposits suggests that the hydrological regime after the abrupt climate change remained lower over the last 4000 years in comparison to the one during period II. This difference might be related to geomorphic changes in the upper catchment basin, as well as to modification of the rainfall pattern.

Deposition of an approximately one meter thick unit of calcareous silt in the summit position above soil unit III (Fig. 6) indicates that dust accession has remained an active process over the last 4000 years, although at a lower rate in comparison to period II. Dust accession does not appear to have recently increased in response to modern agricultural intensification, although the open micro-aggregated fabric of surface horizons, which results from an annual cycle of plant growth, might have increased the sensitivity to aeolian erosion (Table 2). This supports consideration of dust accession to be a reliable indicator of the regional climate trend not modified by human exploitation. The increase of fine sand in top soil unit IV and its homogeneity throughout the different landform components suggest a modification in the source of alluvium. The occurrence of slaking crusts and sandy infillings in the uppermost soil horizon (IVa, table 2) reflects an increase of surface crusting by modern intensive agriculture. Bioturbation, however, has prevented soil degradation.

The difference in morphology and vertical distribution of calcitic features in soil unit IV, and the increased dissolution of the fine mass in comparison to period II soils, indicate higher water percolation and lower evapotranspiration over the last 4000 years (Table 2). This might signal more efficient rainfall because of lower temperatures during the wet season. Lack of a precise soil stratigraphy does not allow us to identify minor climatic fluctuations that affected the region since the abrupt climate change.

Conclusions

A multi-scalar soil proxy record, from microsamples to stratigraphic units to landforms, collected from various Tell Leilan environments, documents the qualitative aspects of environmental dynamics and interactions between natural and societal forces.

The preservation of the soil stratigraphic record in buried soils and archaeological sequences allows us to reconstruct the succession of environmental conditions responsible for accumulation of deposits and their transformation by pedogenic processes. Interpretation of soil attributes with respect to climate variables has been facilitated by the region's location at the transition between various climate influences, the low range variability of geological materials and landform monotony.

In addition, the high sensitivity of the soil-water system to environmental variables has facilitated detection and interpretation of the subtle changes under which soil features formed. Soil evidence documents the resilience of pre-collapse soil landscapes and their capacity to absorb effects of human exploitation without suffering degradation.

Fragility did not increase during the pre-collapse period. The environmental deterioration at ca. 2200 B.C. was, therefore, the result of a new and exceptional force created within a few years. Our new stratigraphic sections provide additional evidence for the nature, rate and amplitude of the phenomena that generated a major shift from normal climate boundary conditions. Initiation of the abrupt climate change appears to have been directly linked to a rapid increase of aerosol fallout rich in glass and other unusual particles, of wind intensity and of soil moisture deficit. The maintenance of unstable air mass dynamics over a ca. 300 year period suggests that the abrupt climate change cannot be the consequence of the explosive volcanic event. The historical and modern record of explosive volcanic eruption shows that climate effects do not last more than 18 months (Sigurdsson 1990). Further investigations will help identify the source of the dust fallout and its regional variability.

We have provided additional evidence to establish a direct causal linkage between the 2200-1900 BC abrupt climate change and collapse of the Leilan IIb Akkadian imperialized settlement. In addition to a considerable and prolonged decrease in rainfall, the sudden deterioration of air quality, and the cloud reduction of solar radiation appear to have been the new constraints on agricultural production. Tracing the dispersion of the dust fallout will help identify the stratigraphic evidence of the abrupt climate change from surrounding regions and the diversity of local environmental effects. At this preliminary research stage, our conclusions stimulate the pursuit of a multidisciplinary effort to elucidate the complex anthropological and environmental aspects of the 2200-1900 B.C. abrupt climate change and its regional impact.

Acknowledgments. We are indebted to the Directorate-General of Antiquities,Damascus, for support, advice and assistance during the course of Tell Leilan fieldwork. The Tell Leilan research and analyses reported here were made possible by grants from the National Science Foundation, National Endowment for the Humanities, Leon Levy, Raymond and Beverly Sackler, Malcolm Wiener and the Centre National de Recherche Scientifique, Paris.

References

Berger A. (1984) Desertification in a changing climate. With a particular attention to the Mediterranean countries. In Desertification in Europe. Proceedings of the Information Symposium in the EEC programme on climatology. Mytilen Greece, pp. 15-34. Birk University Press.

Birkeland P.W., Berry M.E., and Swanson D.K. (1991) Use of soil catena field data for estimating relative ages of moraines. *Geology* 19: 281-283.

Bresson L.-M. and Valentin C. (1994) Soil surface crust formation: contribution of micromorphology. In: A.J. Ringroase-Voase and G. Humphrey (eds.). Soil Micromorphology: Studies in management and genesis. Amsterdam, Elsevier. pp. 737-762.

Bullock P., Fedoroff N., Jongerius A., Stoops G., Tursina T. and Babel U. (1985) Handbook for soil thin section description. Wolverhampton, Waine Research Pub.

Busacca A.J., Nelstead K., McDonald E. and Purser M.D. (1992) Correlation of distal tephra layers in loess in the channeled scabland and Palouse of Washington State. *Quaternary Research*, 37: 281- 303.

Charlson R.J., Schwartz S.E., Hales J.M., Cess R.D., Coakley J.A., Hansen J.E. and Hofmann (1992) Climate forcing by anthropogenic aerosols. *Science* 255: 423-431.

Cheddadi R., Rossignol-Stirck M. and Fontugne M. (1991) Eastern Mediterranean palaeoclimates from 26 to 5 ka B.P. documented by pollen and isotopic analysis of a core in the anoxic Bannock basin. *Marine Geology* 100: 53-66.

Cleuziou S. (1994) La chute de l'empire d'Akkadé: hommes et milieux au Moyen-Orient. Les Nouvelles de l'Archéologie 56: 45-58.

Courty M.A. (1990) Environnements géologiques dans le nord-ouest de l'Inde. Contraintes géodynamiques au peuplement humain protohistorique. Doctorat Es Sciences, Univ. de Bordeaux I, 328 p.

Courty M.A. (1994) Le cadre paléogéographique des occupations humaines dans le bassin du Haut Khabour (Syrie du nord-est). Premiers résultats. Paléorient 20(1): 21-59, CNRS (ed.), Paris.

Courty M.A. and Fedoroff N. (1985) Micromorphology of recent and buried soils in a semi-arid region of North-West India. Geoderma 35: 287-332

Courty M.A., Dhir R.P. and Rhagavan H. (1987) Microfabrics of calcium carbonate accumulations in arid soils of Western India. In: N. Fedoroff, L.M. Bresson. et M.A. Courty, (eds.), Micromorphologie des Sols- Soil Micromorphology, 227-234. AFES, Plaisir, France.

Courty M.A., Goldberg P. and Macphail R.I. (1989) Soils and micromorphology in archaeology. Cambridge Manuals in Archaeology. Cambridge University Press, 344 p.

Courty M.A., Goldberg, P. and Macphail R.I. (1994a) Ancient people - Lifestyles and cultural patterns. In: Transactions 15th Meeting of the International Soil Science Society, Symposium B "Micromorphological indicators of anthropological effects on soils". International Society of Soil Science, Mexico. Vol. 6a: 250-269.

Courty M.A., Marlin C., Dever L., Tremblay P. and Vachier P.(1994b) Morphology, geochemistry and origin of calcitic pendents from the High Arctic (Spitsbergen). Geoderma 61, 71-102.

Courty M.A., Fedoroff N., Jones M.K., Castro P. and MacGlade J. (1994) Chapter 2. Environmental dynamics. In: Understanding natural and anthropogenic causes of desertification in the Mediterranean basin, S. Van Der Leeuw (ed.). Final report of the Archaeomedes Project, Directorate General XII of the Commission of the European Union (Contract EV5V-0021).

Fedoroff, N. and Courty M.A. (1987) Morphology and distribution of textural features in arid and semi-arid regions. In: Fedoroff, N., L.M. Bresson. et M.A. Courty, (eds.), Micromorphologie des Sols- Soil Micromorphology, 213-219. AFES, Plaisir, France.

Fedoroff, N. et Courty M.A. (1989) Indicateurs pédologiques d'aridification. Bull. Soc. géol. France, 8(V-1): 43-53.

Fedoroff N., Courty M.A. and Thompson M.L. (1990) Micromorphological evidence of palaeoenvironmental change in Pleistocene and Holocene paleosols. In: L.A. Douglas (ed.), Soil Micromorphology: a basic and applied science, Elsevier, 653-666.

Gasse F. and Van Campo E. (1994) Abrupt post-glacial events in West Asia and North Africa monsoon domains. *Earth Planet. Sci. Lett.* 126: 435-456.

Glassner J.-J. (1994) La chute de l'empire d'Akkadé, les volcans d'Anatolie et la désertification de la vallée du Habur. Les Nouvelles de l'Archéologie 56: 49-51.

Hassan F. (1986) Holocene lakes and prehistoric settlements of the western Faiyum. Journal of Archaeological Sciences 13.5: 483-501.

Jim C.Y. (1986) Experimental study of soil microfabrics induced by anisotropic stresses of confined swelling and shrinking. Geoderma 37: 91-112.

Kutzbach J.E. (1981) Monsoon climate of the Early Holocene: climate experiment with the Earth's orbital parameters for 9000 years ago. *Science* 214: 59-61.

Macphail R.I., Courty M.A. and Gebhardt A. (1990) Soil Micromorphological evidence of early agriculture in North West Europe. World archaeology 22(1): 53-69.

McFadden L. and Knuepfer P. (1990) Geomorphology: the linkage of pedology and surficial processes. *Geomorphology* 3: 197-205.

Mallowan, Max 1947 Excavations at Brak and Chagar Bazar Iraq 9.

Matthews, Roger 1994 Imperial Catastrophe or Local Incident? An Akkadian Hoard from Tell Brak, Syria Cambridge Archaeological Journal 4.2: 290-316.

Matthews. R.J., W. Matthews, H. McDonald 1994 Excavations at Tell Brak, 1994 Iraq 55: 177-194.

Pedersen H. S. and Hasholt B.N. (1995) Influence of wind speed on rain splash erosion. Catena 24: 39-54.

Penner J., Dickinson R. and O'Neill C. (1992) Effects of aerosols from biomass burning on the global radiation budget. *Science* 256: 1432-1436.

Petit-Maire N., Sanlaville P. and Zhong Whei Y. (1994) Changements globaux et paléomoussons. La zone de transition afro- asiatique au cours des derniers 140 000 ans. *Géochronique* 50: 16-19.

Ponomarev B.Y. and Bivshev A.S. (1966) The geological map of Syria, V.P. Ponikarov (ed.). Sheets J-37-V, XI. Syrian Arab Republic. Ministry of Industry. Department of Geological and Mineralogical Research.

Prentice I.C. and Sarnthein M. (1993) Self-regulatory processes in the biosphere in the face of climate change. In: J.A. Eddy and H. Oeschger (eds), Global changes in the perpective of the past. Wiley, New-York, 31-38.

Retallack G. J. (1994) The environmental factor. Approach to the interpretation of paleosols. In: R. Amundson, J. Harden and M. Singer (eds.): Factors of soil formation: a fiftieth anniversary retrospective. SSSA Special Publication 33, Madson, Wisconsin, USA.

Rind D. (1993) How will future climate changes differ from those of the past. In: J.A. Eddy and H. Oeschger (eds.), Global changes in the perspective of the past. Wiley, New York, pp. 39-49.

Rosen A.M. (1989) Environmental change at the end of early Bronze Age Palestine. In: P. de Miroschedji (ed.), L'urbanisation de la Palestine à l'age du Bronze ancien. BAR Int. Ser. 527 (ii): 247-255.

Rosen A.M. (1995) The social response to environmental change in early Bronze age Canaan. *Journal of Anthropological Archaeology* 14: 26-44.

Scott D.F. and Van Wyk D.B. (1990) The effect of wildfire on soil wettability and hydrological behaviour of an afforested catchment. *Journal of Hydrology* 121: 239-256.

Sigurdsson H. (1990) Evidence of volcanic loading of the atmosphere and climatic response. *Palaeogeography, Palaeoclimatology Palaeoecology* 89: 277-289.

Simonson R.W. (1995). Airborne dust and its significance to soils. Geoderma 65, 1-2: 1-43.

Soil Survey Staff (1975) Soil Taxonomy. A basic system for soil classification for making and interpreting soil surveys. Agricultural Handbook, no. 436, U.S.D.A., Washington D.C.

Street-Perrot. A. and Perrot R.A. (1990) Abrupt climate fluctuations in the tropics: the influence of Atlantic Ocean circulation *Nature* 343 (607): 607-612.

Turco R.P. and Golitsyin G.S. (1988) Global effects of nuclear war: a report status. *Environment* 30 (5): 8-16.

Warburton, D. and J. Eidem in press Tell Brak Regional survey Iraq.

Weiss, Harvey 1996a Causality and Chance: Late Third Millennium Collapse in Southwest Asia in: O. Rouault, ed., La Djezireh et l'Euphrate Syriens. Paris: ERC.

Weiss, H. 1996b Akkade in: E. Meyers, ed., Encyclopedia of the Ancient Near East. Oxford: Oxford University Press.

Weiss H. and Courty M.A. (1993) The genesis and collapse of the Akkadian empire: the accidental refraction of historical law. In: M. Liverani (ed.), Akkad: the first World Empire. Padua: Sargon. pp. 131-156.

Weiss H. and Courty M.A. (1994) Entre droite épigraphique et gauche archéologique, y-a-t-il une place pour la science? Les Nouvelles de l'Archéologie 57: 33-41.

Weiss, H., Courty, M.A., Wetterstrom W., Meadow R., Guichard F., Senior L. and Curnow A. (1993) The origin and collapse of Third Millenium North Mesopotamian Civilization *Science* 261: 995-1004.

Weiss, H., M-A. Courty, and M. Zeder in press Tell Leilan 1993 Annales archéologiques arabes syriennes.

Wilkinson T.J. (1990) Soil development and early land use in the Jazira region, Upper Mesopotamia World Archaeology 22: 87-103.

Wilkinson T.J. (1994a) Review. *Journal of Field Archaeology* 21: 385-390.

Wilkinson T.J. (1994b) The structure and dynamics of dry-farming states in Upper Mesopotamia Current Anthropology 35.5: 483-505.

Cultural Change in the Aegean c.2200BC

Sturt W. MANNING

Department of Archaeology,
University of Reading,
PO Box 218 Whiteknights,
Reading RG6 6AA,
U.K.

Abstract

In the Aegean a wave of collapse and change in the later third millennium BC has long been recognised in the archaeological record (Forsén 1992). In general terms, a date for these events c.2200BC is feasible given conventional interpretations and chronologies (Manning 1995a; Warren and Hankey 1989). This offers the potential hypothesis of contemporaneity with the c.2200BC climate change event in the Near East (Weiss *et al.* 1993). The question is whether, on a detailed and independent analysis, the precise date of c.2200BC is correct for episodes of change in the Aegean? The problem is the lack of high-resolution data. Modern, high-quality, radiocarbon series are rare (dendrochronology does not yet apply), and connections with the historical Near East and Egypt are problematic, or non-existent, at this time. Further, we must consider the major anomaly: Crete. There is no similar collapse on Crete, and, indeed, significant developments appear from around this same time on the island, and the state-level Old Palace civilisation emerges c.2000BC. An explanation couched in terms of prestige goods economies, control of resources (monopolisation, exclusivity), and geographic location on the periphery of multiple world-systems is offered (following Manning 1995b).

Introduction

In the Aegean a wave of change in the later third millennium BC has long been recognised in the archaeological record. In Greece, Caskey (1960) proposed an invasion by newcomers to be responsible for the sharp break he observed between the Early Helladic (EH) II and III periods. Renfrew (1972: 116) regarded this break to be 'more marked than any other subsequently seen in Greek prehistory, or any previously documented since the development of farming life'. In the Cycladic

NATO ASI Series, Vol. I 49
Third Millennium BC Climate Change
and Old World Collapse
Edited by H. Nüzhet Dalfes, George Kukla,
and Harvey Weiss

Islands a new 'Anatolianising' assemblage was recognised before what almost seemed to be a complete 'gap' in the archaeological record (Rutter 1979; 1983; 1984). In each case, and in general terms, a date for these events c.2200BC is feasible given conventional interpretations and chronologies (Manning 1995a; Warren and Hankey 1989; Karantzali 1996; Cosmopoulos 1991: 117-124; Cadogan 1986a). It is thus very tempting to wish to correlate these changes in the Aegean with the cultural and climate change event identified c.2200BC in the Near East (Weiss *et al.* 1993). The question is whether, on a detailed and independent analysis, the precise date of c.2200BC is correct for episodes of change in the Aegean? If so, a unified hypothesis is reasonable. Further, we must consider the major anomaly: Crete. There is no similar collapse on Crete, and, indeed, significant developments appear from around this same time on the island, and the state-level Old Palace civilisation emerges c.2000BC.

Mainland Greece

The third millennium BC of the southern Greek mainland is dominated by the EHII period. This is a long-lived period of some three-four hundred years, and is easily visible due to distinctive ceramic and other forms. 'More ceramics identified as Early Helladic II have come from more sites than any other prehistoric period' (Pullen 1995: 19). The period sees the development of a fairly uniform complex society across southern and central Greece with large, even monumental, buildings at a number of sites. There is evidence for a sealing system as part of an administrative hierarchy, and, based on variations in site sizes, settlement hierarchies are also proposed. Clear development is seen within the course of the period. After a gradual start occupying at least half the overall span of the period, there is a rapid development of complexity in the final century or so which sees the building of the monumental Corridor Houses (Shaw 1987) with their sealing systems, and also elaborate fortifications at Lerna, Thebes and probably Kolonna on Aigina. By the late EHII period, southern Greece has an advanced civilisation comparable to Early Bronze Age (EBA) Troy, Poliochni, and other sites linked to the Anatolian world. In comparison, Crete is a relative backwater at this time (see Hägg and Konsola 1986; Themelis 1984; Pullen 1985; 1994; Stewart 1988; Wiencke 1989; Konsola 1990; Forsén 1992; Rutter 1993: 761-774 with many further references).

Then, seemingly at its peak, this EHII civilisation collapses. Some EHII sites are destroyed, there are many fewer succeeding sites (mostly smaller as well), the material cultural is quite changed, and, crucially, the succeeding EHIII period lacks monumental architecture, sealing systems, ceramic uniformity, evidence of settlement hierarchies, and all the indications of a sophisticated and complex society. In southwest Greece no discernible EHIII phase can even be detected (Rutter 1979: 15; 1988: 74; 1993: 772-773; Cavanagh *et al.* 1996: 16). The change is quite dramatic; it affects things even as mundane as stone tools. In later EHII there is evidence for a widespread system of exchange of tool blanks and of specialised production based on the assemblage at Lerna, whereas this exchange system has ceased in EHIII (Runnels 1985). Likewise, the new forms and decorations in EHIII ceramics suggest a changed economy with greater nomadism (Rutter 1988). Caskey (1960), drawing mainly on his excavations at Lerna where there was a burnt destruction marking the end of the EHII phase, saw a widespread destruction horizon across Greece. Further work has of course complicated this simple scenario, and a number of sites (especially Tiryns, Kolonna, Thebes) now have EHII-III continuity, or differing processes of change. Nonetheless, a pattern of change, reorientation, collapse and decline is evident across the southern mainland from both excavations and surveys (Forsén 1992: 157-196). And, in general, whereas EH sites are prevalent (often mainly EHI-II), subsequent Middle Helladic sites are comparatively few (Rutter 1993: 748 Table 1). EHIII is quite unlike EHII, but, given spatial and temporal variations in data for destructions, new introductions, and changes (Forsén 1992), a more complex explanation than a simple invasion/migration is now required.

These events were not occurring in a vacuum. Excavations in the 1960s at Lefkandi on Euboia uncovered below EHIII levels a hitherto unknown assemblage with north Aegean or Anatolian linkages. At sites in central and eastern Greece examples of this Lefkandi assemblage were found associated with late EHII material. Further, some of the new ceramic types of the EHIII period seemed to be explainable in terms of prototypes from the Lefkandi assemblage (Rutter 1979; 1995: 475-477, 648-649). This new assemblage might be thought to provide evidence of new peoples, of migration/invasion. However, on the mainland, the new elements are restricted to drinking cups (especially one handled tankards and the *depas* cup). They are also rare (one or a few examples from any site), and are even imitated in stone (Caskey 1956: 164, fig.4, pl.47.i). The exact relative chronological placement of the Lefkandi assemblage has been a controversial subject, but careful analysis reveals that it belongs to the later EB2 period (and not EB3) in the southern Aegean (Manning 1995a: 51-62, 66-71). It is contemporary with the rise of the EHII Corridor Houses of southern Greece and the high point of the EHII civilisation ('chiefdoms': see Wiencke 1989). These new artefact types therefore are revealed more as exotic types linked both to the greater worlds of the northern and western Aegean via the 'voyaging' Early Cycladic trader world (Broodbank 1993; Manning 1994: 226-229), and to the feasting activities of the élite during the high-point of Renfrew's 'International Spirit' of late EB2 in the Aegean (1972: 451-455; Cosmopoulos 1991). The collapse comes after this final EB2 phase.

The most important recent evidence on the EHII collapse derives from geoarchaeology. Studies of the Argolid, home to several of the major EHII sites, have shown that there was a major episode of soil erosion in the mid-third millennium BC (Zangger 1993; 1992; van Andel *et al.* 1986; 1990; see also Jahns 1990). This is dated by the presence of EHII sherds in the alluvium, and by a single radiocarbon date (SMU 1643, 4027±155BP, calibrated calendar age range: 2900-2350BC at 1σ confidence from Ramsey 1995 employing the 1986 bi-decadal calibration data [cal10.dta]). In general, these studies show a correlation of the few major erosional episodes with periods of intense human settlement and exploitation of the landscape (Jameson *et al.* 1994). It is notable that areas such as the Skourta plain which were not intensively settled in the EBA have no such erosional history. The general hypothesis is that excessive land clearance onto hillslopes, and largescale intense arable farming, created the circumstances where slope denudation through unusual heavy rain-storms was possible.

Evidence for the use of the plough in the EHII period has been found (Pullen 1992). It may reasonably be hypothesised that the plough was an élite technology given equipment and especially animal costs and maintenance. Plough agriculture is designed for extensive, large field, agriculture. Systems of sealing associated with the main large (élite, or élite controlled) structures from later EHII suggest control of significant resources by some group. A large probable granary built in later EHII at Tiryns (Kilian 1986) further indicates large-scale farming and mobilisation of produce. Since dense settlement is attested from survey in the main regions of EHII settlement, we may assume that land was probably an increasingly scarce resource during the course of the EHII period in the main regions as an élite founded on control of land, plough agriculture, and the mobilisation of such resources, developed.

Data on climate change through this period is as yet rudimentary in Greece. Some evidence of an increase in temperature/aridity has been suggested (see Forsén 1992:

241-242, 245-246, 259-260), but the pollen evidence for alleged forest decline is unsatisfactorily dated. How much to make of finds of bird remains suited more to drier soils at Lerna from EHIII onwards is also not clear. The silting of Lake Lerna might be attributed to either (or both) climatic or anthropogenic factors.

What is the date of the end of the EHII period? Unfortuantely, the available chronological data are poorly constrained. EH Greece has few foreign connections, and none directly to the historical cultures of the Near East. Although a number of EH radiocarbon dates exist, most were either run many years ago, or appear to have been made on inappropriate samples. No short-lived sample material directly relevant to the close of EHII has been dated (and in fact one determination on a grain sample from EHIII Kolonna is the nearest we have to evidence: Manning 1995a: 152 and fig.18.a). Data from long-lived sample material (charcoal) from the late EHII phases at Lerna offer dates in the mid-third millennium BC (Manning 1995a: 149-150, figs.11, 12), and these may probably be assumed to date the construction of Lerna IIIC-D buildings. The evidence for the succeeding EHIII period (Manning 1995a: 151-153) suggests that the EHIII period began after c.2300-2200BC (again with no precision). In combination with evidence from Crete and elsewhere related by archaeological synchronisms, a date c.2200BC is therefore often posited for the close of the EHII period, but the quality of this dating is poor. New high-quality radiocarbon data are sorely needed.

The Cyclades

The EBA sees the rise of the Early Cycladic civilisation made archaeologically famous in recent times by Renfrew (1972; 1991; see also Cosmopoulos 1991; Karantzali 1996). Although based on small islands and consisting of small settlements, this civilisation had an important impact in the EBA Aegean. The high-point was during the EB2 period. At this time Early Cycladic objects, including the well-known Fold Arm Figurines in marble and ceramic sauceboats, are found in a wide distribution around the Aegean (Cosmopoulos 1991). A set of larger sites in the islands appear to have dominated or controlled trade within and especially beyond the islands to the surrounding mainland and Crete. Representations of 'longboats' and maritime and celestial imagery suggest that voyaging and trade were central to power within the Cycladic Islands (Broodbank 1989; 1993).

The Cycladic EB2 period consists of two phases. The first is often referred to as the Keros-Syros phase, and this relates to earlier EHII in southern Greece and EMIIA on

Crete. The second phase is marked by elements of a new north-Aegean/Anatolianising assemblage. This is the same basic assemblage as found at Lefkandi and in central Greece in later EHII (Manning 1995a: 51-62, 66-71, 81-86). In the Cyclades this phase is referred to as the Kastri phase after the site where it was identified (on Syros: Bossert 1967). The Kastri phase is poorly understood in the absence to date of fully published settlement excavations (relevant publications of excavations at Markiani on Amorgos are awaited, for example). At Ayia Irini on Kea it provides only about 10% of the ceramic material in use at the site (Wilson and Eliot 1984; Davis 1992: 708). Kastri, and Panormos on Naxos, have often been said to have entirely Kastri phase material (and so be settlements of invaders/immigrants), but neither site has been fully published (nor excavated). The Kastri phase ceramics are characterised by drinking and eating forms (tankard, depas cup, lentoid jug, wheel-made plate). The shapes and often black/grey/dark glossed surfaces also suggest the influence or imitation of metal forms. Interestingly, prestige metal objects (weapons) from Kastri have an Anatolian provenance (Stos-Gale *et al.* 1984).

As on the mainland, the appearance and development of the north Aegean/Anatolianising assemblage provides no clear evidence of invasion or migration. Instead, it belongs with the final high-point of the Early Cycladic civilisation. At this time the Early Cycladic world is communicating, and sharing artefact links, with most of the Aegean world (Renfrew 1972: 451-455; Cosmopoulos 1991; Karantzali 1996). Collapse then follows this phase, and we have almost no idea of what is going on in the Cyclades until around 2000BC (Rutter 1983; 1984; see Manning 1995a: 51-62, 66-71). As Broodbank (1993) suggests, it is tempting to see a prestige goods economy in operation in the EB Cyclades, with an inflationary cycle of prestige acquisition and competition. Fashions from the much greater (regional 'core') world of Anatolia are acquired by trade (see also Davis 1987: 32) and exploited by the Aegean élite groups as prestige and status symbols (with the Cycladic groups perhaps playing the key intermediary role with regard to mainland Greece). Such economies are almost inherently liable to boom-and-bust cycles (Friedman 1982; Allen 1984). Deliberate discard of Fold Arm Figurines at Keros may indicate attempts to control circulation of value goods. Increased evidence of fortifications in later EB2 might also indicate increased levels of competition. In addition to such endogenous logic, since the Cycladic élite depended on trade/contact with the large-scale civilisations on the surrounding mainland (in western Anatolia, in Greece), and if these economies collapsed through problems with agriculture (etc.), then the raison d'être of Cycladic trading and voyaging would be fatally undermined.

The key problem is the date of the end of the Cycladic EB2 period. Published radiocarbon dates relevant to this question are simply not available from the Cyclades. Extrapolations must be made via Anatolia (Manning 1995a: 79-86, 155-160, 170-172). However, since Manning (1995a) was written, major new radiocarbon data and analysis relevant to Troy have appeared (Korfmann and Kromer 1993). It is evident from these that the Troy I period occupied the interval from during the 30th century BC through to c.2500BC (Korfmann and Kromer 1993: Abb.21; Weninger 1995; consistent data exists also from dendrochronology [Korfmann and Kromer 1993: 155-157] and TL [Wagner and Lorenz 1992]). Similar results are available also from two radiocarbon determinations from Thermi I on Lesbos (Manning 1995a: 28), and two new dates from Poliochni period Blue (V. Tiné, pers. comm.).

The problem is the chronology of the critical Troy II phase linked with the Aegean later EB2. Only five relevant data exist, all from Troy IIg find contexts (Korfmann and Kromer 1993: 147-149; note, the dates previously published as from Troy IIg are now re-assigned: Korfmann and Kromer 1993: 161-164). Three of the Troy II data are on 'beams' ('Balken') from the Troy II megaron buildings (HD 14008, HD 14573 and HD 14561). Radiocarbon measurements on wood from large beams (likely to include hundreds of years of tree-rings [and one may compare the 226 year specimen from Troy I and other long sequences from the relevant Troy II megaron buildings: Korfmann and Kromer 1993: 155-157]) may be expected to offer dates much older than the construction of the relevant buildings, yet alone the close of the Troy II cultural phase. Not surprisingly, these three data offer calibrated ages in the first half of the third millennium BC (**Figure 1**). The two other relevant data are HD 13930 from the burnt city (i.e. Troy IIg) and HD 14572 on charcoal. Since the sample matter is still long-lived, i.e. charcoal (just not known to be potentially *very* long-lived), the dates derived are still *termini post quos* for the cultural context. The calibrated ages are around 2500BC (**see Figure 1**). The end of Troy IIg is later.

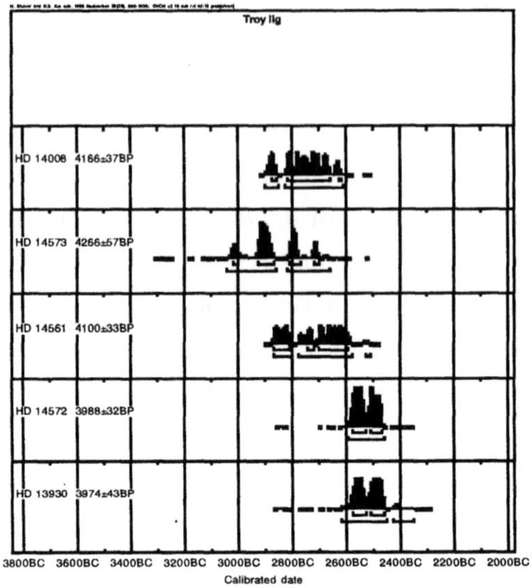

Figure 1. Probabilistic calibrated calendar age ranges for radiocarbon determinations from Troy IIg (Ramsey 1995, Cal10.dta; data from Korfmann and Kromer 1993: 148 and Abb.5).

Korfmann and Kromer (1993: 165-168) reach a contrary conclusion, however. They point out that the radiocarbon dates from later Troy I and late Troy II overlap (which they do), and so suggest that later Troy I and Troy II are in fact contemporary, with Troy II being the élite culture for the surrounding later Troy I town. Apart from this scenario running against all previous analyses of the stratigraphy and sequence at Troy, it fails to note that the three 'early' Troy IIg radiocarbon determinations on 'beams' are likely to be up to a couple of centuries earlier than the date of the Troy II cultural phase. The other two data offer the reasonable and likely *terminus post quem* for the late Troy II period, around or after c.2500BC. Oddly enough, this follows on nicely from the Troy I sequence of radiocarbon data.

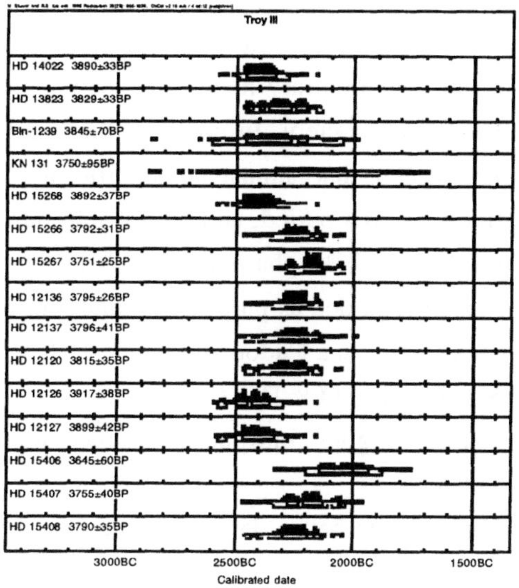

Figure 2. Probabilistic calibrated calendar age ranges for radiocarbon determinations from Troy III (Ramsey 1995, Cal10.dta; data from Korfmann and Kromer 1993: 147-148 and Abb.5, 161)

The other constraint on the date of the close of Troy II is offered by a large group of dates from Troy III (Korfmann and Kromer 1993: Abb.5, p.161) (see **Figure 2**). Again, these are on long-lived, charcoal, samples, and so offer a *terminus post quem* for the Troy III period. The calibrated age ranges for the samples lie nicely in the centuries after 2500BC, with the majority centering around the interval c.2300-2200BC. There is no need for the hiatus c.2500-2400BC posited by Korfmann and Kromer (1993: Abb. 23); instead, Troy II is revealed as lying in the period c.2500-2300/2200BC, with the Troy III cultural phase following on afterwards. The evidence from Troy suggests that the Kastri phase in the Aegean (correlated with Troy IIc/d to IIg especially, but also probably into Troy III: Manning 1995a: 81-86) thus runs from about the 25th century BC through to about 2200BC. However, the accuracy and precision on this date is, as with the end of EHII on mainland Greece, of poor quality. (Three determinations among a dozen as yet unpublished radiocarbon determinations from Markiani [C. Renfrew, pers. comm.] provide dates for material from the Kastri phase which yield calibrated ages consistent with the above scenario.)

Crete

The middle of the third millennium BC on Crete (the Early Minoan [EM] II period) is notably different to the contemporary mainland. The mainland exhibits every sign of a complex 'chiefdom' level society. In contrast, there is little such evidence on Crete, and it has been argued that almost no significant evidence for a complex society exists until after the EMII period, and so not until after the time of the EHII period, and the general Aegean EB2 phase elsewhere in the Aegean (Cherry 1983; Watrous 1994). Crete is not regarded as particularly integrated into the 'international spirit' of the Aegean later EB2 (see Cadogan 1986b: 157; Rutter and Zerner 1984; Cosmopoulos 1991; Karantzali 1996). This situation is of considerable interest, since, in the sequel, the mainland goes into cultural 'recession', while a state-level civilisation emerges on Crete.

The EMII period is only a little better dated than the contemporary EHII period. A few debated linkages do exist between Crete and Egypt in the form of some stone vases, and these offer a general synchronism of EMII with some part of the Old Kingdom (Manning 1995a: 104-107). Otherwise, there is one set of radiocarbon determinations run over 25 years ago from the small site of Myrtos in southern Crete. Although the precision of these dates is not up to modern standards, the excellent recording and publication of the excavation by Warren (1972) permit reasonably useful and precise conclusions to be drawn (Manning 1995a: 145-149). The dates from the EMIIB phase show that this lasted from around the 25th century BC to c.2200BC (see also **Figure 3**).

What happens in the succeeding EMIII period is hotly disputed. In contrast to southern Greece and the Cyclades (and elsewhere further east), no great wave of change occurs at the close of the EMIIB period. There are abandonments or destructions at a couple of sites, and some EMIII sites appear to be fortified, and in defensive locations (Alexiou 1979; Warren 1987: 52-53 and n.31), but the major site of Knossos continues and develops. However, it is also true that the EMIII period is difficult to define, and difficult to find on the ground in large areas of Crete (e.g. south-central Crete: Watrous *et al*. 1993). In a provocative recent synthesis-review of pre-palatial Crete, Watrous (1994) tries in effect to define the EMIII period out of existence, arguing that there was no evidence for significant complexity in EMII, and that the next major phase, and evidence for changes, followed only in MMIA (around 2000BC).

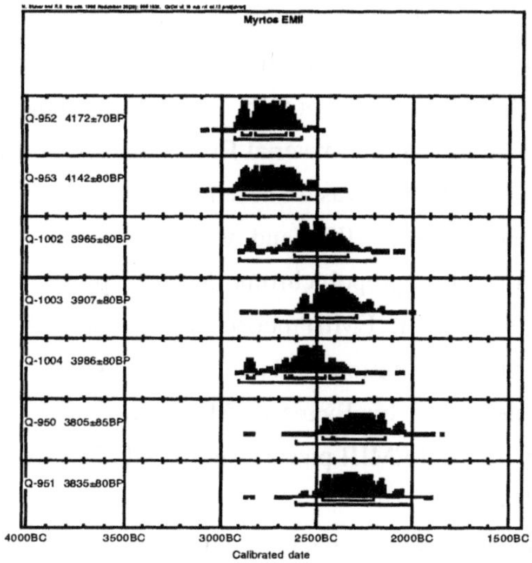

Figure 3. Probabilistic calibrated calendar age ranges for radiocarbon determinations from EMII Myrtos (Ramsey 1995, Cal10.dta; data from Warren 1972: 344-345; Manning 1995a: 145-149, 175, figs.8-10). Q-952 and Q-953 were from samples from large rooms and may be assumed to be have been reasonably large timbers from beams and the roof. This probably explains their 'old' ages. Q-1002, Q-1003 and Q-1004 come from construction timbers relating to the building of the EMIIB phase at the site. These dates offer a *terminus post quem* for the start of the EMIIB period around the 25th century BC. Q-950 and Q-951 are thought to derive from a ladder in use at the end of the settlement in later EMIIB. These dates should offer a *terminus post quem* for later EMIIB after sometime within the range c.2400-2200BC.

It is correct that for many years the definition (even the existence) of an EMIII phase outside east Crete, and its relation to MMIA, was highly problematic, and evidence for regionalism was all too clear (Warren 1965; Zois 1968; Alexiou 1978; Manning 1995a: 63-65). The cause of this EMIII controversy was Evans' (1921: 103-126, esp.108-113, figs. 76-80) problematic definition of EMIII, based on stray finds at Knossos, and in fact almost exclusively on unstratified (or poorly understood) deposits from east Crete. Thus, because EMIII pottery – as defined by Evans – was found only in east Crete, many scholars over the next two generations questioned the existence of this pottery as either a unified style or a chronological phase elsewhere on Crete. In contrast, Evans' 'definition' of MMIA as a phase in which polychrome decoration is a prominent feature was, apparently, regarded by scholars as less problematic – although the evidence to support this definition was equally sparse in the deposits employed by Evans (Andreou 1978; Momigliano 1991).

However, Hood's excavations at Knossos during the late 1950s to early 1960s (Hood 1961-1962; 1966) remedied the situation. Hood identified deposits stratified directly above EMIIB in which the pottery did not resemble EMIII as defined by Evans. It was, in fact, more like the material Evans had often ascribed to MMIA, but with the key difference that there was an absence of polychrome decoration. Yet, this material was stratified above EMIIB (and was before the MM sequence). Hood therefore suggested that this 'prepolychrome MMIA' phase in fact represented the (now stratigraphically defined) EMIII period at Knossos. A new rigorous set of stratigraphic and stylistic definitions was thus established. A re-examination of a number of deposits described by Evans as MMIA revealed that they were in fact similar to the now recognised EMIII phase (Andreou 1978; Momigliano 1991; Wilson 1994; Cadogan *et al.* 1993).

Whereas Watrous (1994) downplays the EMIII period, and developments on Crete pre-2000BC, other scholars report interesting developments at this time. New finds of EMIII pottery and architecture have recently been reported from Malia (Pelon 1989; Touchais 1989), and other major buildings may be associated with the EMIII-MMI period (e.g. Spence 1990; Poursat 1988: 71); a large EMIII/MMIA building is reported from Mochlos (French 1994: 81-82); monumental 'royal' tombs appear at Malia and Archanes (Soles 1992 – Watrous 1994 dates these to MMIA); and at Knossos there are indications of major building work beginning in EMIII (Wilson 1994: 44) and of rapid and substantial settlement expansion (Momigliano 1991). Since this is a time of recession in many areas of the Aegean, how is this to be explained?

A review of the social evolution of Crete through the third millennium BC is instructive. The EMIIA period (earlier EHII, early EB2 in the Cyclades) sees the development of 'emergent' complex society on Crete. The evidence includes: instances of differential mortuary practices at Malia, Gournia and Mochlos which are assumed to reflect social ranking (Soles 1988; 1992; Whitelaw 1983: 337-340); a large building at Knossos with finds of 'prestige' and/or imported wares (Wilson 1994: 36-37, 39-40; 1985); and major levelling and building activity in central Knossos which points to a 'large-scale reorganisation of the settlement in EMIIA' (Wilson 1994: 36). Contacts with the earlier EB2 Cyclades are evident (Stucynski 1982), and, unique in the Aegean, contacts exist with Egypt and the Near East (Warren 1995: 1-2). The following EMIIB period sees further developments on Crete, but no major changes. The period is most notable for the cessation of the overseas contacts characteristic of

EMIIA (especially the Aegean contacts). Imported material from EMIIB find-contexts reasonably arrived in EMIIA. Crete appears uninvolved in the later EB2 developments in both mainland Greece and the Cyclades. It has no part in the Anatolianising fashions of the Cyclades and central mainland. Dramatic changes from the EMII pre-palatial pattern occur only during EMIII. At Knossos, there is evidence for the start of major monumental building activities at the site, beginning with the EMIII terrace wall and fill (Wilson 1994: 37-38), and the likely large or important buildings for which it formed the foundations (Wilson 1994: 38 and n.101). Monumental tombs appear at Archanes and Malia. Crucially, some of the value objects in these are imitations of, or heirloom, earlier EB2 imports (e.g. at Tholos C at Archanes: Sakellarakis 1977).

It is tempting to argue that whereas a boom-and-bust cycle typical of an inflationary prestige goods-based economy occurred in the rest of the Aegean at the close of later EB2 despite widening contacts, the introduction of Anatolian fashions, and so on, because the élite could not maintain exclusivity and monopolisation of value, that in contrast in Crete the absence of overseas Aegean contacts in EMIIB (as trade in the rest of the Aegean focussed east to Anatolia) and EMIII permitted the maintenance, and in fact development, of strategies of exclusivity and monopolisation of a now largely static quantity of value goods by the élite on Crete (just some east Mediterranean imports: Warren and Hankey 1989: 125-127; Wilson 1994: 41). Competition centred around the existing (and only slightly increasing) supply of prestige goods, and so their value increased through time making investment in increased socio-political complexity feasible. Crete's marginal position both on the edge of the Aegean world, and the edge of the east-Mediterranean world system, was the key. Some access to a wide range of exotica occurred, but an enforced scarcity also existed. These two factors complemented each other. (One may compare the case of 'remote' Kiriwina Island in the Trobriands, and the conclusion that: 'A position in the mainstream of the flow of goods brings about an *egalitarian* local structure. The function of the valuables as a basis for the growth and rank of an elite group is undermined as their abundance does not allow exclusive control. A position at the periphery of exchanges, on the contrary, generates a *hierarchical* local structure. Thanks to their scarcity, the goods may be monopolized to serve as a chiefly fund of power' [Persson 1983: 40; see also Peregrine 1992: 27-46].)

A Cretan evolutionary trajectory through EMII-MMI along the following lines might be proposed (for a detailed analysis, see Manning 1995b). Contact with the rest of the Aegean, and technological change across this region, led to internal competition reaching new levels in the earlier EBA, as both the internal subsistence economy was intensified, and external exchange provided an additional prestige goods economy. The best, prime, areas of Crete developed preferentially, following a core-periphery model within Crete, and would have attracted additional population. Competition within these core areas would have been fierce, and threats from 'jealous' groups from neighbouring areas may also have existed. Both factors promoted selection for larger groups within the core areas in order to be part of a successful grouping (Kosse 1994), while external threats may have provided opportunities for successful leaders to enjoy an evolution of chiefly authority (Spencer 1993).

However, rather than this emergent situation developing via inflation, and then collapsing as a boom-and-bust cycle, Crete found itself isolated from the Aegean in EMIIB-EMIII. Competition and intensification around the monopolisation of limited valuables led to a more complex and (with time – since there was no boom-and-bust cycle) a more permanent hierarchy (Friedman 1982; Peregrine 1992: 27-46). This new intensified monopolisation competition again favoured selection for membership of larger and larger groups (and so greater complexity, and there are hints of larger buildings/structures from EMIIB [e.g. a ramp leading to the top, probably main, élite, area of the settlement at Knossos may date to EMIIB: Wilson 1994: 37-38; and the fragments of large walls of EMIIB date (MacGillivray and Driessen 1990: 399) at Palaikastro; there are also a variety of other walls or larger building fragments of more general EMII date: Pernier 1935: 117, 119, pl.VI; Levi 1961-1962: 429, figs. 82-83; 1976: 478-479; Laviosa 1973; Poursat 1988: 78; French 1991: 73]), especially since this competition for restricted resources probably involved 'direct' competition via warfare (and some EMIIB destructions are known). Hence there was nucleation of settlement, and the regionalisation of ceramic sequences. The leaders from the EMIIA emergence, through the EMIIB-EMIII period of competition and monopolisation-intensification, established themselves as the direct (rather than just better) source of sacred and secular authority. They were buried in the élite house tombs (Soles 1992) (and their contemporaries in the Mesara).

To be functional, and so successful, large groups require further complexity; moreover, nucleation aids processes of political centralisation through maximising communication opportunities and costs. Further hierarchy thus developed during EMIII. This hierarchy was now becoming historically permanent. Many generations of chiefs or élite stood behind any current claimant(s). The bones of ancient chiefs were revered (Soles 1992). The hierarchy had become the normal state of things. The crisis perhaps came in the century after c.2200 BC, with the problems in the Near East (e.g. Bell 1971; Weiss *et al.* 1993). The remaining prestige economy (links to the east Mediterranean) was under threat. Intensified monopoly-based competition might be anticipated at each of the larger settlements and within regions, perhaps assisting in the sorting out of several leading families into one dominant family or lineage. However, the situation was also more complex. The new larger, centralised, communities required ever increasing 'incomes' to make them viable. But, no further internal subsistence intensification was on offer in Crete. Expansion into the external arena was the only possibility other than collapse. Moreover, the nature of the competition on Crete changed in EMIII-MMIA. By this time entrenched sets of regional élites had developed on Crete, and relations, distinction, and competition at a new 'élite' level between these regional élites might be expected. This offers the beginnings of a peer-polity model (Renfrew and Cherry 1986). Again, it was the external world, and in particular the core eastern Mediterranean, which offered the potential source of much of the currency of such élite competition.

Expansion into the east Mediterranean was the dramatic and decisive step; it catapulted Crete into a new trajectory of evolution, and a new order of magnitude. Building upon the established EMII-III hierarchy, and empowered through 'peer-polity' competition between a plurality of site or regionally-based élites, emergent state-level societies began to form from late EMIII, and especially during MMIA. Crete developed and expanded into the Aegean interregional core civilisation. The time interval between this key expansion and the first palaces is sadly unknown at present (and cases could be made for it being either: part of EMIII and MMIA, or largely just MMIA, or even only the later part of MMIA), and this affects our view of what occurred (a couple of new radiocarbon dates from Knossos confirm the basic chronology, but do not offer refinement: D. Wilson and N. Momigliano, pers. coms.). If a longer interval, it suggests that although the leaders of the initial expansion undoubtedly achieved great kudos, there was a significant period of emergent state development and naturalisation as the superstructure of potential state power formed over at least a century, whereas, if a shorter timescale is envisaged, the two processes may be conflated and compounded within a few generations. As in the case of 'state' formation on Hawaii (Saxe 1977), the final evolution to a full 'state' probably all happened in a short space of time, perhaps under one leader, or at most a couple of generations of leaders. The Old Palaces/states of Crete probably formed within a couple of generations at the close of MMIA.

Whether Crete made contact with the east Mediterranean, or the east Mediterranean re-activated trade with Crete, trade and exchange with the east Mediterranean (in EMIII/MMIA) saved the élite on Crete, who by the close of the third millennium BC would have been facing a situation of diminishing returns threatening collapse. Without some new resource arena and impetus, emergent EMIIB-III Crete would have stalled or collapsed, having reached its local, internal, ceiling of development. Ambitious and skilful leaders seized the new, exotic, infinite, and value-laden world of the Near East to finance a new round of internal intensification enshrined in monumentalisation. They became 'kings'. Crete expanded into the Aegean (from MMIA), becoming the core of a new Aegean core-periphery world system. The Minoan state was born.

Conclusion

It is evident that major changes occur around 2200BC in the Aegean. The advanced EHII civilisation collapses, along with the Early Cycladic EB2 civilisation which had become increasingly oriented towards western Anatolia by this time. A refined chronology is not, however, available from present data. New radiocarbon series as yet unpublished, and also others currently in progress, will assist to clarify the situation. On the mainland, environmental damage linked to over-exploitation of the landscape under a rapidly developing socio-political hierarchy based around extensive plough agriculture might be linked to climate change. The situation on Crete is very different. Whereas EHII settlements of several to several tens of hectares are known (Konsola 1990), the largest contemporary site on Crete (Knossos) is about 5ha. and the few other 'large' sites are less than half this size (Whitelaw 1983: 339-340). Significant regional settlement hierarchies and evidence for large-scale, extensive, plough agriculture, bureaucracy, and so on, are not to be found. Crete does not appear to have been complex, or large-scale, enough to be seriously undermined by climate change c.2200BC. (one can particularly contrast the Near Eastern mid-third millennium BC civilisations reliant on a sophisticated superstructure: Weiss *et al.* 1993; Weiss 1986). Instead, its 'emergent' character offered the potential for adaptation in the face of change. Its special marginal position in two greater EB2 world systems (Aegean-Anatolian, east Mediterranean), and the re-orientation of these, and then collapse c.2200BC, provided circumstances whereby a sustainable process of social evolution was possible on Crete (and so a route towards an answer to the question posed by Lewthwaite 1983). Again, however, the details of chronology of these developments are embarrassingly poorly known. In view of high-quality data now available from Anatolia and the Near East, programmes of modern chronometric research into the third millennium BC Aegean are urgently required.

Notes to Figures

The calibrated calendar age ranges were calculated by the OxCal computer programme (version 2.18) of Ramsey (1995). The 1986 bi-decadal calibration curve was employed (Cal10.dta). This is probably the most suitable and representative dataset currently available for long-lived sample material in view of acknowledged problems with the 1993 dataset from Belfast published in *Radiocarbon* 35 (1993) (see Kalin *et al.* 1995; McCormac *et al.* 1995). The distributions show total probability. The horizontal lines underneath show (upper) 1σ (68.2%) and (lower) 2σ (95.4%) probability regions within these overall distributions.

References

Alexiou S (1979) Teikhe kai akropoleis ste minoike Krete. *Kretologia* 8: 41-56.

Allen J (1984) Pots and poor princes: a multidimensional approach to the role of pottery trading in coastal Papua. In van der Leeuw SE & Pritchard AC (eds) *The many dimensions of pottery: ceramics in archaeology and anthropology:* 407-463. Cingula VII. Universiteit van Amsterdam, Albert Egges van Giffen Instituut voor Prae- en Protohistorie Amsterdam.

Andreou S (1978) *Pottery groups of the Old Palace period in Crete.* Ph.D. diss. University of Cincinnati.

Bell B (1971) The dark ages in ancient history I. The first dark age in Egypt. *American Journal of Archaeology* 75: 1-26.

Bossert EM (1967) Kastri auf Syros: Vorbericht über eine Untersuchung der prähistorischen Siedlung. *Arkhaiologikon Deltion* 22: 53-76.

Broodbank C (1989) The longboat and society in the Cyclades in the Keros-Syros culture. *American Journal of Archaeology* 93: 319-337.

Broodbank C (1993) Ulysses without sails: trade, distance, knowledge and power in the early Cyclades. *World Archaeology* 25: 315-331.

Cadogan G (ed) (1986a) *The end of the Early Bronze Age in the Aegean*. E.J. Brill Leiden.

Cadogan G (1986b) Why was Crete different? In Cadogan G (ed) *The end of the Early Bronze Age in the Aegean:* 153-171. E.J. Brill Leiden.

Cadogan G, Day PM, MacDonald, CF, MacGillivray, JA, Momigliano, N, Whitelaw, TM, and Wilson, DE (1993) Early Minoan and Middle Minoan pottery groups at Knossos. *Annual of the British School at Athens* 88: 21-28.

Caskey JL (1956) Excavations at Lerna, 1955. *Hesperia* 25: 147-173.

Caskey JL (1960) The Early Helladic period in the Argolid. *Hesperia* 29: 285-303.

Catling HW (1973) Cyprus in the Early Bronze Age. In Edwards IES, Gadd CJ & Hammond NGL (eds) *The Cambridge Ancient History*. Third edition, vol.I Part 2: 808-823. Cambridge University Press Cambridge.

Cavanagh W, Crouwel J, Catling RWV & Shipley G (1996) *The Laconia survey. Vol. II: archaeological data*. British School at Athens London.

Cherry JF (1983) Evolution, revolution, and the origins of complex society in Minoan Crete. In Krzyszkowska O & Nixon L (eds) *Minoan society. Proceedings of the Cambridge colloquium 1981:* 33-45. Bristol Classical Press Bristol.

Cosmopoulos MB (1991) *The Early Bronze 2 in the Aegean*. Studies in Mediterranean Archaeology 98. Paul Åströms Förlag Jonsered.

Davis (JL) 1987 Perspectives on the prehistoric Cyclades: an archaeological introduction. In Getz-Preziosi P *Early Cycladic art in north American collections:* 4-45. Virginia Museum of Fine Arts Richmond.

Davis JL (1992) The islands of the Aegean. *American Journal of Archaeology* 96: 699-756.

Evans AJ (1921). *The palace of Minos at Knossos,* vol.I. Macmillan London.

Forsén J (1992) *The twilight of the Early Helladics: a study of the disturbances in east-central and southern Greece towards the end of the Early Bronze Age*. Studies in Mediterranean Archaeology and Literature Pocket-book 116. Paul Åströms Förlag Jonsered.

French EB (1991) Archaeology in Greece 1990-91. *Archaeological Reports* 37: 3-78.

French EB (1994) Archaeology in Greece 1993-94. *Archaeological Reports* 40: 3-84.

Friedman J (1982) Catastrophe and continuity in social evolution. In Renfrew C, Rowlands MJ & Segraves BA (eds) *Theory and explanation in archaeology: the Southampton conference:* 175-196. Academic Press New York.

Hägg R & Konsola D (eds) (1986) *Early Helladic architecture and urbanization. Proceedings of a seminar held at the Swedish Institute in Athens, June 8, 1985*. Studies in Mediterranean Archaeology 76. Paul Åströms Förlag Göteborg.

Hood MSF (1961-1962) Stratigraphic excavations at Knossos, 1957-61. *Kretika Chronika* 15-16: 92-98.

Hood S (1966) The Early and Middle Minoan periods at Knossos. *Bulletin of the Institute of Classical Studies London* 13: 110-111.

Jahns S (1990) Preliminary notes on human influence and the history of vegetation in S. Dalmatia and S. Greece. In Bottema S, Entjes-Nieborg G & van Zeist W (eds)

Man's role in the shaping of the eastern Mediterranean landscape: 333-340. A.A. Balkema Rotterdam.

Jameson MH, Runnels CN & van Andel TH (1994) *A Greek countryside: the southern Argolid from prehistory to present day*. Stanford University Press Stanford.

Kalin RM, McCormac FG, Damon PE, Eastoe CJ & Long A (1995) Intercomparison of high-precision ^{14}C measurements at the University of Arizona and the Queen's University of Belfast radiocarbon laboratories. *Radiocarbon* 37: 33-38.

Karantzali E (1996) *Le Bronze Ancien dans les Cyclades et en Crète*. BAR International Series 631. Tempus Reparatum Oxford.

Kilian K (1986) The circular building at Tiryns. In Hägg R & Konsola D (eds) *Early Helladic architecture and urbanization. Proceedings of a seminar held at the Swedish Institute in Athens, June 8, 1985*: 65-71. Studies in Mediterranean Archaeology 76. Paul Åströms Förlag Göteborg.

Konsola D (1990) Settlement size and the beginning of urbanization. In Darcque P & Treuil R (eds) *L'habitat Égéen préhistorique*: 463-471. Bulletin de Correspondance Hellénique Supplément XIX. Boccard Paris.

Korfmann M & Kromer B (1993) Demircihüyük, Besik-Tepe, Troia – Eine Zwischenbilanz zur Chronologie dreier Orte in Westanatolien. *Studia Troica* 3: 135-171.

Kosse K (1994) The evolution of large, complex groups: a hypothesis. *Journal of Anthropological Archaeology* 13: 35-50.

Laviosa C (1973) L'abitato Pre-Palaziale di Hagia Triadha. *Annuario della Scuola Archeologica di Atene e delle Missioni Italiane in Oriente* 50/51: 503-513.

Levi D (1961-1962) Gli scavi a Festò negli anni 1958-60. *Annuario della Scuola Archeologica di Atene e delle Missioni Italiane in Oriente* 39-40: 377-504.

Levi D (1976) *Festòs e la civilta' Minoica*. I*. Rome: Instituto per gli studi Micenei ed egeo-anatolici.

Lewthwaite J (1983) Why did civilization not emerge more often? A comparative approach to the development of Minoan Crete. In Krzyszkowska O & Nixon L (eds.), *Minoan society. Proceedings of the Cambridge colloquium 1981:* 171-183. Bristol Classical Press Bristol.

McCormac FG, Baillie MGL, Pilcher JR & Kalin RM (1995) Location-dependent differences in the ^{14}C content of wood. *Radiocarbon* 37: 395-407.

MacGillivray JA & Driessen JM (1990) Minoan settlement at Palaikastro. In Darcque P & Treuil R (eds) *L'habitat Égéen préhistorique:* 395-412. Bulletin de Correspondance Hellénique Supplément XIX. Boccard Paris.

Manning SW (1994) The emergence of divergence: development and decline on Bronze Age Crete and the Cyclades. In Mathers C & Stoddart S (eds) *Development and decline in the Mediterranean Bronze Age:* 221-270. John Collis Sheffield.

Manning SW (1995a) *The absolute chronology of the Aegean Early Bronze Age: archaeology, radiocarbon and history*. Sheffield Academic Press Sheffield.

Manning SW (1995b) *Before Daidalos: the origins of complex society, and the genesis of the state on Crete*. Ph.D. dissertation, Cambridge.

Momigliano N (1991) MMIA pottery from Evans' excavations at Knossos: a reassessment. *Annual of the British School at Athens* 86: 149-271.

Pelon O (1989) Le palais. *Bulletin de Correspondance Hellénique* 113: 771-785.

Peregrine, PN (1992) *Mississippian evolution: a world-system perspective*. Monograohs in World Archaeology No.9. Prehistory Press Madison.

Pernier L (1935) *Il Palazzo Minoico di Festòs*, vol. I. La Libreria Dello Stato Rome.

Persson J (1983) Cyclical change and circular exchange: a re-examination of the Kula Ring. *Oceania* 54: 32-47.

Poursat JC (1988) La ville minoenne de Malia: recherches et publications récentes. *Revue Archéologiques* 1: 61-82.

Pullen DJ (1985) *Social organization in Early Bronze Age Greece: a multidimensional approach*. Ph.D. Dissertation, Indiana University. University Microfilms International Ann Arbor.

Pullen DJ (1992) Ox and plow in the Early Bronze Age Aegean. *American Journal of Archaeology* 96: 45-54.

Pullen DJ (1994) A lead seal from Tsoungiza, ancient Nemea, and Early Bronze Age Aegean sealing systems. *American Journal of Archaeology* 98: 35-52.

Pullen DJ (1995) The pottery of the Neolithic, Early Helladic I, and Early Helladic II periods. In Runnels C, Pullen DJ & Langdon S (eds) *Artifact and Assemblage: the finds from a regional survey of the southern Argolid, Greece. Volume I: the prehistoric and Early Iron Age pottery and the lithic artifacts:* 6-42. Stanford University Press Stanford.

Ramsey CB (1995) Radiocarbon calibration and analysis of stratigraphy: the OxCal program. *Radiocarbon* 37: 425-430.

Renfrew C (1972) *The emergence of civilization. The Cyclades and the Aegean in the third millennium B.C.* Methuen London.

Renfrew C (1991) *The Cycladic spirit: masterpieces from the Nicholas P. Goulandris collection*. Thames and Hudson London.

Renfrew C & Cherry JF (eds) (1986) *Peer polity interaction and socio-political change*. Cambridge University Press Cambridge.

Runnels CN (1985) The Bronze Age flaked-stone industries from Lerna: a preliminary report. *Hesperia* 54: 357-391.

Rutter JB (1979) *Ceramic change in the Aegean Early Bronze Age*. Occasional Paper 5. Institute of Archaeology, University of California Los Angeles.

Rutter JB (1983) Some observations on the Cyclades in the later third and early second millennia BC. *American Journal of Archaeology* 87: 69-76.

Rutter JB (1984) The Early Cycladic III gap: what it is and how to go about filling it without making it go away. In MacGillivray JA & Barber RLN (eds) *The prehistoric Cyclades. Contributions to a workshop on Cycladic chronology*: 95-107. Department of Classical Archaeology, University of Edinburgh Edinburgh.

Rutter JB (1988) Early Helladic III vase painting, ceramic regionalism, and the influence of basketry. In French EB & Wardle KA (eds) *Problems in Greek prehistory:* 73-89. Bristol Classical Press Bristol.

Rutter JB (1993) Review of Aegean prehistory II: the prepalatial Bronze Age of the southern and central Greek mainland. *American Journal of Archaeology* 97: 745-797.

Rutter JB (1995) *Lerna. A preclassical site in the Argolid. Volume III: the pottery of Lerna IV.* The American School of Classical Studies at Athens Princeton.

Rutter JB & Zerner CW (1984) Early Hellado-Minoan contacts. In Hägg R & Marinatos N (eds) *The Minoan thalassocracy: myth and reality. Proceedings of the third international symposium at the Swedish Institute in Athens, 31 May-5 June, 1982*: 75-83. Skrifter Utgivna av Svenska Institutet i Athen, 4°, XXXII. Paul Åströms Förlag Stockholm.

Sakellarakis J (1977) The Cyclades and Crete. In Thimme J (ed) *Art and culture of the Cyclades:* 145-154. Trans. Getz-Preziosi P. The University of Chicago Press Chicago.

Saxe AA (1977) On the origin of evolutionary processes. State formation in the Sandwich Islands: a systemic approach. In Hill JN (ed) *Explanations of prehistoric change:* 105-151. University of New Mexico Press Albuquerque.

Shaw JW (1987) The Early Helladic II Corridor House: development and form. *American Journal of Archaeology* 91: 59-79.

Soles JS (1988) Social ranking in prepalatial Cemetries. In French EB & Wardle KA (eds) *Problems in Greek prehistory. Papers presented at the centenary conference of the British School of Archaeology at Athens, Manchester April 1986*: 49-61. Bristol Classical Press Bristol.

Soles JS (1992) *The prepalatial cemeteries at Mochlos and Gournia and the house tombs of Bronze Age Crete.* Hesperia Supplement 24. American School of Classical Studies at Athens Princeton.

Spence Y (1990) Was there a guarded southern entrance way to the first palace at Mallia? *Annual of the British School at Athens* 85: 369-374.

Spencer CS (1993) Human agency, biased transmission, and the cultural evolution of chiefly authority. *Journal of Anthropological Archaeology* 12: 41-74.

Stewart ST (1988) Bureaucracy and packaging at Lerna: evidence from the clay sealings in the House of Tiles. *American Journal of Archaeology* 92: 253.

Stos-Gale ZA, Gale N & Gilmore GR (1984) Early Bronze Age Trojan metal sources and Anatolians in the Cyclades. *Oxford Journal of Archaeology* 3: 23-43.

Stucynski SL (1982) Cycladic imports in Crete: a brief survey. In *Temple University Aegean Symposium* 7: 50-59. Philadelphia.

Themelis P (1984) Early Helladic monumental architecture. *Mitteilungen des Deutschen Archäologischen Instituts. Athenische Abteilung* 9: 335-351.

Touchais G (1989) Chronique des fouilles et découvertes archéologiques en Grèce en 1988. *Bulletin de Correspondance Hellénique* 113: 581-700.

van Andel TH, Runnels CN & Pope KO (1986) 5000 years of land use and abuse in the southern Argolid, Greece. *Hesperia* 55: 103-128.

van Andel TH, Zangger E & Demitrack A (1990) Land use and soil erosion in prehistoric and historical Greece. *Journal of Field Archaeology* 17: 379-396.

Wagner GA & Lorenz IB (1992) Thermolumineszenz-Datierungen an Keramik vom Besik-Yassıtepe. *Studia Troica* 2: 147-156.

Warren P (1965) The first Minoan stone vases and Early Minoan chronology. *Kretika Chronika* 19: 7-43.

Warren P (1972) *Myrtos. An Early Bronze Age settlement in Crete*. The British School of Archaeology at Athens/Thames and Hudson London.

Warren PM (1987) The genesis of the Minoan palace. In Hägg R & Marinatos N (eds) *The function of the Minoan palaces. Proceedings of the fourth international symposium at the Swedish Institute in Athens, 10-16 June, 1984:* 47-56. Skrifter Utgivna av Svenska Institutet i Athen, 4°, XXXV. Paul Åströms Förlag Stockholm.

Warren P (1995) Minoan Crete and Pharaonic Egypt. In Davies WV & Schofield L (eds) *Egypt, the Aegean and the Levant: interconnections in the second millennium BC:* 1-18. British Museum Publications London.

Warren P & Hankey V (1989) *Aegean Bronze Age chronology*. Bristol Classical Press Bristol.

Watrous LV (1994) Review of Aegean prehistory III: Crete from earliest prehistory through the Protopalatial period. *American Journal of Archaeology* 98: 695-753.

Watrous LV *et al.* (1993) A survey of the western Mesara palain in Crete: preliminary report of the 1984, 1986, and 1987 field seasons. *Hesperia* 62: 191-248.

Weiss H (1986) (ed) *The origins of cities in dry-farming Syria and Mesopotamia in the third millennium B.C.* Four Quarters Publishing Guilford.

Weiss H, Courty MA, Wetterstrom W, Guichard F, Senior L, Meadow R & Curnow A (1993) The genesis and collapse of third millennium north Mesopotamian civilization. *Science* 261: 995-1004.

Weninger B (1995) Stratified ^{14}C dates and ceramic chronologies: case studies for the Early Bronze Age at Troy (Turkey) and Ezero (Bulgaria). *Radiocarbon* 37: 443-456.

Whitelaw TM (1983) The settlement at Fournou Korifi Myrtos and aspects of Early Minoan social organization. In Krzyszkowska O & Nixon L (eds) *Minoan society. Proceedings of the Cambridge colloquium 1981:* 323-345. Bristol Classical Press Bristol.

Wiencke MH (1989) Change in Early Helladic II. *American Journal of Archaeology* 93: 495-509.

Wilson DE (1985) The pottery and architecture of the EMIIA West Court House at Knossos. *Annual of the British School at Athens* 80: 281-364.

Wilson DE (1994) Knossos before the palaces: an overview of the Early Bronze Age (EMI-III). In Evely D, Hughes-Brock H & Momigliano N (eds) *Knossos: a labyrinth of history. Papers in honour of Sinclair Hood:* 23-44. The British School at Athens Oxford.

Wilson DE & Eliot M (1984) Ayia Irini III: the last phase of occupation at the EBA settlement. In MacGillivray JA & Barber RLN (eds) *The prehistoric Cyclades*.

Contributions to a workshop on Cycladic chronology: 78-87. Department of Classical Archaeology, University of Edinburgh Edinburgh.

Zangger E (1992) Neolithic to present soil erosion in Greece. In Boardman J & Bell M (eds) *Past and present soil erosion:* 133-147. Oxbow Oxford.

Zangger E (1993) *The geoarchaeology of the Argolid.* Mann Berlin.

Zois A (1968) Hyparkhei PM III epokhe. In *Pepragmena tou B' Diethnous Kretologikou Synedriou,* vol.A': 141-156. Athens.

The Third Millennium BC in Iberia: Chronometric Evidence for Settlement Histories and Socio-cultural Change

Katina T. LILLIOS

Department of Anthropology and Sociology
Ripon College
Ripon, WI
USA

Abstract

The late third millennium BC was a period of increased socio-political flux not only in eastern Mediterranean, but also among the complex societies of the western Mediterranean. Contemporary with the widespread abandonment of settlements and population transfers in the Nile Valley (First Intermediate Period), Mesopotamia (Akkadian-Ur III), and the Aegean (LHIII), the Iberian Peninsula underwent one of its most dynamic cultural phases. Between about 2500 and 1900 BC, during the transition from the Late Copper Age to the Early Bronze Age, many settlements throughout the peninsula were abandoned or newly established. Although most settlements experienced temporary short-term abandonment episodes during their occupation span, the population dislocations at the end of the Copper Age were of a more permanent nature. Sixteen sites richly dated by radiocarbon show that the settlements in eastern Iberia tended to last shorter and get abandoned earlier than in the more humid west. The difference may have to do more with environmental and climatic factors than with cultural diffusion.

Introduction

The late third millennium BC was a period of socio-political flux not only among the state-level polities of the eastern Mediterranean but also among the complex societies of the western Mediterranean. Contemporary with widespread settlement abandonment and population transfers in the Nile Valley (First Intermediate Period), Mesopotamia (Akkadian-Ur III), and the Aegean (LHIII), the Iberian Peninsula underwent one of its most dynamic cultural phases. At approximately 2200 BC, during the Late Copper Age/Early Bronze Age transition, many settlements throughout the Peninsula were abandoned. These settlements, in general, had been

NATO ASI Series, Vol. I 49
Third Millennium BC Climate Change
and Old World Collapse
Edited by H. Nüzhet Dalfes, George Kukla,
and Harvey Weiss

occupied for about one thousand years, and although many had experienced short-term abandonment phases, the population dislocations at the end of the Copper Age were of a more permanent nature. Some of those settlements that were newly established at this time, particularly in southern Portugal, show evidence for socio-economic decentralization (Gilman 1987; Lillios 1991, 1993), while others, particularly in southeast Spain, show evidence of consolidation and hierarchization (Gilman 1987; Chapman 1990). Thus, what we see at approximately 2200 BC in Iberia are widespread socio-political transformations - either centralization or fragmentation - and, at some sites, abandonment.

No attempt has ever been made to determine whether the cultural changes of late third millennium BC Iberia were impacted by the climatological changes discussed in this volume. To do so would require the demonstration of the following:

1) the determination of the the existence and nature of a climatological change that impacted the Iberian Peninsula at this time,

2) an evaluation of the archaeological evidence, with accurate dating, for the period in question,

3) the establishment of hypothetical linkages between any climatological changes that occurred and the socio-cultural changes that transpired, and

4) the development of highly focused problem-oriented geoarchaeological studies that test the relationship between climate change and social change.

Such a proof is outside the scope of this paper. Only with a coordinated effort by prehistorians and palaeoecologists working in Iberia can a correlation between the palaeoenvironmental and culture historical records be made. In this paper, I hope to contribute to this end by synthesizing the available absolute chronologies for settlements in Iberia occupied in the third millennium and to assess whether there are regional patterns which might be linked to global environmental changes.

The Environmental Zones of Iberia

The Iberian Peninsula, incorporating the nations of Portugal and Spain, is a land of extreme environmental contrasts. The climatic and vegetational diversity characteristic of Iberia is largely a result of the differing air currents of the Atlantic and Mediterranean which pass over the complex system of interior and coastal mountain ranges (Linés Escardó 1970). Generally speaking, Iberia can be divided into two zones: an Atlantic zone and a Mediterranean zone. The Atlantic zone, located in the northern and western portions of the Peninsula, experiences abundant rainfall (1000-2500mm per annum) (Linés Escardó 1970: Fig. 19) and is typified by a deciduous forest cover. The more arid Mediterranean zone, to the south and east, typically experiences 200-1000mm of precipitation per year (Linés Escardó 1970:Fig. 19). Here, the natural forest cover has largely disappeared, and the landscape is one of wheat fields, vineyards, olive groves, and scrub. There are portions of 'dry' Iberia where soil erosion and other degradation processes have left barren, lunar-like 'badlands' (e.g., at Tabernas [prov. Almería]) so similar to the American Southwest that they have been used for the filming of many 'spaghetti' Westerns. Whether the climatic and environmental conditions that exist now, such as the distinction between the Atlantic and Mediterranean zones, are comparable to those that existed in the third and second millennia BC is unclear. There does, however, seem to be some agreement that the environmental conditions in later prehistory are comparable to those of today (Gilman and Thornes 1985:12-15; Chapman 1990:97-112).

Even more important is what, if any, environmental or climatic change occurred between the third and second millennia BC. There are good discussions of the current state of thinking on the palaeoenvironmental evidence from the Spanish Southeast in Gilman and Thornes (1985:12-15) and Chapman (1990:108-112). The evidence for Portugal is summarized by Lillios (1991:43-47). In general, however, the palynological, geomorphological, archaeobotanical, and archaeofaunal data from Portugal and Spain are somewhat contradictory and do not allow us to say much, with precision, regarding the changes between the third and second millennia. Any changes that are documented in the palynological record have been argued to be to be the result of human intervention, not climate change. For example, cores taken from the Padul bog, near Granada, Spain (Florschütz et al. 1971), Alpiarça in central Portugal (Leewaarden and Janssen 1985), and the Serra da Estrela in northern Portugal (Brink and Janssen 1985) show decreases in arboreal pollen (principally Quercus) during the third millennium BC, with concomitant increases of domestic Vitis, Olea, and cereals. Clearly, however, regionally specific and chronologically refined climatic histories for the later prehistory of Iberia do not exist at this time.

Archaeological Background

The archaeological record for the Copper and Bronze Age of Iberia has been traditionally weighted to the Southeast (Andalucia, Murcia, and Valencia, in Spain) and the Southwest (Estremadura, Alentejo, and the Algarve, in Portugal). The true diversity of cultural adaptations in the late third/early second millennium BC throughout the entire Peninsula is only now being realized with the fluorescence of regional archaeological projects throughout Spain and Portugal (Lillios 1995, Díaz-Andreu 1995). Until this new evidence is synthesized and chronological sequences integrated, our understanding of the complex histories of Iberia during its later prehistory will remain tentative.

In general, however, it can be said with some certainty that the Copper Age (3000-2200 BC) of Iberia was a period of widespread social transformations (Gilman and Thornes 1985; Chapman 1990). For the first time in Iberia, the archaeological record points to a departure from egalitarian to non-egalitarian societies. Numerous walled, hilltop settlements concentrated goods acquired by long-distance exchange (copper, ostrich eggshells, and ivory from North Africa) (Harrison and Gilman 1977) and produced by craft specialists (flint daggers, Beaker ceramics, copper items). Monumental tombs, housing collective burials, show marked variability in their associated luxury good assemblages. Items such as slate plaques, limestone figurines, Beaker ceramics (Spindler 1981:fig. 49), and copper objects are differentially distributed among Copper Age burials, which include megaliths, caves and rock shelters, and corbel-vaulted tombs.

Traditional explanations for the notable increase in social complexity during the Copper Age invoked contacts with 'more civilized' peoples (e.g., traders, missionaries) from the Eastern Mediterranean (Siret and Siret 1887; Blance 1961). The application of radiocarbon dating and its calibration to the prehistoric record of Portugal has demonstrated, however, that the changes in society during third millennium BC pre-date or are roughly contemporary with comparable developments in the Eastern Mediterranean (Renfrew 1973) and, thus, must be seen principally in terms of local processes.

The Early Bronze Age (2200-1500 BC) signaled another phase of distinct social change. During this phase, however, there are greater regional differences in evolutionary trajectories (Gilman 1987). In many parts of Iberia, particularly the southern half of Portugal, the early second millennium BC is period of widespread settlement abandonment, a decline in goods acquired via long-distance exchange or produced by craft specialists, and population dispersal, with new unwalled settlements, such as Agroal (Lillios 1991), established.

In much of Spain, however, and the Southeast, in particular, there is fairly good evidence, principally from the intramural burials characteristic of this period (also known as the Argaric), for an increase in wealth concentration, craft specialization, and agricultural intensification (Chapman 1990:179-210, Gilman 1987). Some settlements continued to be occupied through the Bronze Age, or the early second millennium BC (e.g., Moncín, El Castillo, Cerro de la Virgen). Abandonments also occur, however, but the newly established Argaric sites show similar patterns in their material assemblage to those Argaric settlements occupied throughout the Copper Age.

Methods

There are hundreds of third millennium sites known in the Iberian Peninsula. Unfortunately, however, only a very small portion of these are settlements that have been excavated and have well-established chronological sequences dated by radiocarbon means. For the discussion of third millennium BC chronologies in this paper, sixteen sites were selected (Figure 1). In order to be included in this discussion, each site had to fulfill the following criteria:

1) Settlement sites were chosen as more sensitive gauges of population movements than burials.

2) These settlements have multiple radiocarbon dates per cultural phase (Copper Age, Bronze Age).

3) All the radiocarbon dates have been published in association with their stratigraphic contexts and/or the chronological sequence of the site.

What became apparent while assembling these dates was how problematic most dates were for the purposes of this study. While there are approximately five hundred dates from Copper Age or Bronze Age levels from Iberian settlements, their stratigraphic associations have often not been clearly articulated in publications. Furthermore, few seem to have been collected with specific cultural or stratigraphic questions in mind, such as when the site was initially occupied or abandoned. The 'old wood' effect is rarely discussed as a possible complication to the date (Dean 1978). This may be a particular problem in those sites with multiple phases of occupation, such as Cerro de la Virgen, in which wood from previous occupations may have been scavenged for new structures. Without a doubt, the uncritical use of a site's radiocarbon dates to answer questions not of interest to the excavators of the

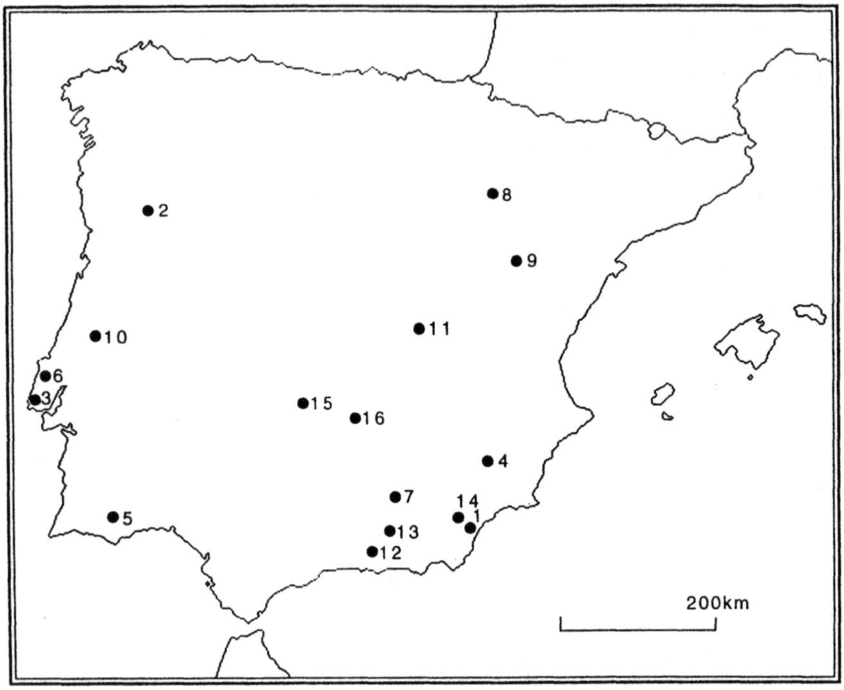

Location of Sites Discussed
1. Almizaraque; 2. Castelo de Aguiar; 3. Liceia; 4. El Prado; 5. Castelo de Santa Justa; 6. Zambujal; 7. Cerro de la Virgen; 8. Moncín; 9. El Castillo; 10. Agroal; 11. Castillejo; 12. Cerro de la Encina; 13. Cuesta del Negro; 14. Fuente Alamo; 15. Los Romeros; 16. Quintanar

Figure 1

site is a dangerous pursuit. Thus, while the number of sites presented in this paper are few, they can be considered the best that exist for addressing the relevant questions.

For such a systematic study, it was necessary to have access to a reasonably complete body of uncalibrated radiocarbon dates. Gilman's (1992) chapter in Chronologies in Old World Archaeology provided the starting point in assembling these dates. For the more recent dates, I consulted the regularly updated database which Gilman maintains. An on-line clearinghouse of all radiocarbon dates (searchable by region and dates) would have clearly facilitated this portion of the research, and it is hoped that, in the near future, this will be available.

For calibrations, the most recent version of CALIB (Version 3.0.3c; Stuiver and Reimer 1993; Stuiver and Pearson 1993; Pearson and Stuiver 1993) was used. All dates were calculated to one standard deviation using probability distributions (Method B).

In order to provide an estimate for the span of occupation, the earliest and latest dates for each cultural phase (Copper Age, Bronze Age) for each site were chosen. For an establishment date, the earliest date from the lowest stratigraphic level was chosen (even if the date was later than a stratigraphically superior level). For an abandonment date, the latest date from the uppermost stratigraphic level was chosen (even if it was earlier than a stratigraphically inferior level). Thus, stratigraphy, primarily, provided the basis for date selection.

Site Histories

The sites included in this study are located in a range of environmental settings on the Peninsula, from the arid badlands of the Spanish Southeast to the temperate forests of northern Portugal (Figure 1). What follows below is a short discussion of the chronologies of the principal settlements in the Iberian Peninsula for which there are fairly secure third millennium BC dates (summarized in Figure 2). The dates indicate the earliest date (for an approximation of the date of establishment) and the most recent date (for approximate abandonment date). When there are two cultural phases at the site (Copper Age and Bronze Age), four dates are given - two for each phase.

Sites with only Copper Age occupation (dated)

(1) Almizaraque (Almería, Spain) Thirteen radiocarbon dates exist for this Copper Age settlement (Delibes et al. 1986a, b; dates cited below in González Gómez et al. 1986b:1202). Five phases of occupation have been identified, which span pre-Beaker and Beaker levels. Neolithic levels may underlie the Copper Age levels, although they have not been adequately documented (Delibes et al. 1986a). No dates associated with the earliest occupation of the site (Phase 1) have been produced, nor dates for the end of Phase 5 (Beaker Copper Age).

UGRA 1634120±200 BP, end of Phase 2 (pre-Beaker Copper Age)

UGRA 1643950±100 BP, end of Phase 3 (pre-Beaker Copper Age)

(2) Castelo de Aguiar (Vila Real, Portugal) Three radiocarbon dates exist for this Copper Age settlement in northern Portugal (Jorge 1985, 1986; dates cited below in González Gómez et al. 1987:386). Six archaeological levels were identified at the site. No dates exist for the terminal phase of occupation of the site.

SITE HISTORIES OF THIRD MILLENNIUM SETTLEMENTS IN IBERIA

Site (# on map)	Level/ Phase	Culture Phase	Significance to Site History (estab/aband)	Sample #	BP	Cal BC (1 SD)
Agroal (10)	Phase 1	EBA	estab	GX 15390G	3560±145	2122-1688
Almizaraque (1)	Phase 2	CA-preBkr	*estab*	UGRA 163	4120±100	2871-2508
Almizaraque (1)	Phase 3	CA-preBkr	*aband*	UGRA 164	3950±100	2576-2286
Castelo de Aguiar (2)	Level 6	LN/CA	estab	UGRA 179	4700±100	3625-3363
Castelo de Aguiar (2)	Level 4-s	CA	*aband*	UGRA 181	3730±140	2325-1928
Castelo de Santa Justa (5)	Phase 1	CA	estab	UGRA 131	4390±150	3332-2884
Castelo de Santa Justa (5)	Phase 2	CA	aband	UGRA 85	3890±130	2557-2143
Castillejo (11)	Level 4	EBA	estab	I 11118	3740±170	2400-1914
Castillejo (11)	Level 2	MBA?	*aband*	I 11120	3230±110	1671-1329
El Castillo (9)	Phase IA	CA	estab	GrN 17550	3685±35	2132-1981
El Castillo (9)	Phase IB	CA		GrN 17551	3660±30	2114-1971
El Castillo (9)	Phase IIB	BA		GrN 17575	3595±20	1951-1895
El Castillo (9)	Phase IIC	BA	aband	GrN 17577	3560±25	1931-1788
Cerro de la Encina (12)	Phase 1b	EBA-Argar B	*estab*	GrN 6634	3625±40	2030-1916
Cerro de la Encina (12)	surface	EBA/LBA	aband	Ly 2656	3350±100	1741-1517
Cerro de la Virgen (7)	Phase 2	CA-Bkr	*estab*	GrN 5593	3890±40	2454-2319
Cerro de la Virgen (7)	Phase 2	CA-Bkr		GrN 5598	3835±35	2326-2201
Cerro de la Virgen (7)	Phase 3	EBA-Argar A		GrN 5594	3735±55	2198-2033
Cerro de la Virgen (7)	Phase 3	EBA-Argar B	aband	GrN 5595	3865±50	2452-2207
Cuesta del Negro (13)	Phase 1b	EBA	*estab*	GrN 7286	3620±35	2024-1915
Cuesta del Negro (13)	Phase 2	LBA	aband	GrN 7284	3095±35	1404-1313
Fuente Alamo (14)	Phase 1	EBA	estab	?	3970±90	2584-2320
Fuente Alamo (14)	Phase 5	LBA	aband	?	3160±90	1518-1312
Liceia (3)	Phase 1	LN/CA	estab	ICEN 312	4530±100	3363-3044
Liceia (3)	Phase 3	CA-Bkr	aband	ICEN 315	3730±170	2397-1892
Moncín (8)	Phase IVB	CA-Bkr	*estab*	BM 2477	3900±40	2456-2334
Moncín (8)	Phase IVB	CA-Bkr		BM 2479	3730±40	2091-2037
Moncín (8)	Phase IIE	EBA		BM 2476	3620±40	2027-1908
Moncín (8)	Phase IIA2	LBA	aband	BM 2006	3050±50	1390-1222
El Prado (4)	Level 5	CA	estab	Beta 7072	4350±50	3031-2901
El Prado (4)	surface	CA	aband	HAR 146	4080±130	2868-2469
Los Romeros (15)	Phase 1	EBA	estab	CSIC 78	3600±120	2131-1768
Los Romeros (15)	Phase 3	MBA?	*aband*	CSIC 77	3290±120	1731-1425
Quintanar (16)	Phase 1	EBA	estab	UGRA 166	3780±110	2395-2033
Quintanar (16)	Phase 3	MBA?	aband	CSIC 491	3290±50	1628-1469
Zambujal (6)	Phase 2	CA	*estab*	GrN 7009	4200±40	2882-2696
Zambujal (6)	Phase 4	CA	*aband*	GrN 6668	3625±65	2113-1887

UGRA 1794700±100 BP, Level 6 (Copper Age, beginning of occupation)

UGRA 1813730±140 BP, Level 4-top (Copper Age)

(3) Liceia (Lisboa, Portugal) Fourteen radiocarbon dates, which span three occupational phases, exist for this Copper Age settlement (Cardoso 1994). Abandonment occurred between Phases 1 and 2 and between Phases 2 and 3. Final abandonment occurred after Phase 3.

ICEN 3124530±100 BP, Phase 1 (Late Neolithic and pre-Beaker Copper Age) ICEN 315 3730±170 BP, Phase 3 (Beaker Copper Age)

(4) El Prado (Murcia, Spain) Fourteen radiocarbon dates from Copper Age levels (1-5) exist for this settlement (Walker 1985; Otlet and Slade 1974:186). Beta 70724350±50 BP, Level 5 (pre-Beaker Copper Age)

HAR 1464080±130 BP (collected from surface of site and, thus, without precise context)

(5) Castelo de Santa Justa (Faro, Portugal) There are eight radiocarbon dates for this Copper Age settlement located in the Alentejo (Gonçalves 1980, 1981a, 1981b, 1982, 1983-4; dates cited below from González Gómez 1985). Two occupational phases, separated by a fire, characterize this site.

UGRA 1314390±150 BP, Phase 1 (Copper Age)

UGRA 853890±130 BP, Phase 2 (Copper Age)

(6) Zambujal (Lisboa, Portugal) There are sixteen radiocarbon dates for this settlement, which spans the Copper and Early Bronze Ages (Sangmeister and Schubart 1981; Schubart and Sangmeister 1983-4). Five building phases (I-V) were identified, although no dates exist for Phases I and V (establishment and final abandonment of site).

GrN 70094200±40 BP, Phase II (Copper Age)

GrB 66683625±65 BP, Phase IV (Copper Age)

Sites with Copper and Bronze Age occupations (dated)

(7) Cerro de la Virgen (Granada, Spain) Seven radiocarbon dates, spanning the Copper Age (Beaker phase) and Early Bronze Age (Argar A and B) exist for this settlement, which is associated with intramural burials (Schüle 1980; Vogel and Waterbolk 1972:74-5; Almagro 1972). The dates for the Argaric burials do not seem to be consistent with the phasing of the period and do not overlap when calibrated to their first standard deviation. Argar A, which has traditionally been thought to precede Argar B (Blance 1964), has produced a somewhat later date. Lull (1983:385)

suggests this is due to the contamination of the Argar B date, although he also argues that the 'classic' Argar A-B sequence may be more an artifact of social evolution than chronology. For this paper, I have simply reported the dates as they are. There are no dates for the earliest occupation of the site (Phase 1- pre-Beaker Copper Age).

GrN 55933890±40 BP, Phase 2 (Beaker Copper Age)

GrN 55983835±35 BP, Phase 2 (Beaker Copper Age)

GrN 55943735±55 BP, Phase 3 (Early Bronze Age, Argar A)

GrN 55953865±50 BP, Phase 3 (Early Bronze Age, Argar B)

(8) Moncín (Zaragoza, Spain) Sixteen radiocarbon dates exist for this settlement occupied from the Late Copper Age through the Late Bronze Age (Phases IIA-V) (Harrison, et al. 1987; Harrison 1993; Ambers et al. 1989:29-30, 1991:65). No dates exist for Phase V, the initial occupational phase.

BM 24773900±40 BP, Phase IV (Copper Age, Beaker)

BM 24793730±40 BP, Phase IV (Copper Age, Beaker)

BM 24763620±40 BP, Phase IIE (Early Bronze Age)

BM 26063050±50 BP, Phase IIA2 (Late Bronze Age)

(9) El Castillo (Teruel, Spain) Six radiocarbon dates exist for this settlement occupied in two phases during the Copper and Bronze Ages (Harrison and Wainwright 1991).

GrN 175503685±35, Phase IA (Copper Age)

GrN 175513660±30, Phase IB (Copper Age)

GrN 175753595±20, Phase IIB (Bronze Age)

GrN 175773560±25, Phase IIC (Bronze Age)

Sites with only Bronze Age occupation (dated)

(10) Agroal (Santarém, Portugal) Two radiocarbon dates of this Bronze Age settlement, located in central Portugal, have been obtained (Lillios 1991). Although there are no dates for the final occupation of Agroal (Late Bronze Age, based on pottery form), it was included in this study as it is the only radiometrically dated Early Bronze Age settlement in Portugal.

GX 15390G3560±145 BP, Phase 1 (Early Bronze Age)

(11) Castillejo (Cuenca, Spain) Three dates exist for this Bronze Age settlement (Martínez Navarrete and Valiente Cánovas 1983:149). Level 4 corresponds to the earliest phase of occupation of the site. There are no dates for the terminal phase of occupation (Level 1).

I 111183740±170 BP, Level 4 (Early Bronze Age)

I 111203230±110 BP, Level 2 (Middle Bronze Age?)

(12) Cerro de la Encina (Granada, Spain) Nine radiocarbon dates, which span the Early Bronze Age (Argar B) and Late Bronze Age, exist for this settlement (Arribas et al. 1974; Aguayo de Hoyos 1986; Evin et al. 1985:426). There are no dates for the earliest occupation of the site.

GrN 66343625±40 BP, Phase 1b (Early Bronze Age, Argar B)

Ly 16563350±100 BP (mixed Argaric/LBA level)

(13) Cuesta del Negro (Granada, Spain) Four radiocarbon dates, from the Early and Late Bronze Ages, exist for this settlement (Molina González and Pareja López 1975; Aguayo de Hoyos 1977:263). No dates exist for the earliest phase of occupation of the site (Phase 1a).

GrN 72863620±35 BP, Phase 1b (Early Bronze Age, Argar B)

GrN 72843095±35 BP, Phase 2 (Late Bronze Age, Argar C)

(14) Fuente Alamo (Almería, Spain) 23 radiocarbon dates from Early Bronze Age (Argar A and B) and Late Bronze Age levels exist for this site (Schubart and Arteaga 1978, 1980, 1983, 1986). The dates included in the study correspond to the earliest and final phases of occupation.

no lab#3970±90 BP, Phase 1 (Early Bronze Age, Argar A)

no lab#3160±90 BP, Phase 5 (Late Bronze Age)

(15) Los Romeros (Ciudad Real, Spain) Three radiocarbon dates were deemed by the excavators to be the only valid ones for this Bronze Age *motilla* (García Pérez 1987, 1988; Alonso et al. 1978). No dates correspond to the terminal phase of occupation.

CSIC 783600±120 BP, Phase 1 (level 9, Bronze Age)

CSIC 773290±120 BP, Phase 3 (level 6, Bronze Age)

(16) Quintanar (Albacete, Spain) There are 26 dates for this Bronze Age *morra* (Martín Morales 1983, 1984; González Gómez et al. 1986:1203). Dates for the earliest and terminal phases of occupation are:

UGRA 1663780±110 BP, Phase 1 (Early Bronze Age)

CSIC 4913290±50 BP, Phase 3 (Middle Bronze Age?)

Discussion

The sixteen sites discussed above are those sites in Iberia whose settlement histories, as known through radiocarbon dates, are most richly documented. There are a total of over 150 radiocarbon determinations for these sites' chronologies. Nonetheless, these sites represent a small percentage of the total number of sites known for the third millennium BC in Iberia. Thus, all conclusions regarding contemporaneity and regional patterning must be considered highly tentative. At best, they should be used as testable hypotheses for future research.

A conservative approach has been taken throughout this analysis. The radiocarbon dates, although abundant, have been inconsistently collected, and thus, sophisticated statistical analyses such as those promoted by Dewar (1991) and Shott (1992) were not thought appropriate here.

One question has dominated the analysis of these dates: is there regional or temporal patterning in the abandonment and establishment of settlements in the Iberian Peninsula during the late third/early second millennium BC that would suggest the effects of a global climatological event? More specifically, are there significant differences in the settlement chronologies between the Atlantic façade and the arid Mediterranean coast? The assumption I make here is that, if there was a phase of aridity in the late third millennium BC, the Mediterranean East would be impacted more acutely and, thus, cultural realignments would occur earlier there than in the more temperate North and West.

Figure 3 shows the duration of occupation of the sites included in this study. Occupational spans were calculated by determining the difference between the end points of the establishment date and abandonment dates (in cal BC). For example, at Fuente Alamo the earliest Early Bronze Age date is 2584-2320 BC, and the latest Late Bronze Age date is 1518-1312 BC. For the occupational span, the difference between 2584 and 1312 was calculated (approx. 1300). Two points must be understood with regard to this map. First, because in many cases the settlements did not have dates for their initial and terminal phases, the estimates for occupational spans may be on the low side. On the other hand, temporary abandonment phases within the sites' histories were not taken into account. These length of these abandonment phases were, for the most part, imprecisely known. It is important, however, to recognize that abandonments are a 'normal' part of the history of most

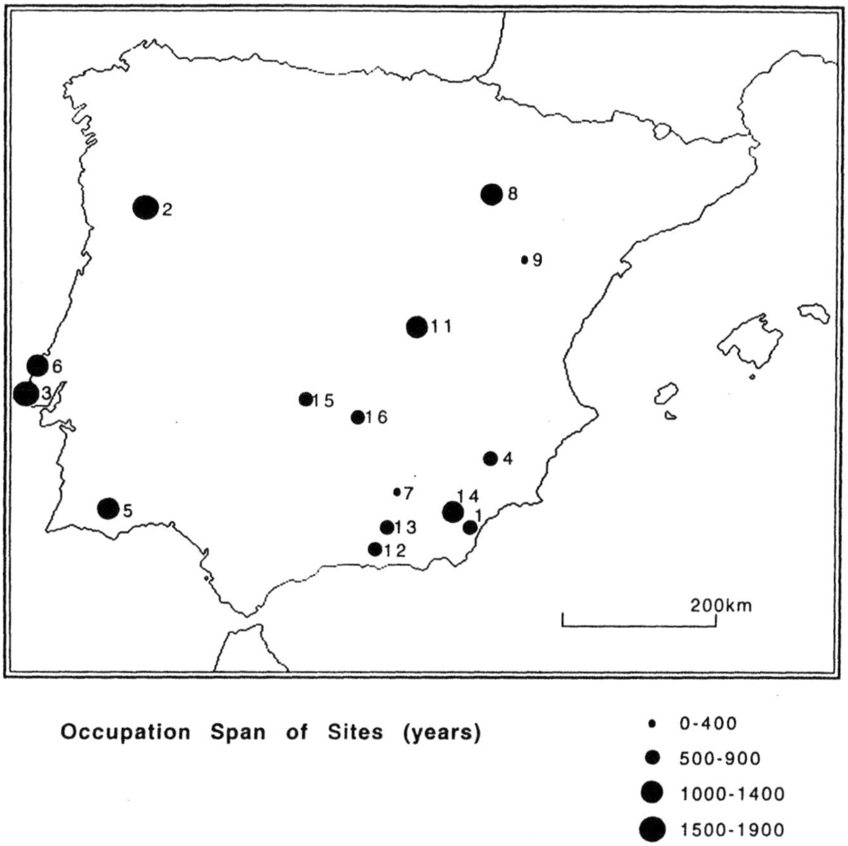

Figure 3

settlements (Cameron and Tomka 1993); that is, they are usually deliberate and planned and generally not the result of a catastrophic, Pompeii-like event.

Briefly put, there appear to be some regional differences in the occupational spans of late prehistoric Iberian settlements. Those settlements with the longest periods of occupation (over 1000 years) tend to be located in western Iberia, and those with briefer spans (less than 1000 years), in the more arid zone. Taken into account with Figure 2, we can see that these western sites were also the earliest occupied of the Copper Age settlements.

Figure 4 plots the establishment and abandonment events that occurred between 2500-1500 BC. Once again, there appear to be regional distinctions in the abandonment of Copper Age settlements. The earliest abandonments occurred along

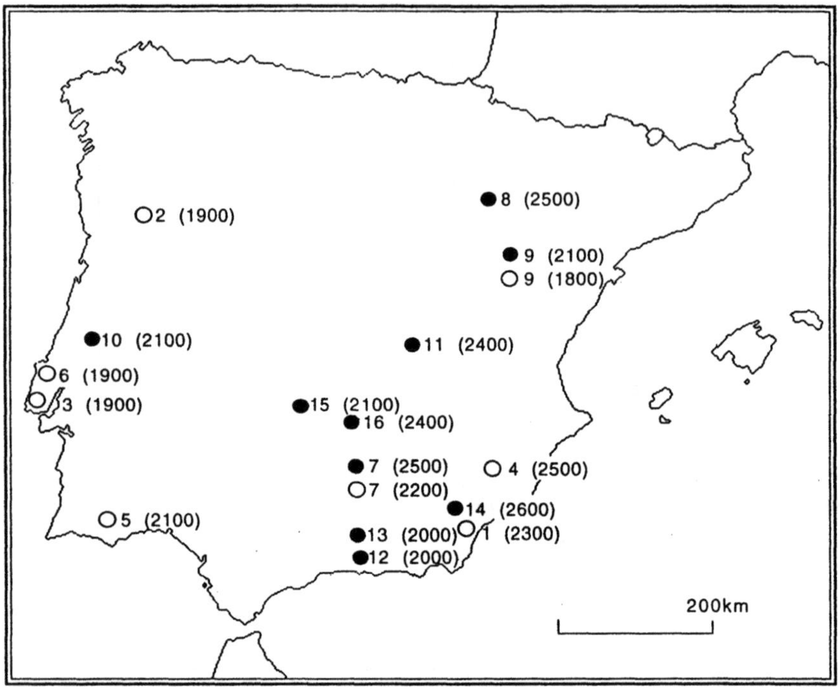

Settlement Establishment and Abandonment: 2500-1500 BC

● Establishment Approximate Year (BC) of
O Abandonment Establishment/Abandonment (2000)

Figure 4

the eastern half of Iberia, and the latest in the west. Thus, the Early Bronze Age of Spain (Argaric) seems to begin earlier than the Early Bronze Age of Portugal (2100 BC). Interestingly, it was traditionally thought that the Argaric 'culture' influenced other areas of Iberia, such as southern Portugal (Blance 1964:130). The chronometric data suggest, however, that differences in the dates of Bronze Age cultures of Iberia may have more to do with environmental or climatic factors than cultural diffusion. Whether this variability between eastern and western Iberia is due to multiple climatic events or one event that resulted in multiple cultural trajectories is uncertain. Only more refined determinations of cultural and environmental sequences will provide us with an answer to this question.

If this patterning reflects a genuine trend and is not an artifact of site preservation, then what are we to make of it? These data suggest the possibility that environmental or climatic factors were partially influencing the behavior of human populations in late prehistoric Iberia. Certainly, the constraints of Atlantic and Mediterranean ecosystems on human behavior, such as farming practices, mobility, political organization, are quite different. For example, the deeper soils of the more humid zones of Iberia would have allowed for longer periods of continuous occupation without the abandonment, either temporary or permanent, of a settlement and without the intensification of agricultural practices.

If a climatic shift was in effect during the late third millennium BC, it did not have the same impact on communities in eastern and western Iberia, nor was the impact 'perceived' at the same time. Local environmental constraints as well as cultural factors must have played an important role in community decisions regarding the abandonment and establishment of settlements.

Conclusion

The social, cultural, and natural landscapes upon which climatic change operates are usually highly variegated. Even if a global climatic event is found to have occurred in the western Mediterranean, archaeologists must be careful in integrating such an event into their models of culture process. Appropriate and comparable scales of analysis must be used when articulating climate change with human behavior. Local social conditions and political agendas of leaders are often the overriding forces in the face of global change-witness the current factionalism and resistance to environmentalists' attempts to curb the depletion of the earth's ozone layer. To paraphrase an oft-used slogan in North America: archaeologists should think globally but keep in mind that their 'people' acted locally.

Acknowledgment

I am deeply indebted to Antonio Gilman, who provided critical comments to drafts of this paper and shared with me his unpublished database of Iberian radiocarbon dates.

Bibliography

Aguayo de Hoyos, P. (1986) La transición de la Edad del Cobre a la Edad del Bronce en la provincia de Granada. Homenaje a Luis Siret (1934-1984). pp. 262-269.

Almagro Gorbea, M. (1972) Nuevas fechas para la prehistoria y la arqueologia peninsular. Trabajos de Prehistoria 29:232-236.

Alonso, F., V. Cabrera, T. Chapa, and M. Fernández-Miranda (1978) Indice de fechas arqueológicos de C-14 en España y Portugal. Fundación Juan March, Serie Universitaria 77:155-183.

Ambers, J., K. Matthews, and R. Burleigh (1989) British Museum natural radiocarbon measurements XXI. Radiocarbon 31:15-32.

Arribas, A., E. Pareja López, F. Molina González, O. Arteaga Matute, and F. Molina Fajardo (1974) Excavaciones en al poblado de la Edad del Bronce 'Cerro de la Encina', Monachil (Granada): el corte estratigráfico núm. 3. Excavaciones Arqueológicos en España 81.

Blance, B. (1961) Early Bronze Age colonists in Iberia. Antiquity 35:192-202.

Blance, B. (1964) The argaric Bronze Age in Iberia. Revista de Guimarães 74:129-142.

Brink, L.M. van der and C.R. Janssen (1985) The effect of human activities during cultural phases on the development of montane vegetation in the Serra da Estrela, Portugal. Review of Palaeobotany and Palynology 44:193-215.

Cameron, C.M. and S.A. Tomka, eds. (1993) Abandonment of Settlements and Regions: Ethnoarchaeological and Archaeological Approaches. Cambridge: Cambridge University Press.

Cardoso, J. (1994) Leceia 1983-1993: Escavações do Povoado Fortificado Pré-histórico. Oeiras: Câmara Municipal de Oeiras.

Chapman, R. (1990) Emerging Complexity: The Later Prehistory of South-East Spain, Iberia and the West Mediterranean. Cambridge: Cambridge University Press.

Dean, J.S. (1978) Independent dating in archaeological analysis. In Advances in Archaeological Method and Theory, vol. 1, ed. by M.B. Schiffer. New York: Academic Press, pp. 223-255.

Delibes, G., M. Fernández-Miranda, M. D. Fernández-Posse, and C. Martín (1986a) El poblado de Almizaraque. Homenaje a Luis Siret (1934-1984). pp. 167-177.

Delibes, G., M. Fernández-Miranda, M. D. Fernández-Posse, and C. Martín (1986b) Die kupferzeitliche Siedlung von Almizaraque (Cuevas del Almanzora, prov. Almería). Madrider Mitteilungen 27:11-26.

Dewar, R.E. (1991) Incorporating variation in occupation span into settlement-pattern analysis. American Antiquity 56(4):604-620.

Díaz-Andreu, M. (1995) Complex societies in Copper and Bronze Age Iberia: a reappraisal. Oxford Journal of Archaeology 14(1):23-39.

Evin, J., J. Maréchal, and G. Marien (1985) Lyon natural radiocarbon measurements X. Radiocarbon 27:386-454.

Florschütz, F., J. Menéndez, and T.A. Wijmstra (1971) Palynology of a thick Quaternary succession in southern Spain. Palaeogeography, Palaeoclimatology, Palaeoecology 10:233-264.

García Perez, T. (1987) La motilla de los Romeros, Alcázar de San Juan (Ciudad Real). Oretum 3:109-65.

García Perez, T. (1988) La motilla de los Romeros, Alcázar de San Juan (Ciudad Real). Actas, I. Congreso de Historia de Castilla-La Mancha 3:13-19.

Gilman, A. and J.B. Thornes (1985) Land-use and Prehistory in South-East Spain. London: George Allen and Unwin.

Gilman, A. (1987) Unequal development in Copper Age Iberia. In Specialization, Exchange, and Complex Societies, ed. by E.M. Brumfiel and T. Earle. Cambridge: Cambridge University Press, pp. 22-29.

Gilman, A. (1992) Iberia. In Chronologies in Old World Archaeology, ed. by R.W. Ehrich. Chicago: University of Chicago Press. pp. 238-255.

Gonçalves, V. (1980) Cerro do Castelo de Santa Justa (Alcoutim), escavações de 1979: extractos do caderno de campo. Clio 2:133-149.

Gonçalves, V. (1981a) Cerro do Castelo de Santa Justa (Alcoutim): campanha 2(80). Clio 3:165-170.

Gonçalves, V. (1981b) Cerro do Castelo de Santa Justa (Alcoutim): campanha 3(81). Clio 3:171-176.

Gonçalves, V. (1982) Cerro do Castelo de Santa Justa: um povoado calcolítico fortificado no Alto Algarve oriental. Arqueologia 6:42-48.

Gonçalves, V. (1983-4) Doze datas 14C para o povoamento calcolítico do Cerro do Castelo de Santa Justa (Alcoutim): comentarios e contextos específicos. Clio/Arqueologia 1:81-92.

González Gómez, C. P. Sánchez Sánchez, and M. Domingo García (1985) University of Granada radiocarbon dates II. Radiocarbon 27:610-615.

González Gómez, C. P. Sánchez Sánchez, and E. Villafranca Sánchez (1986) University of Granada radiocarbon dates III. Radiocarbon 28:1200-1205.

González Gómez, C. P. Sánchez Sánchez, and E. Villafranca Sánchez (1987) University of Granada radiocarbon dates IV. Radiocarbon 29:381-388.

Harrison, R.J. and A. Gilman (1977) Trade in the second and third millennia BC between the Maghreb and Iberia. In Ancient Europe and the Mediterranean, ed. by V. Markotic. Warminster, Aris and Phillips, pp. 89-104.

Harrison, R.J. (1993) A Bronze Age cultural sequence from Moncín, Borja (prov. Zaragoza). Empúries 48-50:384-391.

Harrison, R.J., G. Moreno López, and A.J. Legge (1987) Moncín: poblado prehistórico de la Edad del Bronce (I). Noticiario Arqueológico Hispánico 29:7-102.

Harrison, R.J. and J, Wainwright (1991) Dating the Bronze Age in Spain: a refined chronology for the high-altitude settlement of El Castillo (Frías de Albarracín, prov. Teruel). Oxford Journal of Archaeology 10(3):261-268.

Jorge, S.O. (1985) Datas de carbono-14 para a pré-história recente do norte de Portugal: os dados e os problemas. Arqueologia 12:154-183.

Jorge, S.O. (1986) Povoados da Pré-História Recente (III-Inícios do II Milénios A.C.) da Região de Chaves - Vila Pouca de Aguiar (Trás-os-Montes Ocidental). Porto: Instituto de Arqueologia da Faculdade de Letras do Porto.

Leewaarden, W. van and C.R. Janssen (1985) A preliminary palynological study of peat deposits near an oppidum in the lower Tagus Valley, Portugal. I Reunião do Quaternário Ibérico, Lisboa, 1985. vol. II, pp. 225-236.

Lillios, K.T. (1991) Competition to Fission: The Copper to Bronze Age Transition in the Lowlands of West-Central Portugal (3000-1000 BC). Ph.D. Dissertation, Yale University.

Lillios, K.T. (1993) Regional settlement abandonment at the end of the Copper Age in the lowlands of west-central Portugal. In Abandonment of Settlements and Regions: Ethnoarchaeological and Archaeological Approaches, ed. by C.M. Cameron and S.A. Tomka, Cambridge: Cambridge University Press, pp. 110-120.

Lillios, K.T., ed. (1995) The Origins of Complex Societies in Late Prehistoric Iberia. Ann Arbor, MI: International Monographs in Prehistory.

Linés Escardó, A. (1970) The climate of the Iberian Peninsula. In Climates of Northern and Western Europe, ed. by C.C. Wallén, pp. 195-239.

Lull, V. (1983) La 'Cultura" de El Argar. Madrid: Akal.

Martín Morales, C. (1983) Las fechas del Quintanar (Munera, Albacete) y la cronología absoluta de la Meseta sur. Homenaje al Prof. Martín Almagro Basch, vol. 2:23-35. Madrid: Ministerio de Cultura.

Martín, C., M. Fernández-Miranda, M.D. Fernández-Posse, and A. Gilman (1993) The Bronze Age of La Mancha. Antiquity 67(254):23-45.

Martín Morales, C. (1984) La morra del Quintanar. Al-Basit 15:57-73.

Martínez Navarrete, M.I. and S. Valiente Cánovas (1983) El Cerro del Castillejo (La Parra de las Vegas, Cuenca). Noticiario Arqueológico Hispánico 16:59-223.

Molina González, F. and E. Pareja López (1975) Excavaciones en la Cuesta del Negro (Purullena, Granada). Excavaciones Arqueológicas en España 86.

Otlet, R.L. and B.S. Slade (1974) Harwell radiocarbon measurements I. Radiocarbon 16:178-191.

Pearson, G.W. and M. Stuiver (1993) High-precision bidecadel calibration of the radiocarbon time scale 500-2500 BC. Radiocarbon 35:25-33.

Renfrew, C. (1973) Before Civilization. London: Methuen.

Sangmeister, E. and H. Schubart (1981) Zambujal: die Grabunden 1964 bis 1973. Madrider Beiträge 5(1).

Sangmeister, E. and H. Schubart (1987) Estações, monumentos. Castro de Zambujal, povoados fortificados de idade do cobre. Arqueologia 16:135-137.

Schubart, H. and O. Arteaga (1978) Fuente Alamo: Vorbericht über die Grabung 1977 in der bronzezeitlichen Hohensiedlung. Madrider Mitteilungen 19:23-51.

Schubart, H. and O. Arteaga (1980) Fuente Alamo: Vorbericht über die Grabung 1979 in der bronzezeitlichen Hohensiedlung. Madrider Mitteilungen 21:45-61.

Schubart, H. and O. Arteaga (1983) La cultura de 'El Argar' y las excavaciones en Fuente Alamo. Revista de Arqueologia 24:16-27; 25:54-63; 26:56-63.

Schubart, H. and O. Arteaga (1986) Fundamentos arqueológicos para el estudio socio-económico y cultural del área de El Argar. Homenaje a Luis Siret (1934-1984). pp. 284-307.

Schubart, H. and E. Sangmeister (1983-4) A cronologia absoluta (datações C14) de Zambujal. Clio/Arqueologia 1. pp. 31-40.

Schüle, W. (1980) Orce und Galera: zwei Siedlungen aus der 3. bis 1. Jahrtausend v. Chr. im Südosten der Iberischen Halbinsel. Mainz am Rhein: Verlag Philipp von Zabern.

Shott, M.J. (1992) Radiocarbon dating as a probabilistic technique: the Childers site and Late Woodland Occupation in the Ohio Valley. American Antiquity 57(2):202-23.

Siret, H. and L. Siret (1887) Les premiers ages du métal dans le sud-est de l'Espagne. Anvers.

Spindler, K. (1981) Cova da Moura. Madrider Beiträge 7.

Stuiver, M. and G.W. Pearson (1993) High-precision bidecadel calibration of the radiocarbon time scale, AD 1950-500 BC and 2599-6000 BC. Radiocarbon 35:1-23.

Stuiver, M. and P.J. Reimer (1993) Extended 14C database and revised CALIB radiocarbon calibration program. Radiocarbon 35:215-230.

Vogel, J.C. and H.T. Waterbolk (1972) Groningen radiocarbon dates X. Radiocarbon 14:6-110.

Walker, M.J. (1985) El Prado and the southeastern Spanish Chalcolithic. Research Reports of the National Geographic Society 20:799-834.

Climate and the Eclipse of the Ancient Cities of the Indus

Gregory L. POSSEHL

The University of Pennsylvania Museum,
Philadelphia, PA 19104-6324

Abstract

No great amount of systematic, scientific research has been conducted on the mid and late Holocene climate of the Indian Subcontinent, but there is a growing body of data that can be brought to the attention of the non-specialist. This work will be reviewed here, along with a short history of ideas concerning the eclipse of the ancient cities of the Indus, the chronology for this period of change. It will conclude with thoughts on the current state of knowledge concerning the "end of the ancient cities of the Indus."

The Holocene Climate of the Indian Subcontinent

Introduction

The climate of the Subcontinent does not seem to have been entirely stable since the close of the last great glacial period. There is reasonably good evidence for some significant shifts during the early millennia of this period, but the climate seems to have stabilized. During the Late Holocene, or the time during which the great urban centers of Mohenjo-daro and Harappa flourished as part of the Mature Urban Harappan Civilization, the climate appears to have been much as it is today. Pending further research, it can be proposed that there is no conclusive evidence to support the proposition that the annual rainfall of Mature Harappan times, the fourth through the second millennia, was not significantly different from today. This statement covers the total moisture budget and includes the rain that comes to the South Asian landmass, during the summer from the southwest monsoon, during the winter from the northeast monsoon, and the winter westerlies that are so important for agriculture and pastoralism in Baluchistan, the Northwest Frontier Province and the Punjab. The same seems to be true for solar energy and the temperature regime of the Subcontinent: there is no evidence for a significant difference between "Harappan times" and the present. The

NATO ASI Series, Vol. I 49
Third Millennium BC Climate Change
and Old World Collapse
Edited by H. Nüzhet Dalfes, George Kukla,
and Harvey Weiss

word "significantly" in this context implies a sufficient difference in rainfall that would have a measurable effect on crop yields and the production of grasslands for pasture. Modern weather records, widely available in the Subcontinent since 1901, with many stations maintaining much longer records, have been examined in some detail, using reasonably sophisticated statistics (Krishnan and Thanvi 1982). The results of these examinations are unequivocal: "Rainfall variability in the Indian arid zone does not indicate any secular trend" (Ramakrishna 1988: 99).

Third Millennium Climate in South Asia

There is a discussion of the climate of the third millennium in South Asia in the older archaeological literature. The flourishing village farming communities of Baluchistan in prehistoric eras seemed to indicate a more convivial environment. In his discussion of Kolwa, a broad, fertile valley of southern Baluchistan where the famous site of Kulli is located, Sir Aurel Stein said:

> The open and for the most part drainageless tract of Kolwa forms for some 80 miles a natural continuation of the Key Valley proper, from which it is separated by an almost imperceptible watershed. It contains by far the greatest dry-crop area of Makran, and its export of barley to other parts in years of good rainfall is considerable. Yet how rare such rainfall is and how precarious its cultivation in Kolwa, is shown by the very scanty population of the tract being practically all nomadic. All the more interesting is the evidence which the explorations to be described below have revealed as to Kolwa having in pre-historic times been the seat of a large and thoroughly settled population. The contrast between massively built stone structures traceable at more than one ancient site of Kolwa and the wretched huts of palm-matting which house practically the whole of even the settled population of the Kej Valley as elsewhere in Makran, is striking and illustrated how great the change which has come over this whole region (Stein 1931: 9).

Sir John Marshall presented this position on climate during the period of the Harappan Civilization (2500-2000 BC) in some detail in the opening chapter of Mohenjo-daro and the Indus Civilization (Marshall 1931b: 2-4) (Figure 1, Sites of the Harappan Civilization and Places Mentioned in the Text). The basic elements of his proposition included the observation that the high density of prehistoric villages in Baluchistan can be accounted for only by a more productive environment, now severely constrained by the lack of water: Stein's point. Furthermore, elephants, tigers and rhinoceri are all depicted on Indus stamp seals and are animals that prefer a wet habitat (Figure 2, Stamp seals of the Harappan Civilization). The lion, a dry land animal, is conspicuous by its absence from Indus imagery. Finally, Marshall proposed that the baked bricks of Mohenjo-daro were used as protection from rain, the sun-dried variety being susceptible to erosion in climates with heavy rainfall. Moreover, the large drains at Mohenjo-daro must have been there to handle something more than the scanty rainfall of today's environment. The acceptance of this position seems to be one of the few points of agreement shared by Wheeler and Marshall as noted in Wheeler (1953: 4-8; see also Piggott 1950: 135).

The "higher rainfall hypothesis" was thoroughly critiqued in the 1960s, first in a classic article by Raikes and Dyson (1961). Walter Fairservis (1961) also examined the proposal, as part of his long-standing interest in the environment, and found it wanting. Farzand Durrani did, also. (1965).

The Early Holocene Environment of the Northwestern Subcontinent

The Thar, or Great Indian Desert, has received some attention from those interested in environmental reconstruction. The research of an Indo-British team working there in the 1960s and 70s, and many Indian researchers made effective use of both archaeological and environmental data. The Early Holocene environmental sequence in this area seems at least to have been outlined (Allchin and Goudie 1978, Allchin, Goudie and Hegde 1978, Singh, *et al.* 1974, Meher-Homji 1980).

The sequence begins with a period of large rolling sand dunes and pronounced aridity. These dunes were later weathered, as evidenced by soil formation processes within them. Further evidence indicates a period of moving dunes following this period, that can be associated with the western Indian Upper Palaeolithic and some microliths. The dunes are fossilized by a rainfall regime higher than the period during which they were mobile. This sequence is summarized in Table 1.

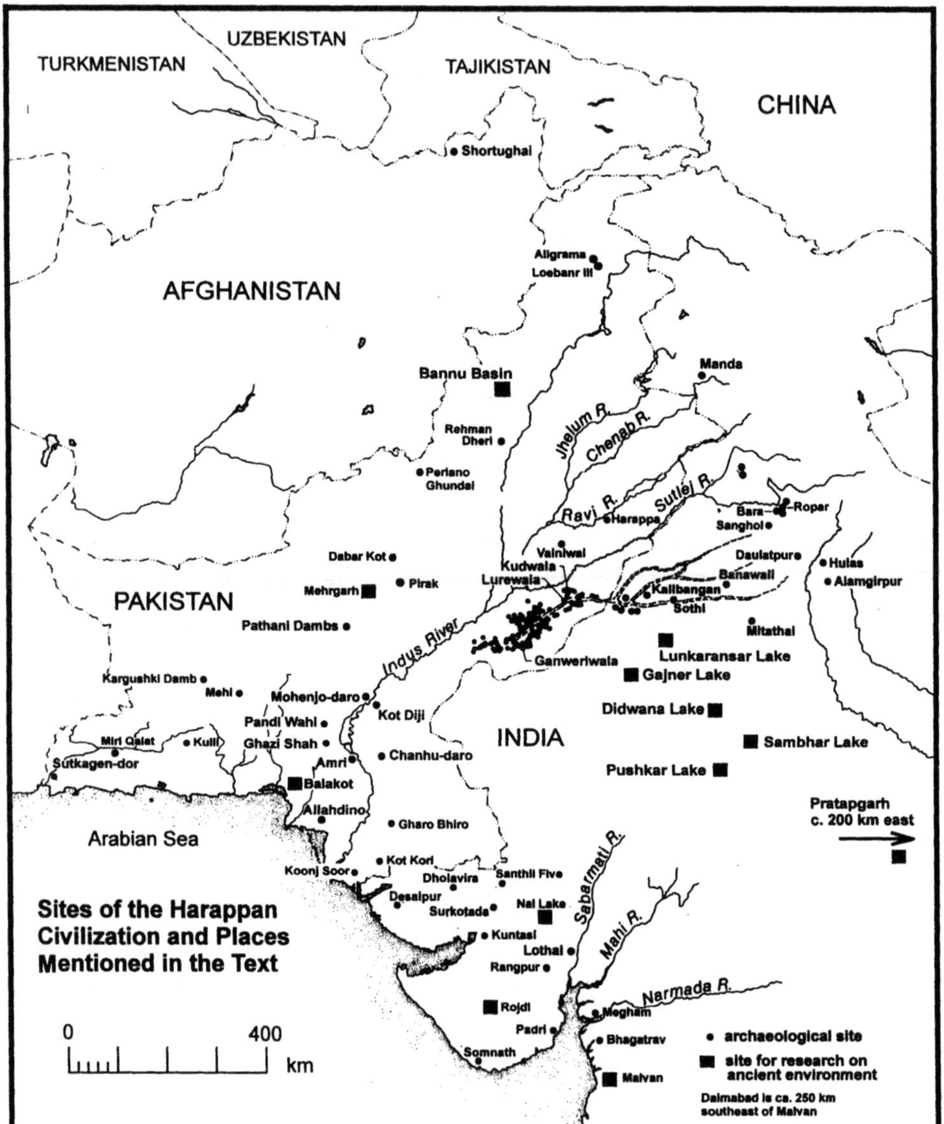

UZBEKISTAN
TURKMENISTAN
TAJIKISTAN
CHINA
Shortughai
Aligrama
Loebanr III
AFGHANISTAN
Bannu Basin
Manda
Rehman Dheri
Jhelum R.
Chenab R.
Periano Ghundai
Ravi R.
Sutlej R.
Harappa
Bara
Ropar
Sanghol
Vainiwal
Dabar Kot
Kudwala
Daulatpur
Hulas
Lurewala
Banawali
Alamgirpur
Pirak
Mehrgarh
Kalibangan
PAKISTAN
Sothi
Pathani Dambs
Mitathal
Indus River
Ganweriwala
Lunkaransar Lake
Gajner Lake
Kargushki Damb
Mehi
Mohenjo-daro
Kot Diji
Didwana Lake
Pandi Wahi
Miri Qalat
Kulli
Ghazi Shah
INDIA
Sambhar Lake
Sutkagen-dor
Amri
Chanhu-daro
Pushkar Lake
Balakot
Allahdino
Pratapgarh c. 200 km east
Arabian Sea
Gharo Bhiro
Kot Kori
Koonj Soor
Dholavira
Santhli Five
Desalpur
Sabarmati R.
Sites of the Harappan Civilization and Places Mentioned in the Text
Surkotada
Nal Lake
Kuntasi
Mahi R.
Lothal
Rangpur
Narmada R.
Rojdi
Megham
0
400
km
Padri
Bhagatrav
archaeological site
Somnath
site for research on ancient environment
Malvan
Dalmabad is ca. 250 km southeast of Malvan

Figure 1

TABLE 1
Early Holocene Climate

ca. 7500 BC	Indication of large scale use of fire by man. Probable beginning of cultivation with 1974 Cerealia-type pollen.		Singh et. al.
8000-7500 BC	Dunes stabilized, with habitation remains within them ca. 240 mm more rain than present	Moist phase	Allchin et. al. 1978
Upper Palaeolithic Pre-8000 BC	Dunes cover lake basins human activity on the edges of the desert	Severe Dry Phase	Allchin et. al. 1978
Middle Stone Age	Deep weathering of soils. Luni River flowing through to Kutch with coarse load	Humid Phase	Allchin et. al. 1978
Early Holocene	Huge sand shields deposited to the windward side of hills	Aridity	Verstappen 1970

See Meher-Homji 1980 for later portions of this sequence, which will be critiqued.

This climatic sequence for the early Holocene has been widely accepted, and it appears to be founded on very sound work. The notion that the farther back in time we go, the more certain our statements about past environments become is somewhat perplexing, but this reconstruction can be accepted, provisionally. But, after the eighth millennium, nearer in time than this widely accepted reconstruction, when there is (or should be) more data to assist in research on the ancient environemnt, the situation is much less certain and presents a vexing problem.

The Higher Rainfall Hypothesis Revived

The hypothesis that there was higher rainfall in the northwestern regions of the Subcontinent during the third millennium BC was revived in some ways by C. Ramaswamy (1968). Using historical records of the monsoon and a reconstruction offered by Lamb, Lewis and Woodruff (1966), he proposed that:

> "...during the south-west monsoon season around 2000-500 BC deep troughs in the upper westerlies must have extended into west Pakistan far more frequently than they do now, inducing monsoon activity and

also causing any monsoon depressions from the Bay of Bengal moving toward longitude 78 °E, to curve to the north or north-east. These developments would have caused frequent active monsoon conditions over the entire Indus Valley..." (Ramaswamy 1968: 629).

Ramaswamy's hypothesis would have affected the latest portion of the Mature, Urban Phase Harappan, and implies a higher rainfall for the Post-urban Phase. His proposal did not receive widespread notice, but in 1971 Gurdip Singh published a paper based on palynological research at four lakes in Rajasthan which served to further the revival of the higher rainfall hypothesis for the Mature, Urban Phase Harappan (Singh 1971, Singh, Joshi and Singh 1972, Singh, Joshi, Chopra and Singh 1974). In addition, Singh proposed that a period of rather marked aridity began at ca. 2000 B.C. and that by implication, the eclipse of the ancient cities of the Indus was caused by a decrease in mean annual precipitation.

The lakes of Rajasthan support the most active field programs researching the Holocene environment of the Subcontinent and a review of this work follows. Research on the Lakes of Rajasthan

The lakes of Rajasthan have received considerable attention from geologists for over a century; in part due to the curiosity that the presence of lakes in a desert brings and also because some of them are now very salty and others are fresh. An Indian custom of long standing is that salt was controlled and taxed by the central government; with revenue at stake the Inland Customs Office's interests in the history of the salt lakes was substantial and legitimate and they caused the Geological Survey of India to begin working on the question as early as 1867). There are a number of general reviews of the literature on these lakes, which give some background to the problem (see especially Sinha 1977; Biswas, Chattopadhyay and Sinha 1982).

The Palynological Research at the Salt Lakes in Rajasthan

A research team headed by G. Singh has worked at the salt lakes of Rajasthan on two occasions: the late 1960s and again in 1980 (Singh, Wasson and Agrawal 1990; Wasson, Smith and Agrawal 1984). The second research venture was more sophisticated than the first and produced additional data that is an important contribution to the study of the palaeoclimate of the northwestern regions of South Asia.

Figure 2

Singh's original research took place at Sambhar, Didwana and Lunkaransar and Pushkar Lakes (see Figure 1). The first three lakes are salty. Pushkar is filled with fresh water. The relevant pollen samples were taken from the salt lakes (Singh 1971; Singh, Joshi and Singh 1972; Singh, Joshi, Chopra and Singh 1974), gathered by digging pits in the dry lake beds. The resulting profiles were sampled at 2.5 to 10 cm intervals. Samples for radiocarbon dating were taken from the same face as the pollen samples, in the form of short monoliths of material, 10 to 15 cm in thickness (Singh, Joshi, Chopra and Singh 1974: 471). The same basic method was used on the second visit to Didwana (Singh, Wasson and Agrawal 1990: 351-52).

According to records with the India Meteorological Department (1962) Sambhar Lake (26° 55' N - 75°11' E) received an average of 494 mm of mean annual precipitation between 1901 to 1950. Rainfall at Didwana Lake (27°24' N - 74° 35' E) averaged 357 mm for the same period. At Lunkaransar Lake (28°30' N - 73°45' E) the average was 233 mm. Surface drainage plays an important role in supplying these lakes with water. Sambhar, for example is now fed by four rivers, the Menda, Rupnagar, Kharian and Khandel which, with some smaller streams, have a catchment area of ca. 6,500 square kilometers (Holland and Christie 1909: 154). Sambhar Lake is also quite large, fluctuating between 200 and 300 square kilometers and always retaining some water. It has a maximum depth of about three meters. Didwana, on the other hand, is only 10 to 15 square kilometers in area and reaches a maximum depth about a meter. It completely dries up during the summer (Biswas, Chattopadhyay and Singh 1982: 69).

A summary of the study resulting from the palynological work work of the first Singh team is given in Table 2.

Table 2
Results of Palynological Research at Four Thar Desert Lakes

Estimated Date	Lake	Environment
1000 AD-today	Rajasthan lakes saline	Present aridity
1000 BC	Lakes begin to turn saline	Beginning of present aridity
1400-1000 BC	Lakes contain fresh water	Aridity ameliorates
1800-1400 BC	Sambhar Lake begins to turn saline; increase in continues halophytes	Aridity and spreads east
ca. 1800 BC	Lunkaransar Lake begins to dry up.	Onset of aridity
3000-1800 BC	Grassy steppe savannah with good tree growth. Obstacle Dunes parabolic dunes developed from the sand washed away from sand shields.	Moist Phase with ca. 500 mm more rain than today
7500-3000 BC	Poor plant cover	Dry Phase; but not as dry as today
ca. 7500 BC	Indication of large scale use of fire by man. Probable beginning of cultivation with cerealia-type pollen.	
8000-7500 BC	Pollen indicates fresh water lakes	Moist phase with ca. 240 mm more rain than today

This sequence of events is in partial agreement with the work of a cooperative venture in palaeoclimatology called COHMAP (Cooperative Holocene Mapping Project), a multi-institutional consortium of scientists studying the Quaternary and Neogene climate of the earth. Their model and data show that the northern tropics, especially Asia and Africa had an orbitally increased amount of summer solar radiation between 10,000 and 4000 B. C. This enhanced the contrast between the sea and land masses and produced strong summer (southwest) monsoons, which accounts for the high lake levels (COHMAP 1988: 1043). The pollen data from the Rajasthan lakes indicates a pronounced wet phase beginning about 1000 years after the end of the COHMAP wet phase and cannot be discounted as a significant point of disagreement.

A second point of disagreement comes from Ramaswamy who proposed that from about 2000-500 B. C. deep troughs in the upper westerlies extended into Pakistan far more often than they do now, inducing frequent active monsoon conditions over the entire Indus Valley (Ramaswamy 1968: 629). This model suggests that there was more rainfall in the period during which the Singh team proposes aridity.

A more general kind of correlation has been advocated by D. P. Agrawal, at least for the glacial and early Holocene times: A comparison of $^{18}O/^{16}O$ measurements on *Globorotalia ruber* and *G. menardil* from the Undermine Sea and the Arabian Sea cores shows that during the last glacial maxima (c. 18,000 B.P.) and at c.10,000 B.P. the southwest monsoon was weaker than today but the northeast monsoon was stronger. During this time the Persian Gulf region, Rajasthan, and the Zagros mountains were drier than today.

Van Campo et. al's. (1982) comparison of oxygen isotopic and pollen data shows three distinct glacial stages at c. 18,000, c. 70,000 and c. 120,000 yrs B.P. Both benthic and planktonic foraminifera show ^{18}O values. There is a positive correlation between 18O and Chenopodiaceae pollen values. Thus Van Campo et. al. (1982) have correlated littoral aridity with sea level fluctuations and oxygen isotopic stratigraphy. Agrawal and Guzder (1978) had established a marine transgression at c. 120,000, 30,000 and 500 B. P. on the Saurashtra coast. The 5,000 B.P. marine transgression seems to coincide with the mid-Holocene warming up. On rechecking, the 30,000 B.P. eustatic rise does not seem to be so well authenticated. The 120,000 transgression appears to coincide with the onset of the last glaciation and therefore a bit difficult to explain. Perhaps this is due to large errors in uranium series dating and may represent the interglacial eustatic rise of the sea preceding the last glaciation and therefore should be older than 120,000 B.P. (Agrawal 1987: 1-4)

The second phase of work at Didwana produced a sequence that can be summarized as follows in Table 3.

Table 3

Results of G. Singh's Palynological Research at Didwana

Estimated Date	Lake	Vegetation
younger than 2230 BC	Ephemeral lake	Semi-arid savannah grassland
2230-4420 BC	Moderately deep fresh water	Sub-humid
4420-5510 BC	Fresh water followed fluctuating fresh and salt	Savannah grassland
5510-12000 BC	Fluctuating fresh and salt water	Shrub, savannah grassland
Older than 12000 BC	Hypersaline	Arid steppe

(adapted from Singh, Wasson and Agrawal 1990: 353)

The second phase of work at Didwana is important because a more substantial attempt was made to use terrestrial plants to reconstruct the environment around the lake. These included Aerva and Calligonum which grow in dry areas, Artemisia (sagebrush), a bush that thrives under conditions of winter rainfall, and Prosopis and Oldenlandia. The effort is important because it shifts the burden of evidence from assessing the aquatic flora within the lakes themselves, to a correlation of environments within the lakes with the surrounding vegetation.

A general observation on the chronology is important if this research is to be used for the reconstruction of culture history. The two sequences have to rely on just a few radiocarbon determinations: eleven for the first work (TF-887, TF-738, TF-739, TF-883, TF-884, TF-886, TF-887, WIS-386, WIS-387, WIS-405, WIS-415) and seven for the Didwana work (PRL-644 through 650); (Singh, Wasson and Agrawal 1990: 354). It is a weakness in the entire scheme, and no one should believe that it is possible to use the chronological schemes offered in publication with confidence.

An examination of the chart presented in (Singh, Wasson and Agrawal 1990: 353) is important for two reasons. First, the column which reflects the water level of Didwana Lake over long periods of time, derived from (Wasson, Smith and Agrawal 1984), moves back and forth in the time period ca. 4000 to 5500 BC. Significant changes are also evidenced in the millennia prior to this period. One wonders if the water level in these lakes would change so quickly and erratically if long term variations in rainfall (e.g. climatic change) were the only factors influencing this variable. It is also interesting, that after this period the water level curve is very stable. An examination of the terrestrial pollen does not seem to indicate climatic change. Aerva starts out as an abundant pollen and then goes flat. Oldenlandia moves around, but does not reveal much, and the same could be said for Prosopis.

The fluctuations in the pollen sequences at the lakes studied by G. Singh and his team are significant because they may have important inferences concerning the total amount of mean annual rainfall that came to Rajasthan in the Holocene. These lakes are complex bodies of water. At times two lakes, one salty the other fresh, are in close geographical proximity as witnessed by freshwater Pushkar's proximity to salty Didwana, Sambhar and Lunkaransar. Over the past century a number of hypotheses have been advanced to account for the salt in some of these lakes and there is a small bibliography on this topic that has been summarized in Sinha (1977). Basically, there are four hypotheses on the origin of these lacustrine salts: 1) it is marine and results from the regression of the Tethys Sea or raised sea levels during the Pleistocene, 2) the salt comes from brine springs in the region, 3) it is wind-borne, from the Arabian Sea and the Ranns of Kutch, and 4) it results from the concentration of salt through the evaporation of ground water with dissolved salts in it.

The Maritime Hypothesis

The first discussion of this problem came from A. O. Hume of the Inland Customs Department in 1867-68 (cited by Holland and Christie 1909: 155-56). Hume attributed the presence of salt in this region to the drying up of an inland sea, a topic of concern to Indian administrators and scientists of the day (Blanford 1876: 95-6).

This position was supported by the Geological Survey of India as late as 1975 (see Godbole 1952, 1972 and Mathur, Verma and Mehra 1975), in spite of a very sensible argument against it which was proposed in 1952 by the same organization. It was at this time that J. B. Auden noted: "..if the salt lakes of Rajasthan are regarded as remnants of the dwindling Arabian or Tethys Sea, they must have been preserved in isolated basins left by the regressing waters from as long ago as the early Tertiary. It is difficult to suppose that such residual salt lakes, cut off from the dwindling sea, could have remained intact through the whole fluviate period, when the region must have been traversed by major silt-laden rivers and rainfall was by inference high" (Auden 1952: 57). The fresh water conditions found in Sambhar Lake at ca. 8,000 B.C. are strong evidence against the marine hypothesis (Singh, Joshi, Chopra and Singh 1974). It is also true that the salts of the lakes are significantly different in composition from sea salt. Those of Sambhar Lake have an excess of carbonate and are deficient in calcium, magnesium and potassium (Sinha 1977: 149). It is reasonably clear that this hypothesis has little supporting evidence and is probably not worth pursuing.

The inland sea hypothesis has been revived to some degree, with speculation concerning the presence of a maritime environment north of the Ranns of Kutch during the Pleistocene (Mathur, Verma and Mehra 1982). There are raised beaches along the coast of Saurashtra with dead coral reefs and sea shells (Gupta 1977; Murty, Kulkarni and Gupta 1969), of a few meters, five or ten at most. Sambhar Lake is approximately 600 meters above modern mean sea level and it is difficult to reconcile this difference.

The Brine Spring Hypothesis

The thought that the salt water of some Rajasthan Lakes is derived from underground brine springs was first advanced in an 1869-70 report on Sambhar Lake by R. M. Adam, another member of the Indian Inland Customs Department (cited in Holland and Christie 1909: 156). Also noted in Holland and Christie (1909: 156) is the fact that Prof. Fritz Noetling in 1902 suggested that that there was a hidden fault plain below the deep desert sands which admitted the waters of a saline spring into the lake bed. Holland and Christie debunk these ideas, not based on any particular field observations but because they run counter to the general geology of the Aravalli system, of which the Rajasthan lakes are a part.

There is some indication of salt deposits lying below some of these lakes. They have been found in Bikaner at minus 553.3 to 556.4 meters (Sinha, Singh, Bakhohi and Bhatnagar 1973). The halites found at minus 7.4 meters at Didwana (Khandelwal 1975) are very localized and do not pertain to the discussion of this problem (Sinha 1977: 149; Biswas, Chattopadhyay and Singh 1982: 73; see also Bhattacharya, Gupta, Rishi, Mitra and Kashyap 1982).

The brine spring hypothesis has not been refuted and although it might not be based on our present understanding of Aravalli geology, it should not be discounted. In fact, the entire issue concerning the subsurface movement of water needs a more complete investigation if the issue of the source of salt in these lakes is to be understood. The brine spring action might be most vigorous during periods of highest water availability thus yielding a correlation between salty conditions and periods of high rainfall; exactly the opposite from the usual interpretations of the lake pollen profiles.

The Aeolian Hypothesis

The thought that the salt of the Rajasthan Lakes is due to its aerial transportation from the Arabian Sea was, once again, discussed by Holland and Christie (1909: 163-66). It is not a viable idea and has been severely critiqued (Godbole 1952, 1972, Auden 1952).

The most important critical comments come from Pascoe (1964: 2005) who observes that while a certain amount of airborne salt can be expected from the Ranns of Kutch, it would tend to move progessively north, even reaching India's northern mountains. There is no evidence for this. Moreover, the analysis of wind direction in this region undertaken by Pramanik and Ramkrishnan (1954) has cast serious doubts on the applicability of this agency providing more than small, even trivial, amounts of salt to the Rajasthan lakes. Finally, it will be recalled that the composition of the salt from Sambhar Lake, at least, is significantly different in composition from that of sea salt.

G. Singh's Evaporation Hypothesis

The proposition that the salt water lakes of Rajasthan result from the long-term evaporation of water, and the resulting concentration of ground water minerals, is an old and attractive idea. It was first postulated by T. D. LaTouche (1902) in his memoir on the geology of western Rajasthan. He proposed that the salt lakes were all closed (endorheic) inland basins without drainage outlets. The water that comes into them therefore has no escape. Generally high temperatures and low humidity led to the evaporation of water, but left behind the minerals once dissolved in it. Over long periods of time this led to the concentration of salt in the lakes, turning them into bodies of briney water. It is also known that some of the streams entering the Rajasthan lakes have accumulated dissolved salts from their environment (Mishra 1982).

By extension it can be proposed that the lakes that are not salty have some outlet, either on the surface or through ground water, which creates an open system and the salts left behind as a part of the evaporation process can escape before being concentrated as brine. Pushkar Lake near Ajmer is a fresh water lake without any surface outlet. R. N. Ghosh (1975) has offered some evidence that there is a subsurface flow of water out of the lake toward the north and the now dry Sarasvati Riverine system.

If this hypothesis is to explain the salt concentration, then changes must be accounted for in the amount of water coming into these basins and the method by which it gets there. For example, there is contemporary evidence that water enters the lakes from local rainfall and drainage, as well as in the form of ground water, extending the argument advanced for the Pushkar Lake. There is further evidence that water leaves these basins through evaporation and subsurface drainage. The salinity of the lake water is the result of a balance between rainfall and surface drainage on the one hand and subsurface conditions on the other. A change in 1) total average annual rainfall, 2) surface drainage or 3) subsurface drainage will therefore change the halite content of these bodies of water.

Changes in Salt Content of the Lakes due to Subsurface Drainage Changing surface and subsurface drainage patterns in this region generally results from large-scale tectonic activity resulting from the collision that the Indian land mass has undergone with the larger Asian land mass. This point is of particular importance to the present assessment of the pollen data because there appears to be a change in lake salinity at ca. 2000 B. C. Singh and his collaborators interpret its increase as a decline in mean annual rainfall. This is, without doubt, one phenomenon that might lead to increasing salinity. But, it is also true that at ca. 2000 B. C. there is geological evidence for major changes in surface, and probably subsurface, drainage in the northern Rajasthan, the eastern Punjab, Haryana and the western Uttar Pradesh. This event led to the desiccation of the Sarasvati Riverine system, the rejuvenation of the Sutlej River and the creation of the Yamuna River.

Perhaps the greatest support for using subsurface drainage to account for variations in salt levels in these lakes comes from a simple observation. All four of them (Sambhar, Didwana and Lunkaransar and Pushkar) are in the same part of Rajasthan, where there is one rainfall regime. Yet, it is an incontrovertible fact that three are salty and one is fresh. The same was true in Bikaner State at the turn of the century were there were two salt lakes: (Chhapar, Lunkaransar) and one with sweet water (Lake Gajner; Erskine 1909: 310). The unpublished report from the Geological Survey of India maintains that the underground drainage outlet for Pushkar Lake has been found (Ghosh 1975).

It would be remiss not to mention a review of the Holocene environment of northwestern India and Pakistan by V. N. Misra (1984). It is possible to reach much the same conclusion as his: that there is basic climatic stability between today and the Mature, Urban Phase Harappan. The work by G. Singh and his team has also been critically evaluated by Vishnu-Mittre (1974).

Settlement Patterns, Subsistence and Climate

There is one more piece of evidence that suggests that the rainfall of northwestern India and Pakistan did not drop during the early second millennium. The Punjab,

Haryana, northern Rajasthan and western Uttar Pradesh is dry farming land, at least prior to the nineteenth century irrigation systems that were dug through the region. Because there is no evidence for canal irrigation during the Mature or Post-urban Harappan periods (2500-1500 B.C.), most archaeologists are content to believe that the village farming communities in this vast region depended on dry farming for their agriculture. There are 217 Mature Harappan sites (2500-2000 BC) in this region, compared to 854 in Post-urban times (2000-1500 BC), just at the time when G. Singh's data are proposed to indicate a drop in rainfall. As a comparison, the figures for Sindh and Kutch at the same times indicates the presence of 140 Mature Harappan sites and only 18 in Post-urban times (Figure 3, Histogram of Site Counts for North India and Sindh/Kutch). An analysis of site sizes does play a role in this interpretation. In Sindh and Kutch the average site size in Mature Harappan times is 8.7 hectares and drops to 6.6 in the Post-urban Phase. In Punjab, Haryana, northern Rajasthan and western Uttar Pradesh the situation is different. The average site size during the Mature Harappan is 10.9 hectares versus only 3.6 for the Post-urban. This kind of decentralization accommodated a large population. Nevertheless, one can project a population increase in these North Indian States during the Post-urban Phase, a trend counter to that of Sindh and Kutch. The increase in the number of sites is also substantial. These observations cannot be easily reconciled with Singh's conclusions and, given the other points taken here, seem to indicate that his implications for his data may not be correct. Other Notices of the Late Holocene Environment of South Asia During Protohistoric Times

Notice should be given to data from four other sources of palaeoenvironmental reconstruction, three based on pollen. They are not entirely consistent among themselves but do tend to indicate that the environment of the third millennium was not so different from the present. There is also an implication that the science of reconstructing the ancient environment of the Greater Indus Region is very young and a great deal of work needs to be done before consistent patterns of climatic stability and change begin to emerge over this vast area.

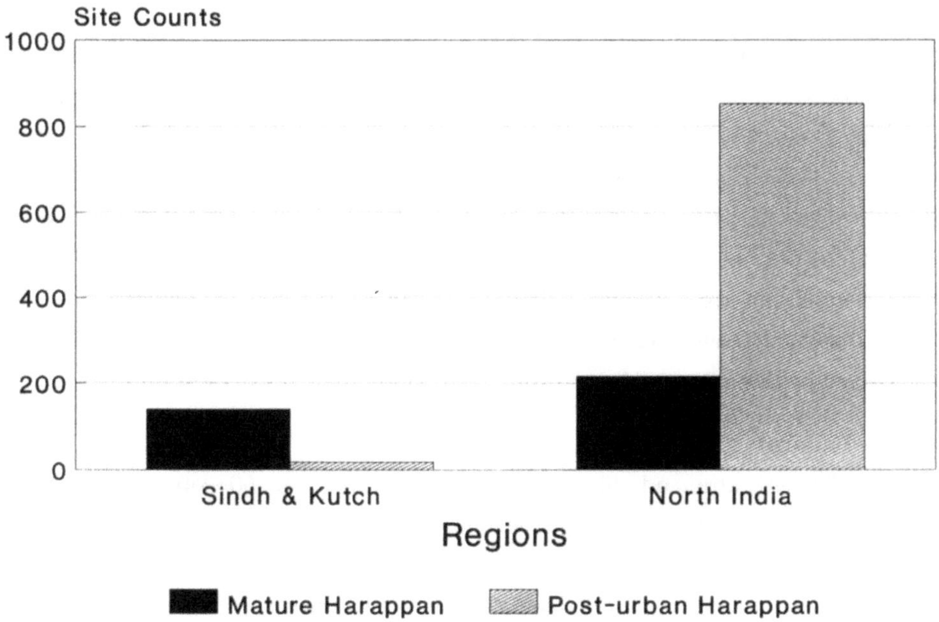

Figure 3

Balakot

The first of these reconstructions comes from Balakot near Sonmiani Bay in Pakistan where Margaret McKean undertook some palynological research aimed at reconstructing the ancient environment. Her conclusions from a comparison of ancient and modern pollen samples covering the Early Harappan and the Mature, Urban Phase Harappan is: "...occupation during both cultural periods existed under essentially modern climatic conditions" (McKean 1983: v).

Malvan

The second study was undertaken as part of the archaeological excavations at Malvan, in South Gujarat, by J. P. Joshi and F. R. Allchin (F. R. Allchin and J. P. Joshi 1970; J. P. Joshi and F. R. Allchin 1972). The palynological research conducted by Vishnu-Mittre and Chayya Sharma at an oxbow lake in the vicinity of the Bronze Age archaeological site has shown considerable environmental stability. The entire pollen sample revealed that an open, grassland of Cheno-Amaranths, along with several other herbaceous elements, surrounded the lake. Distant arboreal forests are also suggested. Of some interest is the fact that there is good evidence for the occasional drying up of the lake, for reasons that are not fully understood. Over all, the pollen sequence suggests little environmental change. However, this is undated and there is no sense of the time depth covered by this sample (Vishnu-Mittre and Sharma 1973).

Nal Lake, Gujarat

Finally, Vishnu-Mittre and Sharma tested Nal Lake ca. 60 km southwest of Ahmedabad. Their reconstruction there indicates the following:

ca. 1819 AD	Brackish water, possibly from marine transgression caused by an earthquake.
1500 BC-AD	Dry lake, no lacustrine pollen at the sampling site.
2500-2200 BC	More tree pollen than non-tree pollen.
3000-1500 BC	Flooding of Nal Lake; abundant fresh water. A grassland-chenopod savannah. The chenopod decline from ca. 8000 B. C. to about 1500 B. C. Pollen from a largespreading tree of the dry deciduous forest (Holoptelea integrifolia) increases until ca. 2000 B. C.
4000-3000 BC	Brackish lake waters
4000 BC	More tree pollen than non-tree pollen.
Mid-Holocene	The vegetational history of Nal Lake
	begins with indications of an open, almost treeless stretch of
	grasslands populated by various Chenopods occurring in
	saline and alkaline areas. (Vishnu-Mittre and Sharma 1979).

Mehrgarh

Stephanie Thiebault undertook the examination of wood charcoal from Mehrgarh (Periods VI and VII) and Lal Shah, the Period VII kiln site (Thiebault 1989, 1992). The species she identified are still endemic to the region, although Juniper, for example, would only be found at higher elevations in the Quetta Valley. There is little, or nothing, to suggest higher rainfall in the past from this data set. The analysis of macrobotanical analysis at Mehrgarh lead to the same tentative conclusion (Costantini and Biasini 1985).

Sindh Kohistan

Michael Harvey and Louis Flam (1993) have an interesting and important study of prehistoric irrigation and land use at the end of the Kirthar Range in western Sindh, near the archaeological site of Ghazi Shah. This is a study of five gabarbands and the associated sources of water around the Kot Dijian sites of Phang, and its neighbor, Kohtras Buthi. While the focus of this study is water management they conclude with the following observation: "Association of both modern and prehistoric sites with springs suggests that there has not been appreciable climatic change in Sindh Kohistan in the last 5000 years" (Harvey and Flam 1993: 109).

Rojdi

The joint team from the Gujarat State Department of Archaeology and The University Museum of The University of Pennsylvania undertook an intensive program of palaeobotanical research (Weber 1991). While palaeoenvironmental reconstruction was not the principal aim of this research it was a topic that received some attention. From an analysis of thousands of seeds from about 100 different plants there is nothing present at Rojdi to suggest that the environment during the third and early second millennia was significantly different from today.

The Bannu Basin

There is another view on the climate of the Indus Age that

deserves notice. Some attention has been paid to the ancient environment of the Bannu Basin in Pakistan as part of the joint archaeological work being undertaken there by the University of Peshawar and a team from the United Kingdom. Although it is clear that the drainage of this valley has undergone significant changes, they seem to be late glacial or early Holocene in age. (Thomas 1986: 22). Helen Rendell suggests that: "Several lines of evidence point to the existence of moister conditions within the basin 3000 to 5000 years ago. In addition to an increase in aridity since that time, there is evidence that the major phase of incision within the basin occurred within the last 300 years..." (Rendell 1986). Thus, just when the G. Singh team is describing a period of aridity (3000-5000 BP), Rendell and her colleagues on the Bannu Basin Project believe there may have been greater rainfall. But, all of this is taken as a very tentative suggestion, as indicated by the following from K. P. Thomas:

> Any climatic shifts which might have led to a more even distribution of rainfall through the year, or to an increase in the amount of winter rain, would have had a distinctly beneficial effect on past arable agriculture. All this is speculation: it must be stressed that we have no unambiguous evidence that late Holocene climates were markedly different from the climate of the Bannu Basin today. This is not, of course, to say that climates have remained the same through much of the later Holocene time in this area; we just do not know (Thomas 1986: 22-3).

The Environmental Reconstruction

There still may be limited merit to interpreting the pollen data from the lakes of Sambhar, Lunkaransar and Didwana as changes in average annual rainfall; certainly not the only possible cause for the changes in pollen. The ultimate origin of the salt in these bodies of water is complex and little understood. Sinha (1977) has noted that the entire subject has been an odd, and unfortunate, mix of armchair speculation, with just enough field research to keep the debate alive. What is needed is a vigorous program of problem oriented field study to resolve the more or less empirical gaps in our knowledge. It is only with the completion of a program of field study that the interpretation of pollen data, such as contributed by G. Singh and his team, will be secure.

The position taken here is that there was no significant difference in the rainfall regime in the period from ca. 7000 BC to 1000 BC as compared to today's long term pattern. Rainfall is a dynamic, like other features of the environment and this position admits differences in yearly patterns, as well as longer term "cycles" of drought and rainfall abundance, in the range of scores or even a hundred years or so. The conclusion reached from the data which are presently available, is that changes in rainfall, or climate, should not be seen as an explanation for major historical events, such as the eclipse of the ancient cities of the Indus.

The relationship between rainfall and a larger environmental reconstruction is more complex and difficult to deal with. We know from a host of reliable sources, including archaeological excavations, historical observations and data collected by men of science that today's landscape in the Greater Indus Region is significantly different from that of even the nineteenth century. The immense irrigation works of Sindh and the Punjab are good examples. The large and interesting literature on Colonial period hunting in India is indicative of an abundance of game, both mammals and birds, on the plains of the Greater Indus Region, especially the Punjab, Rajasthan and Gujarat. The situation by some accounts was not so different from the vast herds of animals on the savannahs of Africa. Intensive hunting over more than a century, using sophisticated weapons technology and knowledgeable guides led to the slaughter of wildlife on a scale that was so vast it is difficult to imagine. In some ways this was simply the acceleration of a process that began much earlier with the introduction of firearms into the Subcontinent, but taken as a whole it can be said with confidence that the density and distribution of game animals is a mere shadow of what it would have been in the prehistoric period. Examples of this, involving animals that were important during Mature Harappan times are abundant. In Pakistan alone, there was once a large population of lions, tigers, elephants, rhinocerii, onager *(Equus hemionus)*, nilgai *(Boselaphus tragocamelus)* and blackbuck *(Antelope cervicapra)*. Now these animals are either extinct there, or limited to a few individuals who stray across the modern border with India in Thar Parkar or along the bed of the ancient Sarasvati (Roberts 1977).

Reducing the number and distribution of wild mammals has had an effect on the availability of pasture but the growing human population and their husbanding of domesticated mammals has more than compromised this as a conservation measure. The presence of relatively large human populations in the Greater Indus Region since ca. 7000 B.C., accompanied by their cattle, sheep, goats, water buffalo and later camels and horses has had the effect of generally, but slowly degrading the landscape of its grass resources. This should not be seen in a catastrophic way, since many plants respond well to grazing, browsing and periodic burning. The presence of these domesticated animals must be seen in concert with man's search for fire wood and timber for wood working and architecture. Clearing land and draining swamps for agriculture, along with the harboring of other land for villages, towns and cities are involved as well. Taken as a whole, it can be seen that we have good reason to imagine that over nine millennia the Greater Indus Region is different today than it was during the prehistoric era under consideration. On a somewhat larger scale, there is evidence that the regional drainage pattern in the northeastern sector of the Greater Indus Region was different from that which we see today. The Yamuna River seems to have been very small, if a river at all, prior to about 2000 BC. The now dry Sarasvati and Drishadvati Rivers, were active, at least early in the time, with some evidence that the Drishadvati dried up before the Sarasvati. The Sutlej also seems to have had a different alignment, flowing into the Sarasvati, rather than the Indus as it does today. The Eastern Nara appears to have continued south, parallel to the Indus, and may have entered Kutch as an independent river. The Indus River in Sindh also had a different alignment than the present one. All of these topics are investigated in detail within the context of describing the ancient Domains of the Harappan Civilization. For the present they offer even more reason to see the ancient landscape of the Indus Age as different from the one evidenced today.

The environmental reconstruction of the Greater Indus Region from ca. 7000 B.C. to 1000 BC has received very little systematic attention. Therefore, it is not yet possible to offer a reconstruction of this landscape based on anything other than a kind of reasoned speculation. Assuming that the rainfall regime from 7000 B.C. to 1000 B.C. was pretty much as it is today, we have the ability to hold this important variable constant; within the parameters of its own dynamic. The massive irrigation facilities and the agricultural areas that they foster should also be removed. A reconstructed regional drainage pattern has to be managed, but some progress has been made there. Many more game animals are implied, and the distribution pattern for many (most?) of them expanded into areas where they are now extinct. Finally, in the absence of more particular data, the density of tree and grass cover has to be increased, with the environment generally "fluffed-up."

Most of this is unacceptable as a reconstruction of the ancient environment of the Greater Indus Region. But for now there is little else that can be done and rectifying the situation represents a challenge for those who take a serious interest in the study of prehistoric life in South Asia.

The Chronology for the Post-Urban Phase

Dating the period of transition from the Mature Harappan to the Post-urban Phase is difficult. There is only one date from Sindh and the West Punjab that pertains to the case, and that seems to be the area that is the most important in the process that brought an end to Urban life in the Greater Indus Region. Just when this process began is also open for discussion, and may well come down to an arbitrary decision. The urban eclipse was a process that took some time, several hundred years at least, and the abandonment of the Great Bath at Mohenjo-daro could be seen as a "starting point."The Great Bath has been assigned to the middle "Intermediate" Period at Mohenjo-daro (Mackay 1937-38: 20), which is corroborated by the German Research Team headed by Michael Jansen. Their analysis of the Great Bath found that it: "...must have been abandoned as such while Harappans were still in occupation of the site. This also applies to the `Granary' to the west" (Ardeleanu-Jansen, Franke and Jansen 1983: 54). There was a great deal of unsorted trash in the Bath when it was excavated, and no coherent architecture in the vicinity. There is also Sir Mortimer Wheeler's testimony concerning the granary: "To a height of 30 feet or more, the tall podium of the Great Granary on the western side of the citadel was engulfed by rising structures of poorer and poorer quality (Wheeler 1968: 127). If this is the case, then these buildings were out of use during the Late Period of Mohenjo-daro, when most of the buildings that have been excavated in the Lower Town were inhabited during the "Late Period." It can be concluded that there were more than a few years between the abandonment of these two key facilities on the Mound of the Great Bath and the final abandonment of the city. This seems to be close to 200 years, as a first approximation, pending the presentation of radiocarbon or other absolute dates.

The date of the Mature Harappan and the transition to what I have called the Post-urban Phase (Possehl 1977) of the Harappan Civilization is based on a growing number of radiocarbon determinations (Possehl 1989). The initial chronology for Mohenjo-daro and Harappa came from the observed similarity of seals from the first two or three seasons of excavation at Mohenjo-daro and Harappa and those found at both Ur and Susa (Sayce 1924) as well as through a consideration of some architectural features (Gadd and Smith 1924: 616). These scholars, and Sir John Marshall of the Archaeological Survey of India, agreed that this evidence suggested that the Mature Harappan Phase could be dated as early as 3000 B.C. and that the civilization lasted for about 1000 years. The first edition of Wheeler's, The Indus Civilization, appeared in 1953 and with revisions to the Mesopotamian chronology these dates had been shifted downward to ca. 2500-1500 B.C. (Wheeler 1953: 84-93). This pre-radiocarbon chronology was now more broadly based on the presence of a wide range of Harappan objects in the west (see Dales 1968, Lamberg-Karlovsky 1972, Ratnagar 1981, Possehl 1996a for general reviews). The most convincing materials are still the stamp seals with Indus script (Wheeler 1953: 1985-86; Chakrabarti 1976, Possehl 1996b), cubical Harappan weights (Ratnagar 1981: 184-86, Possehl 1996b) and etched carnelian beads (During Caspers 1972, Reade 1979). Other kinds of artifacts have been found as well; inlays, several types of beads, cosmetic implements and some pottery. These data focus on the Akkadian Dynasty (2334-2154 BC), with some indications of slightly earlier contact during the Early Dynastic; as early as the "Royal Graves of Ur," ca. 2550 BC (Chakrabarti 1982).

In 1956 Walter A. Fairservis, Jr. developed an anthropological approach to the Harappan Civilization and noted the virtual absence of internal evolution within the Mature Harappan. This suggested to him that the period of time encompassed by these remains was not likely to be as long as a millennium and he proposed that we halve the time that the ancient cities of the Indus were functioning communities. He retained the center of the chronology on the Akkadian period and came up with ca. 2500 to 2000 BC as his best estimate.

This conclusion was supported in some measure by D. P. Agrawal in 1964. His analysis of the radiocarbon dates from Kalibangan and Lothal, as well as a few samples from Niai Buthi, Kot Diji and Mohenjo-daro suggested 2300 to 1750 bc (Agrawal 1964: 950). Agrawal's chronology was generally accepted by archaeologists (e.g. Wheeler 1968, Allchin and Allchin 1968, Fairservis 1971), in spite of the fact that it left the Mature

Harappan disconcertingly late with respect to the Akkadian Dynasty, and the signs of contact with the Early Dynastic.

With the development of radiocarbon calibration curves for the correction of "raw" dates, pioneered by the "MASCA" correction curve from The University Museum (Ralph, Michael and Han 1973), Agrawal's chronology was amended and pushed back into closer conformity with the Fairservis hypothesis (see also Dales 1965d, 1973, Brunswig 1973, 1975). There is nothing wrong with the thought of the Mature Harappan spanning the second half of the third millennium, as long as it is remembered that it almost certainly lacks precision, and gives the impression of a linear sequence, with the Mature Harappan appearing everywhere at once, seeming to go into decline all at the same time. The processes that led to the eclipse of urban life in western South Asia were more complex than they imply and 2500-2000 B.C. are dates of convenience.

Sindh

Mohenjo-daro

As was noted earlier, there are only three dates from Sindh and the West Punjab that bear directly on the establishment of a chronology for the emergence of the Post-urban Phase. The "Jhukar Culture" follows the Mature, Urban Phase Harappan stratigraphically in Sindh but there are no radiocarbon dates from these levels at any of the sites where it appears (e. g. Mohenjo-daro, Chanhu-daro, Amri, Jhukar)

The first radiocarbon dates come from Mohenjo-daro.

Table 4
Radiocarbon Dates From Mohenjo-daro

Site/Lab No.	5568 Half-life bp	5730 Half-life bc	Calibration***
Mohenjo-daro, Upper Levels			
P-1177	3985+64 bp	2155+65 bc	2556, 2546, 2493 cal. BC
P-1179	3913+64 bp	2080+65 bc	2460 cal. BC
P-1180	3828+61 bp	1995+65 bc	2297 cal. BC
P-1178A	3802+59 bp	1965+60 bc	2279, 2231, 2210 cal. BC
P-1176	3801+59 bp	1965+60 bc	2279, 2232, 2209 cal. BC
P-1182A	3702+63 bp	1865+65 bc	2133, 2064, 2048 cal. BC
TF-75	3600+110 bp	1760+115 bc	1961 cal. BC

****All calibrations in this paper are based on the CALIB-2 program*

Pirak

There is one other small bit of information connected with the chronology of the changes in the Harappan Civilization at about 2000 BC. There are two radiocarbon dates attributed to Pirak Period I. The Pirak material is very different from the Mature, Urban Phase Harappan and it extends into the Early Iron Age, but the two dates for Period I seem to indicate that the site was first settled around the beginning of the second millennium BC; although there is an uncomfortable spread between the two determinations.

Table 5
Radiocarbon Dates From Pirak

Site/Lab No.	5568 Half-life bp	5730 Half-life bc	Calibration
Pirak, Period I			
LY-1641	4080+290 bp	2250+300 bc	2250, 2245, 2652, 2647, 2617 cal. BC
LY-1640	3410+140 bp	1560+145 bc	1737, 1713, 1706 cal. BC

Gujarat

There is a long Bronze Age occupation in Gujarat. Sindhi style Mature, Urban Phase Harappan sites are found in Kutch and extend as far southeast as Lothal, just above the head of the Gulf of Cambay. Rojdi and the other Sorath Harappan sites, found mostly on the Saurashtran peninsula are now thought of as a regional aspect of the Mature, Urban Phase Harappan, particularly when compared with the Sindhi Harappan (Possehl and Herman 1990) The culture historical processes in Saurashtra may have been quite different from those that characterize Sindh.

Lothal

There is a series of radiocarbon determinations from Lothal that covers the transition between the Mature, Urban Phase Harappan and Lothal B, the Post-urban Phase at the site.

Table 6
Radiocarbon Dates From Lothal

Site/Lab No.	5568 Half-life bp	5730 Half-life bc	Calibration
Lothal A, Mature Harappan			
TF-13	3915+130 bp	2080+135 bc	2461 cal. BC
TF-22	3845+110 bp	2010+115 bc	2328 cal. BC
TF-27	3840+110 bp	2005+115 bc	2315 cal. BC
TF-26	3830+120 bp	1995+125 bc	2299 cal. BC
TF-133 3	740+110 bp	1900+115 bc	2182, 2166, 2142 cal. BC
TF-29	3740+110 bp	1900+115 bc	2181, 2166, 2142 cal. BC
TF-135	3405+125 bp	1555+130 bc	1735, 1717, 1701 cal. BC
Lothal B, Post-urban Phase			
TF-23	3705+105 bp	1865+110 bc	2134, 2059, 2048 cal. BC
TF-19	3650+135 bp	1810+140 bc	2034 cal. BC

The overlap in these dates again prevents a precise estimate of the time period being dealt with, but it does seem to center on about 2000 B.C., just as in Sindh.

Surkotada

Three radiocarbon dates are available from Surkotada IC, the final phase of occupation at this Mature, Urban Phase Sindhi Harappan site in Kutch. It is the last of three Mature Harappan occupations at the site and seems to come to an end just about the same time that Lothal A is terminated. Again, 2000 B.C. is as good an estimate of this as any other specific year.

Table 7
Radiocarbon Dates From Surkotada

Site/Lab No.	5568 Half-life bp	5730 Half-life bc	Calibration
Surkotada IC, Latest Mature Harappan			
TF-1297	3635+95 bp	1795+100 bc	2030, 1990 cal. BC
TF-1294	3620+95 bp	1780+100 bc	2018, 2002, 1980 cal. BC
TF-1307	3510+105 bp	1665+110 bc	1880, 1830, 1829 cal. BC

Rojdi

Rojdi, in central Saurashtra has a continuous occupation between the second and third millennia. The middle occupation of the site, called "Rojdi B" has six dates, which cross the 2000 B.C. marker, but these dates are stratigraphically consistent within the context of Rojdi B. Of the two dates for Rojdi C, the Late Sorath Harappan occupation there, one (PRL-1081) has to be discarded. The other date, calibrated to ca. 2100 BC is in general agreement with the data from Lothal, Surkotada and the Sindhi/Punjabi sites already discussed.

Table 8
Radiocarbon Dates From Rojdi C

Site/Lab No.	5568 Half-life bp	5730 Half-life bc	Calibration
Rojdi B, Final Sorath Harappan			
PRL-1083	3875+120 bp	2040+125 bc	2452, 2427, 2395, 2374, 2366 cal. BC
TF-200	3810+110 bp	1975+115 bc	2283 cal. BC
PRL-1088	3770+120 bp	1935+125 bc	2198, 2151, 2149 cal. BC
TF-199	3590+100 bp	1750+105 bc	1947 cal. BC
PRL-1281	3520+110 bp	1680+115 bc	1883 cal. BC
PRL-1282	3470+140 bp	1620+145 bc	1866, 1846, 1772 cal. BC
Rojdi C, Early Post-urban Phase			
PRL-1084	3700+150 bp	1860+155 bc	2133, 2067, 2047 cal. BC
BETA-61767	3570+60 bp	1730+60 bc	1973-1872, 1840-1811, 1809-1780 cal. BC
BETA-61768	3520+60 bp	1675+60 bc	1913-1745 cal. BC

Summary

A date of ca. 2000 B.C. can be sustained for an approximation of the transition between the Mature and Post-urban Phase Harappan in Gujarat. Although it lacks precision and is normative in that it simply approximates a point in time during which a complex process was under way, it is the current state of knowledge on the chronology for this topic in the southeasternmost domain of the Harappan civilization.Daimabad: Outside the Harappan Domains Proper

Before examining the northern areas of the Harappan Civilization notice should be taken of Period II at the site of Daimabad on the Pravara River of Maharashtra (Sali 1986). This area is very much a part of the Deccan Plateau and Peninsular India, but there is an occupation at Daimabad with some pottery that is, in some respects, like that of Rojdi C. There is also a small terracotta "seal" with the Harappan "jar" sign on it and a short inscription on the rim of a pot, reinforcing the idea that this period of occupation at

Daimabad had something to do with an expansion of Sorath Harappans out of Gujarat, up onto the Deccan Plateau. But, the period is not well understood, and the neighboring site of Kaothe (Dhavalikar, Shinde, Atre 1990) has a ceramic corpus much like Daimabad II and is now thought to be more within the Malwa-Jorwe fold than the Harappan. There is a lot to learn about the complex culture history of the northwestern Deccan, but the dates from Daimabad II should not be suppressed, even if it is not yet possible to view them in a totally critical way.

Table 9
Radiocarbon Dates from Daimabad II

Site/Lab No.	5568 Half-life bp	5730 Half-life bc	Calibration
PRL-426	3600+150 bp	1760+155 bc	1961 cal. BC
BS-180	3390+100 bp	1540+105 bc	1730, 1729, 1689 cal. BC
PRL-657	3140+100 bp	1285+105 bc	1424 cal. BC
PRL-420	1410+140 bp	500±+45 ad	642 cal. AD

These dates reinforce the thought that the Post-urban Phase of the Harappan Civilization is within the first half of the second millennium, even if they do little to add precision to understanding this phenomenon.

Harappa And The Western Sarasvati Drainage

Harappa

There are two radiocarbon dates from the most recent work at Harappa which are assigned to the Upper levels of the Mature, Urban Phase Harappan. They are very early if judged against the presumed 2000 B.C. date for the transition between the Urban and Post Urban Phases. More work is under way at the site and more dates will be forthcoming. It might even be possible that several radiocarbon dates will be available to give some chronological control over the important Post-urban Phase Cemetery H material from this site.

Table 10
Radiocarbon Dates From Harappa

Site/Lab No.	5568 Half-life bp	5730 Half-life bc	Calibration
Harappa, Mature Harappan, Late Levels			
WIS-2043	3770+70 bp	1930+70 bc	2268, 2263, 2203, 2147, 2146 cal. BC
WIS-2053	3920+210 bp	2090+215 bc	2469 cal. BC

Kalibangan

The ancient cultural affiliations of the Mature, Urban Phase Harappan at Kalibangan are much closer to Harappa and Mohenjo-daro than they are to the site's neighbors farther east and to the north, at places like Banawali and Ropar. It will therefore be considered in the present geographical context. There is a long series of dates from Kalibangan, not all of which are in consistent stratigraphic order or "believable" in terms of developing a realistic chronology for the site. Most of the "bad" dates come from the Early Harappan of Kalibangan I. There is no Post-urban Phase occupation there, and the site seems to have been abandoned at, or near the end of the Mature, Urban Phase Harappan. The dates for the Mature, Urban Phase Harappan at Kalibangan are as follows:

Table 11
Radiocarbon Dates From Kalibangan II

Site/Lab No.	5568 Half-life bp	5730 Half-life bc	Calibration
TF-160	4060+100 bp	2230+105 bc	2587 cal. BC
TF-942	4055+110 bp	2225+115 bc	2586 cal. BC
TF-607	3930+120 bp	2100+125 bc	2464 cal. BC
TF-25	3930+110 bp	2100+115 bc	2464 cal. BC
TF-163 (BS)	3925+125 bp	2090+130 bc	2462 cal. BC
TF-153	3910+110 bp	2075+115 bc	2459 cal. BC
TF-608	3910+110 bp	2075+115 bc	2459 cal. BC
TF-163	3910+100 bp	2075+105 bc	2459 cal. BC
TF-145	3895+100 bp	2060+105 bc	2456, 2412, 2408 cal. BC
P-481	3879+72 bp	2045+75 bc	2453, 2424, 2398 cal. BC
TF-147	3865+100 bp	2030+105 bc	2450, 2348 cal. BC
TF-948	3815+100 bp	1980+105 bc	2286 cal. BC
TF-605	3810+105 bp	1975+110 bc	2283 cal. BC
TF-151	3800+100 bp	1965+105 bc	2278, 2234, 2209 cal. BC

TF-139	3775+100 bp	1940+105 bc	2200 cal. BC
TF-947	3765+85 bp	1930+90 bc	2197, 2154, 2148 cal. BC
TF-150	3740+100 bp	1900+105 bc	2181, 2166, 2142 cal. BC
TF-141	3705+110 bp	1865+115 bc	2134, 2059, 2048 cal. BC
TF-149	3675+140 bp	1835+145 bc	2118, 2083, 2041 cal. BC
TF-142	3635+100 bp	1795+105 bc	2030, 1990 cal. BC
TF-152	3615+85 bp	1775+90 bc	2014, 2006, 1976 cal. BC
TF-946	3605+100 bp	1765+105 bc	1968 cal. BC
TF-149 (BS)	3585+120 bp	1745+125 bc	1942 cal. BC
TF-152 (BS)	3570+125 bp	1725+130 bc	1923 cal. BC
TF-143	3510+110 bp	1665+115 bc	1880, 1830, 1829 cal. BC
TF-244	3250+90 bp	1400+95 bc	1521 cal. BC
TF-138	3075+100 bp	1215+105 bc	1391, 1334, 1327 cal. BC

The report of the Kalibangan excavations is under preparation and it is not possible to place these dates in their true stratigraphic order. But, arranged in chronological order, calibrations for 21 of the 27 dates fall within the third millennium, and all but the last two are perfectly consistent with the notion that 2000 B.C. was a kind of hinge point for the transition between the Mature, Urban Phase Harappan and the Post-urban Phase.

The Eastern Domain Of The Indus Civilization

Radiocarbon dates are available from Bara, Mitathal, Sanghol, Banawali and Hulas. These are presented in Tables 12 through 16.

Table 12
Radiocarbon Dates From Bara

Site/Lab No.	5568 Half-life bp	5730 Half-life bc	Calibration
TF-1205	3730+95 bp	1890+100 bc	2140 cal. BC
TF-1204	3690+150 bp	1850+155 bc	2130, 2074, 2045 cal.
TF-1207	3490+85 bp	1645+90 bc	1875, 1838, 1818, 1800, 1782 cal.
TF-1206	3135+100 bp	1280+105 bc	1422 cal. BC

Table 13
Radiocarbon Dates From Mitathal, Bara Ware

Site/Lab No.	5568 Half-life bp	5730 Half-life bc	Calibration
PRL-292	4210+210 bp	2385+215 bc	2883, 2796, 2784 cal. BC

Table 14
Radiocarbon Dates From Sanghol

Site/Lab No.	5568 Half-life bp	5730 Half-life bc	Calibration
PRL-511	3740+160 bp	1900+165 bc	2181, 2166, 2142 cal. BC
PRL-509	3570+150 bp	1725+155 bc	1923 cal. BC
PRL-510	3550+150 bp	1705+155 bc	1895 cal. BC
PRL-513	3540+150 bp	1695+155 bc	1888 cal. BC
PRL-512	3350+110 bp	1500+115 bc	1673 cal. BC

The thermoluminescence dates from Sanghol do not seem to add clarity to dating this phase.

Table 15

Radiocarbon Dates From Banawali

Site/Lab No.	5568 Half-life bp	5730 Half-life bc	Calibration
PRL-205	3810+180 bp	1975+185 bc	2283 cal. BC
PRL-203	3800+150 bp	1965+155 bc	2278, 2234, 2209 cal. BC
PRL-204	3260+120 bp	1410+125 bc	1523 cal. BC
PRL-207	3100+100 bp	1245+105 bc	1406 cal. BC

Table 16
Radiocarbon Dates From Hulas

Site/Lab No.	5568 Half-life bp	5730 Half-life bc	Calibration
PRL-1032	4380+150 bp	2560+155 bc	3028, 2985, 2930 cal. BC
PRL-1031	3840+110 bp	2005+115 bc	2315 cal. BC

There is a fairly wide divergence of dates from these sites. The four dates from Bara are consistent with the working hypotheses for the date of the Post-urban Phase. The same is true for the dates from Sanghol. But the Mitathal and Hulas dates are very early and those from Banawali could be described as "split."

The culture history of the Eastern Domain has sometimes been thought of as lagging behind Sindh and the Punjab (e.g. S. P. Gupta 1972-73; Thapar 1982). There is no evidence from these radiocarbon dates to support this for the emergence of the Post-urban Phase.

Swat Chalcolithic

Swat is a high valley in the southern outliers of the Hindu Kush. An Italian archaeological team working there since the 1960s has organized the Protohistoric sequence there into two phases, both of which can be dated to the second millennium BC and have material closely related to the Harappan Civilization. These dates are as follows.

Table 17
Radiocarbon Dates From Swat

Site/Lab No.	5568 Half-life bp	5730 Half-life bc	Calibration
Aligrama, Swat Period V (Protohistoric)			
P-2150	3090+40 bp	1230+40 bc	1400 cal. BC
PRL-246	3080+170 bp	1220+175 b	1394, 1331, 1329 cal. BC
PRL-186	3070+230 bp	1210+235 bc	1388, 1336, 1325 cal. BC
PRL-243	2900+110 bp	1035+115 bc	1092 cal. BC
PRL-244	2660+130 bp	790+35 bc	818 cal. BC
Loebanr III, Swat Period IV (Chalcolithic)			
P-256	3360+60 bp	1510+60 bc	1677 cal.
BC P-253	3280+90 bp	1430+95 bc	1570, 1528 cal.
BC P-255	3250+60 bp	1400+60 bc	1521 cal. BC
P-254	3140+60 bp	1285+60 bc	1424 cal. BC

**Two additional dates from Aligrama for Swat Period IV are consistent with those from Loebanr III*

Conclusions On The Date Of The Harappan Post-Urban Phase

The working hypothesis that the transition between the Mature Harappan and the Post-urban Phase is 2000-1900 B.C. is still the best for these data. It is clear, however, that there is a wide range of variation in the radiocarbon dates for this transition, and much more chronological precision is called for.The causes for the abandonment of Mohenjo-daro and many other sites in Sindh and the West Punjab are still not understood. It is important, however, to give some consideration to what the archaeological record indicates did happen at 2000-1900 B.C. in the northwestern part of the Subcontinent.

The Historical Nature Of The Harappan Eclipse

Before attempting to explain an historical "event" it is necessary to develop a sound chronology, just as has been attempted here. Then a sense of what really happened is needed, what it is, precisely, that archaeologists are attempting to explain. The following points in Possehl and Raval (1989: 24-6) are some of the culture historical "facts" that pertain to the transition from the Mature Harappan to the Post-urban Phase of the Harappan Civilization:

1. The abandonment of the cities.

The earliest excavators of Mohenjo-daro and Harappa noted that these two great metropolitan centers were largely abandoned at the end of the Mature, Urban Phase of the Harappan period. This is true for a number of other sites in the Harappan area (e.g. Kot Diji, Balakot, Allahdino, Kalibangan, Ropar, Surkotada, Desalpur). Many other sites have been found that link the Urban Phase of the Harappans to later periods [e.g. Chanhu-daro, Rangpur, Lothal, Bhagwanpura, Dadheri as well as a host of Post-urban Phase sites in the Eastern and Lothal Domains (Possehl 1982: Figure 3) of the Harappan area], and there is transitional or Post-urban materials at each of the two cities [e.g. Jhukar at Mohenjo-daro (see Dales and Kenoyer 1986: 57-9) and Cemetery H at Harappa (Vats 1940: 208 ff. and plates XLVII-LXIV). This abandonment of the metropolitan centers and reorientation of settlement structure, with signs of continuity within smaller, less diversified settlements, is taken as an index of change at a basic level of the Harappan sociocultural system.

2. The disruption to the economy.

During the Urban Phase the Harappan economy has abundant evidence for long distance trade and craft production of a wide range of special materials, many of which seem to have been luxury items. Movement of goods and/or craftsmen over long distances can be inferred for the Urban Phase for 500-600 years. Although it is impossible to quantify such materials from the archaeological record, the impression that comes from the knowledge of Mohenjo-daro, Chanhu-daro, Kalibangan, Lothal and other places, is that the Urban Phase was a time of general economic prosperity for the Harappan peoples. During the Post-urban Phase evidence for economic activity is much changed. Signs of long distance trade are virtually absent, except for possible trade in metals. The range of products, and hence specialized production, is much reduced. The mosaic of stylistic traits which emerges in the Post-urban is indicative of shortened ranges for trade and craftsman intercourse. The impression is not necessarily one of universal economic depression in the Post-urban Phase. This economy is, however, not one which measures up to the dynamism, scale and prosperity of the Urban Phase phenomenon.

3. Shifts in the system of settlement and subsistence.

One of the most striking changes associated with the Post-urban Harappan is the reorientation of the Harappan settlement grid. What happened in Gujarat is still not clear, but further archival work, investigating collections on reserve, that is a part of the overall Rojdi project, will change that. The Post-urban Phase may also mark the expansion of Harappan "pioneers" onto the Deccan Plateau at places like Daimabad (Sali 1982, 1986) and into Dhulia District at sites like Kaothe (Dhavalikar, personal communication and 1984). Site counts in Sindh, on the other hand, seem to drop remarkably. Not fewer than four sites in the vicinity of Karachi are abandoned, for example. The principal Jhukar sites in this region are: Jhukar, Chanhu-daro, Amri and the scattered material found at Mohenjo-daro itself. In Bahawalpur and along the Ghaggar, the settlement trends found by Mughal have already been referred to, and other explorations on the Indian side of the border support his findings (Dalal 1980). On the other hand, in the Punjab, Haryana and Uttar Pradesh the Post-urban settlement pattern is rich and varied, with a marked increase in the number of settlements, many of which seem to represent year round, agriculturally based communities. An overview of this is available in (Joshi, Madhu Bala and Ram 1984).

Shifts in the subsistence regime also need investigation. For example, the integration of millets into the Gujarat subsistence economy needs to be clarified, since it no longer appears that it was coincident with Rangpur II B-C times (Possehl 1986). It is also becoming clear that not all of the millets may be significant. Jowar and bajra are the cultivars of choice today for the summer growing season. Their integration into the subsistence regime of monsoon India may well far outweigh the other millets. The term "millet" is generic, simply denoting a group of unrelated plants that bear relatively large, round seeds. There is no biological reason for it to be suggested that millets as a class of plants might have the same impact on a human subsistence regime. Not enough has been done with subsistence economics of the Post-urban Phase in other regions for comparable statements to be made about them, and this represents a challenge for scholars working there. But, it would surprise no one to learn of interesting developments in this area of understanding once systematic, problem oriented work has been completed.

4. The disappearance of Harappan style items of material culture (see Fairservis 1971: 299). There is a long list of Harappan style artifacts which do not carry over into a Post-urban cultural context; for instance, the distinctive Harappan stamp seals. There are also a number of ceramic vessel forms and designs which fall into this category. Copper based metal tools and weapons also change (Herman 1984). The artifactual inventory of the Urban and Post-urban Phases are distinctly different and indicative of significant change in material culture. It is also true that no systematic work has been done on this kind of inventory, region by region. Information of this kind would be useful.

5. The shift in scale of the stylistic zones of material culture

The Post-urban Phase witnesses a return to a cultural mosaic not unlike the one found during the Pre-urban Phase (Possehl 1982). The very widespread, easily recognized Harappan ceramic corpus although diverse, nonetheless had stylistic coherence. The Harappan black on red ware style is a good example. The boundaries of the Post-urban style zones approximate the Pre-urban "mosaic" with remarkable coherence and may offer an insight into boundaries which were in place during Urban Phase times.

6. The continuity of human occupation within a cultural framework that is demonstrably Harappan. In spite of the noted points of change, some of which seem to be of fundamental significance to an understanding of the Harappan Civilization, there is a strong line of continuity throughout the Harappan region. This links the earliest neolithic settlers of the Indus Plains with the beginnings of the Early Historic Period and has been discussed in Possehl (1977). Seen from this perspective the emergence of Post-urban Phase life is but one moment of change in a continuous process which spans millennia. Thus, those who are seeking a full understanding of the "eclipse" of the Indus Civilization must accommodate the continuities in this cultural tradition, as well as the points of change which have been very briefly noted here.

Theories on the Eclipse of the Ancient Cities of the Indus1

Sir John Marshall, the first great synthesizer of Harappan material, has almost nothing to say about the abandonment of Mohenjo-daro. His colleague, Ernest J. H. Mackay believed that Mohenjo-daro might have been abandoned because of an avulsion of the Indus River (1935: 16), which was reiterated by his wife in another context (1945).

The Post-Urban Harappan: The Raikes/Dales Hypothesis

In recent years the Indus River has reemerged as a dynamic factor in the process of Harappan culture change, due largely to the work of Robert Raikes (1964, 1965, 1967a, 1967b, 1968, 1979); Raikes and Dales (1977, 1986); Dales (1965a, 1965b, 1965c, 1966) and Dales and Raikes (1968). They hypothesize that the waters of the Indus River were impounded by a natural dam across the river in the vicinity of Sehwan (Figure 3) and that Mohenjo-daro, and other sites in the vicinity, were so disrupted that it led to, actually "caused," the abandonment of the city and the eclipse of the Mature, Urban Phase. To support this proposition they claim that the "normal behavior of a big river simply cannot account for the evidence of massive flooding at Mohenjo-daro" (Raikes and Dales 1986:33). Others, notably the geographer H. T. Lambrick (1967) and the geomorphologist R. J. Wasson (1984, 1987), dispute this claim, even following the presentation of long called-for data on the sediments (Raikes and Dales 1986; Wasson 1987). While there is some evidence to support the Raikes/Dales hypothesis, most in local geological history, there is no compelling geophysical evidence from the city of Mohenjo-daro, or the Indus Valley itself to support their position. Some suggestions on what kind of geological evidence might be looked for have been offered (Possehl 1967). The second point of disagreement lies in the consideration of historical process, and what archaeologists should take as an explanation. A discussion on the philosophy of science or history is not necessary here, but it is reasonable to suggest that even if the waters of the Indus River had been impounded as Raikes and Dales propose, the historical consequences would have been different from those they suggest. Moreover, if the Harappans succumbed to new and unpredicted riverine forces the explanation sought by historians and social scientists, lies not directly with geomorphological matters, but rather with the internal structure of the Harappan sociocultural system. The "flaw" that would have led to such catastrophic sociocultural change is not to be found within the natural world of geomorphology but within the human context of the Harappan Civilization; its society and culture. (Discussed in more detail in (Possehl 1967: 37-9). This hypothesis has also been critiqued as in Agrawal (1971: 215-20 and 1982a: 188-90).

The Post-Urban Harappan

Wheeler's Historical Explanation Another "theory" concerning the eclipse of the Harappans was ably stated by Sir Mortimer Wheeler (1947: 78-83). He held that the Rig Veda could be read in part as an historical document and that the conflicts described there were between what he thought of as newly arrived Aryan warriors and the indigenous Harappans. This can be objected to on a number of grounds: chronological, cultural, historical. For example, the chronology of the Aryans in South Asia is far from settled (Allchin 1980). Therefore, no one should presume that they were, or were not, active participants. Also, Dales (1964) convincingly argues that the skeletons found in the upper layers of Mohenjo-daro are actually hasty interments, not the remains of victims of a massacre. Finally, the historical content of the Rig Veda has not even been outlined, let alone convincingly defined. While there is a reasonable supposition that there is important historical information in these otherwise liturgical texts, just where within them this is to be found and how significant this history might be, are still unsettled matters.

In the end, it is fair to say that there is no single "theory" to explain the eclipse of the ancient cities of the Indus. This does not mean that there are not interesting data which might have some bearing on the problem.

The Post-Urban Harappan:

New Data From the Old Sarasvati Archaeological research along the now dry beds of the ancient Sarasvati and Drishadvati Rivers (Ghosh 1952, 1953; Mughal 1980, 1982) have led to the discovery of a large number of archaeological sites. M. R. Mughal's findings on the Pakistan side of the border represent one of the most impressive archaeological feats of this century in the Subcontinent. When placed within the context of the distribution of Harappan sites generally they lead one to suspect that Bahawalpur may well have been the Harappan "bread basket." Mughal's data are the most amenable to quantification at the moment and a long term trend in settlement counts can be illustrated by using his materials. The data are given in Table 18.

Table 18
Bahawalpur Sites by Period

ca. 4000-3000 BC	Hakra Wares	32 sites
ca. 3000-2500 BC	Early Harappan	41 sites
ca. 2500-2000 BC	Mature Harappan	166 sites
ca. 2000-1000 BC	'Late' Harappan (Cemetery H)	72 sites
ca. 1000-500 BC	Painted GreyWare	10 sites

(from Mughal 1980: 94)

The chronology for Mughal's succession of periods is largely in line with the most recent synthesis of material (Shaffer 1992; Possehl and Rissman 1992).During all but the period in which the Painted Grey Ware was in use the settlements were located on the desert scarp above the entrenched river. By Painted Grey Ware times the villages are found within the entrenchment, on the bed of the river, certain indication that the system was very dry. Thus, we have important evidence, and a general chronology for the drying up of this riverine system.The fluvial sequence apparently has little, if anything to do with broad trends in climatic change, but is rather to be explained by tectonics as outlined in Agrawal and Sood (1982). Over what seems to have been the course of the second millennium B.C., waters which drained out of the Himalayan watershed were gradually captured by streams that flowed to the east, into the Bay of Bengal, at the expense of the greater Indus system. This process apparently led to the creation of the Yamuna River, a very young stream, and the drying up of the Sarasvati and Drishadvati Rivers (see also Dikshit 1979; Suraj Bhan 1972 and Pande 1977).

We are not proposing that this stream capture was directly causal to the eclipse of the ancient cities of the Indus. We would seek a sociocultural cause here, just as we would for the Raikes/Dales hypothesis. Over the entire course of the third and second millennia, when the Sarasvati seems to have dried-up, there was plenty of water and arable land within the larger Harappan domain to which these people could have turned. We believe that it should be presumed that the ancient inhabitants of Bahawalpur had options; that they did not have to wither there as their river progressively dried up. But, if this was the case, if somehow the ancient Harappans of modern Bahawalpur did not have options to move, or if they failed to perceive a series of alternatives to their situation and "went down with the ship," we may be closer to the human context for an explanation than ever before. But, we must seek the nature of these sociocultural factors. It should not be assumed that an explanation of the eclipse of the Harappan Civilization is in hand just because there is evidence for the drying up of the Sarasvati, and that this is largely coincident in time with the emergence of the Post-urban Phase. Genuine, compelling historical explanation is far more difficult to formulate than that. The data on the Sarasvati/Drishadvati river history are important, however. It may well be that in some way it bears on the problem at hand, even if it does not explain the events in question.

The Indo-French Archaeological Project has also found many sites along the ancient Sarasvati, but on the Indian side of the border. Their findings apparently conform to Mughal's data, including a significant number of Kushan Period sites with the well known, easily recognized Red Polished Ware (Francfort 1986). They say of the stream capture:

This shift would have occurred either in the still badly defined Late Harappan period or in the PGW period, or gradually during both since Late Harappan (Cemetery H) and PGW sites in the state of Bahawalpur in Pakistan are found on one of the paleochannels known locally as Hakra. However, the adoption of this `hypothesis' poses an interesting methodological problem. If we accept this hypothesis, we must logically accept as well that another tectonic upheaval would have made the rivers revert to the Kushan (Rang Mahal) period since sites from the latter are again found in the region (Francfort 1986:98, original emphasis).

This is an important statement which Francfort intends to

use to sharpen the problem formulation of his team and their colleagues in India. They are seeking answers to relevant questions which are stated precisely, and some of this work is already being pursued (Gentelle 1986; Courty 1986). We are not certain however, that we agree with the literal meaning of the point made above. Is it really true that we `must' accept the same explanation for both data sets; the protohistoric and the early historic? Wars and migrations come about for many reasons, the same is surely true for the regeneration of a settlement grid. Technologies unavailable, or unused, in prehistoric times might have been part of the Kushan agricultural regime. For example, well irrigation and the use of bunds for impounding rain water and soil seem to emerge as important new technological features in western India at about the time of the Kushans (Possehl 1975). This may have allowed renewed settlement in northern Rajasthan, and have been independent and unrelated to tectonics.

An Avulsion of the Indus River

The Indus River in Sindh is a mature stream which frequently changes it course. There has been much river training done on the Indus in modern times, so these course shifts are not as apparent today as they were up to the nineteenth century; but they are a powerful part of the history of Sindh. A number of archaeologists have suggested that a dramatic shift in the course of the Indus led to the abandonment of Mohenjo-daro (Mackay 1935: 16; Lambrick 1967; Mughal 1990).

An avulsion of the Indus River might explain the abandonment of Mohenjo-daro, but there was a good deal more happening at the turn of the third millennium than just the abandonment of Mohenjo-daro. This kind of theory generally reverts to some form of "domino effect" to account for the "eclipse" of the rest of the Harappan Civilization: "as went Mohenjo-daro; so went the Harappan Civilization." This places a great deal of importance on the place of Mohenjo-daro within the fabric of this civilization; something that may not be justified. Moreover, it seems that Mohenjo-daro was undergoing a slow process of change and decline, if we take the data on the abandonment of the Great Bath and Granary into account and this is not a good match with the avulsion hypothesis.

Concluding Statement

We still do not understand the reasons behind the eclipse of the Harappan Civilization. Eclipse may not be the best concept that we should be working with. There is little sound data on which we can reconstruct the environment of the northwestern regions of the Indian Subcontinent. There are some internal inconsistencies in the data sets that seem to apply to this reconstruction. Moreover, tectonic movement of the surface crust in this region, some on a relatively small scale, may well have been a powerful force in shaping the drainage patterns of rivers, streams and lakes there during the Holocene. This observation calls into question the climatic reconstruction derived from the palynological research done on the lakes of Rajasthan. So too do the observations that come from archaeological exploration.

Archaeological exploration in the Punjab, Haryana, northern Rajasthan and western Uttar Pradesh has produced good evidence for human habitation in these areas during the opening centuries of the second millennium B.C. These are dry farming areas dependent on rainfall. The exploration records indicate a rather dramatic increase of habitation at this time; yet this is the period of the so-called "eclipse" of the Harappan Civilization and a period proposed to have been "arid" in this region. Lake Lunkaransar which played a role in the "desiccation hypothesis" is ca. 90 kilometers south of the ancient Sarasvati River, and therefore within, or at least on the borders, of this settlement area. The kinds of climatic changes proposed to have taken place around this body of water can be assumed to have pertained to the Punjab, Haryana, the rest of northern Rajasthan and western Uttar Pradesh. There is an historical awkwardness here on two accounts: first there is a period of "eclipse" with a growth of human habitation; second a proposed "aridity" at a time when archaeological data indicate wide spread dry cropping.

There are other interesting things happening at the beginning of the second millennium, to the south, in Gujarat. The story there is somewhat different from the one unfolding in the Punjab, Haryana, northern Rajasthan and western Uttar Pradesh, but the conclusions that are emerging are similar to those reached in this paper. There is little evidence for climatic change, let alone climatic change that led to culture change. It is also clear that the history of the Harappan Civilization in Sindh and the West Punjab at the beginning of the second millennium, is different from that which took place in Gujarat, especially on the peninsula of Saurashtra. There are few signs of "collapse" or "eclipse" there and at some places, like Rojdi, they are busy with the expansion and rebuilding of their settlements, just as Mohenjo-daro is being abandoned.

What is being learned about the "end of the ancient cities of the Indus" is that it was a good deal more complicated than was thought as recently as only 10 years ago. There is much heterogeneity in the culture history of the domains of the Harappan Civilization. What happened in Sindh and the West Punjab is not necessarily replicated elsewhere. In fact, the very thought of an "eclipse" or "end" as the concept around which our thoughts should focus can be called into question

NOTE:

1. Some of this material has been taken from Possehl and Raval 1989.

Bibliography

Abdul Ali, A. F. M. (1935) Earthquakes in India. Indian Culture, 1(3): 467-76.

Agrawal, D. P. (1964) Harappa culture: new evidence for a shorter chronology. Science, 3609: 950-52.

Agrawal, D. P. (1971) The Copper Bronze Age in India. Delhi: Munshiram Manoharlal: 270 pp.

Agrawal, D. P. (1982) The Archaeology of India. Copenhagen: Scandinavian Institute of Asian Studies Monograph Series No. 46: 294 pp.

Agrawal, D. P. (1987) Environmental changes in India during the last 4 million years. Journal of the Palaeontological Society of India, 32: 1-4.

Agrawal, D. P. and R. K. Sood (1982) Ecological factors and the Harappan Civilization. In, Gregory L. Possehl, ed., Harappan Civilization: A contemporary perspective. Delhi: Oxford & IBH and the American Institute of Indian Studies: 223-31.

Allchin, Bridget and Andrew Goudie (1978) Climatic change in the Indian desert and North-west India during the Late Pleistocene and Early Holocene. In, William C. Brice, ed., The Environmental History of the Near and Middle East Since the Last Ice Age.New York: Academic Press: 307-18.

Allchin, Bridget and Raymond (1968)The Birth of Indian Civilization. Baltimore: Penguin Books.

Allchin, Bridget, Andrew Goudie and Karunarkara Hegde (1978) The Prehistory and Palaeogeography of the Great Indian Desert. New York: Academic Press.

Allchin, F. R. (1980) Archaeological and language historical evidence for the movement of Indo-Aryan speaking peoples into India and Pakistan. Journal of the K. R. Cama Oriental Institute, 48: 68-102.

Allchin, F. R. (1982) The legacy of the Indus Civilization. In, Gregory L. Possehl, ed., Harappan Civilization: A contemporary perspective. Delhi: Oxford & IBH and the American Institute of Indian Studies: 325-33.

Allchin, F. R. (1990a) Indo-Aryan and Aryan: language, culture and ethnicity. Ancient Ceylon, 10: 13-23.

Allchin, F. R. (1990b) The end of the Harappan Urban Phase and its aftermath. Ancient Ceylon, 10: 25-40.

Allchin, F. R. and J. P. Joshi (1970) Malvan--further light on the southern extension of the Indus Civilization. Journal of the Royal Asiatic Society of Great Britain and Ireland: 20-8.

Ardeleanu-Jansen, Alexandra, Ute Franke and Michael Jansen (1983) An approach toward the replacement of artifacts into the architectural context of the Great Bath at Mohenjo-daro. In, G. Urban and M. Jansen, eds., Forschungsprojekt DFG Mohenjodaro. Aachen: Reinische-Westfalischen Technischen Hockschule: 43-6.

Auden, J. B. (1952) Some geological and chemical aspects of the Rajasthan salt problem. Bulletin of the National Institute of Sciences in India, 1:

Bhattacharya, B. N., S. K. Gupta, B. P. Rishi, A. K. Mitra and V. K. Kashyap (1982) Seasonal variation in the composition of Sambhar Lake brine and possible origin of its salinity. In, Proceedings of the Workshop on the Problems of the Deserts of India.Delhi: Geological Survey of India, Special Publication No. 49: 80-91.

Biswas, R. K., G. S. Chattopadhyay and Subrata Sinha (1982) Some observations on the salinity problems of the inland lakes of Rajasthan. In, Proceedings of the Workshop on the Problems of the Deserts of India. Delhi: Geological Survey of India, SpecialPublication No. 49: 68-79.

Brunswig, Robert H.(1973) Prospective tree-ring calibration of the Indus Civilization radiocarbon chronology. Man, 8: 543-54.

Brunswig, Robert H. (1975) Radiocarbon dating and the Indus Civilization: calibration and chronology. East and West, 25(1-2): 111-45.

Chakrabarti, Dilip K. (1976) Seals as an evidence of Indus-West Asia interrelations. In, Debiprasad Chattopadhyaya, ed., History and Society: Essays in Honor of Professor Niharranjan Ray. Calcutta: K. P. Bagchi: 93-116. (First published 1978).

Chakrabarti, Dilip K. (1982) `Long barrel-cylinder' beads and the issue of Pre-Sargonic contect between the Harappan Civilization and Mesopotamia. In, Gregory L. Possehl, ed., Harappan Civilization: A contemporary perspective. Delhi: Oxford & IBH and theAmerican Institute of Indian Studies: 265-70.

Costantini, Lorenzo and Loredana Costantini Biasini (1985) Agriculture in Baluchistan between the 7th and 3rd Millennium BC Newsletter of Baluchistan Studies, 2: 16-30.

Courty, M. A.(1986) Geoarchaeological approach of Holocene paleoenvironments in the Ghaggar Plains. Man and Environment, 10: 111-15.

Dalal, Katy Feroze (1980) A short history of archaeological exploration in Bikanir and Bahawalpur along the `Lost' Sarasvati River. Indica, 17(1): 3-40.

Dales, George F. (1964) The mythical massacre at Mohenjo-daro. Expedition, 6(3): 36-43.

Dales, George F. (1965a) Civilization and floods in the Indus Valley. Expedition, 7(2): 2-10.

Dales, George F. (1965b) New investigations at Mohenjo-daro. Archaeology, 18: 145-50.

Dales, George F. (1965c) Re-opening the Mohenjo-daro excavations. Illustrated London News, May 29: 25-7.

Dales, George F. (1965d) A suggested chronology for Afghanistan, Baluchistan and the Indus Valley. In, Robert Ehrich, ed., Chronologies in Old World Archaeology. Chicago: University of Chicago Press: 257-84.

Dales, George F. (1966) The decline of the Harappans. Scientific American, 214(5): 93-100.

Dales, George F. (1968) Of dice and men. In, William W. Hallow, ed., Essays in Memory of E. A. Speiser. Journal of the American Oriental Society, 88(1): 14-23.

Dales, George F. (1973) Archaeological and radiocarbon chronologies for protohistoric South Asia. In, Norman Hammond, ed., South Asian Archaeology. Park Ridge: Noyes Press: 157-69.

Dales, George F. and J. Mark Kenoyer (1986) Excavations at Mohenjo Daro, Pakistan: The pottery. Philadelphia: The University Museum, University of Pennsylvania.

Dales, George F. and Robert L. Raikes (1968) The Mohenjo-daro floods--a rejoinder. American Anthropologist, 70(5): 957-61.

Dhavalikar, M. K., V. Shinde and S. Atre (1990) Excavations at Kaothe. Poone: Deccan College Post-Graduate and Research Institute.

During Caspers, Elisabeth C. L.(1972) Etched Carnelian Beads. Bulletin of the Institute of Archaeology, London University, No. 10.

Durrani, Farzand A. (1965) Climate of the lower Indus in ancient times. Journal of the University of Peshawar, 10: 33-7.

Dikshit, K. N. (1979) Old channels of Ghaggar in Rajasthan - revisited. Man and Environment, 3: 105-6.

Erdosy, George (1985) The economic organization of early historic states in the Ganges Valley. In, J. Schotsmans and M. Taddei, eds., South Asian Archaeology 1983. Naples: Instituto Universitario Orientale, Dipartimento di Studi Asiatici, Series Minor 23: 509-22.

Erskine, K. D. (1909) The Western Rajputana States Residency and Bikaner Agency. Rajputana Gazetteers, Vol. 3A. Allahabad: The Pioneer Press: 399 pp.

Fairservis, Walter A., Jr. (1961) The Harappan Civilization: new evidence and more theory. New York: American Museum of Natural History, Novitates, No. 2055: 35 pp.

Fairservis, Walter A., Jr. (1971) The Roots of Ancient India. New York: Macmillan: 482 pp.

Francfort, Henri-Paul (1986) Preliminary report (1983-1984): archaeological and environmental researches in the Ghaggar (Saraswati) Plains. Man and Environment, 10: 97-100.

Gadd, C. J. and Sidney Smith (1924) The new links between Indian and Babylonian Civilization. Illustrated London News, October 4: 614-16.

Gentelle, Pierre (1986) Landscapes, environment and irrigation: hypotheses for the study of the 3rd and 2nd millenniums. Man and Environment, 10: 101-10.

Ghosh, A. (1952) The Rajputana desert: its archaeological aspect. Bulletin of the National Institute of Sciences in India, 1: 37-42.

Ghosh, A. (1953) Exploration in Bikanir. East and West, 4(1): 31-4.

Ghosh, R. N. (1975) Report on the basin and aeolian studies of the upper reaches of Luni River in parts of Ajmer, Nagaur and Pali Districts, Rajasthan. Geological Survey of India, Unpublished manuscript.

Godbole, N. N. (1952) The salinity of Sambhar Lake. Bulletin of the National Institute of Sciences in India, 1:

Godbole, N. N. (1972) Theories on the origin of salt lakes of Rajasthan, India. Proceedings of the 24th Session of the International Geological Congress. Canada.

Gupta, S. K. (1977) Quaternary sea-level changes on the Saurashtra coast. In, D. P. Agrawal and B. M. Pande, eds., Ecology and Archaeology of Western India. Delhi: Concept Publishing: 181-94.

Gupta, S. P. (1972-73) A model for understanding the first urbanization in India. Puratattva, 6: 42-50.

Herman, Charles Frank (1984) De Harappa-Cultuur: De koperen en bronzen wapens en werktuigen. 3 Vols. Master of Arts Paper, Departement archaeologie en Kunstwetenschap, Katholieke universiteit Leuven, Belgium: 121 + 369 pp.

Harvey, Michael D. and Louis Flam (1993) Prehistoric soil and water detention structures (gabarbands) at Phang, Sindh Kohistan, Pakistan: an adaptation to environemntal change. Geoarchaeology, 8(2): 109-26.

Holland, Sir Thomas H. and W. A. K. Christie (1909) The origin of the salt deposits in Rajputana. Records of the Geological Survey of India, 5(38): 154-86.

Johnston, Arch C. and Lisa R. Kanter (1990) Earthquakes in stable continental shelf. Scientific American, 262(3): 68-75.

Joshi, Jagat Pati and F. R. Allchin (1972) Malvan. In, S. B. Deo, ed., Archaeological Congress and Seminar Papers. Nagpur: Nagpur University: 36-42.

Joshi, Jagat Pati, Madhu Bala and Jassu Ram (1984) The Indus Civilization: a reconsideration on the basis of distribution maps. In, B. B. Lal and S. P. Gupta, eds., Frontiers of the Indus Civilization. Delhi: Books and Books: 511-30.

Khandelwal, N. M. (1975) On the occurrence of halite in Didwana salt lake area. Current Science, 44: 13-4.

Krishnan, A. and K. P. Thanvi (1982) Study on the variability of rainfall and evaluation of rainfall types in western Rajasthan. In, Proceedings of the Workshop on the Problems of the Deserts of India. Delhi: Geological Survey of India, Special Publication No. 146 49: 156-60.

Lamb, H. H., R. P. W. Lewis and A. Woodruff (1966) World climate from 8000 to 0 B.C. Proceedings of the International Meterological Symposium. London: Royal Meterological Service.

Lamberg-Karlovsky, Carl C. (1972) Trade mechanisms in Indus-Mesopotamian interrelations. Journal of the American Oriental Society, 92(1): 222-29.

Lambrick, H. T. (1967) The Indus flood-plain and the `Indus Civilization.' The Geographical Journal, 133: 483-94.

Mackay, Dorothy (1945) Ancient river beds and dead cities. Antiquity, 19: 135-44.

Mackay, Ernest J. H. (1935) The Indus Civilization. London: Luzac & Co.

Mackay, Ernest J. H. (1937-38) Further Excavations at Mohenjo-daro. 2 Vols. Delhi: Government of India.

Marshall, Sir John, editor (1931a) Mohenjo-daro and the Indus Civilization. 3 Vols. London: Arthur Probsthain.

Marshall, Sir John (1931b) The country, climate and rivers. In, Sir John Marshall, ed., Mohenjo-daro and the Indus Civilization, 3 Vols. London: Arthur Probsthain: 1-7.

Mathur, U. B., K. K. Verma and S. Mehra (1975) Quaternary glacio-isostasy in relation to the origin of Rajasthan salt lakes and desert. Proceedings of the Seminar on Problems of Deserts in India. Jaipur: Central Arid Zone Research Institute.

Mathur, U. B., K. K. Verma and S. Mehra (1982) Quaternary glacio-eustacy in relation to the origin of Rajasthan salt lakes and desert--an evaluation. In, Proceedings of the Workshop on the Problems of the Deserts of India. Delhi: Geological Survey of India, Special Publication No. 49: 59-67.

McKean, Margaret Bernard (1983) The Palynology of Balakot, a Preharappan and Harappan Age Site in Las Bela, Pakistan. PhD Dissertation, Department of Anthropology, Southern Methodist University.

Meher-Homji, V. M. (1980) The Thar desert: its climatic history. Man and Environment, 4: 1-7.

Mishra, Shiva Poojan (1982) Geochemical evolution of Sambhar Lake, Jaipur and Nagaur District, Rajasthan. In, Proceedings of the Workshop on the Problems of the Deserts of India. Delhi: Geological Survey of India, Special Publication No. 49: 92-8.

Misra, V. N. (1984) Climate, a factor in the rise and fall of the Indus Civilization--evidence from Rajasthan and beyond. In, B. B. Lal and S. P. Gupta, eds., Frontiers of the Indus Civilization. Delhi: Books and Books: 461-89.

Mughal, M. R. (1980) New archaeological evidence from Bahawalpur. Man and Environment, 4: 93-8.

Mughal, M. R. (1982) Recent archaeological research in the Cholistan desert. In, Gregory L. Possehl, ed., Harappan Civilization: A contemporary perspective. Delhi: Oxford & IBH and theAmerican Institute of Indian Studies: 85-95.

Mughal, M. R. (199) The decline of the Indus Civilization and the Late Harappan Period in the Indus Valley. Lahore Museum Bulletin, 3(2): 1-22.

Murty, Y. G. K, G. R. Kulkarni and S. K. Gupta (1969) Interim report on the rate of silting in the Rann of Kutch and Quaternary movements of Saurashtra coast line. Geological Survey of India, Unpublished manuscript.

Pande, B. M. (1977) Archaeological remains along the ancient Sarasvati. In, D. P. Agrawal and B. M. Pande, eds., Ecology and Archaeology of Western India. Delhi: Concept Publishing: 55-60.

Pascoe, E. H. (1964) A Manual of Geology of India and Burma. Calcutta: Government Press.

Piggott, Stuart (1950) Prehistoric India to 1000 B.C. Baltimore: Penguin Books.

Possehl, Gregory L. (1967) The Mohenjo daro floods: a reply. American Anthropologist, 69(1): 32-4.

Possehl, Gregory L. (1975) The chronology of gabarbands and palas in western South Asia. Expedition, 17(2): 33-37.

Possehl, Gregory L. (1977) The end of a state and the continuity of a tradition. In,

Richard Fox, ed., Realm and Region in Traditional India. Durham: Carolina Academic Press: 234-54.

Possehl, Gregory L. (1982) The Harappan Civilization: a contemporary perspective. In, Gregory L. Possehl, ed., Harappan Civilization: A contemporary perspective. Delhi: Oxford & IBH and the American Institute of Indian Studies: 15-28.

Possehl, Gregory L. (1986) African millets in South Asian prehistory. In, Jerome Jacobson, ed., Studies in the Archaeology of India and Pakistan. Delhi: Oxford & IBH and the American Institute of Indian Studies: 237-56.

Possehl, Gregory L. (1989) Radiocarbon Dates for South Asian Archaeology. Philadelphia: The University Museum, An Occasional Publication of the Asian Section.

Possehl, Gregory L. (1996a) Meluhha. In, Julian E. Reade, ed., The Indian Ocean in Antiquity. London: Kegan Paul International in Cooperation with the British Museum: 133-208.

Possehl, Gregory L. (1996b) Indus Age: The writing system. Philadelphia: University

of Pennsylvania Press.

Possehl, Gregory L. and M. H. Raval

1989 Harappan Civilization and Rojdi. Delhi: Oxford & IBH.

Possehl, Gregory L. and Charles Frank Herman (1990) The Sorath Harappan: a new regional manifestation of the Indus Urban Phase. In, Maurizio Taddei, ed., South Asian Archaeology 1987. Roma: Instituto Italiano per il Medio ed Estremo Oriente, Serie Orientale Roma, 66(1): 295-320.

Raikes, Robert L. (1964) The end of the ancient cities of the Indus. American Anthropologist, 66(2): 284-99.

Raikes, Robert L. (1965) The Mohenjo-daro floods. Antiquity, 39: 196-203.

Raikes, Robert L. (1967a) Water, Weather and Prehistory. New York: Humanities Press.

Raikes, Robert L. (1967bThe Mohenjo-daro floods--reposte. Antiquity, 41: 309-10.

Raikes, Robert L. (1967c) The Mohenjo-daro floods--further notes. Antiquity, 41: 64-6.

Raikes, Robert L. (1968 Kalibangan: death from natural causes. Antiquity, 42: 286-91.

Raikes, Robert L. (1979)The Mohenjo-daro floods: the debate continues. In, M. Taddei, ed., South Asian Archaeology 1977. Naples: Instituto Universitario Orientale, Seminario di Studi Asiatici, Series Minor 6: 561-66.

Raikes, Robert L. and George F. Dales (1977) The Mohenjo-daro floods reconsidered. Journal of the Palaeontological Society of India, 20: 251-60.

Raikes, Robert L. and George F. Dales (1986) Reposte to Wasson's sedimentological basis of the Mohenjo-Daro flood hypothesis. Man and Environment, 10: 33-44.

Raikes, Robert L. and Robert H. Dyson, Jr. (1961) The prehistoric climate of Baluchistan and the Indus Valley. American Anthropologist, 63: 265-81.

Ralph, Elizabeth K., Henry N. Michael and Mark Han (1973) Radiocarbon dates and reality. MASCA Newsletter, 9(1): 1-20.

Ramakrishna, V. S. (1988) Climatic changes in relation to desertification in Indian arid zone. In, A. K. Tiwari, ed., Desertification: Moritoring and Control. Jodhpur: Scientific Publishers: 99-115.

Ramaswamy, C. (1968) Monsoon over the Indus Valley during the Harappan period. Nature, 217(5129): 628-29.

Ratnagar, Shereen (1981)Encounters: The westerly trade of the Harappa Civilization. Delhi: Oxford University Press.

Reade, Julian E. (1979) Early Etched Beads and the Indus-Mesopotamia Trade. London: British Museum, Department of Western Asiatic Antiquities, No. 1.

Rendell, Helen (1981) A preliminary investigation of the sedimentary history of the Bannu Basin in the Late Holocene. In, H. Hartel, ed., South Asian Archaeology 1979. Berlin: Dietrich Reimer Verlag: 219-25.

Sali, S. A. (1982) The Harappans of Daimabad. In, Gregory L. Possehl, ed., Harappan Civilization: A Contemporary Persepctive. Delhi: Oxford & IBH and the American Institute of Indian Studies: 175-84.

Sali, S. A. (1986) Daimabad, 1976-79. Memoirs of the Archaeological Survey of India, No. 83.

Sayce, A. H. (1924) Remarkable discoveries in India. Illustrated London News, September 27: 566.

Shaffer, Jim G. (1984)The Indo-Aryan invasions: cultural myth and archaeological reality. In, John R. Lukacs, eds.,The People of South Asia: The biological anthropology of India, Pakistan and Nepal. New York: Plenum Press: 77-90.

Shaffer, Jim G. (1992) The Indus Valley, Baluchistan and Helmand traditions: Neolithic through Bronze Age. In, Robert W. Ehrich, ed., Chronologies in Old World Archaeology. 3rd ed. 2 Vols. Chicago: University of Chicago Press: 441-64 & 425-46.

Singh, Gurdip (1971) The Indus Valley Culture seen in the context of Post-glacial climatic and ecological studies in northwest India. Archaeology and Physical Anthropology in Oceania, 6(2): 177-89.

Singh, Gurdip, J. R. Wasson and D. P. Agrawal (1990) Vegetational and seasonal climatic changes since the last full glacial in the Thar Desert, northwestern India. Review of Palaeobotany and Palynology, 64: 351-58.

Singh, Gurdip, R. D. Joshi and A. B. Singh (1972) Stratigraphic and radiocarbon evidence for the age and development of three salt lake deposits in Rajasthan, India. Quaternary Research, 2(4): 496-505.

Singh, Gurdip, R. D. Joshi, S. K. Chopra and A. B. Sing (1974) Late Quaternary history of vegetation and climate of the Rajasthan Desert, India. Philosophical Transactions of the Royal Society of London, Biological Sciences, 267(889): 467-501.

Sinha, B. P., G. S. Singh, A. R. Bakhohi and G. C. Bhatnagar (1973) A note on the occurrence of halite in the Bikaner area, Rajasthan. Current Science, 42: 874.

Sinha, Subrata (1977) Origin of salinity in Rajasthan salt lakes. In, D. P. Agrawal and B. M. Pande, eds., Ecology and Archaeology of Western India. Delhi: Concept Publishing: 147-56.

Suraj Bhan (1972) Changes in the course of the Yamuna and their bearing on the proto-historic cultures of Haryana. In, S. B. Deo, ed., Archaeological Congress and Seminar Papers. Nagpur: Nagpur University: 125-28.

Thapar, B. K. (1982) The Harappan Civilization: some reflections on its environments and resources and their exploitation. In, Gregory L. Possehl, ed., Harappan Civilization: A contemporary perspective. Delhi: Oxford & IBH and the American Institute of Indian Studies: 3-13.

Thiebault, Stephanie (1989) A note on the ancient vegetation of Baluchistan based on charcoal analysis of the latest periods from Mehrgarh, Pakistan. In, Karen Frifelt and Per Sorensen, eds., South Asian Archaeology1985. Scandinavian Institute of Asian Studies, Occasional Papers No. 4. London: Curzon Press: 186-88.

Thiebault, Stephanie (1992) Complementary results in anthracological analysis from sites in Baluchistan. In, Catherine Jarrige, ed., South Asian Archaeology1989. Madison: Prehistory Press: 271-76.

Thomas, Kenneth D. (1986) Environment and subsistence in the Bannu Basin. In, F. R. Allchin, B. Allchin F. A. Durrani and M. Farid Khan, eds., Lewan and the Bannu Basin. Oxford: British Archaeological Reports International Series, 310: 13-33.

La Touche, T. D. (1902) Geology of western Rajputana. Memoirs of the Geological Survey of India, 5(35), Pt. 1.

Vats, M. S. (1940) Excavations at Harappa. 2 Vols. Delhi: Government of India: 488 pp.

Vishnu-Mittre (1974) Climate vs. biotic factor: pollen evidence in the postglacial history of northwest India. In, Ashok K. Ghosh, ed., Perspectives in Palaeoanthropology: Professor D. Sen Festschrift. Calcutta: Firma K. L. Mukhopadhyay: 25-31.

Vishnu-Mittre and Chhaya Sharma (1973) Pollen analysis of the salt flat at Malvan, Gujarat. The Palaeobotanist, 22(2): 118-23.

Wasson, R. J. (1984) The sedimentological basis of the Mohenjo-daro flood hypothesis. Man and Environment, 8: 88-90.

Wasson, R. J. (1987) The sedimentological basis of the Mohenjo-daro flood hypothesis–a further comment. Man and Environment, 11: 122-3.

Wasson, R. J., G. I. Smith and D. P. Agrawal (1984) Late Quaternary sediments, minerals and inferred geochemical history of Didwana Lake, Thar Desert, India. Palaeogrography, Palaeoclimatology, Palaeoecology, 46: 345-72.

Weber, Steven A. (1991) Plants and Harappan Subsistence: An example of stability and change from Rojdi. Delhi: Oxford & IBH and the American Institute of Indian Studies.

Wheeler, R. E. M. (1947) Harappa 1946: the defenses and cemetery R-37. Ancient India, 3: 58-130.

Wheeler, Sir Mortimer (1953) The Indus Civilization. 1st edition. Supplementary Volume to the Cambridge Ancient History of India. Cambridge: At the University Press.

Wheeler, Sir Mortimer (1968) The Indus Civilization. 3rd edition. Supplementary Volume to the Cambridge Ancient History of India. Cambridge: At the University Press.

Sociopolitical Discontinuity in the Near East C. 2200 B.C.E.: Scenarios from Palestine and Egypt

Karl W. BUTZER

The University of Texas at Austin

Abstract

Between about 2300 and 2000 B.C.E. there was widespread sociopolitical devolution in the Near East. Tell Leilan, the particular case in point, was abandoned after deposition of volcanic tephra. Analogs from the Mt. St. Helen's eruption of 1980 suggest that a volcanic dust mantle would have serious biological repercussions, but that recovery would take only years, at most decades. There is no case for a global event, let alone for a "volcanic winter." The local evidence for an incisive and protracted "dry" anomaly is inconclusive, and examination of a wide range of paleoclimatic proxy records fails to support an abrupt climatic shift to greater aridity affecting the Near East and Aegean world between 2400 and 1900 B.C.E. Settlement records to the east, west, and northwest of Leilan are inconsistent; in two cases the settlement network expanded rapidly after the Leilan collapse, in another, sites were abandoned at about the same time. This is incompatible with a major climatic shift and demands exploration of more complex, alternative hypotheses to explain sociopolitical devolution.

The main thrust of this paper is an in-depth examination of devolution in Old Kingdom Egypt and Early Bronze Palestine, regions closely linked by a contentious complementarity. New evidence from Egypt and East Africa shows that the Nile floods were relatively low after 2900 B.C.E., with a brief minimum around 2200 B.C.E.; thereafter floods were exceptionally high 2150-1900 B.C.E. Examination of the historical record suggests that Old Kingdom collapse was unrelated to Nile failure, but preconditioned by decentralization, dynastic weakness, and a shift of wealth and power to several provincial centers during Dynasty 6 (c. 2380-2230 B.C.E.); royal power, anchored in part on the trade monopoly with Syria (e.g., timber, wine, olive oil imports), was devastatingly undercut c. 2300 B.C.E. by Akkadian conquest of the Egyptian entrepot Byblos, with anarchy and then civil wars prevailing in Egypt for almost two centuries after c. 2240 B.C.E.

NATO ASI Series, Vol. I 49
Third Millennium BC Climate Change
and Old World Collapse
Edited by H. Nüzhet Dalfes, George Kukla,
and Harvey Weiss

In Palestine, Early Bronze developments saw a string of prosperous, fortified towns emerge along the hilly margins of the coastal plain. Their wealth was based on wine and oil production, much of it exported to Egypt. Parts of the coastal plain began to be abandoned c. 2700 B.C.E., compensated by clusters of larger towns elsewhere; general depopulation was underway two centuries later, and the last towns fell about 2400 B.C.E., some of them patently destroyed. Two Egyptian military campaigns against this part of Palestine are actually recorded c. 2370-2340 B.C.E., the second specifying large-scale destruction. About 2300 B.C.E. parts of Syria were also ravaged, perhaps by Sargon of Akkad. The ensuing Early Bronze IV of Palestine (c. 2300-2000 B.C.E.) was marked by de-urbanization and agricultural dis-intensification, but contrary to older assumptions, there is little or no evidence for a shift to pastoralism, let alone for settlement by invading pastoralists. Analogs from elsewhere in the Mediterranean region show how transhumant pastoralism was traditionally practiced from farming villages, as a complementary strategy; when villagers were forced to seek refuge in defensible hill sites, they continued to cultivate vineyards and olive groves on a smaller scale, despite a lack of urban demand. The Early Bronze IV bioarchaeological evidence now accumulating shows that household pig-raising and plow-oxen remained as prominent as before, that sheep/goat-herding did not increase, and that grapes and olives continued to be cultivated -- all demonstrating agronomic continuity in a ruralized setting, until the Middle Bronze urban revival.

In conclusion, it is argued that the Early Bronze Age represented a politico-economic network ("world-economy") that functioned as a set of subsystems, the prosperity of the whole dependent on unimpeded flows of energy and information from Spain to the Indus Valley. Rampant militarism eventually weakened the whole, impoverishing its component parts, and allowing political simplification. As the system was restructured, renewed urbanization and intensification are evident. Climatic change can indeed be examined as on potential variable in such a multi-dimensional, systemic network. But climatic change cannot be treated in isolation, divorced from its historical, structural, and regional context.

1.1 The Problem of Sociopolitical Discontinuity

The rise and fall of "high civilizations" have long intrigued the human imagination. Already the ancient Egyptians recognized the continuity of their 2500-year history, while at the same time grasping its cyclic evolution, when the priest Manetho (c. 300 B.C.E.) divided Egyptian history into 30 dynasties, arranged into three "books"--representing the crests of political power that we now know as the Old, Middle and New Kingdoms (see Waddell 1956). When Edward Gibbon published the first volume of *The History of the Decline and Fall of the Roman Empire* in 1776, he touched a nerve among Western thinkers who began to confront the instability of their own civilization (Bowersock 1996). Borrowing from the social darwinism of Spencer (1876), Oswald Spengler (1926) viewed societies as organisms with an inevitable life cycle of youth, maturity and decay. In a less deterministic manner, Toynbee (1934) emphasized the role of creative individuals in the growth of civilizations, and attributed breakdown and disintegration to the failure of creative power among the dominant elite, to alienation of the masses, and to pressures of foreign peoples. Similar themes are echoed by Cowgill (1988) in his analysis of collapse.

Clarke (1968) reformulated the model of birth, growth and death as "culture system ontogeny," with formative, coherent, and postcoherent stages, recognizing that a society is an interdependent system of specialized parts. But the analogy has severe limitations, since sociopolitical systems are far less stable than cultural ones and, unlike the inevitable senescence implicit to ontogenetic models, systems are simplified through equilibrium shifts, related to chance concatenations of inputs and complex feedback loops (Butzer 1980). Without belaboring systemic analogs, it bears noting that human ecosystems (Butzer 1982) are "open", liable to both external inputs and internal (evolutionary or disjunctive) change.

1.2 The Tell Leilan Collapse

One of the more intriguing questions of macro-historical study is the apparently synchronous nature of political collapse across large areas. The earliest, recognized phenomenon of this type dates between perhaps 2300 and 2000 B.C.E. The Aegean world and much of the Near East were engulfed in a "time of troubles" that terminated the Early Bronze Age. Many if not most urban sites from the Balkans to Mesopotamia and Palestine were abandoned, destroyed, or much reduced in size. Early states such as the Akkadian Empire and Old Kingdom Egypt collapsed c. 2230 B.C.E., Troy II was destroyed, and the Indus Valley civilization came to an end. Mysterious invaders, such as Indo-Europeans or Semitic pastoralists, have long been conjured up (in two waves, where necessary) to explain these widespread discontinuities, with some sort of climatic deterioration suspected or invoked to explain their apparently synchronous nature (Bell 1971).

The problem is that the late third millennium is barely within the pale of historical records in Egypt and Akkad, while elsewhere we are entirely dependent on archaeology. Yet cause-and-effect relationships are complex, and multicausal explanations are difficult enough in well-documented, contemporary contexts. For historical examples, such arguments become increasingly inferential, since external triggering mechanisms may be more conspicuous than a complex web of underlying socioeconomic problems, typically accentuated by political factors or events. Direct, contemporaneous testimony is unavailable in prehistoric settings, and one can rarely do more than correlate a potential explanatory factor with population shifts or political disjunctions; possible concatenations of variables can seldom be deciphered, so that "conclusions" are little more than models, generally not amenable to satisfactory testing.

Recently, Harvey Weiss and Marie-Agnes Courty (Weiss et al., 1993; Courty and Weiss, this volume) have proposed that an abrupt increase in aridity c.2200 B.C.E. caused not only abandonment of the North Mesopotamian Plain, but also the fall of the Akkadian Empire and synchronous political collapse in adjacent regions. The argument can be summarized as follows. It is based on excavation of the large urban mound of Tell Leilan and examination of other sites on the alluvial plain of a tributary to the Khabur River, in northeastern Syria. Severe and persistent drought is invoked--as an abrupt climatic change--to explain the abandonment of Leilan and adjacent urban tells, located in what is now a marginal environment for rain-fed agriculture. Abandonment of these large towns, forming part of the Akkadian imperial state, signaled agricultural ruin of a key province, to cut state revenues at a time of declining Euphrates discharge in the Akkadian heartland (see Kay and Johnson 1981). The state crumbled and "nomadic" groups moved in. The presence of wind-borne volcanic glass shards at Leilan and adjacent sites is attributed to the cataclysmic eruption of a volcano in Anatolia, possibly causing a "volcanic winter" that destabilized global climate (Weiss et al. 1993; Courty 1994). This is the prime mover invoked to explain political devolution in Syria-Mesopotamia, "synchronous with climate change and collapse in the Aegean, Egypt, Palestine, and the Indus" (Weiss et al. 1993: 1002).

Presented in logical steps and with care, this argument must be taken seriously. Although circumstantial hypotheses for causation cannot be "proven" to be correct, it may be possible to test the validity of supporting data. Specifically in the case to point: (1) Are the empirical climatic data valid? (2) If so, was the climatic anomaly effective and synchronous across the area in question? and (3) Did sociopolitical unrest or collapse everywhere follow upon climatic change, in a form consonant with greater aridity?

If the answers to these questions are unsatisfactory, the next logical question is whether archaeological and historical sources provide demographic, socioeconomic and political information to support alternative hypotheses. After concluding that there is no evidence for synchronous and incisive climatic change during the late third millennium, the present study examines the many threads of devolution in Egypt and Palestine in seeking a more complex understanding of the crisis of the late third millenium.

2.1 Palaeo-Environmental Evidence: the Khabur Plains

Local palaeoclimatic evidence to support the "Leilan Collapse" is based primarily on pedogenetic properties and changes over time, in floodplain exposures of the upper Khabur River. The background study relies on successive archaeological associations rather than radiometric dates to identify a sequence of sedimentary units (Courty 1994). Sedimentological relationships and properties are downplayed in favor of selected microscopic analysis of sedimentary particles and post-depositional alteration. For some centuries prior to the collapse, fluvial dynamics were already marked by semiarid aggradation of a braided channel, incorporating wind-derived materials (pelletal sands), the sediments then altered by subarid soil processes. As I understand it, the silt-sized volcanic glass (5 to 25 microns) on top of third millennium settlement strata at Leilan are also found in mixed eolian-colluvial strata that rest on the previous soil, in other floodplain exposures representing similar stratigraphic positions (Courty 1994). Some 30cm of fine sands and coarse pelletal sands, mixed with fine glass shards and silt, begin the standard sequence; they suggest alternations of blowing sand and reworking by heavy rain spells amid degraded vegetation. The next 30cm include much finer materials of similar composition, representing eolian dust, also reworked by heavy rains. The final 40cm are once more coarse-grained and include gypsum particles, attributed to local deflation by wind.

The settlement hiatus at Leilan is therefore linked with devegetation and biological aridity. Modern analogs can be sought 200 or 400 km downwind of the more familiar, ash-spewing eruption of Mt. St. Helens in 1980, when 2 to 3.5 cm of volcanic dustfall of similar grain size blanketed or choked vegetation along the main dispersal axis in eastern Washington state (Carey and Sigurdsson 1982). Of course, the parameters of distal tephra accumulation vary widely from one volcanic explosion another (Carey and Sparks 1986), but this example provides a sense of what is involved. The impacts of Mt. St. Helens on agriculture varied according to crop and the stage of growth (Rosenfeld and Cooke 1982: 74-77; Lyons 1986), and plant recolonization was rapid (Franklin et al. 1985). Biological recovery in the semiarid climate of the Khabur Plains would be slower, so that wind and water could be expected to rework loose sediment on floodplain surfaces for years and perhaps decades--but not centuries. The post-Leilan strata appear to represent a volcanic and edaphic horizon, documenting a catastrophic biological event, rather than an abrupt climatic change in the strict sense of the word. Stream behavior immediately after that event (see Courty 1994: Figs. 10, 13-15) did not change notably, except for a much greater sediment supply. Critical is that a thin mantle of volcanic silt, spread out over the farm lands of a marginal environment, might destroy a staple crop, although the psychological trauma of an unfamiliar natural calamity may have been far more effective in provoking abandonment than actual agricultural losses.

The scenario of a "volcanic winter" finds no support. There is no evidence of a globally effective eruption in this general time range from the published ice corings from Greenland (Hammer et al. 1980), and deep-sea cores in the eastern Mediterranean fail to record volcanic materials from this period (Guichard and Courty, this volume). The Leilan volcanic glass represents a regional, not a global event; its tangible environmental impact would have been limited to the direct biological repercussions of a dust and ash mantle, rather than an indirect response to stratospheric aerosols, such as those released in more recent times by Tambora, Krakatoa, or Pinatubo. A better picture of the impacts of a regional volcano may eventually emerge from reexamination of the potential effects of the eruption Thera (Santorini) in the Aegean c. 1628 B.C.E. on proxy records for Near Eastern vegetation (see Roberts et al., this volume; Bottema, this volume).

The regional settlement records immediately east and west of the Khabur Plains are also not consistent with that around Leilan, as might be expected if either climate or vulcanism was a prime, forcing agency. An archaeological survey to the east, in the northern Jazira of Iraq, with an identical ecology, showed a growth of urban centers at the expense of smaller settlements during the course of the third millennium (early to mid-Early Bronze Age [EBA]); subsequently these towns continued to be occupied but the number of satellite villages expanded (late Early to early Middle Bronze [MB]) (Wilkinson 1990). To the west, the survey of the ecologically comparable Carchemish-Birecik sector of the middle Euphrates, north of the Turkish-Syrian border, reveals that settlement actually reached a nadir during the early to mid-EBA, contemporary with exponential growth at Leilan, rebounding strongly during the late EBA to early MB; few of these sites are located on floodplains, but the riverine town of Carchemish in northern Syria expanded greatly after the collapse of Leilan (Algaze et al. 1994). A little upstream along the Euphrates, in the more humid area of the Atatürk Reservoir, the picture is reversed, with growth and decline paralleling that at Leilan (Algaze et al. 1992). Similar inconsistencies are apparent in Syria west of the Euphrates River (Matthiae 1993). Without the benefit of site-specific excavations, archaeological surveys are too coarse and generalizing to provide hard answers. But such contrasting patterns of demographic and urban growth or decline do not support an environmental hypothesis. Instead they suggest a productive avenue of exploration into shifting politico-economic constellations, complemented by micro-studies of socioeconomic change (e.g., Wattenmaker 1994).

2.2 The Palaeoclimatic Record c. 2400-1900 B.C.E.: the Near East

The most intriguing evidence for accentuated aridity in the wider area comes from the Anatolian headwaters of the Tigris and Euphrates rivers. The bottom sediments of Van Gölü, an endorrheic lake, have been studied through a series of disconnected ("floating") cores (Degens et al. 1984). The cross-correlations appear to be good, judging by the "best" match between vegetation trends documented in a radiocarbon-dated pollen core from an adjacent mountain bog (at Sögütlü, Van Zeist and Bottema 1991: Fig. 18) with both the pollen record and ^{18}O isotope trace from the varve-dated "floating" cores taken in the lake (see Van Zeist and Bottema 1991: Fig. 17; Schoell 1978). Enrichment of ^{18}O in Lake Van implies reduced dilution of lake water by precipitation, presumably accompanied by shrinking lake volume (Lemcke, this volume). The isotope trace indicates a pronounced dry spell c. 2350-2075 B.C.E., with positive ^{18}O deviations of up to 1.40, compared with negative deviations before

2500 and after 2075 B.C.E. (Shoell 1978: Table 1). That indicates an initially slow trend, culminating c. 2220 B.C.E., then reversing within about 150 years, in short, not an abrupt change.

Assuming that the Lake Van trace is valid, modern climatic analogs in the region (Kay and Johnson 1981) imply greater aridity across eastern Anatolia and the adjacent Khabur Plains, as well as reduced Euphrates discharge in Mesopotamia. That may help explain increasing salinization and declining crop productivity in Lower Mesopotamia 2350-1850 B.C.E., but the temporal overlap in only partial, and this particular problem has been satisfactorily attributed to intensified cropping without benefit of fallow, a perennial problem in the area (Adams 1981: 151-52).

An arid climatic trend such as that implied by the isotopic chemistry change in Lake Van should leave an imprint on regional vegetation, for which a reasonably tight grid of pollen cores provides proxy records. Curiously enough, that is not the case (Bottema, in this volume).

The key examples can be discussed briefly, by calibrating ^{14}C years to calendar dates (after Stuiver and Becker 1993) to calculate sedimentation rates between the limited numbers of radiometric dates, and then to interpolate or extrapolate to identify the pollen traces for c. 2400-1900 B.C.E. For Lake Zeribar in the Iranian Zagros there was an abrupt decline of oak and other tree pollen (by 40%), compensated for by grass and other herbaceous pollen, reversing a long-term trend of forest colonization by some 500 years (see Van Zeist and Bottema 1991: Fig. 14). But the switch can only be dated approximately, after 2200 and before 1950 B.C.E. In the case of the cores from Lake Van and nearby Sögütlü (Van Zeist and Bottema 1991: Figs. 17-18) there is no distinctive trend or oscillation in our selected time range, which is surprising, to say the least. To the west, there is the Beyşehir I profile, in south-central Anatolia, where there is no evidence of vegetation change, just as in the Ghab III coring, northwestern Syria (Van Zeist and Bottema 1991: Figs. 23 and 36).

In a more southerly direction, the Lake Huleh core in northern Palestine shows an abrupt decline of arboreal pollen, but at an interpolated date of roughly 2000 B.C.E. (Van Zeist and Bottema 1991: Fig. 37). The Lake Kinnereth core records a modest increase of deciduous oak at the expense of olive, in the time frame c. 2450-1900 B.C.E. (Baruch 1990), but comparison of the oxygen and carbon isotope curves with those from tightly dated speleothems in Upper Galilee (Issar, Geyh, et al. 1992. Fig. 2) suggests that this part of the pollen profile should be some 400 years older. The speleothem isotopes imply a drying or warming climate c. 2450-1900 B.C.E., followed by a sharp reversal (Geyh 1994), but the level of the Dead Sea was relatively low throughout the period c. 2500-1400 B.C.E., following a high stand during the first half of the Early Bronze (Frumkin et al. 1994; see also Donahue et al. n.d.).

The ring of palynological control points can be amplified from further afield, in other areas affected by sociopolitical discontinuity. Notably, pollen cores from northwestern Anatolia (Beug 1967), Macedonia and Thessaly (Bottema 1985), Epirus (Willis 1992), Dalmatia (Jahns 1991), and the Peloponnese (Jahns 1993) yield no evidence for climatic variation at the end of the third millennium.

In effect, the substantial body of palynological data does *not* support an abrupt climatic change, with an amplitude of several centuries, and of regional significance. Lakes Zeribar and Huleh are the only examples documenting greater aridity, but the dates are "soft" and point more towards a shift two centuries later than the Leilan Collapse horizon. Particularly troublesome is the lack of corroboration for the Lake Van isotopic record, itself not "robust" because of the "floating" varve chronology. Finally, even if we accept that isotopic record and the Zeribar pollen core as significant records coeval with the Leilan Collapse, the aridity implied would have been circumscribed by the Tigris-Euphrates drainage basin. There was, then, no abrupt climatic shift to greater aridity affecting the larger region--the Near East and the Aegean world--between c. 2400 and 1900 B.C.E.

The spacing of pollen samples in most cores is sufficiently wide to allow for brief anomalies, measured in decades. Nile flood variations in Egypt, discussed below, were of such a scale. But such short trends are less likely to affect biotic stability, whereas the abrupt climatic change that has been proposed is assigned a duration of no less than three centuries (Weiss et al. 1993; Courty 1994).

3.1 The End of the Egyptian Old Kingdom

The Egypt of the Old Kingdom (c. 2730-2230 B.C.E.) was a complex society and territorial state. It extended along the Nile from the mouth of its Delta well upstream into Nubia, exerting some measure of control over the adjacent deserts, and commercially linked to the maritime entrepot of Byblos (modern Jubail, Lebanon), where it enjoyed some degree of hegemony. Agricultural productivity was entirely dependent upon the Nile floods, derived from tropical rainfall in the Ethiopian highlands, far removed from the mid-latitude westerly circulation that dominated the Aegean world and western Asia. The aridification of the Sahara during the fourth and third millennia affected very small populations in already precarious micro-environments (Kroepelin, this volume), and had little significance for Old Kingdom Egypt.

At the end of the Dynasty 6 (c. 2420-2230), Egypt underwent political simplification and decentralization. The next two centuries, known as the First Intermediate Period, were remembered as a time of anarchy, causally linked to Nile failure, starvation, and political impotence (see sources in Butzer 1984). Various textual elaborations on famine, abandoned farm lands, and dislocated people seem to imply that the agricultural system lost its ability to sustain previous levels of productivity. Food redistribution in times of need as well as the restoration of agricultural production became political virtues. The difficulty of interpreting this "lamentation" literature, much of it written shortly after the political re-unification of Egypt, lies in sorting out fact from ideological interpretation.

Three agricultural disasters can indeed be recognized. The first took place while an Ankhtifi was a provincial governor in upper Egypt, c. 2210-2185; he prided himself on being the first ruler to distribute famine supplies. The second relates to another governor, Djehuti, c. 2100, who was obliged to find food to supply the main temple in Thebes during certain famine years. The third comes from a collection of letters by the owner of a network of estates, Hekanakht, written in 2002.

For the period in question, the Egyptian information base is unrivaled, so that Old Kingdom collapse merits critical examination. The issues will be discussed in two parts. Firstly, did the First Intermediate Period coincide with several centuries of low Nile floods, or were the three episodes of Nile failures nothing but short-term anomalies of monsoonal rainfall in East Africa? Secondly, are there more convincing, alternative explanation for Old Kingdom devolution?

3.2 Nile Flood Behavior during the Late Third Millennium

Three proxy records are now available to facilitate more objective resolution of Nile history, two from East Africa and one from the Nile Delta. These are summarized here, with individual ^{14}C dates or overall sedimentation rates recalibrated to calendar years according to the program of Stuiver and Reimer (1993). Equally important has been research in Egypt's Fayum Depression, an adjunct of the Nile Valley, that helps unravel a surprisingly complex flood history in the context of the other proxy records. The new Fayum sequence is analyzed and synthesized further below.

Although mainly located in Kenya, Lake Turkana (Rudolf) now derives 80 or 90% of its water from the Omo River basin, in the western highlands of Ethiopia. Although rainfall anomalies in the Blue Nile catchment are not always in phase with those of the Atbara or Sobat rivers (northeast and west of the Blue Nile), Holocene trends in Nile volume are broadly similar to fluctuations in the level of Lake Turkana. That lake was moderately high c. 3200-3000 B.C.E., at an intermediate level c. 2750-2200, rising abruptly to its overflow level 2150-2050, then dipping back briefly to its intermediate stage c. 1700, before returning to overflow level c. 1600-1500 (Butzer 1980b; Owen et al. 1982; Harvey 1982). This segment of the lake level trace, approximately fixed by 15 ^{14}C dates, defines the major trends but has a relatively coarse resolution of ±100 years. It mainly identifies periodicities in the order of several centuries.

Cores taken from the bed of Lake Turkana now complement the lake's recent history with higher resolution. Taken in the northern part of the lake, Core LT 84-ZP provides a measure for the influx of Omo River sediment, anchored on ^{14}C dates and interpolated for intervals 50 to 100 years in length (Halfman et al. 1992). There are strong fluctuations at this scale, with high influxes c. 3350-3075 and 2875 B.C.E., intermediate influxes c. 2600-2300 and 2125-2050, and low influxes c. 2975, 2200, and after 1900. The match with the lake level trace is good, except that there is no correspondence with the high lake stand c. 1500 B.C.E.

That event does however find support in cores from the northeastern Nile Delta, analyzed by neutron activation for rare earth elements in the silt fraction (Hamroush and Stanley 1990; MEDIBA 1991), using ratios of lanthanum to lutetium and of chromium to scandium. Lower values of the first imply greater Blue Nile discharge; higher values of the second infer either greater White Nile and/or Atbara discharge or reduced Blue Nile influx (for database see: Kammar et al. 1992). The results from three, differentially compressed cores (S-7, 22 and 36) are difficult to fully reconcile and have only coarse resolution, but they closely replicate the Lake Turkana trace: moderately high Blue Nile influx c. 3100-2900 B.C.E., mainly low after 2900, to a minimum at c. 2250, then rising rapidly to a maximum c. 1750-1500 B.C.E. This justifies the use of Lake Turkana as a proxy for Nile behavior.

Direct geoarchaeological associations are given in the Fayum Depression, linked to the nearby Nile floodplain by a secondary channel. Here high Holocene lakes were supported by influx of Nile flood waters, with recessions during periods of lower floods. The chronology of early historical lakes is established near the western margin of the residual Birket Qarun, around the small temple of Qasr el-Sagha. Publication, however, is so fragmented as to require detailed delineation. The datum is given by a humic soil, developed on lacustrine White Silts, with Neolithic and terminal Paleolithic sites; that soil marks the beginning of a recession and is dated c. 3900 B.C.E. (Ginter et al. 1982b). The archaeological referents are the temple, attributed to Sesostris II (1897-1878 B.C.E.); a quay or breakwater, linked to a causeway of diabase blocks, with 5th or 6th dynasty pottery (c. 2525-2230 B.C.E.); and two workmen's settlements, related to episodic quarrying up on the plateau, the primary occupation of which has pottery suggesting the period of Amenemhet III (1843-1797 B.C.E.), while a secondary, re-occupation has 13th dynasty pottery (1790-1674 B.C.E.) (Arnold and Arnold 1979). Two ^{14}C dates of about 1690 and 1740 B.C.E. (Dagan-Ginter et al. 1984) confirm the age of these occupations in the Eastern Settlement, while the foundation of the western counterpart has an approximate date of 1880-1920 B.C.E. However, charcoal in an oven pit under the Eastern Settlement is dated 2340 ± 70 B.C.E., i.e. to Dynasty 6 (see Ginter et al. 1982b: 109, 115, Fig. 57). There is then a definite late Old Kingdom presence, in addition to a variety of Middle Kingdom structures related to the two main episodes of royal building activity in the eastern Fayum, under Sesostris II and Amenemhet III.

The sedimentary sequence consists of four Brown Sands, representing transgressive beach deposits up to 20-24m (above modern sea level), separated by recessional stages, lower by at least 8-10m (Ginter et al. 1982a: especially Figs. 6-7; Ginter et al. 1982b). Brown Sand 1 (20m lake) rests on the fossil soil and predates construction of the quay in late Old Kingdom times. The quay was built during the first recession, and was designed to load boats in an elevation range of 12-20m. Brown Sand 2 (22m) is embanked against this quay. The temple was built during the second recession, the workmen settlements 50 or so years later. Wave action accompanying Brown Sand 3 (22 m) undermined the perimeter wall foundations of the Western Settlement (at 22m) and reduced it to 30cm of debris. The site was reused, in part by a small garrison, during the third recession some 50-100 years later, only to be finally destroyed by the highest transgression, Brown Sand 4 (24m). The stratigraphy of Ginter et al. (1982a, 1982b) is corroborated by a profile of Said, Albritton et al. (1970: 48, 50, 52).

Interpretation of the four episodes of exceptionally high, Nile floods that surged into the Fayum is facilitated by the trend of specific Nile flood levels recorded for dynasties 1 to 5 (Bell 1970), and the inscription of catastrophically high Nile floods c. 1842-1770 B.C.E. at the Semna Cataract in Nubia (Bell 1975; De Putter 1994). The oldest (Brown Sand 1) predates Dynasty 6, and can be assigned to Dynasty 1 (c. 3050-2950 B.C.E.), when flood levels averaged a meter higher than during dynasties 3, 4, and 5; one meter higher represents an increase of about 30% in volume (Butzer 1976: 28). The second episode (Brown Sand 2) was 2m higher still, and fits between 2200 and 1900 B.C.E., but cannot be fixed more precisely. The third episode (Brown Sand 3) was of similar magnitude, and is recorded after 1842 B.C.E. at Semna, which indicates a catastrophic flood every third year or so; peak flood volumes were three times those of the ten highest floods of the 19th century (Bell 1975), implying a flood crest 2-4m higher in Egypt (Butzer 1976: 29). The last episode (Brown Sand 4) was the highest, and post-dates the mid-1600s.

Combined with the proxy records from Lake Turkana and the Nile Delta, the Qasr el-Sagha sequence completes this synthetic overview of Egyptian flood history between 3000 and 1500 B.C.E.; it has a range of uncertainty of less that ± 50 years. Nile floods were high c. 3100-2900, then lower, with a brief minimum around 2200. Exceptionally high floods were again common c. 2150-1900, 1840-1770, and 1600-1500. The complex arguments necessary to arrive at these conclusions illustrate the need for detailed resolution and convergent lines of evidence in paleoclimatology, so as to leave little ambiguity as to the direction of change. Until dendroclimatological sequences become available in the Near East, this is the best level of resolution currently possible in the region for the third millennium.

Applied to the sociopolitical history of Egypt, an image of drought at the end of the Old Kingdom was impressed in ancient Egyptian memory by recurrent Nile failure in the time of Ankhtifi, c. 2210-2185 B.C.E. In turn, three documented Nile failures between 2200 and 2000 gave modern researchers (including this author) the misleading impression that this was a time of generally low Niles and dearth, in part based on the didactic Egyptian "lamentation" literature. The fact is that Egyptians mainly had to cope with destructively high floods, over a span of about 600 years. There is a lesson here, not to assume too much from scattered bits of information.

The devolution of the Egyptian state into a number of petty principalities was already far advanced when Ankhtifi proclaimed his magnanimity as a provincial ruler. Put differently, the Old Kingdom had collapsed before the low-Nile anomaly of the 8th and 9th dynasties. That will have compounded the sociopolitical turmoil, but the roots of political decentralization go back two centuries prior to Old Kingdom collapse. The anarchy spawned by devolution subsequently set in train a number of social adjustments. These generalizations can now be examined.

3.3 Old Kingdom Devolution: Decentralization

The authority of the pharaoh was initially uncontested, but construction of the great pyramids came at a price, and also set a precedent for permanent mortuary cults, supported by tax-exempt land endowments. Both such institutional estates and the "pious foundations", i.e., land donations to temples, were authorized on an increasingly large scale during Dynasty 5 (c. 2420-2380 B.C.E.). This, withdrew prime agricultural lands and their produce from the revenue base (Kemp 1983: 85, 104-107), and allowed the larger temples to emerge as strong competitors for power. At the same time, key bureaucrats and provincial magnates began to build sumptuous tombs, and de facto governors combined all branches of administration in their provinces (Baer 1960: 281-286). Pepi I (c. 2360-2330 B.C.E.) formed a close dynastic alliance with a prominent family from Upper Egypt, where provincial autonomy was gradually being consolidated, perhaps with the pharaoh's support, as a misguided effort to counterbalance the excessive power of the high-level bureaucracy in the capital (Baer 1960: 301-302). On his death, that family practically ruled the country in the name of his underage sons, Merenre (died within a year) and Pepi II (c. 2330-2240 B.C.E.), expediting the process of decentralization (Baer 1960: 297-298). A futile attempt to reassert central authority may be indicated by a sudden drop in the ranking of titles for provincial governors after about his 40th regnal year. Whatever the competence of Pepi II as a mature monarch, after he had outlived his dominating uncles, his ability to rule effectively in his old age will have been severely curtailed. At his death, the Egyptian state promptly disintegrated into a suite of regional principalities (O'Connor 1974) or city states (Kemp 1983: 177), with power devolving to powerful local families, as the culmination of a two-century, accelerating decline of central authority.

Table 1. Tentative Correlation of Third Millennium Archaeological Sequences. The approximate dates between time units are inferred from the Egyptian chronology.

Egypt	Palestine	Byblos	Ras Shamra	Amuq	Hama	Ebla	Mesopotamia
Dyn. 12-13	MB II	H		K	H5	III	Isin-Larsa
(2000 B.C.E.)		====	====	====		====	
Dyn. 10-11	↑	JII	IIIA3	J	J1-4	↑	Ur III
(2200 B.C.E.)				====	====		
Dyn. 7-9	EBIV	JI	↑	Late I	J5	IIB2	(Late) Akkad
(2270 B.C.E.)	====	====			====	====	
Dyn. 5-6	EBIIIb	KIV	IIIA2	Early I	J6-7	IIB1	Early Dyn. III
(2520 B.C.E.)	====						
Dyn. 3-4	EBIIb-IIIa	KIII	IIIA1	H	K1-5	IIA	Early Dyn. II
(2730 B.C.E.)	====	====	====	====			
Dyn. 1-2	EBIIa	KI-II	IIIB	G			Early Dyn. I
(3100 B.C.E.)							

==== Destruction Levels

Sources: (Egypt) Kemp 1983, Hartung 1994, and others; (Palestine) Callaway (1980), Seger (1989, 1995), Brandl (1992), Stager (1992), Joffe (1993), and others; (Byblos) Saghieh (1983); (Ras Shamra, Amuq, Hama) Kühne (1976), de Contenson (1989), Conti and Persiani (1993), and others; (Ebla) Matthiae (1982), Matthiae et al. (1985), Oates (1985), Schwartz and Weiss (1992); (Mesopotamia) Gadd (1971), Porada et al. (1992).

Old Kingdom prosperity and, indirectly royal power, was also linked to the Byblos trade, a royal monopoly that probably was applied to redistribution and even credit at the pharaoh's disposal. Fortunately, the problematic stratigraphy and implications of this site have now been resolved by Saghieh (1983). The town was first founded as an unwalled settlement, at the time of Dynasty I, and its early temple has architectural elements of Mesopotamian origin (level KII). Egyptian influence appears immediately after the subsequent burning and destruction of public buildings: a new temple was built in plainly Egyptian style (level KIII), and Byblos was now fortified. KIII pottery types are found in Egyptian tombs of dynasties 3 and 4 (c. 2730-2520 B.C.E.) (Table 1), suggesting that the political shift was related to events (? Egyptian military action) at the end of Dynasty 2 or during Dynasty 3. With level KIV that temple was modified and enlarged, to incorporate spatial arrangements pioneered by Dynasty 5. But the KIV town was systematically destroyed during the reign of Pepi II, as dated by goods inscribed with his cartouche. When Byblos was rebuilt, in Akkadian times, its public buildings emulated

Mesopotamian and Anatolian models: Egyptian hegemony had been replaced by Mesopotamian indirect control.

Byblos represented *the* pharaonic commercial entrepot in Asia, regardless of whether or not there was an Egyptian military presence. The town was probably destroyed by Sargon of Akkad or one of his immediate predecessors of the late Early Dynastic III (see below), thirty-some years into the reign of Pepi II. Regardless of who sacked Byblos, the impact of its destruction for the Egyptian state will have been catastrophic--by putting an abrupt end to the pharaonic trade monopoly and its lucrative profits. That will have left Pepi II devoid of the currency of power, at a time when the faltering Egyptian monarchy desperately needed it. Finally, after the death of the king, the respect probably still accorded to his person no longer extended to the kingship.

There also appears to have been a progressive economic decline during the reign of Pepi II, judging by an overall reduction in the size and opulence of funerary architecture among both low and high-echelon officials, in the capital as well as in the provinces (Kanawati 1977). That generalization appears to be contradicted by closer study of a very large sample of cemeteries and rock-cut tombs at Qau, the dating of which was tightly controlled by pottery seriation (Seidlmayer 1987). Here both the number of pots placed in graves and the volume of the burial chambers increased steadily through Dynasty 6 and the First Intermediate Period, until the reunification of Egypt, declining with equal regularity thereafter. This is attributed to greater economic prosperity.

During Dynasty 6, what is today called "capital" shifted from Memphis to the provinces, continuing to accumulate in local centers until re-unification, after which the provincial centers were once again impoverished. Kanawati's (1977) argument is based mainly on rock-cut, elite tombs and on provincial centers in Middle rather than Upper Egypt. In conjunction, the two studies seem to imply a greater equalization of wealth across the class spectrum, but an uneven equalization of capital among different centers along the Nile Valley. The elites, now dispersed away from the capital, were less wealthy than before, but they were not impoverished either, judging by gold objects and beads from Byblos found in tombs at Qau belonging to dynasties 8 to 10.

The funerary architecture speaks for decentralization, reduced elite wealth, some equalization of wealth across the class spectrum, and capital accumulation around new nodes of provincial power between c. 2300 and 2130 B.C.E. That is incompatible with repetitive agricultural disasters, which should instead lead to rural impoverishment and depopulation, something precluded by the Qau cemetery record. But it is consonant with political devolution, social readjustments, and reduced economic productivity--a typical outcome of weakened politico-economic integration (see Butzer 1990, 1994).

By 2150 B.C.E. an increasing fraction of the male population was buried collectively in the rock-cut tombs of powerful men, matching a shift in representational art, which once had shown "big men" in a mix of bucolic and domestic contexts, but now depicted them surrounded by squads of retainers rather than family members, with a novel emphasis on armed men (Seidlmayer 1987). This suggests a change from a society organized around family units, to one focused on patron-client coteries, perhaps analogous to the proto-feudal organization of Merovingian Gaul around "big men" and their retinues. Such trends normally stem from prevailing insecurity and violence. If these inferences are correct, then the underlying theme of anarchy in the "lamentation" literature becomes appropriate: political devolution would have brought a breakdown of law and order, that was confronted by a militarization of society at the local level. The practice was ended under Sesostris III (1878-1842 B.C.E.), who seems to have neutralized the local or provincial "war-lords" and thus completed the return to autocratic central authority.

This analysis of Old Kingdom collapse suggests an alternative scenario or model, unrelated to ecological disaster. It emphasizes elements of internal weakness that, amplified by feedback loops and accentuated by external crisis, led to political devolution and subsequently provoked adaptive, social change. The Egyptian "insider's" view of these events is eloquently rendered by the controversial "admonitions" of Ipuwer, a litany of social chaos, probably addressed to a new pretendent to the throne after the fall of the Old Kingdom (see text in Faulkner 1965). Most of the passages represent a dialectic around the theme that the high-born and the wealthy were deprived of their property or dignity, while simple people or robbers basked in undeserved glory or ill-gotten gains. This didactic device served to affirm the traditional values of the old social order. More significant are comments on the prevalence of violence, land laid waste, hunger, and towns destroyed--possibly with reference to civil war and a breakdown of law and order. The living king was deposed by the rabble, and the tombs of dead kings looted, implying disrespect for the kingships and the previous dynasties. The palace and the temple stood empty, perhaps because Memphis had been abandoned as the capital. Lastly, the archives and land records had been thrown out or destroyed, perhaps a symbolic note inserted by Ipuwer, but possibly also an allusion to rampaging mobs during the last days of Memphis.

But the end of the Old Kingdom had nothing to do with a social revolution, which itself is a modern concept. The successors of Pepi II were disempowered when rival provincial magnates gradually withdrew their support for the kingship, and subsequently engaged in numerous military raids or minor wars in a vain struggle for supremacy that dragged on for two centuries (Hayes 1971). Even so, decentralization was a continuing process, until a new, Theban dynasty began to consolidate power and ultimately restored the kingdom. There was no "revolution" or systemic implosion, as the incremental changes in burial patterns and grave goods verify (Seidlmayer 1987). But the periodic warfare supported by varying alliances of war lords was brutal, including the plundering of cemeteries by troops (Hayes 1971: 466), and insecurity or desolation must have descended on some contested areas for decades at a time.

Much like the "lamentation" literature, other Egyptian records of the Old Kingdom are introspective. They never reveal the degree to which Egypt was interlinked with the outside world, nor the economic interdependency that bound the societies of the Near East during the third millennium. Thus we must turn to the archaeological record of Palestine for insight into processes of change different than those in Egypt, that draw attention to the importance of commercial networking and military confrontation between the two countries.

4.1 Early Bronze Palestine: Ecological and Ethnic Dichotomies

The Early to Middle Bronze Age archaeology of Palestine drives home the point that Old Kingdom Egypt approximated what is now a nation state. By contrast, even at its apex of third millennium urbanism, Palestine was a melange of small, probably autonomous towns and nascent city-states, possibly interconnected by rival confederacies or ephemeral principalities that controlled several towns and their hinterland. Its socio-political organization was less complex than that of Egypt and, in default of written records, its bureaucratic institutions less elaborate.

In economic terms, Palestine resembled Syria and the broader Aegean world, with regional clusters of nodal points engaged in complex networks of short and long-distance interaction. How this translates into political terms--on the spectrum of independence, autonomy, alliances, hegemony, or direct control--is largely obscure, because of a lack of familiar modern precedents for loosely-structured commercial centers along the axes of a major trading network. Such clusters or centers fostered the exchange of ideas, and did not represent a commercial and cultural periphery. In the Bronze Age, eastern Mediterranean world, such configurations were common and apparently stable for centuries at a time, but also prone to abrupt and synchronous simplification, followed by renewed elaboration and commercial reintegration after a century or two, thus at the end of the Early Bronze or the Late Bronze Age.

A second contrast between mid-third millennium Egypt and Palestine was in regard to agroecology and what that implies in the economic and social spheres. Largely surrounded by hyperarid deserts, Egypt was practically homogeneous in ecological terms, composed almost exclusively of farmers practicing flood or irrigation agriculture, regardless of elite stockraising in marginal floodplain environments. But Palestine, in the prevailing archaeological conception, was ecologically riven between groups practicing agriculture versus those engaged in pastoralism, as well as spatially between sedentary and nomadic forms of settlement. Some archaeologists have created a polarity between sedentary agriculturalists and nomadic pastoralists--as distinct ethnic rather than socioeconomic groups, perhaps even belonging to different linguistic families. These dichotomies tend to be implicitly superimposed on evolutionary schemata--of dispersed settlements consolidating into nucleated towns or vice versa--so as to introduce an additional, processual dialectic. Such implicit models have a variety of roots: deep-seated misconceptions about the practices of Mediterranean agriculture, the old dichotomy of "the desert and the sown," notions of "black tent nomadism" such as that of the Rwala Beduin, or oversimplified views of the Arab expansion in early Islamic history (S. Rosen and Avni 1993). As a result, urban collapse towards 2200 B.C.E. is commonly linked to the ascendancy of pastoral groups.

Pastoralism is noted for the versatility of its adaptations today (Galaty and Johnson 1992), and there is reason to believe that it also evolved rapidly over time, just as specialized forms of agriculture have. Yet its prehistory is poorly understood, because archaeological resolution has lagged and remains difficult (Bar-Yosef and Khazanov 1991). Several points bear emphasis in regard to interpretation of the Palestinian Early Bronze Age, in support of the generalizations offered above:

a) Historically and ethnographically, agro-pastoralism in subhumid Mediterranean environments was a unified, complementary enterprise (Stager 1985b; Spurr 1986; Butzer 1988, 1994a, 1996; Whittaker 1988). Farmers worked their fields and also kept livestock for milk products, meat, and fiber. During the cool season, sheep or goats were grazed on fallow or stubble, where their manure provided the fertilizer necessary for long-term productivity. During the warm season, animals were herded on the rougher terrain of the local commons, or driven beyond a town's jurisdiction, to mountain pastures used through leasing arrangements. The latter involved professional shepherds, rarely accompanied by family members. The switch from local to medium or long-distance movements was dictated by the size of herds, and large herds tended to reflect wealthy or corporate owners, i.e. a commercial component.

In the early Middle Ages, Christian or Muslim herders drove sheep or cattle back and forth across the shifting frontier zones of Spain, in long-distance treks, in part to sell at market; later, under the designation Mesta, the herds of large cities, grandees, or the king were driven across the length of the peninsula (Butzer 1988). Except for the scale, there were similar state-owned herds in Ur III and Old Babylonian Mesopotamia (Zeder 1994; Kraus 1966), and the contracts binding shepherds in the Spanish Mesta recall those of the Code of Hammurabi (T. J. Meek in Pritchard, 1969: Laws 261-267); such conventions are therefore deeply grounded in Mediterranean common law, back at least to the time of Hammurabi, i.e., to beyond 1750 B.C.E. Related, ecological patterns of pasture management are apparent in Spain by 4000 B.C.E. (Stevenson and Harrison 1992). The hallmark of such variable forms of *transhumance* is that the herds belong to sedentary villagers or town folk, who practice full-time cultivation. Above all, pastoralism and cultivation were complementary strategies to minimize subsistence-risk, a critical fact overlooked by Wilkinson (1994) in his emphasis on vulnerability to climactic anomalies.

b) By contrast, *nomadism* refers to the seasonal rounds of herds that are accompanied by their owners, their families, and some or all of their belongings. There are several degrees of nomadism, and such activities have not been confined to semiarid, desert margins. In the southern Balkans, for at least a millennium prior to 1912, minority Vlachs of Latin tongue lived seasonally in permanent mountain villages, following their herds into the Greek, Slavic, Albanian, or Turkish-speaking lowlands, where they occupied seasonal settlements during the cool season (Beuermann 1967: 120-195). These patterns of the Aramani Vlachs are accentuated among their Sarakátsani brethren, who originally lacked permanent settlements, although their seasonal dwellings and animal sheds were elaborate, and would leave a complex archaeological record of post holes and rock-wall rubble. In recent decades, the Sarakátsani have moved into abandoned mountain villages, or settled down on the lowland margins. Where such semi-nomadic peoples had access to cultivable land, they practiced supplementary agriculture, but in general they filled the ecological and spatial interstices of sedentary societies, as a marginalized socioeconomic class or as different ethnicities. Distinguishing such semi-nomadic groups from agro-pastoral societies in the archaeological record would probably require extensive excavations in an adequate number of small sites, selected in distinctive ecological contexts. For the moment, that leaves considerable uncertainty in regard to the interpretation of small Early Bronze sites in Palestine.

c) The camel nomads of the "ethnographic present" (e.g., Musil 1928) did not exist in the third millennium, when the camel and horse had not yet been domesticated. Camels, raised for their milk products by nomadic beduin, can subsist on ephemeral desert pastures inadequate for other livestock, and can carry the baggage of whole encampments. Horses not only reinforced that mobility, but made swift, mounted pastoralists a military force to be reckoned with. The Biblical image of destructive Amorite raiders belongs to a later era, as do the horsed beduin who terrorized Palestine during the Ottoman era (see Esse 1991: 19-25). Such historical dichotomies did not exist 2200 B.C.E. Without such military advantages, the small contingents of interstitial or marginal pastoralists of Early Bronze Palestine can have had little more than a nuisance effect through petty thievery for the much larger sedentary populations implied by the very scale of agricultural labor. The instructions of the Egyptian ruler Merikare (c. 2100 B.C.E.) argue the point explicitly: the nomadic pastoralist does not conquer, yet cannot be conquered; he may rob a single person, but he cannot seize a populous town (J. A. Wilson, in Pritchard, 1969: 416). Occupying specific economic niches and complementing each other through the exchange of produce, meat, milk products, and dung, the sedentary and non-sedentary peoples of Palestine most probably formed symbiotic relationships. It seems implausible that pastoral groups living in or around Palestine had either means or motive to destroy urbanized civilization in Palestine in the third millenium.

4.2. Early Bronze Palestine: Settlement, Economy, Time

The Early Bronze Age of Palestine is subdivided into time frames that parallel events in Egypt, reflecting the close interactions between the two regions (Stager 1992; Amiran and Gophna 1992; Finkelstein and Gophna 1993; Hartung 1994).

During EBIa (spanning Nagada IIa-IIb and Maadi), Egyptian-style pottery--mainly household wares--was being made in several settlements of southwestern Palestine, suggesting at least some Egyptian craftsmen in residence c. 3750-3550 B.C.E. (Brandl 1992). The Egyptian presence broke off at the beginning of EBIb (equivalent to Nagada IIc-IIIc, c. 3550-3150 B.C.E.), while Canaanite wares appeared in increasing quantity at Upper Egyptian sites. By about 3350 B.C.E. these included great quantities of imported wine jugs in royal tombs. Not long thereafter, Egyptian way stations were established along the north Sinai coasts and Egyptian potters were reestablished in adjacent parts of Palestine (EBIc, Seger 1995). Egyptian as well as local, clay seal-impressions suggest the presence of some bureaucratic institutions. Large, mainly Egyptian storage vats and tall, cylindrical vessels also appear, presumably for olive oil carried back to Egypt by donkey caravans.

This intensive overland trade seems to have broken off late in Dynasty 1 (c. 3000 B.C.E.). That marks the beginning of EBII, when Egypt's Near Eastern trade began to switch to large ships, plying the open sea between the Nile Delta and Byblos. The EBII appears to span dynasties 1 to 3, judging by final ^{14}C dates of c. 2600 B.C.E. (Arad: Stager 1992: II, 51). During EBIII times (dynasties 4 to 6, c. 2660-2230 B.C.E.), Egypt's main commercial links were with Byblos. This maritime trade (Drower and Bottéro 1971: 343-351) terminated abuptly with Pepi II, when Byblos was destroyed (Saghieh 1983). EBIII ended at about this time, to be replaced by the EBIV (also Middle Bronze I or Intermediate Bronze), notable for a fundamental change in settlement.

The Egyptians depended on Syria and Palestine for raw materials such as timber (especially for long cedar beams or planks), fir resin or Dead Sea asphalt (as adhesives), and copper (for metallurgy), but especially for wine and olive oil, for cult purposes and elite use (Stager 1985b). Wine is verified as an import by grape pips in jars dated c. 3350 B.C.E. (Abydos cemeteries, of Nagada IIIa1 age: Hartung 1994). Olive oil is trickier to prove, since olive pits or wood are difficult to distinguish as to wild or cultivated. The oldest possible wine press in Palestine dates to EBIb, but presses only become common in the EBIII (Esse 1991: 119-125). However the oil lamps of many EBII sites suggest olive oil already was common, at the same time that clay pastes resistant to acid and suitable for long-term oil storage became standard for large pottery vessels (Miroschedji 1989). This supports the impression that olive oil was exported to Egypt in quantity somewhat before Dynasty 1, reflecting commercial production of wine and olive in Palestine by the end of EBIb.

Significant agroeconomic changes are apparent amid the foothills debouching on the coastal plain, where a string of large villages appeared along gently-dissected stream valleys or near large springs at the head of small catchments. The population of this ecotonal belt, with 300-500 mm rainfall today, increased several fold between EBIa and EBIb (Gophna and Portugali 1988), but the northern third (the Plain of Sharon) lost half its population by EBII times, whereas some of the larger villages further south (the Shephelah) took on a more urban character and were surrounded by stout walls. Scattered small towns also appeared in small basins of the central hill country, as well as near permanent water sources in the Upper Jordan catchment (Finkelstein and Gophna 1993; Esse 1991).

Towards the end of EBII, a number of towns were abandoned and the northern Shephelah lost a third of its population, while that of the southern Shephelah increased by a half (Gophna and Portugali 1988, and other sources). That trend continued during the early EBIII. These drastic realignments probably reflect more complex processes than local shifts in hegemonic control. That impression is heightened by more detailed analysis of urban histories during the course of EBII and III (Callaway 1980, Seger 1989, 1995). Of nine key Shephelah towns, four date back to EBIIa or earlier, and one was abandoned at the end of EBIIb and three more by the beginning of EBIIIb1. Only two towns remained by the beginning of EBIIIb2, one of these surviving well into EBIV. The pattern in other parts of Palestine appears to be similar, with a general depopulation under way by the end of EBIIIa (perhaps c. 2500 B.C.E.), and settlement discontinuity almost complete by the end of EBIIIb1 (perhaps 2400 B.C.E.) (Seger 1989). Whatever the nature of the crisis, urban devolution of EBIII Palestine was progressive and essentially complete a century before the EBIV becomes archaeologically visible, i.e. well before the critical date of c. 2200 B.C.E.

Assuming 200 people per hectare within a compact, fortified settlement, the EBII/III towns had populations in the order of 1000 to 5000 people. There were some public buildings, primarily linked to religious functions, although elite architecture was modest. Authority may have been wielded by a hereditary priesthood. But the importance of walls and their implementation according to a sophisticated and standardized plan (oval ramparts with rounded bastions, above a long, sloping glacis) implies a professional, military caste (Miroschedji 1989). Fulltime craftsmen were part of these town fabrics, perhaps living in special quarters and supplied through redistributed foodstuffs. Children suffered from poor health and malnutrition, judging by depressed stature and dental hypoplasias, while life expectancy was about 30 years (Smith 1989). A case for sociopolitical stratification seems convincing, at first glance.

The function of the EBII-III towns is partly elucidated by their location. All are found adjacent to or within potentially productive lands, some are sited with respect to defense, while others still are either found along trade routes or at the intersection of commercial arteries. The Shephelah towns, for example, are situated at the terminus of the overland route from Egypt. Even though Egyptian objects are uncommon, trade may have emphasized basic commodities rather than elite goods, their transport in the hands of pastoral caravaneers.

Yet the basic economy was agricultural, with subsistence largely dependent on wheat, barley, and legumes, and arboriculture focused on olive groves, vineyards, and fig-trees (Liphschitz 1989). Although cattle were prominent, oxen and donkeys were most commonly used for plow yokes (Davis 1988), so that sheep and goats assumed primary importance in supplying both meat and secondary products (Horwitz and Tchernov 1989; Horwitz 1989b). Reduced cortical thickness and bone mass implies that the high ratio of older, female goats and sheep was selected for milking (Smith and Horwitz 1984), suggesting that at least part of the ovicaprines were not engaged in seasonal movements, but stall-fed or grazed directly around the towns. The composite picture speaks for high sedentary, commercialized Mediterranean-style polyculture, with peripheral exploitation constrained by security concerns, and probably limited to the 5 or 6 km perimeter dictated by two hours of walking in the course of daily work rounds. That would have included both valley-floor cultivation and arboriculture on the adjacent hills formed by Tertiary strata. A mean intersite distance of 10 km supports such an interpretation and, in the absence of smaller sites, speaks more for autonomous economic units than for a functionally-differentiated but integrated settlement hierarchy in the Shephelah.

Agricultural land use upon the tableland topography of the central hill country (southern Samaria and Judea) appears to have been discontinuous, and focused on broad valleys or swales, where soils were thicker and water was near the surface, amid smooth but stony uplands of hard limestone. At least locally, rock-faced terraces were constructed during Early Bronze times (Gibson et al. 1991; Finkelstein and Gophna 1993), another trait of intensified Mediterranean polyculture. But there also is reason to believe that most of the hill country remained forested until the early Iron Age, when terracing became more common and iron tools effective to fell large, live oaks became available (Stager 1985a, Liphschitz et al. 1989; Gibson et al. 1991). Only then is it probable that semi-pastoral groups began to insinuate themselves into the interstices between older, agricultural settlements (Stager 1985a).

The settlement ecologies of the hill country and the Shephelah were starkly different during the period of EBIII decline. In the Shephelah, urbanized towns were tightly nested within an "orchard" landscape that now receives 300-500 mm rainfall and was dependent on large springs or upland catchments for waters that made outfield cultivation possible, presumably through irrigation. In the hill country, widely dispersed villages and scattered towns demarcated patches of arboriculture, embedded within a wooded landscape that received adequate rainfall for crop cultivation. Nonetheless, both areas were depopulated, in stages across some 250 years. How can this be explained?

4.3 Early Bronze Palestine: Environmental and Systemic Factors

Potential explanatory modes must be examined in both local and regional contexts, and in possible conjunction with other variables.

Local climatic anomalies or trends could readily affect the Shephelah, located on the climatic margins of rain-fed agriculture. However cultivation was probably dependent on stream or spring flow, derived in some good part from aquifers recharged by highland precipitation, particularly if the limestone hill country remained largely forested at the time. Wine and olive production would be reduced by drought years, but olive trees are more commonly damaged by frosts than by droughts, and even if vine stalks are destroyed by persistent drought, they are readily replaced and productive within two or three years.

The floodplains of the Shephelah were very broad during the 4th millennium, channels shallow, and discharge perennial, and much less torrential than today (A. Rosen, 1991). That would have allowed flood agriculture, perhaps complemented by small-scale irrigation, and organic, waterlogged soils suggest that these "wet" floodplain conditions persisted during most of the third millennium probably until EBIV times, when a flush of initially fine, then increasingly coarse sediment accumulated on top of the floodplain near Tell Lachish (A. Rosen, 1986: 57-69). That indicates a more episodic flow, i.e. a different hydrological regime, reflecting either a more arid climate or devegetation, with soil erosion in the upper catchment, rather than on local slopes. This helps explain thriving EBIb to EBIII land use in the area, but offers no direct insight as to EBIII devolution, since A. Rosen's (1995) argument for the collapse of floodwater farming seems to apply to either the EBIV or Middle Bronze.

Quite apart from the inconclusive paleoclimatic proxies (see above), what most excludes climatic change as a significant factor is that the pattern of incremental EBIII settlement abandonment affected all of Palestine, including those areas with over 600 mm precipitation today, i.e. outside the zone of contemporary agricultural stress. Environmental degradation relates to local land-use histories and also would not affect a whole region simultaneously. There simply is no case to attribute collapse of metastable EBII-III sociopolitical structures to a "climatic shift," "desiccation," or other environmental change (contra Esse 1989).

An alternative scenario emphasizes deviation-amplifying feedback loops that lead to system failure without major recourse to external variables (Dever 1989), a sort of system implosion. A number of internal dynamics are suggested: (i) Exponential population growth leads to overcrowding, heightened social inequalities, and increasingly arbitrary efforts to impose control; (ii) Population growth also outstrips the effective carrying capacity of the environment, exacerbating ecological degradation, further curtailing surpluses in agricultural commodities, and increasing shortages to the point of famine; (iii) Subsistence crises would intensify pressures to increase yields, aggravating social tensions and economic chaos; (iv) The end of trade with Egypt, for whatever reason, would isolate Palestine from its closest markets; (v) Local, complex redistributive institutions, operated by a poorly-developed, inexpert, and over-centralized bureaucracy would be unable to cope with mounting crisis; and (vi) Increasing social in equality in times of shortage would provoke unrest, ultimately sundering the social fabric through progressive distintegration rather than revolution (Dever 1989). External inputs--a "slight" climatic fluctuation and nomadic incursions--are then invoked as triggering mechanisms for the final breakdown of an already fragile system (Dever 1989: 238).

After castigating proponents of the same external factors, Dever's neo-Marxian scenario seems to want to have its cake and eat it too. Internal systemic collapse is indeed possible, with declining efficiency of energy and information pathways, particularly when several negative trends reinforce a complex crisis in times of ineffective government (Butzer 1980a). But such crises are specific to a particular context; that is, they can effect one integrated system and at a particular time. In the case of EBIII Palestine, "collapse" effected a host of spatially disjunct and politically autonomous units, progressively and at different times over several centuries. Even within the Shephelah, the continuing disappearance of towns can hardly be attributed to internal "systemic collapse," even if efforts to achieve economic autarky were maladaptive and power was abused or misdirected. That would ignore the prominent evidence for insecurity and warfare, and would imply that ruling elites could not learn from one social implosion how to adapt and quell further disruptions before they in turn became disastrous.

4.4. Early Bronze Palestine: Warfare and Urban Destruction

The elaborate town walls or defensive sites in EBII-III Palestine must be seen against the settlement history of the coastal plain. First there was extensive abandonment in the north, then in the center; eventually almost three quarters of the population was crowded together in the southern Shephelah. All smaller sites were abandoned, some medium centers disappeared, but few new ones were formed. Finally, the remaining towns disappear one by one in a grim "until there were none" scenario. Within 200 years or less, the Shephelah lost all but one twentieth of its sedentary population, a catastrophic ratio. Although the majority of EBIII towns appear to have been simply abandoned, there also is substantial evidence for fire and destruction, e.g., two horizons at Erani, one at Hesi, and three at Halif (Seger 1989, 1995), or two at Ai in Samaria and one at Numeira, on the slopes above the Dead Sea (Callaway 1980; Stager 1992: II, 50-52). The Early Bronze towns will have been intensely competitive, but a convincing case for internecine warfare--with abandonment as a result of forced dislocation or destruction--has to confront the virtual certainty that Egyptian campaigns ravaged the Shephelah. Finally, there is always the lingering question why the very last Shephelah town was eventually abandoned.

Apart from perfunctory Old Kingdom references to successful hostilities with the Asiatics ("sand-dwellers") of Palestine, at least two specific records leave no doubt about serious Egyptian campaigns (Kemp 1983: 142-143). The tomb reliefs of Inti (early Dynasty 6) shows a standard, oval town-wall with rounded bastions under siege. The second is an inscription of Uni, an officer of Pepi I (c. 2340 B.C.E.), who boasts of five land and one seaborne campaign against the rebellious Asiatics, one of which involved conscripts from every district of Egypt, together with Nubian mercenaries, i.e. a large army, Uni threw down the Asiatics' walls, set fire to their settlements, killed immense numbers in battle ("many ten-thousands"), cut down their fig-trees and vines, and brought back countless captives (Drower and Bottéro, 1971: 351-362; J.A. Wilson, in Pritchard, 1969: 227-228). Fig-trees may be a misnomer for olive groves--unknown in Egypt, but in any event the reference to Palestinian polyculture is unambiguous. The attack by sea may relate to an area just beyond the Carmel promontory. These facts imply considerable, direct engagement by Egypt along the coastal plain, and possibly a strongly-disputed claim to hegemony.

In the absence of archaeological evidence to the contrary, the towns of the Shephelah will not have been under Egyptian control, but some sort of commercial dependency is plausible. The presence of moderately large EBIII towns at Paran and Gerisa, where wadis cut through the cordon of coastal dunes and kurkar ridges to the sea, suggests anchorages from which olive oil and wine could have been exported by boat to Egypt. Their townspeople may have thrived as Egyptian allies. Allowing for typical exaggeration in Uni's account, the sort of campaign described still suggests that Egyptian intrusion for whatever cause, could have disastrous demographic and ecological implications, sufficient to wipe one or more towns of several thousand inhabitants off the map. Some sort of Egyptian involvement in Ai, located on a trade route from Gerisa to Jericho, also cannot be dismissed (see Callaway 1980).

Whatever the logic of Egyptian aggression, this scenario helps illuminate several aspects of the Palestinian EBII-III. The peaceable role that Egypt enjoyed in the Shephelah, be it commercial or administrative, ended early in Dynasty I. This marked the onset of the EBII, with most larger settlements throughout Palestine being walled, beginning a process of "urban" nucleation that also served defensive purposes. Coastal sites from Meghanim to Lod were abandoned, the Plain of Sharon partly depopulated, and large agglomerations began to assemble in a tightly -nested cluster in the Shephelah, while smaller settlements disappeared. Then even the largest towns were knocked off, one by one. When Dynasty 6 went under, only the small foothill-town of Beit Mirsim and a squatter (?) settlement at Halif remained. Projecting Egyptian warfare as a prime, but not exclusive agency, this archaeological progression of settlement regression makes good sense. It would, however, cast a very different light on the roots of militarization, centralization, and greater sociopolitical complexity. Contrary to the shopworn, Marxist model of emerging elites grasping for greater wealth and power, at the expense of common people, local "urbanization" would be reinterpreted as an increasingly desperate effort by a beleaguered people to survive with some level of group dignity. The emerging elites and priestly caste then become heroes instead of villains.

Military campaigns may also have led to the destruction of other towns in Syria. The cuneiform library of Palace G in Ebla (Mardikh IIB1) was destroyed during or just after the first 30 years of the reign of Pepi II (i.e., c. 2330-2300 B.C.E.), and at the time of the same king, Byblos suffered massive destruction, as did Hama and towns in the Amuq Plain (Table 1) (Matthiae 1982; Matthiae et al. 1985; Schwartz and Weiss, 1992: I, 236-240; Drower, 1971: 335-340, 345-351; Posener et al., 1971: 587-589; Kühne 1976). That seems to place these events in the reign of Sargon of Akkad (2293-2237 B.C.E., according to the "middle chronology"), who campaigned far to the northwest, to Mari, Ebla, the Cedar Mountain, and a place called Iarmuti, probably on the Mediterranean coast, perhaps in his 11th regnal year (Gadd, 1971: 424-426; Schwartz and Weiss, 1992: 238-240; but see also Cooper 1986 for alternative possibilities). Allowing for the uncertainties in correlating the Mesopotamian and Egyptian chronologies, the Mesopotamian dynast who destroyed Ebla, also appears to have burned down Byblos--the richest city on the coast--which switched from Egyptian to Mesopotamian hegemony at this time (level J1, Saghieh 1983). Roughly a century later Ebla was reconquered by Sargon's grandson Naram-Sin (c. 2291-2255 B.C.E.), as was Mari, an Armanum (? Aleppo), and a coastal town called Ulusum (Drower and Bottéro, 1971: 325-327; Gadd, 1971: 441-442). Destruction horizons are found at Ebla (end of level IIB2) and Ras Shamra (Ugarit), on the coast (end of level

IIIA3) (Table 1); although their regional or absolute dating is not firmly fixed (Oates 1985; Schwartz and Weiss, 1992: 240; Posener et al., 1971: 340; de Contenson 1989), this and similar problems of correlation or linkage should ultimately be amenable to pottery seriation, AMS dating, or dendrochronology.

Large scale, organized warfare is currently not popular as an explanatory factor among anthropologists enamored with internal, evolutionary dynamics in what are, in effect, closed-system models. But the historical record leaves no doubt as to its recurrent role and impacts.

4.5. Early Bronze Palestine: EBIV Disintensification

It is now common for archaeologists to attribute the Palestinian EBIV to a process of "ruralization." Indeed, the hallmark of the EBIV is a changing settlement archaeology: disappearance of EBII-III towns, with total abandonment of some areas, and new, small settlements in others. Some sites were compact and defensive (Richard 1986), whereas in the north larger sites sprawled across earlier sites with little nucleation (Prag 1974; Esse 1991). Although EBIV settlements were substantially smaller than those of the EBI, EBII or MBII, rank-size analyses show a modest level of integration that did not change appreciably over time (Falconer 1994; Fall et al. 1996). That argues for "ruralization" rather than "pastoralization."

Although the Shephelah remainly thinly populated, small villages reappeared along the coastal plain for the first time since EBI times; but, including Beit Mirsim, there were only three towns with 1000 inhabitants or more (Gophna and Portugali 1988). There is a string of small, isolated cemeteries in the Judean Hills, but only a handful of occupation sites, mainly small; further north in Samaria, there is an approximately equal number of cemeteries and sites, several of them large, and most located east of the watershed (Finkelstein 1991). Settlement density was highest in the Beth Shan Valley and Lower Galilee, where site number and size suggest little decline of population, when prorated per century (Esse 1991: Table 6 and Site List). East of the Jordan, population size along some tributaries was greater than at any time since the EBI (Helms 1992).

The geography of settlement begs ecological interpretation, because the densest patterns are found in drier parts of the country, by itself an argument against desiccation. (For an exemplary study relating Early Bronze occupation to landscape geoecology in Lebanon, see Marfoe, 1979). Although EBIV sites remained concentrated near springs or running water, they were commonly located above deeply incised valleys or channels but rarely found on floodplains or alluvial surfaces, liable to floodwater farming or simple irrigation (e.g., Mabry and Palumbo, 1992). Yet there appears to be little change in the proportions of different domestic animals from the increasing number of zoo-archaeological studies for EBIV (e.g., Prag, 1974; Horwitz 1989a, 1989b), with pig and cattle remaining prominent, implying household animals and plow-oxen, with at least some use of secondary products (e.g., milk) from the dominant herds of sheep and goat. This implies the diversified stockraising of sedentary communities. Only at Bir Rassisim in the Negev do EBIV livestock take on a distinctive pastoralist cast (see Dever 1983). Similarly, at sites east of the Jordan River, percentages of cattle, sheep and goat remained almost unchanged between the EBIV and renewed, Middle Bronze agglomeration; olives, grapes, and figs were prominent in EBIV flotation samples, with a decline in grapes and an increase in olives during the Middle Bronze (Fall et al. 1996), i.e. greater investment in commercial crops that require a longer start-up time and imply a greater sense of security or long-term stability. Mediterranean polyculture was by no means extinct, although the Lake Kinnereth pollen core suggests that olive cultivation declined overall (Baruch 1990).

The archaeological data do not support settlement by a new group of people. The pottery and, to a degree, the metalwork show: (a) basic continuity from EBII-III traditions; (b) a trend to regional diversification, running against the earlier tide of regional standardization and suggesting greater isolation or local reaggregation according to kin-group relationships; and (c) limited innovations derived from northern Syria and the middle Euphrates, perhaps via refugee groups fleeing Akkadian military destruction (e.g., Prag 1974, 1991; Helms 1989; Dever 1991; Esse 1991; Palumbo and Peterman 1993). The depopulation of the Shephelah and the western, central hill country stands out, and it seems reasonable to suggest that the last Egyptian campaigns, with the ruthlessness of that of Uni, destroyed that urban network and almost emptied the area. People may have fled to the east and north, with only lagging resettlement of the coastal plain during the EBIV, after Old Kingdom devolution.

Site excavations, that might support the persistent hypothesis of an expansion of pastoralism at the expense of sedentary agriculture (Dever 1991; Finkelstein 1991), are lacking. The case now hinges largely on hill cemeteries away from any local settlements, as in the central hill country. The idea is that these were sacred sites of itinerant pastoral groups, possibly shared by unrelated pastoral groups across centuries or even millennia. Most of the skeletal remains are disarticulated, suggesting ritual reburial. But disarticulation was already common in EBI times, and extended transport of human remains required months of prior deputrefaction and seems impractical for mobile groups (Hanbury-Tenison 1986: 245-250). The isolated hill cemeteries may well represent a cultural tradition with ideological overtones, rather than socioeconomic implications.

The matter of ideology needs considerably more attention. Joffe (1993) makes the important point that the strong trend to urbanization during the Early Bronze may have created a dialectic between two competing social forms, between the older, non-urban way of life and its more recent urban counterpart. These would entail different social structures and ideological visions, representing deep-seated contradictions and tensions within society. By extension, "urban collapse" would have discredited the new way, thus favoring the return to a more rural way of life and its values, symbols, and ideology. It would therefore be interesting to examine more closely the possible links of the hill cemeteries with the "megalithic" tombs found on convex topographies along both flanks of the Ghor, northwards to the Orontes Basin (Zohar 1991). Near Jerash, such tombs cluster near a string of EBI settlements (Sapin 1992: 171), but elsewhere they may continue into the Middle Bronze. In addition to real dolmens, built of large rock slabs placed horizontally over smaller, vertical ones, such cemeteries include tumuli and rectangular, built cists, such as are found in EBIV burial complexes (Prag 1974).

Where were the sedentary sites related to the hill cemeteries? First and foremost it requires reemphasis that small EBIV sedentary sites have been identified in the central hill country, as in the Refaim Valley near Jerusalem (Horowitz 1989a, 1989b). But a fresh angle is put on the issue by Dahr Mirzbaneh, a steep-sided hilltop located 17 km northwest of Jericho and explored by Finkelstein (1991). Walls enclose an area of 70 by 140m, dominated by a prominent rectangular tumulus at the highest end, and enclosing some stone foundations. This is a classic refuge site, analogous to Monte Mollet in eastern Spain (Bazzana and Guichard 1978), where the coastal plain was abandoned because of insecurity c. 400-1000 C.E. (Butzer et al. 1983). A more recent, Greek analog can be suggested to infer the function of Monte Mollet. In the Ottoman Peloponnese, the villages of the coastal plain were abandoned in favor of adjacent defensive mountain sites; although some cultivation continued in the lowlands, using temporary housing, or on terraced hillsides near permanent villages, much of the population moved up and down the mountains with their livestock on a seasonal schedule (Beuermann 1967: 92-119). Mediterranean polyculture continued, but the people became mobile and switched their emphasis to herding. These examples illuminate the likely mechanisms of depopulation, de-urbanization, and disintensification of agriculture in late Roman--Visigothic Spain (Butzer et al. 1985). Similar processes are apparent in Italy, where lowland settlements were abandoned c. 500-700 C.E. in favor of hilltop towns (Potter 1979). Mediterranean polyculture was resilient and could be maintained even when villagers were forced to shift their livelihood to mobile pastoralism. Obliged to adopt a new sociocultural role, they experimented with a different combination of familiar economic strategies. But this disintensification did not involve "retribalization," because traditional, sedentary lifeways were promptly resumed when conditions changed, as they did during Middle Bronze "re-urbanization" in Palestine.

Dahr Mirzbaneh appears to be an important anomaly, that contrasts with settlement continuity at most sites in the north of Palestine and east of the Jordan, and in at least some central locations such as the Refaim Valley. If the interpretation offered -- a defensive refuge site -- proves to be tenable, it would imply that for part of the EBIV time span, security was a fundamental problem in some areas. Just who represented the major threat remains obscure, but the apparently drastic nature of the solution suggests something far more serious that sporadic, pastoralist raids. More likely is the presence of strong hostile forces, within some proximity, for at least some decades. The small size of the walled enclosure suggests a community of only modest size which -- in its specific archaeological context -- implies a crisis situation not at the very beginning of the EBIV, but after the larger towns had already disappeared and population levels had been downsized.

4.6. Early Bronze Palestine: The Nature of Sociopolitical Devolution

That the EBIV of Palestine represented troubled times, with ruralization and partial or local depopulation is not disputed. But the tide of evidence has begun to turn strongly against the assumption that there was an increase of mobility and herding. The bioarchaeological evidence in particular has failed to support either a shift from agriculture to pastoralism, or from household pigs and plow-oxen to potentially mobile sheep or goat herds. Furthermore, even though the markets for commercial products had evaporated and the commercial network was decentralized, arboriculture -- always a multi-year investment in the future -- continued to be practiced. But did pastoral groups infiltrate the area, and eventually become sedentary? That would imply displacement and discontinuity. Here the archaeological record becomes paramount, to trace family and kin organization, through excavation within dwellings and by uncovering spatial relationships between adjacent houses and among house clusters. Ideally, both multicomponent and smaller, single-component sites should be tested in future excavations, to look for synchronic variabilty and to compare with "domestic" archaeology of earlier phases of the Early Bronze.

The much broader issue is whether EBIII breakdown represented *sociopolitical devolution,* that is, structural simplification of hierarchical institutions, or *socioeconomic collapse*. The latter implies a reduction of social complexity, from a stratified to a segmentary society, according to Esse (1989: 92-93; 1991: Fig. 33). Esse equates degrees of social complexity with settlement size and level of urbanization; from this he ascribes a ranked society ("chiefdoms") to the proto-urban EBIb, a stratified society ("state") to the "urban" EBII-III, and a segmentary society (roughly, "clan/tribe") to the "hamlets" of the EBIV. Since the late Chalcolithic Beersheva culture (Levy, 1992) may have been a ranked society, Esse's demotion of the EBIV to a segmentary/tribal society would represent a 2000 year retrogression on the local evolutionary time scale! Yet Esse's argument simply explicates what is implicit in Dever (1991), who uses "retribalization," or in those authors who infer that much of Palestine was given over to "nomadic pastoralists."

Anthropological typologies such as "tribe," "chiefdom," or "stratified society" should not be applied in lieu of explanation. They are heuristic devices to illustrate evolutionary theory, that cannot be applied in the real world where societies shift across a spectrum of available economic strategies and sociocultural roles (Marfoe, 1979). For a nuanced and more appropriate assessment of social systems in Early Bronze Greece, see Renfrew (1972: chap. 18).

Early Bronze urbanism in Palestine began with population growth during EBIb, as reflected in both settlement number and size, at a time of active trade with Egypt. That trade gradually shifted from raw materials to agricultural products, presumably stimulating surplus production, market exchange at several levels, and some form of managing agricultural credits. By creating incentives and buffering subsistence-risk, that might lead to population growth, especially in centers where market functions would encourage a degree of specialization in craft products. That may have allowed the formation of a small elite group, whose multiple roles included not only the maintenance of order and regulation of markets, but also redistribution of part of the crop yield in times of subsistence crisis, or as credits to underwrite investments in the expansion of staple production, vineyards, or olive groves. On modern risk and rural credits, see Marfoe (1979: 5-8). Sooner or later, military leadership would be required, presumably strengthening the role of a community-anchored cult that legitimized the existing social order and served to rally social solidarity in times of crisis. From these hypothetical processes some sort of amalgam would emerge that combined bureaucratic, military, and cult roles. The functional details and the intertwining of powers among administrative elites, privileged lineages, and priestly castes remain poorly understood, because the distinction of secular vs. religious or of public vs. private are cultural artifices. They cannot be deciphered with the perceptions and preconceptions of Western cultural traditions.

In historical times, population growth has generally been linked to market integration, urban nucleation, and agricultural expansion or intensification. The multiple and overlapping jurisdictions of the High Middle Ages in Europe suggest that political centralization was not indispensible. Far more difficult to comprehend are the social arrangements (the subsistence ethic or "moral economy") and ideological underpinnings that facilitated these processes. The goals, justification, degree, and rigidity of social stratification must therefore remain conjectural, despite the self-assurance of current investigative paradigms.

As the relationship between Egypt and Palestine became more contentious, the palace-temple elites of the Palestinian towns adopted strong defensive measures, compressing their populations within elaborate walls and clustering their settlements in the Shephelah. Although Egypt's major trade artery was anchored in Byblos, there were periodic attempts to exert hegemony in Palestine, resulting in destructive military campaigns that may have culminated in early Dynasty 6, with the eradication of almost all the towns in the southern Shephelah. Whether these clustered, large towns of the Early Bronze represented city-states with stratified socioeconomic classes is open to dispute. Compared with the evidence of concentrated wealth and structural specialization at Ebla, or the planned Middle Kingdom towns of Egypt, the towns of the Shephelah were proto-urban and their societies more probably ranked than stratified, judging by the modest architectural differentiation. Nonetheless, centuries of cramped living and enforced cooperation within the walled confines of a defensive site implies adherence to a social contract diverging from that upheld by the inhabitants of dispersed villages or by mobile peoples who moved with their herds.

In the wake of EBIII demise and ruin, the social contract that bound town-dwellers in Palestine became discredited, presumably reaffirming older ideological systems. In that sense, the EBIV may have represented an incisive, sociocultural break with the past, provoking deliberate institutional simplification, with perhaps unexpected, negative repercussions for market exchange, economic integration, and agricultural intensification. Subsequently, the collapse of Egypt and Akkad, two bastions of the regional exchange system, will have deepened and prolonged the economic recession for as much as two centuries. But regardless of how EBIV society was organized, many of the key institutional concepts will have been preserved in skeletal form. Indeed, in the context of renewed, Middle Bronze market integration, agricultural expansion, and urbanization, they were rapidly re-implemented.

Sociopolitical devolution, as so defined, helps to elucidate the processes of change without categorizing the attributes of a flexible social system, in constant flux, by deductive typologies that obscure rather than illuminate the issues.

5. Political Devolution and the Near Eastern Economic Network

In studying sociopolitical devolution, understanding connections may be more productive than trying to establish causes. The key elements of a societal system are energy flows (primary production, redistribution, trade) and information pathways (technology, structural nodes, institutions) that explicitly function in response to human cognition and decisions. As these flows and pathways expand, the system grows in complexity, when they retract, it declines and simplifies.

Specific events, trends, or changes are only intelligible in their particular context. Local or minor inputs, whether cognitive or not, can spiral into systemic perturbations through unanticipated, "positive" feedback loops. Such a chain of events or repercussions is likely to be unique. Similarly, major inputs may be absorbed by the system through buffering, "negative" feedbacks, resulting in little or no change. Possible outcomes depend on the rate and efficiency of energy flows and information pathways--at a particular moment in time--that are imperfectly regulated or steered by human institutions and decisions, that may falter or turn maladaptive. Actual events or sustained trends are therefore not only unpredictable but also to some degree stochastic.

In a larger perspective, the cyclic expansion and retraction of societal systems can be compared with steady-state readjustments, possibly superimposed on long-term, net growth. Whatever kind of proxy data category is used--demography, area of political control, economic statistics--such cyclic behavior is characteristic of systems of all kinds, which are, contrary to common assumption, not homeostatic.

Different societies within a complex region function as "open" subsystems, within a much larger, interdependent whole. For instance, shifts or changes in Early Bronze Palestine depended to some degree on trends or events in Egypt, Syria, and Mesopotamia, and may well be affected by what was going on in the more distant Aegean world or even in the copper-producing areas of Spain. Such connectivities should stimulate comparative studies of interlinked subsystems, individually as well as in conjunction. That is why this study has dealt with both Egypt and Palestine. Examining their devolution, in isolation as well as together, offers complementary insights into what were complex processes, ultimately intelligible in an interactive context that includes Syria, Mesopotamia, and areas even further afield.

The Egyptian pharaohs apparently consolidated and expanded their power through their monopolistic control over the foreign import trade, which probably stimulated redistribution and surplus productivity in Egypt, in turn placing a premium on internal security and risk limitation, thus also favoring population growth. Power became highly concentrated and was predicated on centralized authority. Palestine responded to the economic demands of its neighbor by increasing production of the desired commodities, not only in the Shephelah but also by expansion of settlement and intensification of agriculture in the highlands. As market centers grew at strategic points, administrative structures had to be created to handle intraregional exchange, local redistribution, and extraregional commerce. That favored population growth and nucleation around the best networked market centers. Similar processes presumably affected western Syria, and indirectly stimulated trade in the Aegean world, with repercussions as far afield as Spain. The presumed Akkadian thrust into Syria and the Persian Gulf region would imply that the new dynasts in Mesopotamia wanted a larger stake in this network.

Collapse of the Near Eastern commercial network and the ensuing political devolution appear to have been a consequence of militarism. Based on circumstantial evidence, Egypt seems to have decided to enforce its hegemony over Palestine at all costs, leading to catastrophic destruction of its trade, its towns, and its essential productivity. Depopulated and devoid of export "capital," that region ceased to be productive, weakening the system, regardless of whether Palestinian oil and wine had mainly flowed to Egypt, Syria, or Mesopotamia. Sargon's strategy--to extend Mesopotamian hegemony to Syria and the Mediterranean coast--anticipated that of the New Assyrian Empire, and was based on "windfall" profit-taking: the destruction and looting of towns of their accumulated capital, leaving behind precariously weakened market centers that remained dependent, but were not administered directly, thus short in terms of "re-investment. That would have left the Akkadian economy incapable of growth and therefore fragile. For Egypt, the destruction of Byblos probably proved fatal cutting Egypt off from the Near Eastern exchange network.

With Palestine reduced to a subsistence economy, Egypt spiralling into decentralized chaos, and Akkad unable to generate capital, the Near Eastern politico-economic system collapsed. How this affected the Aegean world is obscure. Even a speculative scenario is precluded by the chronological confusion in the Aegean world, where the abandonment or destruction of towns on the Greek mainland can only be dated with respect to a relative archaeological stratigraphy, that is not supported by overlapping clusters of ^{14}C dates, and that cannot be correlated to the destruction of Troy II (compare Coleman 1992; Forsén 1992; Mellink 1992). Despite some similarities of settlement patterns with Palestine, "towns" in Greece and western Anatolia were, with notable exceptions, modest and little more than villages (see Renfrew 1972: Fig. 14.7). A survey in Thessaly shows that half the Early Bronze sites were still occupied in Middle Bronze times, and there is no change in ecological preference (Halstead 1977; see also Renfrew 1972: 245). That appears to support the ceramic evidence of both continuity and discontinuity in the Peloponnese, arguing against the invasion theory (Forsén 1992). At least in these two regions, disjunction towards the end of the Early Bronze appears limited compared with that in Palestine.

It is now appropriate to examine the Near Eastern exchange network of the third millennium as a possible "world-economy" (as distinct from a global, world economy). That concept, due originally to Wallerstein (1974), is elaborated by

Braudel (1979: 21-22) to describe an economic "theater"--a world unto itself, so to speak. At one level, this appears to describe the Near East of the third millennium, but its pre-capitalist structures operated without coinage or a formal banking system, and it did not have a logistic center, let alone a single "core." Instead there were multiple nuclei, representing subsystems with different degrees of "openness" and sociopolitical complexity: (1) Egypt and, at times, Mesopotamia were archaic states, in which external trade essentially was a government monopoly focused mainly on the importing of raw materials or luxury items destined for elite consumption. (2) The Aegean region, and possibly western Anatolia, included a mix of polities with variable levels of independence and social complexity, ranging from micro-states (Minoan Crete) to autonomous towns, probably linked in shifting confederacies or alliances, with none holding a monopoly on trade; although there will have been elite imports and some trans-shipment across the region, most exchange probably was internal. (3) Syria and Palestine were internally as variegated as the Aegean world, but the region also was a hub, a nexus of many interconnecting threads; external trade with Egypt and Mesopotamia, and to a lesser extent, Anatolia and the Aegean, was critical, while transshipment undoubtedly was an important component. The coastal and island entrepots of the Persian Gulf present a similar case in point, between Mesopotamia and the Harappan towns of western India. As defined by different forms of sociopolitical organization, monopolistic control over exchange, and ratios of internal to external trade, these three categories can provisionally be ascribed by value-neutral descriptors as a (1) major nucleus, (2) minor nucleus, or (3) nexus.

The advantage of such distinctions is not to create artificial boundaries in what was a territorially fluid and sociopolitically changing matrix, but to propose a sliding scale of complementary, functional roles. That in turn allows an emphasis on the multiplicity of long-distance interchanges representing energy and information exchange, predicated on heterogeneous natural and human resources. The "world economy" conceptualized here is a network model (Fig. 1), that is more explicitly dynamic and practicable than the core-periphery premise underlying the Wallersteinian world system model. It has closer analogs to the loosely organized "circuits" interlinked within the 13th century C.E. "world system," ably described but imperfectly articulated by Abu-Lughod (1989: Fig. 1) Each nucleus or nexus to some measure exploited the raw materials of its own "periphery," sometimes in a dependent relationship (e.g., Egypt in Nubia), more often perhaps via intermediaries or direct but sporadic contacts (e.g., Egyptian expeditions to Punt). But such "peripheries" were not necessarily dependent nor disadvantaged, although there is no pragmatic and generally acceptable measure for what might be considered advantageous, since most external interactions eventually have social or economic repercussions.

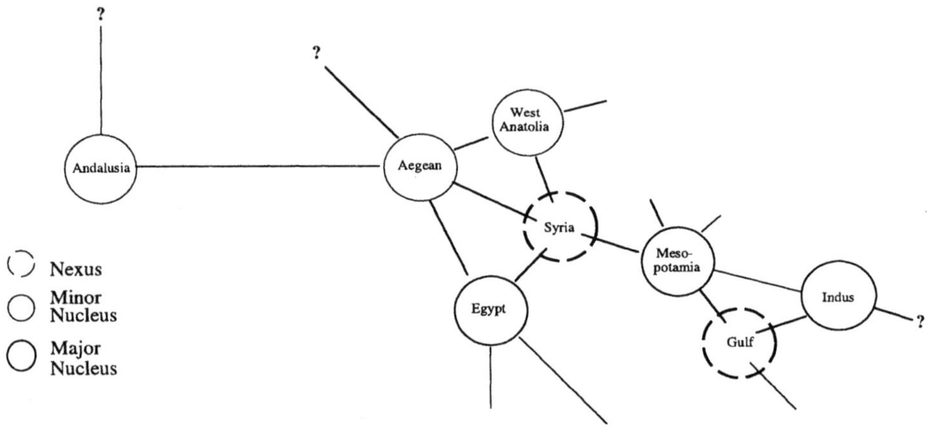

Figure 1. A Network Model for the Early Bronze World-Economy

Whatever we agree to call it, the Near Eastern commercial network and its unequal participants as far afield as Andalusia and the Indus Valley had empirical reality. As a heuristic device it can illuminate the vital role of exchange networks in generating "capital," by virtue of the interdependence of productivity and demand across a diversified spatial matrix. The details of that exchange vary widely from one context to another, but two common threads deserve emphasis: (1) the insatiable demand for copper and, eventually tin -- the indispensible alloy for bronze; and (2) the inevitable need for foodstuffs of symbolic importance (such as wine and olive oil) or in periodic or perennial short supply (grain).

Even in its initial application to elite objects, copper was processed by the ton during the EBA (Edens 1992), placing great strategic importance on commercial, diplomatic, client, or even military relationships not only with respect to sources such as Cypress and Sinai, but also distant areas such as India, Oman, Armenia, the Balkans or Andalusia. For example, Andalusian arsenic copper has been identified in Greek sites of the Chalcolithic and EBA (Sangmeister 1975), and feedbacks are presumed in the changing, sociopolitical organization of southern Spain. The search for tin lead even further afield (Muhly 1973), to places as distant as Cornwall.

As a bulk commodity, the mobilization of grain has been underappreciated, especially in overland transport (e.g., Wilkinson 1994). It bears noting that maize in prehispanic Mexico was transported considerable distances on human backs, making carriage by donkey appear efficient. Maritime shipment was more economical by a factor of ten, and grain exports from Mesopotamia to the Persian Gulf may have been a key commodity by the late third millennium (Edens 1992). It would be an error to simply assume that natural disasters such as a string of famine years could only be buffered by redistribution from the stored, grain surplus of a local sustaining area, even in the case of interior towns. Droughts tend to be localized and, given a profitable reciprocal relationship, customary prohibitions to the export of food staples in a time of crisis might well be waived. Although it is a later example, long distance food relief from Egypt to the Hittite kingdom is actually documented c. 1210 B.C.E. (Faulkner 1975:234). Without minimizing the cost of famine in human terms, it is a fact that such disasters -- even if not mitigated by external assistance -- are usually followed by a rapid return to normal productivity. Critical for recovery is a presumption of continuing "credits." Military disasters are a different matter, because much of the infrastructure is destroyed, fields and orchards ravaged, a substantial part of the population killed or enslaved, and intraregional "credits" likely to be suspended indefinitely. Recovery from such a calamity might lag for generations.

The functioning of such a networked world-economy seems to have been predicated on sustained productivity, conflict resolution, and continuing market exchange with minimal friction. Violence and destruction are not only counterproductive, but threaten the integrity of such a world-economy. The early Near Eastern prototype expanded and dissipated through three cycles, culminating in the Early, Middle, and Late Bronze. Various examples can be cited of conciliatory, trade-oriented policies, e.g., the political restraint of Middle Kingdom Egypt in Asia, the remarkable commercial orientation of the Early Assyrian state (unlike its successors), the international diplomacy of Amarna-age Egypt, or the lasting peace treaty and mutual assistance between Ramessid Egypt and the New Hittite kingdom. There was warfare, of course, but it remained directed to specific goals. Both the Middle and Late Bronze versions of the Near Eastern world-economy (c. 2000-1200 B.C.E.) were brought down not by military adventures of the key powers, but those of emerging peripheral nations or mobile marauders. Throughout the period, Syria, as a non-core region, remained at the intersection of commercial networks steered from politico-economic centers located near the perimeter; as in Medieval times, the role of entrepot cities was mainly a profitable one and their lack of real military power may have been advantageous in the long run.

As a systemic analog, this network model for an early world-economy is neither mechanistic nor anonymous. It presumes a strong cognitive component, and accords a central role to "prime movers" such as imperial ambition, war, invasion or pandemics that figure prominently in the historical record and will have been no less important in proto-historic times. Equally so it can accomodate the range of sociocultural components so splendidly identified by Edens (1992) for the Early Bronze Persian Gulf trade, such as diplomacy, cultural hegemony, and the social contexts of production and composition. But its primary advantage is that it draws the necessary attention to the interconnective strands and feedbacks that make politico-economic units functional components of a much larger whole. Then as now, power cannot be sustained indefinitely without a healthy economic base. Periodic collapse of the Near Eastern world-economy rendered its political structure latently unstable, even without any direct military crisis. Not surprisingly, political collapse tended to sweep the region in domino fashion. Such devolution compounded the problem of economic disintegration, through a set of feedback loops. That in turn greatly retarded the restructuring, without which stable political institutions were unlikely to reemerge.

The scenario proposed here is then based on two hypotheses: First, a cyclical integration and dissipation of the politico-economic network, to a large extent controlled by (i) the rate of commodity exchange and intraregional balancing of key resource or product deficiencies ("energy flows"), and (ii) the functioning of what might be called international "banking" arrangements (see Halstead 1992 for the later Bronze), whereby short and long-term credits and obligations could be managed on a fairly routine basis, to the satisfaction of second and third parties ("information pathways"). Second, a cyclical centralization and decentralization of power within local polities, primarily in response to the changing health of the larger system, but also impacted by various internal processes, trends and problems, as well as external inputs such as military events, attempted migration by force, or natural disasters. More often than not, the political cycles within a particular area would lag behind economic cycles affecting the entire region.

There is room for an impact by climatic perturbations or longer-term trends in such a model, but only as one potential variable within a multi-dimensional, interactive system. Phrased differently, climatic change cannot be treated in isolation; it must be examined in its historical, structural, and regional context.

Acknowledgments.

This paper benefited from discussions with Carlos Córdova (University of Texas), and I am grateful to my former students, Arlene Miller Rosen and Steven Rosen (Beer Scheva University) for information and advice. The manuscript profited from suggestions by Joe D. Seger (Cobb Institue of Archaeology) and Lawrence E. Stager (Harvard University); their encouragement was much appreciated. Charles Rosenfeld (Oregon State University) kindly provided references. The graphic was drafted by Kieke Butzer. Finally, I owe a major debt to that superb historian of Egypt, the late Klaus Baer.

References

Abu-Lughod, J.L. 1989. Before European Hegemony: The World System A.D. 1250-1350. Oxford: Oxford University Press.

Adams, R. McC. 1981. Heartland of Cities: Surveys of ancient Settlement and Land Use on the Central Floodplain of the Euphrates. Chicago: University of Chicago Press.

Algaze, G., Misir, A. and Wilkinson, T. 1992. Sanliurfa Museum/University of California excavations and surveys at Titris Höyük, 1991: A preliminary report. *Anatolica* 18: 33-60.

Algaze, G., Breuninger, R. and Knudstad, J. 1994. The Tigris-Euphrates archaeological reconnaisance project: Final report of the Birecik and Carchemish Dam survey areas. *Anatolica* 20:1-96.

Amiran, R. and Gophna, R. 1992. The correlation between Lower Egypt and southern Canaan during the Early Bronze period. In The Nile Delta in Transition, ed. E. van den Brink, pp. 385-394. Jerusalem: Israel Exploration Fund.

Arnold, Dieter and Arnold, Dorothea 1979. Der Tempel Qasr el-Sagha. Mainz: P. von Zabern (Archäologische Veröffentlichungen 27).

Baer, K. 1960. The Structure of the Egyptian Administration in the Fifth and Sixth Dynasties. Chicago: University of Chicago Press.

Bar-Yosef, O. and Khazanov, A., eds. 1992. Pastoralism in the Levant: Archaeological Materials in Anthropological Perspectives. Madison WI: Prehistory Press (Monographs in World Archaeology 10).

Baruch, U. 1990. Palynological evidence of human impact on the vegetation as recorded in Late Holocene lake sediments in Israel. In Man's Role in the Shaping of the Eastern Mediterranean landscape, eds. S. Bottema, G. Entjes-Nieborg and W. van Zeist, pp. 283-293. Rotterdam: Balkema.

Bazzana, A. and Guichard, P. 1978. Un important site refuge du haut Moyen Age dam la région valencienne. Le *despoblado* du Monte Mollet (Villafamés, Castellón). *Mélanges de la Casa de Velázquez* 14: 1-20.

Bell, B. 1970. The oldest records of the Nile floods. *Geographical Journal* 136: 569-573.

Bell, B. 1971. The Dark Ages in ancient history, I. The first Dark Age in Egypt. *American Journal of Archaeology* 75: 1-26.

Bell, B. 1975. Climate and the history of Egypt: The Middle Kingdom. *American Journal of Archaeology* 79: 223-269.

Beuermann, A. 1967. Fernweidewirtschaft in Südosteuropa. Braunschweig: G. Westermann.

Beug, H.J. 1967. Contributions to the Postglacial vegetational history of northern Turkey. *Quaternary Palaeoecology* 7: 349-356.

Bottema, S. 1985. Palynological investigations in Greece with special reference to pollen as an indicator of human activity. *Palaeohistoria* 24 (1982): 257-289.

Bowersock, G.W. 1996. The vanishing paradigm of the Fall of Rome. *Bulletin, American Academy of Arts and Sciences* 49(8): 29-43.

Brandl, B. 1992. Evidence for Egyptian colonization of the southern Coastal Plain and lowlands of Canaan during the Early Bronze I period. In The Nile Delta in Transition, ed. E. van den Brink, pp. 441-476. Jerusalem: Israel Exploration Fund.

Braudel, F. 1979. Civilization and Capitalism 15th-18th Century. III. The Perspective of the World. New York: Harper and Row.

Butzer, K.W. 1976. Early Hydraulic Civilization in Egypt. Chicago: University of Chicago Press.

Butzer, K.W. 1980a. Civilizations: Organisms or Systems? *American Scientist* 68: 517-523.

Butzer, K.W. 1980b. The Holocene lake plain of North Rudolf. *Physical Geography* 1: 44-58.

Butzer, K.W. 1984. Long-term Nile flood variation and political discontinuities in Pharaonic Egypt. In From Hunters to Farmers, eds. J.D. Clark and S.A. Brandt, pp. 102-112. Berkeley: University of California Press.

Butzer, K.W. 1988. Cattle and sheep from Old to New SpaIn Historical antecedents. *Annals, Association of American Geographers* 78: 29-56.

Butzer, K.W. 1990. The realm of cultural-human ecology: Adaptation and change in historical perspective. In The Earth as Transformed by Human Action, eds. B.L. Turner II and others, pp. 685-701. Cambridge: Cambridge University Press.

Butzer, K.W. 1994a. The Islamic traditions of agro-ecology: Crosscultural experience, ideas and innovations. *Ecumene* 1: 7-50.

Butzer, K.W. 1994b. Toward a new cultural curriculum for the future: A first approximation. In Re-Reading Cultural Geography, ed. K.E. Foote, pp. 409-428. Austin, Tx: University of Texas Press.

Butzer, K.W. 1996. Ecology in the long view: Settlement histories, agrosystemic strategies, and ecological performance. *Journal of Field Archaeology* 23:141-150.

Butzer, K.W., Miralles, I. and Mateu, J.F. 1983. Urban geo-archaeology in Medieval Alzira (Prov. Valencia, Spain). *Journal of Archaeological Science* 10: 333-349.

Butzer, K.W., Mateu, J.F., Butzer, E.K., and Kraus, P. 1985. Irrigation agrosystems in eastern Spaln Roman or Islamic origins? *Annals, Association of American Geographers* 75: 495-522.

Callaway, J.A. 1980. The Early Bronze Age Citadel and Lower City at Ai (El Tell). Cambridge MA: American Schools of Oriental Research.

Carey, S.N. and Sigurdsson, H. 1982. Influence of particle aggregation on deposition of distal tephra from the May 18, 1980, eruption of Mount St. Helens volcano. *Journal of Geophysical Research* 87(B8): 7061-7072.

Carey, S. and Sparks, R.S.J. 1986. Quantitative models of the fallout and dispersal of tephra from volcanic eruption columns. *Bulletin of Volcanology* 48: 109-125.

Clarke, D.L. 1968. Analytical Archaeology. London: Methuen.

Coleman, J.E. 1992. Greece, the Aegean and Cyprus. In Chronologies in Old World Archaeology, ed. R.W. Ehrich, Vol. I, pp. 247-288, Vol. II, pp. 203-221.

Contenson, H. de 1989. Rapports entre la Palestine et Ras Shamra-Ugarit du Bronze ancien. In L'urbanisation de la Palestine à l'âge du Bronze ancien, ed. P. de Miroschedji, pp. 317-330. Oxford: British Archaeological Reports, International Series 527.

Conti, A.M. and Persiani, C. 1993. When worlds collide: Cultural developments in Eastern Anatolia in the Early Bronze Age. In Between the Rivers and over the Mountains, eds. M. Frangipane, H. Hauptmann, M. Liverani, P. Matthiae, and M. Mellinck, pp. 361-414. Rome: Università de Roma "La Sapienza."

Cooper, J.S. 1986. Presargonic Inscriptions. New Haven: American Oriental Society.

Courty, M.A. 1994. Le cadre paléogéographique des occupations humaines dans le bassin du Haut Khabour, Syrie du Nordest. *Paléorient* 20: 21-60.

Cowgill, G.L. 1988. Onward and upward with collapse. In The Collapse of ancient States and Civilizations, eds. N. Yoffee and G.L. Cowgill, pp. 244-276. Tucson: University of Arizona Press.

Dagan-Ginter, A., Ginter, B., Kozlowski, J.K. Pawlikowski, M. and Sliwa, J. 1984. Excavations in the region of Qasr el-Sagha, 1981: Contribution to the Neolithic period, Middle Kingdom settlement and chronological sequences in the northern Fayum Desert. *Mitteilungen, Deutsches Archäologisches Institut, Abteilung Kairo* 40: 33-102.

Davis, S.J. 1988. The mammal bones: Tell Yarmuth 1980-1983. In Yarmouth I, ed. P. de Miroschedji, pp. 143-149. Paris: Editions Recherche sur les civilisations.

Degens, E.T., H.K. Wong, S. Kempe, and F. Kurtman 1984. A geological study of Lake Van, Eastern Turkey. *Geologische Rundschau* 73: 701-734.

De Putter, T. 1993. Les inscriptions de Semna et Koumma (Nubie): Niveaux de crues exceptionnelles ou d'un lac de retenue artificial du Moyen Empire? *Studien zur Altägyptischen Kultur* 20: 255-288.

Dever, W.G. 1983. Be'er Resisim -- a late third millennium B.C.E. settlement. *Qadmoniot* 16(2-3): 52-57.

Dever, W.G. 1992. Pastoralism and the end of the urban Early Bronze Age in Palestine. In Pastoralism in the Levant: Archaeological Materials in Anthropological Perspective, eds. O. Bar-Yosef and A. Khazanov, pp. 83-92. Madison WI: Prehistory Press.

Donahue, J., Peer, B. and Schaub, R.T. n.d. The southeastern Dead Sea PlaIn Changing shorelines and their impact on settlement patterns of the major historical periods of the region. Unpublished manuscript.

Drower, M.S. and Bottéro, J. 1971. Syria before 2200 B.C. In The Cambridge Ancient History, Vol. I, Part 2, pp. 315-362. Cambridge: Cambridge University Press, third. Ed.

Edens, C. 1992. Dynamics of trade in the ancient Mesopotamian "world system." *American Anthropologist* 94: 118-139.

Esse, D.L. 1989. Secondary state formation and collapse in Early Bronze Age Palestine. In L'Urbanisation de la Palestine à l'âge du Bronze ancien, ed.P. de Miroschedji, pp. 81-96. Oxford: British Archaeological Reports, International Series 527.

Esse, D.L. 1991. Subsistence, Trade and Social Change in Early Bronze Age Palestine. Chicago: Oriental Institute of the University of Chicago, Studies in Ancient Oriental Civilization 50.

Falconer, S.E. 1994. The development and decline of Bronze Age civilisation in the southern Levant: A reassessment of urbanism and ruralism. In Development and Decline in the Mediterranean Bronze Age, eds. C. Mathers and S. Stoddart, pp. 305-333. Sheffield: Sheffield Academic Press (Sheffield Archaeological Monographs 8).

Fall, P.L., Lines, L. and Falconer, S.E. 1996. Seeds of civilization: Bronze Age rural economy and ecology in the southern Levant. *Annals, Association of American Geographers* 86: in press.

Faulkner, R.O. 1965. The admonition of an Egyptian sage. *Journal of Egyptian Archaeology* 51: 53-62.

Faulkner, R.O. 1975. Egypt from the inception of the Nineteenth Dynasty to the death of Ramesses III. In the Cambridge Ancient History (third ed.), Vol. II, pt. 2, pp. 217-251.

Finkelstein, I. 1991. The Central Hill Country in the Intermediate Bronze Age. *Israel Exploration Journal* 41: 19-45.

Finkelstein, I. and Gophna, R. 1993. Settlement, demographic, and economic patterns in the highlands of Palestine in the Chalcolithic and Early Bronze periods and the beginnings of urbanism. *Bulletin, American Schools of Oriental Research* 289: 1-22.

Forsén, J. 1992. The Twilight of the Early Helladics. Jonsered, Sweden: P. Aströms.

Franklin, J.F., MacMahon, J.A., Swanson, F.J., and Sedell, J.R. 1985. Ecosystem responses to the eruption of Mount St. Helens. *National Geographic Research* 1: 198-216.

Gadd, C.J. 1971. The dynasty of Agade and the Gutian invasion. In The Cambridge Ancient History, Vol. 1, Part 2, pp. 417-463. Cambridge: Cambridge University Press, third Ed.

Galaty, J.G. and Johnson, D.L., eds. 1992. The World of Pastoralism: Herding Systems in Comparative Perspective. New York: Guilford Press.

Gibson, S., Ibbs, B., and Kloner, A. 1991. The Sataf Project of landscape archaeology in the Judaean Hills. *Levant* 23: 29-54.

Ginter, B., Kozlowski, J.K., Pawlikowski, M. and Sliwa, J. 1982a. El-Tarif and Qasr el-Sagha. *Milteilungen, Deutsches Archäologisches Institut, Abteilung Kairo* 38: 97-129.

Ginter, B. Kozlowski, J.K., Pawlikowski, M., Sliwa, J. and Pazdur, M.F. 1982b. Qasr el-Sagha 1980: Contributions to the Holocene Geology, the Predynastic and Dynastic Settlements in the Northern Fayum Desert. Warsaw-Krakow: Nakladem Uniwersytetu Jagiellonskiego.

Gophna, R. and Portugali, J. 1988. Settlement and demographic processes in Israel's coastal plain from the Chalcolithic to the Middle Bronze Age. *Bulletin, American Schools of Oriental Research* 269: 11-28.

Halfman, J.D., Jacobson, D.F., Cannella, C.M., Haberyan, K.A. and Finney, B.P. 1993. Fossil diatoms and the mid to late Holocene paleolimnology of Lake Turkana, Kenya: A reconnaissance study. *Journal of Paleolimnology* 7:23-35.

Halstead, P. 1977. Prehistoric Thessaly: The submergence of civilisation. In Mycenaean Geography, ed. J. Bintliff, pp. 23-29. Cambridge: British Association for Mycenaean Studies.

Halstead, P. 1992. Agriculture in the Bronze Age: Towards a model of palatial economy. In Agriculture in Ancient Greece, ed. B. Wells, pp. 105-118. Stockholm: Skrifter Utgivna av Svenska Institute i Athen 42.

Hammer, C.U., Clausen, H.B. and Dansgaard, W. 1980. Greenland ice sheet evidence of post-glacial volcanism and its climatic impact. *Nature* 288: 230-235.

Hamroush, H.A. and Stanley, D.J. 1990. Paleoclimatic oscillations in East Africa interpreted by analysis of trace elements in Nile delta sediments. *Episodes* 13: 264-269.

Hanbury-Tenison, J.W. 1986. The Late Chalcolithic to Early Bronze I Transition in Palestine and Transjordan. Oxford: British Archaeological Reports, International Series 311.

Hartung, U. 1994. Bemerkungen zur Chronologie der Beziehungen Agyptens zu Südkanaan in spätprädynastischer Zeit. *Mitteilungen, Deutsches Archäologisches Institut, Abteilung Kairo* 50: 107-113.

Harvey, C.P.D. 1982. The archaeology of the southern Sudan: Environmental context. In Culture History in the Southern Sudan: Archaeology, Linguistics and Ethnohistory, eds. J. Mack and P. Robert-Shaw, pp. 7-18. Nairobi: British Institute in Eastern Africa.

Hayes, W.C. 1971. The Middle Kingdom in Egypt. In The Cambridge Ancient History, (third ed.), Vol. 1, pt. 2, pp. 464-531. Cambridge: Cambridge University Press.

Helms, S. 1989. An EBIV pottery repertoire at Amman, Jordan. *Bulletin, American Schools of Oriental Research* 273: 17-26.

Helms, S. 1992. The 'Zarqa Triangle': A preliminary appraisal of protohistorical settlement patterns and demographic episodes. *Studies in the History and Archaeology of Jordan* 4: 129-136.

Heltzer, M. and Eitam, D., eds. 1987. Olive Oil in Antiquity. Haifa: Culture and Art Division, Ministry of Education and Culture.

Horwitz, L.K. 1989a. Sedentism in the Early Bronze IV: A faunal perspective. *Bulletin, American Schools Oriental Research* 275: 15-26.

Horwitz, L.K. 1989b. Diachronic changes in rural husbandry practices in Bronze Age settlements from the Refaim Valley, Israel. *Palestine Exploration Quarterly* 121: 44-54.

Horwitz, L.K. and Tchernov, E. 1989. Animal exploitation in the Early Bronze Age of the southern Levant. In L'urbanisation de la Palestine à l'âge du Bronze ancien, ed. P. de Miroschedji, pp. 279-296. Oxford: British Archaeological Reports, International Series 527.

Issar, A.S., Govrin, Y., Geyh, M.A. Wakshal, E. and Wolf, M. 1992. Climate changes during the Upper Holocene of Israel. *Israel Journal of Earth Sciences* 40: 219-223.

Jahns, S. 1991. Untersuchungen über die holozane Vegetationsgeschichte von Süddalmatien und Südgriechenland. Doctoral Dissertation, Göttingen University.

Jahns, S. 1993. On the Holocene vegetation of the Argive Plain (Peloponnese, southern Greece). *Vegetation History and Archaeobotany* 2: 187-203.

Joffe, A.H. 1993. Settlement and Society in the Early Bronze Age I and II, Southern Levant: Complementarity and Contradiction in a Small-Scale Complex Society. Sheffield: Sheffield Academic Press (Monographs in Mediterranean Archaeology 4).

El-Kammar, A.A., Phillip, G. and Arafa, E.H. 1992. Geochemistry of Recent Nile sediments from the main course of the Nile in Egypt, and the principal tributaries in Ethiopia and Sudan. In Geology of the Arab World, pp. 527-541. Cairo: Cairo University.

Kanawati, N. 1977. The Egyptian Administration in the Old Kingdom: Evidence on its Economic Decline. Warminster UK: Aris and Phillips.

Kay, P.A. and D.L.Johnson 1981. Estimation of Tigris-Euphrates streamflow from regional paleoenvironmental proxy data. *Climatic Change* 3: 251-263.

Kemp, B.J. 1983. Old Kingdom, Middle Kingdom and Second Intermediate Period in Egypt. In Ancient Egypt: A Social History, eds. B.G. Trigger, B.J. Kemp, D. O'Connor and A.B. Lloyd, pp. 71-182. Cambridge: Cambridge University Press.

Kraus, F.R. 1966. Staatliche Viehhaltung im altbabylonischen Lande Larsa. Amsterdam: K. Nederlandse Akademie van Wetenschappen, Letterkunde 29, No. 5.

Kühne, H. 1976. Die Deramik vom Tell Chuera und ihre Beziehungen zu Funden aus Syrien-Palästina, der Türkei und dem Iraq. Berlin: Mann. (Vorderasiatische Forschungen der Oppenheim Stiftung 1).

Levy, T.E. 1992. Radiocarbon chronology of the Beersheva culture and Predynastic Egypt. In The Nile Delta in Transition, ed. E. van den Brink, pp. 345-356. Jerusalem: Israel Exploration Fund.

Liphschitz, N. 1989. Plant economy and diet in the Early Bronze Age in Israel. In L'urbanisation de la Palestine à l'âge du Bronze ancien, ed. P. de Miroschedji, pp. 268-278. Oxford: British Archaeological Reports, International Series 527.

Liphschitz, N., Gophna, R., and Lev-Yadun, S. 1989a. Plant economy and diet in the Early Bronze Age in Israel. In L'urbanisation de la Palestine à l'âge du Bronze ancien, ed. P. de Miroschedji, pp. 268-278. Oxford: British Archaeological Report, International Series 527.

Liphschitz, N., Gophna, R., and Lev-Yadun, S. 1989b. Man's impact on the vegetational landscape of Israel in the Early Bronze Age II-III. In L'urbanisation de la Palestine à l'âge du Bronze ancien, ed. P. de Miroschedji, pp. 269-278. Oxford: British Archaeological Reports, International Series 527.

Lyons, J.V. 1986. Agricultural impact and adjustment to the Mount St. Helens ashfall. In Mount St. Helens: Five Years Later, ed. S.A.O. Keller, pp. 423-429. Cheney WA: Eastern Washington University Press.

Mabry, J. and Palumbo, G. 1992. Environmental, economic, and political constraints on ancient settlement patterns in the Wadi al-Yabis region. *Studies in the History and Archaeology of Jordan* 4: 67-72.

Marfoe, L. 1979. The integrative transformation: Patterns of socio-political organization in southern Syria. *Bulletin, American Schools of Oriental Research* 234: 1-42.

Matthiae, P. 1982. The problem of the relations between Ebla and Mesopotamia in the time of the royal palace of Mardikh IIB1 (ca. 2400-2250 B.C.). In Mesopotamia und seine Nachbarn, eds. H.J. Nissen and J. Renger, Vol. 1, pp. 11-124. Berlln Berliner Beiträge zum Vorderen Orient.

Matthiae, P. Mazzoni Archi, S., and Scandone Matthiae, G., eds. 1985. Da Ebla a Damasco. Rome: Electa.

Matthiae, P. 1993. On this side of the Euphrates: A note on urban origins in Inner Syria. In Between the Rivers and over the Mountains, eds. M. Frangipane, H. Hauptmann, M. Liverani, P. Matthiae, and M. Mellinck, pp. 523-530. Rome: Università de Roma "La Sapienza."

MEDIBA (Mediterranean Basin Program) 1991. Nile Delta project database listings, Washington: National Museum of Natural History.

Mellink, M.J. 1992. Anatolia. In Chronologies in Old World Archaeology, ed. R.W. Ehrich, Vol. I, pp. 207-220, Vol. II, pp. 171-184.

Muhly, J. 1973. Copper and Tin. Hamden CT: Arzhon Books.

Musil, A. 1928. The Manners and Customs of the Rwala Bedouins. New York: American Geographical Society.

Oates, J. 1985. Tell Brak and chronology: The third millennium. *Mari: Annales de recherches interdisciplinaires* 4: 137-144.

O'Connor, D. 1974. Political systems and archaeological data in Egypt: 2600-1780 B.C. *World Archaeology* 6: 15-38.

Owen, R.B., Barthelme, J.W., Renaut, R.W. and Vincens, A. 1982. Palaeolimnology and archaeology of Holocene deposits north-east of Lake Turkana, Kenya. *Nature* 298:523-529.

Palumbo, G. and Peterman, G. 1993. Early Bronze Age IV ceramic regionalism in central Jordan. *Bulletin, American Schools of Oriental Research* 289: 23-32.

Porada, E., Hansen, D.P., Dunham, S., and Babcock, S.H. 1992. The chronology of Mesopotamia, ca. 7000-1600 B.C. In Chronologies in Old World Archaeology, ed. R.W. Ehrich, Vol. I, pp. 77-121; Vol. II, pp. 90-124. Chicago: University of Chicago Press.

Posener, G., Bottéro, J. and Kenyon, K.M. 1971. Syria and Palestine c. 2160-1780 B.C. In The Cambridge Ancient History, Vol. I, Part 2, pp. 532-594. Cambridge: Cambridge University Press, third Ed.

Potter, T.W. 1979. The Changing Landscape of South Etruria. New York: St. Martin's Press.

Prag, K. 1984. Continuity and migration in the South Levant in the late third millennium. *Palestine Exploration Quarterly* 116: 58-68.

Prag, K. 1991. Preliminary report on the excavations at Tell Iktanu and Tell al-Hammam, Jordan, 1990. *Levant* 23: 55-66.

Pritchard, J.B. ed. 1969. Ancient Near Eastern Texts relating to the Old Testament. Princeton: Princeton University Press, third Ed.

Renfrew, C. 1972. The Emergence of Civilization: The Cyclades and the Aegean in the Third Millennium B.C. London: Methuen.

Richard, S. 1986. Excavations at Khirbet Iskander, Jordan: A glimpse at settled life during the "Dark Age" in Palestinian archaeology. *Expedition* 28:3-12.

Rosen, A.M. 1986. Cities of Clay: The Archaeology of Tells. Chicago: University of Chicago Press.

Rosen, A.M. 1991. Early Bronze Age Tell Erani: An environmental perspective. *Tel Aviv* 18: 192-204.

Rosen, A.M. 1995. The social response to environmental change in Early Bronze Age Canaan. *Journal of Anthropological Archaeology* 14: 26-44.

Rosen, S.A. and Avni, G. 1993. The edge of the empire: The archaeology of pastoral nomads in the southern Negev Highlands in late Antiquity. *Biblical Archaeologist* 56: 189-199.

Rosenfeld, C.E. and Cooke, R.I. 1982. Earthfire: The Eruption of Mount St. Helens. Cambridge MA: MIT Press.

Saghieh, M. 1983. Byblos in the Third Millennium B.C. Warminster: Aris and Phillips.

Said, R. Albritton, C.C., Wendorf, F., Schild R., and Kobusiewicz, M. 1970. A preliminary report on the Holocene geology and archaeology of the Northern Fayum Desert. In Playa Lake Symposium, ed. C.C. Reeves, pp. 41-62. Lubbock, TX: International Center for Arid and Semi-Arid Land Studies 4.

Sangmeister, E. 1975. Die Anfänge der Metallurgie in Europa. In Ausgrabungen in Deutschland, Part 3, 297-299, with maps. Mainz: Römisch-Germanisches Zentralmuseum.

Sapin, J. 1992. De l'occupation à l'utilisation de l'espace à l'aube de l'age du Bronze dans la région de Jérash et sa péripherie Orientale. *Studies in the History and Archaeology of Jordan* 4: 169-174.

Schoell, M. 1978. Oxygen isotope analyses of authigenic carbonates from Lake Van sediments. In The Geology of Lake Van, eds. E.T. Degens and F. Kurtman, pp. 92-97. Ankara: Mineral Research and Exploration Institute of Turkey.

Schwartz, G.M. and Weiss, H. 1992. Syria. In Chronologies in Old World Archaeology, ed. R.W. Ehrich, Vol. I, pp. 221-243, Vol. II, pp. 185-202. Chicago: University of Chicago Press, third Ed..

Seger, J.D., 1989. Some provisional correlations in EB III stratigraphy in southern Palestine. In L'Urbanisation de la Palestine à l'âge du Bronze ancien, ed.P. de Miroschedji, pp. 117-137. Oxford: British Archaeological Reports, International Series 527.

Seger, J.D. 1995. The point-one principle: A case study from the Halif Terrace. In Retrieving the Past: Essays on archaeological Research and in honor of Gus W. Van Beek, ed. J.D. Seger, pp. 245-268. Winona Lake IN Eisenbrauns.

Seidlmayer, S. 1987. Wirtschaftliche und gesellschaftliche Entwicklung in Übergang vom Alten zum Mittleren Reich. In Problems and Priorities in Egyptian Archaeology, eds. J. Assmann, G. Burkard and V. Davies, pp. 175-218. London: Routledge and Kegan Paul.

Smith, P. 1989. The skeletal biology and paleo-pathology of Early Bronze Age populations in the Levant. In L'urbanisation de la Palestine à l'âge du Bronze ancien, ed. P. de Miroschedji, pp. 297-313. Oxford: British Archaeological Reports, International Series 527.

Smith, P. and Horwitz, L.K. 1984. Radiographic evidence for changing patterns of animal exploitation in the southern Levant. *Journal of Archaeological Science* 11: 467-476.

Spencer, H. 1876. The Principles of Sociology. London: Williams and Norgate, Vol. 1.

Spengler, O. 1926-28. The Decline of the West. London. 2 vol.

Spurr, M.S. 1986. Arable Cultivation in Roman Italy c. 200 B.C.--c. A.D. 100. London: Journal of Roman Studies, Monograph 3.

Stager, L.E. 1985a. The archaeology of the family in ancient Israel. *Bulletin, American Schools of Oriental Research* 260: 1-35.

Stager, L.E. 1985b. The first fruits of civilization. In Palestine in the Bronze and Iron Ages, ed. J.N. Tubb, pp. 172-188. London: Institute of Archaeology Monographs.

Stager, L.E. 1992. Palestine. In Chronologies in Old World Archaeology, ed. R.W. Ehrich, Vol. I, pp. 22-41; Vol. II, pp. 17-60. Chicago: University of Chicago Press, third Ed.

Stevenson, A.C. and Harrison, R.J. 1992. Ancient forests in SpaIn A model for land-use and dry forest management in south-west Spain from 4000 B.C. to 1900 A.D. *Proceedings of the Prehistoric Society* 58: 227-247.

Stuiver, M. and Reimer, P.J. 1993. Extended ^{14}C database and revised CALIB 3.0 ^{14}C age calibration program. *Radiocarbon* 35: 212-230.

Toynbee, A.J. 1934. A Study of History. London: Oxford University Press, Vols. 1 and 3.

Van Zeist, W. and Bottema, S. 1991. Late Quaternary Vegetation in the Near East. Wiesbaden: Beihefte zum Tübinger Atlas des Vorderen Orients A18.

Vaux, R. de, 1971. Palestine in the Early Bronze Age. In Cambridge Ancient History (third ed), Vol.1, pt.2, pp. 208-237. Cambridge: Cambridge University Press.

Waddell, W.G. 1950. Manetho, with an English translation. London: Heinemann.

Wallerstein, I. 1974. The Modern World System (I). New York: Academic Press.

Wattenmaker, P. 1994. Political fluctuations and local exchange systems: Evidence from the Early Bronze Age settlements at Kurban Höyük. In Chiefdoms and Early States in the Near East, eds. G. Stein and M.S. Rothman, pp. 193-208. Madison WI: Prehistory Press (Monographs in World Archaeology 18).

Weiss, H., M.A. Courty, W. Wetterstrom, F. Guichard, L. Senior, R. Meadow, A. Curnow 1993. The genesis and collapse of third millennium North Mesopotamian civilization. *Science* 261: 995-1004.

Whittaker, C.R., ed. 1988. Pastoral Economies in Classical Antiquity. Cambridge: Cambridge Philological Society.

Wilkinson, T.J. 1990. The development of settlement in the North Jazira between the 7th and 1st millennia B.C. *Iraq* 52: 49-62.

Wilkinson, T.J. 1994. The structure and dynamics of dry-farming states in Upper Mesopotamia. *Current Anthropology* 35: 483-520.

Willis, K.J. 1992. The late Quaternary vegetational history of northwest Greece. III. A comparative study of two contrasting sites. *New Phytologist* 121: 139-155.

Zeder, M.A. 1994. Of kings and shepherds: Specialized animal economy in Ur III Mesopotamia. In Chiefdoms and Early States in the Near East, eds. G. Stein and M.S. Rothman, pp. 175-191. Madison WI: Prehistory Press (Monographs in World Archaeology 18).

Holocene Climate Change and Man in the Present Day Sahara Desert

Nicole PETITE-MAIRE, L. BEUFORT and N. PAGE

Laboratoire de Géologie du Quaternaire
Luminy, case 907
13288 Marseille, Cédex 9
France

Abstract

About 2,000 radiocarbon dates identify a chronology of multiple proxy indicators of the Holocene climate change in the Sahara. Histograms illustrate progressive shifts in three quite distinctive types of environment: typical Saharan interior lowlands, mountainous highlands and Atlantic littoral. The first displays abundant rains (partly summer monsoons, partly winter westerlies) with associated fauna, flora and human populations, from 9,000-6,000 BP (uncalibrated ^{14}C yr) falling to zero by 2,000 BP. The second shows mixed rainfall peaks before 8,000 BP; 4,000 and 2,000 BP. The third builds slowly to a peak before 3,000 and 2,000 BP. All display complete aridification by 2,000 BP, that is, long before modern overgrazing and political interference.

Introduction

The Sahara is the largest desert in the world, covering more than 9,000,000 km^2 of which half is hyperarid. The boundary between the desert and its marginal semi-arid transitional zones (the Sahel to the south, the Mediterranean region to the north) is identified by the 100-150 mm isohyets, roughly coinciding with the limit of active sand dunes and sedentary human habitats (except for mountainous areas and oases). The precipitation gradient is very steep, decreasing 1 mm/1.4 km from the Sahel to the Tropic of Cancer where the precipitation/evaporation (P/E) ratio is nearly zero and the potential evaporation ranges between 5 and 8 m for an average annual rainfall under 5 mm.

Over the last 20 years, many new observations and analyses of geological records here yielded a great number of data on past environmental and climatic changes

NATO ASI Series, Vol. I 49
Third Millennium BC Climate Change
and Old World Collapse
Edited by H. Nüzhet Dalfes, George Kukla,
and Harvey Weiss

throughout the present-day desert, reaching into its most desolate core along the Tropic of Cancer.

Data

The Holocene has been particularly well studied. Almost 2,000 radiocarbon dates of multiproxy paleoenvironmental samples (Vernet and Aumassip, 1992) and some continuous lacustrine sections (Petit-Maire et al., 1991) ranging from 10,000 to 3,000 yr BP, permit a good analysis of the regional evolution. A major part of this research has been provided by the members of former IGCP 252 "Evolution of Deserts: Past, Present, Future".

The data bank was used for preparing the map "The Sahara in the Holocene" (Petit-Maire, 1993), compiled from literature published over the last 20 years and adapted from earlier compilations (Commelin and Petit-Maire, 1979; Vernet and Aumassip, 1992). It has now been transmitted to the NOAA World Data Center (Paleoclimatology) at Boulder, Colorado, where it is available for consultation.

• Only radiocarbon-dated information has been used, which eliminates a good number of observations, most of which from literature prior to 1970. Data corresponding to some 1890 radiocarbon ages have thus been considered. They cover two types of information:

a) *Proxy markers of humidity,* implying local precipitation or aquifer rise:

• surface water: lakes, swamps (generally dated from shells of freshwater mollusks).

• geochemical deposits: calcaretes, travertines.

• vegetal cover: organic matter and paleosols, pollen, charcoal.

• skeletal remains of aquatic animals:: fish, crocodiles, hippopotamuses.

• skeletal remains of large land mammals (elephants, rhinoceroses, antelopes, etc., implying both fresh-water accessibility and a ground cover of grassland, or even wooded grassland (giraffes)

b) *Prehistoric habitats:* archaeological sites, burials, and artifacts, implying the possibility of a semi-sedentary life, i.e. accessible water and food (game, vegetation).

The climatic significance of these proxies has been widely discussed in literature (Petit-Maire and Riser, eds, 1983; Haynes and Mead, 1987; Neumann, 1989; Schultz, 1991).

Using this bank of data, we have here tried to couple significant Early Holocene climatic changes towards wetter/drier conditions with:

a) the human migrations *into* the Late Pleistocene hyperarid Sahara (Mainguet et al., 1980; Rognon, P., 1980: Talbot, 1984; Ousseini and Morel, 1989; Petit-Maire and Page, 1992) succeeded by the onset of the Neolithic civilization; and then

b) the human migrations *out of* the Early to Mid-Holocene semi-arid Sahara associated with the end of the humid phase.

Results

The paleomap shows the following general pattern :

(1) In the whole of the present Saharan area the Early Holocene warming induced a significant rise of the P/E ratio. Major environmental changes took place as early as 9,500 BP, with fresh-water lakes, Sahelian-type vegetation and fauna appearing even in the most continental areas, the hydrologically isolated lowlands around the latitude of the Tropic of Cancer.

(2) The hydrological optimum was between 8,500 and 6,500 BP. Since 6,000 BP, a renewed expansion of the Sahara took place over semi-arid zone of the Early Holocene; the desert reached its present-day extent between 3000 and 2000 BP. These processes were related with astronomically forced global changes (Berger, 1992) observed both in oceanic (Imbrie et al, 1993) and atmospheric circulation, the latter manifested especially by a strong increase of monsoon activity and range (Petit-Maire, 1994; Yan and Petit-Maire, 1994) following the end of the Younger Dryas. This expanded monsoon was then succeeded by a decrease corresponding to a reduction of insolation after 6,000 BP. No artificial desertification was involved at these latitudes and periods. It is hard to imagine severe environmental intrusion caused by small Neolithic groups probably as careful about protecting the environment they relied upon, as the present-day semi-nomads were before the modern socio-economic and political boundaries forced changes in their traditional ways.

The distribution of data shown in the published paleomap (1993) enables us to make a comparison of climatic evolution and human settlements in the Sahara during the Holocene (Fig. 1). It must be realized that the number of Neolithic sites used in construction of the figure widely underestimates the actual occurrence of relevant evidence throughout the present desert. At least twice as many sites are described in literature, but have not been radiocarbon dated. However, it is clearly evident from the distribution of the geo-biological "wet" indications that the latter preceded the signs of human life by about 1,000 years. The hydrological optimum was between 8,500 and 7,000 BP, whereas the settlement maximum was between 6,500 and 5,000 BP.

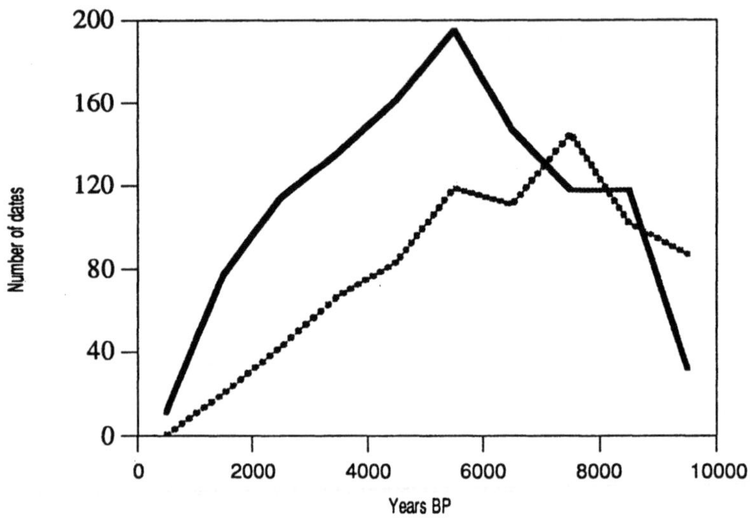

Fig. 1..Compared distribution of humidity evidence (dotted line) and of archaeological sites (full line).

Obviously, after the onset of a positive hydrological balance, a time lag was needed for a full biological response in a region which had been hyperarid during the preceding 10,000 years. In particular, the migratory response of vegetation, of game and man probably following one another in that order in logical sequence. The frequency of sites steadily rises after the humid optimum and decreases once more corresponding to the climatic deterioration at ca 5,500 BP until ca 2,000 BP. Field evidence shows that archaeological sites were located around the former lakes between 9,000 and 5,000 BP, but on the dry lake beds, generally around the centers of topographic depressions, after 5,000 BP.

The development of the Neolithic civilization in the Sahara is clearly related with climatic change. The wide extension of grass cover correlates with the ubiquity of mortars and grinders. The abundance of large fish in shallow lakes is implied by the development of bone hooks and harpoons [115 harpoons were found in a single excavated site of about 200 m^2 at 19°15'N (Decobert and Petit-Maire, 1985)]. The possibility of settling near the clay pans favored the introduction and expansion of rich ceramics industry (Commelin, 1984; Roset, 1988). The variety and quality of pottery decoration and of stone and bone adornments (Camps-Fabrer et al., 1982, Camps-Fabrer, 1983), suggest spare time for refinement, incompatible with a hard struggle for life in a hostile environment.

In order to verify the accuracy of these observations, we have compared the climatic and archaeological changes in three separate units of the Sahara, whose different geographic settings are likely to have influenced both the hydrological balance and human life, namely (a) the wetter mountainous areas (Hoggar, Tibesti), (b) oceanic fringes and (c) the drier continental lowlands which today constitute the hyperarid core of the desert (fig 2)

Following conclusions can be made from the plots:.

(a) Archaeological sites are present all the time in the highlands, although they are scarcer before 9,000 BP and after 3,000 BP.

(b) The Atlantic coast was wet after 9,500 yr BP, but humans were not present until the climatic deterioration began at ca. 6,000 BP. This colonization of the ocean fringes of the desert reached maximum after 4,000 BP and lasted until 2,000 BP or even later. The population there lived mainly on sea food, as testified by thick continuously accumulated shell middens lining the coast from 28° to 17°N (Petite-Maire, 1979, 1980).

(c) The maximal occupation period in the interior lowlands is coeval with the hydrological optimum; then its frequency decreased between 6,000 and 2,000 BP, and only the lower latitudes were occupied after 4,000 BP throughout the year.

It is reasonable to conceive that, when the continental lowlands became too dry for providing people with accessible fresh water, substantial vegetal food and game, there was migration to the nearest areas that were still wet; the latter included the banks of large rivers (Niger, Nile, Senegal), the highlands and the oceanic fringes where atmospheric humidity and fresh-water wells still permitted settlement. It has been shown (Petit-Maire, 1979; Guérin and Faure, 1983) that large game also migrated into those areas followed by hunters.

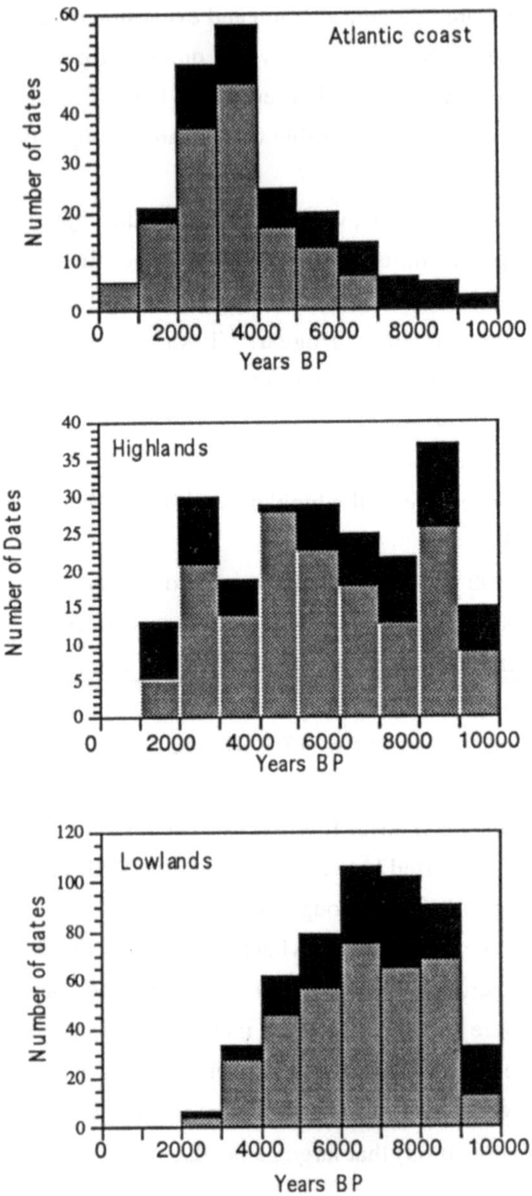

Fig. 2. Number of humidity indicators (in black) and of archaeological sites (in grey) in the Saharan lowlands, the Saharan highlands and along the Atlantic coast.

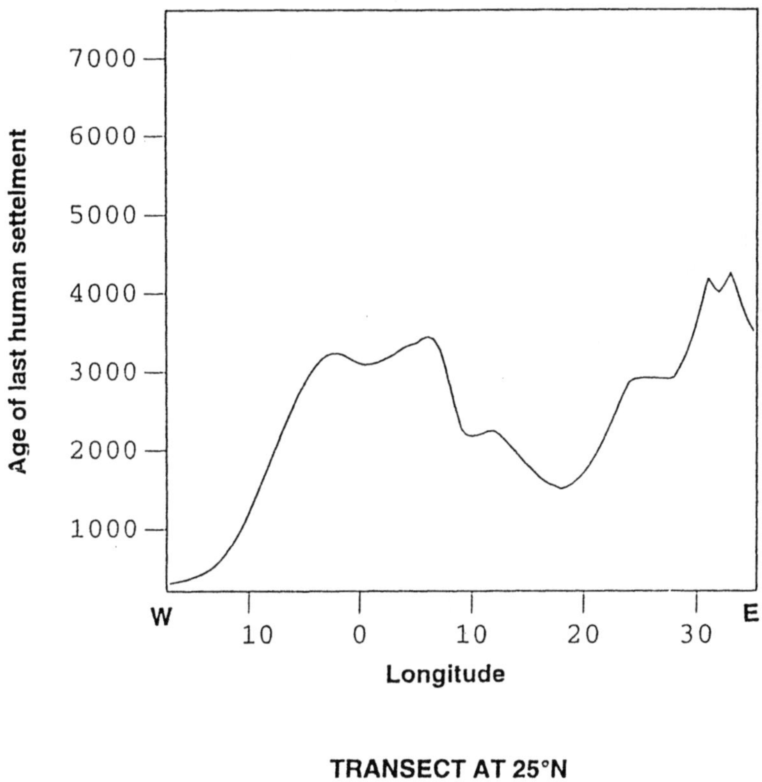

Fig. 3.Smoothed distribution of dates for the latest recorded settlements, according to longitude.

Fig. 3 supports the above observations. The curve shows the date of the last known human settlements, plotted from east to west along the 25 degrees North latitude. The lowest parts of the curve correspond to the Hoggar and Tibesti (10° - 20°E) and to the Atlantic coast (ca 10°W).

The situation in the southern Sahara and its monsoonal changes are particularly well illustrated by Figs. 4a and 4b based on the data from the Sahara in Mali (17° to 24°N), where a large number of multiproxy observations were made between 1980 and 1988 (Petit-Maire, 1986; Petit-Maire and Riser, 1983, 1988). Anthropological studies of numerous well typified human remains from burials between 20° and 16°N and their archaeological context (Petit-Maire and Dutour, 1987; Dutour, 1989; Raimbault and Dutour, 1990; Raimbault, 1994) have shown a southwards migration of the same population after the drying up of the paleolakes at 19°N (Petit-Maire and Gayet, 1984; Petit-Maire and Riser, 1987).

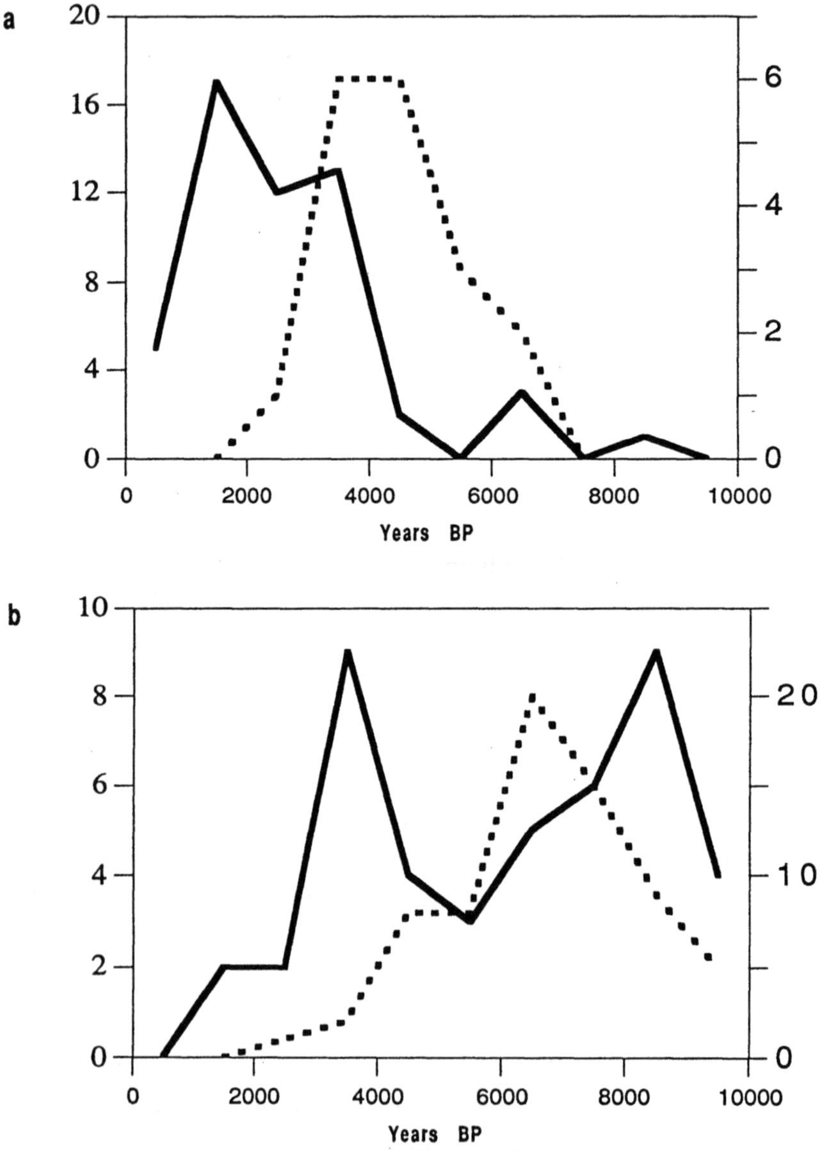

Fig. 4. Distribution of number of data in northern Mali:———17° to 20°N (scale to the left); -------20° to 25°N (scale to the right); a: climatic data: b: archaeological data.

Fig. 5 shows the age of maximum human occupation in relation to latitude. Along the 24th parallel it also applies to the highlands, the Atlantic coast and the Nile valley.

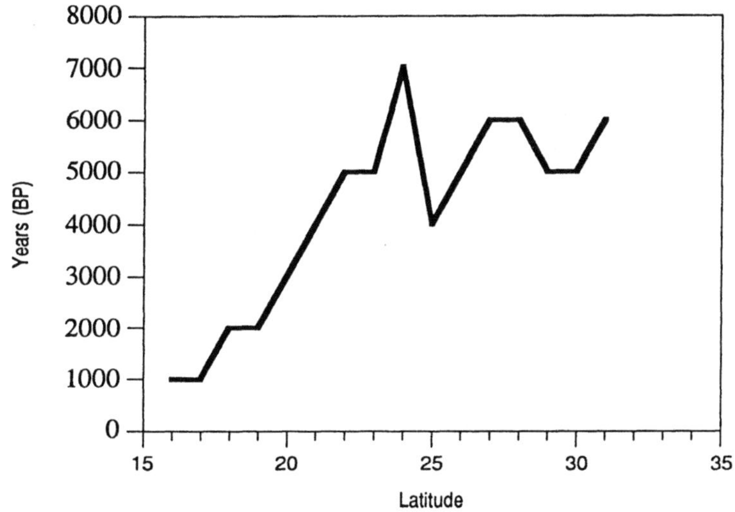

Fig. 5. Age of maximal human occupation in relation to latitude.

Conclusion

Due to the high variability of the monsoon, both in activity and range, on one hand, and due to variable penetration of Atlantic cyclones on the other, the arid/semi-arid belt in northern Africa is prone to extreme environmental changes, closely correlated with global climate variations.

During the Early Holocene warming, the arid belt around the Tropic of Cancer practically disappeared (Petit-Maire et al., 1993) and was replaced in Sahel and Sudan by semi-arid steppes. The distribution in time and space of numerous multiproxy data throughout the presently hyperarid area of the Sahara shows how sensitive human populations have been in responding to the climatic evolution of this region. Both the hydrological optimum of the Early Holocene and its progressive deterioration throughout the Mid-Holocene until about 2,000 years ago have induced wide migrations into or out of a territory covering several million km^2. The African highlands, the Atlantic coast and the riverine shores of the Niger, Senegal and Nile provided a refuge.

It is possible that if either a new man-made or natural global warming (Sadourny, 1994) occurs and reaches +1°C relative to the present, i.e. approximately the same value as during the Holocene optimum, the climatic conditions in northern Africa will again evolve in the same way as they did under the past natural forcing and that the environments there will again become more hospitable to human life. Models of the high CO2 world (Sadourny, 1992) indicate, although to a lesser extent, that corresponding changes in atmospheric circulation and in particular of the monsoon activity would take place. However, one must not forget that a lag of about one thousand years was necessary for a shift from the arid to semi-arid biotops. The desert greening occurred over the interval of several centuries and it is doubtful that modern technologies could speed up such a process to only a few decades or a single century.

References

Berger A (1992) Le Climat de la Terre. Un passé pour quel avenir? De Boeck-Wesmaël, Brussels, (B), 447 p.

Camps-Fabrer H, Commelin D and Petit-Maire N (1982) Objets rares et parures néolithiques du Sahara malien. Travaux du Laboratoire d'Anthropologie des Pays de Méditerranée occidentale. Aix-en Provence ,1-19.

Camps-Fabrer H (1983) Parures et engins de pêches. In: N. Petit-Maire and J. Riser ªeds.), Sahara ou Sahel? Quaternaire récent du Bassin de Taoudenni. Marseille, 367-401. Commelin, C., Petit-Maire, N., 1979. Chronologie isotopique saharienne pour les derniers 10.000 ans. Essai d'inteprétation: migrations humaines et paléoclimatologie. Bulletin du Musée d'Anthropologie préhistorique de Monaco, 23: 37-88.

Commelin D (1984) La céramique néolithique dans le Bassin de Taoudenni (Sahara Malien). Thèse, Université de Marseille.

Decobert M, Petit-Maire N (1985) An early Neolithic midden and necropolis in the Malian Sahara. Nyame Akuma 226: 26-32.

Dutour O (1989) Hommes fossiles du Sahara: peuplement holocene du Mali septentrional. Editions du CNRS, Marseille (FR), 342, p.

Guérin C, Faure M (1983) Mammifères. *In*: N. Petit-Maire and J. Riser (eds.) Sahara our Sahel? Quaternaire récent du Bassin de Taodenni (Mali). Marseille ,239-272.

Haynes CV, Mead AR (1987) Radiocarbon dating and paleoclimatic significance of subfossil Limicolaria in Northwestern Sudan. Quaternary Research 28: 86-99.

Imbrie J, Berger A, Boyle E.A, Clemens SC, Duffy A, Howard WR, Kukla G, Martinson DG, McIntyre A, Mix AC, Molfino B, Morley JJ, Peterson LC, Pisias NG., Prell WL, Raymor MW. Shackleton NJ, Toggweiler JR (1993) On the structure and origin of major glaciation cycles: The 100,000 years cycles. Paleoceanography 86: 699-735.

Le Houérou HN (1989) Classification écoclimatique des zones arides (s.l.) de l'Afrique du Nord. Ecologia Mediterranea, 15, 314:95-144.

Mainguet M, Canon L, Chemin MC (1980) Le Sahara: géomorphologie et paléomorphologie éoliennes, In: William, M. And Faure, H. (eds.), The Sahara and the Nile. Balkema-Rotterdam, (NL): 607 : 17-35.

Neumann K (1989) Holocene vegetation of the eastern Sahra: charcoal prehistoric sites. *African Archaeological Review*, 7: 97-116.

Ouseini I, Morel A (1989) Utilisation de formations alluviales pour l'étude des paléoenvironnements du Pléistocène supérieur et de l'Holocène au Sud du Sahara: l'exemple de la Vallée du fleuve Niger dans le Liptako nigérien. Bulletin de la Société géologique de France (8), 5/1: 85-90.

Petit-Maire N (ed.) (1979) La Sahara atlantique à l'Holocène. Peuplement et écologie. Mémoire. Alger (AL), CRAPE: 340 p.

Petit-Maire N (1980) Holocene biogeographical variations along the northwestern African coast (28° - 19°): Paleoclimatic implications. In: Sarnthein, M., Seibold, E., Rognon, P. (eds.) The Sahara and surrounding seas. Balkema-Rotterdam :365-377.

Petit-Maire N, Riser J (eds) (1983) Quaternaire récent du Bassin di Taoudenni (Mali) IN: N. Petit-Maire and J. Riser (eds.) Sahara ou Sahel? CNRS, Marseille 473 p.

Petit-Maire N, Gayet M (1984) Hydrographie du Niger (Mali) à l'Holocène ancien. Computes Rendus d l'Académie des Sciences, Paris ,II, 298:21-23.

Petit-Maire N (1986) Paleoclimates in the Sahara of Mali: A multidisciplinary study. Episodes 9/1:7-16.

Petit-Maire, N., Dutour, O., (1987) Holocene populations of the Western and Southern Sahara. Mechtoids and paleoclimates. In: Close, A.E. (ed.) Prehistoyr of arid North Africa, Southern Methodist Univeristy, Dallas (LUSA); 259-286.

Petit-Maire, N., Riser, J. (1988) Le Sahara à l'Holocène (Mali): Map 1/1.000.000. CGMW Paris (FR).

Petit-Maire, N., (Ed.) (1991) Paléoenvironnements du Sahara: Lacs holocènes à Taoudenni (Mali) Editions du CNRS, Marseille/Paris (FR):191-194..

Petit-Maire, N., and Page, N., (1992) Remote sensing and past climatic changes in tropical deserts: the case of the Sahara. *Episodes*, 15,2: 113-117.

Petit-Maire, N., Page, N., and Marchand, J. (1993) The Sahara in the Holocene: Map 1/5.000.000.UNESCO, CGMW. Paris (FR).

Petit-Maire, N., (1994) Natural variability of the Asian, Indian and African monsoons over the last 130 ka. IN: Desbois, F. Désalmand (ed.), Global Precipitations and Climate Change, NATO ASI series, I, 26: 3-26.

Raimbault, M., (1994) Sahara malien: environnment, populations et industries préhistoriques. Thèse, Université de Provence, Marseille (FR), 3 vol.

Raimbault, R., Dutour, O., (1990) Découvertes de populations méchtoïdes dans le néolithique du Sahel malien (Gisement lacustre de Kobadi); implications paléoclimatiques et paléoanthropologiques. Comptes Rendus de l'Academie des Sciences, Paris (FR), III, 310: 631-638.

Rognon, P., (1980) Une extension des déserts (Sahara et Moyen Oreint) au cours du tardiglaciare (18,000-10,000 and BP.. Revue de Géologie dynamique et Géographie physique, 22: 313-328.

Roset, J.P. (1988) Paleoclimatic and cultural condition of Neolithic development in the Early Holocene of the Northern Niger (Aïr and Ténéré).. In: Close, A.E. (ed.) Prehistory of Arid North Africa, SMU, Dallas (USA): 211-234.

Sadourny, R., (1991) La modélisation du climat. Revue scientifique et techniqe de la Défense, Paris (FR), 1: 1-22.

Sadourny, R., (1994) L'influence du soleil sur le climat. Comptes Rendus de l'Académie des Sciences, Paris (FR) II, 319, 11:1325-1342.

Schultz, E., (1991) Les pollens des Lacs de Taoudenni-Argorgott. Paléoenvironnements du Sahara: lacs holocènes à Taoudenni (Mali). Editions du CNRS, Marseille/Paris (FR), 143-162.

Talbot, M., (1984) Late Pleistocene rainfall and dune building in the Sahel. Palaeocology of Africa, 16:203-214.

Vernet, R. Aumassip, G., (1992) Le Sahara et ses marges. Paléoenvironnement et occupation humaine à l'Holocène. Inventaire des datations 14C. Editions du CNRS, Mendon/Paris (FR), 107 p.

Yan, Z.W. and Petit-Maire, N., (1994) The last 140 ka in the Afro-Asian climatic transitional zone. *Palaeogeography, Palaeoclimatology, Palaeoecology*, 110: 217-233.

Mid-Holocene Dry Spells in Bohemia, Central Europe

Vaclav CILEK

Geological Institute of the Czech Academy of Sciences
Rozvojova 135
16500 Praha 6
Czech Republic

Abstract

The earliest long lasting dry period that affected the area of the Czech Republic was before the Late Boreal-Atlantic climatic optimum. The Early and Middle Neolithic cave settlements dated from the 6th millennium to the beginning of the 5th millennium BC show that during the most humid period of the Holocene there were short, possibly century long dry spells. In contrast, the relatively dry phases before and after the climatic optimum were long lasting, stable and very mild. Reconstructed mean annual precipitation was at about the modern levels. The younger dry phase started at approximately 4,500 BC and lasted up to about 3,700 BC. Later in the Epiatlantic between 2,000 and 4,000 BC we observe several (probably 3-4) dry spells which lasted, judging from sedimentation rates, about 100-300 years each. The Epiatlantic represents one of the most chaotic cultural period of the entire Holocene, characterized by several waves of the so-called invasion cultures. The discontinuous cultural and/or ethnic migrations from the Balkans and Anatolia took place at the beginning of the Neolithic. Two migration waves are particularly important: one at about 2,500 and the second between about 2,250 and 2,000 BC. They had a profound effect on socio-economic structure in the region. Already some half a century ago they were described as the "First Great Migration" (Hrozny, 1940: Filip, 1948). These migrations were traditionally explained as the result of aridification of the Mid-Russian steppe belt. There are no signs of major climate oscillations affecting the area of the Czech Republic at that time. The most expressed dry spell of the entire Holocene there took place later, in Subboreal (1,250-700 BC). Preferred model of the Holocene climate development is a secular progression toward mope continental regime composed of several (5-7) never repeated phases, switching on and off rather suddenly. These

NATO ASI Series, Vol. I 49
Third Millennium BC Climate Change
and Old World Collapse
Edited by H. Nüzhet Dalfes, George Kukla,
and Harvey Weiss

general phases were interrupted by minor fluctuations on a century-scale level, sometimes accompanied by abrupt changes of precipitation.

Introduction

Bohemia forms the central part of the Czech Republic. Moravia in its eastern margin has different cultural and environmental history due to its position on the periphery of the Pannonian basin. Thus most of the data discussed here come from the semi-enclosed zone of Bohemia. Systematic studies of Holocene climatic and environmental fluctuations in this area are over two decades old. They consist mostly from the palynology of peat bogs and marshes and malacozoology of calcareous sediments (Jankovska 1970, Horacek and Lozek 1988). At least 60 profiles distributed at different elevations and in different environments were studied in detail. Unfortunately the reconstructed succession of landscape changes and climatic oscillations lacks a sufficient supporting framework of radiometric dates. The time resolution is thus much less then 2 to 3 centuries and the ages are often derived from the associated archeological finds.

I will try to summarize here all important information relevant to the dry spells which affected Bohemia in the Holocene, as derived from the studies of my own, Lozek's (Lozek and Cilek, 1955) and others (Bouzek 1993; Buchvaldek et al. 1985). An overall summary of the Holocene climate changes in terms of precipitation and temperature is given in Lozek's contribution in this same volume. One has to bear in mind the limited dating accuracy which makes minor shifts in the timing of individual peaks possible and even probable.

Three lines of evidence

The dry oscillations in the middle latitudes of Central Europe were never so conspicuous as to leave any clearly detectable traces in pollen spectra of marshes and peat bogs (Krippel 1986). Stable water levels there were usually a rule. The variable proportion of steppe elements is usually interpreted as reflecting the man-made changes of the area of arable land. It is very difficult, if not next to impossible to distinguish the climatic and anthropogenic signals (Krippel 1986).

High resolution fossil records in lacustrine sediments are almost non existent in Bohemia. One of such very few potentially suitable areas, the Komorany lake, was destroyed by brown-coal mining. The situation is further complicated by the lack of dendrochronologic records. For these reasons I have to rely on a combination of three lines of evidence in describing the Mid-Holocene climate changes. They are: (1) changes of archeological cultures and related land use, (2) stratigraphy of the floodplain deposits and (3), evolution of slope and cave sediments and fresh water carbonates.

Cultural data

Wenland and Bryson (1974) proposed that the cultural history of Man may be interpreted as a proxy record of environmental change. This is not to suggest that Man fully or solely responds to his environment and its changes, but rather that initial conditions may be established by the environment. However individual prehistoric societies tend to respond to the environmental change in individual ways, basically characterized by agriculture styles (lowland and highland agriculture and agriculture to pasture ratio). Dry spells in Central Europe paradoxically seem to lead to demographic and economic boom because agricultural land is utilized in the more humid hills covering several times larger area than the more fertile but drier chernozem lowlands.

The Mid-Holocene climatic information provided by archaeological evidence may be summarized as follows:

Neolithic. The Central European Neolithic period is environmentally more stable and culturally more homogenous than any later prehistoric stage. Several long-lived cultures (Linearband Keramik, Stichband Keramik, Lengyel complex) are almost evenly distributed and quasi continuously present over more than 2,000 years. The entire Neolithic falls within one large broad interval of Atlantic humid climatic optimum. That is apparently contradicted by numerous Early to Middle Neolithic settlements in nowadays uninhabitable wet caves testifying to drier climate than today. One has to realize that a rainy day in the karst translates into additional 2 to 3 wet days inside the cave because of the long water seepage through the limestone massif. Caves are therefore inhabited only during relatively dry local conditions such as existed during Glacial, Early to Middle Neolithic, Middle Eneolithic and Late Bronze period (Bouzek, 1993). The number of cave settlements in Eastern Slovakia rapidly decreased after the Buekk culture (Lichardus, 1968) after approximately 4,700 years BC. Although the Early Neolithic corresponds to the most humid phase of the Holocene (see Lozek, this volume), the growing evidence from the caves points to short, maybe only a century long arid intervals punctuating this time. Lamb et al.

(1995) refer to the occurrence of arid winters in mid-Holocene, associated with cooler sea surface temperatures in the North Atlantic. Zaitseva et al., (1994) find six lake transgressions and regressions between 5,500 BC and 700 AD in the Dvina basin in northwestern Russia. The Czech faunistic and floral record according to our observations does not indicate existence of any dry episodes in the Neolithic period, but it has a low time resolution. On the contrary the cultural data such as those from the cave settlements make the existence of such short anomalies highly probable.. In any case the dry intervals of Neolithic period had to be less intensive and less frequent than the dry spells of the Eneolithic because the latter but not the former ones left clear traces in the travertine deposits.

Eneolithic The equivalent of the Near East Chalcolitic is known in Europe as “Eneolithic”. This is due to the fact that the copper artifacts are rare but the societal structure, agriculture and craftsmanship reached more advanced levels. The areas inhabited during this phase in Europe were isolated and small. The unifying force of large valleys in Europe is limited and thus no larger empires or urban states came into existence. The Eneolithic of Central Europe can be characterized by the more complex, less stable cultural patterns with several waves of cultural and/or ethnic migrations (Bouzek et al., 1982; Buchvaldek et al.1985). The scarcity of anthropological material does not permit any unambiguous model of ethnic migrations to be developed, but the profound changes at the end of the 3rd millennium BC lead to the idea of the First Great Migration or Völkerwanderung which took place between about 2,250 to 2,000 BC. This migration has been traditionally ascribed to the drying of the central Russian steppe belt in the area around Black, Caspian and Aral Seas (Hrozny, 1940: Filip, 1948). Related scientific paradigm is paradoxical. The traditional model of climate changes as the underlying cause of the collapse of social order is frequently doubted by younger East European archaeologists, while the explanation of internal social roots of the collapse is doubted by many western archaeologists (Gibbons, 1993).

The first indication of the Middle Eneolithic drought appears in Baden culture in Slovakia as early as at 3,500 BC, but it peaked sometimes before the first half of the 3rd millennium BC. Settlements from this time are fortified and sometimes occur at altitudes over 1,000 m a.s.l. This contrasts with all previous and most of the later cultures which seldom cross the altitude of 300-450 m a.s.l. In addition the Baden pottery in the east and the contemporary Rivnác pottery in the west are often found in the caves. To explain the anomalous settlement pattern of the Baden culture, which is believed to be the combination of indigenous European Mesolithic stock and Anatolian-Balkan influences, some specific social or climatic factors are required (Bouzek, 1993). These may very well be precipitation related, since the rains are generally more plentiful at higher elevations.

The second half of the 3rd millennium BC is marked by the presence of the so-called invasion cultures. Pleiner et al.(1978) suggest that the migration of the shaft and ochre graves culture was triggered by the drought in Russia and in a domino-like pattern initiated the westward movement of the corded ware nomadic culture. Similar, somehow later migration is supposed to take place in Iberia, where the Bell Beaker people with round skulls moved to Central Europe at about 2,500 years BC. Unfortunately only few radiocarbon dates from this important episode are available.

The generally favorable conditions during the first half and the middle part of the 3rd millennium BC were probably drier than today. Rapid, previously unheard of cultural and probably also ethnic changes occurred in the vast areas of Europe at that time. The causes are disputed, because no indication of climatically induced changes are observable in fossil record. It is therefore traditionally believed that the profound social changes at the end of the 3rd millennium were triggered by the dry spells outside the Bohemian region (Buchvaldek et al., 1985; Hrozny, 1940).

Bronze Age. The Unetice culture, formerly considered a part of the Early Bronze Age (approximately 2,000-1,700 BC) is now included in the final Eneolithic phase (Bouzek, 1982). It represents the return to more stable agriculture in one of the civilization peaks of the Bohemian prehistory. It was gradually replaced during the general deterioration and agricultural decline of the 1,500-1,200 BC interval.

By far the most important dry period of the Holocene is in Bohemia represented by the Subboreal (1,250-700 BC) or in the archaeological sense by the Late Bronze cultures (Knoviz, Lusatia cultures). Bouzek (1993) presumes that the dry spells preferentially affected the chernozemic lowlands with the bulk of the Late Bronze population moving to the wetter hilly areas where climatic impacts were less severe. The changed land use pattern, mainly the turnover of deciduous woodlands to cultivated steppes led in spite of the droughts to the economic and population boom, comparable to early Middle Ages. The Mediterranean realms were during the droughts more vulnerable to both: the hunger-driven nomads from the East and expanding barbarians from the North. The Subboreal phase is interrupted by only a slightly more humid interval. The settlements at river banks and the character of the fluvial sediments testify to relatively low and stable water tables with absence of major floodings. At the same time however, the Subboreal is the time of intensive talus scree formation. The causes of the dilemma are unclear but one may speculate on the effects of artificial deforestation combined perhaps with intensified winter frosts..

Iron Age to AD. The precipitation regimes of this period are unclear because of the strong anthropogenic overprint. No major climate extremes are indicated by the available evidence. Probably the most significant paleoclimatic change is reflected by the decline of travertine formation between 700 BC. and 1 AD. Even the deep-seated karst springs and reservoirs were loosing capacity to precipitate larger travertine bodies in spite of uninterrupted karst water outflows (Goudie et al., 1993). Bouzek (1993) expects existence of several more dry periods such as in the 3rd and the 6th centuries BC. The ethnic migrations of that time could be possibly related with changing seasonality. More continental weather with hard winters and long snow cover would have stressed the winter pasture of cattle but would have not seriously affected the agrarian societies. The pastoral and agrarian migrations could have been thus influenced not only by social developments such as wars or by major droughts, but also by oceanicity-continentality shifts.

Karst sediments and travertines

The lack of classical Holocene sediments such as lacustrine varves in the area of the Czech Republic led to studies of subaerial sediments. As early as at the beginning of this century, J. Petrbok started to collect malacofauna in calcareous sediments of karst areas, swamps ands in travertines and to interpret past environments from fossil assemblages (Petrbok, 1916). Several hundreds of Petrbok's papers published mostly before World War II provided a sound basis for future refinement of malacostratigraphy described in almost 800 Lozek's papers (see e.g. Lozek, 1964). The review of his results related to the 3rd millennium BC is given elsewhere in this volume. I want to draw attention to three general but often overlooked aspects of Holocene sedimentation reflected in tens of profiles described from various parts of former Czechoslovakia:

(1). The Central European Holocene is divided into 5 - 7 distinct layers which can be correlated over large regions. The macroscopical character, soil development and fossil content is characteristic of each layer and is never repeated. Therefore the layers reflect a basic environmental evolution, which is secular, not cyclical. The low time resolution of karst scree deposits, slope fans and cave sediments provides a natural filter of the data smoothing minor fluctuations of often unknown intensities. The detail is lost in favor of general picture.

(2) Transitional layers are seldom observed. Most of the layers start and end abruptly. When confronted with the field evidence it is difficult to believe in the sinusoidal character of the Holocene environmental and climatic changes. Nor is it possible to accept any simple flip-flop scenario. We suggest existence of several singular unrepeated Holocene climate phases lasting between one to three thousand years, switching on and off rather suddenly.

(3) This type of fossil record does not provide much information on shorter, century-scale dry intervals and their possibly cyclic character. The cultural data suggest the existence of dry intervals even in the Atlantic but 3 or 4 thin horizons of buried soils in calcareous tuff bodies directly point to Epiatlantic dry spells. The duration of each dry spell can be estimated from the average sedimentation rates to some 100 - 300 years.

Flood plain deposits

The intensive downcutting of the Late Glacial rivers caused the deepening of valley bottoms and provided new space for the early Holocene accumulations. The sedimetary history of Czech floodplains has been studied several times by various authors (e.g. Lozek, 1988; Ruzicková and Zeman, 1994). Despite similar general pattern the detailed correlation among various localities is difficult, This is because the sediments of small rivers and the oxbow lakes often reflect local processes, while the sediments of large rivers are often destroyed erosion.

The floodplain of the largest Czech river, Labe (Elbe) was intensively studied during last few years in order to decipher its Holocene history and for conservation purposes. Ruzicková and Zeman (1994) summarized the main stages of the floodplain development as follows:

10,000 years BP. and earlier: dominant erosion of the Pleistocene sandy gravels and sands.

9,500 - 9,000 years BP: accumulations of the sediments of the upper (earlier) terrace, common oak trunks, Mesolithic settlement.

8,000 years BP: fluvial erosion, formation of the upper terrace as a morphological level, the important downcutting event which is also found in travertines and gullies.

7,700 - 6,900 years BP: accumulation of fluvial sands and gravels of the lower (younger) terrace, Neolithic colonization.

3,750 years BP: a sudden accumulation event, with prehistoric settlements along the rivers overlain by fluvial sediments.

3,500-3,200 years BP and later: an equilibrium regime is established. River meandering, deposition of, gradual accumulation of floodplain without any marked hydrological discontinuity continues up to the 12th century AD. when major floods destroy settlements and cause downcutting. The chaotic hydrological regimes combine with anthropogenic impacts. New equilibrium is reached only between 1,800 -1,900 AD.

The dry periods are almost undetectable in the floodplain sediments of larger central European river because the thin layers and or soils from such times are easily eroded. The floodplain development of the 3rd millennium BC displays normal, mild sediment aggradation but with water levels lower than prior to 3,500 years BC. The Neolithic colonization of the floodplain continued up to the Bronze age.

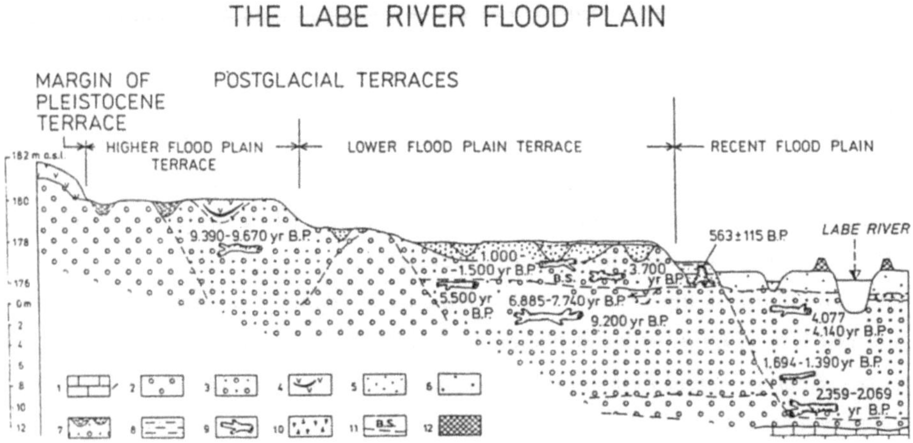

Fig. 1. Scheme of the flood plain in the middle course of the Labe River (Ruzicková and Zeman, 1994). 1 - Upper Cretaceous marlstone; 2 - fluvial sandy gravels of the Upper Pleistocene; 3 - fluvial sand with admixture of gravel; 4 - wind-blown sand, locally with an interlayer of volcanic ash (marked with black lens); 5 - loams and sandy-loams of the Late Holocene; 6 - flood loams; 7 - deposits filling ox-bow lakes (loams and sands); 8 - medium-grained fluvial sand with admixture of gravel; 9 - tree trunks; 10 -fragments of bricks; 11 - geological boundary - buried surface; 12 - flood dam

Synthesis

The Holocene research of the area of the Czech Republic is founded on about 60 profiles studied in detail and irregularly scattered over most local landscape types. The principal methods of Holocene studies have been so far by biostratigraphic methods - mostly palynology, malacozoology and microvertebrate paleontology (Horacek and Lozek 1988; Firbas 1935; Jankovská 1970). Several decades of such studies resulted in well-established relative chronologies and successions of environmental changes. However the time resolution of the fossil record is in most cases low (usually more than 200-400 years). Modern radiocarbon data are scarce and no absolute chronologies are developed. Most palynological profiles are dated by correlation with classical Holocene floral stages (Firbas, 1935) and many terrestrial sediments such as slope and karst sequences are dated only by embedded archaeological finds. The use of cultural data as climatic proxies is hindered because huge amount of archaeological research reports is never published. Several projects were only recently launched to fill the gaps.

I tried to combine several lines of evidence to reach the following synthesis of climate development in the Mid-Holocene with particular focus on the dry spells affecting the Czech Republic:

(1) An important dry period influenced the area of the Czech Republic before the Late Boreal-Atlantic climatic optimum. The abundant cave settlements of the Early and Middle Neolithic (6th millennium to the beginning of the 5th millennium BC) show that even during the most humid Atlantic period there were short, century or so long dry spells. Both drier phases before and after the climatic optimum were long lasting, stable and very mild. The mean annual precipitation lowered to about 50% of peak Atlantic values but it still remained on the values close to modern ones. The younger dry fluctuation started about 4,500 BC and ended at about 3,700 BC.

(2). Several, probably 3 or 4 dry spells, lasting, judging from the sedimentation rates of travertines, some 100-300 years each, occurred between 4,000 and 2,000 BC. At the same time this Epiatlantic period represents one of the most chaotic cultural periods of the entire Holocene characterized by several waves of the so-called invasion cultures. The discontinuous cultural and/or ethnic migrations from the Balkan-Anatolian area took place from the beginning of the Neolithic, but two Eneolithic migrations seem to be particularly important: at about 2,500 and 2,250 - 2,000 BC. They profoundly changed the socio-economic structure of the Late

Eneolithic societies and were described, already some half century ago, as the First Great Migration of Völkerwanderung (Hrozny 1940; Filip 1948). Both migrations have been traditionally explained as the result of recurrent droughts in the Mid-Russian steppe belt. New, partly unpublished data cast doubt on any simple model which would directly link the migrations to climate (Turek, pers. com. 1995).

(3) The most important and striking Holocene "dry" event, reflected in the Czech fossil record took place in Subboreal (1,250 - 700 BC). Its social impacts were less adverse than in the Mediterranean because in Central Europe the rainfall is higher and more evenly distributed over stable, predominantly plain areas with well developed soils. The Central European tribes can adjust to the droughts by moving their land use to hilly regions which occupy several times larger spaces than dry chernozem basins of the originally settled areas. The demographic boom in the North may have increased the stress on the frontiers of Mediterranean empires weakened at the same time by nomadic invasions and sea raiders (Bouzek, 1993) This scenario may have been repeated during the early Middle Ages.

Conclusions

The fossil record in the area of Czech Republic does not reflect any striking climatic changes during the 3rd millennium BC. except for travertine deposits, where 3 to 4 buried soils indicate existence of short (100-300 years long) dry spells. These dry periods were less intense than the abrupt Subboreal dry spell of 1250 to 700 years BC.

The Eneolithic cultural pattern suddenly changed at approximately 2,500 to 2,000 years BC due to the influx of the so-called invasion cultures. This immigration at the end of the 3rd millennium BC, since long ago recognized as the First Great Migration (Völkerwanderung) was ascribed not to the local climate shift, but to the drying of steppes in central Russia (Hrozny,1940; Buchvaldek et al., 1985).

References

Bouzek J (1993) Climatic changes: new archaeological evidence from the Bohemian Karst and other areas. Antiquity 67: 386-393.

Bouzek J et al(1982) Nastin Evropskeho Praveku (The Review of the European Prehistory; in Czech). Praehistorica IX : 1-194 , Charles University, Praha.

Buchvaldek M et al(1985) The Prehistory of Europe (in Czech). Statni Pedagogicke Nakladatelstvi, Praha.

Filip J (1948) Praveke Ceskoslovensko. Uvod do studia dejin praveku. (The Prehistory of Czechoslovakia; in Czech) Stopami Veku, Vol19-22: 1-390, Praha

Filip J (1952) Indoevropska otazka a lid se snurovou keramikou. Arch. Rozhledy IV: 38-96, Praha

Firbas F 1935) Die Vegetationsentwicklung des mitteleuropäischen Spätglazials. Bibliographia Bot. 112: 68.

Horacek I, Lozek V (1988) Paleozoology and the Mid-European Quaternary past: Scope of the approach and selected results. Rozpravy CSAV, Matem. Prirod Ved 98:4 , Academia, Praha.

Hrozny B (1940) O nejstarsim stehovani narodu a problemu civilizace proto-indicke (On the oldest Great Migration and the problem of proto-indian civilization -in Czech) Orientalni Ustav v Praze, 28pp, Praha.

Gibbons A (1993) How the Akkadian Empire was hung out to dry. Science 261:285

Goudie AS et al (1993) The Late-Holocene tuffa decline in Europe. Holocene 3: 181-186.

Jankovská V (1970) Ergebnisse der Pollen- und Grossrestanalyze des Moors "Verlanská cesta" in Südböhmen. Folia Geobot. Phytotaxonom. 5: 43-60.

Krippel E (1986) The postglacial evolution of Slovakian vegetation (in Slovakian). Veda Bratislava 307pp

Lamb HF et al (1995) Relation between century-scale Holocene arid intervals in tropical and temperate zones. Nature 373: 134-137.

Lichardus, J. (1968) Domica Cave, the most important Buekk culture settlement (in Slovakian). Pamätniky Nasej Minulosti 5, Bratislava.

Lozek V (1964) Quartärmollusken der Tschechoslowakei. Rozpravy UUG , 31, Nakladatelstvi CSAV, Praha 374 pp

Lozek V (1988) Nacheiszeitliche Molluskenfaunenfolge in der Aue des Tales Tiché Udolí bei Prag. Malakologische Abhandlungen 19, (12): 110-117, Dresden.

Lozek V Cílek V (1995) The humidity and temperature course of the Mid-European Holocene. In Ruzickova E, Zeman A, eds Manifestation of climate on the Earth surface at the end of Holocene. Proceedings of 1994 PAGES STREAM I meeting in Kolín, . p 79-81, Praha

Lozek V Cílek V. (in print) The Late Weichselian to Holocene transition in terrestrial record. Anthropozoikum Praha.

Petrbok J (1916) On the knowledge of the Holocene molluscs in Bohemia (in Czech). Rozpravy Ceske Akademie XXV, 20:1-15, Praha.

Pleiner R edit (1978) Praveke dejiny Cech (The Bohemian Prehistory -in Czech) Academia, Praha. 870pp

Ruzicková E, Zeman A, eds. (1994) Holocene flood plain of the Labe River. Geological Institute Abliccademy of Sciences of Czech Republic , Report on Grant 31305, 116 pp. Praha.

Wenland WM Bryson RA (1974) Dating climatic episodes of the Holocene. Quat. Res. 4:9-24.

Zaitsava GI et al (1994) The Chronology of the occupation of the region between the rivers Dvina and Lovat in the context of dynamics of environmental changes. 15th Radiocarbon Conference Book of Abstracts, Glasgow.

Geomorphologic Arguments for Mid- to Late Holocene Environmental Change in Central Anatolian (Pluvial) Lake Basins

Oğuz EROL

İ. Ü. Deniz Bilimleri Enst.
Müşküle Sok. 34770
İstanbul Türkey

Abstract

In the interior of the Anatolian Peninsula several shallow swampy lakes occupy the floors of former Pleistocene basins. Traces of emerged abandoned coastlines enable recognition of several developmental phases of the lakes. The Late Pleistocene and Mid-Holocene segments of the basins are covered with fertile farmland, occupied by ancient villages dating mostly from Early Bronze Age. They have concentric fields and radial road patterns centered at old mounds (höyüks). The lake floors of Late Holocene age are occupied by less fertile salty soil and at times flooded. Dunes developed in these sections in Konya-Karapınar and Sultan Lake areas. This author concludes that the different character of the Middle and the Late Holocene deposits indicates a drying phase in the early part of the Late Holocene.

Introduction

Quaternary environmental conditions are the final result of a long continued geomorphological and climatological evolution of the Anatolian peninsula since mid-Tertiary times. For better understanding this natural history one should consider the plate-tectonic, palaeoclimatologic and palaeogeomorphologic conditions of Eurasiatic, Tethyan and Africa-Arabian area and this is made by a separate paper in this NATO-ARW volume (Fairbridge et al. in this volume). Additional environmental changes in the area of Anatolian Peninsula and its coastline were summarized by the author in a correlation table (Erol, 1993). This table was actually a summary of the author's former researches was also made before and concluded therein (Erol, 1978: p. 112): "High water levels should not be automatically attributed to climatic causes, and in their study, the tectonic, hydrographic and geomorphic history of the basins

NATO ASI Series, Vol. I 49
Third Millennium BC Climate Change
and Old World Collapse
Edited by H. Nüzhet Dalfes, George Kukla,
and Harvey Weiss

and their surrounding mountains must also be taken into account". Indeed, the later studies on lake and river basins, which are discussed below have also added several examples to this subject. New radiometric datings in the lake and fluvial sediments and at archeological sites related them, also make interesting contributions to this subject.

Central Anatolian Quaternary Lake Basins

The several lake basins of Quaternary age found in the Anatolian Central Plateau mainly are the continuation of their Neogene antecedents. Since their size, form and water quality, however, has changed repeatedly through time, it is not correct to regard them simply as relics of the Neogene lakes. This is a natural result of the active and dynamic nature of the Anatolian Plate. Indeed, the structure, morphology, climate etc. were in continuous change since the early Tertiary (Fairbridge et al. in this volume). Therefore each Anatolian lake or river basin has its own natural history, which were evolved under the control of local, regional and global factors. In this point, it should be noted that, especially the lake-level changes of completely closed, or outward drained fluvial or karstic lake, or river basins, even close to each other, show considerably different characteristics. This indicates the importance of the local differences, especially in the relief of the country. These phenomena may be more clearly observed if compared for example, with the wide Mesopotamian plains. An abrupt change of climate, during the third Millennium in Northern Mesopotamia may cause a dramatic collapse of civilization, but the influence of the same phenomenon in the various Anatolian lake or river basins may be reflected by considerably different forms. Observing and evaluating these differences is often difficult, and demands comprehensive research. As a geomorphologist believes that careful geomorphic study may be the basis for this kind of work, but it needs several contributions from other related scientific research. This means that, there are still a lot of things to do and therefore only some examples from some of the studied basins are given below.

Western Taurus And South Central Anatolian Lake District

In the Western Taurus and South Central Anatolian Lake district is one of the best areas of Anatolia to study its environmental changes. There, the tectonics, surfacial and underground-karstic hydrology, together with climatic change, all acted together as a whole on the life of vegetation, animals and man. The diversity of each between the basins and mountains led to very varied and even contrasting conditions.

Burdur Lake Basin

Burdur Lake basin is a good example of an intramountain closed basin, which was open to exotic drainage during the Pleistocene and became closed during the Holocene; also it was formerly a fresh water lake, which became brackish after being closed. This basin had been occupied by lakes since the Early Miocene and its history has already been summarized (Atalay, 1977; Erinç, 1978; Erol, 1986; Kazancı, 1987; Kazancı, 1989; Louis, 1938; Sungur, 1978).

In the Quaternary basin there are a series of the Pleistocene and Holocene lake terraces (Erol 1978), the highest of which is at about 1000 m today and probably of earliest Pleistocene age while the recent lake surface is at 854 m. The radiometric datable terraces are today at 925 to 870 m elevation. The C 14 age (Kış et al., 1989) of the older sediments of 900-925m terrace are ranging between 28,444 ± 160 and 20,958 ± 175 years B.P. and the age of the younger ones are 15,365 ± 155 and 13,218 ± 160 years B.P. The youngest samples are of 14,218 ± 170 and 14,850 ± 165 years BP age were collected at an elevation of 870 m at Evtepe, where the Bronze age people were digging holes into the lake sands for burial purposes.

The famous Early Neolithic site of Hacılar (9,000-7,400 B.P.) is about 5 kilometers southeast of the Evtepe locality. According to the author the site of Hacılar is situated on the Late Pleistocene lake terrace sediments. In alluvial sediments in the Suludere valley 5-6 km northeast of Hacılar a piece of pottery of very primitive appearance was found in situ below 6-7 m of sediments (Cohen, 1969, Fig. 9). To the north of all these places, between the villages Yarıköy, Yazıköy and Soğanlı there is a fairly large salty lake floor a few meters above the modern lake. Putting together all these observations, one can conclude the following environmental evolution (Kış et al. 1989).

1. 28-20,000 years B.P., formation of sandy littoral sediments of Soğanlı Lake terrace at 925-920 m,

2. 15,360 years B.P., formation of lake floor sediments of Bozçay Köprüsü at 915 m,

3. 15-14,000 years B.P., formation of Evtepe coastal spits at 870 m.

4. 11-13,000 years B.P., formation of Tozlutepe coastal sandy sediments at 910-905 m,

5. Lowering of the Late Pleistocene lake level and recession of its shorelines towards the north. The Mesolithic site at Baradiz in the north of the Burdur Lake

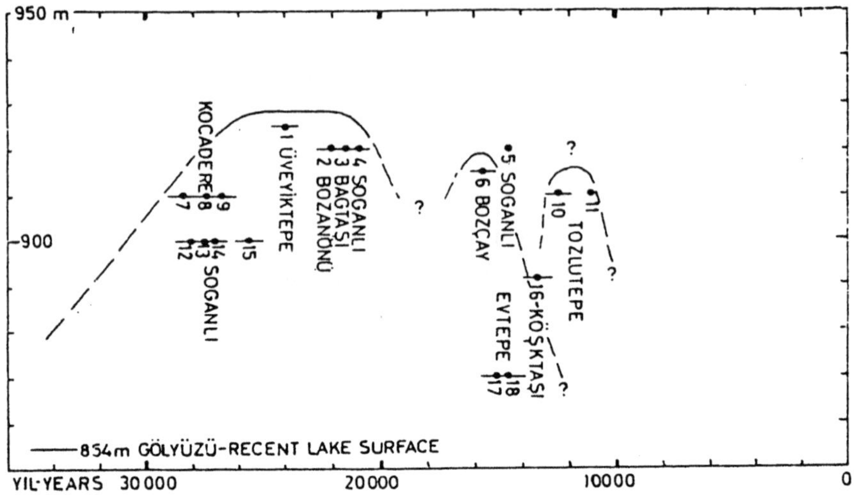

Figure 1. High lake-levels of the Burdur pluvial lake between 28,000 to 10,000 years B.P. (Modified from Kış et al. 1989).

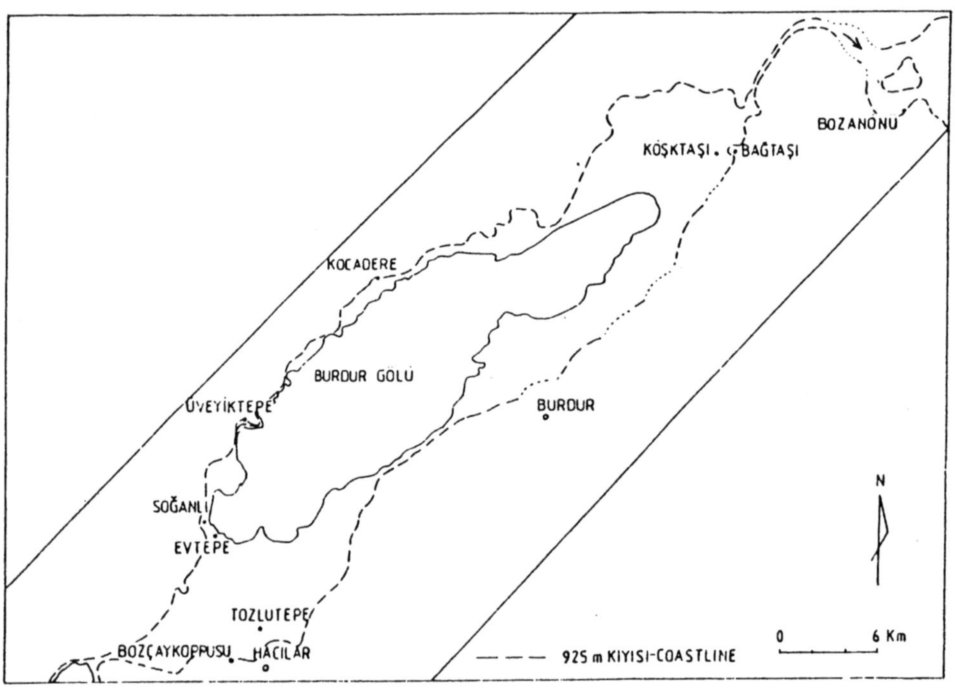

Figure 2. The map of the maximum extent of the Würm pluvial lake. (Modified from Erol, 1972).

may be an indirect indication of this major recession, that is close to the Pleistocene/Holocene transition (Erol, 1978: p. 138),

6. 9,000-7,400 years B.P., Early Holocene, formation of Suludere alluvial sediments at about 880-870 m, exposure of Hacılar site on the surface of dried lake terrace surface, near a karstic spring,

7. 5,000-4,000 years BP, Middle Holocene recession of the lake shore,

8. Late Holocene, formation of salty lake floor of Yarıköy-Yazıköy at 855-854 m,

9. Recent lake level 854 m fluctuating,

With all these characteristics and dating possibilities, the Burdur Lake Basin is a good example for the West Taurus Lake District, and even for Turkey.

The other lake basins of this district partly had the same conditions, but with the local differences among them are also seen. For example, the Acıgöl to the northwest of Burdur Basin, west of Dinar has high shoreline traces and had an outflow during the last pluvial period (Sungur, 1978). Southwest of the Burdur Basin, on the other hand, the small lake basins of Yaraşlı, Tefenni, Salda, Söğüt and Gölhisar, in spite of being partly karstic origin, are also contains some traces of high lake levels evidence of Pleistocene wet climatic conditions. For example the Yaraşlı Lake were connected to Burdur Lake during the last (Würm?) pluvial period, but it became an isolated water body during the Holocene, and contains elevated early Holocene beaches, in the sands of which hundreds of pithos are found a graveyard probably of the Bronze age. On the other hand, the Salda Lake basins near to Yaraşlı basin is one of the deeper, non-karstic hollow of the district. The Atabey Ovası to the NW of Burdur is a small closed basin today, but it was a part of the Pleistocene Burdur Lake and had an outflow towards the south, on the threshold of Isparta towards the Mediterranean, and Paleolithic humans were lived in the Bozanönü Cave (Erol, 1978) on its shore.

The Eğridir Lake Basin, further east of Bozanönü is a deeper tectonic depression, which is almost completely karstified; its southern extension is Lake Kovado. Pluvial shore traces are few and less characteristic. This basin had karstic outflows all during the Pleistocene and it still has these karstic and even surficial outflows. With these characteristics, the Eğridir Lake is an exact opposite example compared with the Burdur Basin. This observation makes it clear that despite the regional climatic trends local neotectonics created the local history of each basin. Thus minor climatic changes might be recorded geomorphologicaly in closed basins, while in the basins with surficial or underground karstic outflows caused a dominant fluvial or fluviokarstic erosion, and should be a subject of detailed karstic researches (Erol, 1991; Güldalı, 1981). Akşehir Lake basin to the NW of the Eğridir basin shows very similar conditions and lake terraces compared to the Burdur basin (Wenzel, 1932; Sanır, 1948; Atalay, 1975).

Beyşehir and Suğla Lake Basins

Beyşehir and Suğla Lake basins are very good examples of drained karstic intramountain basin. The Beyşehir Basin was occupied by an Upper Miocene to Pliocene Lake; during the Pleistocene and Holocene the two basins were occupied by freshwater lakes (Farrand, 1963 and 1965; Erol, 1978; Güldalı, 1981; Roberts, 1991 and 1994). The main basin: Beyşehir Lake dried out from time to time due to accelerated karstic outflow periods (Acatay, 1966) and filled by freshwater during the wet periods of Pleistocene, at times of positive hydrologic conditions the karstic activity and travertine formation was increased (De Planhol, 1956). The fully karstic Suğla Lake basin is actually the extension of the Beyşehir Lake. Under normal conditions, collects the surficial outflow of the main lake, and transports those waters by means of swallow holes into the karstic reservoirs of the Taurus Mountains, which are evidently connected to the sea (Acatay, 1966). During the extra wet periods of the past, however, as the swallow holes could not carry all the water, a surficial overflow river, which is the Çarşamba Çayı became active and transported the excess waters of Beyşehir drainage system into the Konya Basin. Therefore the geomorphic basins of the Beyşehir District were excellent areas for farming and fishing, which were living naturally assured against droughts and floods. There was a special period, which is called "Beyşehir Occupation Phase" in the prehistory of Turkey. This phase begun at or just before 3000 yr BP (1400 BC cal.) continued until 1350 yr BP (600 AD cal.) (Roberts, 1994; Bottema, 1990), that is during the Late Holocene. During that time the people living there had the possibility fishing during the winter and farming during the summer in the same place. There were one or more salinization phases during the late Holocene, for example in Burdur, Karapınar, Tuzgölü and Sultan Sazlığı, but not in Beyşehir-Suğla area. The reason of these contrasts is the local differences among various basins, and therefore the Late Holocene salinization occurred only in those ones, of which geologic and geomorphologic conditions are appropriate to reflect the influence of climatic change.

Konya Pleistocene Lake Basin

Konya Lake Basin is one of the first studied pluvial lake basins of Turkey (Chaput, 1936; Darkot, 1938; Louis, 1938 a, b; Lahn, 1946; Cohen and Erol, 1969; Cohen, 1970) in these works the relatively fresh looking shoreline traces and sediments were studied and were attributed to the last glacial period but later as Erol (1978) continued his work in the Central Anatolian pluvial lake basins, especially in Konya, Suğla, Tuzgölü and Burdur basins, he observed some other shore traces and sediments, which were obviously higher and most probably older. Meanwhile Roberts had started his doctorate thesis in Konya basin and brought the first radiocarbon dates, which range between 35,000 to 10,000 years BP and provided proof for the former estimations; later some other radiometric datings are also made in the Burdur basin (Roberts, personal communication; Kış and Erol, 1989).

Absolute dating of the older lake shorelines (pre Würm) and their sediments is still not possible, but aerial photographs aid field mapping and disclose the presence of a extensive series of older pluvial levels in Tuzgölü (Erol, 1970), Burdur (Erol, 1972) and Konya (Erol, 1972 and 1978) basins. Erol (1991) concluded that in the north eastern part of the Konya basin a set of multiple shore generations are connected to well-developed sinkhole (in Turkish obruk). The generations indicate Pleistocene wet periods, whereas dry valleys and older dune generations (De Meester, 1971; Erol, 1991) indicate dry. All these are sequentially related to each other and lower from the periphery to the central part of the basin. These observations should indicate, according to the author, a continuous subsidence in the central part of the basin where there must always have been a shallow lake, which this lake should expanded during the wet periods and even flow out by karstic underground water and even by surface rivers into the Tuzgölü basin to the north and towards the south into the swallow holes at the foot of the Taurus Mountains.

These observations show that the Konya basin was received huge amounts of waters and sediments from the Taurus Mountains during the late Pliocene to Pleistocene wet periods . At those times, the excess waters were drained towards the north into Tuzgölü basin, but the sediments were principally deposited in the Konya basin itself. Thus, the Konya depression was subsided and accumulated more than 200-250 m sediments. During the fieldwork it was observed that the local people were digging deep and wide holes in the clayey floor of the lake bed near Sarıcalar village and were taking sand and gravel for the city of Konya. It appears that this was a remnant of several shallow lakes, which were present at least since the late Pliocene-Early Pleistocene. This lake during all this time, the paleolake kept its general appearance, but showed several fluctuations,

However, there is another important environmental factor in the Konya basin which should be considered. The tranquil Plioquaternary history of the pluvial lakes had always angry neighbors: volcanoes. They lived together all the time by influencing each other, but not making war. Volcanoes in the lakes were making from time to time some disturbance, but lakes were keeping the environment more constant. Trees, grasses, wild animals and ancient man lived there in one of the centers of domestication (Roberts, 1989; Atalay, 1992). The wall paint at Çatal Höyük with houses, cattle and volcanoes may indicate this environmental harmony.

The detailed Holocene history of these individual lakes and swamps are now studied by Roberts (from 1981 to 1994) and Kuzucuoğlu (1992-1994). The summary of their results are indicated in the tables attached to this text, in order to make a comparison with the other neighboring basins. Therefore the comparison between Konya and Sultan Lake basins will be made in following chapters.

In this respect, however, it should be noted here:

Study of aerial photographs here by the writer, he observed on the low terrace, which is about 4-8 meters above the floor of the recent basin and on fertile agricultural ground between the city of Konya and Çumra, that is in the west of Çatalhöyük (Todd, 1976; Roberts, 1982a) some mounds (höyüks, tells) with a clear pattern of a "radial roads and concentric fields". This pattern has no relation to the recent villages. In some places this field and road pattern has remained undisturbed, roads and fields are later, but the hüyük mounds have remained empty. As this observation compared with "the maps of coastline traces" of the author (Figures 3 to 5) and with the map of "the distribution of höyüks" of French (1970) one can conclude that the pattern of radial roads-concentric fields coincides principally with hüyüks which are older than the first millennium and must be of Neolithic-Bronze Age and they are always observed above the 1002 m coastline of the ancient lake. Some of these concentric patterns are indicated in Figures 5, 6 and 10. Additionally, if this observation, compared with Wilkinson's publications (from 1978 to 1993) one can conclude that there must be here a fertile and really organized agricultural environment during the Bronze Age.

Such changes are not limited only to this basin and this period. According to the author's observations, similar changes had been occurred in the other pluvial basins before, during the passage from Pleistocene to Holocene and even during the older phases of Pleistocene. According to him, the well developed coastal spits are the obvious geomorphological indicators of a medium deep lake and the direction of dominant and probably prevailing winds for a given period (Figures 4, 5 and 6). Coastal sandy beaches were also developed to the land side of these lake sand bars. When a climatic change happened, the lake level may drop slightly and the former beach sands converted to coastal dunes. Following this phase at along the lowered coastline a second beach belt may develop again (Figures 4 and 5). But some times, between two successive phase the dominant-prevailing wind directions may also change, in this case the two successive spits develop in different directions opposite to each other (Figures 4 and 7). According to the observations of the author in Central Anatolia the prevailing northerly cold or cool winds were dominated usually during the wet periods, and prevailing southerly warm winds were dominated during the dry periods, therefore the opposite directed coastal spits (Figures 4 and 7) should indicate a climatic change from wet to dry and even cold-cool to warm conditions. As the result of this the shallow lakes should have disappeared, and then fairly expensive dune generations developed. For example in the Konya basin there are Pleistocene (De Meester, 1971; Erol, 1991) and Holocene (Erinç, 1978; Erol, 1978) dune areas. Because of extensive evaporation during dry periods some playa-lake, swampy-salty basin floors were also developed. Even in some cases, like İznik Lake basin in the north, during the periods excessive evaporation beachrocks had also developed about 3-2000 years before, then a recent fresh water lake had replaced that basin again (Kayan, 1993).

But in Anatolia, and of course in the Konya pluvial basin, because of highly fluctuating climatic conditions in some places much complicated coastal spits were also developed (Figure 8). Therefore, depending on several local landscape units, it is not possible to make similar observations everywhere. For this reason the author hesitates to make a generalization at the scale of the whole country. Only during detailed geomorphological mapping this kind of phenomena is indicated where it is observed.

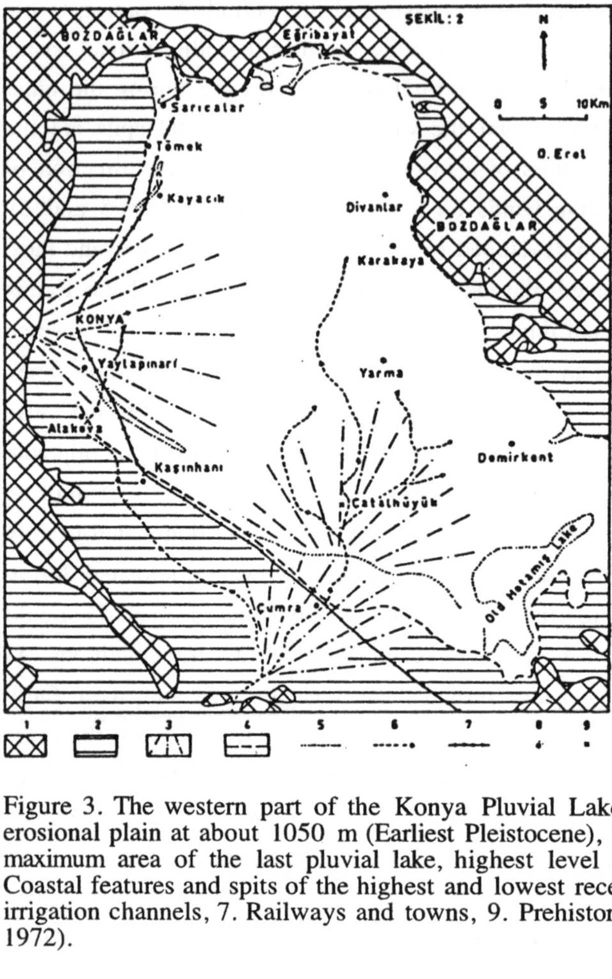

Figure 3. The western part of the Konya Pluvial Lake. 1. Uplands, 2. Undulating erosional plain at about 1050 m (Earliest Pleistocene), 3. Large alluvial fans, 4. The maximum area of the last pluvial lake, highest level indicated by broken line, 5. Coastal features and spits of the highest and lowest recessional phases, 6. Rivers and irrigation channels, 7. Railways and towns, 9. Prehistoric sites. (Modified from Erol, 1972).

Sultansazlığı Pleistocene-Holocene Lake Basin

The Sultan Lake Basin in some hundreds kilometers to the east shows a somewhat similar appearance to the Konya basin. That is, the Konya and Sultan Basins are receiving much sediments and water from the Taurus Mountains

Yeşilhisar Ovası, in which the beautiful Lake of Sultansazlığı, one of the bird paradises of Turkey in the east Central Anatolia and it is another tectonic basin at the foot of the Taurus Mountains. As it is indicated above, this depression is principally in the same nature with the Konya basin, and in a broader sense it is the most eastern member of the same tectonic basin chain along the northern margin of the Middle

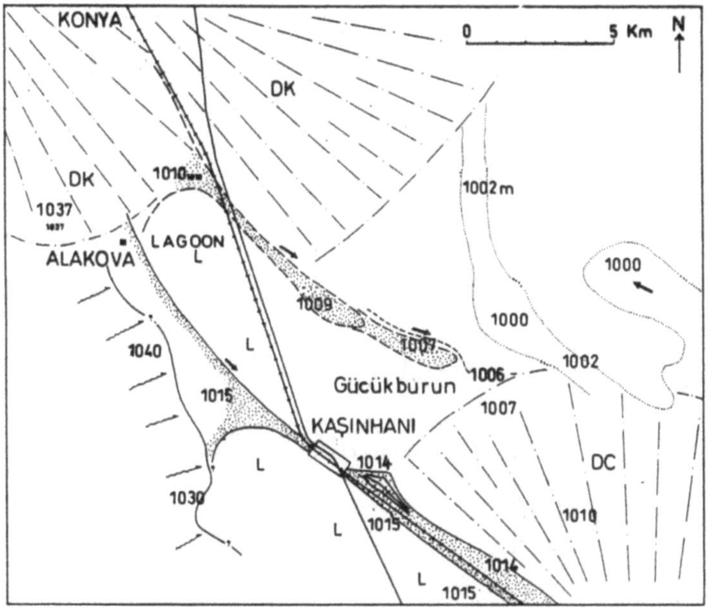

Figure 4. Geomorphological map of the spits between Konya and Kaşınhanı. DK Dejection cone of Konya, DC Dejection cone of Çumra-Çarşamba river. Note the NW-SE directed 1015 m Kaşınhanı, 1007-9m Gücükburun spits and SW-NE directed 1000-1002m spits, that is the opposite directions indicating changes in the coastal current directions. (Modified from Erol, 1972).

Taurus Mountains. Because of being at the northern end of the "Ecemiş Tectonic Trench" which is the result of a left lateral horizontal fault belt crossing the Middle Taurus Mountain Chain in a SW-NE direction, the Sultansazlığı depression, or Sultan Ova, is a round shaped deep hollow. It was developed at the crossing point of two main fault zones of SW-NE directed Ecemiş Trench and E-W directed Cappadocian fault belt between the Taurus Mountain Chain and Central Anatolian Plateaus since the mid-Tertiary. A continuous volcanic belt, starting in the west of Konya Basin at Erenler Dağı, extends in the north of Konya basin with huge Karadağ, Karacadağ, Hasandağ, Melendizdağ volcanic cones and hundreds of smaller ones, which rise above an extensive ignimbritic plateau. The easternmost of this volcanic cones is the Erciyes Mountain to the north of Sultan Ova. This South Central Anatolian lava and ignimbritic plateau was developed principally during the period of middle to late Miocene, and started to break down during the Pliocene. Local volcanic activities and basin formation continued until the end of Pleistocene. In some places, for

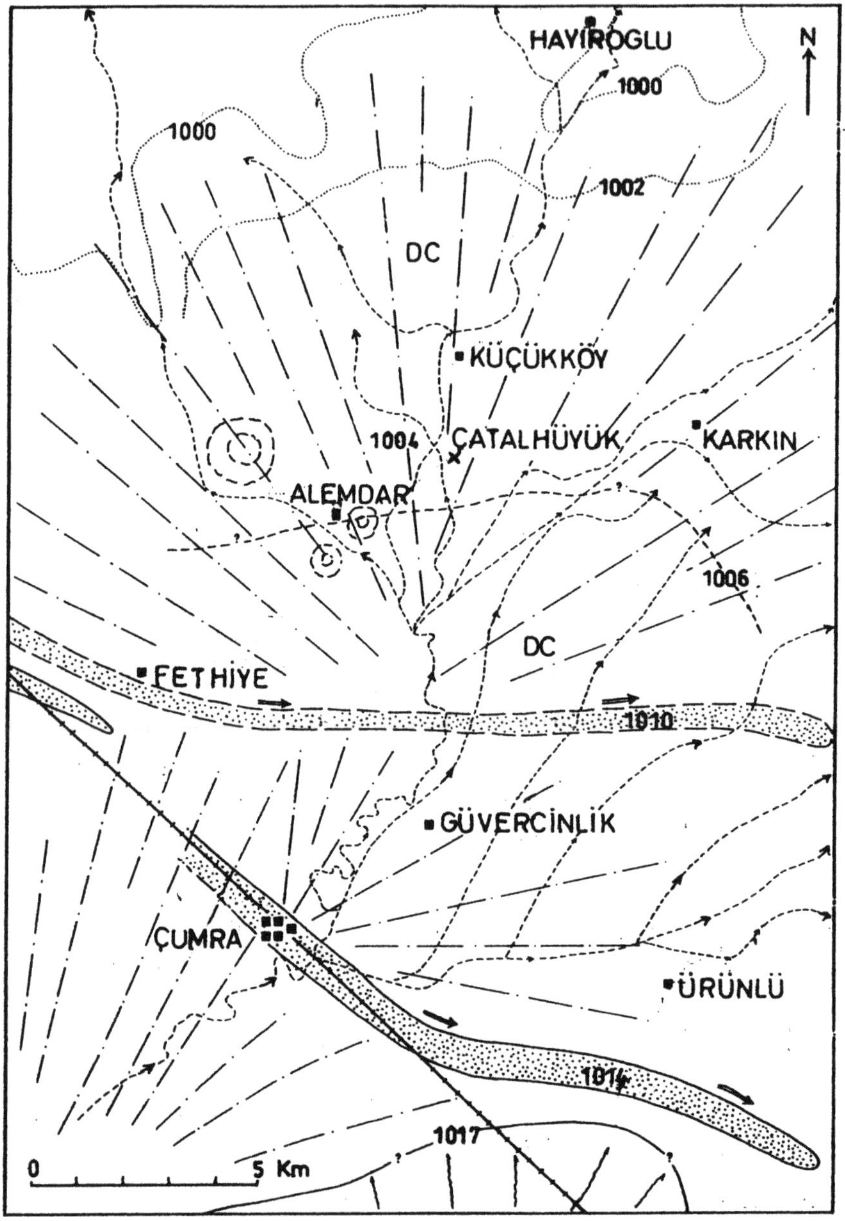

Figure 5. Geomorphological map of the spits and alluvial fans around Çumra and Çatalhüyük. Circles are indicating concentric fields and radial road pattern around the Alemdar village, in the west of Çatalhüyük, they must be younger than 1006m and older than 1002m coastlines. Note the 3 different cones related to former coastal spits. (Modified from Erol, 1972).

example at the foot of the Erciyes Mountain some minor volcanic activities had been observed even during historic times.

Erciyes Volcanic Cone, starting at late Miocene, had principally grown in a tectonic basin at the northern end of Ecemiş Fault belt together with Sultan Ova basin to the south of it and Kayseri basin to the north. That is, the Erciyes volcanic cone has grown up in a larger collapsing tectonic basin and today this basin includes one huge volcanic cone and two basins to the north and south of it. Therefore the geologic and geomorphic history of the Sultan Ova has an intimate dependence on the Erciyes mountain. Kayseri Ovası, the northern part of this main tectonic basin is opened to the outer drainage by the tributaries of the Kızılırmak, the largest river of Central Anatolia. The Sultan Ova to the south of the Erciyes Volcano, was from time to time open to outer drainage (İzbırak, 1953). During the Pliocene, it became closed as northwest directed huge lava flows of the Erciyes volcano dammed the ancient Incesu valley, that is when the Aliboran lava threshold developed, the basin became closed again during the Middle Pleistocene and the first pluvial lake, which is called by the author as Aliboran Lake in the Sultan Ova had developed. Let us consider now, following the earlier 2 denudational-fluvial stages, the development phases of this Upper Pleistocene pluvial lake.

We consider now phase 3 a fluvial interval preceeding this Late Pleistocene pluvial lake first pluvial lake shore extends along the foot of an inclined "glacias type" terrace of probably early Pleistocene age (indicated as 3 in the map). The relics of this terrace surface are fairly extensive to the southwest and northeast of the basin, and extend also towards the Incesu town at along the foot of the Hodul Mountain in the north (See map) and indicate the last fluvial phase before the pluvial lake developed.

During the wet phase (4.1) following the fluvial phase 3, a pluvial lake formed, named Aliboran Lake by the writer. It had developed during this wet phase the lake should have an outflow on the lava threshold towards the north, and at that time the huge recent dejection cone of Yeşilhisar in the south should have been started as a small delta growing at the coast of the Aliboran Lake (See the map, Figure 9).

The maximum elevation of Aliboran pluvial lake-level is estimated as 1155 m. This level is the boundary between the Arapbağı sand dune belt (D4) and an eroded terrace surface (indicated as 4.1 on the Map, Figure 9) at the foot of Hodul Mountain and it corresponds to the maximum level of the lava threshold in the north of the study area, about 5 km from Incesu town (See the Map, Figure 9).

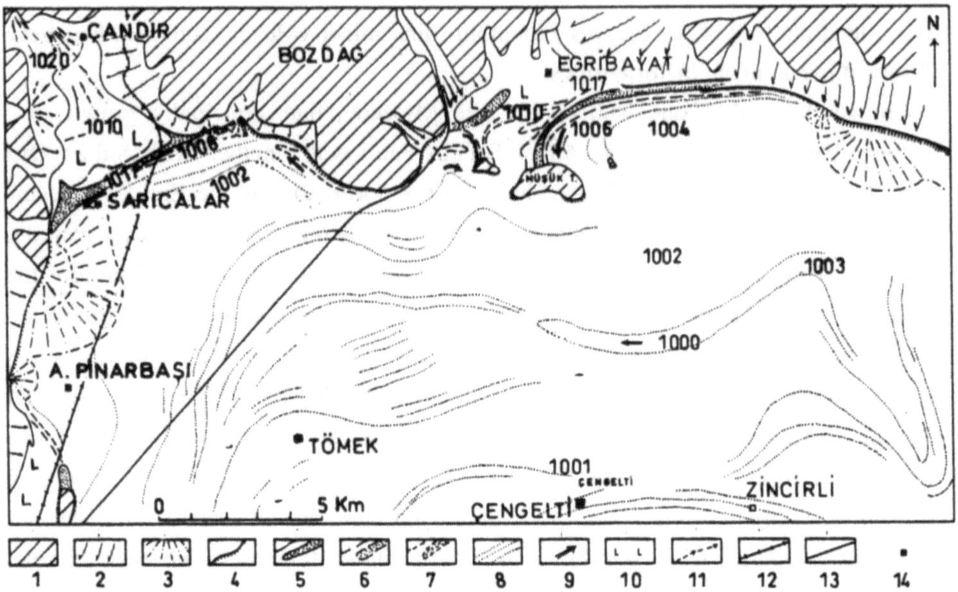

Figure 6. Geomorphological map of the coastal features around Sarıcalar-Eğribayat villages. 1. Highlands, 2. Erosional and depositional surfaces developed on lacustrine limestones (Neogene, mainly Pliocene) and red-colored detritic deposits (Plioquaternary), 3. Dejection cones, 4. Old (pluvial) coastal cliffs, 5. Coastline traces and spits belonging to 1017m level, 6. to 1010m level, 7. to 1006m level, 8. to 1002 and 1000 m levels, 9. Probably directions of longshore currents, 10. Old lagoons, 11. Recent rivers and channels, 12. Railway, 13. Main Roads, 14. Villages. (Modified from Erol, 1972).

As the climatic conditions changed and probably due to continuing subsidence in the central part of the basin its spillway dried and the lake-level behind it dropped. This was the reason for the development of a fairly extensive D4 Arapbağı dune area at the foot of Hodul dağ, in the northwest of the basin. As the relation of dune belt and estimated coastline regarded these dunes should be developed under the influence of southeasterly winds, and they could not be regarded as the result of dominant northerly winds of today.

The second phase (4.2) of the Pleistocene lake is named by the author as Çalbama Lake and its level is estimated as maximum 1125 m and it had also a spillway on the lava threshold near the village of Çalbama and the lowest level of this terrace is 1110 m. During this time the Yeşilhisar delta seem relatively more developed into the lake (See the map). At the end of this phase some of the Arapbağı dune sands should have been washed or blown down on the terrace surfaces of this generation. The phases of

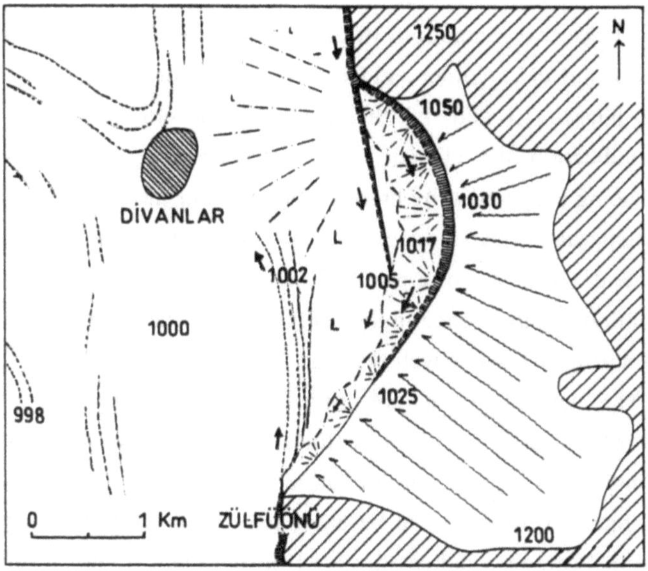

Figure 7. Geomorphological map of the coastal features of the pluvial lake in the east of the village of Divanlar, in the Konya pluvial lake basin. Note the opposite directions of 1005m and 1002m coastal spits. (Modified from Erol, 1972).

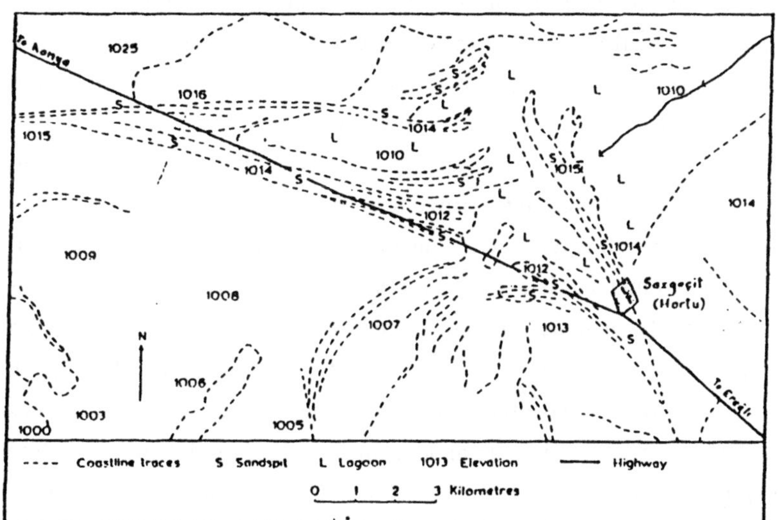

Figure 8. Geomorphological map of the composite coastal spits, on which the Sazgeçit village and Konya Karapınar highway were built. (Field maps of Oğuz Erol).

the Sultan Lake giving outflow to open drainage are interpreted as the result of wet period of last glacial age, because two generations of glaciers were recognized on the Erciyes Mountain (Erinç 1952 and 1953; Güner and Emre, 1983; Kurter, 1991).

Phase 5. Following the phases 4.1 and 4.2 the basin became closed to open drainage. This may be the indication of a climatic change towards the transition to early Holocene. In spite of being surrounded by high mountains, and the presence of several springs bringing karstic and artesian water into the basin it remained closed afterwards. This may indicate the influence of a gradually increased drought towards the recent times, but the continuing subsidence in the central part of the basin should also be considered. Indeed during the water well drillings at least 250-300 m detritic material (clay, sand, gravel) are cut and they are considered of Plioquaternary age.

During the earlier phases of Holocene the lake coastlines were running close to the basin margins, which may indicate there was still a relatively deep lake in the middle of the basin. The coastline traces of this lake can be obviously followed in the aerial photographs and therefore they are also indicated by 5.1 on the map. A very fertile soil which is covered by continuous cereal fields are observed on the aerial photographs and even on the satellite images. But at the end of this phase the coastline traces seem to have shifted towards the floor of the basin and several minor dejection cones spread on the edge of the lake floor, which is indicated by 5.2 (see the map). Therefore the coastline of 1080 m is suggested as the transition between Early to Middle Holocene Kebirkaşı lowest terraces and Late Holocene Cibik and Sultan Lake floors. Çayırözü Dunes (D5) were developed on lowest terraces which developed on the basaltic lava flows near to the Çayırözü village, as the lake had recessed back from the Early Holocene Çayırözü terrace.

Phase 6. Below the 1080 m level there is an extensive swampy-salty lake floor. Here there are no field and this part is a good grazing ground during the dry seasons of the year and covered by marshes or high vegetation, such as reed and rushes during the wet season. Dry ground is covered by a salty soil, but the ponds of fresh water springs are covered by intensive reed and rushes, therefore this lake is called as "Sultansazlığı" in Turkish. These rushes are an ideal area for birds and therefore Sultan Sazlığı is now a well-known bird paradise. Since Late Holocene and recent shallow swampy lake water coastlines fluctuating back and forewords on the flat bottom and 3 different minor sand dune generations, that is Kopçu (D6), Cibik (D7) and Yığıntı dunes (D8) could be separated from each other. They are accepted by the author as

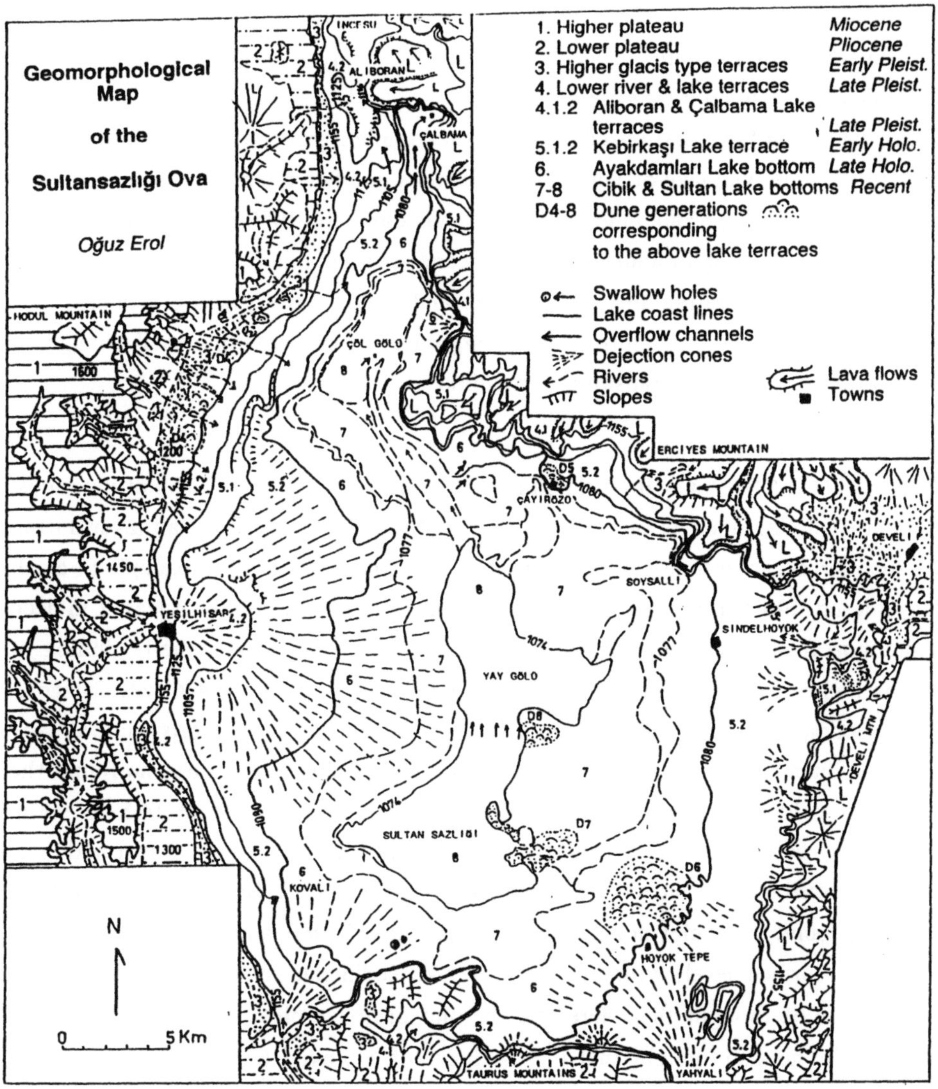

Figure 9. Geomorphological Map of the Yeşilhisar-Sultansazlığı Ova. (Modified from Erol, 1994).

indicator of minor dry phases. It is interesting to note that the low hills of Cibik dunes are invaded partly by the fresh water of Sultan Sazlığı Lake and provided a very convenient grazing ground for cattle, because the sandy bottom of shallow water is convenient to walk on it (see the map).

So, the Sultan Lake basin became a tiny but complete example of a late Pleistocene and Holocene pluvial lake of Turkey. Here there is another interesting observation to note: All the villages around the Sultan Lake are above the 1080 m coastline, which is separating the early and middle to late Holocene terraces, this is the boundary drawn by the nature separating the arable lands and sandy salty grazing grounds, and therefore it is another geomorphological argument of the environmental change of the middle/late Holocene transition in Anatolia.

Tuzgölü Pluvial Basin

Tuzgölü (The Salt Lake in Turkish) is another interesting pluvial basin of innermost Central Anatolia. Since its lake floor is covered by a 30-50 cm thick salt layer during the summer, at earlier times, in the literature it was given as an ideal example of a salt playa desert. But during the winter the salt layer become covered by at least 1 m deep and almost fresh rain water. Therefore the author believed that since the beginning of his researches it is not a simple playa lake. Therefore he started in 1966 with Tuzgölü, and for comparison he went to Konya basin on which the first reports were already printed (Chaput, 1936; Darkot, 1938; Louis, 1938a,b; Lahn, 1946). Later he found the similar fresh looking coastline traces in the Tuzgölü basin at about 905 to 930m which may correspond to the 996 to 1017 m coastlines of Konya Basin and may probably attributed to last (Würm?) pluvial age (Erol, 1970). But additional to these traces he found much elevated deltaic sediments at 955, 980, and 1015 meters and an extensive surface level, without apparent lake sediments at 1050 m the Tuzgölü basin. Than he thought that there must be pluvial traces older than Würm age, and additionally he also thought that the maximum of a pluvial level should not simply coincides with the maximum of a glacial period at higher latitudes, simply saying the high levels are not simply of climatic origin, and they must be the result of a hydrologic balance of wet and dry periods unit and local tectonics. Later he followed the 1050 m surface towards the Konya basin and put forward that this surface is an early Pleistocene level extending in larger areas of Central Anatolia (Erol, 1991). Additionally he found that the salt layer at the floor of the lake, originating from the

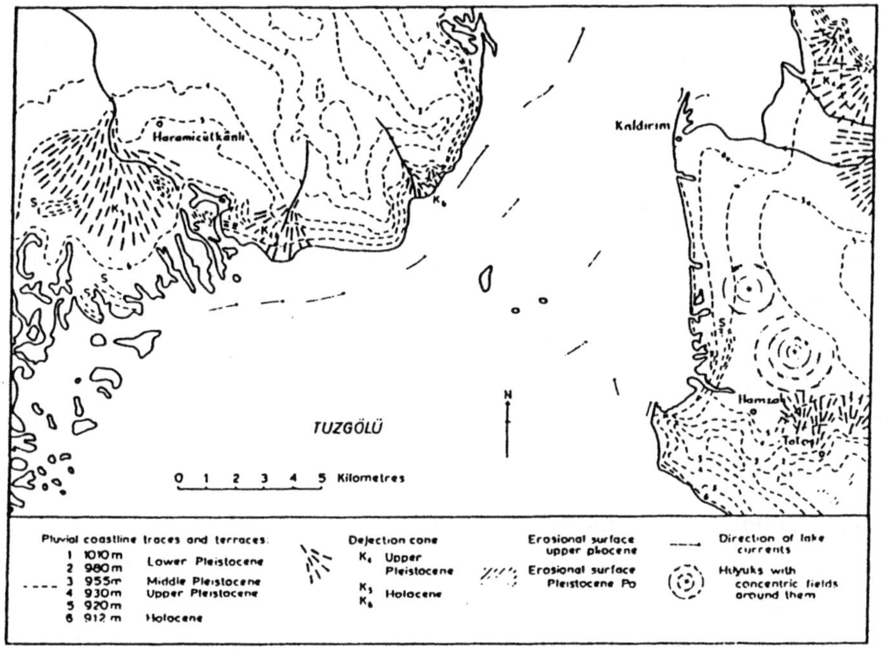

Figure 10. Geomorphological Map of the Middle part of Tuzgölü. Note to the coastal spits directed to north and circles indicating concentric field and radial road patterns. (Field maps of Oğuz Erol).

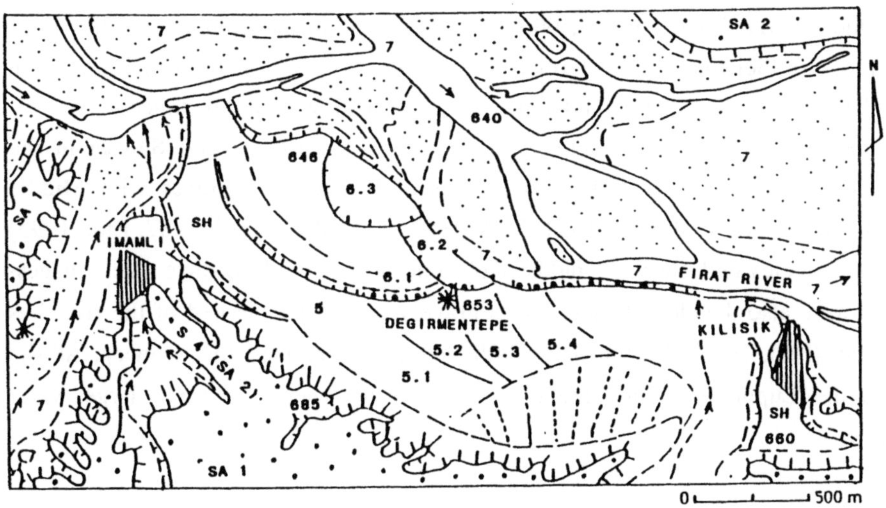

Figure 11. Geomorphological Map of the Değirmentepe area. (Field maps of Oğuz Erol).

deeper and older salt deposits of at least 900 000 years old (Bagge and Willkomm, 1966) and coming up by natural artesian springs, that is simply Tuzgölü is not a playa lake.

For the Holocene question there is again an observation. On the detailed geomorphological maps of the lake coastlines (Figure 10) the subrecent spits at the lake coastline are directed towards the north, and some old höyüks, even with radial roads and concentric fields are found immediately at the salty coastline, along which fertile cereal fields are expending. This may indicate a conclusion:

1. Probably since the Bronze age immediately the coastal belt was arable,

2. The subrecent coastal currents were directed towards the north, while the recent dominant winds are blowing from north steadily during the summer months, therefore the last phase of lake-level lowering may be the result of southerly winds.

Euphrates Valley In The Eastern Anatolia

Similar to the pluvial lake basins in Central Anatolia, 2-5 m elevated Early to Middle Holocene lowest terraces with a fertile soil cover and old (Bronze Age) villages and gravely Upper Holocene valley floors are also to be seen in the Central and East Anatolian river valleys. As examples of this group, the conditions around Değirmentepe (Figure 11) and Samsat (Figure 12) areas two maps are given here. Shortly saying that Değirmentepe site near to the city of Malatya was established during the Obeid culture (about 6,500-6,550 BP) on the former Euphrates valley floor at the foot of a Pleistocene river terrace (Erol, 1995). Afterwards, the Euphrates have cut its bed about 4-6 m deeper and a gravely recent valley floor developed.

On the other hand in the south of the city of Adıyaman around Samsat in the Euphrates valley (Figure 12), there is a series of Pleistocene terraces with Middle Palaeolithic artifacts (Albrecht, 1984; Völk, 1987; Yalçınkaya et al. 1987; Erol, 1995) and a 2-5 m elevated early to middle Holocene terrace with Bronze age villages-höyüks were also developed there and a gravely late Holocene bed of the river has cut it later (Erol, 1995). Wilkinson (1990b) has also described this scenery. All the sites in the Euphrates valley are now covered by waters of three large water reservoir.

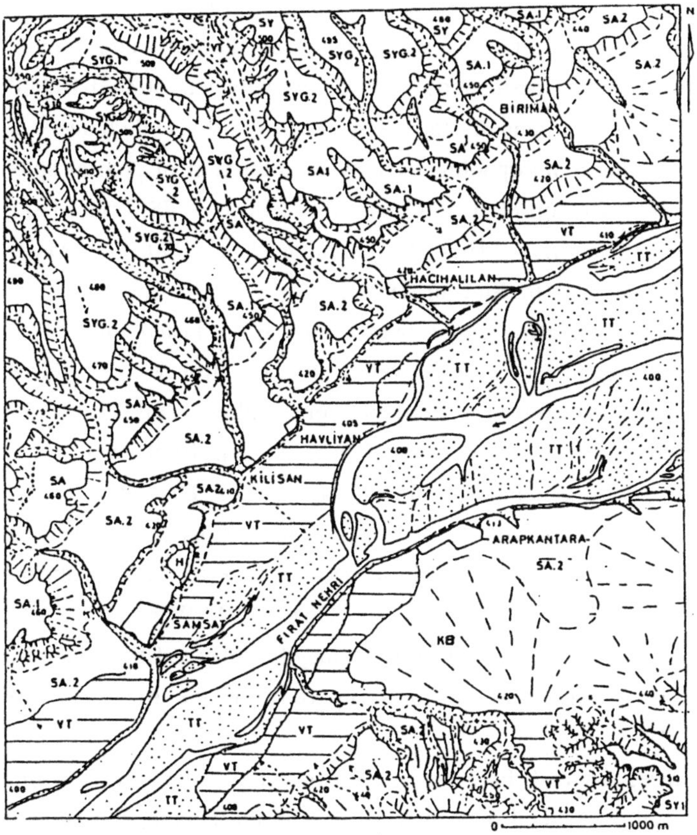

Figure 12. Geomorphological map of the Samsat area. SY. and SA. are the Pleistocene terraces, KB. Dejection cones, VT. Early Holocene lowest terrace on which the Bronze age hüyüks and roads are found, TT. gravely recent valley floor and riverbeds.

Conclusion

According to the geomorphic arguments wet (or pluvial) conditions in Anatolia were indicated by high lake-levels, increasing travertine formation (De Planhol, 1956; Darkot and Erinç, 1951; Vita-Finzi, 1969b). The transitions into dry conditions were marked by lake-level lowering, cutting of terraces and halite formation on the valley and lake floors (Erol, 1978; Rosen, 1994; Schuldenrein, 1994), deforestation (Brice, 1978; Erinç, 1978, Eisma, 1978; Atalay, 1992) soil erosion and cutting of terraces, increased alluviation and deltaic advance (Darkot, 1938; Russel, 1954; Erinç, 1955; Vita-Finzi, 1969 b, c, Eisma, 1974 and 1978; Kayan, 1991 and 1994)

These phenomena have been repeated several times during the Early and Middle Pleistocene, but only the last wet period could be dated between 30,000-10,000 years B.P. These changes may be principally caused by climatic shifts, and changes in wind directions are observable geomorphologically by the direction of coastal spits of pluvial lakes (Atalay, 1992; Bryson et al. 1992 and 1994; Brice, 1978; Erinç, 1978 and 1989, Erol, 1972; 1978; Fairbridge, 1994; La Fontaine et al. 1990, Petit-Maire 1994 a, b; Roberts 1989, Rognon, 1976, Sinclair, 1994).

Optimum climatic conditions starting about 8000 years B.P. lasted until 5000 years B.P. and may correspond to some lake-level rises when Bronze age villages established on a lowest terrace which was covered by a fertile alluvial soil (Erol, 1972 and 1978; Todd, 1971; French, 1970; Roberts, 1982 a, b; Wilkinson, 1978-1993).

There was an unstable period generally at 5000 to 4000 years. Dissection of the lowest terrace in the river valleys, drying up of lakes in the basins may indicate a marked environmental change. But this change seems not to reflect as "an abrupt collapse" due to climate change as in Mesopotamia (Hole, 1994; Weiss, 1993 and 1994; Foley, 1994; Kukla, 1994). This kind of drought may be tolerated by mountain rivers therefore such disasters may usually of local importance in the Anatolian fine geomorphic fabric.

There may be minor three dry and three wet oscillations since 4000 B.P. The dry periods are usually indicated by several dune generations on the Konya, Burdur and Sultan lake basins. Today is the latest relatively wet period.

References

Acatay,T. (1966) The relation between Beyşehir Lake and Manavgat River. D.S.İ. Teknik Bülten 9: 1-24. Ankara. (In Turkish).

Albrecht, G., Engelhardt, H., Müller-Beck, H.J., Unrath. G., Yalçınkaya, I. (1984) Vorbericht über die Untersuchungen an der Faustkeilstation Şehremuz in der südöstlichen Türkei. Eiszeit alter und Gegenwart 34: 43-86. Stuttgart.

Atalay, İ. (1975) Quaternary deposits and Geomorphology of Akşehir, Eber and Karamuk Basins. Congress of Earth Sciences on the Occasion if the 50th Anniversary of Turkish Republic, 17-19 December 1973 Papers: 365-385. Publ of MTA Inst. Ankara.

Atalay, İ. (1977) Geomorphological evolution of the Burdur Basin and its surroundings. *Jeomorfoloji Dergisi* 6: 93-110.Ankara (In Turkish, English summary)

Atalay, İ. (1992) The paleogeography of the Near East. From late Pleistocene to early Holocene. 38 pp. İzmir.

Bagge, E. and Willkomm,H. (1966) Geologische Altersbestimmung mit 36 Cl. Atomkernenergie 11: 176-184.

Blackburn, M. and Fortin, M. (1994) Geomorphology of Tell'Atij, Northern Syria. Geoarcheology 9/1: 57-74. Wiley.

Bottema, S. (1978) The Late Glacial in the Eastern Mediterranean and the Near East. In Brice W.C. (ed.) The Environmental History of the Near and Middle East Since the Last Ice Age: 15-28. Academic Press.

Bottema, S. (1986) Late Quaternary and modern distribution of forest and some tree taxa in Turkey. Proceedings of Royal Soc. of Edinburgh 89 B: 103-111

Bottema, S. and Zeist, W.v. (1990) Vorderer Orient. Vegetation im Frühholozån (ca 8000 B.P.).Tavo A VI 2. 2.1 Nordwesten. Palynologisch bearbeitet. A VI 3 Vorderer Orient. Vegetation im Spåt-PleistozÑn und Holozån. Beispiele 3.1 Südliche Levante, 3.2 NW Anatolien, 3.3 Nordwstlicher Zagros. Middle East. Early Holocene Vegetation (c. 8000 B.P.) Tavo A VI 2. 2.1 North West. Palynologically Investigated. Hers. von SFG 19. Tübinger Atlas des vorderen Orients (TAVO) der UniversitÑt Tübingen Ludwig Reichert Verlag.

Bottema, S. (1994) Vegetation and climate history of the Near East based upon palynological investigations. NATO-ARW on Third Millennium BC Abrupt Climate Change and Old World Social Collapse, 19-24 September 1994 Kemer Turkey.

Bottema, S. and Zeist, Wv. (1989) Vorderer Orient. Vegetation im Spåt-Pleistozån und Holozån. Beispiele 3.1. Südliche Levante, 3.2. Nordwest Anatolien, 3.3.Nordwestlicher Zagros.TAVO Karte AVI3.

Brice, W.C. (1978a) (ed.). The Environmental History of the Near and Middle East Since the Last Ice Age. 384 pp. Academic Press.

Brice, W.C. (1978b) The Desiccation of Anatolia. In Brice W C Env. Hist. of the Near and Middle East.: 141-148.

Bryson, R. A. (1992) A macrophysical model of the Holocene Intertropical Convergence and jetstream positions and rainfall for the Saharan Region. *Meteor. and Atmos. Phys.* 47: 247-258

Bryson, R. A. (1994) Proxy indications of Holocene winter rains in Southwest Asia compared with simulated rainfall. NATO-ARW on Third Millennium BC Abrupt Climate Change and Old World Social Collapse, 19-24 September 1994 Kemer Turkey.

Bryson, R. A. and Bryson, R. U. (1994) High Resolution Simulations of regional Holocene climate: North Africa and the northwestern United States. NATO-ARW on Third Millennium BC Abrupt Climate Change and Old World Social Collapse, 19-24 September 1994 Kemer Turkey.

Bukowski, G. von (1890-92) Geologische Forschungen im westlichen Kleinasien. Verh. Kais. K. Geol. Reichsamt. 5: 134-141

Chaput, E. (1936) Voyages d'etudes geologiques et geomorphogeniques en Turquie. Mem. de l'İnst. Fr. Archeologie de Stamboul II, VIII. Paris.

Cohen, H.R. (1970) The palaeoecology of south Central Anatolia at the end of the Pleistocene and the Beginning of the Holocene. *Anatolian Studies* 20: 121-137

Cohen, H. R. and Erol, O. (1969) Aspects of Paleogeography of Central Anatolia. *The Geographical Journal.* 135. 3: 388 -398.

Darkot, B. (1938) Türkiye'de yakın iklim değişiklikleri ile ilgili bazı morfoloji müşahedeleri. Quelques observations morphologiques en relation avec les variations recentes du climat en Turquie (Resume). Coğrafi araştırmalar ı: 15-27, 89-93. Istanbul Üniv. Coğrafya Enst Neür. 4.

Darkot, B. (1938) Ege haliçlerinin menşe ve tekamülü. Sur l'origine et l'evolution morphologique des estuaires de la Mer d'Egee (Resume). Coğrafi Araştırmalar ı: 29-52, 93-95. İstanbul Üniv. Coğrafya Enst. Neür.4.

Darkot, B. and Erinç,S. (1951) Aksu batısında Antalya traverten taraçaları. Terrasses de Travertin d'Antalya a l'ouest d'Aksu. *İst. Üniv. Coğ. Enst. Derg.* 1. 2: 56-65.

Degens, E.T. and Kurtaman, F. (eds.) (1978) The Geology of Lake Van. MTA Institute Publ. No. 169. Ankara.

Degens, E.T., Wong, H.K., Kempe, S. and Kurtman, F. (1984) A geological study of Lake Van, Eastern Turkey. Geologische Rundschau 73.2: 701-743. Stuttgart.

De Meester,T. (1971) Highly Calcareous Lacustrine Soils of the Great Konya Basin, Turkey. PUDOC Wageningen.

De Planhol, X. (1956) Contribution a la etude geomorphologique du Taurus occidental et de ses plaines bordieres. Rev. Geogr. Alpine 44: 609-685.

Eisma, D. (1962) Beach ridges near Selçuk (Ephesus), Turkey. Tijdschr. Kon. Ned. Aardr. Gen. 79. 3: 234-246.

Eisma, D. (1964) Stream deposition in the Mediterranean area in historical times. *Nature* 203. 4949: 1061.

Eisma, D. (1978) Stream deposition and erosion by the eastern Shore of the Aegean. In W C Brice (ed) Env. Hist. Middle and Near East: 67-79.

El-Moslimany, A.P. (1982) The Late Quaternary vegetational history of the Zagros and Taurus mountains in the regions of Lake Mirabad, Lake Zaribar and Lake Van. A reappraisal. In Bintliff J L, Zeist Wv(Eds) Palaeoclimates, Palaeoenvironments and Human communities in the Eastern Mediterranean Region in Later Prehistory: 343-351. British Archeological Reports Int. Ser 133. Oxford.

Erinç, S. (1949a) Eiszeitliche Formen und gegenwårtige Vergletscherung im nordostanatolischen Randgebirge.*Geologische Rundschau* 37.

Erinç, S. (1949b) Bathimetric map of the Sapanca Lake and its morphometry. Türk Coğrafya Dergisi 6-8 / 11-12: 139-140. İstanbul. (In Turkish).

Erinç, S. (1951) The glacier of Erciyes in Pleistocene and post-Glacial Epoches. İstanbul Üniv. Coğrafya Enst. Derg 1/2: 82-90. Istanbul. (In Turkish).

Erinç, S. (1952) Glacial evidences of the climatic variations in Turkey. Geogr. Annaler 34: 89-98. Stockholm.

Erinç, S. (1952) The present glaciation in Turkey. Proceed. 17th Intern. Geogr. Congress, Washington: 326 -330.

Erinç, S. (1953) Geography of the Eastern Anatolia. Publ. of the Geographical Institute of the İstanbul Univ. No 15 (in Turkish)

Erinç, S. (1978) Changes in the Physical Environment in Turkey Since the End of the Last Glacial. In Brice W.C. (Ed.) The environmental History of the Near and Middle East since the last Ice Age: 87-108. Academic Press.

Erol, O. (1963) Die Geomorphologie des Orontes-Deltas und der anschliessenden pleistozaenen Strand- und Flussterrassen, Provinz Hatay, Türkei. Publ. of Ankara Univerståt, Inst.für Geographie, No.148. 110 pp.

Erol, O. (1969) Geology and Geomorphology of the Tuzgölü Basin. A research from the viewpoint of neotectonics, pluvial lake terraces and potas-salt formations. 334 pp. Report for Turkish Scientific and Technical Council. Ankara. In Turkish, unpublished.

Erol, O. (1970) Les hauts niveaux pleistocenes du Tuzgölü (Lac Sale) en Anatolie Centrale, Turquie. Ann. de Geogr. 79: 39-50. Paris.

Erol, O. (1972) Geomorphological evidence of the recessional phases of the pluvial lakes in the Konya, Tuzgölü and Burdur Basins in Anatolia. Coğrafya Araştırmaları Dergisi 3-4: 13-52. Ankara. In Turkish with English summary.

Erol, O. (1976) The palaeoecology of some neolithic sites on South Central Anatolia (Abstract). UİSPP 9e Congres Resume de Communications: 363. Nice.

Erol, O. (1978) The Quaternary history of the lake basins of Central and southern Anatolia. In Brice W.C. (Ed.) The environmental History of the Near and Middle East since the last Ice Age: 111-139. Academic Press.

Erol, O. (1981a) Quaternary pluvial and interpluvial conditions in Anatolia and environmental changes in south-central Anatolia since the last glaciation. Frey, W. and Uerpmann, H.P. (Hrsg) Beitråge zur Umweltgeschichte des Vorderen Orients. Beihefte zum Tübinger Atlas des orderen Orients. Reihe A. Nr 8: 101-109. Wiesbaden.

Erol, O. (1981b) Neotectonic and Geomorphologic evolution of Turkey In Fairbridge R.W. (Ed.) Neotectonics. Zeitschr. für Geom. Suppl. Bd. 40: 193-211.

Erol, O. (1984) Geomorphology and neotectonics of the pluvial lake basins in the Taurus Belt and South Central Anatolia. In Tekeli, O. and Göncüoğlu, C. (Eds) Geology of the Taurus Belt. International Symposium: 119 -124. Ankara.

Erol, O. (1986) Plioquaternary evolution of the Burdur Lake basin as an example of geomorphologic system analysis in neotectonic research. 4. Müh. Haftası, 5-9 May 1986, Isparta. Abstract: 60-61.

Erol, O. (1991a) The relationship between the phases of the development of the Konya-Karapınar obruks and the Pleistocene Tuzgölü and Konya pluvial lakes, Turkey (Abstract and Summary). Bulletin of the Inst. of Marine Sci. and Geography, University of İstanbul. No.7: 5-49. İstanbul.

Erol, O. (1991b) Geomorphological evolution of the Taurus Mountains, Turkey. Zeitschr. für Geom. N.F. Supp. Bd 82: 99-109.

Erol, O. (1993) Quaternary Chronology of Turkey according to the geomorphologic arguments. Workshop on the Quaternary of Turkey, November 17-19, 1993 İstanbul. Abstracts: 54-59. Publ. of the Technical University of İstanbul, Faculty of Mining, Dep. of Geological Engineering and TÜBİTAK. İstanbul. In Turkish.

Erol, O. (1994) Geomorphology of Sultan Sazlığı and its environ. 72 pp. Unpublished report for Ministry of Forestry, Division of National Parks of Turkey.

Erol, O. (1995) Archeometric and Geomorphological Research in Turkey. In print in the Festschrift for Prof.X de Planhol, University of Paris-Sorbonne, Geography Department.

Erol, O., Akkan, E., Elibüyük, M. and Doğu, A.F. (1987) The Present and Quaternary natural environmental conditions in the Lower Euphrates region. Lower Euphrates Project 1978-1979 activities. METU Lower Euphrates Project Publications. Series 1. No 3: 15-23. Levha-Plate 1. Ankara.

Erol, O., and Pirazzoli P.A. (1992) Seleucia Pieria: an ancient harbor submitted to two successive uplifts. *The Inter. Jour. of Nautical Archeology* 21. 4: 317-327.

Erol, O., Şenel, S. and Kış, M. (1986) Preliminary results of C14 dating of the Burdur Lake coastal deposits. Arkeometri Ünitesi Bilimsel Toplantı Bildirileri, İstanbul 6: 81-88.1986. In Turkish.

Erol, O. and Tüfekçi, K. (1993) Ilgın Gölü dolayının jeomorfolojisi. The geomorphology of the surroundings of the Ilgın Lake (Abstract). *Jeomorfoloji Dergisi* 20: 1 -12. Ankara.

Fairbridge, R.W. (1994) Climate and Chronology of the Holocene Sub- boreal (Bronze Age). NATO-ARW on Third Millennium BC Abrupt Climate Change and Old World Social Collapse, 19-24 September 1994 Kemer Turkey.

Fairbridge, R. W., Erol, O., Karaca, M. and Yılmaz, Y. (in this volume) Background to Mid-Holocene Climate Change in Anatolia and Adjacent Regions. In print in NATO/ARW volume "Third Millennium Abrupt Climate Change and Old World Social Collapse".

Farrand, W. R. (1965) Geology and physiography of the Beyşehir-Suğla depression, Western Taurus Lake District, Turkey. Türk Arkeol. Dergisi. 13: 149-154.

Farrand, W.R., Solecki, R.S. (1963) An archeological reconnaissance in the Beyşehir-Suğla Area of Southwestern Turkey. Report of the Columbia University Archeological Exp. to Turkey.

Foley, J. A. (1994) Climate and Terrestrial Biosphere Interactions during the Middle Holocene. NATO-ARW on Third Millennium BC Abrupt Climate Change and Old World Social Collapse, 19-24 September 1994 Kemer Turkey.

French, D. H. (1970) Notes on site distribution on the Çumra area. Anatolaian Studies 20: 139-148.

Güldalı, N. (1981) Karstic hydrogeology of Suğla Plain and the problem of Suğla Lake. *Jeomorfoloji Dergisi*, 10: 33-58. (In Turkish).

Güner, Y. and Emre, Ö. (1983) Pleistocene glaciation on Mt. Erciyes and its relation to volcanism (Abstract). *Jeomorfoloji Dergisi*, 11: 23-34. Ankara. (In Turkish).

Higgs, E. S., Fagg, A. E., Vita-Finzi, C. and Harris, D. R. (1967) The Climate, Environment and Industries of Stone Age Greece: Part III. Proceedings of the Prehistoric Society Vol 33: 1-29. Appendix: Bottema S.A Late Quaternary pollen diagram from Ionnia.

Hole, F. (1994) Evidence for Mid-Holocene environmental change in the Western Habur drainage, northeastern Syria. NATO-ARW on Third Millennium BC

Abrupt Climate Change and Old World Social Collapse, 19-24 September 1994 Kemer Turkey.

İzbırak, R. (1953) Develi Plain and its economic development. Publ. of Faculty of Letters, University of Ankara (In Turkish).

Kayan, I. (1991) Holocene geomorphic evolution of the Beşik Plain and changing environment of Ancient Man. Studia Troica Bd 1: 79-92.

Kayan, I. (1994) An investigation on the regression and sedimentlogical change through the Western Anatolian coastal plains during the Bronze Age. NATO-ARW Third Millennium BC Abrupt Climate Change and the Old World Social Collapse. September 19-23, 1994 Kemer Turkey.

Kazancı, N., Özkan, H. M., Alkan. A. and Erol, O. (1986) Grain surface features of Pleistocene deltaic quartz sands from Burdur basin, Turkey: a SAM application. Doğa 10/3: 255-266. Ankara.

Kazancı, N. and Erol, O. (1987) Sedimentary characteristics of a Pleistocene fan-delta complex from the Burdur basin, Turkey. Z. für Geomorphologie N.F. 31/3: 261-275. Berlin.

Kazancı, N., Erol, O. and Tüfekçi, K. (1989) Interrelations between synsedimentary tectonism, basin-margin morphology and fan delta formation with examples from Pleistocene -Holocene Burdur Basin of Turkey. Conference on Geomorphology, Man and Natural Resources. March 27-31, 1989 Ankara. Abstracts: 44-45. Ankara. (Abstract).

Kış, M., Erol, O., Şenel, S. and Ergin, M. (1989) Preliminary Results of Radiocarbon Dating of Coastal Deposits of the Pleistocene Pluvial Lake of Burdur, Turkey. Journal of Islamic Academy of Sciences 2: 37-40.

Kukla, G. (1994) Paleoclimatic data from the second and third millennium BC: overview and problems. NATO-ARW Third Millennium BC Abrupt Climate Change and the Old World Social Collapse. September 19-23, 1994 Kemer Turkey.

Kurter, A. (1991) Glaciers of Turkey. In Williams R. S., Ferrigno, J. G.(eds.) Glaciers of the Middle East and Africa. Satellite Image Atlas of the World. US Geological Survey Professional Paper 1386 -G-1: 1-29.

Kutluk, H. (1994) Holocene pollens of Haliç estuary, İstanbul, Turkey Doctorate Thesis. Institute of Marine Sciences, University of Istanbul. (unpublished).

Kuzucuoğlu, C., Karabıyıkoğlu, M., Pastre, J. F., Fontugne, M. and Ercan, T. (1994) Tephra layers and environmental indicators in Pleistocene and Holocene lake sediment sequences in Inner Anatolia (Karapınar-Konya, Turkey). NATO-ARW Third Millennium BC Abrupt Climate Change and the Old World Social Collapse. September 19-23, 1994 Kemer Turkey.

La Fontaine, C. V., Bryson, R. A., Wendland, W. M. (1990) Airstream regions of North Africa and the Mediterranean. *Jour. Climate* 3/3: 366-372.

Lahn, E. (1946) Konya-Burdur bölgelerindeki Pliyosen ve Dördüncü zaman tabakaları. Les depots Pliocenes et Quaternaires de la region de Konya-Burdur. İstanbul Üniv. Fen Fak. Mecm. B 11. 2: 85-106.

Lahn, E. (1948) Türkiye göllerinin jeolojisi ve jeomorfolojisi hakkında bir etüd. Contribution a l'etude geologique et geomorphologique des lacs de la Turquie. MTA Yayını B12. 178 pp. Ankara.

Louis, H. (1938) Eiszeitliche Seen in Anatolien. Z. Ges. Erdkunde Berlin 7/8: 267-285.

Louis, H. (1938) öber alte HochstÅnde anatolischer Seen. Comptes Rend. Congr. Int. Geogr. Amsterdam 1938, Tome II, Sect.A-F: 325- 335. Leiden

Petit-Maire, N., Sanlaville,P. and Yan, Z.W. (1994a) Changements globaoux et paleomoussons: la zone de transition afro-asiatique au course des derniers 140 000 ans. Geochronique 50: 16-19.

Petit-Maire, N. and Page, N. (1994b) Holocene climatic evolution in the current hyperarid Sahara. NATO-ARW on Third Millennium BC Abrupt Climate Change and Old World Social Collapse, 19-24 September 1994 Kemer Turkey, Abstracts.

Pirazzoli, P. A., Laborel, J., Saliege, J.F., Erol, O., Kayan, I. and Person A. (1991) Holocene raised shorelines on the Hatay coasts (Turkey): Palaeocological and tectonic implications. *Marine Geology* 96: 295-311.

Roberts, N. (1981) Late Quaternary Geomorphology and Palaeoecology of the Konya Basin, Turkey. Doctorate Thesis. Dep. of Geography Univ. Coll. London. 296 pp.

Roberts, N., (1981) Radiocarbon Dates Turkey, University of Miami.

Roberts, N. (1982a) A note on the geomorphological environment of Çatalhüyük, Turkey. *Journal of Archeological Science*: 341-348

Roberts, N. (1982b) Forest re-advance and the Anatolian Neolithic. In Bell, M., Limbrey, S. (eds) Archeological aspects of wood-land ecology. Oxford British Archeological Reports, International Series 146: 231-46.

Roberts, N. (1982c) Lake-levels as an indicator of Near Eastern palaeoclimates: a preliminary appraisal. In Bintliff, J.L., Zeist W.v (eds.) Palaeoclimates, Palaeoenvironments and Human Communities in the Eastern Mediterranean Region in Later Prehistory: 234-265

Roberts, N. (1983) Age, palaeoenvironments and climatic significance of late Pleistocene Konya lake, Turkey. *Quaternary Research* 19: 154-171.

Roberts, N. (1984) Reply to comments by Ann P. El-Moslimany. Quaternary Research 21: 117-120

Roberts, N. (1989) The Holocene. An Environmental History. 227 pp. Blackwell publ.

Roberts, N. (1990) Human-induced landscape change in south and southwest Turkey during the later Holocene. In Bottema, S. et al (eds) Man's role in the sapping of the eastern Mediterranean landscape: 53-67. Balkema publ.

Roberts, N. (1991) Late Quaternary geomorphological change and the origins of the agriculture in South Central Turkey. *Geoarcheology* 6 / 1: 1-26.

Roberts, N. (1992) Response of alluvial fan regimes to climatic forcing during the Late Quaternary in the Konya Basin, south central Turkey. Mediterranean Quaternary River Environments, an International Symp. Peterhouse College, Univ. of Cambridge September 1992. Balkema publ.

Roberts, N., Eastwood, W., Lamb, H. and Tibby, J. (1994) Volcanism, climate change and human impact: separating cause from effect in the palaeoenvironmental record of SW Turkey. NATO-ARW on Third Millennium BC Abrupt Climate Change and Old World Social Collapse, 19-24 September 1994 Kemer Turkey, Abstracts.

Roberts, N., Erol, O., De Meester, T. and Uerpmann, H.P. (1979) Radiocarbon chronology of late Pleistocene Konya lake, Turkey. Nature 281. 5733: 662-664.

Roberts, N. and Wright jr, H. E. (1994) Vegetation, lake-level and climatic history of the Near East and Southwest Asia. In Wright jr, H. E. et al. (eds) Global Climates Since the Last Glacial Maximum, Minneapolis, University of Minnesota Press, 569 pp.

Rognon, P. (1976) Essai d'interpretation des variations climatiques au Sahara depuis 40 000 ans. Revue de Geographie Physique et de Geologie dynamique 17/2-3: 251-282.

Rosen, A. M. (1994) Environmental change and human adaptational failure at the end of the Early Bronze Age in the Southwestern Levant. NATO-ARW on Third Millennium BC Abrupt Climate Change and Old World Social Collapse, 19-24 September 1994 Kemer Turkey, Abstracts.

Russel, R. J. (1954) Alluvial Morphology of Anatolian rivers. Ann. Assoc. Amer. Geogr. 44: 363-391.

Sanır, F. (1948) Sultan Dağları, Sakarya Basin and Akşehir Lake. 183 pp. In Turkish.

Schuldenrein, J. and Clark, G. A. (1994) Landscape and Prehistoric chronology of West-Central Jordan. *Geoarcheology,* 9/1: 31-56. Wiley.

Schweizer, G. (1975) Untersuchungen zur Physiographie von Ostanatolien und Nordwestiran. Geomorphologische, klima- und hydrographische Studien im Van See- und Rezaiyeh See Gebiet. 145 s. Tübinger Geographische Studien 60.

Sinclair, L. and Bryson, R. (1994) Preliminary results of joint historical and climatological high resolution modeling study of E. Mediterranean Area, Part I: Egypt, Mesopotamia and Palestine. NATO-ARW on Third Millennium BC Abrupt Climate Change and Old World Social Collapse, 19-24 September 1994, Kemer Turkey, Abstracts.

Sungur, K. A. (1978) Physical Geography of Burdur, Tefenni, Acıgöl basins. 126 pp. Publ. of İstanbul Univ. Geographical Inst.No 95. In Turkish.

Todd, I. A. (1971) The Neolithic Period in Central Anatolia. 8th Congrès International des Sciences Pre- et Protohistoriques, Beograd, September 1971.

Todd, I. A. (1976) Çatalhüyük in Perspective. Menlo Park, California: Cummings.

Vita-Finzi, C. (1966) The Hasa Formation: an alluvial deposition in Jordan. Man 1.3: 386-390.

Vita-Finzi, C. (1969a) Late Quaternary alluvial chronology of Iran. *Geol. Rundschau* 58: 951-973.

Vita-Finzi, C. (1969b) Late Quaternary continental deposits of central and western Turkey. Man 4. 4: 605-619.

Vita-Finzi, C. (1969c) The Mediterranean Valleys: Geological Changes in Historical Times. Cambridge University Press.

Voigt, M. M. (1985) Village on the Euphrates. Excavations at Neolithic Gritille in Turkey. Expedition, 27.1, Univ. Museum Magazine of Archeology/Anthropology, University of Pennsylvania Philadelphia: 10-24.

Voigt, M. N. and Ellis, R. S. (1981) Excavations at Gritille, Turkey. Paleorient 7/2: 87-100, CNRS Paris.

Völk, H. R., Bischoff, R. and Geier, B (1987) JungpleistozÅne Terrassen und Bodenentwicklung am Euphrat bei Samsat, Südosttürkei. Geographisches Institut UniversitÅt Heidelberg.

Weiss, H. (1994) Late Third Millennium Mesopotamian socio-economic collapse. NATO-ARW on Third Millennium BC Abrupt Climate Change and Old World Social Collapse, 19-24 September 1994 Kemer Turkey, Abstracts.

Weiss, H., Courty, M.A., Wetterstrom, W., Guichard, F., Senior L, Meadow, R. and Curnow, A. (1993) The genesis and collapse of Third millennium north Mesopotamian Civilization. *Science* 261: 995-1004.

Wenzel, H. (1932) Sultan Dağ und Akşehir Ova. Schriften Geograph. Inst. Univ. Kiel.

Wenzel, H. (1935) Forschungen in Inneranatolien. I.Aufbau und Formen der Lykaonischen Steppe. Schriften Geogr. Institut Univ. Kiel 542 s.

Wilkinson, T. J. (1978) Erosion and the Sedimentation Along the Euphrates Valley in Northern Syria . In Brice W C (Ed.) The Environmental History of the Near and Middle East since the Last Ice Age: 215-226.

Wilkinson, T. J. (1990a) The development of settlement in the North Jazira between the 7th and 1st millennium BC. Iraq 52: 49-62.

Wilkinson, T. J. (1990b) Town and country in southeast Anatolia 1: Settlement and landuse at Kurban Hüyük and other sites in the Lower Karababa basin. Chicago (IL), Oriental Institute Publications 109.

Wilkinson,T. J. (1992) Off-site archeology. Land-use diversity in Bronze Age Upper Mesopotamia. *National Geographic Research and Exploration* 8. 2: 196-207.

Wilkinson, T. J. (1993) Linear hollows in the Jazira, Upper Mesopotamia. *Antiquity* 67: 548-562.

Yalçınkaya, I., Müller-Beck, H. and Albrecht G (1987) Beobactungen zum Palåolithikum des Euphrattales bei Adıyaman-Samsat und Malatya-Kuruçay 1979. Lower Euphrates Project 1978-1979 Activities, Middle East Technical University, Lower Euphrates Project Publications, Series 1, No 3. Ankara.

Zeist, W. van and Bottema, S. (1982) Vegetational history of the eastern Mediterranean and the Near East during the last 20,000 years. In Bintliff, J.L., Zeist, W.van (Eds) Palaeoclimates, paleoenvironments and human communities in the eastern Mediterranean region in later prehistory. Oxford: British Archeological Reports International Series, pp 277-321.

Zeist, W. van, Bottema, S. (1991) Late Quaternary Vegetation of the Near East. Beihefte zum TAVO, Reihe A, Nr 18. zur Karte A VI 2. Reichert Verlag.

Zeist, W. van, Woldring, H. and Stapert, D. (1975) Late Quaternary vegetation and climate of the southwestern Turkey. *Palaeohistoria* 17: 53-143.

The Late Holocene Environmental and Climate Shift in Russia and Surrounding Lands

Constantin V. KREMENTSKI

Institute of Geography
Russian Academy of Sciences
Staromonetny 29
109017 Moscow
Russia

Abstract

The period between circa 6000-4500 BP was marked by a climate which was considerably less continental than that of today. Broad-leaved trees expanded farther to the north and east as compared to their modern limit in Eastern Europe and West Siberia. Northern treeline reached. The share of broad-leaved trees was largest in forests of Eastern Europe. Steppes and semi-desert belts of today were relatively wet then. Forested area expanded, and soils were less continental in steppes of Ukraine, Southern Russia and Kazakhstan.

Between circa 4500-3500 BP climate was colder and more continental in the boreal belt, and drier in the arid belt. Area of broad-leaved forests were reduced. Ranges of *Tilia* and *Alnus* glutinosa in West Siberia were changed. Northern limits of trees and shrubs in boreal belt retreated to the present day position. Drier climate in the steppes and desert belt of Eurasia has been reconstructed based on pollen, lake levels and soil data.

All existing data prove that serious and relatively rapid climate shift took place ca 4500-4300 BP. It was accompanied by changes in human economy, which was detected mainly in Eurasian steppes and forest-steppes. After 4500-4200 BP nomadic groups penetrated in some parts of forest belt in East Europe and Siberia. At the same time agricultural communities collapsed in south-west Ukraine and Moldova, to be replaced by nomadic cultures.

Between circa 3400/3300 and 2800/2700 BP a new phase of more favorable climate is indicated in southern regions of Eastern Europe, Kazakhstan and Siberia. The Late

NATO ASI Series, Vol. I 49
Third Millennium BC Climate Change
and Old World Collapse
Edited by H. Nüzhet Dalfes, George Kukla,
and Harvey Weiss

Holocene vegetation and climate shift was more pronounced in the northern and southern margins of forest belt than elsewhere.

Introduction

Russia and surrounding states of Eastern Europe and northern Asia occupy a large part of Eurasia. This region is characterized by continental climate with very pronounced contrast between cold and warm, wet and dry seasons. Serious changes in vegetation cover caused by climate oscillation took place during transition from late Atlantic optimum (ca 6000-5500 BP) to the Subboreal after 4500 BP (in calibrated radiocarbon years). Late Holocene climatic shift was reflected in the vegetation structure and in moisture and lake level oscillations in arid regions. Environmental changes seriously influenced human societies. Reaction of human communities in various regions was corresponding to the nature of environmental challenge.

Forest belt of East Europe

Broad-leaved trees were the most widespread in forests in the central European Russia, Ukraine and Bielorussia ca 5000 BP. After 4500 BP the climate became colder in eastern Europe (Nikiforova, 1982; Bezusko et al., 1988; Khotinsky et al., 1994; Velichko et al., 1994).

Within 4600-4200 BP the area of broad-leaved forests was reduced. Birch-pine and pine forests expanded in Bielorussia (Yelovicheva et al., 1988). In Ukraine during 4500-3500 BP the area of broad-leaved forests was less then in earlier periods. After 3500 BP an area of broad-leaved forests expanded in Ukraine.

After 4500 BP an increase in *Picea* pollen is registered in eastern part of the European Russia boreal forests (taiga) belt notably in the Krasnokamsk peat-bog section (58° 03' N, 55° 44' E) (Kokarovtsev, 1992). In Bielorussia spruce expanded in forests within 4000-2500 BP. Ca 3300 BP the share of *Picea* pollen increased in pollen spectra in many regions of Eastern Europe. An expansion of spruce in forests. was due to more a cold and humid climate.

Within 6000-5000 BP the northern limits of ranges of many trees and shrubs in the north of Russia were situated farther north as compared to the present day. Figure 1 shows the position of northern limit of Siberian spruce *Picea obovata* Ledeb (Fig. 1 A), larch *Larix* (Fig. 1 B), and birch-tree *Betula* (Fig. 1 C), ca 6000 BP. That position is reconstructed based on results of radiocarbon dating of macrofossil remains of wood, bark etc. in present tundra. These data are available now for many parts of

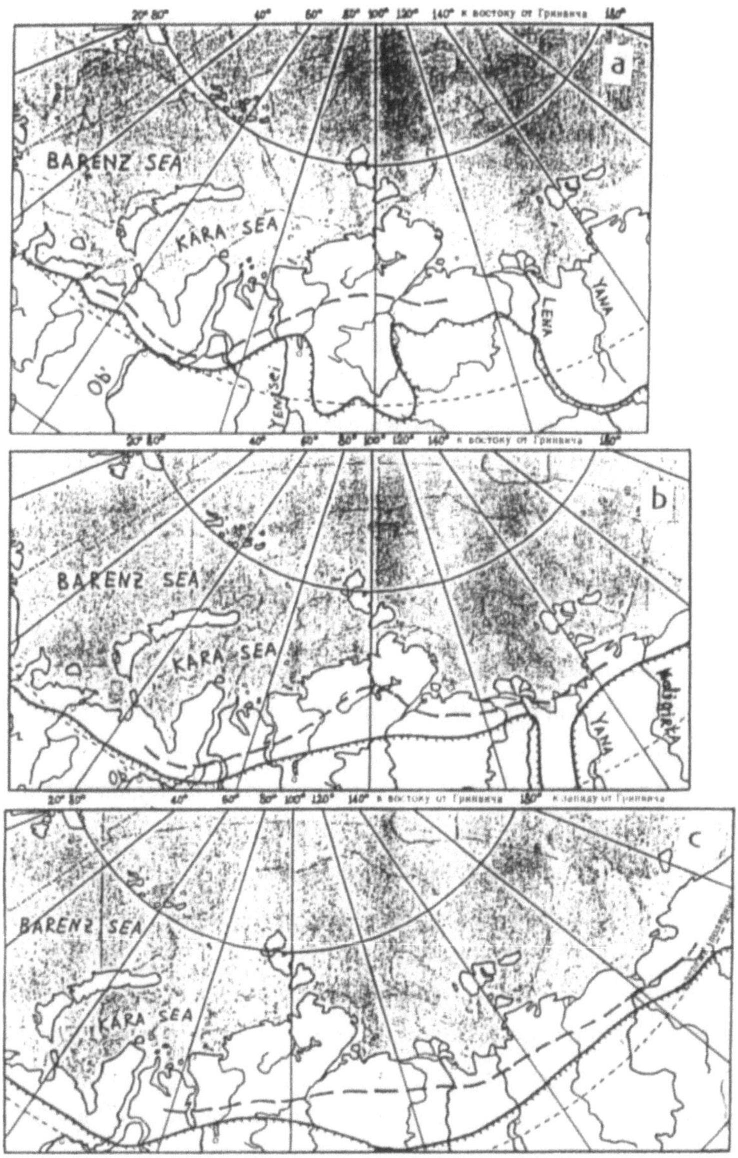

Fig. 1. Position of northern limits of ranges of some trees 6000-5000 BP. Scale 1:30,000,000

1 A). - Siberian spruce *Picea obovata Ledeb.*

1 B). - Larch *Larix*

1 C). - Birch-tree *Betula*

┬┬┬┬ - present northern limit of range

- - - - - northern limit of range 6000-5000 BP.

Siberia. After 5000 BP the northern limits of trees and shrubs moved southward and by 4500-4300 BP the northern limits of trees reached roughly the present day position with the exception only in some parts of Siberia in valleys of great rivers (Yenissey, Lena). By 4500 BP in a major part of Siberia the modern type of forest cover was just formed. Common pine, Siberian pine, larch dominated in these forests.

South of European Russia and Ukraine

Climate was less continental than at present ca 6000-4500 BP. In steppe belt January t° was 1°C higher than today, July t° was 2° lower, annual precipitation was 100-150 mm higher as compared to the modern level.

Floodplain and valley forests were the most widespread in the Dniestr, Dniepr, southern Bug, Don and Volga River valleys. Broad-leaved trees penetrated farther to the south as compared to their modern southern limits (see figures 2,3). Pollen of *Carpinus betulus* is registered in the Volga River ox-bow section near Solenoye Zaymische village (47°54' N, 46°09' E) (Bolykhovskaya, 1990)(figure 3). That fact proves that the hornbeam area expanded in the south of Eastern Europe. Steppe cover became more mesophytic, an amount of meadow elements in steppes increased. Forest area expanded also in the forest-steppe belt of Ukraine and Russia (Artiushenko, 1970; Gerasimenko, 1993; Spiridonova, 1991).

Pollen analysis of the Kharabuluk peatland in the north of Kalmuk region (47°46' N, 44°26' E) shows that 6000-4700 BP was the time of increased forest area in the Yergeni lowhills (47° N, 45° E) in north of Kalmuk region.

Sharp climatic changes occurred also during Subboreal period. After 4500 BP the climate became more arid and continental. In south-western Ukraine a share of Ulmaceae in forests seriously decreased. Thermophilous submediterranean species of oak and linden disappeared from the southern part of Volyn Lowhills in west Ukraine. Forest area decreased and steppe areas increased.

The area of ravine woodlands markedly decreased in Kalmuk region ca. 4700-4500 BP. After 4500 BP some decrease of valley forest is evident for the lower part of Volga River valley (Bolikhovskaya, 1990). Within 4200 and 3700 BP was a maximal aridity of climate. In the steppe zone at that epoch there was 50 mm less precipitation than nowadays (Kremenetski, 1991). Grass steppes replaced grass-meadow steppes in the low and middle part of the Don River basin; and the area of valley forests considerably reduced.

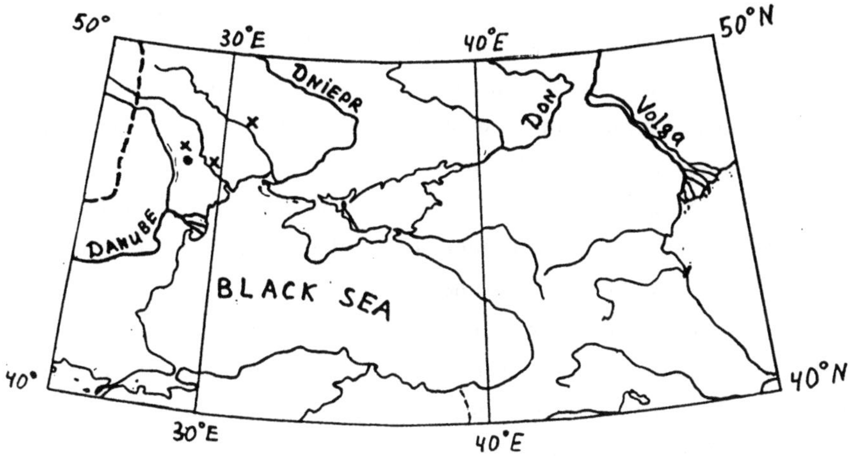

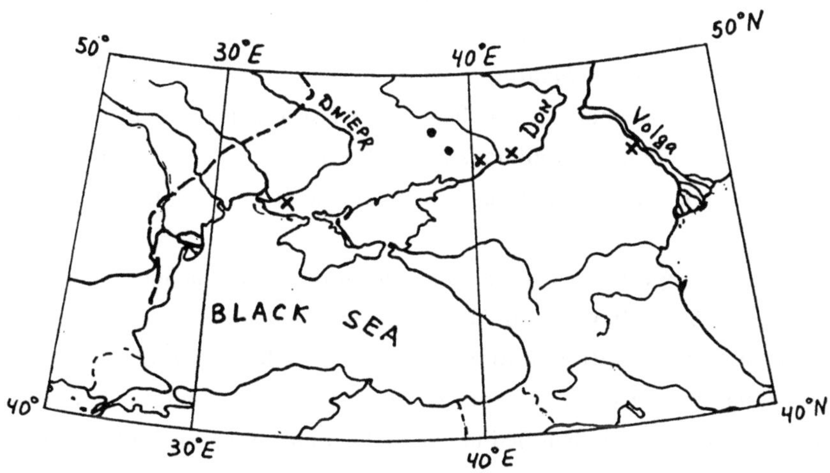

Fig. 2. Late Holocene finds of beech pollen outside limits of the modern range of beech *Fagus sylvatica L.* in West Ukraine and Moldova

Scale 1:17,500,000

---- - modern limit of the beech *Fagus sylvatica* L.

• - isolated site of beech *Fagus sylvatica* L.

x - finds of beech *Fagus sylvatica* L. pollen

Fig. 3. Late Holocene finds of hornbeam pollen outside limits of the modern range of hornbeam *Carpinus betulus* L. in Ukraine and Southern Russia

Scale 1:17,500,000

---- - modern limit of the hornbeam *Carpinus betulus* L.

• - isolated sites of hornbeam *Carpinus betulus* L.

x - finds of hornbeam *Carpinus betulus* L. pollen

A new phase of wetter climate occurred in 3400/3300 and 2800 BP. In lower part of Volga River basin the age of this phase is estimated at 3500-2700 BP (Bolikhovskaya, 1990). Summer temperature was 2° higher that modern, and there was 100 mm precipitation more than at present. In the west part of the forest-steppe belt in Ukraine the January temperature was 1° lower than now.

Climate of the late Subboreal wet phase was favorable for tree growth. Ca 3300-2800/2700 BP the woodland area in Ergeni hills in the Kalmuk region and in the southern part of Volyn-Podol hills in Ukraine. In contrast to the period of 6000-4500 BP in the Volyn-Podol hills less thermophilous *Ulmus glabra* predominated and the share of *Ulmus campestris* was minimal. *Ulmus laevis*, *Ulmus glabra*, *U. campestris* dominated in the floodplain forests in the lower part of the Volga River valley (Bolykhovskaya, 1990). About 3300 BP in western Ukraine, mainly in and near the Carpathian region, the share of spruce in forests increased (Artushenko et al., 1982).

Forest area in floodplain in the lower part of the Dniepr and Don River valleys increased. Hornbeam *Carpinus betulus L.* still grew in these forests. Later it disappeared. At the same time hornbeam pollen for the last time is registered if lower part of Volga River floodplain (Bolikhovskaya, 1990). After 4000 BP hornbeam expanded in forest-steppe area in Eastern Ukraine (Bezusko et al., 1988; Gerasimenko, 1993).

Modern structure of vegetation cover in southern European Russia and in Ukraine was established after 2500 BP. Ca 2500 BP pine disappeared from the lower part of Don Valley, and ca 1500/1200 BP pine disappeared from the lower part of the Dniepr River valley. After 2500 BP human impact upon vegetation increased. Pollen indicators of human activity are identified in pollen diagrams in the south of Moldova and in Western Ukraine (Kremenetski, 1991).

West Siberia and Kazakhstan

In the second half of the Atlantic stage birch-poplar forest-steppe expanded in the south of West Siberian Lowland up to the southern part of the Kazakh Lowhills. Shifts in vegetation cover show an increase of climate moisture.

Ca 4500-3600 BP climate became more continental and arid, and climatic conditions were close to modern ones. Forest area was reduced, while steppes became more xerotic. Forest-steppes of Kazakhstan Lowhills were separated from the forest steppe belt of West Siberian Lowland by a steppe belt. Ranges of broad-leaved trees were seriously reduced. Broad-leaved trees persisted in the higher parts of granite mountains in Kazakhstan Lowhills (Kremenetski et al., 1994).

A phase of wetter climate occurred ca 3300-2800 BP. At that time the ranges of *Quercus*, *Ulmus* and *Tilia* expanded, steppe cover became more mesophitic.

Ca 2500-2100 BP climate was more continental, close to the modern. Ca 2000-1800 and ca 100-800 BP occurred phases of more humid and less continental climate. After 800 BP climate and vegetation structure in the south of the West-Siberian Lowland and in the Kazakhstan Lowhills became like the modern. Ca 1800-1500 BP the distribution of pine forests in the southern limit of *Pinus sylvestris* in Kazakhstan reached its modern position (Kremenetski et al., 1994).

Shifts in ranges of some trees.

Ca 6000-4500 BP ranges of some broad-leaved trees (*Quercus*, *Ulmus laevis*, *Tilia*) and also of black alder (*Alnus glutinosa*) expanded in West Siberia and Kazakhstan (Volkova, Belova, 1980; Kremenetski et al., 1994). Broad-leaved trees did not form homogenous forests, but participated as first and second level of trees in birch-poplar and pine forests. Such a situation is typical now in southern and south-east margin of ranges of broad-leaved trees in the West Siberian Lowland.

Studies of modern flora of isolated pine forests in Kazakhstan Lowhills (Gorchakovski, 1987) agree with results of palynological investigations and confirm that broad-leaved trees penetrated in Kazakhstan Lowhills.

Linden-tree (*Tilia cordata* Mill., *T. sibirica* Fisch.)(figure 4).

Among other broad-leaved trees linden-tree penetrates farther to the East in West Siberia. Linden is less sensitive to the continental climate as compared with other broad-leaved trees. In south-east part of West Siberia there is a linden-tree which is considered as separate species (or subspecies) - *Tilia sibirica* Fisch. (Krylov, 1935). There are some isolated sites of linden to the south, south-east and east of its modern range (see figure 4). The existence of isolated sites shows us that the range of *Tilia* in Siberia in the past was greater than the modern one.

Tilia pollen is shown in pollen diagrams in many sections outside limits of the linden modern range in the southern part of the West Siberian Lowland (see figure 4). In all these sites there is a very small amount of *Tilia* pollen, including soil profiles, which are situated in the margins of *Tilia* range in West Siberia. Pollen data show that ca 5000 BP in southern part of West Siberia *Tilia* range was continuous in area

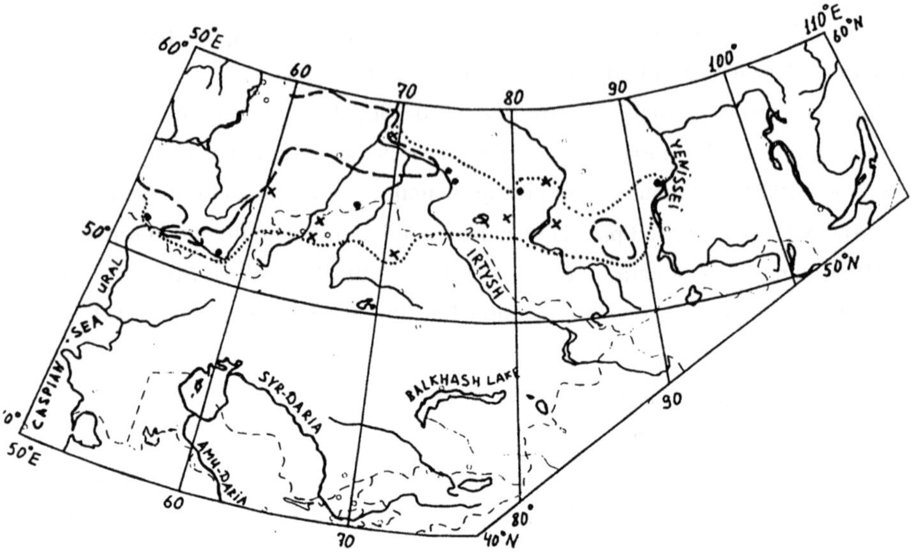

Fig. 4. Late Holocene distribution of linden-tree *Tilia cordata Mill.*, *Tilia sibirica Fisch.* in West Siberia

Scale 1:25,000,000

---- - modern limit of *Tilia* range

• - isolated sites of *Tilia*

x - finds of *Tilia* pollen

.... - limits of *Tilia* range 5000 BP

between the southern Ural mountains and northern foothills of the Kuznetski Alatau Mountains (see figure 4).

Within 4500-3500 BP *Tilia* pollen does not occur in diagrams in West Siberia. By that time we can date a disjunction of modern range of *Tilia cordata* and *Tilia sibirica*, and at that time sites of *Tilia* in upper part of Om River basin (55°40' N, 81°00' E) and near the Medvezhie Lake (55°10' N, 68°02' E) were isolated from the main range of *Tilia*. *Tilia* pollen occurs outside the modern range of *Tilia* about 3400/3300-2800 BP.

Southern expansion of *Tilia* in West Siberia is limited by high summer temperature and by low precipitation. Northern expansion of *Tilia* in West Siberia is limited by low summer and winter temperatures. In south-east of West Siberia the range of *Tilia* is limited by low winter temperature and low precipitation.

The *Tilia* presence in the north of Kazakhstan was possible when annual precipitation was higher than modern by 100-120 mm, and when mean July temperature was lower than modern by 1.5-2°C.

We can draw the same conclusions when we analyze modern climatic parameters of isolated sites of *Tilia cordata* in the Ural River valley (see table 1). Expansion of the major range of *Tilia* in this part of Ural River valley was possible if July temperature was 1°C lower and if precipitation increased by 100 mm .

The presence of *Tilia* in Barabinsk forest-steppe in the eastern West Siberia was possible if January temperature was 1.5-2°C higher than modern and when summer temperature was close to the modern. Precipitation was 100-120 mm higher than modern. With an increase of January temperature by 1.5°C and with near to the modern precipitation *Tilia* could grow in Ob River valley near Novosibirsk city. The penetration of *Tilia* in more northern regions was possible when January temperature was 2°C higher and when July temperature was 1°C higher then modern with a modern level of precipitation.

An absence of *Tilia* in near the Tobol City part of Irtysh River during the major part of Subboreal means that January temperature was 1.5°C lower; July temperature was 1°C lower. Precipitation was at near the modern level.

Palaeoclimatic reconstruction can explain the age of an isolated site of a shrub form of *Tilia sibirica* near the Krasnoyarsk city ("Mansky linden island") (55°58' N, 92°32' E) (Ilyin, 1934). Modern climate of this area is characterized by a meteorological station of Krasnoyarsk city (see table 2). In this table for comparison there are data for some other meteorological stations in this part of the Yenissey River basin. The present growth of *Tilia* is explained by local climatic conditions. If January temperature is 2-2.5°C higher and if precipitation is near the modern level *Tilia* could ca 6000-5000 BP reach the Yenissey River valley near Krasnoyarsk city.

This palaeoclimatic estimation correlates well with a palaeoclimatic reconstruction for the southern part of West Siberia for 6000-4500 BP. An increase of January temperature for ca 6000-5000 BP by 2°C is estimated by Koshkarova (1986) for the Yenisseysk city region (58°30' N).

Palaeoclimatic reconstruction makes it possible for us to presume a late Atlantic (6000-4500 BP) age of the *Tilia* site near Krasnoyarsk. The shift of this site from the main range of *Tilia sibirica* in south-west Siberia in Kuznetski Alatau foothills probably occurred after 4500 BP during a stage of more continental climate.

Black alder (*Alnus glutinosa* Gaertn)(figure 5)

Black alder (*Alnus glutinosa* Gaertn) has now a very limited range in the west margin of the West Siberian lowland (see figure 5). There are some isolated sites of

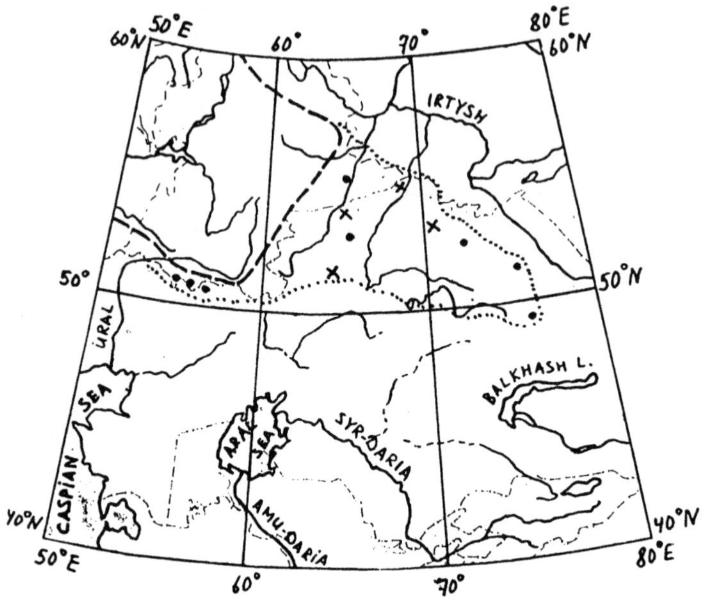

Fig. 5. Late Holocene distribution of black alder *Alnus glutinosa* Gaertn. in West Siberia

Scale 1:25,000,000

---- - modern limit of black alder *Alnus glutinosa* Gaertn range

• - isolated sites of black alder *Alnus glutinosa* Gaertn

x - finds of black alder *Alnus glutinosa* Gaertn pollen

.... - limits of black alder range 5000 BP

black alder in West Siberia and Kazakhstan (Gorchakovsky, 1987) (see figure 5). Pollen of *Alnus* is identified in some sections in the south of West Siberia and Kazakhstan (Krupenikov, 1941; Kremenetski et al., 1994; Nurmambetov et al., 1989) (see figure 5). Period 6000-4500 BP was the most favorable for the black alder growth. Figure shows the limit of black alder range ca 5000 BP.

During the arid phase after 4500 BP black alder persisted in isolated sites, where it is growing up to the present day. A shift of black alder range in West Siberia and Kazakhstan occurred within 4500-3500 BP during a phase of arid climate. Within 3300-2700 BP when climate became more humid the role of black alder in forests increased in wet places.

Low precipitation and high summer temperature limits black alder range in its south-east part. Northern limit of black alder in West Siberia is determined by low winter temperature.

When black alder grew in Naurzum forest in Kazakhstan (51°30' N, 64°30' E) precipitation in this region was 130-150 mm higher than now. An increase of precipitation could be accompanied by some decrease in July temperature. When black alder grew in the Tobol River valley near Petropavlovsk city (54°50' N, 69°09' E) January temperature was at least 2°C higher than the modern.

Climatic shifts. South of West Siberia and Kazakhstan.

Palaeoclimatic reconstruction using the Iversen (1944) method permitted to draw a map of deviations of main climatic parameters from their modern meaning for southern part of West Siberia and Kazakhstan for ca 5500 BP (see figure 6a, b).

Ca 5500-5000 BP climate was less continental than now. Maximal deviations of mean January temperature from the modern meaning +2°C are reconstructed for south part of West Siberia within 59° N and 55-54° N. In the southern margin of the West Siberian lowland and in a major part of the Kazakhstan Lowhills the mean January temperature was 1° higher than now. In the southern part of Kazakhstan Lowhills there was no increase of mean January temperature.

At ca 57° N mean July temperature was 1° higher than modern. More to the south in the belt from Tobolsk City to Novosibirsk City July temperature was at the modern level. Farther to the south July temperature was lower than at present. At ca 55° N July temperature was 1°C lower than the present. In the southern margin of the West Siberian Lowland and in the Kazakhstan Lowhills July temperature was 2°C lower than at present.

Ca 5500-5000 BP in the lower part of the forest belt in West Siberia precipitation was near the modern level (see figure 6c). Farther to the south, in forest-steppe and steppe belts in the southern part of West Siberian Lowland there was 100 mm higher precipitation than now, and in the Kazakhstan Lowhills and in Irtysh River valley near the Semipalatinsk City there was 150 mm higher precipitation than at present. Main features of deviations of climatic parameters from their modern ranges were preserved later through Late Holocene.

Ca 4500-3600/3500 BP the climate became more continental and dry in the south of West Siberia and Kazakhstan. All climatic parameters seems to be close to modern meanings. Only for the region of Irtysh River near Tobolsk City within 4000-2400 BP the January temperature was 1.5°C lower and the July temperature was 1.5°C lower than at present. Ca 3300-2800/2700 BP the climate was less continental. In the Ob River valley near Novosibirsk City January temperature was 2.5°C higher, which is

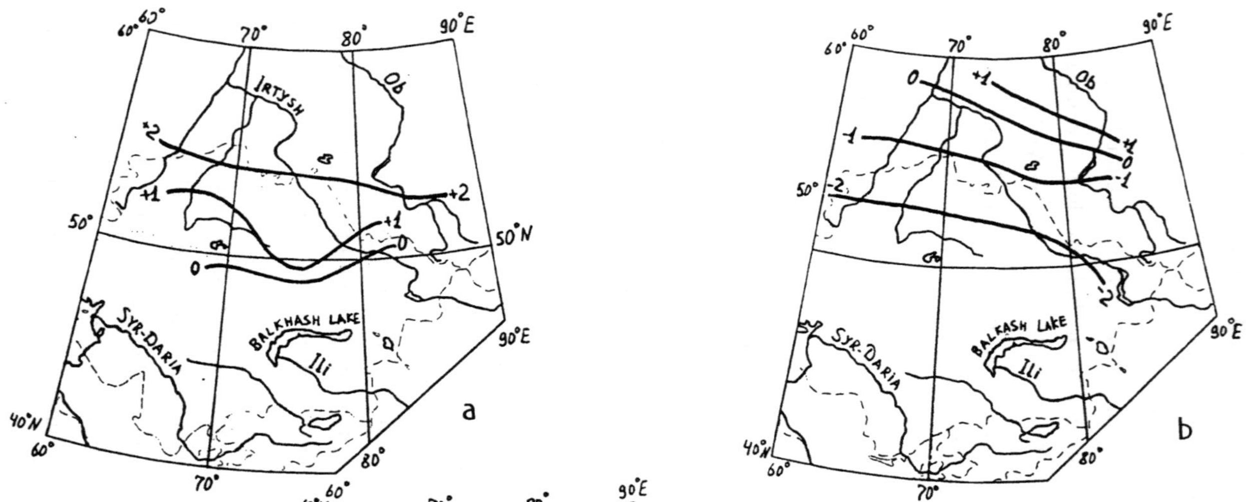

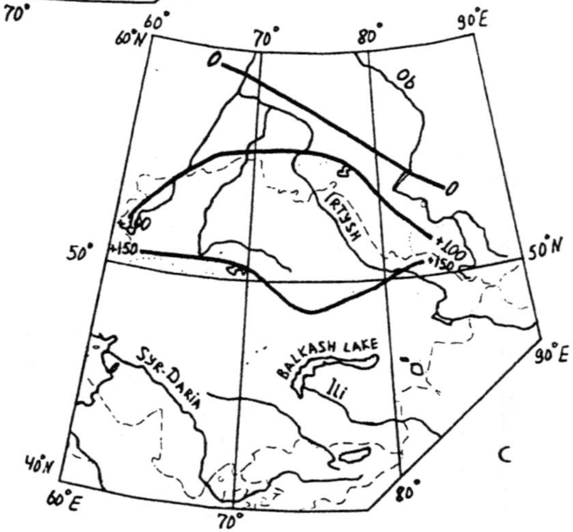

Fig. 6. Deviations of climatic parameters from modern values in south of West Siberia and in Kazakhstan Low mountains for 5000 BP.
Scale 1:25,000,000
A) - mean January temperature °C
B) - mean July temperature °C
C) - mean annual precipitation, mm

equal to the increase of temperature reconstructed for the period ca 5500-5000 BP. Deviations of climatic parameters in the Kazakhstan Lowhills and in the Tobol River basin were same as ca 5500-5000 BP.

South of Middle Siberia and Mongolia

In mountains and in intermountain valleys in Tuva region ca 5500-4500 an area of birch forests increased and steppe area decreased. That means an increasing climate moisture (Ponomareva, Seliverstov, 1991). More wet climate for the second half of the Atlantic is reconstructed by the lake sediments studies in north of Mongolia Republic (Vipper et al., 1989).

Ca 6000-4500 BP in regions to the west of Baikal Lake forest area was maximal. After 4500 BP in this area climate became more arid, with maximal dry conditions ca 4100-3800 BP (Vorobieva et al., 1992). Pollen data show that ca 4000 BP a considerable dry phase occurred in northern part of Mongolia Republic (Vipper et al., 1989).

Ca 3300-2800 BP in regions to the west of Baikal Lake a warmer climate is reconstructed. Ca 2500 BP climate deteriorated, and after 1800 BP modern climatic and vegetation pattern was established (Vorobieva et al., 1992).

Reaction of lakes on climatic shifts.

Results of lakes investigation in center of Asia show that for the last 6000 years lake level oscillations were in major part controlled by climatic factor. History of lakes correlates well with a sequence of vegetation and climate history.

<u>Aral Sea</u>.

In 5500 and 5000 BP the area of the Aral sea was much larger, and salinity was much less (Maev et al., 1983). Since ca 5000 BP the Aral Sea existed near to its modern dimensions. Upper layers of lake-shore transgressive sediments of Novoaral transgression are dated by mollusc shells 2860±80 BP (Tln-104) (Veinbergs, Stelle, 1980). This date reflects the end of that transgression.

<u>Lake Balkhash</u>. (See figure 7)

A transgressive phase of Lake Balkhash history continued up to the end of Atlantic stage (ca 5000/4500 BP). A Balkhash regression is dated 4380±150 BP (Khrustalev, Chernousov, 1992). According to B.G. Venus (1985) the age of buried peats related with the Balkhash regression is 3860±120 BP. During this regression the Balkhash Lake level decreased by 2-3 m as compared with the modern (pre-industrial) level.

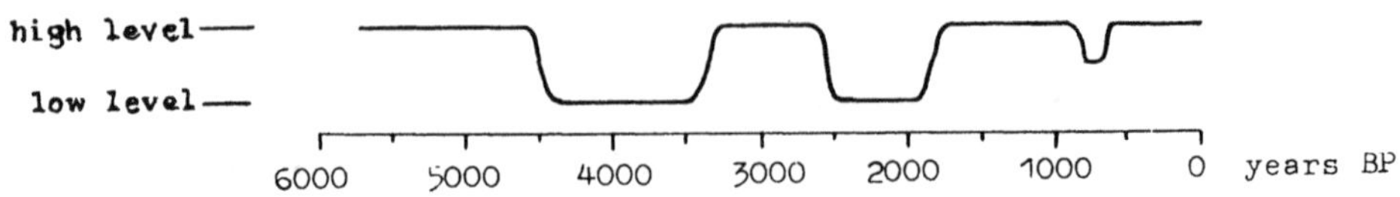

Fig. 7. Scheme of sequence of transgressive and regressive phases of Balkhash Lake since 6000 BP (after Krustalev, Chernousov, 1992; Venus, 1983, 1985).

After 3580±120 BP a novobalkhash transgression occurred, when the lake level was 2-3 m higher than now. After 2690±120 BP a novobalkhash regression began. The age of a modern transgressive layer is 1920±100 BP. A regressive episode occurred ca 800 BP (Venus, 1985).

Regressions of Chany Lake (54°40' N, 72°30' E) in the south of West Siberia and of Shalkar Lake (53°11' N, 68°28' E) in the north of Kazakhstan occurred ca 2500-2100 BP and ca 800 BP (Orlova, 1990; Shnitnikov, 1975). Regression of Sarykamysh Lake (42° N, 57°35' E) in Turkmenia is dated about the same time (Drenova, 1985).

Stages of an arid climate in the interior Eurasian temperate belt are simultaneous with low sedimentation rate in fresh-water lakes in the Kazakhstan Lowhills-Pashenno (49°22' N, 75°24' E), Karasye (53°02' N, 70°13' E) and in Mokhovoye Lake and bog in northern Kazakhstan (53°46' N, 64°15' E) (Kremenetski et al., 1994).

Palaeosol studies

Climate shifts around 5000 and 3000 BP influenced soil processes. Results of studies of buried soils in Eurasian steppes correlate well with reconstructions based on pollen data.

Studies of soils, buried under barrows (human burial mounds) in south of Ukraine and in Kazakhstan ca 5000 BP demonstrated that these soils were formed during more favorable and less continental climate. Climate amelioration was very well pronounced in the region north to the Caspian sea in the Ryn-sands area (desert and semi-desert, ca 48° N, 49° E). Ca 5000 BP the Ryn-sands environment was close to the modern one the south of steppe belt (Ivanov, 1992). That means that ca 5000 BP precipitation in Ryn-sands was ca 100-150 mm higher than at present.

Within 4500-3600/3500 BP soils buried under barrows in the south of Ukraine, of European Russia and in Kazakhstan show patterns of a drier climate. At that time chernozem soils in the north of Kazakhstan were replaced by dark-chestnut colored soils with features of salinity. According to soil studies a maximum phase of aridity in the south of Russia occurred 4200-3700 BP. Increasing soil salinity is reconstructed for a time between 4500 and 3500 BP for the region of Elton saline lake (49° N, 47° E) in a semi-desert belt north to the Caspian sea (Ivanov, 1992). In the semi-desert and desert belt north to the Caspian Sea 4000-3700 BP eolian processes were active. Later on, within 3700 and 3300 BP eolian activity was much less (Lavrushin et al., 1991).

Table 1. Climatic parameters from meteorological stations, situated near the isolated sites of linden-tree *Tilia cordata Mill.* in Ural River valley.

Meteorological stations	Coordinates		Mean temperature °C		Annual precipitation, mm
	Lat. N	Long. E	January	July	
Khalilovo	51°28'	57°55'	-16,5	17,3	371
Orsk	51°15'	58°36'	-16,8	19,5	303
Andreevka	52°19'	51°55'	-15,3	18,9	375
Teplyi	51°32'	51°32'	-14,6	19,8	282

Table 2. Climatic parameters from meteorological stations, situated in the region of isolated site of linden-tree *Tilia sibirica Fisch.* near Krasnoyarsk City in Yenissey River valley.

Meteorological stations	Coordinates		Mean temperature °C		Annual precipitation, mm
	Lat. N	Long. E	January	July	
Krasnoyarsk	56°02'	92°45'	-16,8	18,2	380
Legostaevo	55°08'	91°02'	-20,6	17,7	359
Balakhta	55°23'	91°40'	-20,5	17,1	337

C.V. Kremenetski. The late holocene environmental and climate shift...

Evolution of human communities

Within 5000-3000 BP major changes in human economy and social structure occurred. Ca 5000 BP agricultural and herding archaeological cultures were developed in the south of the East-European plain. In the forest belt hunter and gatherer cultures dominated. In west Ukraine and Moldova a Tripolian Eneolithic culture existed. The technique of copper work became known in this region (Masson et al., 1982). Climate and vegetation situation at that time had a positive influence upon the economic activity of Eneolithic tribes. An increasing human pressure upon the vegetation cover is reflected in pollen diagrams of Neolithic and Eneolithic archaeological sites in Ukraine and Moldova (Kremenetski, 1991).

By 5000/4500 BP archaeological cultures of herders were developed in steppe and forest-steppe belts of West Siberia (Zdanovich, Zaibert, 1989). Bones of wild animals, including forest animals (*Alces* L., *Capreolus pygagrus* Pall., *Ursus arctos* L.) were identified from archaeological sites. These finds prove the persistence of hunting in the economy of late neolithic cultures. Hunting was also very important during Bronze age in West Siberia. Archaeozoological data correlates well with results of pollen investigations (Kremenetski et al., 1994).

Bronze age began ca 4500/4300 BP. At that time some agricultural cultures disappeared in forest-steppe and steppe zones of Eastern Europe.

Herders cultures were widespread, and even penetrated in forest belt. Nomadic cultures were formed in the Eurasia steppe belt after 4500 BP. Specialized cultures of herders were formed in Siberia later than in Eastern Europe. In north Kazakhstan this type of economy was finally established only ca 3500 BP, although the first find of domestic horse bones in this region is dated 4340±120 (IGAN-432) (Zdanovich, Zaibert, 1989). Hunting preserved a major role in forest-steppe, but not in the steppe zone (Nurumov, Makarova, 1988).

Agriculture remained one of the branches of the economy during the Bronze age (Pashkevich, 1991). In the south of West Siberia, by archaeological data, the first evidence of agriculture is dated ca 3500 BP (Zdanovich, Zaibert, 1989).

Specialized settlements of bronze-melters appeared in bronze age in Eurasian steppes. The major region of concentration of such settlements in Ukraine is region of the lower Dniepr sands (Berezanskaya, Sharafutdinova, 1985). Age of these settlements is 3500-2900 years BP. The location of these settlements was very good from palaeogeographical viewpoint, because here was the biggest forest in steppe belt of Ukraine. Decrease of forest area in low part of Dniepr valley reconstructed from pollen data recently was partly in relation with forest clearing by bronze manufacturers, because charcoal was used for smelting metal (Kremenetski, 1994).

Kazakhstan Lowhills in the Bronze age were a major source of copper for bronze production for the whole Eurasian steppe region. Numerous remains of Bronze Age mines and melting workshops were discovered in this area since late 19th century (Kuznetsova, 1988). Besides the presence of rich copper ores a vicinity of extensive pine forests was crucial for the location of bronzes melting workshops. The volume of ancient production of bronze was high according to archaeological reconstructions (Kuznetsova, 1988), but despite this there are no pollen indicators of human induced vegetation changes in pollen diagrams from this area (Kremenetski et al., 1994). Pollen data indicate that although the volume of wood use for industrial needs during bronze age (and later on during the Iron age) in Kazakhstan was relatively high, it did not exceed the natural volume of wood growth, and, in contrast to the situation in Ukraine, the human impact on vegetation did not lead to a general decrease of forest area in West Siberia and Kazakhstan.

Conclusion

Different types of palaeoecological evidence: pollen analysis and palaeobotanic reconstructions, macrofossil dating, palaeosol studies, lake-level investigations, archaeological studies - allprove that within 6000-5000/4500 BP climate conditions were very favorable. Ratio of temperature and precipitation was optimal. Circa 4500 BP serious climatic shift took place. The climate became more colder and wetter in boreal belt and drier and hotter in arid belt.

Climate change was accompanied by a southward retreat of northern treeline. Present day vegetation communities were established in the boreal belt. Lake levels decreased in the arid belt. Features of soil aridization are reconstructed. Such a change was quick but gradual. It provoked some changes in human economy. In Ukraine and Moldova it terminated some agricultural communities and they were replaced by nomads. Nomad economy dominated the steppe belt thanks to domestication of the horse. Nomadic tribes penetrated even into the southern margin of the forest belt. Transition to the Bronze age was accompanied by an increase of human pressure on the vegetation cover.

References

Artushenko A. T. (1970). Vegetation of steppe and foreststeppe of Ukraine in Quaternary. "Naukova dumka", Kiev, 174 p. (in Russian).

Artushenko A. T., Arap R.Ya., Bezusko L.G. (1982) Vegetation history of Vestern Ukraine in Quaternary. "Naukova dumka", Kiev, 136 p. (in Russian).

Berezanskaya S. S., Sharafutdinova I.N. 1985. Sabatinovska culture. In: Archaeology of Ukrainian SSR, v. 1, "Naukova dumka", Kiev, pp. 489-499 (in Russian).

Bezusko L.G., Klimanov V. A., Y.R. (1988) Climate change in Ukraine in Lateglacial and Holocene. In: Holocene palaeoclimates of the European territory of the USSR. Moscow, pp. 125-135 (in Russian).

Bolikhovskaya N. S. (1990) Palaeoiondicators of environmental changes in the low part of the Volga River basin for the last 10,000 years. // Caspian sea. Problems of geology and geomorphology. Moscow. Nauka. P. 52-68. (in Russian).

Drenova A.N (1985). Wet phases in the Sarykamysh region according to radiocarbon data. // Izvestia AN SSSR. Ser. geography 4: 89-93. (in Russian).

Gerasimenko N. P. (1993) Evolution of environment in Donetsk region in Holocene. Ukrainian Geographical Journal 4: 31-35.

Gorchakovski P. L. (1987) Forest oasis of the Kazakhstan Lowhills. Moscow. Nauka, 159 p. (in Russian).

Ilyin M. M. (1934). On Tilia near Krasnoyarsk city. // Botanicheski journal V. 19. 4: 385-392. (in Russian).

Ivanov I. V. (1992) Evolution of soils in steppe zone at Holocene Nauka Moscow, 143 p. (in Russian).

Iversen J. (1944) *Viscum*, *Hedera*, *Ilex* as climate indicators. A contribution to the study of the post-glacial temperature climate. // Geol. Foren. i Stockholm Forhandl. V. 66. P. 463-483.

Khotinsky N. A., Bezusko L.G., Cherkinsky A.E. (1994) Changes in vegetation in central and western regions of the Russian plain. // Palaeogeographical basis of the modern landscapes. Moscow. Nauka. P. 111-118. (in Russian).

Khrustalev Yu. P., Chernousov S.Ya. (1992) On the holocene history of the Balkhash lake. // *Izvestia of Russian Geographical Society*. V. 124. 2: 164-171. (in Russian).

Kokarovtsev V. K. (1992) Resourses and geology of the holocene lake carbonates in the Perm region. Ekaterinburg 215 p. (in Russian).

Koshkarova V. L. (1986) Seeds florae of Siberian peatlands. Novosibirsk. Nauka. 121 p. (in Russian).

Kremenetski C.V. (1991) Palaeoecology of earliest agricultural tribes of the Russian plain. Moscow, 193 p. (in Russian).

Kremenetski C.V. (1994) Holocene vegetation and climate history of the southwestern Ukraine. // *Review of Palaeobotany and Palynology*, (in press).

Kremenetski C.V., Tarasov P.E. (1992) Holocene climate and lakes history of Kazakhstan. // Symposium on global continental palaeohidrology. Ed. L. Starkel Krakow 3p.

Kremenetski C. V., Tarasov P. E., Cherkinski A. E. (1994) A history of the Kazakhstan "island" pine forests in Holocene. // *Botanicheski journal* V. 79. 3: 13-29. (in Russian).

Krupenikov I. A. (1941) On the history of the island forests of the Kustanai region. // Reports of the USSR Academy of Sciences. V. 30. 7: 664-665. (in Russian).

Krylov P. N. (1935) Flora of West Siberia. Tomsk, V. 8. P. 1819-2087. (in Russian).

Kuznetsova E. F. (1988) Ancient mining and metal production in the Central Kazakhstan in Bronze and early Iron Ages. // Problems of palaeoeconomy of Kazakhstan according to archaeological data. Alma-Ata. Nauka KazSSR. P. 50-62. (in Russian).

Lavrushin Yu. A., Spiridonova, E. A., Sulerzhitsky L. D. (1991) Geological and palaeoecological events in the north of the arid belt during last 10000 years. // Quaternary geology and palaeoecology. Moscow.: P. 87-104. (in Russian).

Maev E.G., Maeva S.A., Nikolaev S.D., Parunin O.V. (1983). New data on the Aral sea holocene history. // Cenozoic palaeogeography of the Caspian and Aral seas. Moscow. Moscow University publishing house. Part 2. P. 97-104. (in Russian).

Masson V.M., Merpert N.Ya., Munchaev R.M., Chernysh E.K. (1982) Eneolithic of the USSR. Moscow. Nauka. 359 p. (in Russian).

Nikiforova L. D. (1982) Dynamics of landscape belts in Holocene in the North-East of the USSR European part. // Environment evolution of the USSR territory in Late Pleistocene and Holocene. Moscow. Nauka P. 154-162. (in Russian)

Nurmambetov E. I., Chupina L. N., Akiyanova F. Zh. (1989) On the Late-Pleistocene and Holocene history of the Northern Kazakhstan. // Palaeoclimates of the Lateglacial and Holocene. Moscow. Nauka. P. 102-109. (in Russian)

Nurumov T. N., Makarova L. A. (1988) Domestic and wild animals in Neolithic and Bronze Ages in the Central and Northern Kazakhstan. // Problems of palaeoeconomy of Kazakhstan according to archaeological data. Alma-Ata. Nauka KazSSR. P. 7-36. (in Russian).

Orlova L. A. (1990) Holocene of the Baraba region. Stratigraphy and radiocarbon chronology. Novosibirsk. Nauka. 127 p. (in Russian).

Pashkevich G. A. (1991) Palaeoethnobotanic results in Ukraine (Neolithic - Bronze Age). Kiev 48 p. (in Russian).

Ponomareva D. P. and Seliverstov Yu. P. (1991) Some features of the Ulughem valley (Tuva) holocene palaeogeography. // Vestnik of the Leningrad University. Ser. 7, geology, geography. 3: 86-90. (in Russian).

Shnitnikov A. V. (1975) On the history of lakes in the Northern Kazakhstan. // Lakes of Kazakhstan and Kirgizia and their history. Leningrad. Nauka. P. 5-27. (in Russian).

Spiridonova E.A. (1991) Evolution of the vegetation cover in the Don River basin in Upper Pleistocene and Holocene. Moscow. Nauka. 221 p. (in Russian).

Veinbergs I. G., Stelle V. Ya. (1975) Aral Sea late pleistocene and holocene history. // History of the arid belt lakes and interior seas. Leningrad. Geographical society publishers. P. 53-64. (in Russian).

Velichko A. A., Andreev A. A., Klimanov V. A.(1994) Vegetation and climate dinamics during Late Glacial and Holocene in tundra and forest zones of Northern Eurasia. // Short-term and sharp environmental changes for last 15000 years. Mpscow. P. 4-60 (in Russian).

Venus B. G. (1985) Features of the lake basin evolution in the humid and arid belts // Palaeolimnology in the humid and arid belts. Moscow. Nauka. P. 5-29. (in Russian).

Vipper P. B., Dorofeiuk N. I., Meteltseva E.P., Sokolovskaya V.T. (1989) Holocene landscape and climate changes in Central Mongolia. // Palaeoclimates of the Lateglacial and Holocene. Moscow. Nauka. P. 160-167. (in Russian).

Volkova V. S., Belova V.A. (1980) The role of broadleaved trees in the Holocene vegetation of Siberia. // Palaeopalynology of Siberia. Moscow. Nauka P. 112-117. (in Russian).

Vorobieva G. A., Goriunova O. I., Saveliev N. A. (1992) Holocene chronology and palaeogeography of the Middle Siberia south. // Quaternary geochronology. Moscow. Nauka, P. 174-181 (in Russian).

Yelovicheva Ya. K., Bogdel I. I., Zernitskaya V. N., Klimanov V. A. (1988) Holocene climatic reconstructions of Bielorussia based on pollen data. // Holocene palaeoclimates of the European territory of the USSR. Moscow, pp. 86-94 (in Russian).

Zdanovich G. B., Zaibert V. F. (1989) Main features of the agriculture emergence in the Ural-Kazakhstan steppes. // Emergence and evolution of agriculture in Ural region. Sverdlovsk. P. 70-83. (in Russian).

Environmental and Climatic Changes Between 3 and 5ka BP in Southeastern Ukraine.

Natalia P. GERASIMENKO

National Ukrainian Academy of Sciences
44 Volodymyrska St.
Kyiv 252003
Ukraine

Abstract

Pedological and palynological data from Eneolithic and Bronze Age settlements in the steppe zone of eastern Ukraine show that the transition from the mild climate of the Atlantic period (6,500-5,500 BP) to the more continental climate of the Subboreal was marked by considerable impoverishment of the floristic composition and by reduction or disappearance of several mesophilic and thermophilic taxa. Within this general background, there were also rhythmic oscillations between wet climate stages marked by expansion of forest-steppe, with dry climate stages marked by expansion of steppes. Five such shifts have occurred between 5,500 and 2,500 BP. The strongest aridification occurred between about 4,100 and 3,500 BP. It is marked by the decrease of humus accumulation, biogenic activity and chemical weathering in the soils, which became rather loess-like and contain deep desiccation fissures. Pollen data show sharp reduction of forest areas, significant xerophytization of steppe vegetation, and drying of flood plain swamps. The forest-steppe landscapes of the Early Subboreal were replaced by Artemisia-Gramineae steppes representing the shift through three phytogeographic subzones. None of the Holocene climatic fluctuations was as sharp and contrasting when compared with the preceding and following stages. In studied regions, no major paleoenvironmental changes were observed within that interval that could be clearly ascribed to human impact. The Mid-Subboreal aridification is therefore considered to be of natural origin.

Introduction

The present paper is a report on palynological and pedological studies of the Middle-to-Late Holocene deposits at the archeological sites in the steppe zone of eastern Ukraine.

NATO ASI Series, Vol. I 49
Third Millennium BC Climate Change
and Old World Collapse
Edited by H. Nüzhet Dalfes, George Kukla,
and Harvey Weiss

The Holocene deposits have been palynologically studied in Ukraine since the early fifties (Zerov, 1950; Artyushenko, 1970; Artyushenko et al., 1982; Artyushenko and Arap, 1988; Bezus'ko et al., 1988) but the studied sites are located mainly in the forest belt, partly in the forest-steppe zones. The spread of more mesophilic genera and moistening of climate after 3300 yr BP has been detected. Further data of Bezus'ko et al (1988) reveal several fluctuations of we-cold and dry-warm climate between 5000 and 2500 yr BP.

More than 40% of the Ukrainian territory is occupied by the steppe zone. Studies of the vegetation history is complicated there by limited presence of paludal and lacustrine sediments. The most complete vegetation and climate reconstruction for the southern part of the Russian plain has been made by Kremenetski (1991). Using the palynological data from peat bogs of Dovzhok and Kardashinskoye in the Lower Dnieper region as well as from the soil profiles of the Eneolithic-Bronze age sites of Moldova and the Lower Don region, he has identified the Late Atlantic climatic optimum, the Early Subboreal drying period and the Late Subboreal moistening and cooling. The palynologically investigated sites of the Eastern Ukraine are the bog of the Andreyevsky Liman (Artyushenko1970) and soil profiles in the "Streltsovskaya steppe" preserve (Isayeva-Petrova 1976). The results give evidence of drier and cooler climate of the Late Holocene as compared with the Mid-Holocene.

The Holocene soil history of Ukraine has been revealed further from the studies of soils beneath burial sites of various ages in the Middle-Lower Dnieper region (Zolotun, 1975, Ivanov et al., 1986; Ivanov, 1992). The latest data indicate that the most arid climate in the Subboreal was between 2700 and 1700 BC.. The Late Atlantic and especially the Subatlantic periods were much wetter..

The region under investigation (Fig. 1) including the Donets highlands, the Donets plain and the Azov lowland is situated at present in the northern subzone of steppes characterized by spread of Herbetum mixtum - Gramineae communities. The soils are chernozems of ordinary subtype. Pine forests grow on sand terraces of the Donets river valley while oak groves with lime, elm and ash trees occupy some slopes and valleys of the Donets highland. To the south, they are replaced by shrub associations, and by grass vegetation which becomes more xerophytic as well. The northern steppe is a transitional subzone between forest-steppes and typical Gramineae steppes. The highest and more humid northern part of the Donets highland with its watershed forests and gray forest soils is regarded by some researchers as the forest-steppe

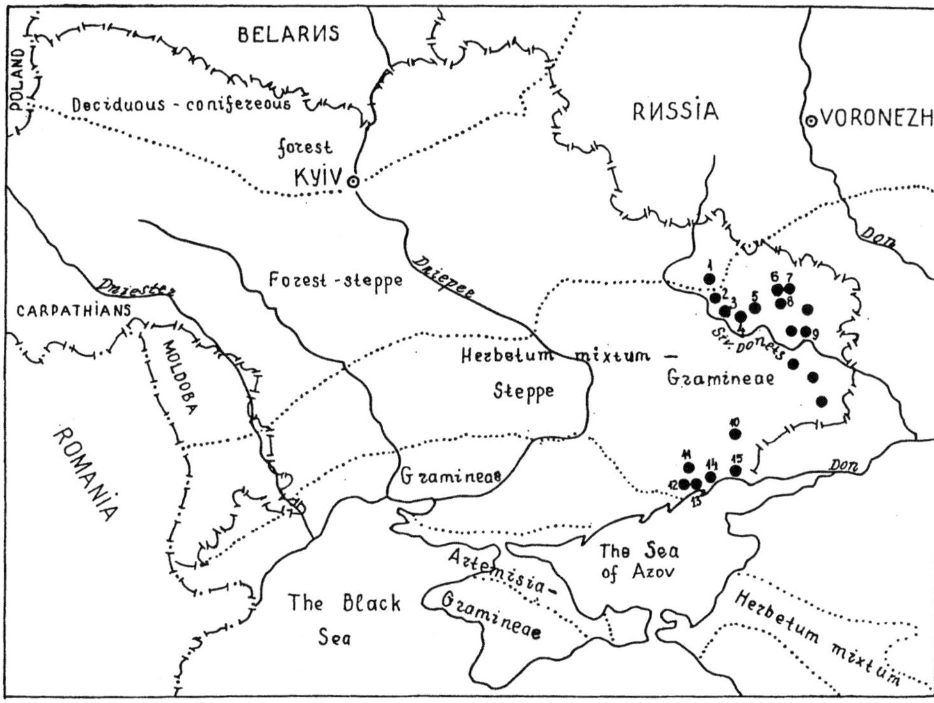

Fig. 1 The location of sites studied and the boundaries of the modern landscape regions in Ukraine. Localities mentioned in the paper: 1-Novosyolovka 1, 1, 2-Moskvino, 3-Illichevka 1, 4-Bakai (Illichevka 2), 5-Glubokoye 1,2 6-Podgorovka, 7-Novosyolovka-Ajdar, 8 Starobelsk, 9-Rogalik, 10-Amvrosievka, 11-Kamyshevataya balka, 12-Buylovataya balka, 13-Fokina balka, 14-Lyapinskaya balka, 15-Bezymennoye.

province. Transitional zones are vulnerable and sensitive to climatic oscillations, so studies of environmental proxy should be informative as paleoclimatic indicators.

Approaches and methods

The main subject of our investigations are soils and soil deposits at archaeological sites. Both pedological and palynological methods were applied in order to obtain paleoenvironental information. The principal pedologic work was done by Veklich and Sirenko (Veklich et al, 1979, Veklich, 1987). They described different rates of soil formation and sedimentation on the plateaus and in depressions. On a plateau, owing to the low rate of dust accretion, soil processes rework the underlying layers and form polygenetic soil profiles. In accumulation zones the ratio between the rate of deposition and soil formation is more favorable and registers different phases of soil formation in the accumulating sequence of soil sediments. More stable components of soil horizons such as morphological features, granulometry, chemical and mineralogical composition or relative content of humus are most appropriate for paleogeographic reconstructions.

Carbonates as more mobile element can form secondary post sedimentary accumulations even in soils that are synchronous with sedimentation. In the soil deposits of topographic depressions, the paleoecological information derived from pedology and from spore-pollen data is in general well correlated. Some discrepancies are observed in the layers with strongly pronounced signs of eluvial-illuvial processes of forest soil formation. Though soils in depression grow upwards, in some cases intensive eluvial-illuvial migration of organic and mineral compounds and clay particles dominates and causes post genetic reworking of older horizons with preserved original spore-pollen composition (exemplified by several figures of this paper). Some pollen can be relocated along the soil fissures and pores as well, but the possibility of unrecognized significant introduction of pollen from the upper soil horizon was reduced by various dedicated methodical studies. These include experiments devoted to the problem of formation of palynological spectra in soils of different zones (Alexandrovsky et al., 1981; Berezina et al, 1969; Isayeva-Petrova, 1979).

Another reason for discrepancies between the sporo-pollen and pedological data can be a lower threshold of sensibility of soils as a climatic indicator in comparison with pollen spectra. The latter provide more generalized information about environmental changes within a whole region, while soils represent a local situation, that does not always reflect regional changes.

Within the accumulation zones of the relief at investigated archaeological sites, the cultural layers are embedded mostly "in situ" in the corresponding genetic horizon. The presence of cultural layer of different age is usually connected with the appearance of a new soil genetic horizon (exemplified by figures of this paper). On the elevated parts of the plateau, the same cultural layer can be embedded in different genetic soil horizon (Alexandrovsky, 1988; Gerasimenko, 1994). Thus the adequate information on paleoecological settings of ancient humans can be derived from the study of soil horizons with corresponding cultural layers

Most of the investigated sections presented in this paper are situated in the depressions or on gentle slopes such as ravines and gully bottoms, in terraces and flood plains .The location of the studied sites is shown in figure 1. The known age of cultural layers combined with few available direct radiocarbon dates has been used for dating of the stages of the Holocene environmental development (see figures)). Archeological dating of the Bronze age sites has been summarized by Gershkovich (1988), Gorbov (1993) and Tsymidanov; dating of Eneolithic sites by Gurin (1992) and Kotova (1993); of the Neolithic, Mesolithic and Late Paleolithic sites Gorelik (1986) and Degermenszhy.

Pedological reconstructions are based on the results of soil morphological studies in the field and on applied analytical methods (granulometric, bulk chemical analyses, definitions of humus content, and of carbonate content). In order to investigate soil variations in catenas, samples were analyzed along cross sections reaching from local watershed to talwegs near the location of archaeological sites (cf., e.g. fig 7).

The pollen samples were processed following methodology of the Saint Petersburg university (1978). This involves treatment with 10% HCl, 10% KOH, cold treatment of HF, disintegration in a solution of pyrophosphate ($Na_4 P_2 O_7$) and separation in heavy liquid (KdI_2, KI) of specific gravity 2.2. The pollen grains were mostly well preserved. However the pollen content was not always high enough and in some cases a sufficient number of pollen grains could be calculated only from 100-200 g of sediment. In the presented spore-pollen diagrams, the microfossil basic includes pollen of arboreous plants, herbs and spores. The percentage of different spore taxa has been calculated in relation to the basic sum while the percentage of different arboreous (or herbaceous) taxa is related to the sum of trees or herbs assumed equal to 100%.

Special tables which depict the ratios of the spectra components in modern natural zones and subzones (Grichuk and Zaklinskaya, 1948: Bolikhovskaya, 1976: Dinesman, 1977) have been applied in the interpretation of pollen diagrams. The correction factors between vegetation and the resulting palynospectra for steppe and forest-steppe zones (Arap, 1976: Bezus'ko, 1992) have been used as well.

The results of pollen analysis of surface samples from the virgin soils are available for the region under investigation (Zubets, 1971). Palynospectra of the steppe type include 80-88% NAP and about 12-20% AP. *Pinus* (50-75%), *Betula, Alnus, Corylus, Acer* (2-4% each), *Quercus* (2-20% are represented in the AP group. In the NAP group, predominance of Herbetum mixtum, especially *Asteraceae (Compositae)* and *Chenopodiaceae* has been observed. High values of ruderal plants are caused probably by increased human impact on the environment.

The Blytt-Sernander scheme of the Holocene in the modification of Hotinsky (1977, 1982) has been used for stratigraphic classification as well as the scheme of Holocene stages published by Veklich (1987). The latter one based on generalization of literary data depicts the global alternation of warm and cool stages (some 0.3-0.5 Kyr long each). The warm stages are designated by odd indices (for example hl_b2^3, the cool stages by even indices (hl_b2^4).

The ^{14}C dating has been carried out by Kovalukh (Institute of Geochmistry and Physics of Minerals of the Ukrainian Academy of Sciences), thermoluminescence dating by Chlukov and Lyashenko (Moscow University).

Results

This study resulted in distinguishing of several soil layers and corresponding pollen zones that characterize the climatic rhythms of the Middle and Late Holocene. In the soil sections of the investigated region, the Late Atlantic period is represented by buried humus horizons. Their upper layers are archaeologically dated to 3800-4100 BC at the Eneolithic sites of Podgorovka, Novoselovka-Ajdar (Gurin, 1992; Kotova, 1993) or give an absolute age (the site of Rogalik-2, Fig. 2) of about 5000 BP. Humus horizons (up to 0.5 m thick) are marked by dark gray color, granular structure and loose porous consistency. They are mostly leached.

The physical-chemical records (from 3 sections) show higher values of humus, of alkaline metal oxides, especially CaO, lower values of silt and clay particles and sesquioxides (R_2O_3) than the Early Subboreal and Early Atlantic soils (cf. Fig. 2). Both morphology and physical-chemical indices of the Late Atlantic humus horizons are most similar to the Late Subatlantic ones, providing evidence for well comparable natural conditions of their formation. Lower values of humus in buried horizon is due to the destruction of organic matter with age.

At the Late Atlantic stage, the soil formation was determined by humus and CaO accumulation accompanied by leaching of carbonates. This reflects a more arid climate as compared to the Early Atlantic time when textural horizons formed in forest soils (dated by enclosed Neolithic remains, cf. figure 10). Development of humus accumulation processes has been shown up to the southern part of the forest zone (Velichko and Morozova, 1986; Alexandrovsky, 1983).

The pollen records from 8 sites show lower values of AP in the Late Atlantic humus horizon than in the Early Atlantic layers. Pollen spectra of the upper and lower parts of

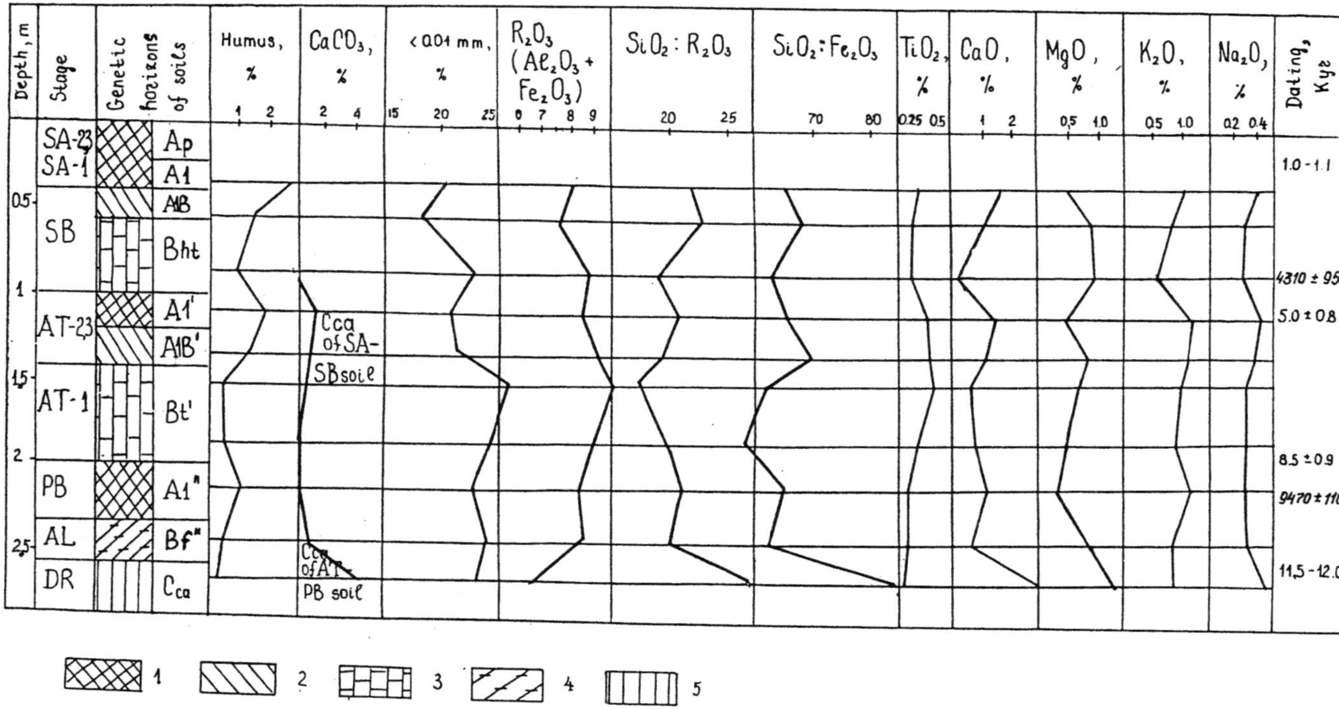

Fig. 2. The physical-chemical characteristics of the Holocene deposits at the Rogalik-12 site, located on the mild slope of the river Yovsug valley. 1- humus horizons Ap, A1 A1', A1", dark-grey (', " - signs of buried horizons), 2- humus-transitional horizon AB, AB', grey, 3- textural-illuvial horizons Bt', Bht, brown, 4- ferric-illuvial horizon Bf", light brown, 5- loess, Cca (carbonate horizon), pale-yellow. The TL-dates are extrapoloated from the corresponding layers of the nearest Rogalik-2 site and given in thousands of years. Radiometric C14 samples in years in right margin)
1900 BC) by H.

the horizon differ by values of pollen from broad-leaved species . Percentage of the mesophytic grass pollen is higher in the lower layers (see Fig. 3). The composition of AP is richer and more diverse as well. *Carpinus, Quercus, Ulmus, Fraxinus, Acer, Corylus, Euonymus, Cornus,* and *Oleaceae* are found. In the upper layers, only *Quercus* and *Tilia* are represented in significant quantities; other trees and shrubs mentioned above have been replaced by taxa of more open environment. The pollen of xerophytes increased as well, indicating a drier climate from the beginning to the end of Late Atlantic. The lower parts of the humus horizon are the last level where *Carpinus betulus* was observed in the investigated region. Today, the eastern boundary of the natural range of *Carpinus betulus* is significantly further west. The pollen spectra of the upper layers are very similar to those from the surface of modern soils. Climate inferences based on pollen spectra correspond well with the pedogenic information showing close resemblance of Late Atlantic soils with the present ones.

In the Holocene sections of the northern part of this region, the Late Atlantic humus horizons are overlain by brown-colored textural horizons of the Early Subboreal soils. Their lower layers are 14C dated to 4310 +/95 BP. These horizons which are up to 0,6 m thick are leached characterized by a rather hard consistency and strong nutty prismatic structure. The structure may in part result from the sandy composition of slope loams. The results of pedoanalytical studies show the enrichment by sesquioxides and silt and clay particles. CaO, K_2O, Na_2O and humus are impoverished compared with the underlying and overlying layers. (Fig. 2). Both morphological and physical- chemical indices give evidence of weaker humus accumulation and stronger argillization and decalcification in the Early Subboreal. These processes are not typical in steppe soils, but are more common under forests. Probably in the Early Subboreal forest soil-forming processes replaced humus accumulation in low-lying sites. The early Subboreal turf-podzolic soils on top of Late Atlantic humus horizons formed in the forest and forest-steppe zones (Alexandrovsky, 1983; Veklich and Gerasimenko, 1993).

Pollen characteristics of the Early Subboreal has been obtained form the dated layers of the Rogalik-12 site (fig. 3) and from the eluvial-illuvial (A2B) horizon of the forest soil (fig. 4). The humous elluvial horizon of the latter is archaeologically dated to Late Subboreal and the illuvial horizon to the Atlantic (the Neolithic culture according to Degermendgy) In both sites, pollen spectra of the Late Atlantic-Subboreal deposits change in similar succession (especially the pollen of broad-leaved

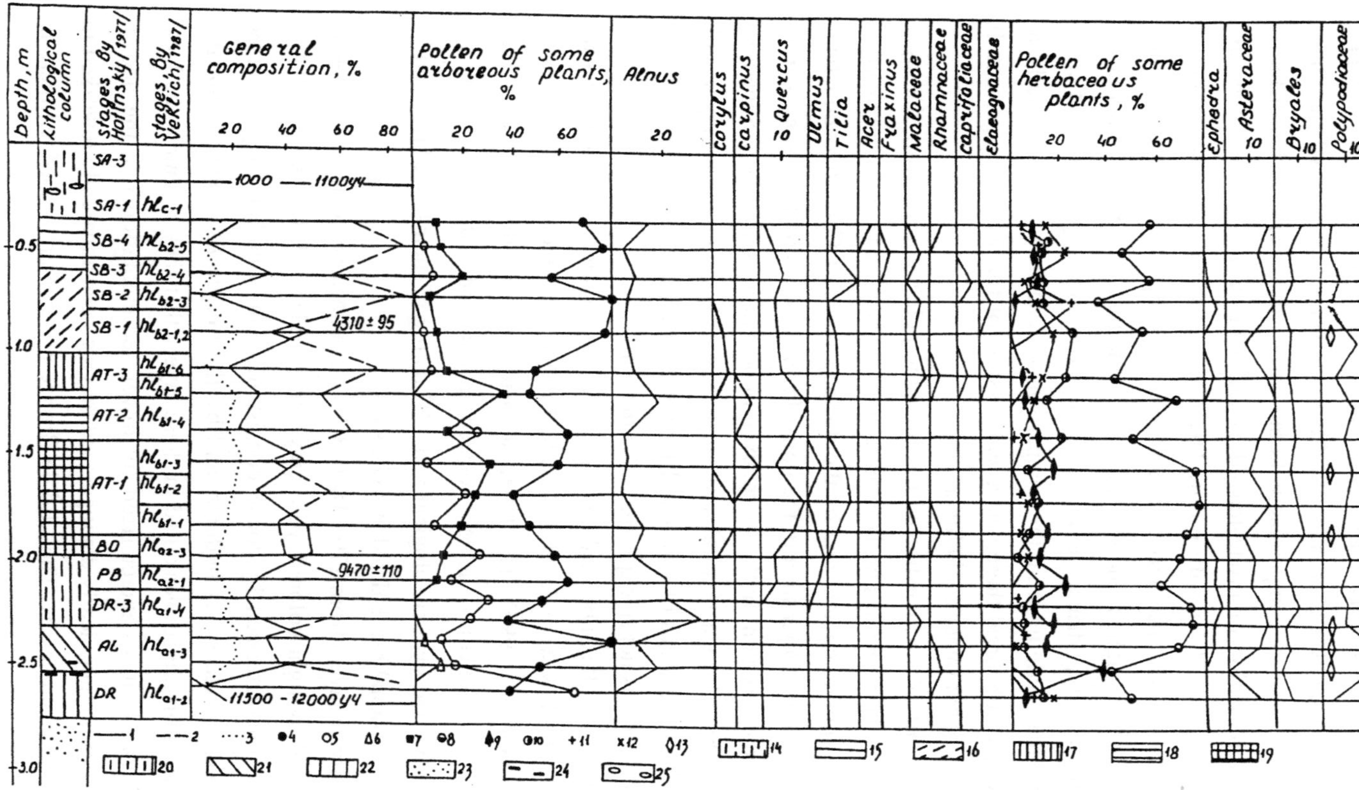

Fig. 3. Spore-pollen diagram of the Rogalik 12 section. At present it is located under Herbetum mixtum-Gramineae communities. Percentage of microfossils: 1-AP, 2-NAP, 3-spores, 4-*Pinus*, 5- *Betula*, 6-*Picea*, 7-sum of broad-leaved trees, 8-Herbetum mixtum, including Astercaeae, 10- Gramineae, 11-*Artemisia*, 12-Chenopodiaceae, 13- Lycopodiaceae. Soil horizons: 14-humus, 15-humus-transitional, 16-textural, 17-buried humus, 18-buried humus-transitional, 19-buried textural, 20-buried humus, 21-ferric-illuvial, 22-loess, 23-sands, 24-the Late Paleolithic cultural layer, 25-the Saltov cultural layer. 1900 BC) by H.

plants), providing a possibility correlation. The spectra represent a forest-steppe type and are characterized by higher values of AP and lower values of NAP than those of the Late Atlantic . They are comparable with the Early Atlantic spectra but differ by poorer floristic composition of broad-leaved plants. *Quercus, Tilia,* and *Acer* are present, but *Carpinus* and *Ulmus* drop out. The *Ulmus* disappearance according to Hotinsky (1977) is the most characteristic feature of the threshold between the Atlantic and Subboreal period throughout the Russian plain. The high values of spores in the spectra, especially of Polypodiacae demonstrate the mesophilic type of ground cover in the Early Subboreal. Judging from the comparison with surface of modern soils, at the time of the Early Subboreal there was a wider extent of forests than at present. It was the stage of a wet but rather cool climate.

In the southern part of the investigated territory, the pollen spectra from the layer overlying the Late Atlantic humus horizon are also characterized by a sharp reduction of the broad-leaved pollen content and by the disappearance of *Carpinus* and *Ulmus* (fig. 5). The AP value is higher than in the upper layers at the site, but lower than in the sites of the northern region. In the south, even in depressions, the Early Subboreal is represented not by a brown textural soil horizon, but by a light- gray layer with leached humus. No clear signs of forest soil formation are present. To the south, the moistening of climate was not so expressed as to the north..

The most ancient Middle Subboreal deposits have been studied at the sites near the Sea of Azov (fig. 6). They were dated by Middle Bronze Age archeological context to 1700-1800 BC (Gorbov, 1993). The latter deposits are embedded in transitional horizons BCca of the Late Holocene soils located on slopes of gullies. The young age of these soils is revealed by archeological and ^{14}C dating of their basal layer and by the characteristic erosional break dated at some sites.

The Middle Bronze layers from 1700-1800 BC with radiocarbon date of 3,720 + 90 BP are represented by light colored, pale-gray, structureless silt loams up to 0.4 m thick. These are characterized by loose consistency, an abundance of dispersed carbonate, secondary calcareous nodules, and by the tongue like lower boundary. Signs of soil formation are so weakened in these layers that they resemble the loess deposits. At elevated sites characterized by lower sedimentation rate the above-mentioned horizons are not clearly distinguished, having been apparently reworked in the later soil-forming processes. The Middle Bronze remains are embedded there in the lower part of the B(C) horizons, which are more structural and more brightly colored (cf. Fig . 6).

According to the analytical data (exemplified by fig. 8), the BC_{ca} horizons are distinguished by the lowest value of humus and very low indices of chemical weathering (content of R_2O_3, TiO_2 and clay particles). On the other hand, it is marked by the high carbonate content, by the highest values of alkalis and earth alkaline metal oxides. As evidenced from this characteristics, the horizon was formed under extremely unfavorable conditions for humus accumulation and soil formation on the whole. It shows attenuation of clay weathering, of any leaching, of structure formation but instead accumulation of loess-like silt. Such conditions point to strong aridification, an interpretation which is also supported by the tongue like base of the horizon. The tonguing is usually caused by deep desiccation fissures developed in expressed continental climate.

In the Russian plain the pedogenic studies of soils buried under mounds dated to 2.500-1,700 BC proved much more arid climate than the modern one (Ivanov et al., 1986, Ivanov, 1992, Dyomkin et al., 1985, Pesochina et al. 1982). The weak humus accumulation and desiccation fissures were found to be accompanied by soil salinization and strong carbonate enrichment.

Pollen records obtained from the Middle Bronze Age layers (fig. 9) show the lowest values of AP and spores and the highest values of NAP found on the sites of the investigated region. The percentage of xerophytes (Chenopodiaceae, *Artemisia, Ephedra*, and so on) is at its maximum as well. The spectra are similar to those from the underlying loess, but differ in occurrence of small amount of broad-leaved plants (sporadic *Quercus* and *Tilia.*). Other tree pollen are also only few. They include Elaeagnaceae, Rhamnaceae and Malaceae. Evidently most of the dominant *Pinus* pollen is blown from a distance. The lowest share of Herbetum mixtum, absence of Cyperaceae and Polypodiaceae provide evidence of sharp xerophytization as well. Only in the bottom of the ravines, the presence of xerophytic herb communities was less extensive (fig. 5).

In the pollen diagrams of the more northern sites (figs. 3 and 4), the spectra with minimum AP and maximum NAP are fixed between the Early Subboreal pollen zone and the one dated to 1300-1400 BC. Both are characterized by a high share of AP. Similar stratigraphic position and pollen composition provides the basis of the interregional correlation of this horizon.. The data demonstrate that the deforestation

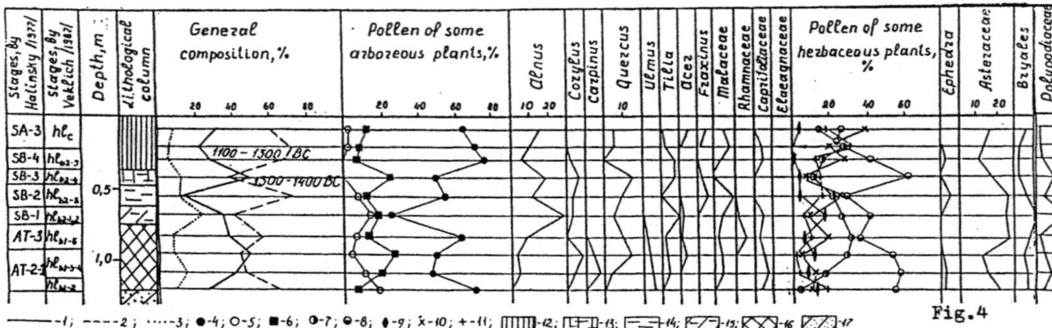

Fig.4

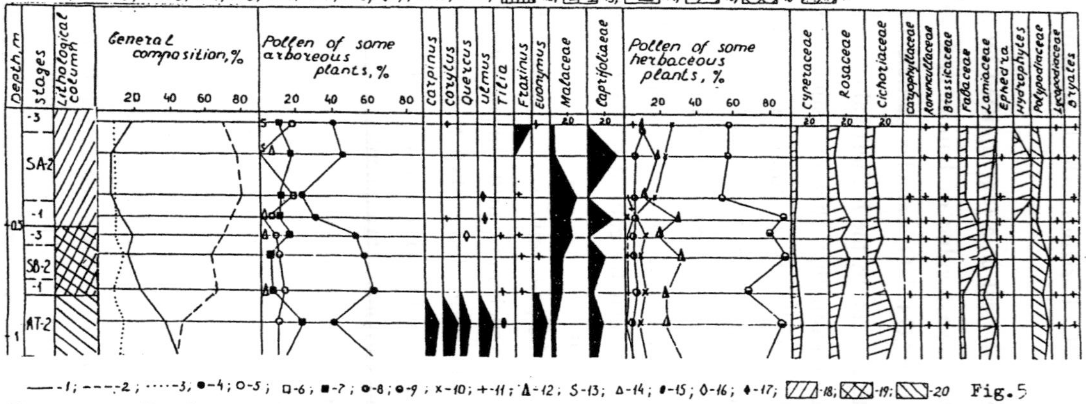

Fig.5

Fig. 4. Spore-pollen diagram of the Glubokoye-2 section (profile -8 in Fig. 10). The section is located on slope of sandy terrace of the Siversky Donets river (gentle, covered by pine forest to the flood plain which is flat and covered by alder forest) Archeological dating by I. Gershkovich. Percentage of microfossils: 1-AP, 2-NAP, 3- spores, 4-*Pinus*, 5-*Betula*, 6-sum of broad-leaved plants, 7-Gramineae, 8-Herbetum mixtum (including Astereaceae), 9-Cyperaceae, 10-Chenopodiaceae, 11-*Artemisia*. Soil horizons: 12-humus A1, 13-humus-eluvial A1A2, 14-eluvial A2, 15-eluvial-illuvial A2B, 16-textural-illuvial Bt, illuvial-transitional B_tC.

Fig. 5. Spore-pollen diagram of the Amvrosievka section . The section is located at the bottom of a ravine, at present covered by mesophytic herbage (shrubs on slopes)> Percentage of microfossils: 1-AP, 2-NAP, 3-spores, 4-*Pinus*, 5-*Betula*, 6-*Alnus*, 7-sum of broad-leaved plants, 8-Gramineae, 9-Herbetum mixtum (including Asteraceae), 10-Chenopodiaceae, 11-*Artemisia*, 12-Asteraceae. Single pollen grains: 13-*Salix*, 14-*Acer*, 15-Oleaceae, 16-Cornaceae, 17-Rhamnaceae. Soil horizons: 18-humus A1, 19-humus-transitional A1B, 20-buried humus A1.

and strong xerophytization of the herb cover reached into the northern part of the territory as well. At the low-lying sites (fig. 4), the amount of AP is not as low, partly because of the occurrence of hygrophytic pollen (*Alnus glutinosa* Gaertn., *Betula pubescens* Ehrh). On the whole, the values of *Artemisia* pollen sharply increased while that of Herbetum mixtum and broad-leaved plants (only *Quercus* is represented) decreased, indicating the advent of arid spell. The stages of significant xerophytization of plant cover have been revealed in the steppe and forest-steppe regions of the Russian plain between about 4200 and 3400 BP (Kremenetski, 1991; Spiridonova, 1991; Klimanov and Serebryannaya, 1986; Bolikhovskaya, 1990).

The layers documenting this dry event are overlain by soils archaeologically dated to 1300-1400 BC in the northern part of the investigated region (Gershkovich et al., 1988) and to 1300-1500 BC) in its southern part (Gorbov, 1993). They contain cultural layers of the early stages of the Srubnaya (Timber-grave) culture of the Late Bronze Age. The pedogenic and pollen records from these layers give evidence of climate amelioration all over the region. In the north (figs. 4 and 10), they are represented by humous elluvia of forest soils indicative of wetter climate,

Ashy gray A2A1 horizon and darker A1A2 horizon, both structureless, loose and sandy loams, are coupled with brown and rusty-brown prismatic illuvia. Studies of the stratigraphic position of the cultural layer disposition in the soil catena at the Glubokoye-2 site (fig. 10) demonstrated that the eluvial-illuvial processes are more intense in the low lying sites. On the surface of the terrace a thin crypto-podzolic soil is developed while down along the slope, the A1A2 and B horizons make their appearance. Farther down even the A2A1 and A2B horizons show up. In this way the pseudogleying processes contributed to the differentiation of soil profiles. The stagnation of superficial waters could be caused by the underlying clay and dense illuvial horizons having been formed during the Atlantic and Early Subboreal. Development of gleying and lessivage processes in the soils of depressions in the Subboreal (after 4000 BP) has already been established in the center of the Russian plain (Velichko et al., 1986).

In the southern part of the region under investigation (Figs. 6 and 7) the layers formed at 3100-3300 BP are represented by dark-brown and gray-brown clay loams (up to 0.6 m thick) distinguished by rather hard consistency and prismatic structure, but without organic-mineral cutans on ped surfaces. They are regarded as Bm (in some cases Bca) horizons of the Late Holocene soils. As for the physical-chemical indices

(Fig. 8), they are characterized by medium values of humus and CaO but are marked out by high content of clay fraction, of aluminum and iron sequioxides and rather low content of alkali metals oxides. These signs of indicate much more favorable conditions for humus accumulation and especially for chemical weathering than during the BCca horizon formation. Furthermore, the clay weathering evidently reached more significant intensity than at the later Holocene stages.

The pollen spectra from the B horizon dated to 1300-1500 BC (Fig. 9,A) contain not so abundant but diverse pollen of arboreous plants. *Alnus, Quercus, Tilia, Fraxinus, Corylus* and a variety of shrubs are represented while the share of *Pinus* pollen is reduced. As compared with spectra from the underlying layers, the values of NAP decrease as well as of xerophytes (Chenopodiaceae, *Artemisia),* and the percentage of Herbetum mixtum sharply increases.

Much higher values of AP, of broad-leaved pollen are found in the corresponding layers of the northern sites (Fig. 3, 4) though the florisitic composition of AP is the same there. These spectra are attributed to a forest-steppe type. It is the evidence for a reduction of forest growth conditions in the north, while predominance of Herbetum mixtum proves the widespread mesophitic grass communities. The top surface tests of modern soils show that at the time of the Early Srubnaya culture, precipitation was stronger than at present and more so than in the preceding Middle Subboreal stage. The increase in moisture and forest expansion after 3400 BP has also been revealed in the steppe and forest-steppe regions of the Russian plain (Kremenetski, 1991; Spiridonova, 1991; Bolikhovskaya, 1990).

Rather continuous records of environmental fluctuations during the Late Bronze age are available both for northern and the southern parts of the investigated region. In many sites (Fig. 4, 6-10), the Early Srubnaya culture layers (1500-1300 BC) are overlain by the layers containing the Later Srubnaya culture, 1300-1000 BC (3100-2800 BP), by Gershkovich (1988), V. Gorbov (1993). These layers differ noticeably from the underlying ones. In the northern sites, they are represented by humus horizons (buried ones in depressions), dark grey, loose, crumb structured, leached (in depressions with dispersal-carbonate horizon in the underlying layers). The humus accumulation, carbonate leaching and their removal to depressions contributed to soil formation while eluvial-illuvial processes ceased. The foundation pits of a 1500-1300 BC dwelling were filled up with the humus-eluvial horizon material while those of a 1100-1300 BC dwelling by the humus horizon (Fig. 10, D-2, A-2').

In the southern sites, the corresponding layers are dark-grey but more lighter colored than the Subatlantic humus horizons and are regarded as A1B horizons of the Late Holocene soils. They differ from B horizon by strong crump-granular structure, by loose earthworm consistency, by lower share of clay particles; all these provide evidence for intensive humus accumulation and weakening of chemical weathering in this time span. Nevertheless, the conditions were not so favorable for humus-forming processes as during the Subatlantic. The above mentioned signs of changes of soil-forming processes indicate the replacement of a relatively wet climatic stage by more arid one.

Further increase of climatic aridity has been detected from the pedogenic characteristics of the soil layers, dated to 800-1000 BC (Gorbov, 1993). They are observed in the sites of the southern part of the region investigated (Fig. 6) and are distinguished by lighter grey color, weaker structure, more silty and less clayey composition than the including layers. They are less porous and earthworm structured as well. The tongued lower boundary of the horizon documents the ground cracking under the severe continental climate. The analytical data (Fig. 8) show low values of humus, low indices of chemical weathering (<0.001mm fraction, R_2O_3 content, SiO_2: R_2O_3) as compared both with the Sublatlantic and Early Srubnaya culture layers. Nevertheless, the horizon is not so low in humius and in TiO_2 and sesquioxides as the Early Bronze layers. Deterioration of soil-forming conditions was strong but not so significant as that of the Middle Subboreal. The xerothermic depression is established near the Black Sea region at the 1200-700 BC time (Zolotun, 1975).

Pollen spectra of the described layers are notable for low values of AP of broad-leaved plants, and high of NAP, of *Pinus* (evidently blown in from some distance) (Fig. 9). As for these indices they are similar to spectra of the Early Bronze Age layers, but differ from them by the lower share of xerophytic pollen and higher of Herbetum mixtum. The spectra indicate xerophytization of plant cover as compared with the Srubnaya culture time span and with the present, but it was not so severe as that of the Early Bronze culture time .

In the northern part of the territory, pollen spectra are only available from layers dated to 1100-1300 BC and those correlated with them (Fig. 3, 4). They demonstrate the significant decrease of AP, especially the broad-leaved plant pollen, and increase of NAP but with medium values of Herbetum mixtum. This characteristic proves the distinct change to more continental climate after 1300 BP.

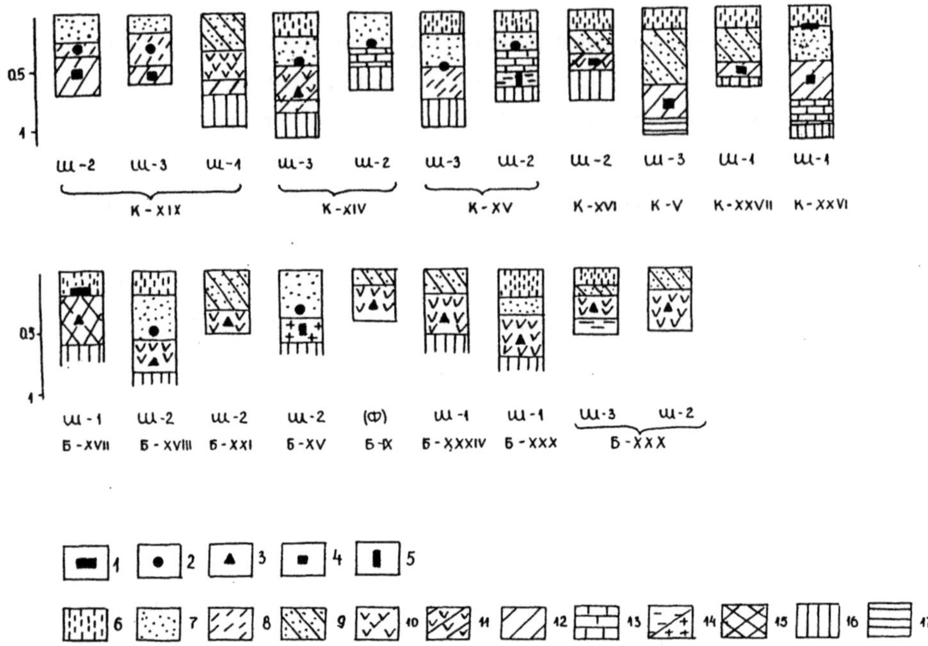

Fig. 6. Genetic profiles of soils at the Bronze Age archaeological sites near the Sea of Azov . The sites are located on slopes and terraces of the Kamyshevataya (K) ravine, the Buylovataya (b) and Fokina (Φ) ravines. The archaeological dating is by Gorbov. The cultural layers of different age: 1-the Saltov culture, 1000-1100 BP; the Bronze Age: 2-800-1000 BC (2600-2800 BP), 3- 1300-1500 BC (3100-3300 BP), 4-1700-1800 BC (3450-3550 BP), 5- the Mesolith. Soil horizons; 8-humus A1 (SA-3), grey; 7-humus A1'(SA-2,1), dark-grey; 8-humus-transitional A1B (SB-4), light-grey; 9-inseparable A1' -A1B (SA-2-SB-4), grey; 10-textural Bm (SB-3), brown, prismatic; 11-transitional B (SB-3,2), grey-brown; 12-transitional BC_{ca} (SB-2), pale-grey; 13-textural Bm' (AT-1), bright brown, prismatic; 14-transitional BC' (BO-PB), light brown; 15-the filling of dwelling pit (SB-3), light grey; 16-loess; 17-the Pleistocene fossil soil.

Paleoecology

Rhythmic changes of vegetation and of soil-forming processes have been observed between 5 and 3 kyr BP. The pollen assemblages and pedogenic signs of the Mid-to-Late Holocene deposits show a large measure of agreement in their paleoenvironmental inferences. It enables one to reconstruct landscape and climate conditions and to retrace the stages of their development (Fig. 11).

In the Late Atlantic, 5000-5400 BP, the South-Eastern Ukraine was located in the Herbetum mixtum-Gramineae subzone of steppes that corresponds to its present physical-geographical situation. Oak, lime, ash tree grew in the gully forests only, as

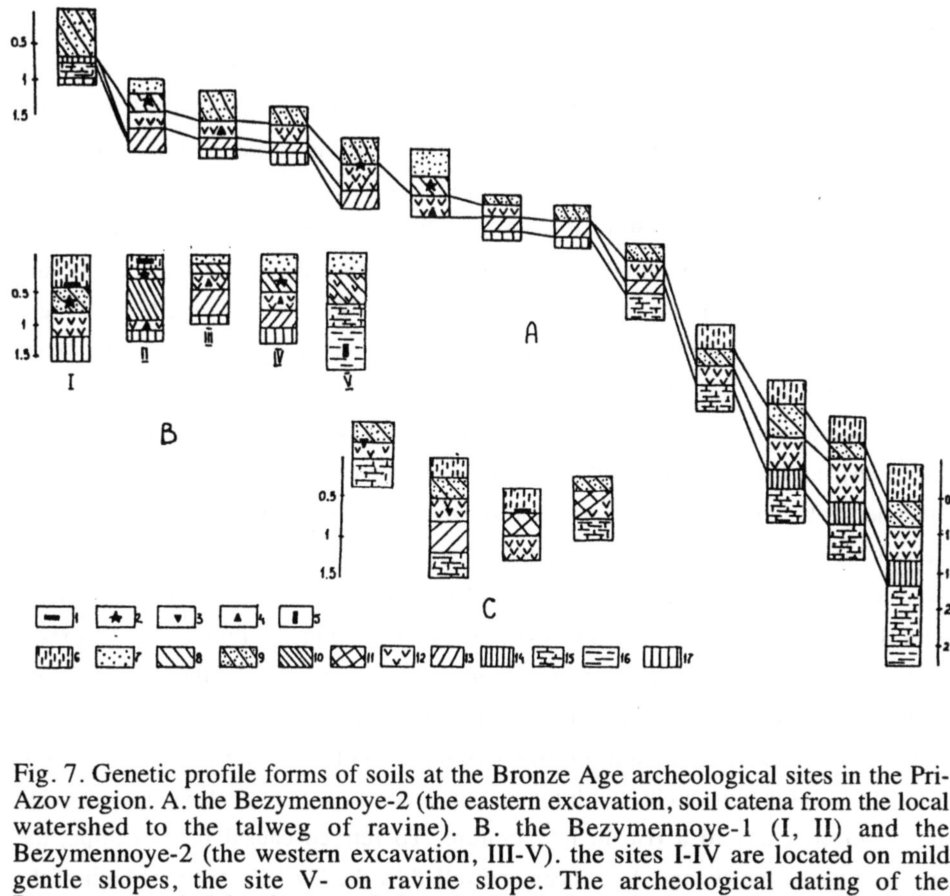

Fig. 7. Genetic profile forms of soils at the Bronze Age archeological sites in the Pri-Azov region. A. the Bezymennoye-2 (the eastern excavation, soil catena from the local watershed to the talweg of ravine). B. the Bezymennoye-1 (I, II) and the Bezymennoye-2 (the western excavation, III-V). the sites I-IV are located on mild gentle slopes, the site V- on ravine slope. The archeological dating of the Bezymennoye 1, 2 has been carried out by V. Gorbov, the Lyapinskaya nalka- by V. Tsimidanov.
The cultural layers of different ages: 1-the Saltov culture, 1000-1100 BP; the Bronze Age: 2-1000-1200 BC; 3-1200-1400 BC, 4- 1300-1500, 5-the Mesolith. Soil horizons: 6-humus A1 (SA-3), grey; 7-humus A1' (SA-2,1), dark grey; 8-humus-transitional A1B (SB-4) more light, grey; 9-inseparable A1'-A1B (SA-2-SB-4), grey; 10-the filling of dwelling pit 1400-1500 BC, high in humic material; 11-the filling of dwelling pit 1200-1300 BC low in humic material; 12-textural Bm (SB-3), brown prismatic; 13-transitional BCca (SB-2) pale-grey; 14-humus A1" (AT-2), dark-grey; 15-textural Bmt" (AT-1) bright brown, prismatic; 16-transitional BC" (BO-PB), light brown; 17-loess, pale-yellow.

well as shrub communities. The soil formation was dominated by chernozem humus-accumulative processes, CaO accumulation, by relatively deep leaching of carbonates - the processes most characteristic also of the latest (the Late Subatlantic) soil formation.

Much milder climatic conditions have been revealed based on the data from the lowermost layers of the Late Atlantic humus horizon. Forest-steppe landscapes (meadow steppes, mezophyllic *Carpinus Betulus* L. in the forest composition) were widespread then. This time span is evidently coeval with the end of the Holocene climatic optimum dated to 5500 BP (4200 BC) (Bezus'ko et al, 1988, Klimanov et al., 1986). After 5500 BP the xerophytization of steppes, the reduction of forest areas, and the disappearance of more ecologically vulnerable *Carpinus betulus* L. - all these provide evidence for the advent of a stage of a more continental climate. In the 4800-5200 BP time span, in the more eastern regions of the Russian plain, the Gramineae steppes spread into the modern forest-steppe zone (Spiridonova, 1991; Klimanov et al., 1986), and humus accumulative processes developed in the southern forest subzone (Velichko et al., 1986, Alexandrovsky, 1983), while in the western regions, steppes were characterized by more mesophytic composition than the modern one (Kremenetski, 1991), and the conditions of chernozem soils formation were rather mild (Ivanov, 1991). It can be supposed that climatic continentality increased strongly to the east. Pedogenic indicators from steppe sites, as well as pollen data that are characterized by a low share of AP and thus also by broad-leaved pollen too, reflect mainly the changes of climatic moistening and do not elucidate their reasons. For example, aridification can be conditioned by a decrease of precipitation or an increase of evaporation due to a rise in temperature. To retrace causal relationships, the observed landscape changes have been controlled by the level fluctuations both of the Black Sea (as an indicator of global temperature changes (Ostrovsky et al., 1977, Balabanov et al., 1988), and of the Caspian Sea as an indicator of continental-wide moistening (Varushchenko et al, 1987). The Black Sea transition from the trangressive stage (around 5200 BP-3800 BC) to the regression one (around 4800 BP -3400 BC) is coeval with the Caspian regression (after 5400 BP-4000 BC). Thus, being arid, the climate changed from warm (before 5100 BP) to more cool. In the global climatic schemes of the Holocene (Zubakov, 1986; Veklich, 1987), the time span 5300-4800/4600 BP is regarded as a cool climate stage. In the Eastern Ukraine, the fall in

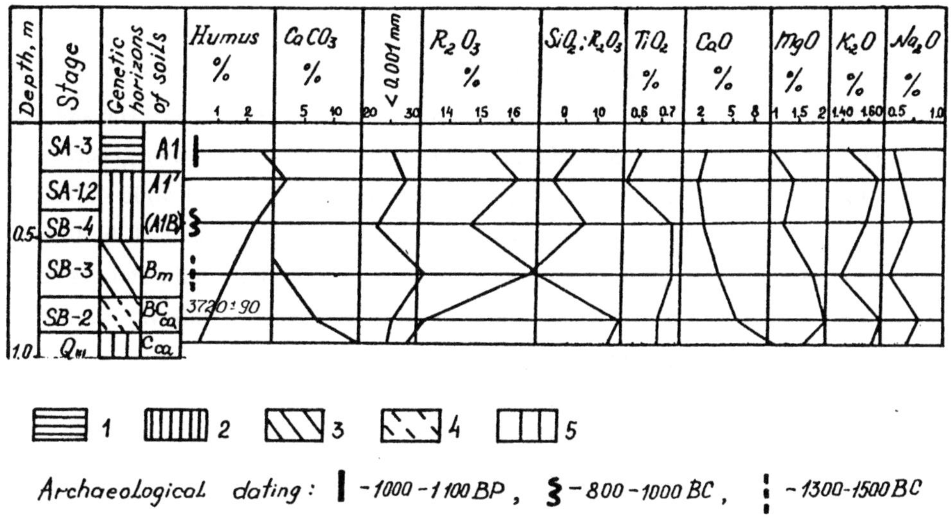

Fig. 8. They physical-chemical characteristics of the Holocene deposits at the Kanyshevataya XIV site (III-3). The site is located on the flattening gentle ravine slope. Archaeological dating by V. Gorbov. 1-humus horizon A1 (SA-3), grey; 2-buried humus horizon A1' (Sa-2-SB-4), dark grey; 3-textural horizon Bm (SB-3), brown, prismatic; 4-transitional horizon BCca (SB-2), pale-grey; 5-loess, pale yellow.

temperature is likely to begin after 5000 BP while the sharp increase of continentality of climate - after 5400 BP.

The Early Subboreal landscapes (4500-4100 BP; 2800-3400 BC) of the investigated regions were dominated by forest-steppes. On slopes and terraces, forest soil-forming processes developed providing textural differentiation of soil profiles, leaching and decalcification. The forests of the different composition grew there while the meadow (Gramineae-Herbetum mixtum) steppes occupied the plateau. As compared with the Atlantic, the broad-leaved forests areas were reduced, and the composition of broad-leaved flora became poorer because of lack of more thermophyllic and mesophylleous species - *Caprinus betulus* L., *Ulmus laevis* Pall. Pine birch and alder forests prevailed. This expansion of the boreal floristic elements provides evidence for the climate cooling around 43 10± 95 BP (2610 BC). The widespread distribution of alder formations, of sedge and spore plant communities indicates even excessive moistening of the northern part of the investigated territory. In the southern regions, the afforestation was not so extensive, but the forests areas were wider than at the later stages of the Holocene. Being rather cool, climatic conditions were characterized by more humidity than now.

It corroborates Veklich's conclusion (1987) about a cool climate stage at 4500-4200 BP time span while Zubankov (1986) regards this period as a warm one . The slightly regressive stage of the Black Sea between 4400-4100 BP has been shown (Ostrovsky et al., 1977), but the Caspian was in a transgressive stage around 4000-4300 BP (Varushchenko, et al, 1987). In the Russian plain, the maximum extent of the forest-steppe zone to the south and the favorable forest growth conditions in the steppes have been established for the period 4800/4100 BP (Hotinksy, 1977; Bolikhovskaya, 1990; Spiridonova, 1991, Kremenetski, 1991).

The landscape of the next Middle Subboreal stage (SB-2) contrasted considerably with the Early Subboreal ones. In the time span 3450-3750 BP (1700-2000 BC), the South-Eastern Ukraine was occupied by the dry steppe vegetation: Gramineae steppes of the Donets plain were replaced south by Gramineae-Artemisia ones. This means a shift over three geographical subzones as compared with the Early Subboreal forest-steppe landscapes. The swamp areas were significantly reduced in the flood-plains (data from the Podgorovka site). The sharp depression of arboreal vegetation led to the almost complete deforestation of the territory. In the sporadic refuges preserved only in valleys and ravines an isolated oak, lime and ash tree occurred among the steppe shrubs. The presence of broad-leaved trees makes it possible to interpret the climatic deterioration as a result of the extreme aridity of climate but not of the significant fall in temperature.

In the more arid southern regions, soil-forming processes, both humus accumulation and chemical weathering, were so weakened that the Middle Subboreal deposits on slopes were rather similar to loess. Low biogenic activity of soil microfauna, intensive accumulation of calcium and alkaline metals compounds, the desiccation fissure formation were the characteristic phenomena of the arid zones. The reduction of depth of humus horizons, development of soil salinization in the time span 2500-1700 BC have been shown (Ivanov, 1992; Dyomkin et al, 1985, Akhtyrtsev, 1986, Zolotun, 1975). The coeval stage of humus- accumulative processes development within the modern south forest subzone has been observed (Alexandrovsky, 1983; Veklich, Gerasimenko, 1993).

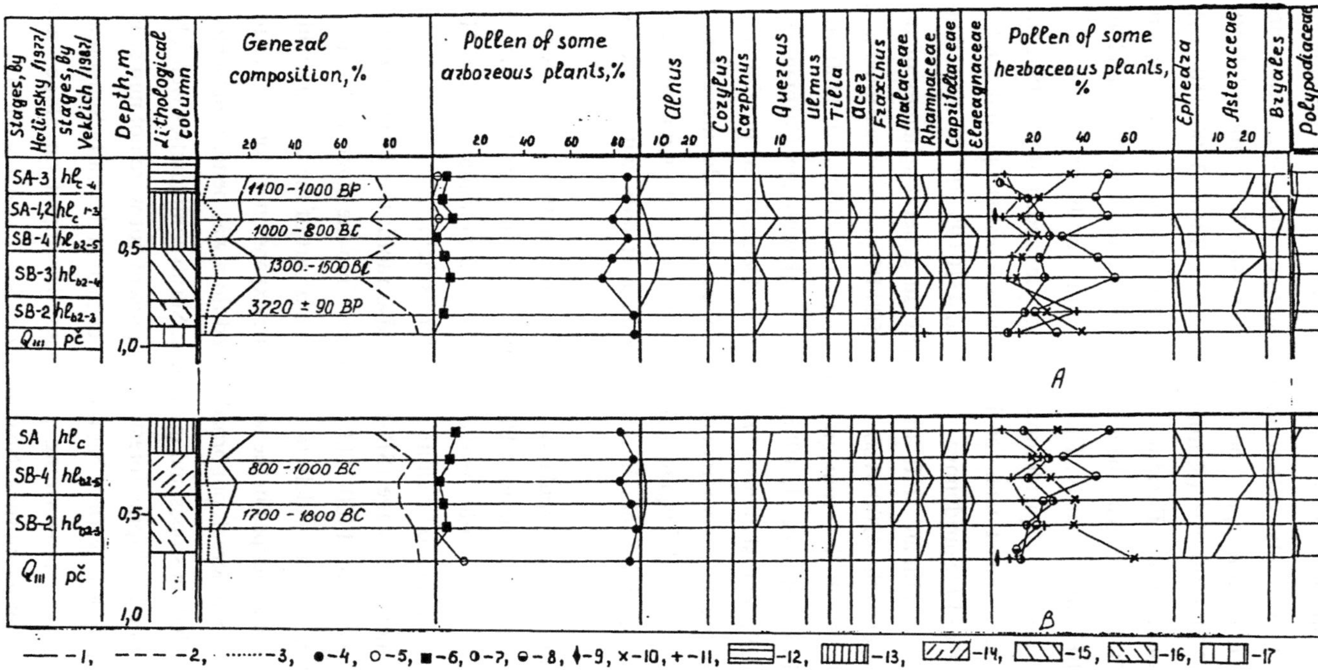

Fig. 9. Spore-pollen diagrams of the deposits of the archeological sites near the Sea of Azov regions. A-the Kamyshevataya-XIV site (III-3), B- the Kamyshevataya-XIX site (III-2). At present, the sites are located on slopes under steppe Herbetum mixtum-Gramineae communities. Archeological dating by V. Gorbov. Percentage of microfossils: 1-AP, 2-NAP, 3-spores, 4-*Pinus*, 5-*Betula*, 6-sum of broad-leaved plants, 7-Gramineae, 8-Herbetum mixutm (including Asteraceae), 9- Cyperaceae, 10-Chenopodiaceae, 11-*Artemisia*. Soil horizons: 12-humus A1(SA-3), grey; 13-buried humus A1' (SA-2,1), dark grey, in some case inseparable with A1B (SB-4); 14-humus-transitional A1B (SB-4), light grey; 15-textural BM (SB-3), brown, prismatic; 16-transitional, BCca (SB-2), pale grey, 17-loess.

The strong shift of Gramineae-Artemisia steppes to the north up to the modern forest-steppe zone, sharp xerophytization of the steppe zone vegetation are reconstructed for the various time spans between 4200-3400 BP, 4200-3700 BP (Kremenetski, 1991); 4200-3700 BP (Klimanov et al., 1986), 3700-3500 BP (Spiridonova, 1991). Though in the Eastern Ukraine, the indicators of aridity are available for the 3450-3750 BP time span, the above evidence allows one to extend the drought episode to 4100 BP approximately. The earliest dates for the dry steppe stage are 4170±100, and 4130±40 BP (Spiridonova, 1991; Alexandrovsky, 1983, the adjacent Don region), while the latest dates for the previous forest-steppe stage in the Don-Donets region is 4300±140 (Spiridonova, 1991),and 4310±95 BP. No more than 200 years for the significant climate shift are needed. In general schemes of the Holocene (Hotinsky, 1977; etc.) the 4200/4100 - 3800/3500 BP period is accepted as a warm climate stage. The Black Sea transgression (between 4100 and 3400 BP, Ostrovsky et al, 1977, Balabanov et al, 1988) was coeval with the Caspian Sea Regression (4000-3500 BP, Varushchenko *et al.*, 1987.) Decrease of precipitation could be combined with the intensive moisture evaporation producing extreme aridification. The Middle Subboreal Caspian regression was the most important one in the whole Middle and Late Holocene. It can be compared only with the Early Holocene Boreal one.

Archaeological data show that the population of the nomadic settlement mainly was not numerous but mainly existed in the East Ukrainian steppes at 4000-3300 BP. No paleoenvironmental evidence is recorded that can be ascribed to human impact. The aridification is therefore considered to be entirely of natural origin

The transition into the next wet stage (SB-3) was also rapid: the 1700-1800 BC span falls into the arid stage while 1400-1500 BC span - into the stage with forest-steppe landscape. The latter extends widely in the northern part of the territory while Herbetum mixtum - Gramineae steppes in the south. The expansion of forests was not so extensive as in the Early Subboreal, but broad-leaved flora was better represented. The oak-lime forests prevailed on slopes and in valleys. Shrub communities were also widespread in the south while birch-pine forests were present but were not numerous over the territory. The climate was warmer but not wetter than in the Early Subboreal . In the Donets region, forest soil-forming processes developed on slopes and pseudogleying processes in low-lying sites. Near the Azov Sea region, chemical

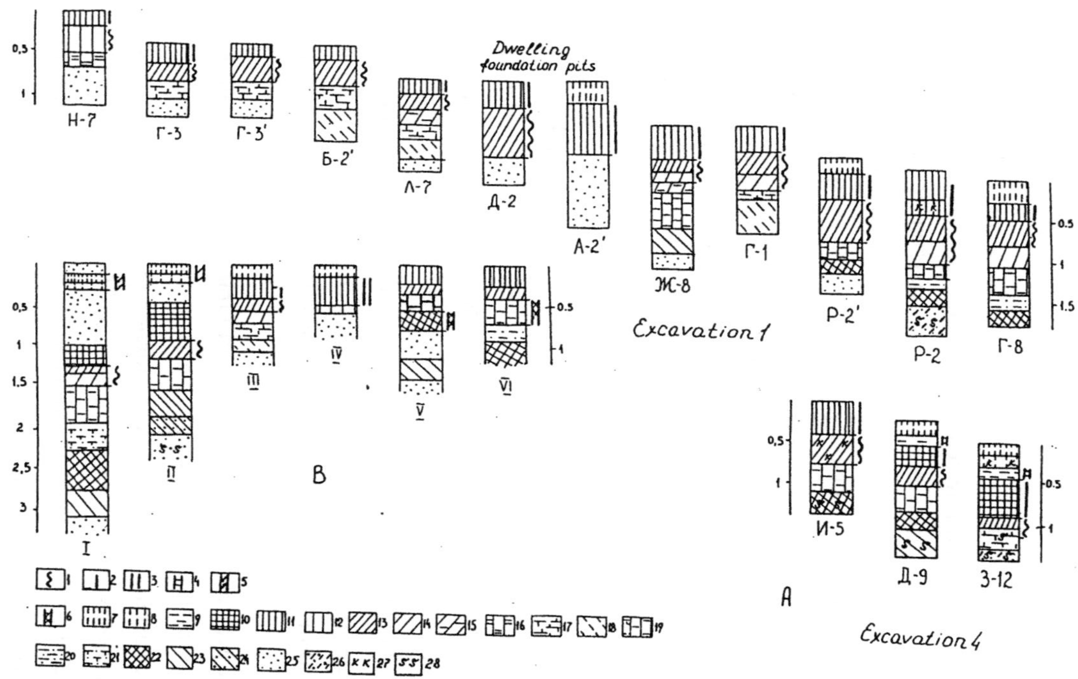

Fig. 10. Genetic profiles of soils of the archeological sites in the Near-Donets river region. A-the Glubokoye-2 site. Excavation 1 - soil catena from the top of sand terrace to the flood-plain excavation; 4 - the high flood plain soils at the foot of the slope. Archeological dating by Y. Gershkovish. B-soils of the Donets river sand terrace at the sites: 1-Novoselovka 2, II-Novoselovka 1, III-Illichevka 1, IY-Glubokoye 1, Y- Moskvino, VI-Bakai (Illichevka 2). The Bronze Age cultural layers have been dated by V. Tsimidanov, the Neolithic layer by S. Degermendgy.

The cultural layers of different age: 1-the Bronze Age, 1300-1400 BC (3100-3200 BP), 2-the Bronze Age, 1100-1300 BC (2900-3100 BP), 3- the Bronze Age, 800-1000 BC (2600-2800 BP), 4- the Scythian age, 500-600 BC (3200-2400 BP), 5- the Saltov culture, 1000-1100 BP, 6-the Neolithic cultures, 5000-5300 BC (6400-6700 BP); The soil horizons: 7-humus A1(SA3,2) 8-humus-transitional A1B (SA-3,2), (-humus-eluvial A1A2 (SA-1), 10-humus-illuvial A1'Bt (SA-1-SB-4), 11-humus A1' (SB-4), 12- humus A1" (SB-3), 13-humus-eluvial A1"A2" (SB-3), 14-eluvial A2" (SB-3), 15-eluvial-illuvial A2"Bt" (SB-3), 16-humus-cryptoilluvial A1"B"(SB-3), 17-illuvial Bt" (SB-3)-SB1?), 18-illuvial-transitional Bt"C (SB-3-SB-1), 19-humus-illuvial A"'Bt (SB-3-AT3/At-1), 20-eluvial A2"' (AT-1), 21- eluvial-illuvial A2"'Bt"' (AT-1), 22-illuvial Bt"'(AT-1), 23 illuvial-transitional Bt"'C(At-1), 24-transitional B"'C(AT-1), 25-sands C, 26-sandy loams C, 27-carbonates presence, 28-gleying signs.

weathering in soils exhibited a significant increase that provided evidence for rather mild climatic conditions.

From the beginning of this stage, the nomadic settlements were replaced by the sedentary settlement system and the farming economy of the Srubnaya (Timber-grave) culture. The wet stage corresponds to its earlier phases - 1500-1300 BC. The climatic amelioration in the period of Srubnaya culture was revealed by pedogenic and pollen studies (Ivanov, 1992; Dymokin et al., 1985, Kremenetski, 1991). The long lasting forest expansion in the forest-steppe and steppe zones of the Russian plain took from about 3500/3400 BP up to 2700-2800 BP (Spiridonova, 1991; Bolikhovskaya, 1990), and the beginning of the Late Subboreal (3200-2800 BP) is accepted as a period of climatic moistening (Hotinksy, 1977; Kremenetski, 1991). But in the Eastern Ukraine, the signs of dessication appeared after 1300 BC (3100 BP).

The later phases of the Srubnaya culture (1300-1000 BC) correspond to somehow drier stage than that of 1500-1300 BC. All over the territory, rather favorable conditions existed for humus accumulation, while chemical weathering was weakened, and eluvial-illuvial processes came to an end. The investigated region was located in the Herbetum mixtum - Gramineae steppe subzone. The forest areas were reduced. The environmental conditions were likely to be similar to the present one or to some even drier. The time span after 1000 BC was characterized by a still more arid climate. Thus, the 1300-1000 BC period was the transitional time between wetter and drier stages.

The 700-1000 BC stage was notable for the spread of Gramineae steppes all over the territory, for further reduction of forest areas and of broad-leaved forests in particular. The weak humus accumulation and the formation of desiccation fissures in soils provide evidence of aridity. Only nomadic camps are found on the Ukrainian steppes at this time. However, from pedogenic and pollen records, it follows that this dryi stage was not as severe as during the 4100-3500 BP interval It can be also deduced that the climate became colder.

This corroborates the soil scientist' opinion about the climatic aridification after the Srubnaya culture period (Ivanov, 1991; Dyomkin et al., 1985; Zolotun, 1975). Reduction of forest areas about 3100-2800 BP has been retraced in the modern forest-steppe zone (Klimanov et al, 1986). The Caspian Sea transgression took place between 3500-3000 BP, then it changed into the regressive stage at 3500-3000 BP,next it again changed to a regressive stage at 3000-2400 BP (Varushchenko et al., 1987). In

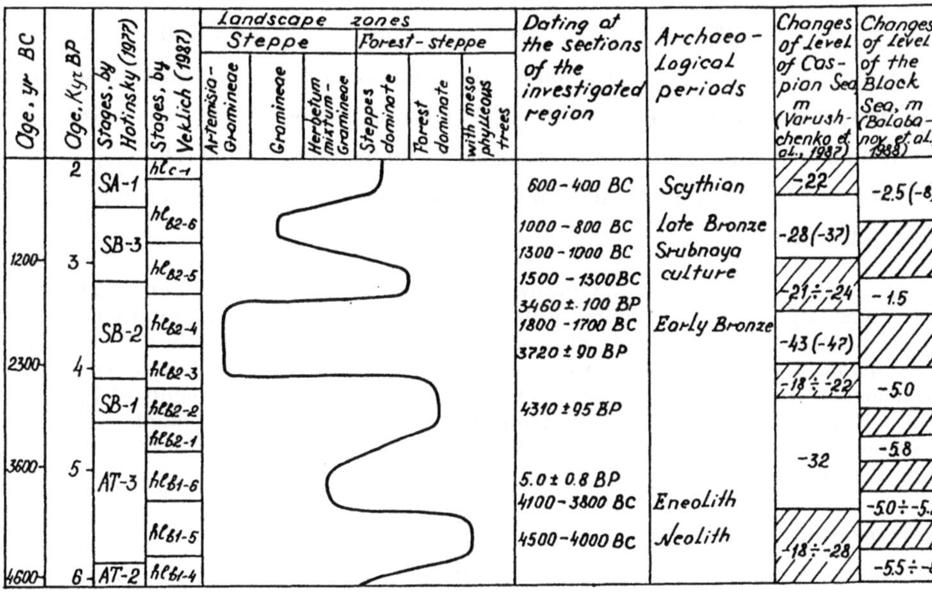

Fig. 11: Schematic reconstruction of the landscape dynamics from 6 - 2 kyr BP in the South-Eastern Ukraine.

complete contrast, the Black Sea regression occurred from 3400 to 3100 BP, and transgression took place between 3100 and 2500 BP (Ostrovsky et al., 1977; Balabanov et al., 1988). Thus just as in the Early-Middle Subboreal, the stages of climate humidity correspond to cooler climatic conditions, and the arid stages to warmer one. Judging by the amplitude of transgressive-regressive movements, the drying (SB-2) and the moistening (SB-1)phases of the Early-Middle Subboreal were both stronger than those (SB-4, SB-3) of the Middle to Late Subboreal. By the soil-pollen records, the cool stage SB-3 was warmer than the cool stage SB-1, while the warm stage SB-4 was cooler than the warm stage SB-2. Between the SB-4 and SB-3 stages, the amplitude of change of climatic elements (temperature, humidity) and of their landscape effects (Fig. 11) were not so great as between SB-1 and SB-2, SB-2 and SB-3 stages. The transitional period between the wet stage 3300-3100 BP and the arid conditions 2600-2800 was also longer and more gradual .

Being in the descending curve from the Atlantic climatic optimum, the general trend of ecologic development was complicated by rhythmic climate and landscape oscillations. Five of them have been observed between 5500 - 2500 BP. The warm and dry climate stages characterized by steppe landscapes alternated with more cool and wet stages, marked out by expansion of forest areas and mesophytization of steppes. The most arid climatic stage of 4100 - 3500 BP was also notable for very severe and contrasting landscape changes when compared with the previous and following stages.

It was the most significant shift of landscape zones in Ukraine in the Middle and Late Holocene.

My thanks go first to Dr. A. Gorelik, Dr. Y. Gershkovich, Dr. Y. Gurin who made the archaeological sites available for the paleogeographic studies and kindly provided archaeological dating. I am also grateful to Dr. N. Kovalukh for the 14C determination, and their calibration in calendar years and to Dr. A. Shlyukov and M. Lyanshenko for TL determinations. The valuable and helpful comments of Prof. M. Veklich on the first draft of the manuscript are gratefully appreciated. Technical assistance from T. Demyanenko is acknowledged. And thanks to Miriam Collwell and Prof. Rhodes Fairbridge for the English language editing.

References

Alexandrovsky A (1983) Evolution of the soils of the East-European plain at the Holocene, Nauka, Moscow 150pp. (in Russian).

Alexandrovsky A, Ovsyannikova S (1881) Palynological sdudies of turf-podzolic soil and chernozem. Soil science 1: 29-39 (in Russian).

Alexandrovsky A, Matskevoy L (1989) Soil and stratigraphical conditions of Mesolithic occurences at the West of Ukraine, in Quaternary Age. Paleontology and Archaeology

Adamenko O and Golbert A Eds, Shtiinitsa, Kishiniv 219-223 (in Russian).

Arap R (1976) Spore-pollen investigation of top surface samples of soils of the nature zones of the plain part of Ukraine, Inst.of Botany of Ac. Sci. of Ukraine Press, Kyiv 25pp. (in Russian).

Artyushenko A (1970) Vegetation of the forest-steppe and steppe of Ukraine at the Quaternary period, Naukova dumka, Kyiv 172pp. (in Russian)

Artyushenko A, Arap R, Bezus'ko L (1982) Vegetation history of western regions of Ukraine at the Quaternary period, Naukova dumka, Kyiv 135pp. (in Russian).

Artyushenko A, Arap R (1988) Climatic optima of the Late Glacial and Holocene of the plain territory of Ukraine, in Holocene Paleoclimates of the European Territory of the USSR, Hotinsky N and Klimanov B Eds, Inst. of Geography of Ac. Sci.of USSR Press, Moscow 136-142 (in Russian).

Ahtyrtsev B, Pryahin A (1988) Paleosoils under burial mounds of the Abashevskaya and Srubnaya cultures of the Bronze age in the Middle Russian forest-steppe. Biological sciences 12: 99-103 (in Russian).

Balabanov I, Izmaylov Y (1988) Changes of the levels and geochemical regime of the Black Sea and the Sea of Azov at the last 20000 yr. Water resources 6: 54-63 (in Russi- an).

Berezina N, Tyuremnov S (1964) Application of spore-pollen analyses for paleophytocoenoses investigation. Bulletin of the Moscow University ser.biol. 3: (in Russian).

Bezus'ko l, Klimanov V, Shelyag-Sosonko Y (1988) Climatic conditions of Ukraine at the Late Glacial and Holocene, in Holocene Paleoclimates of the European Territory of the USSR Hotinsky N and Klimanov V Eds, Inst. of Geography of Ac.Sci. of USSR Press, Moscow 125-135 (in Russian).

Bezus'ko L, Kostylyov C, Popovich S (1992) Representation of the composition of the modern vegetation of Arabatska Strilka in the spore-pollen spectra. Ukrainian botanical journal 49/5: 60-67 (in Ukrainian).

Bolikhovskaya N (1976) Palynology of loesses and fossil soils,in Problems of General Physical Geography and Paleogeography Agadzhanyan A and Dobrodyeev O Eds Moscow University Press, Moscow 257-277 (in Russian).

Bolikhovskaya N (1990) Paleoindication of the landscape changes of the Lower Volga region at the last 10000 yr, in The Caspian Sea Kosarev A Ed, Nauka, Moscow 52-68 (in Russian).

Dinesman L (1977) Biogeocoenoses of steppes at the Holocene, Nauka, Moscow 159pp. (in Russian).

Dyomkin V, Ivanov I (1985) Development of soils of the Near the Caspian Sea Lowland at the Holocene, Sci.Centre of Biol. Research of Ac.Sci. of USSR Press, Pushchino 165pp. (in Russian).

Gerasimenko N (1994) Stages of development of nature of the South-Eastern Ukraine at the Holocene, Inst. of Geography of Ac.Sci. of Ukraine Press, Kyiv 120pp. (in Russian).

Gershkovich Y, Klochko V (1988) Connections of the population of Lower Dnieper region at the Late Bronze epoch, in Intertribal Connections at the Bronze Epoch in the Territory of Ukraine Artemenko I Ed, Naukova dumka, Kyiv 101-114 (in Russian).

Gorbov V (1992) Population complexes of the Near the Sea of Azov region in the system of synchronization of the Western and Eastern circles of the Valikovaya ceramic culture, in New Discoveries and Approaches of the Archaeological Chronology Masson V and Bochkarev V Eds, Inst. of History of Material Cultures of Ac.Sci. of Russia Press, Sankt-Petersburg 90-92 (in Russian).

Gorelik A (1986) The Mesolith of the Siversky Donets basin and of the Near the Sea of Azov region, Inst.of Archaeology of Ac. Sci.of USSR Press, Moscow 16pp. (in Russian).

Grichuk V, Zaklinskaya E (1948) Analysis of fossil pollen and spores and its application in paleogeography, Geografgis, Moscow 91pp. (in Russian).

Gurin Y (1992) The Podgorovka site, in Neolithical Sites of the Steppe of Ukraine Kotova N and Bodyansky A Eds, Inst. of Archaeology of Ac.Sci. of Ukraine Press, Kyiv 15-24 (in Russian).

Hotinsky N (1977) Holocene of the Northern Eurasia, Nauka, Moscow 200pp. (in Russian).

Hotinsky N (1982) Paleogeographical principles of the definition and subdivision of the Holocene of USSR. Striae 16: 95-98

Ivanov I (1992) Evolution of soils of the steppe zone at the Holocene, Nauka, Moscow 144pp. (in Russian).

Ivanov I, Gubin S, Skripnichenko L, Kovalyova I (1986) Development of chernozems of the south of the Russian plain at the Holocene, in Evolution and Age of Soils of the USSR Ivanov Ed, Sci. Centre of Biol. Research of Ac. Sci. of USSR Press,Pushchino 56-64 (in Russian).

Isayeva-Petrova L (1976) Vegetation of the Streltsovskaya steppe preserve at the Holocene. Reports of Ac.Sci. of USSR ser.geogr. 2: 75-87 (in Russian).

Isayeva-Petrova L (1979) On the possibility of palynological studies of chernozem soils. Reports of Ac.Sci. of USSR ser.geogr. 1: 80-87 (in Russian).

Klimanov V, Serebryannaya T (1986) Changes of vegetation and climate in the Middle-Russian Upland at the Holocene. Reports of Ac.Sci. of USSR ser. geogr. 2: 93-102 (in Russian).

Kotova N (1993) Nature-climate conditions and the Neolithical polulation of the steppe of Ukraine, in Environment and ancient population of Ukraine Kryzhytsky S Ed, Inst. of Archaeology of Ac.Sci. of Ukraine Press, Kyiv 23-32 (in Russian).

Kremenetski C (1991) Paleoecology of the past farmers and herdsmen of the Russian plain. Inst. of Geography of Ac.Sci. of USSR Press, Moscow 193pp. (in Russian).

Ostrovsky A, Izmaylov Y, Shcheglov A et al (1977) New data on stratigraphy and geochronology of the Pleistocene marine terraces of the Black Sea coast of the Caucasus and Kerch-Taman' region, in Paleogeography and the Pleistocene Deposits of the Southern Seas of the USSR Kaplin P and Shcherbakov F Eds, Nauka, Moscow 61-68 (in Russian).

Pesochina L, Buylov V (1982) Evolution of soils of the II terrace of the Lower Don river at the Late Holocene, Sci.Centre of Biol.Research of Ac. Sci. of USSR, Pushchino 29pp. (in Russian).

Spiridonova E (1991) Evolution of vegetation cover of the Don basin at the Upper Pleistocene and Holocene, Nauka, Moscow 218pp. (in Russian).

Varushchenko S, Varushchenko A, Klige P (1987) Change of the regime of the Caspian Sea and the closed basins at paleotime, Nauka, Moscow 239pp. (in Russian).

Veklich M (1987) Problems of paleoclimatilogy, Naykova dumka, Kiyv 188pp. (in Russian).

Veklich M, Matviishina Z, Medvedev V, Sirenko N, Fedorov C (1979) Methods of paleopedological investigation, Naukova dumka, Kyiv 271pp. (in Russian).

Veklich M, Gerasimenko N (1993) Stages of the Middle and Late Holocene soil formation in the Desna flood plain. Bulletin of Kyiv University ser.geogr. 40: 85-92 (in Ukrainian).

Velichko A, Morozova T (1986) Paleogeographical basis of the history of the modern soil cover formation, in Evolution and Age of Soils of the USSR Ivanov I Ed, Sci.Centre of Biol.Research of Ac.Sci. of USSR Press, Pushchino 22-36 (in Russian).

Zerov D (1950) The main features of the Late Glacial and Holocene vegetation history in the Ukrainian SSR, in Reports of the Palynological Conference of 1948 Sukachev V Ed, Moscow University Press, Moscow 43-61 (in Russian).

Zolotun V (1974) Development of soils of the South of Ukraine at the last 50-45 centuries, Ukrainian Agricultural Academy Press,Kyiv 74pp. (in Russian).

Zubakov V (1986) Global climatic events of the Pleistocene, Gidrometeoizdat, Leningrad 287 pp. (in Russian).

Zubets R (1971) Spore-pollen studies of the top surface samples of soils of the steppe zone of Ukraine. Ukrainian botanical journal 28/2: 192-198 (in Ukrainian).

Mid- to Late Holocene Changes in Central Europe Climate and Man

Klaus D. JÄGER

Institute of Prehistorical Archäeology
University at Halle
Brandbergweg 23C
D-06099 Halle (Saale)
Germany

Abstract

The upper Holocene of Central Europe is characterized by climatic oscillations reflected in the stratigraphic subdivisions of various sedimentary sequences. Evidence in the inland areas is provided by freshwater limestone deposits alternating with buried humus horizons. In coastal areas the record consists of sequences of alternating transgressional and regressional deposits. Changing geographic pattern of human settlements during different periods supports the link with climate oscillations. Archeological dating of settlements and of associated buried soils enables development of a relatively accurate chronologic framework, and provides evidence of moisture oscillations. Within these bounds the interval around 2000 BC belongs to a markedly dry episode of the Late Holocene, although regionally less expressed than the dry spell between the 13th and 8th centuries BC.

Introduction

Much information concerning Holocene climatic change in Central Europe has been collected in the course of the present century, especially by means of pollen analysis (Firbas 1952) and paleomalacology (Lozek 1964, 1982). The main trends of paleoclimatic development have been proved by these methods. Less attention has been given to actual oscillations of climate in the course of the Holocene. However, we find clear traces in Holocene stratigraphy. This is especially true in the case of freshwater limestone (travertine) deposits (Jäger 1967, 1982).

Notably, this may be demonstrated by one example from Thuringia, in a small valley, the Pennickental, near Jena. It is drained by a minor tributary of the Saale River, where it is flanked by steep slopes of limestone, the Triassic Muschelkalk.

NATO ASI Series, Vol. I 49
Third Millennium BC Climate Change
and Old World Collapse
Edited by H. Nüzhet Dalfes, George Kukla,
and Harvey Weiss

Consequently the water is calcareous and has produced sequences of Holocene travertine deposits in the bottom of the valley. During the early Holocene a continuous sedimentation took place. Later, an repeated interruption of subaquatic sedimentation is proved by buried rendzina soils. Frequently, these humus horizons are datable by archaeological finds. In this case a remarkable example belongs to the Late Bronze Age.

The site provides evidence of several representative facts, including:

a) the continuity of subaquatic calcareous sedimentation during the early Holocene;

b) the alternation of travertine deposits humic paleosols in the top part of the sequences; and

c) the numerous archaeological objects dating the humus horizons.

Other examples in Thuringia include

- Rudolstadt-Schaala, dated to the Hallstatt period, and
- Remda, dated to the centuries about A.D.I.

Also in other regions of Central Europe, such as

- in Bohemia (Sv.Jan p. Skalou),
- in southern Slovakia (Slovakian Karst: Jablonov, site Evetesi Malom),
- in nothern Slovakia (Hozelec near Poprad: Banická),
- in SW Bavaria (Wittislingen).

The sequence of buried paleosols starts with the Central European early Neolithic period, characterized by the so-called linear pottery (in German "Linienbandkeramik"). One example is a site near Wittislingen in SW Bavaria (Seitz 1951, 1956).

The most distinctive time markers are the archaeological datings of buried soils. Each of these time horizons is recorded by different sites and in different regions, but in each case indicating periods of relative dryness. In the same way the calcareous deposits are evidence of intervals of local moist conditions. Local records of an alternation of moist and dry environment can thus be correlated exactly, and this correlation covers an extensive area of Central Europe (at least from S.W. Germany to eastern Slovakia).

In the same way an alternation of lower and higher sealevels, connected with sequences of regressions and transgressions and explicable as eustatic sea-level changes can be observed in the Central European coastal regions of N.W. Germany, Belgium, and the Netherlands (PONS, JELGERSMA, WIGGERS & DE JONG 1963). The regional development of these coastal regions corresponds to the eustatic changes of the ocean level as discovered by FAIRBRIDGE (1965; also FAIRBRIDGE & NEWMAN 1965). Frequently again the regression periods are dated archaeologically by prehistoric sites (e.g. in the Netherlands at Vlaardingen: VAN REGTEREN-ALTENA, BAKKER, CLASON, GLASBERGEN, GROENMAN, VAN WATERINGE & PONS 1962; at Swifterband: VAN DER WAALS & WATERBOLK 1976). The recorded dating of regressions corresponds to the same time-levels as the formation of paleosols within the sequences of travertine deposits in the karstic regions of the Central European inland areas (HAGEMAN & JÄGER 1992). Their widespread occurrence as well as the correlation with eustatic sea-level changes may be interpreted as indicators of climatic cause.

Periods of relative dryness in comparable chronological positions are recorded in the same way in the shores of Central European lakes, as well as in the younger morainic regions of the Baltic region to the north as well as in the prealpine southern areas (JÄGER 1987, JOOS 1982). For example, the regression periods in the coastal regions correspond to periods of low water level in inland lakes. For example, the Late Bronze Age (in the sense of Urnfield Bronze Age: BD to HB according to P. Reinecke 1924) which is well-known from site datings associated with buried soils in the travertine deposits, is frequently observed below the present water level of lakes both in Switzerland and in eastern Germany (territory of Brandenburg: Lake Schwielochsee near Beeskow - CHRISTL, 1985). A similar observation is true for the Slavonic Middle Ages deposits in the northeastern part of Central Europe.

These examples clarify the influence of changing hydrological conditions on the distribution pattern of settlement in the past, both during prehistory and later. This means that observations of archaeology and history on this pattern against a background of regional natural conditions may reflect the contemporaneous palaeoclimatic situation, thus providing a help for its reconstruction. Helpful sources for this purpose provide:

1. the changing distances of sites from the present lake shore in case of beaches of lakes (so-called "Pfahlbauten"; O. PARET 1946; M. JOOS 1982; G. CHRISTL 1985; B PREHN 1985);

2. the displacement in the distribution and selection of prehistoric (and later) sites away from inland water sand their changing criteria chronologically from horizon to horizon (PAVUK, VELIACIK & ROMSAUER 1988; PARET 1961);
3. changing upper limits of settled areas in relation to their altitude (A. CHRISTL 1989);
4. the changing intensity of human occupation and settlement in caves of the karstic regions in Central Europe (WALTER 1989; WEISSMÜLLER 1986);
5. the high frequency of prehistoric fortifications of different ages, especially in dry periods (so-called "Burgenhorizonte" of G. NEUMANN 1954) and as to their cause;
6. archaeological records of human population migrations (e.g. in the Late Bronze Age, the so-called "Fremdgruppenzeit": JÄGER & LOZEK 1978, 1981, 1982; compare in particular KIMMIG 1964).

All these observations provide information concerning:

1. geology (e.g. notably Holocene stratigraphy in case of travertine deposits);
2. geomorphology (notably changing water-levels both of the sea level in coastal regions and of lakes inland);
3. archaeology (notably changes in the distribution patterns of settlement).

These sorts of observations all seem to coincide chronologically with comparable events. They are

a) explicable in terms of common causes and

b) are valid for wide-ranging areas.

Thus, the supposition of a climatological causality gains a high degree of probability.

It may be of interest, to note that periods of relatively dry character (soil formation within freshwater limestone sequences, regression of sea levels and lake levels) are lacking in the early part of the Central European Holocene. Their occurence started just after the first impact of agricultural land use by Man during early Neolithic times (SEITZ 1951, 1956). Therefore, the immigration of early Neolithic farmers to Central Europe cannot be related to any dry time interval. On the other hand, one may question if the first wide-ranging desiccation of Central Europe could be caused by the impact of agriculture. Indeed, at this time a widespread forest cleaning and the succeeding agricultural land use must have led to widespread expansion of open country (LITT 1988, 1992; JÄGER & NEUHÄUSL 1992). Consequently there followed

a) an "anthropogenic continentalization" (HEINRICH & JÄGER 1987) of the regional climate in periods of maximum settlement intensity; and, connected with this phenomenon,

b) an increase of regional albedo.

Real consequences for the regional climatic situation may be expected, and therefore, at this time we try to investigate the different intensities of land clearing and albedo during different time horizons. This may be carried out by regional mapping of all known sites associated by their chronological level. These efforts have been initiated by the authors predecessor at the University of Halle (J. PREUSS, in press).

However, the reliability of correlations between regional desiccation periods and eustatic changes of sealevel (HAGEMAN & JÄGER 1992) contradicts any supposition of regional as opposed to global causes effecting the recorded periods of relative dryness in Central Europe ("Epi-Atlantic" according to JÄGER 1969).

Another observation requires additional field research: The oscillations between relative moisture and dryness during Neolithic as well as Aeneolithic times in Central Europe are characterized by rather short time intervals and probably low amplitudes. During the Bronze Age, i.e. during the second millenium BC, both climatic extremes have increased and reached maximum values: The existing data point to a maximum of moisture between 1600 and 1300 BC (wide distribution of the beech/*Fagus silvatica*) and a very long dry period between 1300 and 750 BC. Another dry period followed about 2000 or 1800 BC, during the earliest Bronze Age, but the evidence up till now seems to indicate less intensive extremes.

A significant decrease of temperature followed at about 800 or 700 BC (so-called "Klimasturz" of SMOLLA 1954). Due to the decrease of temperature at most sites the travertine sedimentation ceased during this period. In some places they are covered by clastic deposits, in other cases only by recent soils.

This change of temperature conditions coincides with the change from the Subboreal period (in the sense of FIRBAS, 1949, 1952) to the Subatlantic time (according to the systems of BLYTT (1876) and SERNANDER (1894). It represents the transition from the Mid-Holocene to the Late Holocene.

REFERENCES

Blytt, A. (1876) Essay on the immigration of the Norwegian flora during alternating rainy and dry periods. Albert Cammeyer, Kristiania.

Christl, A. (1989) Höhengrenzen der urgeschichtlichen siedlung im Erzgebirge und dessen Umland, dargestellt an einem Ausschnitt. Archeol. Rozhledy 41: 386-405.

Christl, G. (1988) Ur- und frühgeschichtliche Fundplätze im Uferbereich des Schwielochsees sowie im nördlich angrenzenden Spreetal. Veröff. Mus. Ur- u. Frühgesch., Potsdam 22: 229-244.

Fairbridge, R.W. (1965) Ancient shorelines and absolute dating. Report of VIth Int. Congr. on Quaternary 1: 653-676.

Fairbridge, R.W., NEWMAN W.S (1965) Sea level and the Holocene boundary in eastern United States. Report of the VI the Int. Congr. on Quaternary 1: 397-418.

Firbas, F. (1949, 1952) Waldgeschichte Mitteleuropas nördlich der Alpen 1, 2. Fischer, Jena.

Hageman, B.P., Jäger, K.D. (1992) Zur stratigraphischen Verknüpfung holozäner Meeresspiegelbewegungen im Küstenraumder Nordsee mit Klimaschwankungen im mitteleuropäischen Binnenland. In: Billwitz K., Jäger KD., Janke W. (eds): Jung-quartäre Landschaftsräume - Aktuelle Forschungen zwischen Atlantik und Tienschan (Springer, Berlin Heidelberg etc.) 157 - 169.

Heinrich, W.D., Jäger, K.D. (1987) Paläoökologische und landschaftsgeschichtliche Interpretation einer klinalen Merkmalsvariation bei rezenten und fossilen Populationen von *Crocidura leucodon* (Mammalia: Soricidae). Petermanns Geogr. Mitt., 131: 93 - 98.

Jäger, K.D. (1987) Holozäne Binnenwasserkalke im Ostteil der Thüringer Triasmulde. In: Jäger K.D., Hrabowski K. (eds), Probleme und Befunde der Holozänstratigraphie in Thüringen, Sachsen und Böhmen (Berlin, Prag): 6 - 31.

Jäger, K.D. (1969) Climatic character and oscillations of the Subboreal period in the dry regions of the Central European Highlands. In: Wright jr. H.E. (ed) Quaternary Geology and climate (Proc. VII Congr. of Int. Assoc. for Quaternary Res. 16): 38 - 42.

Jäger, K.D. (1982) Stratigraphische Belege für Klimawandlungen im mitteleuropäischen Holozän. Z. f. Geol. Wiss. 10: 799 - 809.

Jäger, K.D. (1987) Zum gegenwärtigen Kenntnisstand über nacheiszeitliche Wasserstandsschwankungen an Binnenseen im Jungmoränengebiet des nördlichen Mitteleuropa. Wiss. Z. Univ.Greifswald, Math.-Naturwiss R 36: 44 - 48.

Jäger, K.D., Lozek, V. (1978, 1981) Umweltbedingungen und Landesausbau während der Urnenfelderbronzezeit in Mitteleuropa. In: Coblenz, W., Horst, F. (eds) Mitteleuropäische Bronzezeit - Beiträge zur Archäologie und Geschichte (Akademie Verl. Berlin): 211 - 229.

Jäger, K.D., Lozek, V. (1982) Environmental conditions and land cultivation during the Urnfield Bronze Age in central Europe. In: Harding, A.F. (ed) Climatic change in Later prehistory (Edinburgh Univ Press): 162 - 178.

Jäger, K.D., Neuhäusl, R. (1992) Interactions between natural environment and Neolithic man in Central Europe - an investigation based on comparative studies on vegetation and settlement with special emphasis on the view of natural science. Palaeoclimate Research 8: 75 - 81.

Joos, M. (1982) Swiss Midland-Lakes and climatic changes. In: Harding, A.F. (ed) Climatic change in Later prehistory (Edinburgh Univ Press): 44 - 51.

Krimmig W. (1964) Seevölkerbewegung und Urnenfelderkultur Beih. Bonner Jahrb. 10 (Studien aus Alteuropa) 1: 220 - 283.

Litt, Th. (1988) Stratigraphische Belege für anthropogen ausgelöste Bodenverlagerungen vom Neolithikum bis zur frühen Eisenzeit im circumhercynen Raum. Ethnograph. Archäolog. Z. 29: 129 - 137.

Litt, Th. (1992) Investigations on the extent of the early Neolithic settlement in the Elbe-Saale region and on its influence on the natural environment. Paleoclimate Research 8: 83 - 91.

Lozek V. (1964) Quartärmollusken der Tschechoslowakei (Rozpr. Ustr. ust. geol. 31). Geol. Zentralanst. Praha.

Lozek, V. (1982) Faunengeschichtliche Grundlinien zur spätund nacheiszeitlichen Entwicklung der Molluskenbestände in Mitteleuropa (Rozpr. Cs. ak. ved, r mat. a prir. ved 92/4) Academia Praha.

Neumann, G. (1954) Sieben Gleichbergburgen nach dem Forschungsstande von 1952. Frühe Burgen und Städte - Beiträge zur Burgen- und Stadtkernforschung (Wilhelm Unverzagt zum 60. Geburtstag ...): 7 - 16.

Paret, O. (1946) Das neue Bild der Vorgeschichte. August Schröder, Stuttgart.

Paret, O. (1961) Württemberg in vor- und frühgeschichtlicher Zeit. Kohlhammer, Stuttgart.

Pavúk, J., Veliacik, L., Romsauer, P. - Veränderung von Siedlungsarealen als mögliche Folge klimatischer Schankungen in der Westslowakei. Striae Uppsala: 32

Pons, L.J, Jelgersma, S., Wiggers, A.J., De Jong, J.D. (1963) Evolution of the Netherlands coastal area during the Holocene. Verh. Konigl. Nederlands. Geol. Mijnbouwkund. Genootschap, Geol. Ser. 21-2: 197 - 208.

Prehn, B. (1987) Zu Hinweisen auf Seespiegelschwankungen der Müritz. Wiss. Z. Univ. Greifswald, Math.-Naturwiss. R. 36: 49 - 51.

Preuss, J. (ed., in press) Das Neolithikum in Mitteleuropa: Kulturen - Wirtschaft - Umwelt vom 6. bis 3. Jahrtausend v.u.Z. - Übersichten zum Stand der Forschung.

Reinecke, P. (1924) Zur chronologischen Gliederung der süddeutschen Bronzezeit, Germania 8: 43 - 44.

Seitz, H.J. (1951) Die Süßwasserkalkprofile zu Wittislingen und die Frage des nacheiszeitlichen Klimaablaufs. Berichte Naturforsch. Ges. Augsburg 4: 1 - 132.

Seitz, H.J. (1956) Zur Altersfrage der Bandkeramik und weitere Neuergebnisse aus den Profilen zu Wittislingen (1952-56). Berichte Naturforsch. Ges. Augsburg 7: 5 - 33.

Sernander, R. (1894) Studier öfver den Gotländska vegetationens utvecklingshistoria. Akademisk. afhandl., Uppsala.

Smolla, G. (1954) Der "Klimasturz" um 800 v. Chr. und seine Bedeutung für die Kulturentwicklung in Südwestdeutschland. Festschr. Peter Goessler: 168 - 196.

van der Waals, J.D., Waterbolk, H.T. (1976) Excavations at Swifterband discovery, progress, aims and methods. Helinium 16: 3 - 14.

van Regteren-Altena, J.F., Bakker, J.A., Clason, A.T., Glasbergen, W.; Groenman-van Waateringe, W.; Pons, L.J. (1962, 1963) The Vlaardingen Culture. Helinium 2 (1962): 3-35, 97 - 103, 215-243; 3 (1963); 39-54, 97 - 120.

Walter, D. (1985) Thüringer Höhlen und ihre holozänen Bodenaltertümer. Mus. f. Ur- u. Frühgesch. Weimar.

Weissmüller, W. (1986) Postmesolithische funde aus Höhlen und Abris am Beispiel des südlichen Riesrandgebietes. BAR Inernat. Ser. 279. BAR Oxford.

Westphal, Th. (1995) Die Höhengrenzen der urgeschichtlichen Besiedlung im Thüringer Wald und dessen Vorländern, dargestellt an einem Ausschnitt. Magister thesis, unpubl. (Humboldt-Universität Berlin).

The Age and Causes of Mid-Late Holocene Environmental Change in Southwest Turkey

Neil ROBERTS^, Warren J. EASTWOOD*, Henry F. LAMB* and John C. TIBBY^+

^ Department of of Geography, Loughborough University, U.K.

* Institute of Earth Studies, University of Wales, Aberystwyth, U.K.

+ Department of of Geography and Environmental Sciences, Monash University, Melbourne, Victoria, Australia.

Abstract

Proxy records such as lake sediment sequences provide important data on abrupt environmental changes in the past, but establishing their specific causes from the palaeoenvironmental record can be problematic. Pollen diagrams from southwest Turkey show a mid-late Holocene pollen assemblage zone, designated as the Beyşehir Occupation phase, the onset of which has been ^{14}C dated to ca. 3000 BP (ca. 1250 BC). A second millennium BC date for the start of the Beyşehir Occupation phase can now be confirmed as a result of the discovery of volcanic tephra from the Minoan eruption of Santorini (Thera) in lake sediment cores from the region. Palaeoecological analyses on sediment cores from Gölhisar gölü, a shallow montane lake, indicate that tephra deposition was followed by a sustained response in the aquatic ecosystem, in the form of increased algal productivity. The onset of pollen changes marking the beginning of the Beyşehir Occupation phase was not, on the other hand, precisely coincident with the tephra layer, but rather occurred at least a century later at this site. Despite the paucity of archaeological evidence for Late Bronze Age settlement in the Oro-Mediterranean region of southwest Turkey, it would appear that the second millennium BC saw the start of a period of major human impact on the landscape which continued until the late first millennium AD. The Santorini ash represents an important time-synchronous, stratigraphic marker horizon, but does not appear to have been the immediate cause of the onset of the Beyş ehir Occupation phase.

NATO ASI Series, Vol. I 49
Third Millennium BC Climate Change
and Old World Collapse
Edited by H. Nüzhet Dalfes, George Kukla,
and Harvey Weiss

Introduction

Holocene palaeoecology has provided important insights into long-term environmental dynamics and landscape changes. On the other hand, establishing the causes of these changes may be difficult due to the problem of equifinality, in which similar consequences (e.g. a decline in forest cover inferred from a change in the ratio of arboreal to non-arboreal pollen, or an increased influx of eroded soil) could result from one or more of a number of different causes (e.g. drought, human clearance, ashfall, etc). For example, the mid-Holocene decline in elm across Europe, detected by pollen analysis, has been attributed to disease pathogen attack, prehistoric human impact, climate change, or a combination of these different agencies (Rackham 1980).

This problem of pin-pointing 'cause and effect' relationships is probably greatest for the middle part of the Holocene between ca. 3000 BC to ca AD 1, a time period which witnessed landscape changes for which both human impact and climate change may have been responsible. An important shift in the global climate system took place during the centuries either side of 3000 BC, especially in low latitudes, associated with a weakening of the Indian Ocean monsoon (Street-Perrott and Roberts 1983). This change was ultimately related to the declining receipt of Northern Hemisphere summer radiation, due to orbital forcing (Kutzbach 1987). In addition, it is now known that the mid Holocene was punctuated by a number of 'sub-Milankovitch' abrupt climatic events of century-scale duration (Roberts 1993, Lamb et al. 1995). The last three millennia BC were also marked by significant and increasing human impact associated with the development of the first Old World civilizations. In addition to climate and human impact, other factors - sometimes global, but usually regional - led to environmental change. These included disease, major fires, earthquakes and volcanic eruptions, the last of which may have led to climatic perturbation as well as having direct effects (Porter 1981, White and Humphreys 1994). This range of different and, at times, interacting factors is not easily disaggregated.

It has long been postulated that climate, and specifically desiccation, had an important bearing on the development of Old World cultures (Huntington 1915). In attempting to test hypotheses about the human consequnces of abrupt mid-Holocene climate change, archaeological records have the important advantage of a direct link to cultural history. On the other hand, evidence that a particular site or region was abandoned at a particular time represents a (possible) *consequence* of environmental change, and cannot also be used as an indicator of its cause. An independent test of inferred climatic events may instead be obtained from proxy records, such as lake sediment archives and treerings, so helping to avoid circularity of argument. Unfortunately, these palaeoenvironmental data may not originate from the same location as the archaeology and are not always dated with sufficient precision to demonstrate synchroneity and hence possible causation (Roberts and Butlin 1995). Proxy records commonly form a continuous time series but their temporal resolution is more likely to be decadal than annual. Another problem is the question of radiocarbon reliability, related to uptake of 'old' carbon in lakes receiving hardwater or experiencing catchment erosion during periods of landscape disturbance.

In order to achieve locational correspondence between data sets, archaeological and palaeoenvironmental records should derive from the same landscape unit, preferably as part of integrated studies of landscape change. There have recently been a number of successful multi-disciplinary projects investigating regional landscape evolution, such as the Ystad project in southern Sweden (Berglund 1991) and an investigation of landscape changes in the Outer Hebrides of Scotland (Entwistle et al. 1995). In this study we aim to discover the causal links between human occupation, forest decline, climatic change and volcanism in the montane region of southwest Turkey, with particular reference to the second millennium BC. In order to do this, the archaeological and palaeoenvironmental data that are employed derive from the same landscape unit.

The Beyşehir Occupation phase

Pollen diagrams from southwest Turkey show a major mid-late Holocene disturbance horizon known as the Beyşehir Occupation phase (Bottema and Woldring 1990). ^{14}C dating indicates that this phase began at around 3000 BP (ca. 1250 BC), and it continued until 1350 BP (AD 600), or even later. In diagrams from Beyşehir Gölü (van Zeist et al. 1975, Bottema and Woldring 1984), Pınarbaşı (Bottema and Woldring, 1984), and Söğüt (van Zeist et al. 1975), the Beyşehir Occupation Phase was defined on the basis of the appearance or increase of a number of primary and secondary anthropogenic indicators (sensu Behre 1990, Bottema and Woldring 1990) together with a decline in pollen of *Pinus* at Beyşehir Gölü, *Pinus* and *Cedrus* at Pınarbaşı, and *Pinus* and *Juniperus* at Söğüt. The Beyşehir Occupation phase shows clear evidence of agricultural activity, in particular in the form of arboriculture (walnut, vine, manna ash, olive), although many of the other palynological indicators that are present (e.g. plantains, grasses) cannot be used unambiguously to separate human from natural disturbance of the natural vegetation. The end of this occupation phase appears to have been marked by regrowth of woodland whose species composition and diversity was significantly different from that which had existed in the mid-Holocene.

The occupation phase was evidently widespread, being recorded at pollen sites from Beyş ehir in the east to the Aegean coastal region in the west (Fig. 1). This palaeoecological evidence ought therefore to have a counterpart in the historical and archaeological record of human occupancy in the same area. In the latter part of the first millennium BC, southwest Turkey was home to the Lycians and to allied groups such as the Kabalians and Milyans (Bean 1978, Coulton 1993). Subsequently these independent peoples came to be incorporated in the Roman and Byzantine Empires. Detailed information on the history of human settlement within southwest Turkey has come from a series of archaeological and historical field surveys and excavations. One of the most recent surveys (1985-1993) has focused on the classical city of Balboura, which was founded around 200 BC and continued until at least the sixth century AD (Coulton 1988). Balboura was one of a tetrad of adjacent classical cities known as the Cibyratis; Oenoanda, Bubon and Kibyra being the others (Figure 2). Balboura's territory touched on Söğüt Gölü, the ancient Lake Karalitis, and from which there is a pollen diagram showing a very clear Beyşehir Occupation phase (van Zeist et al. 1975). Systematic field survey of sample areas within Balboura's territory has recorded sites of periods from prehistoric to Byzantine, but by far the most prominant phase of occupation was in the Roman period. At this time, there is archaeological and epigraphic evidence for a flourishing local agricultural economy, with cereal cultivation and viticulture. None the less Balboura, at 1650 m, with valley bottoms down to 1200 m, lies close to the modern altitudinal limit for permanent village settlement, and agriculture must always have been somewhat precarious here. This montane region appears to have been abandoned to permanent settlement during or after the Byzantine period, with seasonal pastoralism probably being the main form of landscape exploitation from the late first millennium AD until the nineteenth and twentieth centuries. Archaeological and historical data consequently appear to be in good agreement with pollen evidence in showing an intensively used and well-

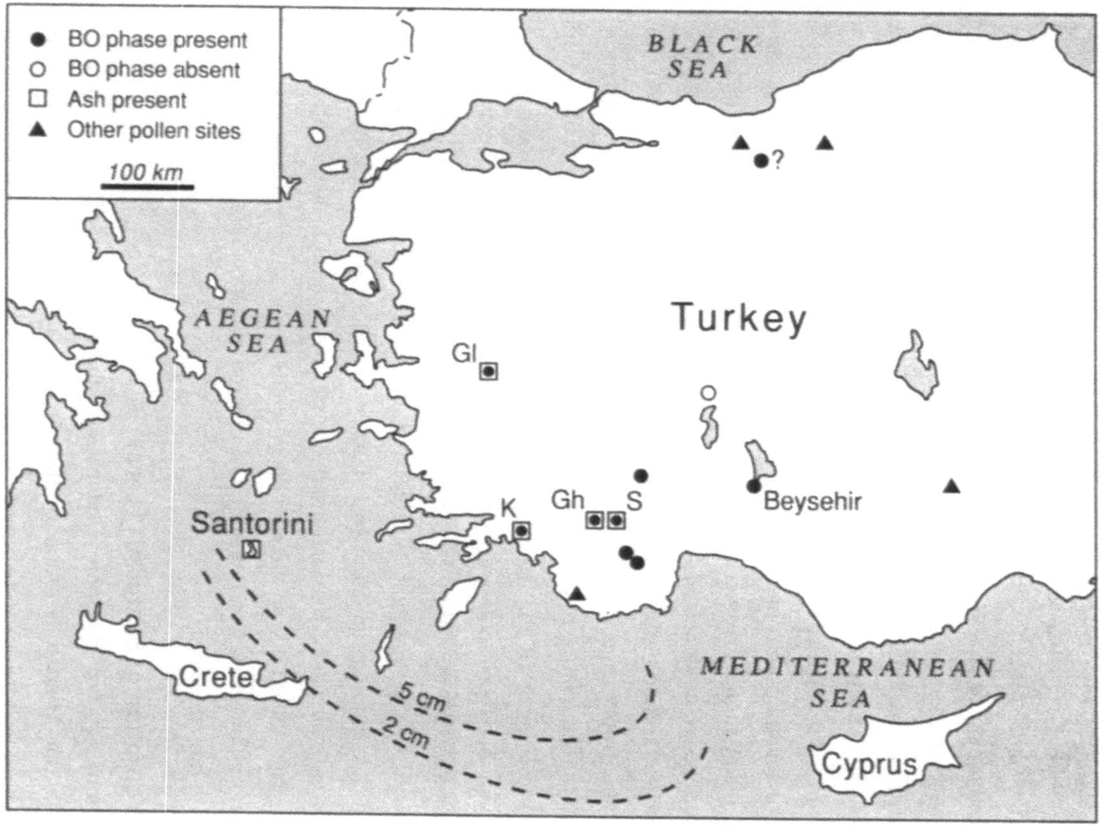

Figure 1: Pollen core sites in southwest Turkey recording the Beyşehir Occupation (BO) phase and/or Santorini tephra. Gl = Gölcük (ash thickness 12 cm), K= Köyceğiz (9 cm), Gh = Gölhisar (4.5 cm), Ŝ = Söğüt (>1 cm), Beyşehir (ash not identified). Isopach lines for tephra thickness in marine cores from Watkins et al. (1978).

organised rural landscape from ca. 200 BC to ca. 600 AD, followed by abandonment and reafforestation.

Palynological and archaeological records are in less good agreement about the onset and early part of the Beyşehir Occupation phase (Roberts 1990). Within Balboura's territory, there is a substantial settlement mound which includes pottery back to the fifth-fourth millennium BC, and a decorated ware which begins in the eighth century BC and is found quite widely in southwest Turkey. No evidence of settlement from the seventh to the third centuries BC was found in the survey, but a number of monumental tombs probably date to this period, and suggest at least some continuity of occupation. However, the Balboura archaeological survey has failed to reveal any conclusive evidence of Middle or Late Bronze settlement. Elsewhere in southwest Turkey, second millennium BC sites are recorded (e.g. Mellaart 1954), but overall the Late Bronze Age is not as well represented in the archaeological record here as it is in the Aegean or in central Anatolia. This would appear to conflict with ^{14}C dates for the onset of the Beyşehir Occupation phase, which indicate that clearance began between 1600 and 1000 BC (see further discussion below). One possible explanation for this might be that the ^{14}C dates for the pollen diagrams have been contaminated by the "hardwater" effect and are too old. Limestone is often the dominant bedrock lithology in southwest Turkey. An alternative might be that the initial mid-second millennium BC landscape disturbance was not caused by human impact at all, but related to another agency such as a major volcanic eruption (although the remainder of the Beyşehir Occupation phase would still have to be associated with human action).

The Minoan eruption of Santorini (Thera)

The beginning of the Beyşehir Occupation phase at ca. 1250 BC broadly coincides with the Minoan eruption of the Aegean volcanic island of Santorini (Thera). There has been active debate over the precise age of this eruption, with ^{14}C dates from Santorini itself lying in a broad time range within the second millennium BC (Manning 1990). Attempts have been made to correlate the Minoan eruption of Thera with acidity peaks in Greenland ice cores (Hammer et al. 1987) and series of exceptional growth rings in timbers from California (LaMarche and Hirschboek 1984), Ireland (Baillie and Munro 1988) and central Anatolia (Kuniholm et al. 1996). These non-archaeological dates have generally placed the Minoan eruption within

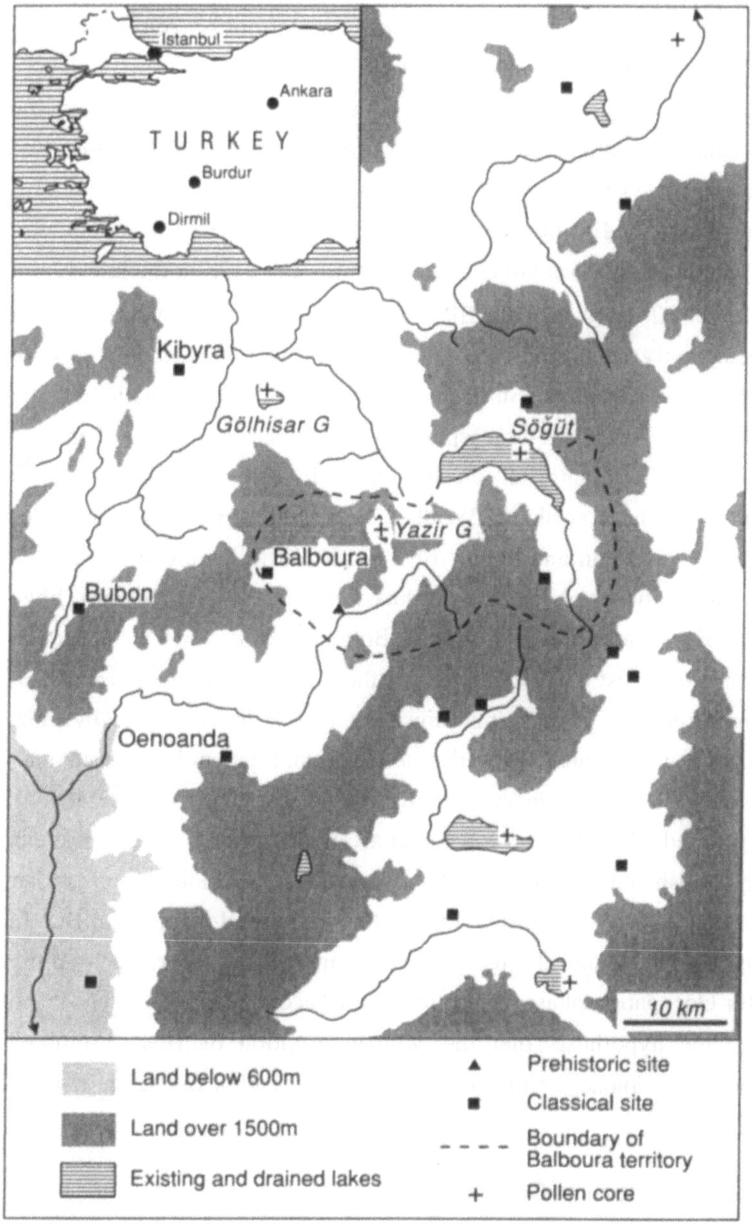

Figure 2: Location of pollen cores and archaeological sites in the central montane district of southwest Turkey.

the time range 1650-1600 BC. In contrast, pottery sequences from Akrotiri have been used to infer a later date for the eruption, ca. 1500 BC.

Santorini ash has been found on Crete and in marine sediment cores from the eastern Mediterranean, suggesting that the main ash plume was to the south-east of the eruption (Watkins et al. 1978)(Fig. 1). In the last decade, Santorini ash has also been found in a number of sediment cores from southwest Turkey. The tephra was first recognised by Sullivan (1988) at Gölcük, where it was ca.12 cm in thickness, and lay below a peat layer ^{14}C-dated to 3100 BP. The Thera ash, approximately 9 cm in thickness, has subsequently been identified in the previously published pollen core from the coastal lagoon at Köyceğiz (Sullivan 1990), while a re-examination of the pollen core from Söğüt (by WJE) shows a tephra layer to have been present there too (Table 1). The presence of the Santorini volcanic tephra in southwest Turkey and in Black Sea sediments (Guichard et al. 1993), albeit in sub-mm thicknesses, provides new information on the direction and extent of the ash fall following the eruption, and shows significant ash dispersal east and northeast of Thera. There may also have been some dispersal of tephra northwest of Thera (Bottema 1980).

The Santorini ash occurs just before the onset of the Beyşehir Occupation phase at several sites, and represents an important stratigraphic marker horizon. At Köyceğiz, the Santorini ash layer immediately precedes a significant palynological change, including a fall in tree pollen and an increase in disturbance indicators (van Zeist et al. 1975). On the other hand, Köyceğiz lies on the coast, not in the Oro-Mediterranean zone like the other Turkish sites (Table 1), and its Beyşehir Occupation phase is rather atypical, lasting - at most - a few centuries during the late second millennium BC. In the Söğüt core, on the other hand, the tephra layer is apparently separated from the start of the Beyşehir Occupation phase. Therefore, these records alone do not seem to be sufficient to test the hypothesis that the Minoan eruption of Thera produced widespread environmental impacts, either directly through ashfall or via short-lived climate perturbation.

The palaeoecological record at Gölhisar

We have investigated the environmental effects of Santorini ash deposition and its relationship to the beginning of the Beyşehir Occupation phase at Gölhisar, a small, shallow montane lake (37° 8' N, 29° 36' E). A preliminary investigation of this site

Table 1: ^{14}C dates for the onset of the Beyşehir Occupation phase and the Santorini tephra in southwest Turkey (see Fig. 6 for stratigraphic positions)

Site	Site elevation	^{14}C date ± 1SD	cal. age range	lab. no.	refs.
Beyşehir	1120 m	3265 ± 35 BP	1590-1510 BC	GrN-6879	1, 2
Köyceğiz	1 m	3070 ± 55 BP	1404-1230 BC	GrN-6451	1, 4
Söğüt	1393 m	2885 ± 35 BP	1117-998 BC	GrN-6452	1
Gölcük	1050 m	3100 ± 160 BP	1518-1124 BC	Beta-6102	3
Gölhisar (GH.E.92-7)	930 m	3225 ± 45 BP	1521-1428 BC	SRR-5188	5
Gölhisar (GH.A.92)	930 m	3330 ± 70 BP	1683-1518 BC	Beta-56673	5
Gölhisar (GH.B.92)	930 m	2830 ± 50 BP	1024-910 BC	SRR-5185	5

References: 1. van Zeist et al. (1975); 2. Bottema and Woldring (1984); 3. Sullivan (1988); 4. Sullivan (1990); 5. this study

Calibration based on Stuiver and Reimer (1993), using 1 SD age range.

had been carried out in 1977, and the resulting undated pollen diagram showed the Beyşehir Occupation phase to have been recorded here (Bottema and Woldring 1984). The sediments from Gölhisar were shown to be relatively organic, unlike many lake deposits in southwest Turkey. This offered the possibility of obtaining a series of ^{14}C dates and achieving a good chronological framework. Significantly, the site also lies in close proximity to a number of important archaeological sites. Gölhisar lies in a fertile plain inside the territory of the classical city of Kibyra, the classical city-state immediately north of Balboura. An important defensive site lies on a peninsula in the lake itself, and which was identified by the late Alan Hall (1994) as the "Sinda" of Livy and other classical authors. Survey of this archaeological site has shown traces of long occupation from Neolithic to Medieval times, but mainly from the eighth century BC through to the Middle Byzantine period (eighth-ninth centuries AD).

Parallel overlapping cores were obtained from exposed mudflats at the southeastern edge of the lake, using a 5 cm diameter Livingstone corer and chain hoist. The longer core (GH.A.92) reached a depth of 813 cm, while the shorter (GH.B.92) was ended at a depth of 532 cm. Core recovery averaged over 80%. The sediments consist of alternating clays and clay-silts, silty marls and peats. An

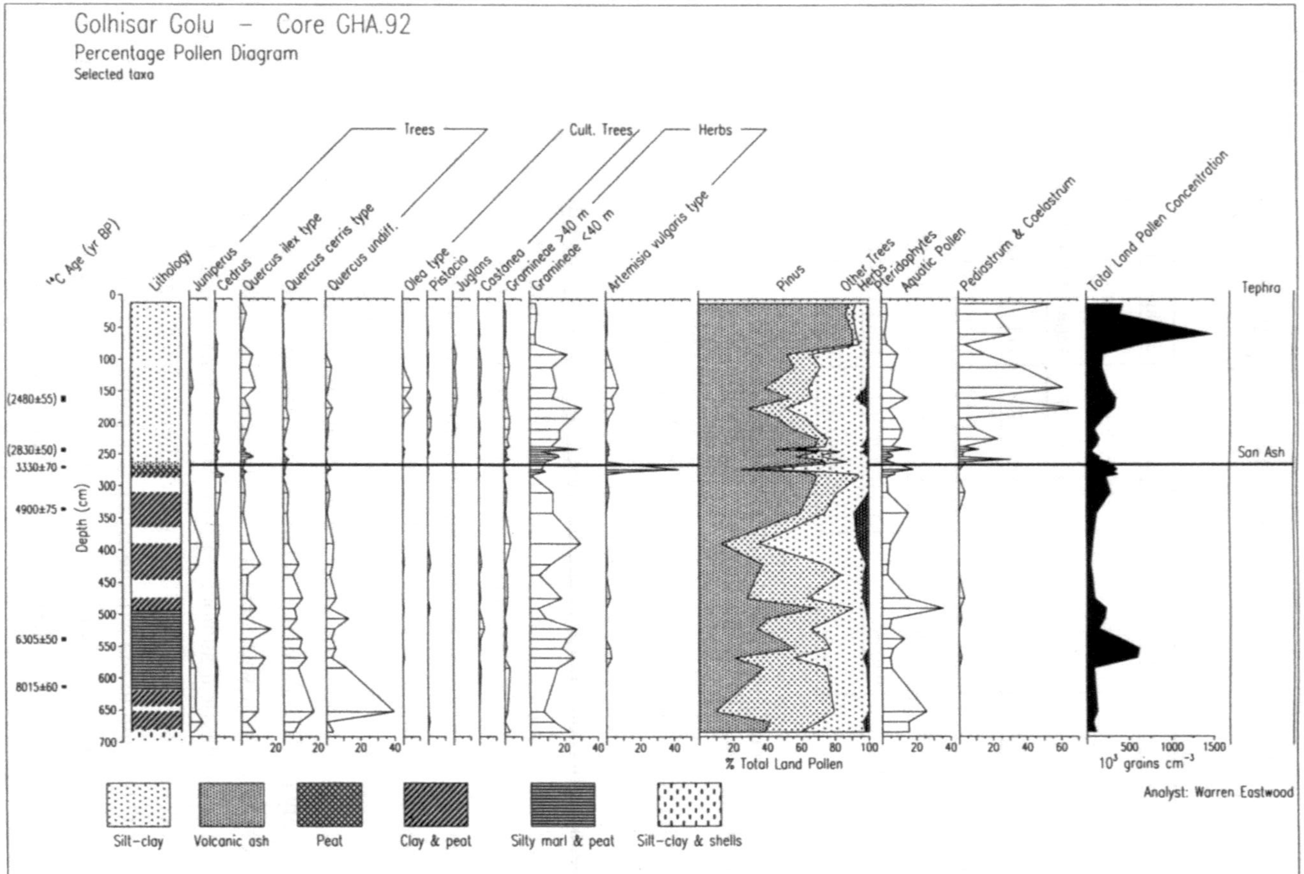

Figure 3: Summary pollen diagram for core GH.A.92 from Gölhisar.

additional series of eight overlapping short cores (<25 cm long) were subsequently taken with a Dachnowsky sampler to provide additional samples for that part of the sediment record which dates to the second millennium BC (GH.E.93 series). The lake water had a *p*H of 8.95 and a conductivity of 1350-1500 μScm^{-1} in July 1992.

Core sediments have been analysed for whole core magnetic susceptibility, Troels-Smith sediment classification, organic and carbonate content via loss on ignition at 550°C and 950°C, whole core x-ray scanning, palynology and diatoms. Magnetic susceptibility shows a very clear set of down-core fluctuations, and permits precise correlation of the parallel cores where no lithostratigraphic changes are visible. Loss-on-ignition values show organic contents up to 65% in core GH.A.92. Carbonate values reach a maximum of 38%. A total of seven conventional ^{14}C dates on total organic matter have been determined on sediment samples from cores GH.A and B, with an eighth date from short core GH.E.93-7. They indicate fairly linear sediment accumulation and with a basal age for core GH.A.92 at around the start of the Holocene.

The cores from Gölhisar contain an ash horizon, at a depth of ca. 275 cm in core GH.B.92. The ash, which was not detectable by either magnetic susceptibllity or by X-ray analyses, displays considerable between core variability in thickness, up to a recorded maximum of 4.5 cm. The volcanic glass shards have been characterised by ICP-MS and electron micro-probe analysis, and these analyses shows a close geochemical correspondence with ash from Thera itself, from deep sea and Turkish lake sediment cores, and from the archaeological site at Knossos (Eastwood et al. in press). There consequently seems little doubt about the provenance of the Gölhisar ash to the Minoan Santorini eruption. Peat immediately below the tephra has been ^{14}C dated to 3330 BP (GH.A.92) and 3225 BP (GH.E.93-7), or between 1683 and 1428 BC according to the calibration of Stuiver and Reimer (1993).

Palaeoecological analyses have been undertaken on core GH.A.92, to provide a full Holocene sequence, and on short-core GH.E.93-6, including fine-interval (0.5-1.0 cm) pollen and diatom analysis spanning the tephra horizon. Pollen and diatoms are not preserved below ca.700 cm in core GH.A.92, and diatoms are also absent between 500 and 350 cm. The summary pollen diagram from GH.A.92 (Fig. 3) covers the last ca. 8400 ^{14}C years, and indicates pine woodland with some oak and juniper prior to the ash horizon, and a Beyşehir Occupation phase with walnut, olive,

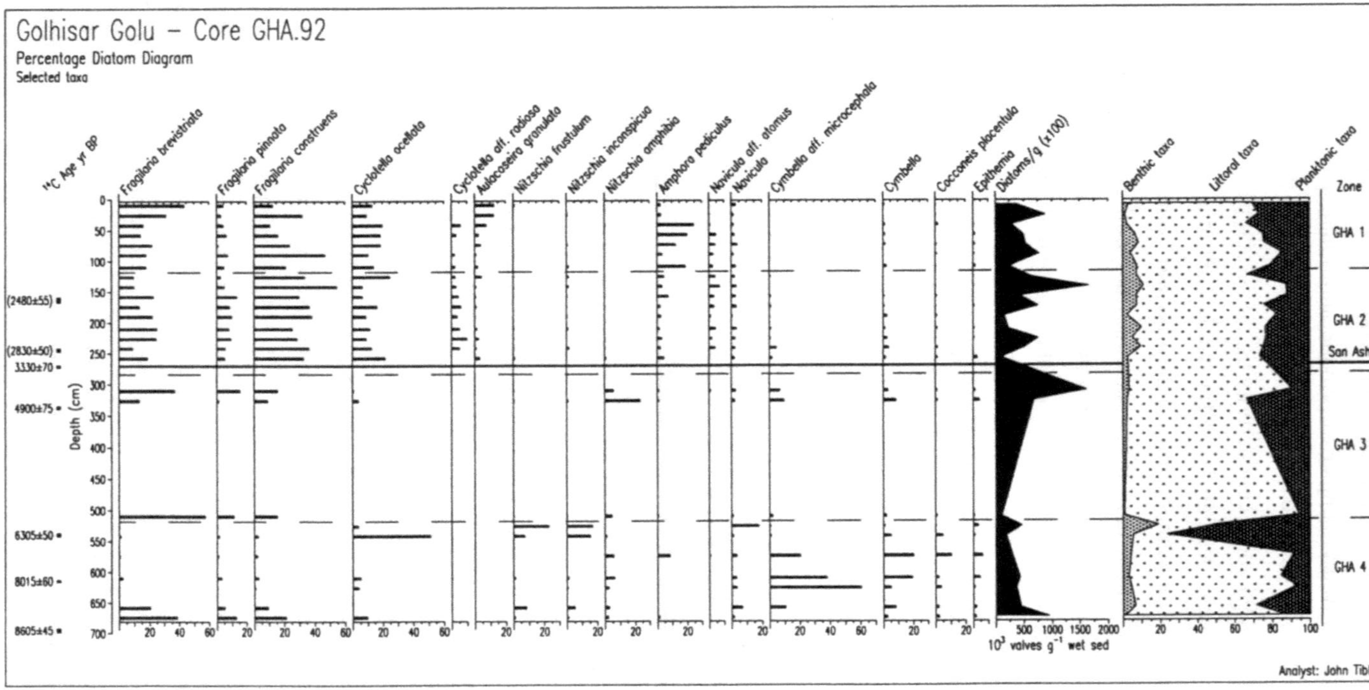

Figure 4: Summary diatom diagram for core GH.A.92 from Gölhisar.

pistachio, *Plantago lanceolata* and "Cerealia"-type pollen, occurring above it. A sharp peak in the herb *Artemisia* occurs slightly below the tephra layer.

A precise definition of the start and end of the Beyşehir Occupation phase is needed to determine its relationship to archaeological evidence for human occupation, and to the ash layer. Juglans (walnut) pollen may be a good indicator, for several reasons. It first occurs in contiguous levels at or just above the base of the Beyşehir Occupation phase in all diagrams, except that of Beyşehir Gölü itself where it is absent from three levels. Similarly, the end of the Beyşehir Occupation phase at all sites is marked by a decrease of Juglans to scattered occurrences of less than 1%, and an increase in arboreal pollen. Juglans is a primary anthropogenic indicator; that is, its pollen is derived from cultivated trees, and it is represented by just one species (J. regia) in Anatolia (Bottema 1980). Other reasons for choice of Juglans include the fact that its pollen is readily identifiable and that it is almost certainly under-represented in pollen diagrams; its presence therefore represents local cultivation. Olea or Fraxinus ornus could also be used as indicators of the Beyşehir Occupation phase, although pollen of these taxa is less clearly of local origin.

If Juglans were to be used as an indicator, the Beyşehir Occupation phase at Gölhisar Gölü would start at a depth of 220cm in core GH.A.92 (ca. 2700 ^{14}C yr BP, ca. 850 BC). On the other hand, other probable cultigens such as Olea and Pistacia appear lower in this pollen diagram (259 cm; ca. 3050 BP, ca. 1300 BC), suggesting that human-induced landscape change started during the later second millennium BC. Significantly, neither of the two levels for the onset of the Beyşehir Occupation phase was precisely coincident with the tephra layer, but rather occurred some time later, by at least a century according to the radiocarbon dates. This phase also includes a ^{14}C date of 2480 ± 55 BP (CAL age) at 158-167 cm. This sample had a low carbon content, and would be particularly susceptible to inwash of "old" soil carbon from cultivated fields surrounding the lake, so its ^{14}C age may be some centuries too old. The Beyş ehir Occupation phase at Gölhisar Gölü ended at a depth of ca. 80cm (ca. 1250 BP, ca. 780 AD, although no ^{14}C dates are directly associated with this level).

The diatom stratigraphy is dominated by freshwater, periphytic or shallow planktonic taxa throughout the Holocene (Fig. 4). In some species (e.g. *Cyclotella ocellata*) the preferred habitat is variable, so firm attribution to particular life form

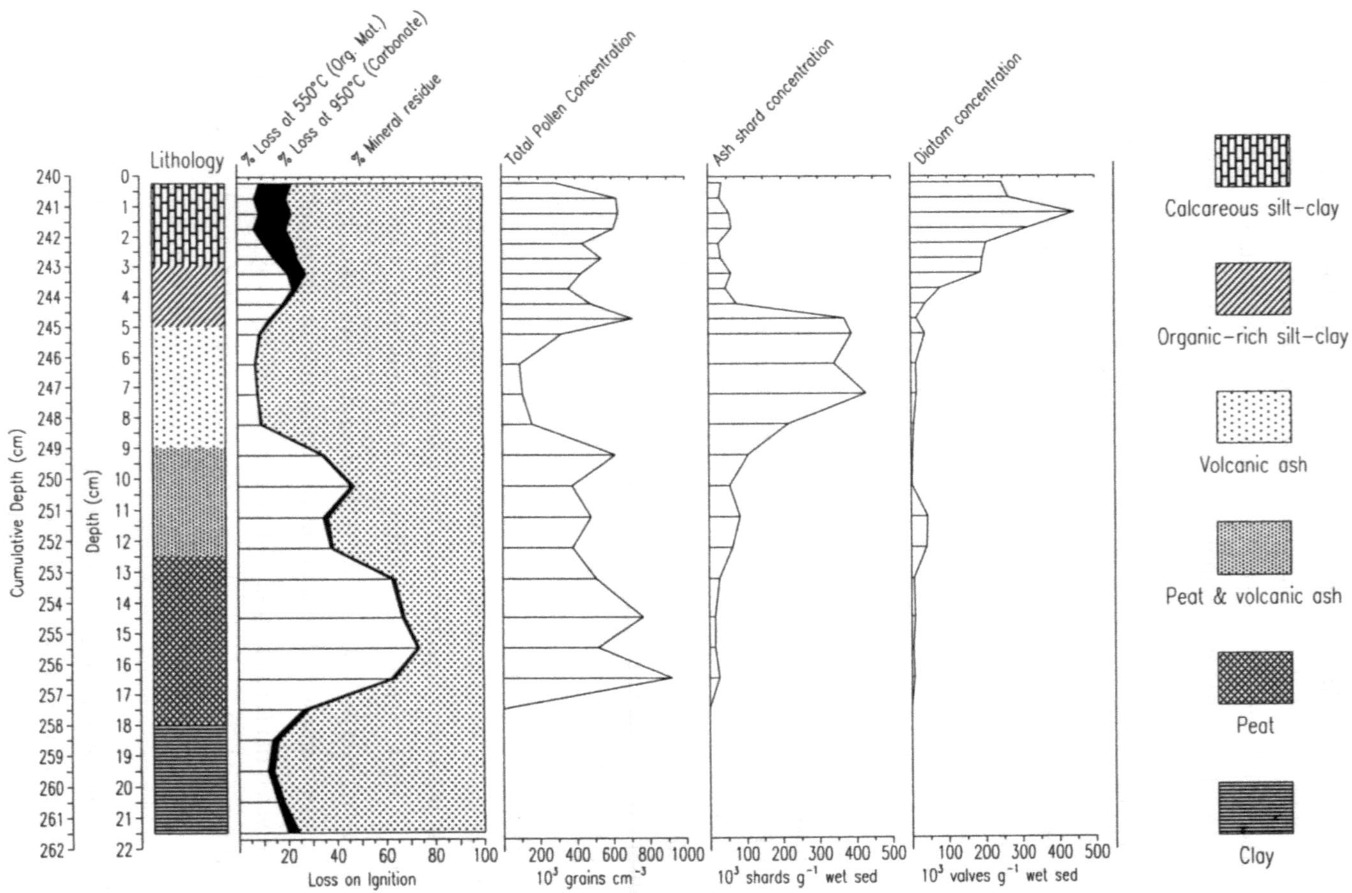

Figure 5: Lithology, Loss-on-Ignition and concentation data for Gölhisar short core GH.E.93-6.

classes is difficult. In general, centric diatoms such as other *Cyclotella* spp. and *Aulacoseira granulata*, are likely to indicate deeper water conditions than *Fragilaria* spp, *Cymbella* spp. and *Amphora pediculus*. In core GH.A.92, a significant stratigraphic change occurs immediately above the Santorini ash layer. Below this, diatom preservation and species composition are both highly variable between levels; above this, diatom preservation is better and species assemblages are more homogeneous. A switch to a *Fragilaria*-dominated assemblage is clearly identified at ca. 275 cm in the CONISS cluster analysis dendrogram. In short core GH.E.93-6 volcanic shard and diatoms were counted from the same slides. The ash shards show a clear peak between 4.0 and 8.5 cm depth representing the main ash layer (Fig. 5). Above this, there is a large increase in overall diatom frustule abundance to values in excess of those found elsewhere in the full Holocene core sequence (>400 x 10^3 valves per cm^3 wet sediment, maximum values elsewhere in core 170 x 10^3 per cm^3, general Holocene range 20-100 x 10^3 per cm^3). A sustained increase in non-silicious aquatic algae (*Pediastrum, Coelastrum*) also occurs at the Santorini ash horizon (Fig. 3).

Discussion

The sequence of events at Gölhisar revealed by palaeoecological analyses can be used to test alternative hypotheses about the causes and consequences of mid-Holocene ecosystem disturbance in southwest Turkey. Specifically, they help to separate the effects of the Santorini eruption from those of human impact. The ash horizon at Gölhisar correlates with a sharp increase in diatom abundance and a change in diatom species composition, suggesting that ash deposition had an effect on the lake ecosystem. The main "beneficiaries" were small *Fragilaria* species, and in particular *F.brevistriata*, which has been found to increase in some other lakes during periods of increased water turbidity and catchment disturbance (Barker et al. 1994, Metcalfe 1995). The fact that the increase in diatom abundance also coincides with a marked expansion in other, non-siliceous algae suggests that ash deposition had a fertilisation effect on the lake ecosystem. Although the greater numbers of diatoms may - in part - be attributable to greater availability of silica, other nutrients must also have been involved. Significantly, algal productivity seems to have been maintained at higher levels throughout the remainder of the core, either because of

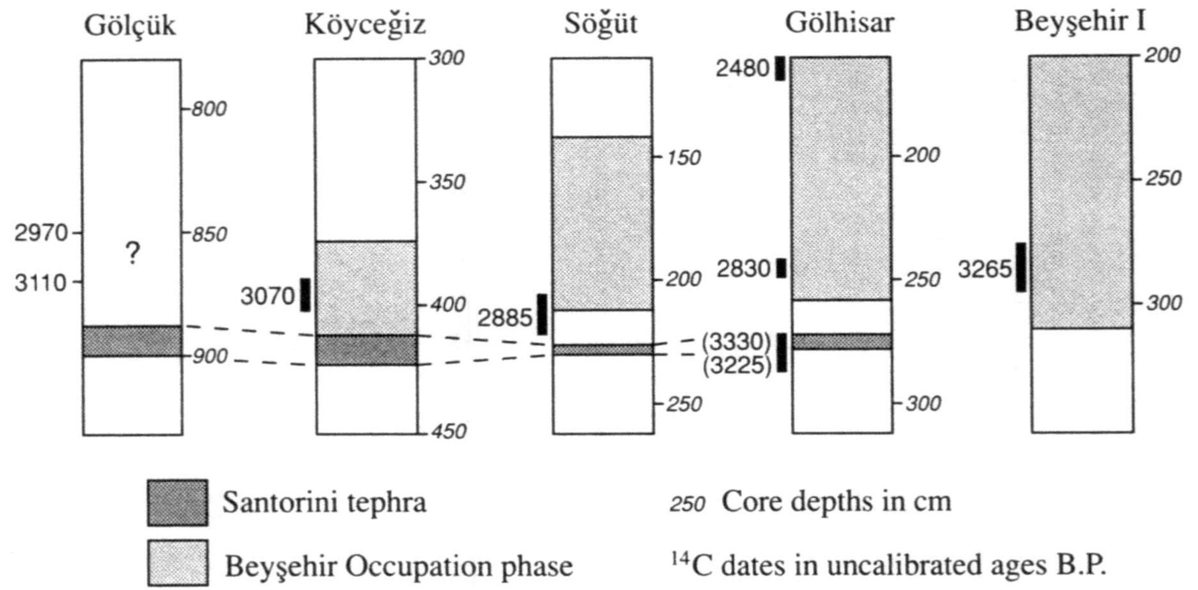

Figure 6: Comparison of ^{14}C dates and stratigraphic relations for the onset of the Beyşehir Occupation phase and the Santorini tephra in southwest Turkey. ^{14}C dates in parentheses are positioned in correct relation to the tephra rather than core depth. See Table 1 for further details.

other catchment-related changes (e.g. associated with the Beyşehir Occupation phase) and/or because the volcanic ash led to a permanent change in the catchment nutrient budget. The presence of *Aulacoseira granulata* above the ash layer supports the hypothesis of nutrient enrichment in the lake.

Our conclusion from the dryland pollen record at Gölhisar, however, is that the tephra horizon resulting from the Santorini eruption was not precisely synchronous with the appearance of disturbance indicators associated with the onset of the Beyşehir Occupation phase. The main evidence of cultural activity in the pollen diagram occurs above 259 cm and, in particular, between 220 and 80 cm, associated with pollen indicative of tree cultivation.

^{14}C dates associated with the onset of the Beyşehir Occupation phase and/or the Santorini ash horizon from Turkish lake sediment cores are shown in Table 1 and Figure 6. These indicate excellent agreement for the age of the tephra at around 3350-3200 BP (1650-1450 BC). Although this is consistent with the date of 1650-1600 BC suggested from ice cores and treerings, the calibrated age range is not sufficiently precise to confirm it. Intriguingly, dates for the start of the Beyşehir Occupation phase show rather less consistency, ranging from before 3265 BP (1590-1510 BC) to at, or after, 2885 BP (1117-998 BC). While these may possibly lie within the range of sampling, counting and calibration error, it also suggests that the start of this occupation phase was not synchronous across southwest Turkey, reinforcing evidence for a cultural cause.

Conclusions

Lake sediment archives provide important data on both human-induced and naturally-occurring palaeoenvironmental change. One such change is the Beyşehir Occupation phase recognised from pollen diagrams in southwest Turkey, whose onset is ^{14}C dated to ca. 3000 BP (1250 BC). This approximate date for the onset of the cultural phase has recently been confirmed by the discovery of volcanic tephra from the Minoan eruption of Santorini in several lake sediment sequences. The Santorini ash occurs just before the onset of the Beyşehir Occupation phase, and forms a time-synchronous, stratigraphic marker horizon. Tephrochronological and radiocarbon dates from five sites in southwest Turkey, however, suggest that that the onset of the Beyşehir Occupation phase may have been time-transgressive, spanning a time period of between 200 and 500 years in the mid-late second millennium BC, that is, during the Late Bronze Age.

Cores from Gölhisar gölü contain an ash horizon up to 4.5 cm thick and ^{14}C dated to 3300-3200 BP (ca. 1650-1450 BC). Characterisation of the volcanic glass shards by ICP-MS and electron micro-probe analysis confirms that this tephra derived from the Santorini eruption. Palynological and diatom analysis, including fine-interval sampling of short cores spanning the tephra horizon, have shown a clear and sustained change in the aquatic ecosystem. The onset of pollen-stratigraphic changes marking the beginning of the Beyşehir Occupation phase was not precisely coincident with the tephra layer, but occurred somewhat later. Gölhisar was inside the territory of the classical city of Kibyra, and a fortified archaeological site (Sinda) lies on a promontory in the lake. This proximity permits comparison of palaeoecological and archaeological data within the same catchment ecosystem. Sinda probably had a main period of occupation from the eighth century BC to the eighth century AD, which matches rather well the main period of cultural disturbance shown in the pollen record from Gölhisar Gölü (estimated ages 850 BC - 780 AD). To judge by evidence from Kibyra and from the adjacent city state of Balboura, agricultural activity may have reached a peak in the Late Hellenistic to Late Roman periods (ca.200 BC to ca.AD 450). Gölhisar lies at a lower elevation than Balboura, however, and the settlement history of the two areas may have been somewhat different. Settled agriculture would have been less marginal around Gölhisar, and the presence of an early Ottoman sanjak in the lake catchment suggests that the area may not have been abandoned completely in the Byzantine period (J.J. Coulton pers. comm.). Archaeological evidence for significant human occupation during the second millennium BC remains elusive, on the other hand. This partial accordance with pollen data suggests that important Bronze Age settlement may be going unrecognised in the archaeological record, perhaps because of cultural overprinting by later, more substantial, occupation phases.

It has been postulated by archaeologists such as Marinatos that the Santorini eruption played a major role in the decline of the Minoan civilization of Crete. However, it is also possible that the disturbance to the landscape and atmospheric influx of nutrients resulting from the eruption were subsequently exploited opportunistically by Late Bronze Age agricultural populations. Intriguingly, there is documentary evidence from Herodotus and from Hittite records for the Lycians (the

"Lukka") originating on Crete and moving into southwest Anatolia around 1450-1250 BC (Bean 1978). In terms of chronology, this would be in good accord with the dating from palaeoecological sequences presented here, although it should be noted that the Beyşehir Occupation phase was not restricted to Lycia, but also extended northwards and eastwards into the adjacent districts of Caria, Pisidia and Lycaonia. Even if precise relationships remain unclear, the Gölhisar data confirm that a major period of socio-economic change and human impact on the landscape occurred in southwest Turkey during the centuries following the Thera eruption.

Acknowledgements

Dr Phil Barker and Dr Carol David were an essential part of the fieldwork programme. We are very grateful to Dr Jim Coulton for providing the opportunity to work alongside the Balboura archaeological survey team, and for helpful comments on this manuscript. Permission for the fieldwork was kindly granted by the General Directorate of Monuments and Museums. The field research programme received the financial support of Royal Society, Royal Geographical Society and Loughborough University grants to NR. WJE acknowledges the support of a University of Wales studentship. Most of the radiocarbon dates from Gölhisar reported here were carried out at the NERC laboratory at East Kilbride, to which we are most grateful. Thanks are also due to Dr Sytze Bottema, and to the British Institute of Archaeology at Ankara, and its then director Dr David French.

References

Baillie, M.G.L. and Munro, M.A.R. (1988) Irish tree rings, Santorini and volcanic dust veils. *Nature* 332: 344-346.

Barker, P.A., Roberts, N., Lamb, H.F., van der Kaars, S. and Benkaddour, A. (1994) Interpretation of lake-level change from diatom life form in Lake Sidi Ali, Morocco. *Journal of Paleolimnology* 12: 223-234.

Bean, G.E. (1978) Lycian Turkey. John Murray, London.

Behre, K.-E. (1990) Some reflections on anthropogenic indicators and the record of prehistoric occupation phases in pollen diagrams from the Near East. In: S. Bottema, G. Entjes-Nieborg and W.van Zeist (eds) Man's role in the shaping of the eastern Mediterranean landscape. Balkema, Rotterdam, 219-30.

Berglund, B.E. (ed.) (1991) The Cultural Landscape during 6000 years in southern Sweden. *Ecological Bulletins* 41

Bottema, S. (1980) On the history of the walnut (Juglans regia L.) in southeastern Europe. *Acta Botanica Neerlandica* 29: 343-349.

Bottema, S. and Woldring, H. (1984) Late Quaternary vegetation and climate of southwestern Turkey II. *Palaeohistoria* 26: 123-149.

Bottema, S. and Woldring, H. (1990) Anthropogenic indicators in the pollen record of the Eastern Mediterranean. In: S. Bottema, G. Entjes-Nieborg and W. van Zeist (eds) Man's role in the shaping of the eastern Mediterranean landscape. A.A.Balkema, Rotterdam, 231-264.

Coulton, J.J. (1988) Balboura survey. Annual report of the British Institute of Archaeology at Ankara 40: 13-15.

Coulton, J.J. (1993) North Lycia before the Romans. In: Borchhardt, J. and Dobesch, G. (eds) Akten des II. Internationalen Lykien-Symposions. Österreichische Akademie der Wissenschaften. 79-85.

Eastwood, W.J., Pearce, N.J.G, Perkins, W.T., Lamb, H.F, Westgate, J. A. and Roberts, C.N. (in press) Geochemistry of Santorini tephra in lake sediments from southwest Turkey. *Quaternary Science Reviews.*

Entwistle, J. A., Lamb, H. F., Abrahams, P.W. and Dodgshon, R. A. (1995) A lake-sediment record of changing land use intensity on the Isle of Lewis, Scotland. In: Butlin, R.A. and Roberts, N. (eds.) Ecological relations in historical times: human impact and adaptation. Blackwell, Oxford, 45-67.

Guichard, F., Carey, S., Arthur, M. A., Sigurdsson, H. and Arnold, M. (1993) Tephra from the Minoan eruption of Santorini in sediments of the Black Sea. *Nature* 363: 610-612.

Hall, A. (1994) Sinda. In: French, D. (ed.) Studies in the history and topography of Lycia and Pisidia. In memoriam. A.S. Hall. British Institute of Archaeology at Ankara Monograph 19, 48-52.

Hammer, C.U., Clausen, H.B., Friedrich, W.L. and Tauber, H. (1987) The Minoan eruption of Santorini in Greece dated to 1645 BC? *Nature* 328: 517-519.

Huntington, E. (1915) Civilization and climate. Yale University Press, New Haven

Kuniholm, P.I., Kromer, B, Manning, S.W., Newton, M., Latini, C.E. and Bruce, M.J. (1996) Anatolian tree rings and the absolute chronology of the eastern Mediterranean, 2270-718 BC. *Nature* 381: 780- 783.

Kutzbach, J.E. (1987) The changing pulse of the monsoon. In Frey, J.S. and Stephens, P.L. (eds) Monsoons. J.Wiley, 247-68.

LaMarche, V.C. and Hirschboek, K.K. (1984) Frost rings in trees as records of major volcanic eruptions. *Nature* 307: 121-126.

Lamb, H.F., Gasse, F., Benkaddour, A., El-Hamouti, N., van der Kaars, S., Perkins, W.T., Pearce, N.J. and Roberts, C.N. (1995) Relation between century-scale Holocene arid intervals in tropical and temperate zones. *Nature* 373: 134-137.

Manning, S. (1990) The eruption of Thera: Date and implications. In: Hardy, D. A. and Renfrew, A. C. (eds.), Thera and the Aegean World III: Volume Three - Chronology. Proceedings of the Third International Congress, Santorini, Greece. London, The Thera Foundation, pp 29-40

Mellaart, J. (1954) Preclassical remains in southern Turkey. *Anatolian Studies* 4: 175-240.

Metcalfe, S.E. (1995) Holocene environmental change in the Zacapu basin, Mexico: a diatom-based record. *The Holocene* 5: 196-208.

Porter, S.C. (1981) Recent glacier variations and volcanic eruptions. *Nature* 291: 139-141.

Rackham, O. (1980) Ancient woodland: its history, vegetation and uses in England. London, Edward Arnold

Roberts, N. (1990) Human-induced landscape change in south and south-west Turkey during the later Holocene. In: S. Bottema, G. Entjes-Nieborg and W. van Zeist (eds) Man's role in the shaping of the Eastern Mediterranean landscape. A.A. Balkema, Rotterdam, 53-67

Roberts, N. (1993) Sub-Milankovitch palaeoclimatic events: their recognition and correlation. *Climatic Change* 24: 175-178.

Roberts, N. and Butlin, R.A. (1995) Ecological relations in historical times: an introduction. In: Butlin, R.A. and Roberts, N. (eds.) Ecological relations in historical times: human impact and adaptation. Blackwell, Oxford, 1-14.

Street-Perrott, F.A. and Roberts, N. (1983) Fluctuations in closed basin lakes as an indicator of past atmospheric circulation patterns. In F.A. Street-Perrott, M. Beran and R.A.S. Ratcliffe (eds.) Variations in the global water budget. Dordrecht, Reidel, 331-345.

Stuiver, M. and Reimer, P.J. (1993) *Radiocarbon* 35: 215-230; CALIB rev. 3.0.3.c (Department of Geological Sciences, University of Washington, Seattle, 1994).

Sullivan, D.G. (1988) The discovery of Santorini Minoan tephra in western Turkey. *Nature* 333: 552-4.

Sullivan, D.G. (1990) Minoan tephra in lake sediments in Western Turkey, dating the eruption and assessing the atmospheric dispersal of the ash. In: Hardy, D. A. and Renfrew, A. C. (eds.), Thera and the Aegean World III: Volume Three - Chronology. Proceedings of the Third International Congress, Santorini, Greece. London, The Thera Foundation, pp 114-118.

Watkins, N.D., Sparks, R.S.J., Sigurdsson, H., Huang, T.C., Federman, A., Carey, S. and Ninkovich, D. (1978) Volume and extent of the Minoan tephra from Santorini Volcano: new evidence from deep-sea cores. *Nature* 271: 122-126.

White, R.S. and Humphreys, C.J. (1994) Famines and cataclysmic volcanism. *Geology Today*, September-October: 181-185.

van Zeist, W., Woldring, H. and Stapert, D. (1975) Late Quaternary vegetation and climate of southwestern Turkey. *Palaeohistoria* 17: 55-143.

Bronze Age Regression and Change of Sedimentation on The Aegean Coastal Plains of Anatolia (Turkey)

İlhan KAYAN

Ege University
Department of Geography
35100 Bornova, İzmir
Turkey

Abstract

Borehole studies indicate three stratigraphical units in the Holocene sediments on the Aegean coastal plains of Anatolia. These are early Holocene shallow marine-estuarine muds at the bottom, fine sandy-silty flood-plain sediments at the top, and a transition layer between them. The transition layer, a few meters thick and a few meters below the present sea-level, generally consists of the coarse sandy sediments of shallow coastal environments. C^{14} datings show that the change in sedimentation occurred during the third millennium BC In addition, relative sea-level fell a few meters during the same period, which accelerated deltaic progression. The transition layer has similar characteristics all along the coastal plains investigated, indicating that the change in sedimentation is a result of normal geomorphological development dependent on sea-level changes; local or regional tectonic movements and sediment compactions are not important here. Climatic-ecstatic sea-level changes may be considered the best explanation for the sedimentation change.

Introduction

Geomorphological-sedimentological data obtained from our long-term research on various alluvial plains of the Aegean coast of Anatolia, Turkey, have been sufficient for some general interpretations. Investigations were mainly concentrated on Troia, Ephesos-Klaros and Old Knidos (Fig. 1). These results were supported by preliminary evidence from less comprehensive studies at Altınova, İzmir and Miletos. Archaeological locations were chosen as research areas in order to use archaeological evidence for chronology and to elucidate human-environment relationships. Geological structures and geomorphological characteristics of the research areas were investigated first. Then, subsurface sedimentation units and their physical

NATO ASI Series, Vol. I 49
Third Millennium BC Climate Change
and Old World Collapse
Edited by H. Nüzhet Dalfes, George Kukla,
and Harvey Weiss

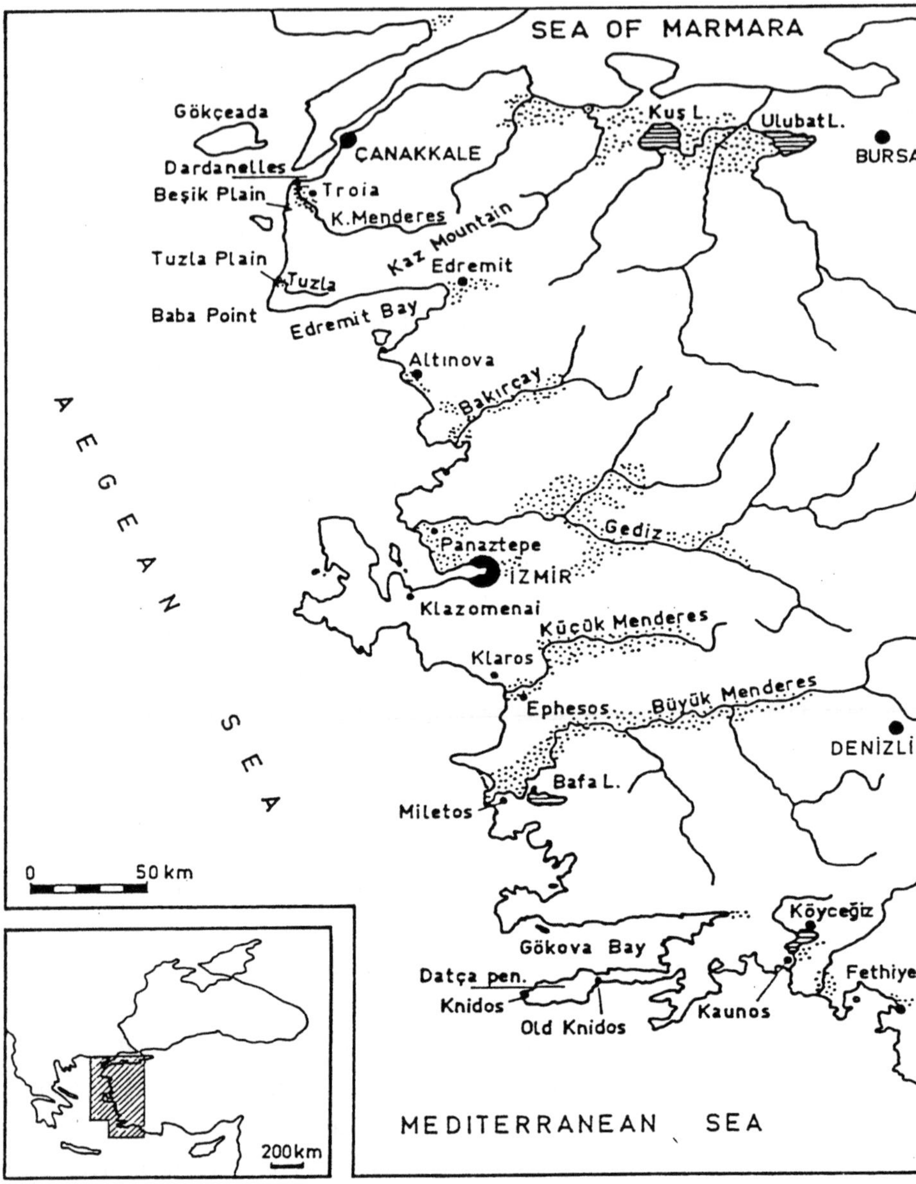

Figure 1. Aegean coastal region of Turkey and locations of research areas. Dotted areas indicate major alluvial plains.

environments were determined by borehole studies on the alluvial coastal plains. These units were dated using radioactive C^{14} datings and archaeological evidence. Thus, paleogeographical reconstruction maps could be drawn of the different stages of the geomorphological evolution of the area.

The most outstanding geomorphological feature of the Aegean coast of Anatolia is its very indented shape (Fig. 1). This is mainly related to the geological structure of the region. Structural units forming Western Anatolia are various. The Menderes massif functioned as a core at the base in the bordering alpine-type mountain belts. In the youthful Neotectonic period, this region was uplifted while the Aegean and Mediterranean basins subsided. During this time, the Menderes massif itself was cut by faults running mostly in an east-west direction, and bordering younger marginal structural units were segmented depending on their own structural lineations. Volcanism which occurred in relation to this break diversified the structure. The Aegean is thus influenced by many fault blocks extending in various directions (west-east in the Menderes massif, the general north-south trend of the Aegean with its NW-SE and NE-SW faults).

Depressions, especially tectonic troughs (grabens) of the Aegean coastal region have become sedimentation basins for the material eroded from the rising mountain blocks (horsts). Clearly the sedimentation in the basins occurred under control of the paleogeographical conditions, and both continuing young tectonic movements and changing climatic conditions formed new landform systems in the inner parts of the basins. For example, there are Neogene lacustrine sediments in most of the depressions. Slightly inclined surfaces and dissected hilly areas of these sediments rise over the alluvial floors of the depressions. The youngest depressions which were formed by newer faults cutting the Neogene formations have been filled by fluvial sediments.

It is generally accepted that the tectonic depressions which had previously formed between mountain blocks were occupied by rivers during the Quaternary period, thus shaping present morphology. This means that the great valleys of the region were not primarily cut by river erosion, but rather were shaped by surface waters flowing into tectonic depressions to form the present drainage systems. Therefore the main rivers of the region pass through various depressions that have different geomorphological developments. Generally each of them has an inner basin occupying a graben in which wide flood plains have formed by alluvial deposition.

These rivers pass through gorges near the sea in the west, and then reach the Aegean Sea over delta plains. The Gediz, Küçük Menderes, Bakırçay and Karamenderes rivers in the north are typical examples of this model. This is an important factor affecting the nature and amount of alluvium deposited over the coastal plains.

Another effect on recent the geomorphological formation of the Aegean coast of Anatolia is the Holocene transgression. Rapidly rising sea-level in the post-glacial period emphasized the zigzag features which already existed. Thus, intrusion of the sea toward the coastal embayments, grabens and the mouths of rivers formed bays and estuaries of different sizes and shapes. Horizontal rates of change of coastline depended on the relation between the topographical inclination, the rate of sea-level rise and the volume of alluvium reaching the coast (Kayan 1988 b).

Although the features originated from the young tectonic movements are prominent in the geomorphology of the Aegean region, there is no clear evidence indicating any notable effect of tectonic movements on the Holocene coastal formations. This is an unexpected characteristic for such a region which is frequently shaken by earthquakes even today. This can be explained in two ways: either amplitude of the Holocene tectonic movements is not great enough to be clearly seen in the morphology in spite of the present expression of earthquakes, or the whole region has been in a continual state of subsidence and old geomorphological traces, for example old coastlines, cannot be seen because they have been drowned by the sea or alluvium. Covering of the Early Holocene coastlines by the sea or alluvium is normal because of eustatic sea-level rise and dependent alluvial drowning. There is sedimentological and C^{14} dating evidence indicating that the rising sea in the post-glacial period first reached its present level 6000 years B.P. In addition, no evidence has so far been found in any place that the Holocene sea-level on the Aegean coast of Anatolia was ever higher than at present. On the contrary, sedimentological and C^{14} dating evidence also indicate that the relative sea-level fell in the period between 5000 and 4000 years B.P. (Bronze Age regression) and it was a few meters below the present (generally about 2 m) in the period between 4000-3000 B.P. (the Third Millennium BC). Later it rose again and was almost at its present level in Roman times, about 2000 years ago (Fig. 2).

Accordingly, the relative sea-level has been very close to the present level for the last 6000 years, that is since Neolithic times, and it has not changed greatly except for a small fall and rise between 5000 and 2000 B.P. However, it is clear that the

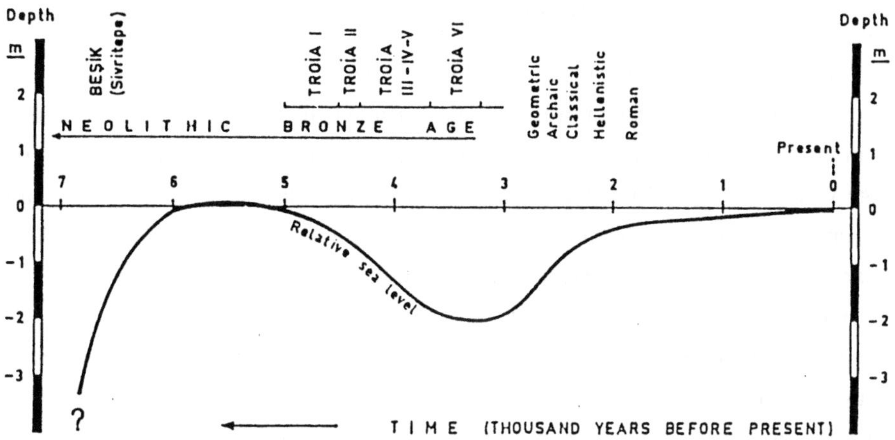

Figure 2. Relative sea level changes along the Aegean coast of Turkey during the last 7000 years (after Kayan 1991). The curve is plotted based on the sedimentological and archaeological evidence, including C14 dates, obtained from the Beşik plain on the coast to the west of Troia.

coastline has changed horizontally over great distances during the last 6000 years and that human activity has been influenced by this change since Neolithic times. These changes have generally been at the mouths of fairly large rivers and depended on alluvial sedimentation. Therefore sedimentological data obtaining by borehole studies over the alluvial coastal plains are very important.

This paper concentrates on the relation between the alluvial sedimentation and coastline changes on the Aegean coastal plains of Anatolia during the last 6000 years. In this development, the small sea-level fall in the Third Millennium BC and its effects on the sedimentation and coastline changes are especially noticeable. There is not yet enough evidence to explain this change by climate changes or to make a discussion on this relation. However, data obtained from about 200 boreholes over the alluvial plains surrounding Troia (Troy) clearly showed that the alluvial sedimentation changed during that period. Therefore, sedimentological and stratigraphical characteristics of the Troian alluvial plains will be introduced below first, then the results will be compared with the other investigation areas along the Aegean coast of Anatolia.

Alluvial stratigraphy of the Karamenderes - Dümrek flood plain surrounding Troia

The Karamenderes River starts from the northern slopes of the Kazdağı (Kaz Mountain, 1774 m) in the south of the Biga peninsula (Fig. 1). It forms a wide alluvial plain in the Bayramiç-Ezine Neogene basin. Then it turns to the north, passes through a gorge which is formed by antecedent river incision on the old geological base, and opens on to a flood-delta plain extending toward the south of the Çanakkale Strait (Dardanelles) (Fig. 3). The Karamenderes flood-delta plain is formed between low ridges about 40-80 m high which consist of various strata of shallow marine Neogene sediments. The plain is separated from the Aegean Sea by a narrow ridge in the west. Although the edges of the plain are indented between the ridges, the average width is 3-4 km, and their length is about 17 km. The Karamenderes River passes through the plain and reaches the sea at the southern end of the Çanakkale Strait. In the lower part, the Dümrek river, which comes from the east as a large tributary, has affected the formation of the delta by depositing coarser sediments (Kayan 1995 a).

Troia is one of the oldest and most extensively examined archaeological sites of the Western Anatolia. Its known history goes back to 5000 years B.P. It is situated to the east of the Karamenderes flood-delta plain and about 4 km south from the present coast. It occupies a platform about 30-35 m high at the western tip of a Neogene ridge on the south side of the Dümrek valley (Fig. 3).

In our first study of the alluvial plains of the Karamenderes and Dümrek rivers surrounding Troia, 7 boreholes were made down to a depth of 35-70 m. The data obtained from these boreholes showed that rapidly rising sea-levels during the post-glacial period intruded into a deeply incised valley at the mouth of the Karamenderes River and started to form an estuarine bay about 10,000 years B.P. The valley bottom and river course of this time were found about 50 m below the present surface in the middle part of the present plain (Fig. 4). Later, the sea extended up to the vicinity of Pınarbaşı in the south and a long estuary formed in the area of the present plain (Kraft et al. 1980, 1982).

Later, a new research project on the alluvial geomorphology of the Beşik plain was made in conjunction with a German archaeological expedition led by Prof. Korfmann. The Beşik plain is formed in a small structural indentation on the Aegean coast, west of the Karamenderes plain (Fig. 3). There is no important stream reaching the sea here, and sedimentation has been dependent mainly on marine dynamics. Therefore

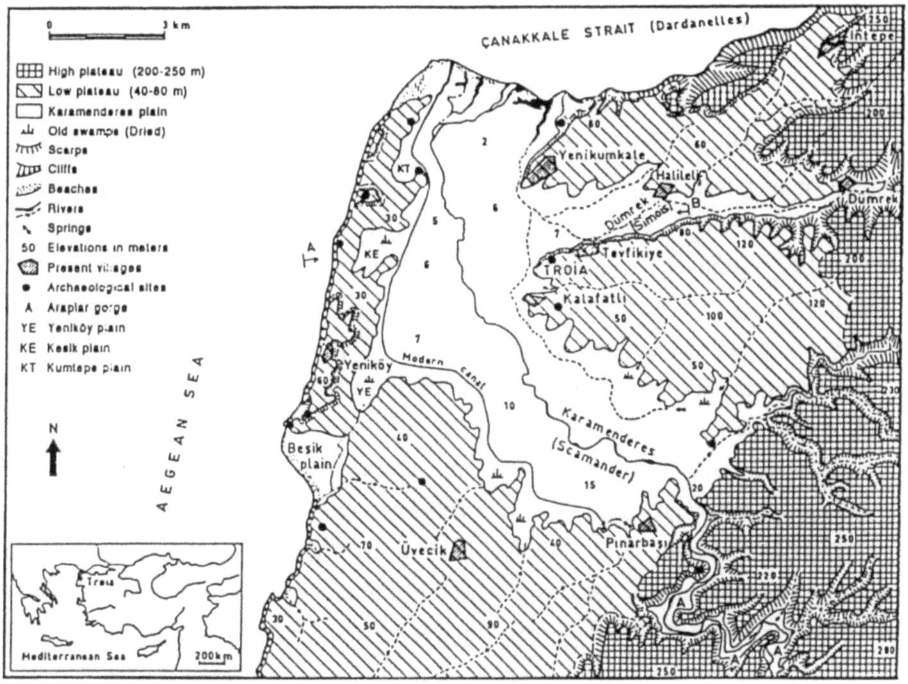

Figure 3. Geomorphological outlines of the Karamenderes-Dümrek flood-delta plain and its environs. **A** and **B** are the end points of cross-section given on Figure 4.

geomorphological development of the plain is rather simple. Hand-operated boreholes (76 in all), about 8 m deep, were made over the plain, which covers only about 3-4 km^2. C^{14} samples taken from appropriate levels were dated by Dr. Kromer at the University of Heidelberg. This research led to some important conclusions about relative sea-level changes in this area during the Holocene. It was shown that the rising sea in the post-glacial period first reached the present level 6000 years B.P. After remaining near that level for about a millennium, it fell about by 2 m between 5000 and 4000 B.P. There is evidence indicating that the relative sea-level again reached its present position around the time of Christ (Fig. 2) (Kayan 1991).

A new archaeological research, excavation and restoration project was started at Troia in 1988 under the leadership of Prof. Korfmann. By joining the project, we were able to make more than 100 boreholes over the Karamenderes and Dümrek plains to examine in detail subsurface sedimentary units down to 20 m. Some new interpretations were made based on sedimentological evidence and chronological data obtained by C^{14} datings and from various archaeological finds (Kayan 1995 a). As a result, it is also confirmed here that the sea, which intruded into the Karamenderes-Dümrek valley, first reached its present level 6000 years ago. The coastline of that time can be recognized by coarse sandy coastal sediments with plenty of shells below the recent alluvium along the present inner edge of the plain. This coastline has never been detected above its present sea-level in any place.

Another noticeable feature of the sedimentation is that the marine muds are regularly encountered a few meters below the present sea-level in the middle part of the present plain. This showed that, the sea retreated from the Karamenderes-Dümrek estuary over a surface about 2 m below the present sea-level, and then the marine sediments were covered by delta-flood plain sediments. There is a coarse sandy transition layer between these two sedimentary units from different environments (Fig. 4). This layer is easily recognizable everywhere. The upper and lower boundaries of the layer are variously clear or gradual from place to place. There are many marine shells and much decayed plant material in the transition layer. This indicates that the layer consists of coastal sediments of the retreating sea. It can be understood that coastal barriers and lagoonal features did not develop on the coast of the retreating sea in the Karamenderes valley. Instead, a swampy, bird's-foot type ("azmak" in Turkish) delta coast was dominant. The very regular flat shape of the upper surface of the coarse sandy transition layer indicates that the formation of this unit occurred in conditions where the water was shallow and had very low wave energy. However, the undulating lower boundary of the coarse sandy transition layer at the mouth of the Dümrek river indicates that some coastal barrier and lagoon formations behind them are possible.

C^{14} dates of the samples taken from transitional coastal sediments which contained organic material showed that the coastline about 6000 B.P. was located on a line from Troia toward the southwest. Later, about 5000 B.P., the coastline was almost to the west of Troia (Fig. 3). This means that Troia was a coastal settlement during those times. On the other hand, coarse sandy alluvium from the Dümrek river was forming a different deltaic sedimentation at its mouth to the north of Troia. In fact, the coarse sandy sediments of the Dümrek River cover the blackish, fine-grained marine sediments about 2 m below the present sea-level (Fig. 4).

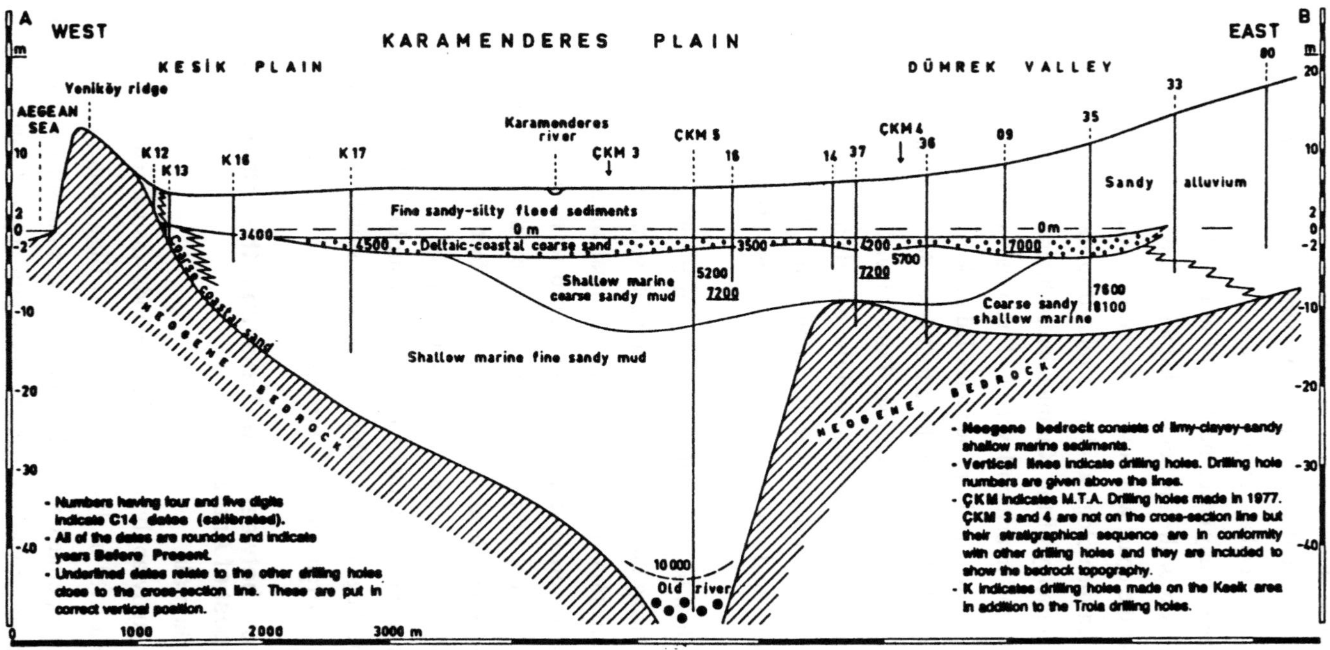

Figure 4. A-B cross-section of the Karamenderes-Dümrek delta-flood plain between the Aegean Sea in the west and the Dümrek valley in the east (see Figure 3).

According to this evidence, the sea, which had risen up to the present level by 6000 B.P., slowly fell about 2 m in the period between 5000 and 4000 B.P. (the Third Millennium BC). Deltaic progradation will have been accelerated by this small sea-level fall when, in any case, the estuary at the joint mouths of the Karamenderes and Dümrek rivers was already being filled with alluvium after sea-level rise had stopped. It seems that the coast was covered by swamps during that time. But in the following period, no sedimentological change related to the small sea-level rise is clear on the alluvial stratigraphy of the Karamenderes-Dümrek delta-flood plain. Flood sediments always cover the transition layer of -2 m. This indicates that, sea water was not able to intrude toward the Karamenderes-Dümrek valley again during the last sea-level rise of 3000-2000 B.P. This is because the sea-level rise was much smaller (about only 2 m) and slower than the first rise. Thus, alluvial deposition along the shallow coast compensated for the small sea-level rise and deltaic progradation continued toward the Çanakkale Strait. However, it can be stated that the small sea-level rise affected the level of ground water and coastal swamps survived for longer and over a wider area.

As a result, it is obvious that the environment changed from marine to fluvial in the Troia area during the Third Millennium BC It seems that the change in sedimentation and the formation of different stratigraphical units below the Karamenderes and Dümrek plains surrounding Troia are related to normal geomorphological processes and accelerated deltaic progradation caused by the small sea-level fall in the Bronze Age. There is no clear evidence of any effect of other factors, such as climatic changes, on the formation and sequence of the different sedimentary units. On the other hand, similar contemporaneous sedimentological changes were found in other plains of the Aegean coast. This may not be a simple coincidence, but may be related to regional environmental occurrences.

Comparison between the Troia and the other coastal plains of the Aegean coast of Anatolia

The Tuzla plain

Another study applying a similar method, was carried out on the flood-delta plain of the Tuzla river in the southwest of the Biga peninsula (Figs. 1 and 5) (Kayan 1994). Hand operated borehole equipment was not sufficient to go deeper than 10 m in this area and it was not possible to obtain any C^{14} dating. Nor was there enough archaeological evidence for dating. However, stratigraphical sequences of the alluvial

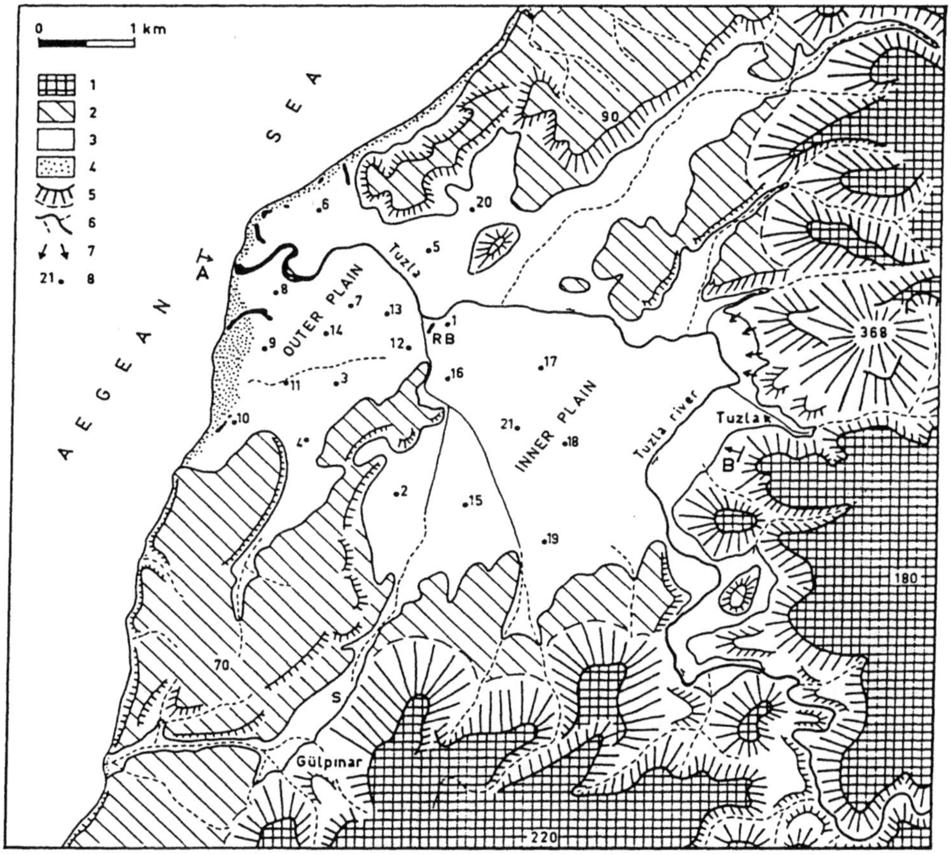

Figure 5. Geomorphological outlines of the Tuzla plain and its environs. **1)** High plateau on the Tertiary volcanic rocks (200-250 m), **2)** Low plateau on the Neogene sedimentary rocks (40-80 m), **3)** Tuzla plain, **4)** Beach and dune cover, **5)** Scarps, **6)** Rivers, **7)** Springs (hot, salty), **8)** Borehole locations and numbers. **S**: Temple to Apollon Smintheus, **RB**: Roman bridge, **A** and **B** are the end points of cross-section given in Figure 6.

deposition were found in the same order as in Troia. A clean, loose and coarse sandy alluvium unit was entered below the surface of the present plain, 2-3 m deep near the coast, and 8-10 m deep about 3-4 km inland (that is, always about present sea-level) (Fig. 6). It is interpreted that the rising sea, which had reached its present level in the middle Holocene, intruded up to the middle part of the plain over this surface. The middle part of the plain is narrow between low ridges which extend from the northern and southern margins (Fig. 5). When the sea advanced inland to its maximum extension, a high coastal barrier developed here which caused a change on the sedimentation over the inner part. In fact, a muddy alluvium a few meters thick was deposited on the clean, loose and coarse sandy alluvium up to the top level of the barrier. The muddy sediments have lagoonal characteristics in the inner (eastern) side of the barrier. The outer (coastal or western) side of the barrier was the area of a new delta formation. Finally, the inner and outer parts of the Tuzla plain were completely covered by fine sandy-silty flood sediments a few meters thick. These recent flood sediments also cover the old coastal barrier.

Although evidence is insufficient to date the stages of geomorphological development of the Tuzla plain, remains of a Roman stone bridge which is situated on the old barrier with its arches rising above the present surface, made it possible to interpret the paleogeography of the area during Roman times. It can be said that the plain was at almost the same level as the old coastal barrier in Roman times, but probably the surface of the plain was not dry enough to cross all the year round because of seasonal swamps caused by the rising sea. Therefore, the Romans used the old barrier because it was a little higher than the plain surface at that time, and it was dry because of its coarse sandy sediments. According to this interpretation, the transition surface at the present sea-level located in the inner plain between the lower coarse sandy unit and covering muddy upper unit can be attributed to the period high sea-level about 6000 years B.P. It is seen that here also that the reason for the sedimentation change is normal geomorphological development related to alluvial deposition.

Ephesos and Klaros

There are some noticeable similarities between the sedimentological-stratigraphical characteristics of the Troia area and Ephesos and Klaros to the south (Fig. 1). Paleontological analyses and C^{14} datings which were to be performed by Dr. Kraft and Dr. Rapp are not yet complete. However, preliminary interpretations of the

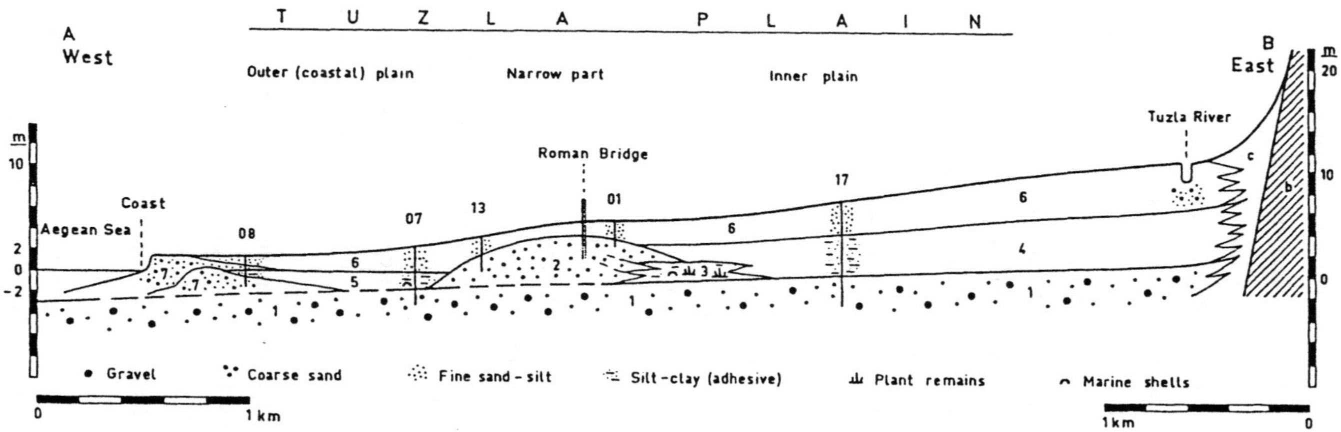

Figure 6. A-B cross-section of the Tuzla plain between the Aegean Sea in the west and the village Tuzla in the east (see Figure 5). **1)** Early Holocene coarse sandy - gravelly alluvium, **2)** Coastal barrier formed following the first sea level rise to its present level in mid-Holocene, **3)** Lagoonal and coastal swamp deposits, **4)** Mid-Holocene muddy alluvium of the inner plain, **5)** Lagoonal and back-swamp deposits to the north of the outer plain, **6)** Last flood-delta plain deposits covering the entire plain, **7)** Coarse sandy coastal beach formations. **b:** Neogene bedrock, **c:** colluvium. Numbers having two digits are borehole numbers (see Figure 5).

sedimentological-stratigraphical data show that the marine sediments lie a few meters below present sea-level in the Küçük Menderes flood-delta plain. Coarse sandy deltaic-coastal sediments form a transition layer a few meters thick over the early Holocene estuarine-marine mud unit at the bottom. Then fine sandy-silty flood plain sediments cover the whole plain at the top. Archaeological finds from the Archaic period are also found a little below the present sea-level in boreholes which were made near the ancient city Ephesos of (for example the oldest cultural layers at the sites of the temple to Artemis and the old harbor. Fig. 7).

The ancient sacred area of Klaros is situated in a small valley about 1600 m from the coastline to the north of the Küçük Menderes delta in the Kuşadası gulf (Fig. 1). Alluvial sediments close to the surface were examined and some interpretations relating to the paleogeographical development were made during the last excavations led by Dr. de la Genière (Kayan, 1995 b). Archaeological remains were found here on a coarse sandy layer 650-730 cm below the present surface of the valley floor (Fig. 8). The coarse sandy layer belongs to an old beach and is partly cemented into beach-rock. There are shallow marine sediments beneath this layer. Then fine sandy-silty old flood sediments from the valley are entered through a gradual boundary about 900 cm below the present valley floor.

The upper surface of the coarse sandy old beach sediment layer, which is 650 cm below the present surface at the site of temple to Apollo, is about 50 cm below present sea-level. There is a sediment unit of a coastal swamp about 100-150 cm thick above the coastal sand. There are many mixed archaeological finds in this dark colored mud such as figurines, shreds, and things relating to fire (ashy mud, charcoal, and burned bones and shells). A piece of firewood from the lower part was dated by C^{14} to the 6th century BC which is in concordance with the dates of the archaeological finds.

The upper surface of the swamp sediments, which is about 500 cm below the present valley floor, was leveled by an artificial fill and a Hellenistic temple to Apollo was constructed on this base. However, the base of an older temple (which might be Archaic according to De La Genière) was found in the swamp sediments below the Hellenistic temple.

According to this evidence, it is understood that the rising sea in the post-glacial period intruded into the valley farther north than the place where Klaros is situated, and then the sea retreated because of alluvial deposition in the valley. The coarse

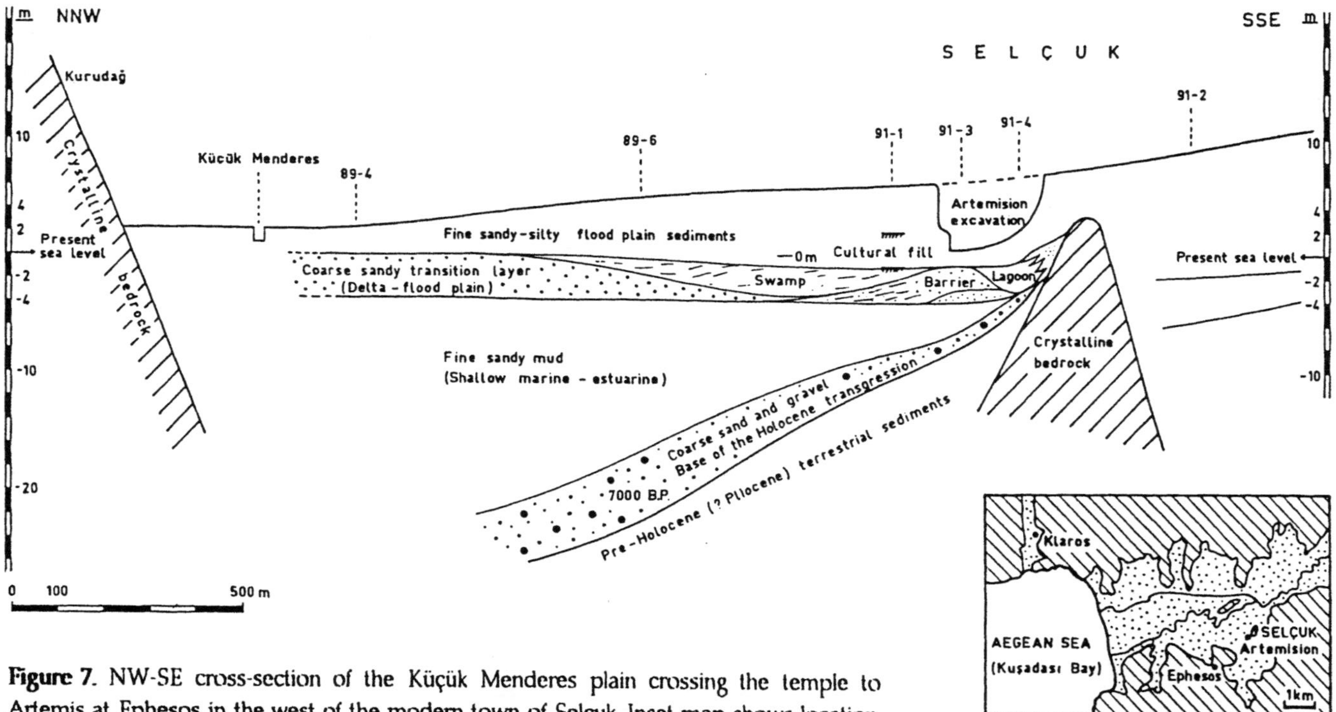

Figure 7. NW-SE cross-section of the Küçük Menderes plain crossing the temple to Artemis at Ephesos in the west of the modern town of Selçuk. Inset map shows location. Dashed areas on this map show mountainous areas; dotted areas represent Küçük Menderes flood-delta plain.

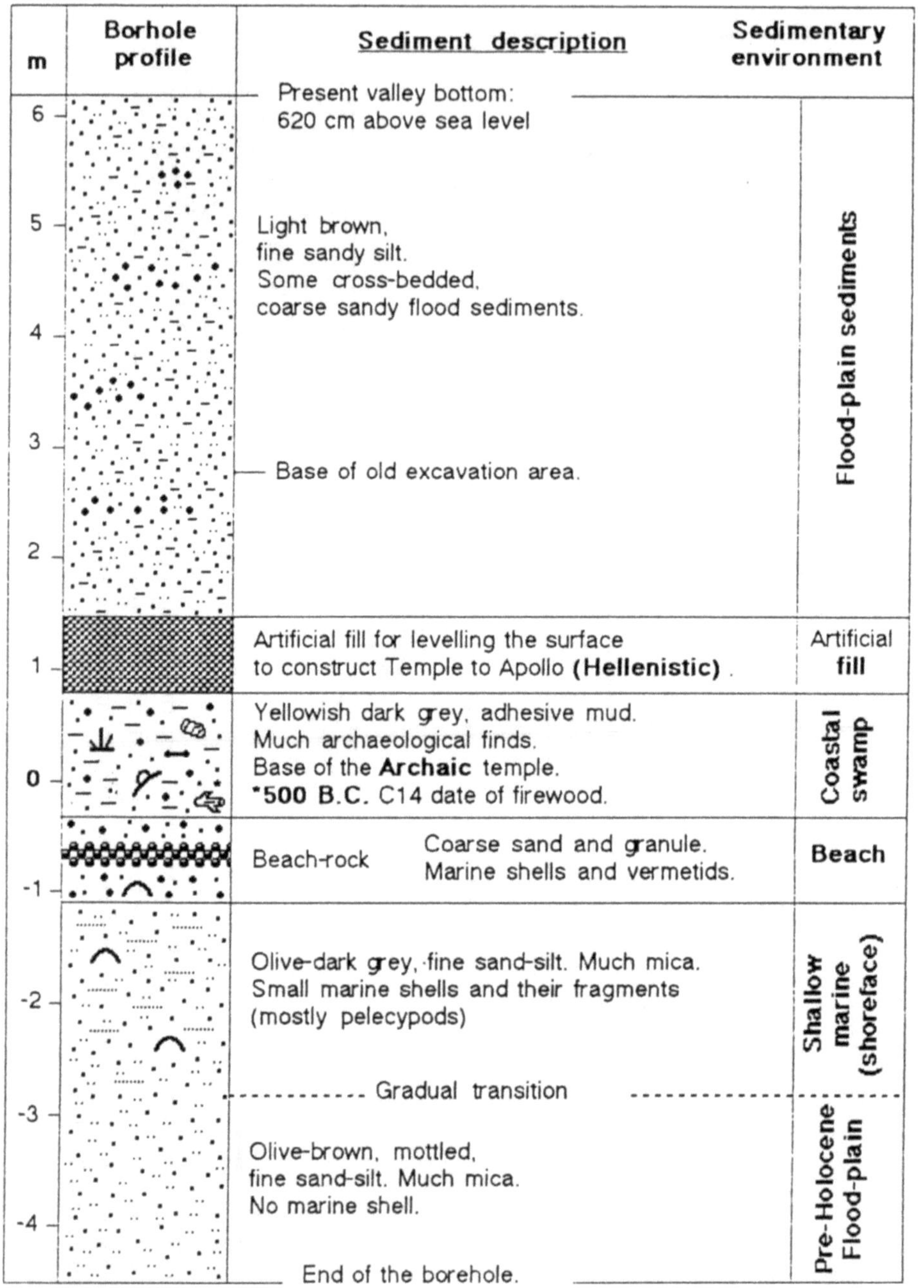

Figure 8. Late Holocene stratigraphical profile of the Klaros area at the site of Temple to Apollo.

sandy beach sediment which is partly cemented as beach-rock layer is noticeable here. It must have been formed when the sea-level was lower than at present. Swamp sediments on the beach sands show that the sea-level rose again and consequently raised the level of ground water behind the coastline. The occurrence of a number of archaeological finds 50-100 cm below the present sea-level which dated to the first half of the first millennium BC implies that the sea-level was still lower than at present, although it had risen a little by that time.

To sum up, alluvial stratigraphy of the Ephesos and Klaros areas indicate that the rising sea in the post-glacial period fell a few meters after first rising to its present level. In this period, coarse sandy sediments along a rapidly progressing delta coast form a transition layer between lower, dark-colored marine mud and upper, fine sandy-silty, light brownish flood sediments. Although muddy sediments of a swamp environment cover the transition layer, which lies a little below the present sea-level, this is a terrestrial-coastal swamp and finds indicate that its surface was at least partly or seasonally dry and firm enough to walk on. Accordingly, it is understood that the falling and retreating sea rose again during the first millennium BC However, the rising sea could not intrude the valleys again, but ground water behind the coastline rose following the sea-level and wider swamp areas developed. This progression is similar to that which is seen in the environs of Troia.

Beach-rocks

Beach-rock formations on the Aegean coast, especially in the south where they are better developed, have important characteristics to interpret sea-level changes and related paleogeography. For example on the coast Datça in the south and west of İzmir, beach rock layers are seen under sea water in some places and these have undergone severe wave erosion (Kayan 1988 b). However, beach-rock formation is never seen at a higher level than the present wave zone in any location in this region. There is no research on the dating of these beach-rocks. Nevertheless, supposing that the beach-rock formations develop along regressive coasts, then it can be said that they were formed by concretion of beaches during the Bronze Age regression of the sea following its first rise in the post glacial period. They were covered by water during the second rise of the sea and broken down by the erosive effects of the waves. There is a noticeable concordance between the beach-rock formations and paleogeographical interpretations of the region which were made according to the alluvial stratigraphy.

Conclusions

In the post-glacial period, the rising sea intruded into embayments along the Aegean coast of Anatolia and formed a very indented coastline. However, these embayments, and especially estuarine bays at the mouths of big rivers which brought a great deal of alluvium from inland, filled rapidly. In such places, shallow marine-estuarine fine grained sediments start a few meters below the present sea-level. They are easily recognizable with blackish dark-colored muds containing various marine shells. This marine sedimentation was bordered by coarse sandy coastal sediments along the edges of the present plains, just at the present sea-level beneath the recent alluvium. C^{14} dates of the shells show the coastline is 6000 years old along the inner parts of the present plains and this date becomes closer to the present as it approaches the present coastline.

In the central parts of the plains various sediments of low coastal environments cover the marine muds with a layer about 1-2 m thick. They are mostly coarse sandy and have marine shells and decayed plant remains. The upper surface of this layer is about 1-2 m below the present sea-level and is filled with fine sandy-silty, light brown flood sediments up to the present surface of the plains.

In this stratigraphical sequence, the coarse sandy layer between the lower marine (estuarine) sediments and upper flood sediments consists of various sediments of low coastal environments such as shallow shoreface, low beaches, inlet (azmak) and coastal swamps. The stratigraphical characteristics of this transition layer and C^{14} dates obtained from its different parts indicate a development order as follows: rising sea in the post glacial period intruded toward the coastal embayments and mouths of the rivers, with the sea first reaching its present level 6000 years B.P. Then the sea-level fell about 2 m in the period between 5000-4000 B.P. (the Third Millennium BC); consequently deltaic progradation accelerated. Following a period of low sea-level between 4000-3000 B.P., the sea-level rose again up to almost its present level in 3000-2000 B.P. However, sea could not intrude inland into river mouths as in its previous rise; on the contrary, deltaic progradation continued, because the sea-level rise was slower and smaller this time.

A coarse sandy transition layer between marine and fluvial sediment units can generally be easily recognized in many places. This change in the sedimentation is a normal result of the silting up of estuarine embayments with alluvium following the ending of the sea-level rise. It is obvious that the change on sedimentation was dependent on sea-level changes. The transition layer has similar characteristics over all the coastal plains investigated from north to south all along the Aegean coast of Anatolia. It always lies on a flat horizontal surface a few meters below the present sea-level. Also there is a coincidence in the time of formation. That is, the change from marine sedimentation to fluvial throughout the Aegean coastal plains of Anatolia occurred in the same sequence and at a similar time (Third Millennium BC).

It might be thought that such a development can be explained by a general tectonic subsidence when the geological structure of the region is taken into account. However, the Aegean region has been formed by tectonic movements of many blocks and it is difficult to suppose that all of the blocks moved in the same order and by the same amount without any distortion or tilting. Therefore, the general configuration of the transition layer cannot be explained by tectonic movements. In addition, no geomorphological feature has ever been found up to now to indicate any tectonic movement of the Aegean blocks during the last 6000 years. For example, no old coastline traces such as coastal terraces, notches, bio-erosional and bio-constructed features are seen on the steep cliffy mountain coasts.

Also, compaction of the sediments might be taken into consideration as a reason for the transition layer being a few meters below the present sea-level. However, this is not a reasonable explanation either because the sediment thickness varies in many alluvial plains depending of the base topography, but the horizontal extension of the transition layer is nevertheless very smooth.

Briefly, according to the present evidence, climatic-eustatic sea-level changes can be taken into consideration as the most reasonable explanation for the change of sedimentation throughout the Aegean coast of Anatolia in the third millennium BC Such climatic changes should have caused some textural and structural changes on the sediment characteristics. More detailed sediment analyses are necessary on the upper levels of the alluvium to interpret such changes. In addition, although no absolute dating is available, beach-rock formations below the present sea-level can be taken into consideration as indicators of a regression of the sea in a hot and dry climatic period with greater evaporation. This period may be supposed to be contemporaneous with the Third Millennium BC sea-level fall.

References

De La Genière, J. (1990) Le sanctuaire d'Apollon à Claros; Nouvelles découvertes. Revue des Études Grecques. CIII, 490-491: 95-110. Paris.

Kayan, İ. (1988a) Datça yarımadasında Eski Knidos yerleşmesini etkileyen doğal çevre özellikleri. *Coğrafya Araştırmaları Dergisi*. 11: 51-70. Ankara Üniversitesi Dil ve Tarih Coğrafya Fakültesi. Ankara.

Kayan, İ. (1988b) Late Holocene sea-level changes on the Western Anatolian coast. Palaeogeography, Palaeoclimatology, Palaeoecology. 68, 2-4: 205-218. Special issue: Quaternary Coastal Changes. Ed. by P. A. Pirazzoli - D. B. Scott. (A selection of papers presented at the IGCP-200 meetings). Elsevier Science Publishers B.V. Amsterdam.

Kayan, İ. (1991) Holocene geomorphic evolution of the Beşik plain and changing environment of ancient man. Studia Troica. Band 1: 79-92. Verlag Philipp von Zabern. Mainz am Rhein.

Kayan, İ. (1994) Tuzla ovasının (Ayvacık-Çanakkale) alüvyal jeomorfolojisi ve Holosen'deki kıyı çizgisi değişmeleri (Summary: Alluvial geomorphology of the Tuzla plain and Holocene coastal changes). Ege Üniversitesi Rektörlüğü Araştırma Fonu Proje No: Edf 1988-027. 74 pp. İzmir. (Unpublished report).

Kayan, İ. (1995a) The Troia bay and potential harbor sites in the Bronze Age. Studia Troica. Band 5. Philipp von Zabern, Mainz am Rhein. Germany. (In print).

Kayan, İ. (1995b) Klaros kazı alanında jeomorfolojik-jeoarkeolojik araştırmalar (Summary: Geomorphological-geoarchaeological investigations on the Klaros excavation area). *Aegean Geographical Journal*. 8: 1-26. Ege University. İzmir. (In print).

Kraft J. C. , Kayan İ. and Erol O. (1980) Geomorphic reconstructions in the environs of ancient Troy. *Science*. American Association for the Advancement of Science. Vol. 209, No. 4458: 776-782.

Kraft, J. C., Kayan İ., and Erol, O. (1982) Geology and Paleogeographic Reconstructions of the Vicinity of Troy, in: Rapp, G. Jr. and Gifford, J. A. (eds.), Troy: The Archaeological Geology, Supplementary Monograph 4, 11-40. Cincinnati.

Environmental Changes In Holocene Lacustrine Sequences From Karapınar In The Konya Plain (Turkey)

C. Kuzucuoğlu[1], M. Karabıyıkoğlu[2], M. Fontugne[3], J-F. Pastre[1], T. Ercan[2].

[1] Laboratoire de Géographie Physique, URA 141 CNRS, 1 Pl. A. Briand, 92195 Meudon cédex, France

[2] MTA General Direction, Department of Geological Studies, 06160 Ankara, Turkey

[3] Centre des Faibles Radioactivités, CEA-CNRS, Domaine du CNRS, 91198 Gif-sur-Yvette cédex, France

Abstract

The Karapınar playa occupies the floor of a lacustrine depression which was part of the Upper Pleistocene (23,000-17,000 yr BP) Konya palaeolake. Four sections opened in the floor of the Karapınar depression permit the study of the stratigraphy of Holocene sequences. From the sedimentological analyses of the sand content, organic carbon content, heavy mineral inputs and molluscan fauna, associated with the sediment observation (marks of aerial weathering and plant roots), several environmental changes can be assessed during Holocene. Underneath a characteristic black marshy layer, a thick sequence of lacustrine clay shows two episodes of cencentrated unbroken shells of Dreissena polymorpha, which is a typical species of the Upper Pleistocene open lacustrine environment in the area. Between these two episodes, stem marks in the clay show that the lake dried during an unknown period, during which soil was formedand carbonatic nodules occurred on its bottom. During Early to Mid-Holocene, the development of a marshy eutrophic environment is testified by an extensive organic and molluscan-rich layer situated in all the sections at about 1 m below the surface. 14C dates obtained both on organic matter and on shells from this marshy layer are mainly distributed between two periods: 13,800-13,400 BP and 6800-6500 BP. The latter period corresponds to the end of the occupation of the Neolithic site of Çatal Hüyük in the Konya plain. Around 6500 BP, the marshes are replaced by a lake which left a 10-30 cm thick clay layer. This Mid-Holocene lacustrine readvance ended with a brutal drought marked by a change in sedimentation, with a 30 to 60 cm thick silty and sandy aeolian deposit covering the lacustrine clay. The Late Holocene is marked by arid to semi-arid processes, with wind

NATO ASI Series, Vol. I 49
Third Millennium BC Climate Change
and Old World Collapse
Edited by H. Nüzhet Dalfes, George Kukla,
and Harvey Weiss

as an important geomorphic agent. During thi uncalib. 14C), we assume that it corresponds to the arid to semi-arid dry period which followed the Mid-Holocene lacustrine stage.

Introduction

The Konya plain, located West of the Cappadocian volcanic region in inner Anatolia, corresponds to the floor of an Upper Pleistocene lake today almost entirely dry. Geomorphological and sedimentological studies of some Pleistocene landforms and sediments in the Konya plain, such as sand pits and beaches (Erol, 1978; Roberts, 1980), lacustrine deposits (de Ridder, 1965; de Meester, 1970; Roberts, 1980, 1982b) and dune systems (Groneman, 1968; Kuzucuoğlu, 1993; Kuzucuoğlu et al., pub.) have shown that the level of the Konya lake changed several times during the Upper Pleistocene (Roberts, 1983, 1993). The beach deposits surrounding the main floor of the lake and corresponding to its maximum extension in Upper Pleistocene have been 14C dated (Roberts et al., 1979) 23,000 BP to 17,000 BP on Dreissena polymorpha shells. We obtained also 14C dates ranging from 18,500 to 17,000 BP from Dreissena shells collected from various sand pits (Gif 10271, 10272, 10274, 10275, 10337). Evidence for Holocene level variations of the Konya lake has been given by both Erol (1979) studying landforms and by Roberts (1980) studying sediments in sections and cores. The chronology of these changes being not yet fixed, the Holocene history of the vanished Konya palaeolake can be studied using the sediment sequences in the bottom of smaller isolated subbasins. One of these is the Karapınar playa, on the northern part of the Konya plain.

Present day environment

Today, the climate of the Karapınar area in the Konya plain is semi-arid with an annual precipitation of less than 300 mm, of which less than 60 mm falls in summer. Rains occur mainly during spring, while some snow falls during winter. Water from the Taurus mountains (maximum altitude = 3553 m), 50 km south of the Konya plain, feeds an abundant underground water reserve which, however, returns to the limestone bedrock throughout sinkholes.

Winds blow throughout the year from the North or the North-North-East, but the strongest and most efficient winds blow from the South-West, mainly during spring and summer. These latter winds are an important agent of morphology, especially on the northern and drier side of the Konya plain where the Karapınar lacustrine depression is situated.

The Karapınar saline lake covers an approximate surface of 120 km^2 at an almost constant altitude of 1000m (figure 1). It occupies a tectonic depression where both subsidence and karstic features influence its hydrologic behaviour. Its eastern edge is limited by a fault line which fringes the Karacadag Pliocene volcanic mountain. Its western and northern edges are cut into Mio-Pliocene lacustrine limestones.

Natural vegetation is formed by steppe or very open shrub vegetation. In the western limestone hills, some trees can be seen near water springs. In the volcanic Karacadag mountains, deciduous oak is not rare, still forming open forests on the northern slopes. A halophytic vegetation grows on the bottom sides of the depression.

Today, the floor of the depression is seasonally flooded, being mostly dry during summer, a time of heavy evaporation and of salt-crust formation. During winter and spring, streams run into the playa from both the eastern volcanic slopes (Karacadag mountains) and the western karstic plateaux (Obruk region).

In summer, the lake is supplied by permanent springs located on the northern and southern edges of the depression. In the South, a drain discharges waste waters from the town of Karapınar.

The Late Pleistocene and Holocene changes in the Konya plain according to previous studies on pollen assemblages and lake levels

According to palynology, the Holocene started in the Anatolian plateau about 11,000 ago (uncal.), with Mediterranean bioclimatic indicators much less extensive at the beginning of the Holocene than they were later (Bottema, 1987; Roberts & Wright, 1993). In the pollen diagrams available from Anatolia, steppe assemblages generally prevailed until 9000 BP and trees were dominant at 6000 BP, with up to 80% of the pollen (Bottema, 1991; van Zeist & Bottema, 1991).

In the area studied, one pollen diagram (´YAdabagY[a]) from the Akgöl subbasin of the Konya plain, at about 50km South of the Karapınar lake, covers the Late Pleistocene and the Early Holocene (Bottema & Woldring, 1984; Bottema, 1987). At the bottom of the diagram (13,000 BP), the tree pollen account for only 10% of the total, while the herbaceous pollen percentages and types show very arid conditions. Between 12,400 BP and 11,600 BP a pollen shift shows a slight increase in precipitation, followed by a period rich in oak until 11,200 BP. Drier conditions

Figure 1 : LOCATION OF THE AREA STUDIED AND OF SAMPLE SITES

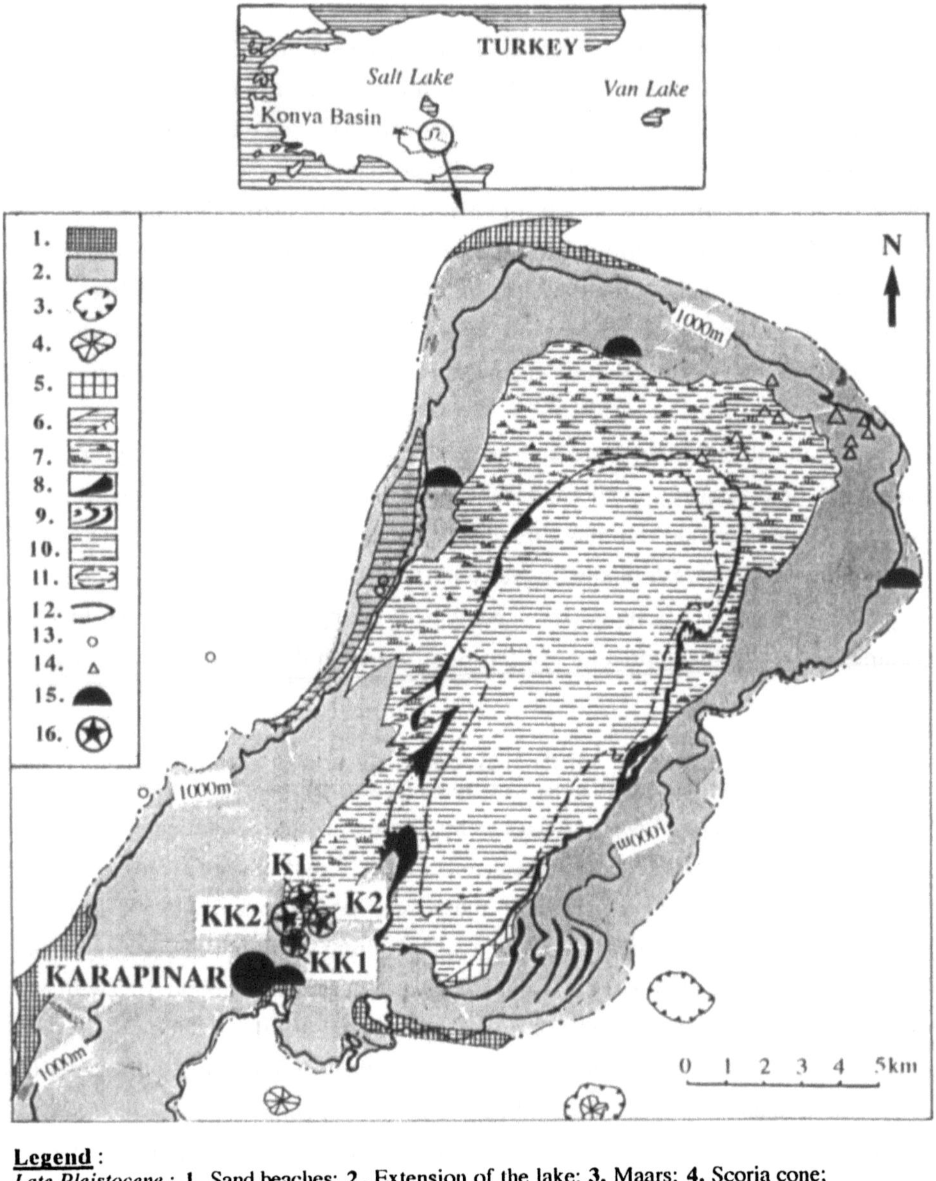

Legend :
Late Pleistocene : **1.** Sand beaches; **2.** Extension of the lake; **3.** Maars; **4.** Scoria cone;
Early Holocene lake : **5.** Sand beaches; **6.** Erosion surface; **7.** Extension of the lake;
Mid-Holocene lake : **8.** Sand beaches; **9.** Contracting sand beaches; **10.** Extension of the lake
Today : **11.** Extension of the playa; **12.** 1000m contour line; **13.** Sink holes; **14.** Springs;
15. Large hüyüks; **16.** Sampling sites;

developed and lasted until about 10,500 BP. Afterwards, forest started to spread with the expansion of birch followed by a rise in decidous oak at about 9000 BP. Pine became the dominant tree after 8000 BP.

From this diagram and others from south-western Turkey, it is concluded that dry conditions were still prevailing at the beginning of the Holocene (12,000-10,500 BP). Afterwards, humidity rose during the first 2 to 3 millenia of the Holocene (until 8000 BP), with climatic conditions still drier than at present.

This chronology illustrates local conditions in rainfall and temperature. For during this same period (the beginning of Holocene), the marine cores of the eastern Mediterranean sea show an episode of increase in fresh water input at 9300 BP (mean value) with the formation of a sapropel layer, ending at 8000 BP (Fontugne et al., 1994). This fresh water input is usually related to freshwater influx via the Nile river. The climatic trend towards drier conditions start, in the marine cores, at about 8000 yr BP, in agreement with the sharp peak in d18O values in the marine cores (Fontugne et al., 1994), with the marine palynological records (Cheddadi et al., 1991) and with observations from continental records (Street & Grove, 1979; Nicholson & Flohn, 1980).

Late Pleistocene and Holocene changes in the levels of the Konya palaeolake and consecutive environmental interpretation are known from a 20 year succession of field researches in the area, with soil analyses and maps (de Ridder, 1965; de Meester, 1970), field observation (Cohen, 1970; Erol, 1979), archaeological stratigraphy related to the Neolithic sites of Çatal Hüyük and Can Hasan in the southern part of the Konya plain (Cohen, 1970; Roberts, 1982a), and core sampling associated with geomorphological and sedimentological surveys (Roberts, 1980). Several syntheses on the Late Pleistocene and Holocene lake level variations in the Konya plain have been published (Roberts et al, 1979; Roberts, 1982a, 1982b, 1983, 1991).

We shall recall here a short summary of the conclusions of these previous researches, mainly available for the western and south-western parts of the plain:

- from 23,000 to 17,000 BP, the Konya lake was a ca 30 m deep freshwater body, extending over 4340 km2 with a salinity below 5 g.l-1;

- from 17,000 to 13,000 BP, the plain was dry, except perhaps in the eastern Akgöl subbasin;

- from ca 12,400 to 11,000 BP, the lake level rose and isolated shallow freshwater lakes appeared in the five subbasins composing the Konya plain (14C dating of mollusca shells from a marshy palaeosoil in the Yarma subbasin) (Roberts, 1983). The marshy layer related to this stage is associated with a rich molluscan fauna;

- from 11,000 to 7000 BP, increasing desiccation led to a complete drying up of the plain. To this period corresponds a fossil soil dated 8040 ± 140 BP (Bottema & Woldring, 1984; Roberts, 1991) and the Neolithic occupation phases of Çatal Hüyük and Can Hasan (8600-6700 BP).

- after 6700 BP, a post-Neolithic readvance of a shallow lake has been evidenced by Erol (1978) and Roberts (1982a; 1983). On the basis of archaeological surveys in the southern part of the Konya plain (French, 1970), the age of this lake renewal, not yet 14C based, might be either 6400-5700 yr BP (Middle Chalcolithic) or 3rd millenium BP, the shift of the sites to the edges of the basin being particularly marked in the latter case (Roberts, 1982a).

- lastly, after the Roman period, increasing drought and summer evaporation have accompanied the disappearance of the main water bodies in the plain.

Periods of dry conditions (climatic or human-induced) since the Late Pleistocene have also led to the construction and activation of sand dune systems, especially on the northern shores of the plain and in the Karapınar area (Kuzucuoğlu, 1993; Kuzucuoğlu et al., pub.).

The Holocene landforms of the Karapınar depression

Several landforms testify that at the time when the Konya palaeolake was at its maximum extension (Upper Pleistocene), the Karapınar lake also was 10 to 15m higher than today: sand beaches on the northern edge, erosion terraces on the western edge, erosion slopes (glacis) on the eastern edge (figure 1).

Some local features show more recent, i.e. Holocene, variations of the lake level. On the western shores, a younger erosion level is clearly distinct from the + 15m Upper Pleistocene highstand; situated at + 3-5m above the bottom of depression, its surface is partly dissolved by karstic sinkholes. On the south-eastern shores, the satellite SPOT imagery shows regressive stages of a contracting lake; the contraction started from a + 5 to 8 m sand beach fringing the volcanic slopes of the Karacadag. Furthermore, three large hüyüks (Neolithic or Bronze Age mounds) are located on the floor of the depression at some distance from its edges (figure 1); their location is closely related to freshwater inputs from the surrounding hills and mountains. The time of their occupation is yet unknown.

These features suggest that after the drying up of the Upper Pleistocene lake, a lake reappeared in the Karapınar depression, with a water body 3 to 5 m deep, occasionnaly constructing beaches on the south-eastern edges and eroding low terraces on the western edges. After this stage, it dried slowly. During its progressive contraction, parallel curved sand beaches were abandoned in its south-eastern corner. At some time during the Mid-Holocene evolution, while the lake had disappeared or was much contracted, man settled in prosperous villages, which he abandoned later.

After the last drying up of the lake, wind erosion activated the sand from the southern sand beaches and dune systems, constructing small aeolian sand dunes moving northwards over the lacustrine floor and invading the southern part of the depression.

The Holocene lacustrine sequences in the bottom of the Karapınar lake

The chronology of the Holocene environmental evolution as testified by lake level changes may be made more precise by the sedimentological study of the upper part of the lacustrine sequences opened in the bottom of the lake.

In a clay quarry North of the town of Karapınar (section KK1, figures 1 and 2), a 10cm thick black organic layer clearly appears at a 0.95 m depth. In two holes dug in the vicinity of the quarry, using a shovel dredger provided by the Karapınar municipality, (sections K1, K2, figures 1 and 2), the same dark layer appears at about a 1.0 to 1.10 m depth. A fourth section (KK2, figures 1 and 2), available in a 3 m deep test-well opened next to the main front of the quarry, shows this organic-rich layer at a 0.90-1.00 m depth. These four sections are grouped in the southern part of the Karapınar depression.

Except for section KK1, which was only 1.10 m high, samples were collected, every 5 cm down to the watertable level, i.e. at about 2.18 m at station K1, 1.95 m at station K2 and 3.1 m at station KK2. Various analyses were performed on samples selected according to the stratigraphy and to the variations in sedimentology along the profiles of the sections. 14C dating was performed on the organic matter and mollusca from the marshy layer of sections KK1 and KK2. 14C dating was unsuccessfully attempted on Dreissena shells from the lacustrine clay underneath the marshy layer of the KK2 section (the amount of shells collected was insufficient for 14C dating).

* The main feature, common to the four sections, is the organic matter-rich black layer containing large and small individuals of a rich freshwater molluscan fauna. This

layer lies at a 1 m depth in all sections. J. Mouthon, from the Cemagref in Lyon, has determined the following molluscan species: small individuals (800 μ - 2 mm) are mainly Planorbis planorbis, Bithynia tentaculata, Valvata piscinalis, Valvata cristata, Limnea radix, Micromelania sp., Gyraulus cronicus, Armiger crista. Very large individuals are mainly Planorbis corneus, Limnea palustris and Limnea radix. Three 14C ages have been measured in large individuals: 6830 ± 230 (Gif 9823) and 13740 ± 110 BP (Gif 9824) in section KK1; 13,470 ± 100 BP (Gif 10081) in section KK2. Two other dates have been measured on the organic matter from the black layer: 9170 ± 70 BP (Gif 9825) in section KK1, and 6520 ± 90 BP (Gif 10082) in section KK2. In the four sections studied, this organic layer covers a 5 cm thick sandy episode rich in rather coarse and reworked material including heavy minerals supplied by the surrounding volcanic slopes (_zecek dagi). This bottom sandy episode lays on top of a thick sequence of older lacustrine marls.

These results must be compared with other dates obtained elsewhere in the Konya plain. In the western part (the Yarma subbasin), a dark organic marl layer containing diverse molluscan fauna, probably equivalent to that described here in the Karapınar subbasin, gave an age of 12,010 ± 65 and 10,950 ± 65 BP (Roberts, 1983). In the eastern part (the Akgöl subbasin), a marshy environment starting at 10,920 ±150 BP is overlaid by a 10 cm thick palaeosol indicating a dry episode dated 8040 ± 140 BP (Bottema & Woldring, 1984) and marking the disappearance of the marshes in the central area of the subbasin.

After that date, in the Adabag core (Bottema & Woldring, 1984) like in the Karapınar sequences, some 1 m thick clay contains indicators of swampy conditions.

Other remarks must be made on the possible deviations in 14C dates obtained in the shells and sediments of the lacustrine sequences of the area. Deviations in 14C dates can occur by hard water effect and recrystallisation or mixing with older and younger carbon from either the limestones of the catchment basin (both lakes have a karstic behaviour) or from the overlying lacustrine marls. The mollusca may have anomalous ages because of uptake of old carbon; the organic matter may have been contaminated by leaching with younger carbon entering the system. Thus, 14C in carbonated environments are frequently subject to contamination. The dates obtained on the mollusca shells and organic matter from the 1 m deep and 10 cm thick marshy layer in the Karapınar subbasin seem to belong to two distinct phases: the oldest one would be dated ca 13800 to 13400 BP, to be compared with the dates (12400-11200

Figure 2 : SECTIONS STUDIED IN THE FLOOR OF THE KARAPINAR LAKE

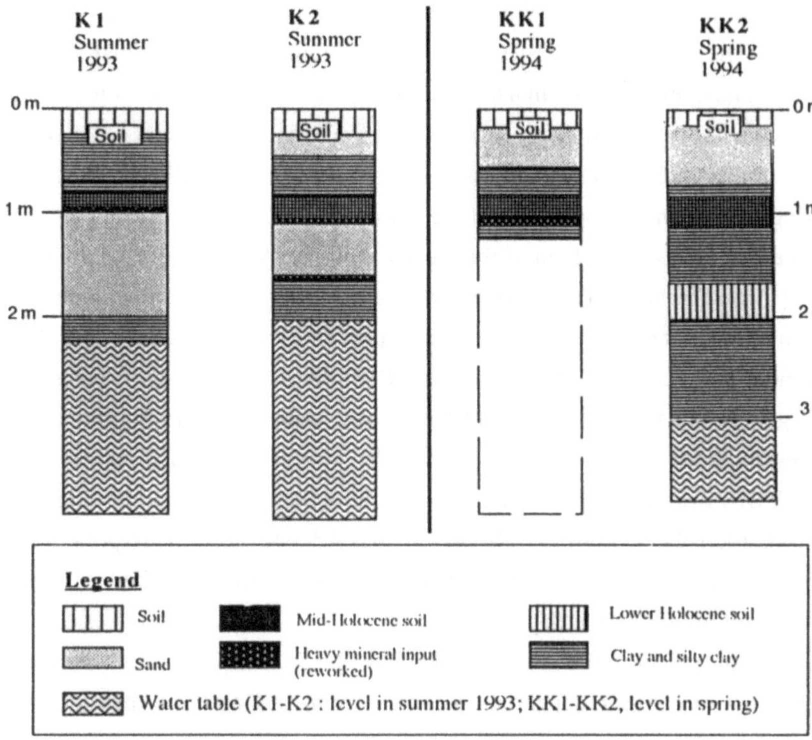

2.2. GRAIN SIZE AND SHELL CONTENT

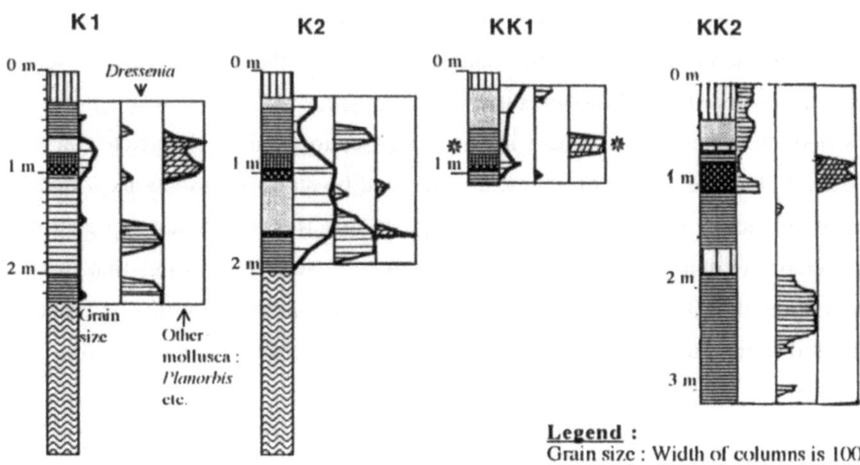

Legend :
Grain size : Width of columns is 100%
Shells : Width of columns is 50 individuals
Stratigraphic log : see previous figure

BP in Roberts, 1983) obtained on freshwater mollusca in the Yarma subbasin; the youngest phase would be dated ca 7000-6500 BP with yet no equivalent with any published dates. The age of 9170 ± 70 might result from one or the other type of contamination. Thus, either the marshy environment evidenced in the Yarma depression may have started earlier (in ca 13,800 BP) and lasted until ca 11,200 BP, or there may be another and older marshy stage (ca 13,800 to 13,400 BP), or there may be a mis-dating by contamination in one of the sites.

In this frame, as well as in the light of the results from the Adabag core (Bottema & Woldring, 1984), and from the Yarma section (Roberts, 1983), we interprete our results in the following way. After the complete drying up of the lakes around 17,000 BP, a very dry period succeeded (17,000 - 13,800 BP) which ended with increasing precipitation generating the appearance of freshwater marshes in the bottom of the subbasins (approximate dates: 13,600 to 11,200 BP), eventually in two stages. Next (11,200 - 7000 BP) is a period increasingly favourable to vegetation, although dry, with a soil formation dated 8040 ± 140 BP in the Adabag core. This period was followed by a more humid period (7000-6500 BP) which is marked by the renewal of the marshes in the bottoms of the depression.

* The period starting ca 6500 BP is marked by the succession of two different environments: first the lake reappeared, second a semi-arid and dry period starts which has lasted until today. In the sections studied, the marshy episode ending in ca 6500 BP is covered unconformably by 20 to 30 cm thick lacustrine clays. This episode may match with the lacustrine episode also evidenced in the Akg–1 core posterior to the soil dated 8100 BP. This clay layer is interrupted by a thicker (up to 50 cm) layer of aeolian silt-mixed fine sands, covered by the present surface soil also enriched in reworked, wind-blown, sand.

This aeolian sequence includes, in the KK2 section, three stem-rich soil levels associated with some carbon enrichment, separated by stem-poor sections which have also been subject to soil formation. This suggests that the environment prevailing in the bottom of the depression in the period which followed the drought that dried the last lake (around 5000-4500 BP ?) was that of a dried lake bottom, maybe seasonally flooded and subject to aeolian processes in arid to semi-arid climatic conditions.

The Mid-Holocene environmental changes in the Karapınar basin and the position of the 3rd millenium BC (= 3500 to 5000 BP uncalib., approx.)

The sedimentological analyses of the Karapınar sections support the evidence for Holocene lake level variations also confirmed by the geomorphological study of the local landforms. The end of the second marshy episode at ca 6500 BP was followed by a renewal of the lacustrine environment in the bottom of the depression, which could be contemporaneous with the end of the occupation of the Neolithic sites in the area, mainly located at the output of the rivers from the surrounding mountains, probably next to the shores of the marshes

After the abandonment of the sites, i.e. after 6700 BP, lake conditions reappeared as shown by all the sections in the Karapınar depression. This stage corresponds probably to that observed by N. Roberts (1982a) over the Çatal Hüyük site and attributed by him to the 3rd millenium BP.

After this Mid-Holocene lake expansion a severe drought occurred, TL dated at about 5000-4500 BP (Kuzucuoğlu et al., pub), followed by seasonal desiccation. Steppic vegetation expanded over the emerged lake floor submitted to both wind erosion and accumulation of aeolian silt and sand.

We may assess that the environment prevailing during the 3rd millenium BC was dry conditions continuing the severe drought (5000-4500 BP TL dated) which had interrupted the Mid-Holocene lacustrine episode. The last strong environmental change occurring in the area and recorded in the sediment sequences is this sudden and severe drought which has been TL dated 5000-4500 BP. After this severely dry phase accompanied by dune construction and motion, environmental conditions allowed the extension of a herbaceous vegetation cover on the emerged floor of the subbasin.

On the contrary, Mid-Holocene seems to have been more humid, with marshes (7000-6500 BP) followed by lake extension (6500-5000 BP). Early Holocene (11,200-7000 BP) was probably drier, as shown by soil formation and fluvial input of sand in the depression. The older dates (Late Pleistocene) obtained in the Karapınar sections belong to a marshy environment in the bottom of the depression with swamps dated 13,800 to 13,400 BP or to 11,200 BP. Most of these results are in accordance with previous data published about the area; they also introduce new data on the chronology of Mid-Holocene to present events.

Acknowledgments

The authors would like to thank for their financial, technical and scientific support, in addition to their own laboratories, the TÜBITAK-CNRS cooperation programme, the CNRS/Volcanism in Anatolia PICS programme, the ITÜ and NATO/Global Change Programme, the Köy Hizmetleri in Karapınar, Konya and Ankara, the Mayor of the Karapınar town, and Jacques Mouthon (Cemagref, Lyon, F.) Special thanks are also addressed to Naim Aydinbelge and his family in Karapınar.

References

Bottema, S. (1987). Chronology and climatic phases in the Near East from 16000 to 10000 BP. in O. Aurenche, J. Evin & P. Hours (eds), Chronologies int the Near East, BAR International Series, 295-310.

Bottema, S. (1991). Développement de la végétation et du climat dans le bassin méditerranéen oriental à la fin du Pléistocène et pendant l'Holocène. L'Anthropologie, 4, 295-310.

Bottema, S. & Woldring, H. (1984). Late Quaternary vegetation and climate of south-western Turkey. Palaeohistoria, 26, 123-149.

Cheddadi R., M. Rossignol-Strick and M. Fontugne (1991). Eastern Mediterranean palaeoclimate from 26 to 5 ka B.P. documented by pollen and isotopic analysis of a core in the anoxic Bannock Basin. Marine Geology,100, 53-66.

Cohen, H.R. (1970). The palaeoecology of south-central Anatolia at the end of the Pleistocene and the beginning of the Holocene. Anatolian Studies, 20, 119-137.

Erol, O. (1978). The Quaternary history of the lake basins of central and southern Anatolia. in W.C. Brice (ed.)The Environmental History of the Near and Middle East since the Last Ice Age, 111-139, London: Academic Press.

Fontugne, M., Arnold, M., Labeyrie, L., Paterne, M., Calvert S.E., Duplessy, J-C. (1994). Paleoenvironment, sapropel chronology and Nile river discharge during the last 20,000 years as indicated by deep-sea sediment records in the eastern Mediterranean. in O. Bar-Yosef & R.S. Kra (eds), Late Quaternary Chronology and Paleoclimates of the Eastern Mediterranean, Radiocarbon, 75-88.

French, D.H. (1970). Notes on site distribution in the Çumra area. Anatolian Studies, 20: 139-148.

Groneman, A.F. (1968). The soils of the wind erosion control camp area, Karapınar, Turkey. Centre of Agricultural Publishing and Documentation, 160 pp., PUDOC: Wageningen.

Kuzucuoğlu, C. (1993). The climatic significance of the Upper Pleistocene and Holocene aeolian sand flats around Karapınar, Konya. Preliminary results.Türkiye Kuvaterneri Worshop Bildiri Özleri, ITÜ, Istanbul, 2-16.

Kuzucuoğlu, C., Parish, R., Karabiyikgolu, M. (accepted). The dune systems of the Konya Plain (Turkey). Their relation to the environmental changes in Central Anatolia during the Late Pleistocene and Holocene. To be published in Geomorphology.

de Meester, T. (ed.) (1970). Soils of the Great Konya Basin, Turkey. Centre of Agricultural Publishing and Documentation, 290 pp., PUDOC: Wageningen.

Nicholson, S.E. & Flohn, H. (1980). African environmental and climatic changes and the general atmospheric circulation in the Late Pleistocene and Holocene. Climatic Change, 2, 313-348.

de Ridder N.A. (1965). Sediments in the Konya basin, central Anatolia, Turkey. Palaeogeography, Palaeoclimatology, Palaeoecology, 1, 225-254.

Roberts, N. (1980). Late Quaternary Geomorphology and Palaeocology of the Konya Basin, Turkey. Unpublished Ph. D. Thesis, 266pp., London University: London.

Roberts, N. (1982a). A note on the geomorphological environment of Çatal Hüyük, Turkey. Journal of Arachaeological ScienceT, 9, 341-348.

Roberts, N. (1982b). Lake levels as an indicator of Near Eastern palaeoclimates: A preliminary appraisal. in J.L. Bintcliff & W. van Zeist (eds), Palaeoclimates, palaeoenvironments and human communities in the Eastern Mediterranean Region in Later Prehistory, BAR Reports, International Series, 133, 235-267.

Roberts, N. (1983). Age, palaeoenvironments and climatic significance of Late Pleistocene Konya lake, Turkey. Quaternary Research, 19, 154-171.

Roberts, N. (1991). Late Quaternary Geomorphological change in South and Southwest Turkey during the later Holocene. in S. Bottema, G. Entjes-Nieborg and W. van Zeist (eds), Man's role in the Shaping of the Eastern Mediterranean Landscape, 53-67, Balkema: Rotterdam.

Roberts, N., Erol, O., de Meester, T., Uerpmann, H.P. (1979). Radiocarbon chronology of late Pleistocene Konya Lake, Turkey. Nature, 281, 662-664.

Roberts, N. & H.E. Wright Jr. (1993). Vegetational, Lake-Level and Climatic History of the Near East and Southwest Asia. in H.E. Wright, Jr, J.E. Kutzbach, T. Webb III, W.F. Ruddiman, F.A. Street-Perrott, P.J. Bartlein (eds), *Global climates since the Last Glacial Maximum*, 194-220, University of Minnesota: Minneapolis

Street, F.A. & Grove, A.T. (1979). Global maps of lake-level fluctuations since 30,000 yr BP. Quaternary Research, 12, 83-118.

van Zeist, W. & Bottema, S. (1991). Late Quaternary of the Near East. Beihefte zum Tübinger Atlas des Vorderen Orients, Reihe A (Naturwissenschaft), 18, Riechert: Wiesbaden.

Note: All dates in this paper are uncalibrated 14 C dates BP.

Proxy Indications of Holocene Winter Rains in Southwest Asia Compared with Simulated Rainfall

Reid A. BRYSON

Center for Climatic Research,
University of Wisconsin,
Madison, WI 53706 USA.

Abstract

Reconstructed history of the monsoon rainfall in Rajasthan shows that the monsoon was weak or absent in latest glacial time. With the advent of Holocene climatic patterns, fresh water lakes formed in dune fields and the pollen rain preserved in these reservoirs provides a basis for the reconstruction of the monsoon history. This history shows maximum monsoon amounts in the early Holocene, with a roughly two-thirds decrease to the present, and salinization in Sub-Boreal time. There appears to have been long intervals of near complete desiccation in the last four millennia. The variability of the winter rains is especially pertinent to the past climate of regions to the far west of Rajasthan.

Introduction

The scientific traveler to northwestern India cannot fail to notice the impressive aerosol loading of the atmosphere in the dry season. This aerosol loading is (a) deflated soil material (Bryson et al., 1964; Bryson, 1972) and (b) significant to the climate of the region (Bryson and Baerreis, 1967). These facts in turn raise questions as to the history of the terrigenous aerosol in terms of its relation to cultural history (since it appears to be associated with human activity).

In response to these questions, a program of paleoclimatic reconstruction for Rajasthan was undertaken through a contract with Dr. Gurdip Singh of the Birbal Sahni Institute of Paleobotany in Lucknow, U.P., India. Some results of this investigation will be described below, placing the reconstructed climate in a broader global context where possible.

NATO ASI Series, Vol. I 49
Third Millennium BC Climate Change
and Old World Collapse
Edited by H. Nüzhet Dalfes, George Kukla,
and Harvey Weiss

Method

The basic data for the study consisted of pollen profiles from sediment monoliths excavated from the bottoms of dry or seasonally dry lake beds. The details of these profiles are given in Singh et al. (1972, 1974). In addition, 114 surface samples of pollen were collected at 22 different sites in northwestern India for which climatic data were available (Singh et al., 1973).

A transfer function for relating present day pollen rain to present day climate was then constructed using the method of Webb and Bryson (1972), and this function was applied to fossil pollen assemblages obtained from the sediment monoliths.

The Lunkaransar Climate Profile

Figure 1 shows the reconstructed history of the rainfall in the Lunkaransar Lake vicinity, 65 km from Bikaner in northwestern Rajasthan at 28 30' N, 73 45' E from the time of lake formation to the present. This lake formed in a dune field at the time of the major transition to Holocene climates about 10,800 yr BP. The rising water table at that time, necessary to produce permanent freshwater lakes in the low areas, implied that significant monsoon rains did not appear prior to about 10,800 yr BP.

The lake appears to have disappeared during periods with annual precipitation less than about 200 mm. This supports the conclusion, obtained by general circulation modeling of the 18,000 yr BP climate, that a northwestern Indian summer monsoon was not characteristic of late Pleistocene conditions (Manabe and Hahn, 1977).

Though fluctuating, the summer monsoon rainfall appears to have averaged around 500 mm per season for the first 6000 yr of the history of the lake. This, plus the winter rains ranging up to 500 mm or so per season, seems to have maintained fresh water in the lake and stabilized the dunes (Singh, 1971, p.184). Indeed, agriculture appears to have arrived very early in the various parts of Rajasthan including the area of Lunkaransar Lake, as shown by the appearance of cultigens in the pollen rain and charcoal fragments from what may be stubble burning (Singh, 1971). Dates of first evidence of cultivation range from 9400 yr BP. at Lunkaransar to about 8000 yr BP. at Sambhar. By 5000 yr BP. the Indus Culture was well established in the region, with a large number of upland occupation sites where dominantly rainfed agriculture would seem to be indicated. Location of such major cities as Harappa and Mohenjo Daro on the Indus floodplain suggests to some that the culture was essentially riverine (Agrawal, 1971). However, most of the sites are not on floodplains. at least in Rajasthan (Ghosh, personal communication, 1966), and the Lunkaransar reconstruction suggests not only far more summer rainfall than the present but maximum winter precipitation as well during the Indus Culture interval of 5000-3500 yr BP.

Using the Lunkaransar pollen profile, Singh (1971) interpreted the 5000-3500 yr BP period as the wettest in the past 10,000 yr. The reconstruction shows total annual rainfall not much different at that time than in the previous centuries. Singh's interpretation was based on an increase in the percentages of several pollen types, including Syzygium, Pinus, and Artemisia, that are normally found in relatively moist environments. He did not note that these taxa have spatial distributions that have high positive correlations with the winter rainfall pattern rather than the summer monsoon (Table 1). The Indus period may have thus had higher precipitation efficiency than earlier or later, due to reduced net loss of water during the winter. Certainly during the intervals of maximum winter rain during Indus time there must have been a net gain during the winter.

Around 3700 yr BP the lake became saline, shown by halophytes replacing Typha and other freshwater aquatic species in the pollen record (Singh, 1971: Singh et al., 1972), and within a few centuries dried up entirely. At this same time the dunes once more became mobile, suggesting a significant reduction of the vegetative cover, and the Indus Culture disappeared (Singh, 1971; Agrawal et al., *f*01964). A number of the old towns were buried under sand dunes as at Osian, near Jodhpur. There appear to be no dated occupation sites during the next 700 yr or so in the whole region. Indeed, the last Indus Culture date is within the uncertainty of radiocarbon dating of the time of desiccation of Lunkaransar and Sambhar Lakes and the remobilization of the dune fields.

The first Painted Grey Ware or Aryan sites were occupied about seven centuries after the last Indus dates and were not generally re-occupation of the Indus sites (Ghosh, 1966, personal communication). Osian, where the main temple was built about 900 BC. on dune sand over a previous town-site, is a case belonging to the minority. Since this appearance of a new culture appears before the re-establishment of a seasonal salt lake about 2000 yr BP, one might reason that the aridity had diminished enough for at least nomadic herdsmen to occupy much of the region.

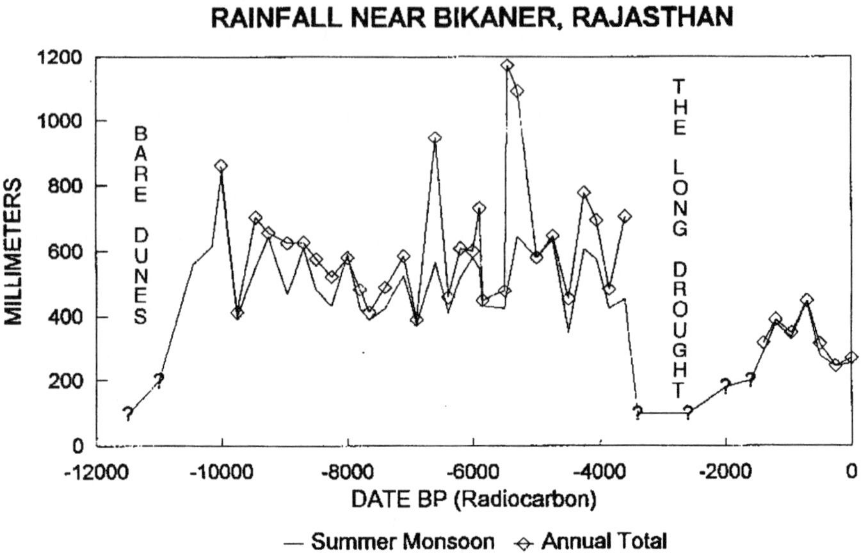

Fig. 1. Summer (June-October) and annual precipitation at Lunkaransar Lake, Rajasthan reconstructed from pollen percentages by the canonical correlation method of Webb and Bryson (1972).

After the deposition of the sterile saline sediments of the "long drought" (ca. 3600-2000 yr BP), the pollen rain into the lake indicates, when translated into seasonal rainfall amounts, a summer monsoon only half as copious as the pre-drought value, and negligible winter rain. Within this era, the highest computed amounts occurred between 700 and 1100 yr BP, plus or minus the error of interpolation between radiocarbon dates.

Implications of the Lunkaransar Profile

There is no in situ data to test the validity of the monsoon reconstruction other than the mobility of the dunes and absence of occupation during the aridity shown by the desiccation of the lake. However, there are some consistency tests that might be applied to the profile. Because of the inter-connectedness of the climatic system, there must be climatic events elsewhere that are related to the monsoon fluctuations.

One qualitative measure has already been mentioned. During late glacial times, i.e., prior to 10,800 yr BP., the monsoon was suppressed. This is consistent with the modern climatic condition in that the monsoon rains are a summer phenomenon and appear in India after the spring contraction of the circumpolar vortex has taken the jet stream north of the Himalayas (Yin, 1949). Certainly the late glacial time was winterlike. It would be reasonable, then, to expect that one should be able to find evidence of high latitude cooling and an expansion of the Arctic climates during the "long drought".

There is such evidence in the paleosols of northern Canada. These fossil soils show a marked expansion of tundra at the expense of boreal forest starting about 3600 yr BP. (Bryson et al., 1965; Sorenson et al., 1971). By 2000 yr BP. the forest had readvanced to near its present position but only half way to the position it had occupied prior to 3600 yr BP. This failure of the Arctic climatic region to shrink to its former size (as indicated by tundra extent) is consistent with the failure of the monsoon to regain its former intensity. A second retreat about 700800 yr BP. followed a readvance of the forest halfway from its present and 2000 yr BP. position to the pre3600 yr BP. position. This would imply scanty monsoon rains at 2000 yr BP. and the present, with larger amounts around 800 yr BP. This is what the Lunkaransar profile in Figure 1 shows. For 2000 yr BP. we must assume less than present day rainfall to explain the complete desiccation of the lake.

The Winter Rains

According to Naqvi (1962), winter rains in northwest India and Pakistan are almost entirely associated with cyclonic storms that originate in the eastern Mediterranean. These storms, as expressed at the surface, move through Mesopotamia, along the Mekran and Las Bela coasts, and wind up in northwest India and Pakistan. Being winter storms, they should produce rains in the lowlands of the "Fertile Crescent" and snows on the Toros and Zagros Mountains, on the headwaters of the Tigris and Euphrates.

Figure 2 shows the winter rains in Rajasthan derived from the pollen data, as a function of date converted to calendar year (using the Stuiver calibration). The mid-Holocene appears to have had the heaviest episodes of heavy winter rain, in some centuries exceeding the amount of the summer monsoon (see Fig. 1). These times must have been times of either numerous or strong storms following the Cyprus-Mesopotamia-Mekran-Rajasthan track. The spacing of the data points does not allow great precision in specifying which decades or century had the very heavy rains.

Of particular interest is the largest peak of winter rainfall about 5000 years ago, or 3000 BCE. It is nearly coincident with the "Piori oscillation" of the Alpine glaciers and the end of the Atlantic climatic episode.

Mesopotamia

The precipitation of Israel, Syria, and Iraq is almost entirely a function of winter storms associated with the westerlies. In fact, the monthly precipitation at Kameshli, in northern Syria correlates 0.95 with the number of cyclonic centers in the area. At Hasseke, Syria, the correlation is 0.93. These storms are associated with the southern branch of the Eurasian jetstream system.

Using the techniques described in Bryson (1992), the relationship between the position of the jetstream and the precipitation, the history of precipitation at several Syrian locations was simulated (Fig. 3).

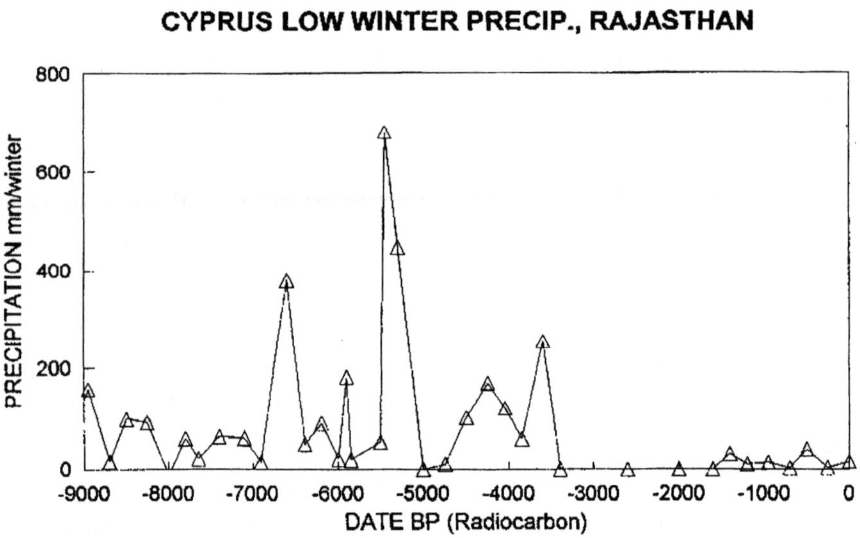

Fig. 2 Winter rains near Bikaner, Rajasthan reconstructed from fossil pollen. This record appears to be related to the number of "Cyprus lows" that pass through Mesopotamia to reach northwest India and Pakistan.

Discussion

The interpretive problem posed by the differences between figures 2 and 3 has not been resolved. It is clear that winter rainfall in northwest India and Pakistan is related to Mediterranean storms, but only to those which do not disappear before reaching the area, or change to a different track. On the other hand, Syrian precipitation responds to essentially all storms which leave the eastern end of the Mediterranean.

Figure 2, the simulated history of winter precipitation in Kameshli, Syria, is based on the calculated latitude of the jetstream. It probably is a good estimate of the number of storms that cross the region, for it is an observational fact that essentially all of the precipitation of this region is with the latitude of the jet south of the latitude of the station at which the precipitation is measured.

Unfortunately, the number of storms crossing Syria is not a good measure of how many reach northwestern India. How to relate the two remains as future research.

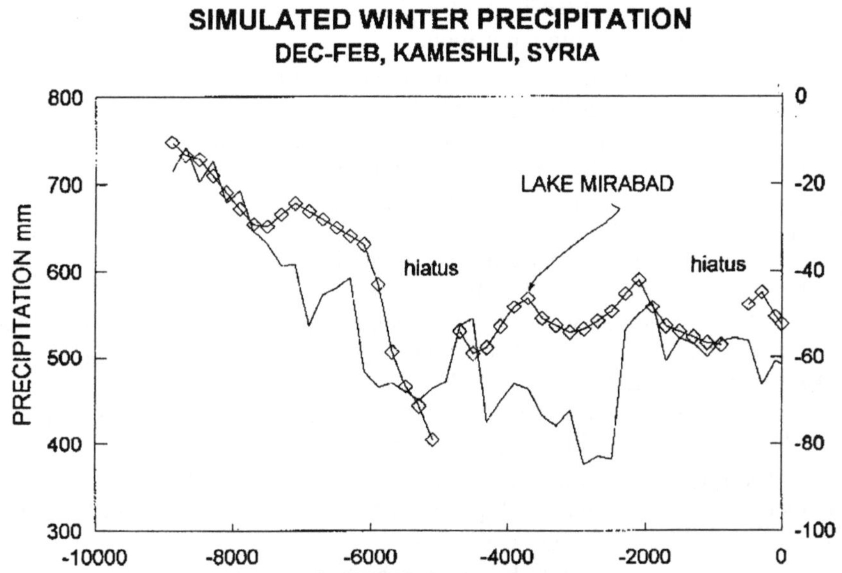

Fig. 3. Simulated winter rainfall at Kameshli, Syria compared to oak pollen at Lake Mirabad in Iran.

References

Agrawal, D. P. (1971). "**The Copper Bronze Age in India**." Munshiram Manoharlal, New Delhi.

Agrawal, D. P.. Kusumgar. S., and Sarna, R. P.Q(1964). Radiocarbon dates of archaeological samples. Current Science **33**, 226.

Bryson, R. A. (1972). Climatic modification by air pollution. In "**The Environmental Future**" (N. Polunin, Ed.), pp. 133-154. Macmillan, New York.

Bryson, R.A., (1992), "A Macrophysical Model of the Holocene Intertropical Convergence and Jetstream Positions and Rainfall for the Saharan Region" Meteor. and Atmos. Phys. **47**, 247-258

Bryson, R. A., and Baerreis, D. S. (1967). Possibilities of major climatic modification and their implications: Northwest India, a case for study. Bulletin of the American Meteorological Society **48**, 136-142.

Bryson, R. A., Irving, W. M., and Larson, J. A. (1965). Radiocarbon and soils evidence of former forest in the southern Canadian tundra. Science **147**, 46-48.

Bryson, R. A., Wilson, C. A., III, and Kuhn, P. M. (1964). Some preliminary results of radiation sonde ascents over India. **Proceedings WMO-IUGG Symposium on Tropical Meteorology, Rotorua, New Zealand, November, 1963**, Wellington, New Zealand Meteorological Service.

Ghosh, A. (1966). Personal communication.

Manabe, S., and Hahn, D. G. (1977). Simulation of the tropical climate of an Ice Age. Journal of Geophysical Research **82**, 3889-3911.

Naqvi, S. N. (1962). Data presented at Conference on the Climate of the Eleventh and Sixteenth Centuries. Aspen. Colorado. June 1624. 1962.

Roberts, N, and H.E. Wright Jr., (1994). "Vegetational, Lake-Level, and Climatic History of the Near East and Southwest Asia", p.194-220 in H.E. Wright Jr. et al., (1994), **Global Climates since the Last Glacial Maximum**, Minneapolis, Univ. of Minnesota Press, 569 pp.

Singh, G. (1971). The Indus Valley Culture seen in the context of postglacial climatic and ecological studies in northwest India. Archaeology and Physical Anthropology in Oceania **6**, 177-189.

Singh, G., Chopra, S. K., and Singh, A. B. (1973). Pollen rain from the vegetation of northwest India. New Phytologist **72**, 191-206.

Singh, G., Joshi, R. D., and Singh, A. B. (1972). Stratigraphic and radiocarbon evidence for the age and development of three salt lake deposits in Rajasthan, India. Quaternary Research **2**, 496-505.

Singh, G., Joshi. R. A., Chopra, S. K., and Singh, A. B. (1974). Late Quaternary history of vegetation and climate of the Rajasthan Desert, India. Philosophical Transactions of the Royal Society of London **B 267**, 467-501.

Sorenson, C. J., Knox, J. C., Larson, J. A., and Bryson, R. A. (1971). Paleosols and the forest border in Keewatin. Quaternary Research **1**, 468-473.

Webb, T., III, and Bryson, R. A. (1972). Late- and post-glacial climatic change in the northern Midwest, USA: Quantitative estimates derived from fossil pollen spectra by multivariate statistical analysis. Quaternary Research **2**: 70-115.

Yin, Maung Tun (1949). A synoptic-aerologic study of the onset of the summer monsoon over India and Burma. Journal of Meteorology **6**, 393-400.

Monsoon Regime Variations During the Late Holocene in the Southwestern India

Ilhem BENTALEB[1], Claude CARATINI[2], Michel FONTUGNE[1], Marie Thérèse MORZADEC-KERFOURN[3], Jean Pierre PASCAL[4], Colette TISSOT[2]

[1]Centre des Faibles Radioactivités, Laboratoire mixte CNRS/CEA, Domaine du CNRS, 91198- Gif-sur-Yvette cedex, France
[2]Institut Français, P.O. Box 33, 605001-Pondicherry, India
[3]Institut de Géologie, Université Rennes I, 35042-Rennes cedex, France
[4]Laboratoire de Biométrie, Génétique et Biologie des Populations, Université Lyon I, 69622-Villeurbanne Cedex, France

Abstract

A marked change in vegetation of the Western Ghats at ca. 3500 yr BP (uncalibrated ^{14}C yr) has been observed from a study of two marine cores taken from the inner shelf off Karwar (Karataka, western India), near the mouth of the Kalinadi River. This change is indicated by decrease in tree pollen, particularly those of evergreen forests, reduction of mangrove derived pollen. and by corresponding increase of savanna grass pollen. At the same time, the stable carbon isotopic ratios of organic matter increased abruptly, indicating a higher contribution of organic matter of marine origin and consequently reduced input of terrestrial organic matter due to reduced flow of the Kalinadi River. All these changes point to a less humid climate and shorter rainy season in the region, and are unlikely to result from human activity. The pattern of drier climate associated with forest reduction, increase of savanna elements, decrease of the mangrove vegetation, lower fresh water runoff of the Kalinadi River, continued up to ca. 2200 yr BP. Since then, the pollen assemblages have remained uniform up to the present day. We conclude that the climate similar to the present one was established in the studied area about 2200 yr ago.

Introduction

Though the general trends of the climatic evolution in the Indian monsoon area are now well documented for the last 150kyr (Duplessy, 1982; Van Campo et al, 1982; Prell and Kutzbach, 1987), few attempts have been made to study the climatic

NATO ASI Series, Vol. I 49
Third Millennium BC Climate Change and Old World Collapse
Edited by H. Nüzhet Dalfes, George Kukla, and Harvey Weiss

fluctuations during the Holocene. Recent studies have shown that short climatic events have occurred over the last 24, 000 years (Sirocko et al., 1993, for the southwest Arabian Sea; Gasse and Van Campo, 1994, for the Holocene in tropical Africa and the mountainous regions of south China). However, for the southwestern India no data have been published. The climate of this region of the Indian peninsula is at present characterized by the southwest summer monsoon regime, with two contrasting seasons, dry and rainy. The rainy season begins at the end of May in the extreme south and progresses steadily northwards up to July. Then, it retreats gradually at a slower pace from September to November. Nevertheless there is a high inter-annual variability in the length of the two seasons, as well as in rainfall (Pascal, 1982), Soman et al. (1988) pointed out a reduction of monsoon rainfall since the beginning of the century. For a better understanding of climatic evolution over the last millennia, a continous paleoclimatic record from about 5000 yr BP to the present with a time resolution less than 100 yr was obtained from a study of palynomorphs and isotopic markers of cores taken from the inner shelf off Karnataka. Our results are derived from three independent types of proxy data, and are compared with the global climatic history of this period in low-latitude regions.

Sites

The Karnataka region, especially its coastal plains, is characterized by numerous short rivers flowing from the western Ghats mountains to the Arabian Sea through a complex system of lagoons and backwaters. The Kalinadi River presents an exception because it flows directly into the sea through a steep-sided narrow valley. The sediments are transported directly to the sea without any complications resulting from the usual meanderings in estuarine plain formations. The topography of the inner shelf is regular, forming a gentle slope covered by clayey silt (Hashimi and Nair, 1981). Thus the marine deposits should record paleoclimatic signals representative of a fairly large region represented by the catchment basin drained by the Kalinadi. This is large, the largest of the central part of the Western Ghats. In this basin, where the dry season average 6-7 months (with extremes of 5-8 months), the mean rainfall ranges from 3300mm in the coastal region to less than 1000mm in the eastern part, on the Karnataka Plateau; values exceed 6000 mm at the highest elevations. This wide range supports several types of vegetation (Pascal, 1988): evergreen forest of the low altitude type is the most widespread in this part of central Ghats, while humid deciduous forest or dry deciduous forest occur in places; mangrove forests are scattered along the coastal estuaries. The wind regime, following the general monsoonal pattern (Rao et al, 1992), suggests that airborne allochtonous pollen should not be expected in significant amounts in the pollen spectra. The influence of marine surface currents is small since the waters within the Bay of Karwar are mainly under the influence of the Kalinadi as indicated by their variable salinity (28 ‰ in summer; 34 ‰ in winter).

Selected Markers

We have selected three different independent markers in order to provide information about the precipitation over the Western Ghats: inland biological markers (evergreen deciduous and mangrove forests, and savanna pollen), marine palynological markers (dinoflagellate cysts) and stable organic carbon isotopes. The selected taxa require precise climatic and environmental conditions for their optimal development and thus can provide reliable clues for paleoclimatic reconstructions.

Data were obtained from core SK 27B/8 taken off Karwar (14° 49' 43N, 73° 59' 31E) at a depth of 22 m close to the mouth of the Kalinadi River and directly downstream of the continuation of the estuary (Figure-1).The radiocarbon datings were obtained on sediment organic matter. The conventional ^{14}C ages are given according to Stuiver and Polach (1977) and are uncalibrated. A second core SK 44/13 (14°43'80N; 74°02'64E, 25 m depth) in the same area provides similar palynological results as those recorded in the core SK 27B/8 (Caratini et al, 1994).

Evergreen forest

The different taxa present in these regions have been described in Caratini et al (1994). The western Ghats are subject to a latitudinal climatic gradient with the length of the dry season regularly increasing northwards and inducing a latitudinal distribution of vegetation. Some taxa such as *Cullenia exarillata Agrostitachys* and *Dipterocarpus* are accurate markers of humid conditions since they are not presently found in the Kalinadi basin Nowadays, they grow only in the more southern areas where rainy season lasts longer than 6-7 months instead of 5-6 months for the Karwar region. Taxa like *Allophyllus cobbe*, *Aporosa indica* characterises forest clearings and margins. Any changes in the ratio between the percentage of evergreen forest and that of clearings may signal a disturbance in natural environment probably resulting from human activity.

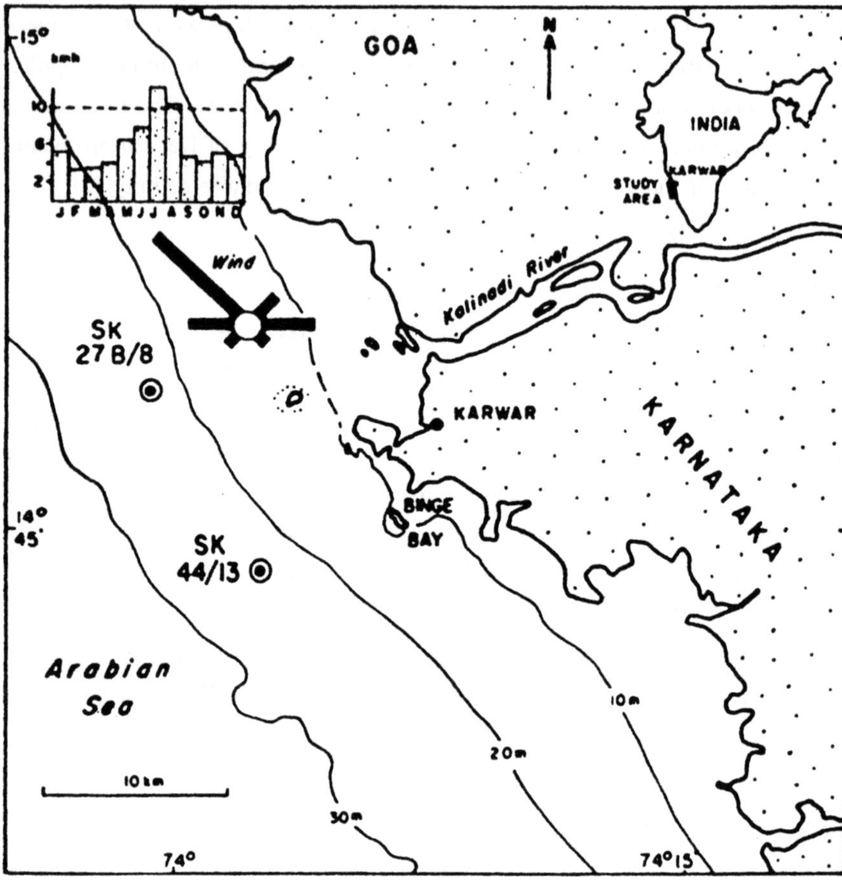

Figure 1. Map of the study area: core location, wind distribution and average monthly intensity.

Agrostitachys and *Dipterocarpus* are accurate markers of humid conditions since they are not presently found in the Kalinadi basin Nowadays, they grow only in the more southern areas where rainy season lasts longer than 6-7 months instead of 5-6 months for the Karwar region. Taxa like *Allophyllus cobbe*, *Aporosa indica* characterises forest clearings and margins. Any changes in the ratio between the percentage of evergreen forest and that of clearings may signal a disturbance in natural environment probably resulting from human activity.

Deciduous forests

As their development is conditioned by the intensity of the rainfall they are also good climatic indicators. However, they can be useful in recording even a small fall in humidity which would lead to a decrease in the area under evergreen forest with a concurrent rise in the extent of deciduous forest.

Savanna

In pollen spectra, Poaceae is the main taxon representative of savannas which are generally grass-dominated environments with scattered trees and shrubs. Though this family may also occur in deciduous forest as is often observed in the southern part of the western Ghats this is not the case for the Kalinadi basin. A larger representation of savanna taxa is interpreted as indicating less precipitation.

Mangrove forests

Mangrove forests grow only in coastal regions influenced by brackish or sea water. Thus, their optimal development occurs during the most humid intervals. As mangroves do not compete with inland formations, variation of their pollen and spores are reliable markers of rainfall (Thanikaimoni, 1987; Van Campo et al., 1982).

Marine palynological markers

In the marine sediments, dinoflagellate cysts provide reliable clues about paleoenvironmental aspects such as depth, distance from the coast (Wall et al., 1977; Caratini at al, 1987) and water chemistry. Recent studies attempt to understand the relationships between the distribution of dinoflagellate cysts and environmental factors (Bradford et Wall, 1984; Morzadec-Kerfourn, 1988a, b). As the sediments of core SK27/B8 have been deposited after the stabilisation of the Holocene transgression (shoreline and depth were fixed), variations in dinoflagellate cyst assemblages would reflect changes of physico-chemical properties of waters mainly resulting from change in the ratio between fresh and marine waters.*Tuberculodinium vancampoae* appears to be a species especially tolerant to low salinity.

Organic carbon stable isotopes

The $^{13}C/ ^{12}C$ ratio ($\partial^{13}C$) of organic matter in marine sediments has been routinely used to infer the relative proportion of marine and terrestrial organic carbon in a given sediment (Sackett and Thompson, 1963). In most sedimentary environments, marine planktonic organic carbon is isotopically heavier than that of the carbon delivered to the ocean from terrestrial environments (Deines, 1980). Solution of the simple mixing equation $\partial^{13}C = F_t * \partial^{13}C_t + F_m * \partial^{13}C_m$, with $F_m + F_t = 1$, where F_t and F_m are the fractions of terrestrial and marine organic matter and $\partial^{13}C_t$ and $\partial^{13}C_m$ are the isotopic compositions of the terrestrial and marine organic carbon sources, respectively, provides a method for estimating the proportions of the two types of organic matter present in a given sample. Making such estimates requires an accurate knowledge of the isotopic compositions of the marine and terrestrial end-members (Fontugne and Duplessy, 1986). Only very small changes in the isotopic composition of terrestrial organic material have been recorded even during glacial to interglacial transition, so that the isotopic composition of this source material is not likely to have varied significantly. Numerous plankton samples were analysed for seasonal variations in carbon isotopic composition over a three-year period. The planktonic $\partial^{13}C$ values do not show significant variations (Fontugne and Duplessy, 1986). During the late Holocene, the carbon isotopic values of marine phytoplankton were rather constant since the temperature and partial pressure of carbon dioxide in the atmosphere and consequently in surface sea waters which govern the variations of this parameter do not show significant variations (Fontugne and Calvert, 1992). Consequently, the carbon isotopic values chosen for marine and terrestrial end members, on the base of these previous studies, are respectively -20 and -26 ‰ versus P.D.B. During the period recorded by the core (4.5 kyr), sea-level variations were less than ± lm and could not induce a significant increase or decrease of terrestrial organic matter input due to erosional variations. Under the observed conditions, variations of terrestrial organic carbon input can only result from river flow variations (Fontugne and Jouanneau, 1987, Bentaleb et al., 1994).

Paleoenvironmental conditions and paleoclimatic trends

The ecological markers (Figure-2) selected and the isotopic signal (Figure-3) show converging trends corresponding to a clear response of the environment to past changes in the climate. Thus, the climatic evolution can be described through three sequence during the last 5000 years BP.

Sequence I (5-3, 5 kyr BP)

This sequence starts with a high content of evergreen and mangrove pollen indicating humid conditions. At 4.3kyr BP, the pollen spectra of inland

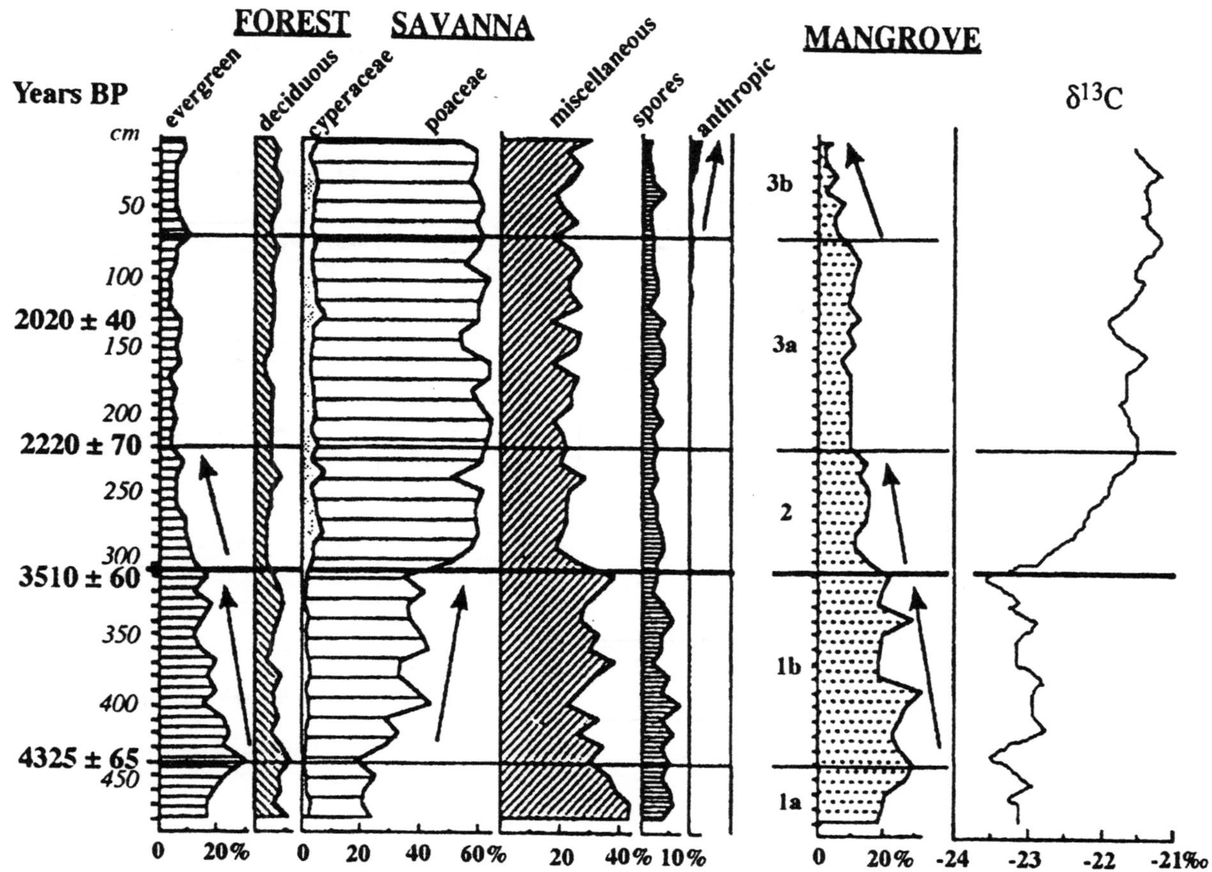

Figure-2: Zonation of core SK27B/8 based on pollen, spores and $\partial^{13}C$ of sedimentary organic matter. Frequencies are expressed as percentage of the sum of inland pollen and spores for inland vegetation (mangrove excluded) and as percentage of the total number of pollen and spores for mangrove vegetation.

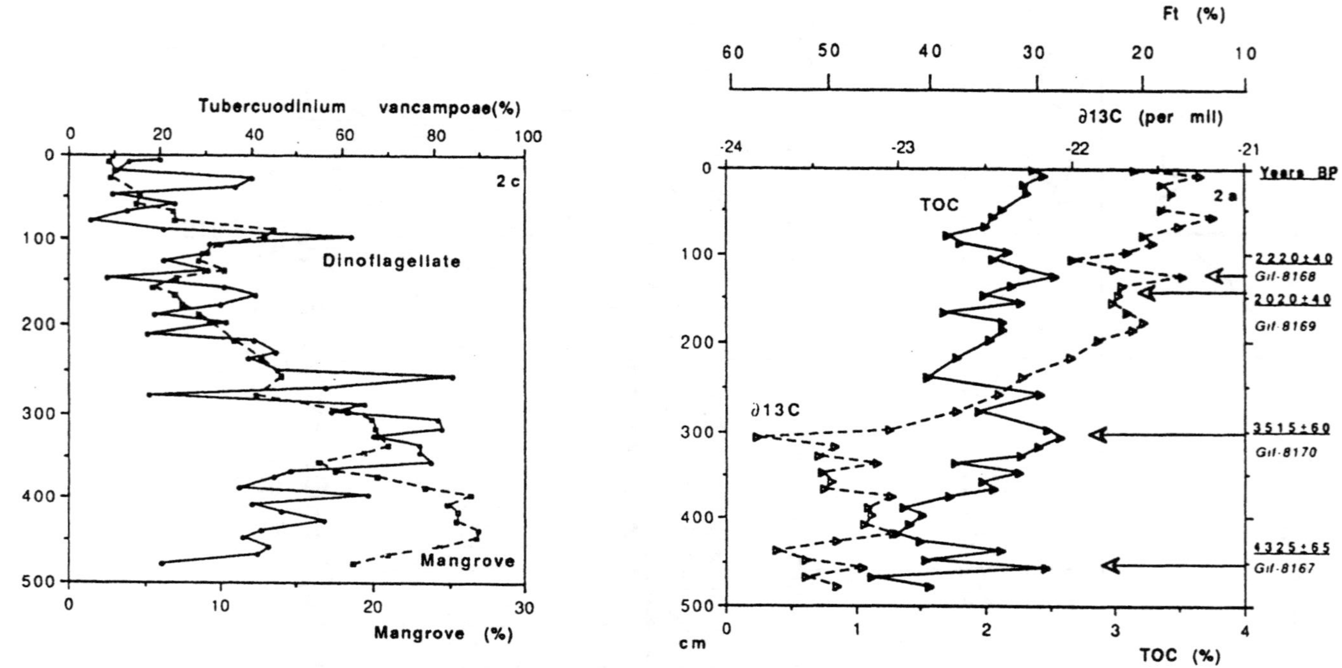

Figure-3 Variations within the core SK27B/8 of a) the organic carbon content and isotopic composition and the terrestrial organic carbon input b) of percentage of the total number of Gonyaulacoid Dinoflagellate cysts (related to autotrophic dinoflagellate) for *Tuberculidinium vancampoe*. and mangrove pollen.

vegetation is dominated by forest markers but the main changes are a substancial increase in Poaceae from 15 to 30 % accompanied by decreasing values for evergreen and mangrove forests. This suggest that humidity had been declining. However these changes in climate probably progressed with only a moderate intensity, dispaying only a 2-3 % fall in both deciduous forest and pteridophytes spores. The $\partial^{13}C$ values vary between -21, 5 to -23, 5 ‰ and show no correlation with organic carbon content. From 4500 to 3500 yr BP, carbon isotopic composition is rather stable around -23, 3 ‰ indicating a slight predominance of the terrestrial fraction.

Abrupt reduction of rainfall around 3500 years BP

At around 3500 years BP, an increase of $\partial^{13}C$ values is recorded indicates a reduction of the terrigenous organic matter input (figure 3a). Such a change can only be related to a decrease of river flow since the export of terrestrial matter is directly linked to the strength of the flood (Fontugne and Jouanneau, 1987) and the decreased land drainage. The rapid changes in vegetation occurring within a short period as shown by sedimentation rate also record reduction of precipitation. The pollen spectra indicate an increase of savanna extent and the corresponding fall in the inland forest both evergreen and deciduous. This results from a climatic change of primary importance since a small reduction in rainfall would have led to a decrease in evergreen forest but an increase in deciduous forest. Mangroves were also strongly affected. This limit, at 3500 yr BP, is also clearly recorded by the decrease of *Tuberculodinium vancampoae*, a dinoflagellate cyst characteristic of littoral zones in tropical areas (Figure-3b). The question whether such a reduction in forest can be attributed to human activity rather than to less precipitation is a subject of debate. Human activity cannot simply be ruled out in this region where human settlements and the beginning of cultivation are generally dated at around 3000 years BP following the introduction of iron smelting (Gadgil and Subash Chandran, 1988). Nevertheless, several palynological grounds can be cited to support climatic change against human activity, the most conspicuous being:

(1) the occurrence of *Cullenia exarillata*, *Agrostitachys* and *Dipterocarpus*, is restricted to Sequence I. As mentioned above, these taxa are not found in the Kalinadi Basin at present but grow only in the southern part of the Western Ghats where the rainy season is longer;

(2) the consistent values for the ratio between core evergreen forest pollen and these of the edges and openings on either side of the limit, these values would have

been modified by any man-induced clearings as can be seen in the upper part of the core;

(3) the quasi-absence of anthropic pollen is another noteworthy feature, with only some rare *Cocos nucifera*, and native species being recorded during this period.

Evidence from carbon isotopic record also supports climatic change.

(1)The tropical savanna grass may photosynthesize via the C4 pathway, yielding $\partial^{13}C$ values around -12 ‰. If this is the case, their contribution to the shelf sediment would lead to an increase of $\partial^{13}C$ values and consequently the shift observed at around 3500 yr BP would not reflect decrease in the river flow. This interpretation has to be discarded since no evidence of the occurrence of C4 plants is available and no correlation between the variations of $\partial^{13}C$ values and savanna pollen abundance is observed. Furthermore, Quay et al. (1992) show that contribution of C4 plants is low because of their preferential oxidation in the river and Mariotti et al. (1991) show that organic carbon isotopic composition of suspended material in the Congo river presents typical C3-plant values in spite of the occurence of large savanna areas covered by C4 plant in the hydrographic basin. Nevertheless, if we postulate a C4 plant signal in the shelf sediment core at 3500 years BP, its ecological significance is an increase of drier conditions and consequently lower precipitation rates and decrease of river flow.

(2) Conversely, the $\partial^{13}C$ increase cannot reflect human activity since deforestation would lead to an increased soil erosion and transport of organic matter with lower $\partial^{13}C$ values (-26 ‰) representative of C3 plant derived soils.

Sequence II (3.5 to 2.2 kyr BP): further reduction in moisture

Sequence II is a transitional zone, the composition of vegetation continues to change in the same direction and the trends marking the primary limit persist but are considerablely weaker. Savannas pollen increase from 50 to 60 % while inland forests and mangroves declines. Since 2220 years BP, $\partial^{13}C$ values stabilize around -21, 5 ‰ indicating that the terrestrial contribution was reduced to about 25 %.

The 2.2 kyr BP limit: establishment of present climate

If we consider the pollen and spores record of inland vegetation and mangrove and the origin of organic matter, the 2200 years BP limit is less defined than the 3500 boundary. Nevertheless, this limit corresponds to a clear climatic change since most of the markers reach a rather constant value that approaches the present values.

Sequence III- (2.2 kyr to present): stability of climatic conditions

The upper part of the pollen spectra present fluctuating values without any relevant change: savanna remaining the principal source area for pollen supply. This pattern marks the establishment of the current status of vegetation over the central part of the western Ghats. This period also corresponds to the stabilization of the marine palynomorph asssemblage which continues until today and shows a greater marine influence (Caratini et al., 1994). Mangrove pollen and spores reduction is observed during this last phase. This change was most probably due to clearing of mangrove forest for land reclamation and cultivation, since in older sequences, when human activity could not have been a factor affecting the vegetation, mangrove and inland forest followed the same evolution. Moreover, if we consider the curve of mangrove pollen and the terrigenous fraction of sedimentary organic carbon the 0.7 m limit is conspicuous. These two curves present good covariation below 0.7 m in older sequences, while the supply of organic matter do not change in the upper part of sequence III, the decrease in quantity of mangrove pollen and spores at that time is another reason to consider this change as a result of human activity. Occurrence of increasing anthropic taxa like *Cocos nucifera* and *Oryza sativa* is noted while recently introduced species such as *Eucalyptus*, *Mimosa pudica* and *Casuarina* are present in the last centimetres at the top of the core.

Conclusion

In the western part of south India, from our data obtained with three independant markers, the evolution of the late Holocene climate can be summarised as follows:

from 4500 to 2200 year BP a gradual reduction in moisture, with an abrupt intensification in this trend at around 3500 yr BP taking place within a short period.

from 2200 BP to the present a continuous record of all markers implies a relative stability in climatic conditions which seem to have undergone no changes of major importance. The actual cause which led to the onset of a drier climate in the Kanara hills cannot be clearly elucided. It could be the result of decrease in rainfall or an increase of the duration of the dry season. Since the current average rainfall is high enough (above 3000mm) to sustain rain forest vegetation, any limitation on forest development is more likely to be due to the length of dry season which may have been shorter than its present 6-7 months length.

Such a pattern in the alteration of the climate is rather well documented in the Rajasthan desert from palynological work of Singh et al (1974, 1990) confirmed by Krishnamurty et al (1981) or Courty (1990) (See also Bryson this volume). Paleoclimatic reconstruction by Bryson and Swain (1981) or Wasson et al (1984) agree that there has been a reduction of precipitation during both winter and summer in the Thar desert. From an archeological point of view, in the northwestern India the 3500yr BP event corresponds to the decline and the end of the Harappan Culture, while the establishment of a present-day type of climate at 2200 yr BP appears to be associated with the Rangmahal Culture which found adequate water supplies and probably obtained a partial resuscitation of the river system. However, both for Rajasthan and Southwest India, paleoenvironmental changes appears to be dependent on climatic change rather than the result of human activities. The increasing aridity has also been described for the same period from several cores taken in the northern Arabian Sea (Caratini et al, 1981 & 1994; Bryson 1989). Similar phenomena which have been described at 3500 year BP in the many parts of the tropical belt (see Caratini et al, 1994; Gasse and Van Campo, 1994) argue for the existence of a global climatic phenomenon. More studies are needed, however, to understand its true significance or its link with the Broecker's salt oscillator theory.

Acknowledgments

We thanks Dr B.N. Desai, Director of the National Institute of Oceanography in Goa, Dr R.R. Nair and Dr R. Nigam who provide sediment samples, S. Martin, N. Noury and M. Rousseau for preparing samples for dating.

References

Bentaleb I, Fontugne M, Caratini C, Morzadec-Kerfourn MT, Tissot C, (1994) River paleoflow evolution in the west Indian coastal region: organic carbon isotopic evidences from marine sediment cores for a drier climate since 3500 years BP. *C.R.Acad.Sci.*,Paris, 317, 713-718.

Bradford MR, Wall DA, (1984) The distribution of recent organic walled dinoflagellate cysts in the Persian Gulf,gulf of Oman and northwestern Arabian Sea. *Paleontographica*, B, 192, 16-84.

Bryson RA, Swain AM, (1981) Holocene variations of monsoonal raifall in Rajasthan. *Quat. Res.*, 16, 135-145.

Bryson R A, (1989) Modelling the NW India monsoon for the last 40 000 years. *Climate Dyn.*, 3, 169-177.

Caratini C, Bentaleb I, Fontugne M, Morzadec-Kerfourn MT, Pascal JP, Tissot C, (1994)- A less humid climate since ca. 3500yr BP from marine cores off Karwar, western India. *Palaegeogr. Palaeoclim. Palaeoecol.*, 109, 371-384.

Caratini C, Bellet J, Tissot C, (1981) Etude microscopique de la matière organique: palynologie et palynofaciès. In: *ORGON IV. Golfe d'Aden. Mer d'Oman.* CNRS. Paris, 265-308.

Courty MA, (1990) Environnements géologiques dans le nord ouest de l'Inde. Contraintes géodynamiques au peuplement protohistorique. (Bassins de la Ghaggar-Saraswati-Chautang). Thesis, Univ. Bordeaux I (France), pp328).

Deines P, (1980) The isotopic composition of reduced carbon. In: *Handbook of Environmental Isotope Geochemistry*, P. Fritz and JC Fontes eds, Elsevier, Amsterdam, 329-406.

Duplessy JC, (1982) Glacial to interglacial contrast in the northern Indian Ocean.*Nature*, 295, 494-498.

Fontugne MR, Jouanneau JM, (1987) Modulation of the particulate organic carbon flux to the ocean by a macrotidal estuary: evidence from measurements of carbon isotopes in organic matter from the Gironde system. *Estuarine Coastal Shelf Sci.*, 24, 377-387.

Fontugne MR, Duplessy JC, (1986) Variations of the monsoon regime during the upper Quaternary: evidence from carbon isotopic record of organic matter in north Indian Ocean sediment cores *Palaegeogr. Palaeoclim. Palaeoecol.*, 56, 69-88.

Fontugne MR, Calvert SE, (1992) Late Pleistocene variability of the organic carbon isotopic composition of organic matter in the eastern Mediterranean: monitor of changes in carbon sources and atmospheric CO_2 concentrations.*Paleoceanography*, 7, 1-20.

Gadgil M, Subash Chandran MD, (1988) On the history of Uttara Kannada forest. In: J., Dargavel, K., Dixon & N., Semple (eds). *Changing tropical forests* Aust. Natl. univ Canberra.

Gasse F, Van Campo E, (1994) Abrupt post glacial climate events in West Asia and North Africa monsoon domains. *Earth Planet. Sci Lett.*, 126, 435-456.

Hashimi NH, Nair RR, (1981) Surficial sedimentsof the continental shelf off Karnataka. *J. Geol. Soc India*, 22, 1981, 266-273.

Krishnamurty RV, Agrawal DP, Misra VN, Rajaguru SN, (1981) Paleoclimatic inferences from the behaviour of radiocarbon dates of carbonates from sand dunes of Rajasthan. *Proc. Indian Acad. Sci. (Earth Planet)*, 90, 155-160.

Mariotti A, Gadel F, Giresse P, Kinga-Mouzeo, (1991) Carbon isotope composition and geochemistry of particulate organic matter in the Congo river. Application to the study of Quaternary sediments of the mouth river. *Chemical Geology (Isotope Geoscience Sect.)*, 86, 345-357.

Morzadec-Kerfourn MT, (1988a) Paléoclimat et paléoenvironnements du Tardiglaciaire au récent en Mediterranée orientale à l'est du Delta du Nil: l'apport des microfossile à membrane organique. *Bull. Centr. Rech. Explor. Elf Aquitaine*, 12, 267-275.

Morzadec-Kerfourn MT, (1988b) Distribution des kistes de dinoflagellés dans les sédiments pléistocènes de la marge guinéenne de l'Afrique (Equamarge I. 1983). *Palaegeogr. Palaeoclim. Palaeoecol.*, 65, 201-216.

Pascal JP, (1982) Bioclimates of the Western Ghats. *Trav Sect. Sci. Techn. Inst. Fr. Pondichery*, 17, 2 maps.

Pascal JP, (1988) Wet evergreen forest of the western Ghats of India. *Trav. Sect. Sci. Techn. Inst Fr. Pondichery*, 20bis, 345pp.

Prell W, Kutzbach JE, (1987) Monsoon variability over the past 150,000 years.*J. Geophy. Res.*, 92 (D7), 1987, 8411-8425.

Quay PD, Wilbur DO, Richey JE, Hedges JI, Devol AH, Victoria R, (1992) Carbon cycling in the Amazon River: implication from the ^{13}C compositions of particles and solutes. *Limnol. Oceanogr.*, 37, 1992, 857-871.

Rao AD, Dube SK, Sinha PC, (1992) Wind driven seasonal ocean circulation around southern Indian peninsula. *Indian J. Mar. Sci.*, 21, 196-200.

Sackett WM, Thompson RR,(1963) Isotopic organic carbon composition of recent continental derived clastic sediments of eastern gulf coast, Gulf of Mexico *Am. Assoc. Pet. Geol.Bull.*, 47, 525-531.

Singh G, Joshi R, Chopra S, Singh AB, (1974) Late Quaternary history of vegetation and climate in the Rajasthan desert, India. *Trans. R. Soc. London,* 267, 467-501.

Singh G, Wasson RJ, Agrawal DP, (1990) Vegetational and seasonal climatic changes since the last full glacial in Thar desert, northwest India. *Rev Paleobot. Palynol.*, 64, 351-358.

Sirocko F, Sarnthein M, Erlenkeuser H, Lange H, Arnold M, Duplessy JC (1993) Century-scale events in monsoonal climate over the past 24,000 years. *Nature*, 364, 322-324.

Stuiver M, Polach H, (1977) Discussion: reporting of ^{14}C data.*Radiocarbon*, 19, 355-363.

Thanikaimoni G, (1987) Mangrove palynology . *Trav. Sect. Sci. Techn. Inst Fr. Pondichery*, 24, 100pp.

Van Campo E, Duplessy JC, Rossignol-Strick M, (1982) Climatic conditions deduced from a 150kyr oxygen isotope-pollen record from the Arabian Sea.*Nature*, 296, 1982, 56-59.

Wasson RJ, Smith BJ, Agrawal DP, (1984) Late Quaternary sedimentary minerals and inferred geochemical history of Didwana Lake, Thar desert India. *Palaegeogr. Palaeoclim. Palaeoecol.,* 46, 345-378.

Third Millennium Climate in the Near East Based upon Pollen Evidence

Sytze BOTTEMA

Biologisch-Archaeologisch Instituut
(Groningen Institute of Archaeology)
Poststraat 6
9712 ER Groningen
The Netherlands

Abstract

Third millennium BC abrupt climatic change and the Old World social collapse, as suggested from archaeological evidence, is traced in pollen diagrams from the Near East. The third millennium pollen assemblages is followed closely in a selection of pollen diagrams from various locations in the Near East. The distribution of the pollen evidence is fairly even, but only one diagram is available from the Mesopotamian area itself.

The third millennium period in general and two dry periods within this era more specifically, are defined chronologically for the pollen diagrams. The given calendar dates were converted to uncalibrated years. The chronology of the pollen diagrams is based upon uncalibrated radiocarbon dates which, together with the sedimentation rate, were used to indicate the suggested drought periods.

The behaviour of plant and tree species in relation to climate is discussed. The concept 'response time' as the period of time necessary to cause a change in the growth curve of a plant species is explained. Long living trees react upon changes with a lower frequency than short-living annuals, for instance certain crops.

Eight pollen diagrams, distributed over the Near East, have been traced for types that might be indicative for abrupt climatic changes. One pollen diagram originates from Mesopotamia, the others come from the adjoining mountains or farther away in the Near East. As the water discharge in the Euphrates and Tigris depends on rain and snow in the Anatolian and Zagros Mountains, drought will also have affected that part of Mesopotamia.

NATO ASI Series, Vol. I 49
Third Millennium BC Climate Change
and Old World Collapse
Edited by H. Nüzhet Dalfes, George Kukla,
and Harvey Weiss

The pollen evidence does not indicate sudden drought, and no support for centuries-long dry periods is found. The only location where a dry period may have occurred is Bouara. Towards the final centuries of the third millennium peat is formed here in an otherwise open water system.

Introduction

For the reconstruction of the paleoclimate, palynology forms an important contributor. Botanical evidence obtained from Near Eastern pollen diagrams covers a period of about 50,000 radiocarbon years. Palynological investigations in the Eastern Mediterranean started in the early sixties and since then a fairly great of pollen diagrams distributed over the area, have been prepared. The area for which we have obtained palynological information is shown in fig. 1. Those parts that receive less than 300 mm precipitation nowadays, produced almost no pollen bearing sediments. Some sites date back to halfway the last glacial period but the majority of the pollen records is of Holocene origin (Van Zeist & Bottema 1991). The pollen archive formed during Late Quaternary times gives rise to a reconstruction of the past vegetation and this reconstruction acts as proxy data for the climate reconstruction. From the glacial period throughout the older half of the Holocene a gradual change in vegetation is concluded from the pollen diagrams. The general vegetation development, concluded from the subfossil pollen rain, points to a change from open, steppic vegetation to a vegetation with trees or forest. The increase in longer lasting (living) woody elements is conspicuous but changes in the herb vegetation under the tree cover and in the steppe took place also.

It is known from observations that special climate types are linked with special vegetation and that trees demand more moisture than steppe vegetation. Thus information on the hydrothermical system can be obtained from the plant cover reconstructed for a particular area. Moisture conditions favouring tree growth, which developed from Late-Glacial time onward, are demonstrated for instance by an increase in deciduous oak pollen in northern Israel at about 15,000 BP (Baruch & Bottema 1991). At that time most of the Near East was still devoid of trees and was characterized by steppe vegetation. Arboreal pollen values increased in northern Turkey along the Black Sea towards the end of the Late Glacial. In the Pisidian Lake district in southwestern Turkey, tree pollen developed modern values from about

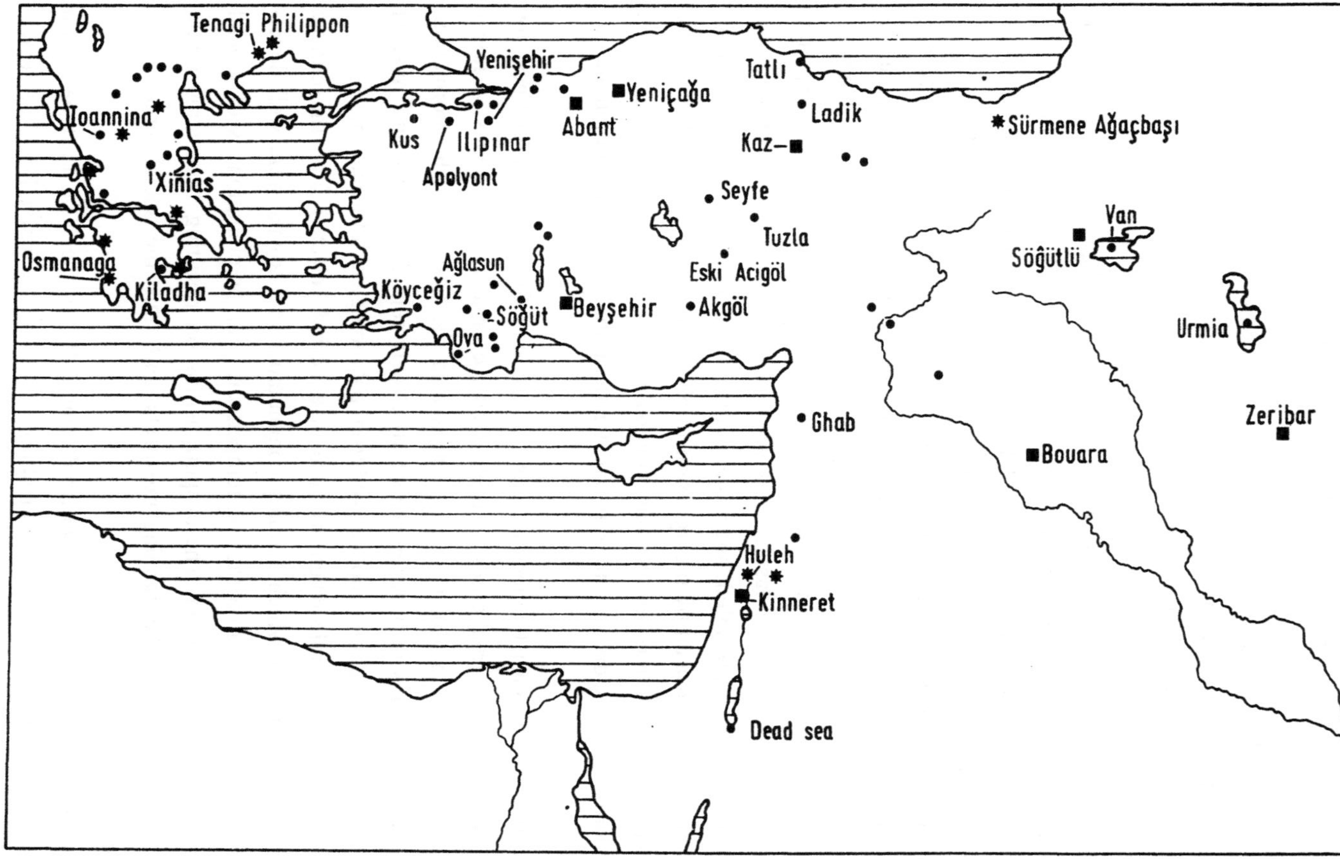

Fig. 1. Map of the Near East showing the core locations from which Late Quaternary pollen diagrams have been produced. The location of the selected eight diagrams are indicated by black squares.

9000 BP onward, whereas eastern Turkey demonstrates such pollen behaviour starting at about 7000 BP. Maximum tree pollen values, in the Zagros mountains in northwestern Iran, had developed during the sixth millennium only. From these pollen data it is concluded that during the Holocene a general increase in moisture gradually moved over de Near East from the west to the east. The moisture increase culminated around 5500-6000 BP in optimal moisture conditions for almost the whole area known as the Near East (Van Zeist & Bottema 1991). There is one exception. Northern Israel witnessed optimal conditions for tree growth from about 15,000-11,500 BP, illustrated by the large amount of Tabor oak pollen. After 11,500 up to the present time Israel witnessed much drier conditions (Baruch & Bottema 1991).

The conditions for plant cover in the Near East are characterized by a set of factors of which among others high evapotranspiration is the result. Not only the amount of annual precipitation is of importance. The distribution of the rainfall and the temperature during the year are crucial. We conclude from the vegetation history that moisture increased during the last 15,000 years. Compared to northern latitudes where low temperatures prevailed, one has to realize that in the Near East temperature must have been compensated by precipitation to allow tree growth especially during the Holocene thermic optimum. We can conclude to general conditions from the pollen records but it is difficult to separate temperature from precipitation in this pattern.

Information on the past temperature can be obtained from other disciplines, for instance isotope records from cores from the land ice or marine sediments. A lower Glacial temperature for the Near East is assumed on the basis of such information. When we think in terms of a difference of 10°C between the glacial and modern temperature, such information is not derived from palynological information.

Several factors play a role when we turn to a period selected from the climate history of the past during which farming communities and towns were well developed. The variation in the two factors, temperature and precipitation, is of crucial importance for crop cultivation. The evapotranspiration resulting from temperature and precipitation defines the farming prospects and thus the social structure of habitation.

In this paper the climate of the third millennium BC in the Near East based on information obtained from the palynological record will be discussed. It will be seen if a change or if changes in climate can be concluded from the pollen record. To do so we first have to consider what the relation of plants or vegetation and climate and especially climate change is. If this relation is established, or at least better understood, it will be investigated whether a relation between changes in pollen precipitation and changes in vegetation and climate does exist.

Climate is one of the decisive factors ruling the presence or absence of plant species c.q. vegetation. Changes in climate provoke a change in vegetation, at least if the change is of a certain magnitude. In this respect it should be stressed that climate is according to agreement the average of 30 years weather record. When periods shorter than 30 years are discussed we still speak about the weather.

To what extent do changes in parameters, that form the weather (and after 30 years the climate), change a vegetation or one or more of its components? The daily weather already knows large fluctuations which are perfectly "normal". The difference between day and night temperature for instance may be 20°C. This is in sharp contrast with the difference between the average annual temperature of a glacial period and that of an interglacial that may be much smaller, for instance 10°C. Still we observe that the vegetation is not visibly harmed by the diurnal fluctuation, but is importantly changed by the glacial/interglacial cycle. It seems reasonable to accept that plants are adapted to fluctuations in day and night temperature and it is obvious that in the case of long-term fluctuations they are not adapted. Somewhere in between there must be a point where the length of time and the dimension of a temperature or precipitation change become decisive.

It is very instructive to observe the course of the temperature during the complete growth cycle of an annual plant, as is shown by Woodward (1987). This course indicates the diurnal and nocturnal temperatures that fluctuate during the 24 hour cycle, but they also show that the average temperature gradually increases from spring to summer. For the annual plant the average temperature during its short life is more decisive than the daily fluctuation. An annual plant that may have been important for the Mesopotamian agricultural economy is for instance summer barley. It is obvious that also in the case of summer barley the diurnal cycle fluctuates more than the seasonal cycle and that the plant is well adapted to both. It turns out that the barley is sensitive to a change in the daily average temperature if this change lasts for a certain time. Any change in temperature with a frequency smaller than 31 days would not evoke a response in summer barley. The shortest response time for annual plant species reported by Woodward turns out to be 16 days for those plants that have a very short life cycle (from germination to seed producing).

Relatively short-lived plants may respond to short-term weather changes in the annual to interannual bands but forests are buffered against most changes on this scale (Prentice 1992: p. 314-316). When we turn to plants with a much longer life cycle, for instance trees, the response time can be up to 15 years. A model by Davis & Botkin (1985, quoted by Prentice 1992) suggests that forested landscapes may be buffered against decadal climate changes but respond (with some lag) to millennial climate change. Frequent disturbance may allow the landscape to respond to higher frequency changes, but only if they are of large enough amplitude.

There is some pollen evidence that vegetation responded to the Little Ice Age, a period that did altogether not have very cold winters so much as well as cold and moist springs and summers. There are spatial variations in the sensitivity and response time of the landscape. Different soils for instance react differently on the same climatic change. Plants on poor soils react weaker than plants on richer soils. Prentice suggests that the full course of an invasion, from the initial pioneers to the slow growing taxa, would take 200-1000 years, caused for instance by a climatic change. This change concerns expanding plants or trees but in the present paper we are interested in those situations where a decrease or retreat of floral elements is concluded from pollen assemblages and we want to know how such a decrease has to be explained.

On one occasion plants can be destroyed by a very short change in conditions, contradictory to the description of the fairly long response time as mentioned above. Such an abrupt change is known as a catastrophe, an event with a rate of change that is shorter than the response time but that still causes a strong reaction, even death of the living organism. Often catastrophes cause death instantly but on other occasions the effect depends upon the tolerance of the respective species. As an example, the tolerance of three different groups of trees is given by Woodward (1987).

1) Boreal forest can stand a variation in temperature of 110°C, a rainfall of 500-1500 mm and a wind speed of 15 m/sec. The Boreal forest cannot stand hard wind because of dry-freezing.

2) For the temperate forest these figures are 80°C, 800-1800 mm precipitation and 25 m/sec wind.

3) Tropical forest tolerates 5-50 °C, 16 m/sec wind and 2000-4000 mm annual precipitation. Tropical forest cannot stand low temperatures which altogether do not occur. To build a resistance against non existing conditions is not economical.

A catastrophe with a rate of change smaller than the response time will be effective if it acts outside the tolerance of a plant or tree species. Examples are fire, exceptional frost or hurricanes which may cause immediate death.

If we agree upon an abrupt climatic change that caused social disrupture, we agree that this supposed climate change acted negatively upon the growth of plants, especially crops. One of the common threats in the Near East would be drought. Drought can be caused by an increase in temperature causing higher evaporisation and transpiration during the growth season that could be fatal. Another possibility, for instance mentioned in texts on clay tablets, is drought that developed from decreased rainfall. Under the ruling Mediterranean climate the rainfall occurred during the winter halfyear. A drop in precipitation may have lasted long enough to exceed the response time of the short living crop plants.

Archaeology and climate

In this paper the possibility of changes in climate during the third millennium is discussed. The study will focus upon examples of possible climatic change as is suggested by excavations. The periods, suggested by the archaeologist are of a magnitude that may be responsible for social collapse at least present in the archaeological sequence as sterile layers indicating absence of habitation.

It is not uncommon to suggest periods of serious drought as the cause of economical and political disorder leading to absence of habitation in Near Eastern sites. Direct proof for such weather or climatic changes is seldom given but indications can be found for instance in Mesopotamian, Hittite or Egyptian texts. The influence of the new (catastrophic) climatic conditions is mostly given in terms of qualitative changes. To explain and to prove the change in conditions, archaeologists should describe modern climate in terms of temperature and precipitation and give estimated values for these parameters for the respective prehistoric period.

Weiss (this volume) thinks two periods of socio-economic collapse to be present in the northern Mesopotamian area. A short period at the beginning of the third millennium, from 3000 to 2900 BC, and a longer-lasting period from 2300 to 1900 BC (Weiss, abstract Causality and chance: The origins and collapse of Third Millennium

Mesopotamian civilizations, Kemer, 1994). In terms of uncalibrated years BP this would be 4360-4260 BP and 3820-3560 BP according to the reference curves of Pearson & Stuiver (1986) and Stuiver & Pearson (1993). The periods suggested are described by Weiss as Leilan IIIa (3000-2900 BC/4360-4260 BP) and Habur hiatus 1: Subir I (2300-1900 BC/3820-3560 BP). In the present study these phases are abbreviated to Leil. IIIa and Hab. I.

One wonders whether a hiatus in habitation was caused by the same constantly bad climatic conditions, or that only initially dry weather happened. It remains the question whether both options can be identified in the pollen records as different from each other.

The first period lasted from about 3000 to 2900 BC. The resolution of most of the pollen diagrams used in this study is less than one sample per hundred years. That means that a drought of one hundred years could go unnoticed in a pollen diagram assuming that the vegetation is completely restored to the preceding situation after the change. In practice this can hardly have been the case because the vegetation cannot have disappeared completely and restored in such a short time. The eight pollen diagrams that will be discussed in this paper must indicate some kind of effect upon the pollen precipitation. It is worth mentioning that some odd thirty diagrams of the Near East have been screened for such events.

The second period lasted from about 2300 to 1900 BC according to Weiss and concluded from the average diagram mentioned in this paper, two spectra would fall in this period. The periods Leil. IIIa and Hab. I have been calculated for each pollen diagram by calculating the calendar year dates back into uncalibrated ^{14}C years before present and assuming a constant sedimentation rate. In the respective figs 2-9 the periods have been indicated and the pollen spectra that represent the period are compared with preceding and following spectra.

The political and economical system in third millennium Mesopotamia was of a level that a well-organized state would survive crop failure several years. However the elasticity in the organized economy did not last endlessly and after 5-6 continuously bad harvests the society could come in trouble (verbal communication NATO-meeting, Kemer).

Climatic changes, in fact a pronounced drought, have been mentioned in connection with cuneiform texts that speak of famine caused by drought. In this respect, proxy data with evidence are brought forward by Kay & Johnson (1981). These

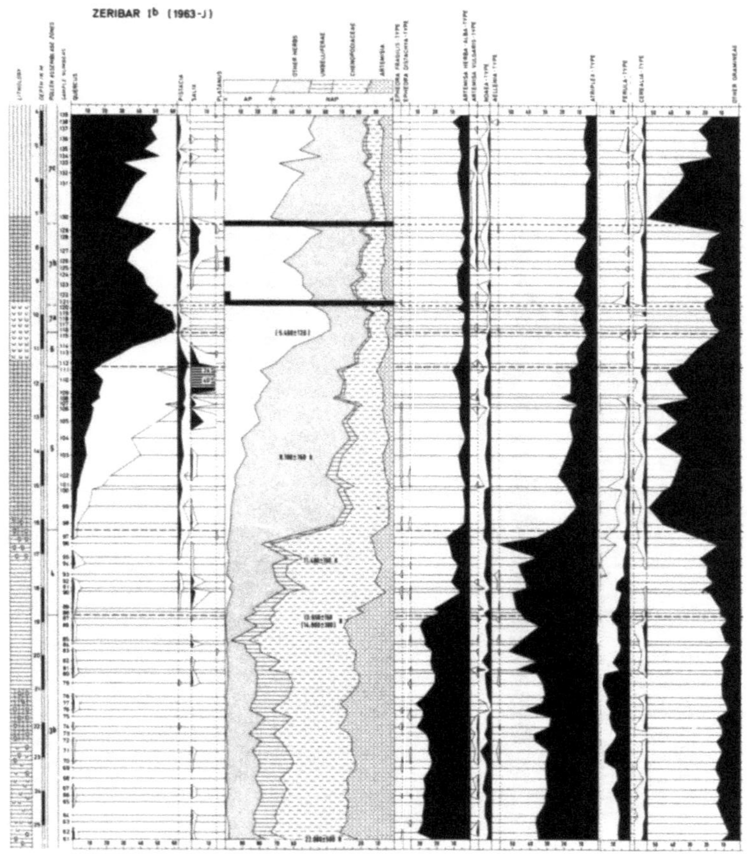

Fig. 2. Part of the diagram Zeribar Ib (Van Zeist & Bottema 1977) with selected pollen curves discussed in this paper. The third millennium BC is indicated by III, the period of Leilan IIIa (3000-2900 BC) by L, and Habur I (2300-1900 BC) by H.

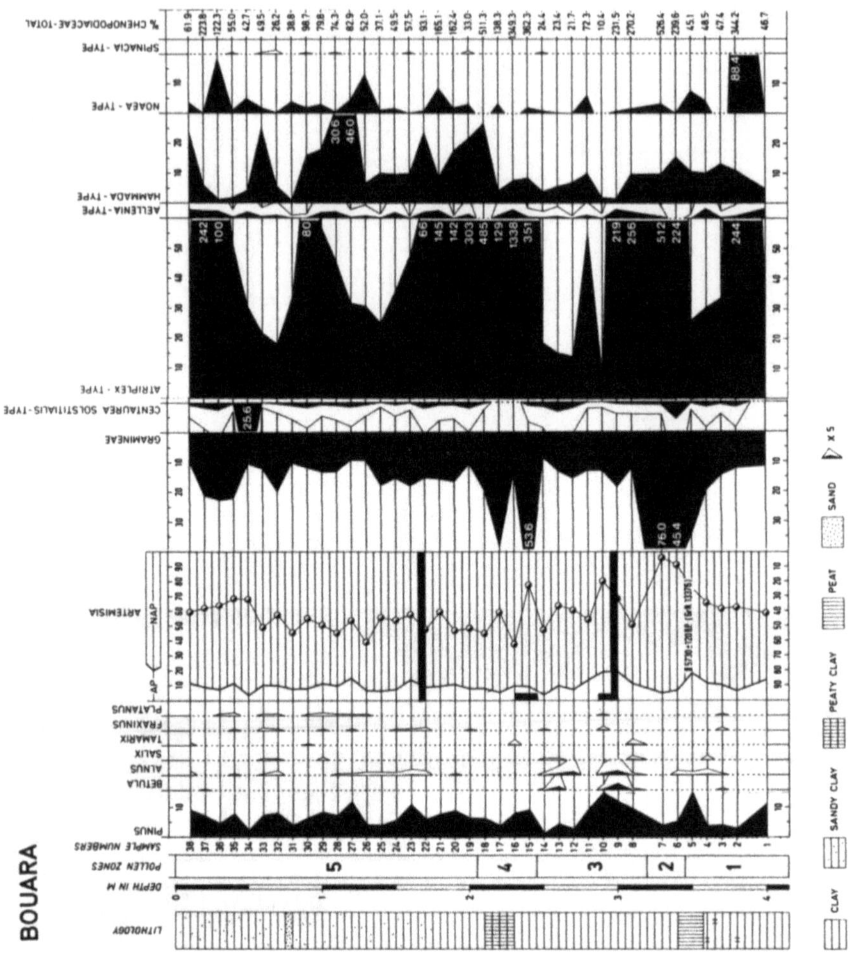

Fig. 3. Part of the Bouara diagram (Gremmen & Bottema 1991). For legend, see fig. 2.

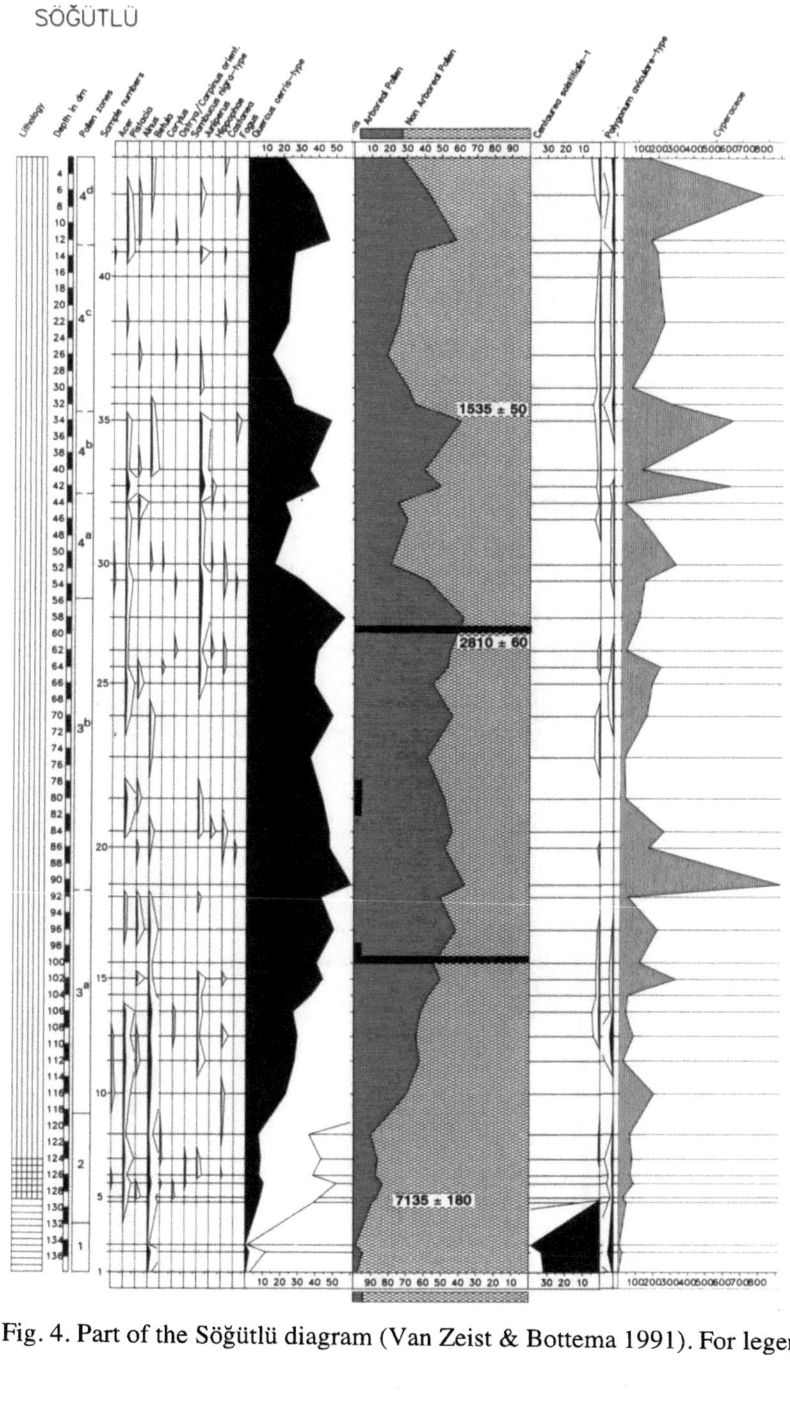

Fig. 4. Part of the Söğütlü diagram (Van Zeist & Bottema 1991). For legend, see fig. 2.

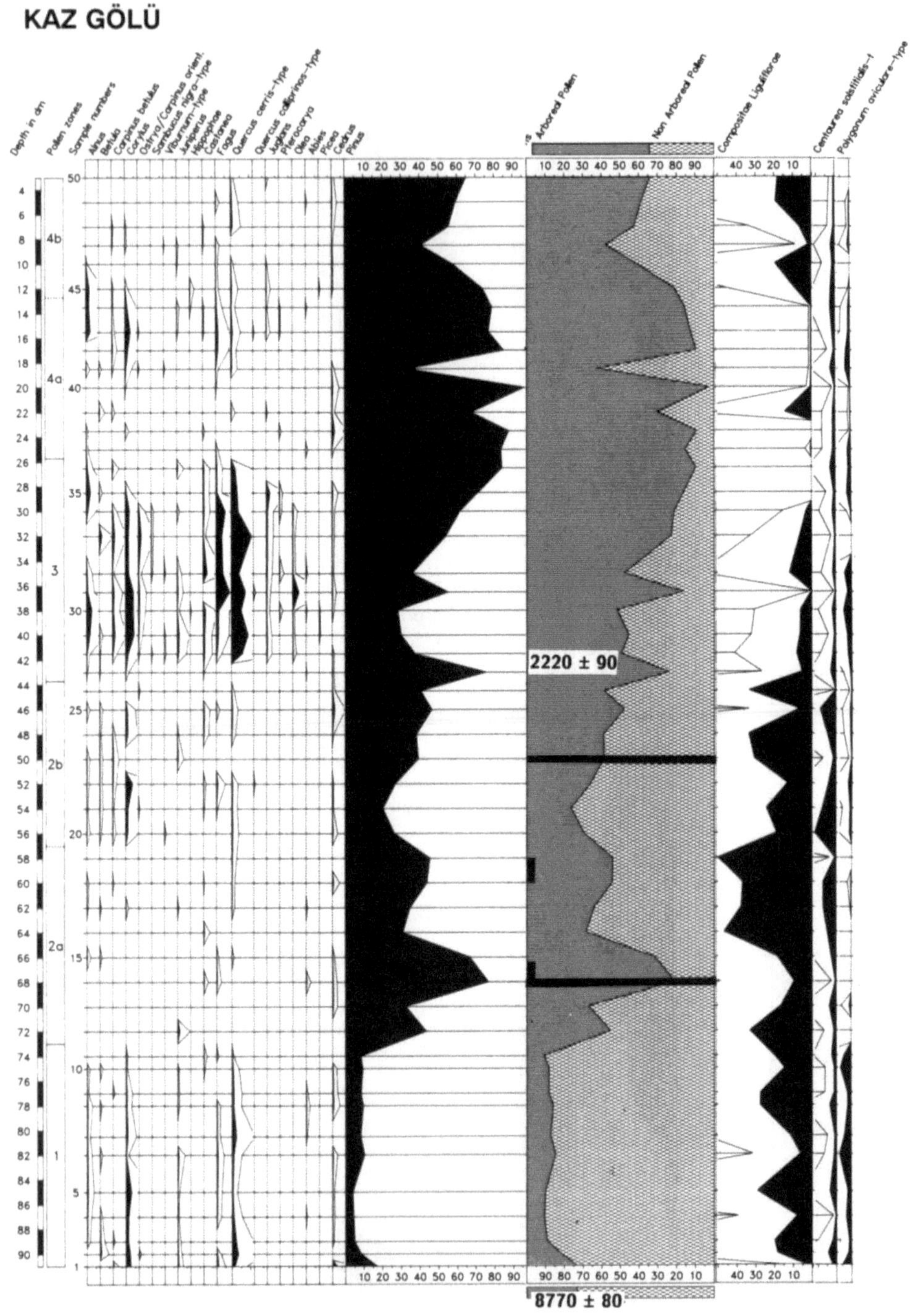

Fig. 5. Part of the Kaz Gölü diagram (Bottema et al. 1995). For legend, see fig. 2.

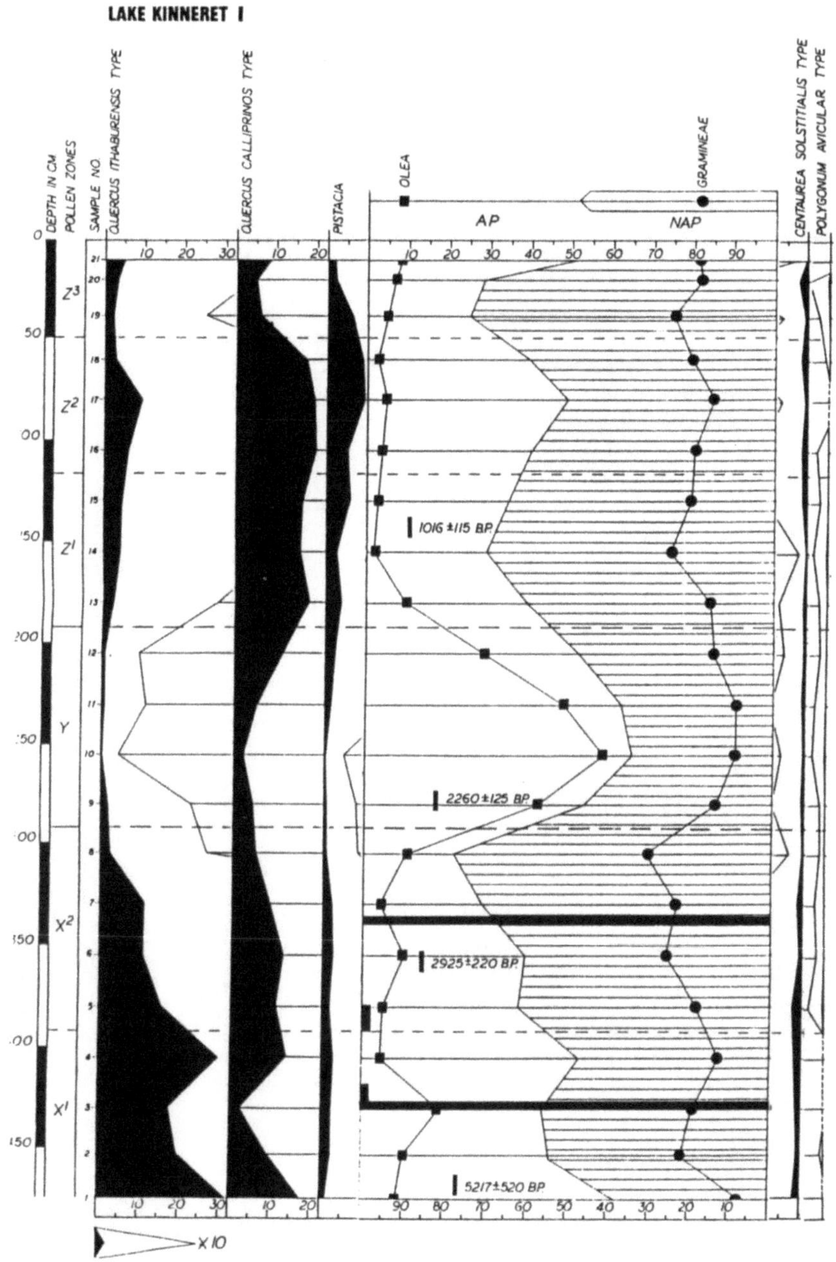

Fig. 6. Part of the Kinneret diagram (Baruch 1986). For legend, see fig. 2.

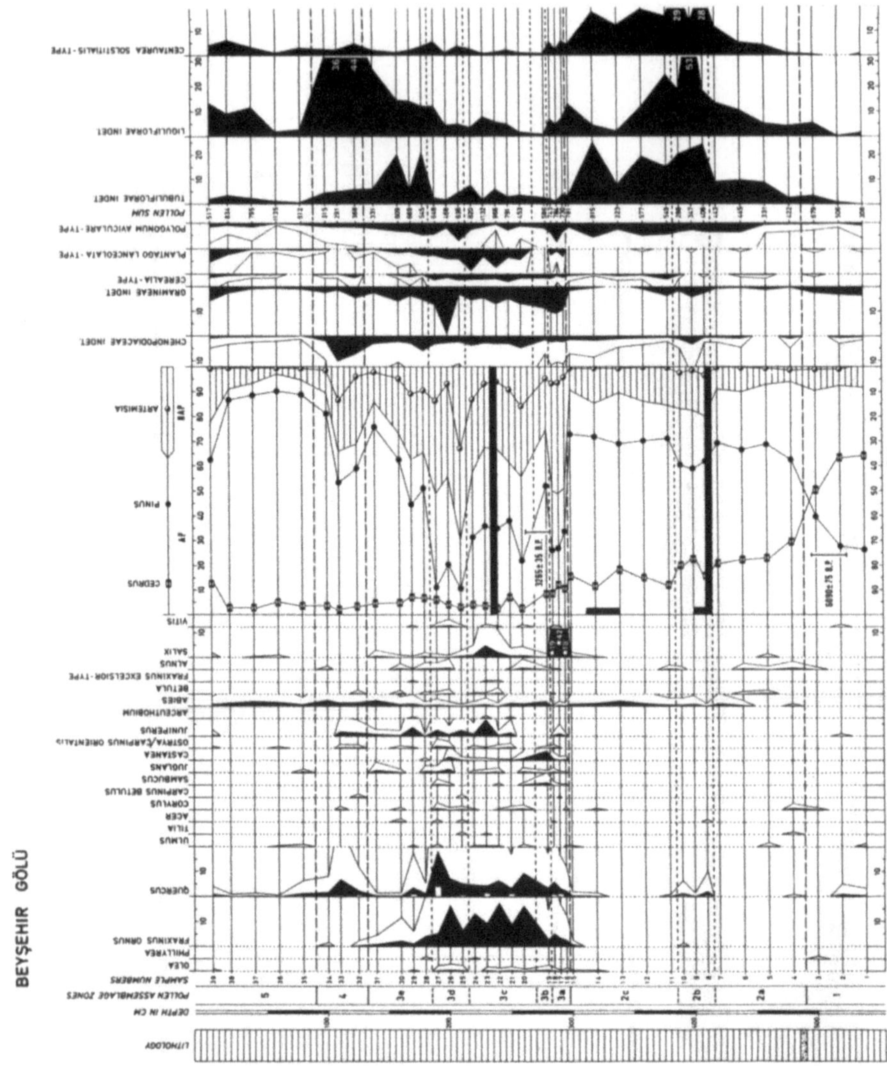

Fig. 7. Part of the Beyşehir diagram (Van Zeist et al. 1975). For legend, see fig. 2.

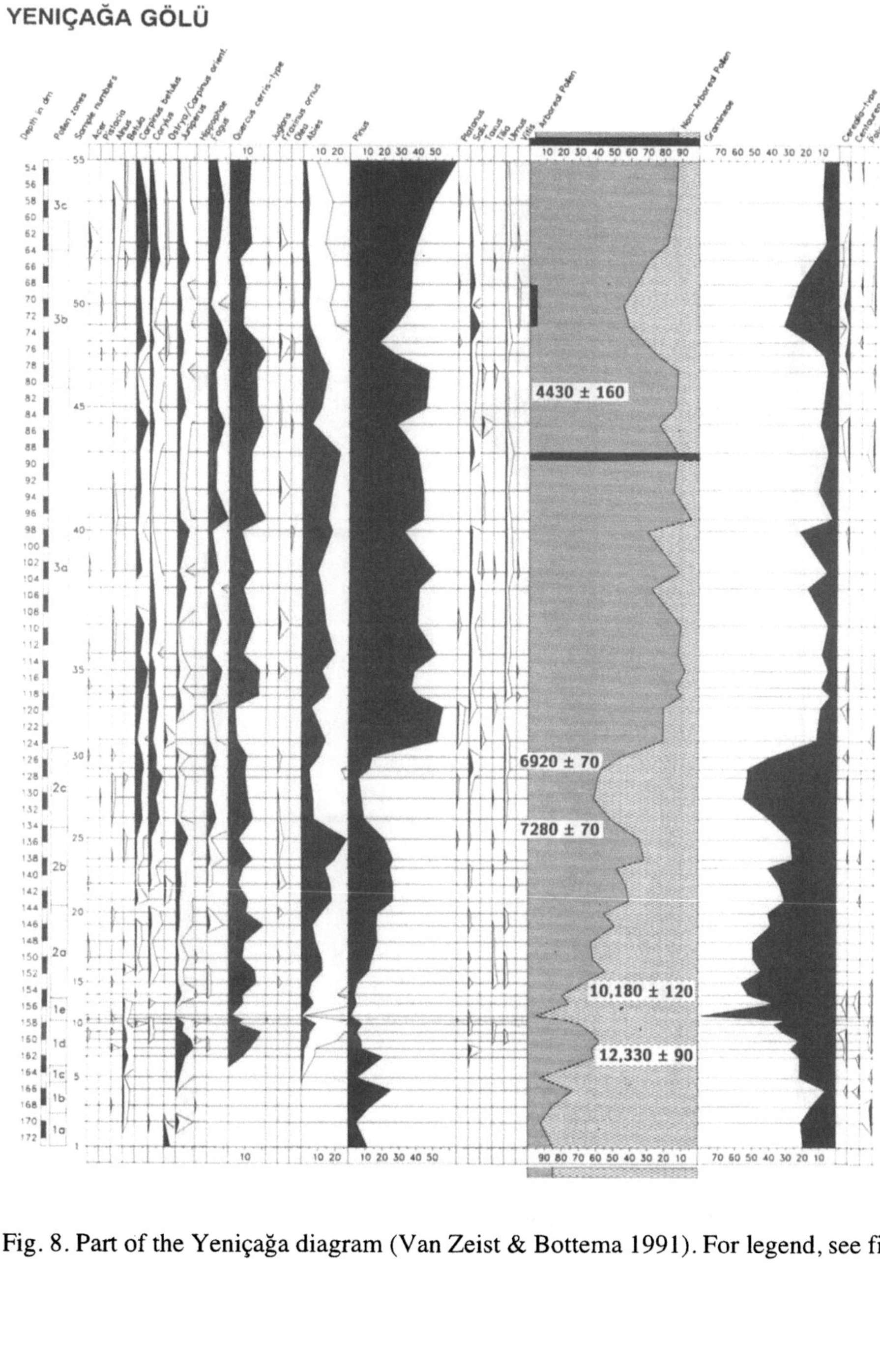

Fig. 8. Part of the Yeniçağa diagram (Van Zeist & Bottema 1991). For legend, see fig. 2.

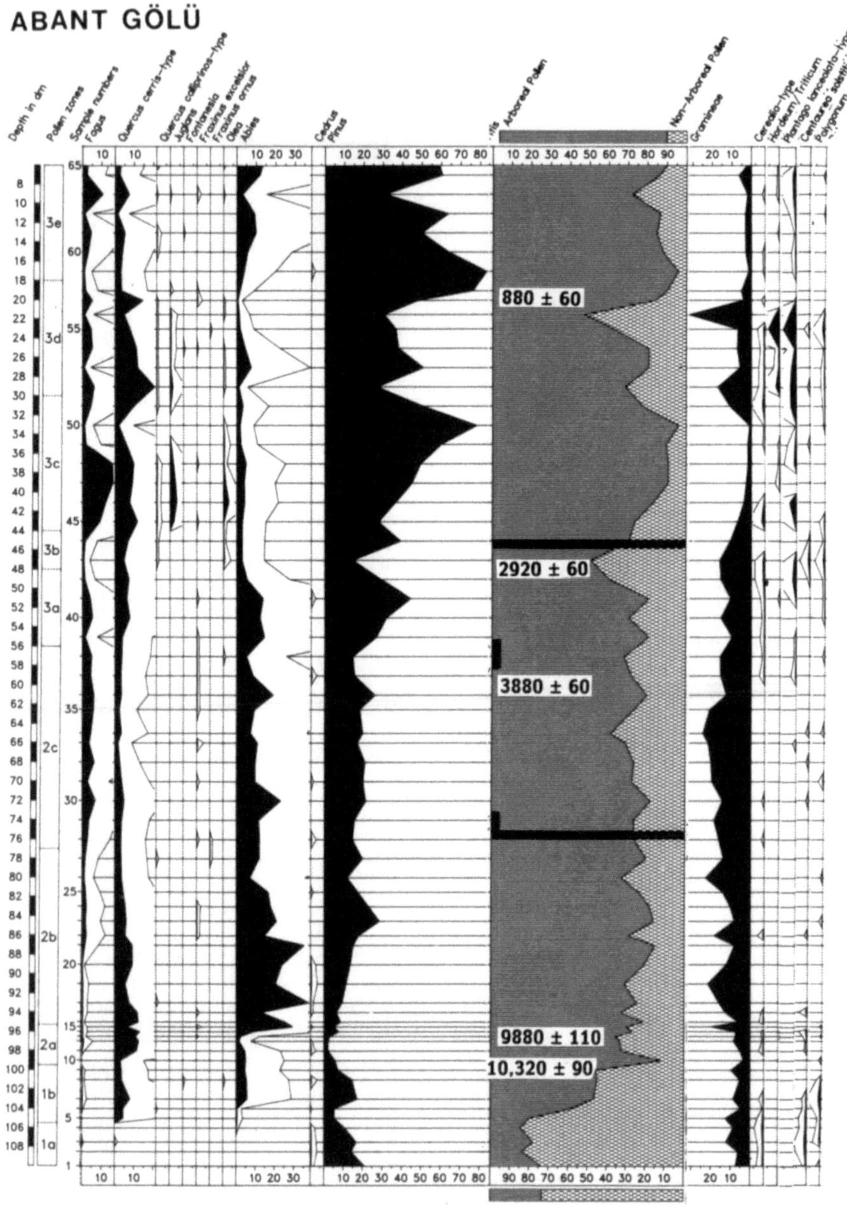

Fig. 9. Part of the Abant diagram (Van Zeist & Bottema 1991). For legend, see fig. 2.

authors indicate a phase of strongly reduced water flow in the Euphrates and the Tigris around 1200 BC, a period during which the Hittite empire collapsed. The low level of the two large rivers was altogether not indicated by direct measurements but was concluded by Kay & Johnson from the pollen curves of Lake Van (Van Zeist & Woldring 1978) which they decided to show a dry period around this time. Kay & Johnson based their conclusions on the chronology of the Van record that was dated by annual laminations (Kempe & Degens 1978). Chronologically the diagram was not matched by either Turkish or Iranian pollen diagrams. Recent evidence from the radiocarbon dated pollen diagram of the peatbog of Söğütlü, on the southwestern slope of the volcano Nemrut Dağ west of Lake Van, demonstrated that the chronology of the Van record was wrong (Van Zeist & Bottema 1991). These authors explain that the laminations are probably annual but that about 1100-1750 years must be missing. Later studies by Lemcke et al. (verbal communication, Kemer) demonstrate that the idea of Van Zeist & Bottema is right and that the discrepancy is even larger in the oldest part towards the Late glacial, which means that a dry period as suggested by Kay & Johnson for Mesopotamia is concluded from a pollen record that in reality runs out of phase one to several millennia. It also illustrates how difficult it is to conclude to climate changes from proxy data.

Nowadays the amount of annual precipitation in northern Mesopotamia is 400-600 mm. In general this is enough for rain-fed agriculture. Rain-fed (dry farming) needs at least 200 mm (Wirth 1971). The predictability of the amount of rain was and is of course crucial. The philosophy behind the existence of large empires, based upon hard grain (Steensberg 1989:147), cannot be the exploitation of marginal land. During the development of large-scale cereal production there must have been prognoses of promising harvests based upon acceptable conditions. It does not even differ much if such an agriculture was based upon rain-fed or upon irrigation farming. In both cases the water was needed and had to be predictable. When a sudden catastrophic drop in precipitation took place in northern Mesopotamia, the irrigation farming area of southern Mesopotamia must have been equally measuring a decrease resulting in a lower level of the river water. However, the question remains to what extent the southerners were using the water supply (assuming that availability was the same also during a dry year). If they used a relatively small part under normal conditions and let most of the water run into the Arabian/Persian Gulf, they possibly could claim the same amount of irrigation water for their farming even from a smaller supply. This would become relevant when archaeological evidence indicates a hiatus in habitation for rain-fed northern Mesopotamia, whereas in southern Mesopotamia where irrigation farming was practised no hiatus occurred.

The impact of drought on pollen precipitation

The identification level of pollen types varies from type to type. Some plant species are identified on the family level, some on the genus and for only a restricted number of plants, identification goes down to a few or one species. The length of time of the drought determines the categories of plant species that will be harmed. Their response time indicates whether they suffer or not. The group that receives the first blows is that of the annuals, one-season plants that includes a number of farm crops. Such plants could in principle indicate whether there was at a certain time a reduction in precipitation given the fact that their identification level is high enough to use them as indicators. There are, however, also other limitations. For the Mesopotamian area only one pollen diagram has been prepared so far, that of Bouara, a salt marsh on the Syrian/Iraqi border (Gremmen & Bottema 1991). Another question is: to what extent are crop plants represented in the pollen precipitation? Some species are self-pollinating, others are insect-pollinated, and both have a very low share in the pollen precipitation. The result is that such types are not found or only occasionally met with in pollen diagrams. This fact, together with the absence of pollen cores from the cultivated part of Mesopotamia, stresses the lack of direct evidence. Still drought periods will also have expressed themselves in diagrams that are not from the cultivated area. For instance, the record of Bouara is situated near the inhabited parts of the Khabur area in Mesopotamia. Other information can be obtained from the adjoining mountains.

The first pollen types to be looked for are the types produced by annuals, because these are the kind of plant species that have the shortest response time and thus will react the quickest. When checking the list of identified types in the pollen diagrams to look for potential annuals, the low identification level becomes clear. Most annuals form part of a genus or a family type and such plants cannot be separated from perennials that are also part of the taxon. Two pollen types are very much restricted to the special species that gave their name to the type. Although the two taxa of *Polygonum aviculare*-type and *Centaurea solstitialis*-type represent more than one species, they are named after the most common representatives and they repre-sent the name-bearers in many cases. In the case of *Centaurea solstitialis*-type pollen, the producer most likely is *Centaurea solstitialis* (Bottema & Woldring 1990).

Pollen evidence

Pollen diagrams from locations on both sides of the Mesopotamian lowland were screened for indications of climatic change during the third millennium BC. The selected pollen diagrams are dated by radiocarbon, and these dates are shown here uncalibrated. It should be stressed that a certain error is always present, indicated by the ∂ value of the radiocarbon date and inferred dates may suffer from possible irregularities in the deposition rate.

The limits of the periods discussed in this paper, have been very seldom precisely indicated by radiocarbon. Radiocarbon dates in Near Eastern pollen diagrams are mostly obtained from rare organic material in cores that generally contain clay sediments. The periods under discussion were calculated by means of the available radiocarbon dates, assuming a constant sedimentation rate. In some diagrams characteristic changes in the pollen assemblages are dated by correlation with nearby paly-nological curves which had been dated by radiocarbon already. The characteristic palynological changes are discussed by Bottema & Woldring (1990).

The pollen diagrams used here are investigated for the periods of Leil. IIIa (3000-2900 BC) and Hab. I (2300-1900 BC) which are suggested to be periods of social collapse and/or desertion connected with desertification (Weiss 1994).

The pollen diagrams

Zeribar Ib (Van Zeist & Bottema 1977)

The site of Zeribar is found at an elevation of 1300 m in the Zagros mountains of Iran (fig. 1). Modern rainfall amounts from 400 to 600 mm. Present day natural vegetation is a deciduous oak woodland (Quercus brantii). From all the diagrams prepared for this site, Zeribar Ib is the one selected and shown in fig. 2.

The period from 3000 to 1000 BC is calculated from the radiocarbon date of 5460±120 (Yale) from a core that was taken nearby. This date was transferred to the core Zeribar Ib on the basis of pollen-floristic features. The pollen core Zeribar II has a radiocarbon date of 5640±70 for the pollen assemblage in spectrum 34 that correlates well with the date and the pollen assemblages of Zeribar Ib. On the basis of these dates the period from the third to the first millennium has been calculated to cover the part of the record from spectrum 121-129. The distance in time of two spectra is on average a little more than 200 years.

The suggested dry period from 3000 to 2900 BC (Leil. IIIa) must be located around spectrum 121. A change in several pollen curves is not restricted to this spectrum but starts already in spectrum 119, where a decrease in deciduous oak and pistachio pollen is visible. In spectrum 121 Centaurea solstitialis-type is absent. Suddenly a few percent of Ferula-type appears in spectrum 121, whereas lower values are found in spectra 120 and 119. Ferula-type pollen played a prominent role during the Full Glacial and becomes rare after about 9000 BP. In the glacial period the type is found together with steppe-indicators. From the local marsh-plants Sparganium-type pollen decreased from over 20% in spectrum 119 to about 5% in spectrum 121. At the same time the pollen percentages of this marsh plant have diminished, values of the waterplant Myriophyllum verticillatum/spicatum increase considerably, suggesting that the level of the lake rose. Thus the pollen evidence could lead to somewhat contradictory conclusions. At the one hand there is a decrease in tree pollen, the disappearing of an annual (Centaurea solstitialis-type), at the other hand there is the appearance of the annual Polygonum aviculare and the increase of waterplants. It is possible that the trees suffered more from the action of prehistoric people than that the decrease in AP values points to a change in climatic conditions. The fact that Cerealia-type pollen grains start again with a continuous curve in spectrum 121 is not only an indication of cereal growing as this curve is also prominent throughout the glacial part of the diagram when no cultivation of cereals took place.

The second period with suggested drought from about 2300 to 1900 BC (Hab. I) is located in the Zeribar Ib diagram around spectrum 125 and 126. Deciduous oak pollen has reached a minimum, from 60% at the maximum at about 5500 BP to c. 30%. Pistacia pollen is even absent in spectrum 126. On the other hand is there a steady increase in the pollen of the moisture demanding willow (Salix). Pollen of Tubuliflorae and Liguliflorae appears during both spectra, together with Sanguisorba minor, an indicator of open, grazed vegetation. Myriophyllum, a local waterplant that peaked in spectrum 121, shows the same behaviour during spectra 125 and 126. The annuals Centaurea solstitialis and Polygonum aviculare reappear during spectra 124 and 125 after a short period during which they were not found in the pollen samples.

It is obvious that a period of climatic drought is difficult to conclude from spectra 125 and 126, the part for which such a period is calculated. Spectrum 124 shows no pollen of annuals, an absence that might have been caused by an extreme drought. Spectra 125 and 126 have low values for deciduous trees together with the appearance of pollen of two moisture demanding taxa, willow and plane tree. Are these two taxa simply the signal of a diminishing lake and is their presence explained by the closing in of the lake side, trees approaching the core location? This explanation is altogether not feasible if we look at two Holocene peaks of oak expansion, accompanied by high values of willow pollen.

Concluding, it can be stated that percentages of deciduous oak-pollen in the Zeribar Ib diagram gradually decreased from about 5400 to 4000 BP pointing at a continuos proces. Some typical drought indicating features accompany the short period calculated for 2300-1900 BC. Some other indicators are found before this period and it is not clear which pollen effect could point to a drastic change. In fact the indications observed so far seem to be rather insignificant.

Bouara (Gremmen & Bottema 1991)

The saltflats of Bouara are located in the Mesopotamian area, on the border of Syria and Iraq at an elevation of c. 300 m (fig. 1). Average annual precipitation amounts to 150 mm.

Local vegetation is scarce, mostly represented by Tamarix and Chenopodiaceae. Tree pollen met with in the samples originates from long-distance transport. Two pollen types dominate the diagram: Chenopodiaceae, local salt plants that are excluded from the basic pollen sum, and Gramineae (grasses). The part of the Bouara diagram dated before 5730 BP, from 5000-4000 BP and after 3350 BP up to the present time, is fairly identical. The diagram shows two short periods, from 5800-5000 and 4000-3350 BP, during which local pollen was relatively important and pollen percentages of extraneous elements including Betula, Alnus, and Pinus, are much lower (fig. 3).

The sediment collected at Bouara consists mainly of clay or a mixture of clay and sand. At two depths peat or other organic deposits were formed. One peat layer stretches from 340-360 cm and is dated 5730±120 BP at a depth of 430-350 cm (GrN-13376). A younger deposit of organic material is found from 210-225 cm and dated by interpolation to about 3350-3765 BP. Both organic sediments correlate with peaks of Gramineae pollen and the grasses may have produced the peaty layers. Compared to the clay sediments that were formed in open water, both organic layers can be related to periods during which the water table was lowered, giving rise to swamp vegetation. The conditions for swamp vegetation to develop, point at shallow water and the drop in water level may have been the result of a dry period that preceded the peaty sediments.

For the Leil. IIIa period there is no change visible in the lithology column. In the pollen record, spectra 9 and 10 show an increased long-distance transport of some tree pollen-types. Pine pollen has it nearest source in the Levant and the Taurus mountains. At that time Betula grew in limited numbers in Anatolia, but its pollen may also originate from Europe (Bottema 1990). The higher values for long-distance tree pollen points to a lower precipitation of local pollen. Such lower precipitation may have been caused by an extension of the water that moved the local vegetation away from the core location.

The Hab. I period is represented by Bouara spectra 15 and 16 which are found just before and at the beginning of the organic layer. One could explain the difference in time by the fact that it had to be dry first to result in drying up of the Bouara basin somewhat later. The Hab. I period in the Bouara diagram shows at least traits that suggest that it was drier than before and afterwards.

Söğütlü (Van Zeist & Bottema 1991)

The core of Söğütlü was taken by H.E. Wright Jr in a peatbog on the southwestern slope of the Nemrud Dağı, west of Lake Van (1646 m), near the small village of Söğütlü (fig. 1). The elevation of the site is about 1500 m and the average annual precipitation amounts to c. 600-800 mm.

The period 3000-1000 BC is calculated to have taken place from spectrum 16-27 (fig. 4). The oldest period to be discussed (3000-2900 BC) coincides with spectrum 16. The second period (2300-1900 BC) is ascribed to the level of spectra 21 and 22. During the part of the Söğütlü diagram in between spectra 16-27, pollen behaviour is rather uniform. The only fluctuations are caused by pollen of the Cyperaceae (sedges) that must have grown in the local marsh.

The two periods that we are interested in, are represented by relatively high oak pollen values and a dry period cannot be concluded from them. The indicative annuals *Centaurea solstitialis* and *Polygonum aviculare* display low values altogether and the latter is absent in spectra 18-23. During the steppic period before 7000 BP these two annual herb types were far more common in samples with very low tree pollen numbers. It is concluded that the Söğütlü pollen diagram does not show an effect that can be ascribed to climatic change for the Leil. IIIa and Hab. I periods.

Kaz Gölü (Bottema et al. 1993)

The lake of Kaz is found in northern central Anatolia at an elevation of c. 500 m (fig. 1) where the average annual precipitation is nowadays about 600-800 mm. The period under discussion has to be calculated from a radiocarbon date of 2220±90 BP (420-430 cm) and palynological correlation with the nearby dated record of Lâdik Gölü. The lower boundary (3000 BC) cannot be defined with great certainty and is most likely found around spectrum 14. The reason for this uncertainty is the irregular sedimentation rate found in the lakes of Kaz and Lâdik. This irregularity is for instance caused by the extensive floating vegetation mat in both lakes. For the same reason the period of 2300-1900 BC, calculated for spectra 18 and 19 (fig. 5), must be considered as the most likely approach.

For both periods (spectrum 14, and spectra 18 and 19) arboreal pollen values show a maximum, which is exclusively caused by pine. During spectrum 19 there is a low value for *Centaurea solstitialis*-type and *Polygonum aviculare*-type is even absent. It must be stressed that the latter type is absent for most of the period covered by spectra 12-20. High values for Liguliflorae point to local conditions that at least had dry hot summers from 8000 to 2200 years BP. Also for the period under discussion, 2300-1900 BC, Liguliflorae show appreciable numbers. During spectra 20 and 21, arboreal pollen values, mainly Pinus, decrease from 80% to 25%. Concluded from other pollen types, for instance increasing Corylus (hazel), this effect seems to be anthropogenic and not caused by climate.

Kinnereth (Baruch 1986)

The lake of Kinnereth (Tiberias) lies in Israel (fig. 1), 210 m below sea level and has an average annual precipitation of 400 mm. The period we are interested in here is represented by the most pronounced arboreal part of the sequence and deciduous oak pollen, very likely *Quercus ithaburensis*, measures up to 30% (fig. 6). The herb types do not indicate dry periods and Centaurea solstitialis-type and *Polygonum aviculare*-type do not fluctuate. One has to consider however the very low resolution of the analysis.

Beyşehir (Van Zeist et al. 1975, Bottema & Woldring [1984]1986)

Beyşehir Lake is found in southwestern Anatolia at an elevation of 1120 m (fig. 1). Precipitation is about 500 mm annually. The lake lies in the Oro-Mediterranean vegetation belt, characterized by *Cedrus libani* and *Juniperus excelsa*. The third to first millennium period covers spectra 8-14 (fig. 7), while the early period 3000-2900 BC is located at spectrum 8 and the second period (2300-1900 BC) would be represented by spectra 12 and 13. Spectrum 8 forms part of a period with low AP values covering spectra 7-10. The spectra with lower AP values lasted about 500 years, considerably longer than the 100 years indicated for the period. What kind of change do these spectra represent? Pine pollen decreases but deciduous oak shows a short recovery. There is a slight increase in grass and Artemisia pollen. It is not clear whether spectra 7-10 represent a man-induced or a climate-induced change.

At the level of spectra 13 and 14 no changes are visible in the pollen assemblages and consequently no dry period is indicated. Important changes take place in somewhat more recent time, above spectrum 15. There the AP/NAP ratio in the pollen assemblages suddenly drops and a series of types appear or increase, indicating farming and horticulture. This phase is described as the Beyşehir Occupation Phase, found in a large part of Turkey and mainland Greece (Bottema & Woldring 1990, 1995). This phase started at about 1600 BC and is a clear anthropogenic action although it may have been triggered by climate that altogether did not become drier.

Yeniçağa (Van Zeist & Bottema 1991; Bottema et al. 1995)

The site of Yeniçağa lies in a valley east of Bolu at an elevation of 975 m (Fig. 1). The core that produced the pollen diagram (fig. 8) was taken in a large peatbog on the west side of the Yeniçağa lake. The Bolu area is reported to have 600-800 mm of precipitation annually (Yeni Türkiye Atlası 1977). The oldest period is located around spectrum 34 where no changes in the pollen record can be observed. The period 2300-1900 BC is calculated for spectra 49-51, a period that shows an important change in the AP/NAP ratio. Coniferous pollen decreases already in spectrum 47, to be followed by that of oak, beech, hornbeam and hazel. Most of the increase of the herbs is caused by the grasses, but there is also a significant rise in Cerealia which played no role before. The latter type suggests that the changes in the pollen diagram are the reflection of human activity and not of climate.

Abant (Van Zeist & Bottema 1991; Woldring et al. 1993)

The Abant Lake is found at about 1300 m in the Elmacık mountains (fig. 1). Precipitation averages about 1000-1250 mm annually (Yeni Türkiye Atlası 1977). Abant lies in the mountain forest zone that is dominated by Nordmann fir (Abies nordmanniana ssp. bornmuelleriana) and Austrian pine (*Pinus nigra*). Beeches (*Fagus orientalis*) are few in numbers (Van Zeist & Bottema 1991).

The period of 3000-1000 BC is attributed to spectra 29-44 (fig. 9). The first century of the third millennium BC, represented by spectrum 29, shows no changes pointing to a change in climate. The part of 2300-1900 BC is represented by spectra 37 and 38. The tree pollen values do not indicate changes of any importance. Pollen of *Centaurea solstitialis* is scarce, whereas the curve for *Polygonum aviculare* starts at the level of spectrum 37, together with that of the *Cerealia* and *Plantago lanceolata*. As is the case in the basin of Yeniçağa, human action might be responsible for the changes after 2300 BC.

Conclusive remarks

Screening the palynological information on the Leilan IIIa and Habur I periods, obtained from eight Near Eastern sites, does not produce much evidence for drought. It is brought forward that plant species with a short response time (so-called annuals) are very likely to suffer first from a period of drought. Among the annuals are many farm crops that may have been threatened. The palynological indication of cultivated or wild annuals is rather poor, either because of their low share in the pollen precipitation or because of the low identification level. The few types that are available were tested in this study. Information from pollen of shrubs and trees with a longer life and a longer reponse time was also screened for evidence on drought. One may assume that a dry period, if felt in northern Mesopotamia, also occurred in the Anatolian and Zagrosian realm. A reduction of the precipitation of northern Mesopotamia can only have occurred on the level of affection of some annuals to some part. The longliving plants or trees were not visibly influenced.

The conclusions of the eight pollen diagrams are given in Table 1. It is clear from this table that no persistent drought in the two given periods Leilan IIIa and Habur I can be concluded from pollen evidence. The absence of such evidence throws doubt upon the statement of such drought.

Table 1. Indications of climate change during 3000-2900 BC and 2300-1900 BC.

Symbols: - indications for drier conditions; + indications for moister conditions; ± no visible change in climate; ? it is not clear whether a change in pollen must be ascribed to climate change or to human impact.

	Leilan IIIa 3000-2900 BC	Habur I 2300-1900 BC
Zeribar 1b	-	±
Bouara	±	-
Söğütlü	±	±
Kaz Gölü	±	±
Kinnereth	+	+
Beyşehir	?	±
Yeniçağa	±	?
Abant	±	?

References

Baruch U (1986) The late Holocene vegetational history of lake Kinnereth (Sea of Galilee), Israel. *Paléorient* 12(2):37-48

Baruch U, Bottema S (1991) Palynological evidence for climatic changes in the Levant ca. 17,000-9,000 B.P. In: Bar-Yosef O, Valla FR (eds) The Natufian Culture in the Levant. International Monographs in Prehistory, Archaeological Series 1, Ann Arbor/Michigan USA, pp. 11-20

Bottema S, Woldring H (1986)) Late Quaternary vegetation and climate of southwestern Turkey. Part II. *Palaeohistoria* 26:123-149

Bottema S, Woldring H (1990) Anthropogenic indicators in the pollen record of the Eastern Mediterranean. In: Bottema S, Entjes-Nieborg G, Van Zeist W (eds) Man's role in the shaping of the Eastern Mediterranean landscape. Balkema Rotterdam Brookfield, pp. 231-264

Bottema S, Woldring H (1995) The environment of Classical Sagalassos. In: Waelkens M, Poblome J (eds) Sagalassos III, Report on the fourth excavation campaign of 1993. Leuven University Press Leuven, pp. 327-340

Bottema S, Woldring H, Aytuğ B (1993) Palynological investigations on the relation between prehistoric man and vegetation in Turkey: The Beyşehir occupation phase. In: Demırız , H, Özhatay, N (eds) OPTIMA, Proceedings of the fifth meeting at Istanbul, 8-15 September 1986. Istanbul, pp. 315-328

Bottema S, Woldring H, Aytu_ B, (1995) Late Quaternary vegetation history of northern Turkey. *Palaeohistoria* 35/36:13-72

Gremmen WHE, Bottema S (1991) Palynological investigations in the Syrian Gazira. In: Kühne H (Hrsg) Die rezente Umwelt van Tall Seh Hamad und Daten zur Umweltrekonstruktion der assyrischen Stadt Dur-katlimmu. Berlin, pp. 105-116

Kay PA, Johnson DL (1981) Estimation of Tigris-Euphrates Streamflow from Regional Paleoenvironmental Proxy Data. *Climatic Changes* 3:251-263.

Kempe S, Degens ET (1978) Lake Van varve record: the last 10,420 years. In: Degens ET, Kurtman F (eds) The Geology of Lake Van. MTA Report 169:6-65.

Pearson GE, Stuiver M (1986) High precision calibration of the radiocarbon time scale 500-2500 BC. *Radiocarbon* 28B:839-852.

Prentice C (1992) Climate change and long-term vegetation dynamics. In: Glenn-Lewin DC, Peet RK, Veblen TT (eds) Plant succession. Theory and prediction. Chapman & Hall London Glasgow New York Tokyo Melbourne Madras, pp. 293-339.

Steensberg A (1989) Hard grains, irrigation, numerals and script in the rise of civilisations. Paul Kristensen Copenhagen.

Stuiver M, Pearson GW (1993) High precision bidecadal calibration of the radiocarbon time scale, AD 1950-500 BC and 2500-6000 BC. In: Stuiver M, Lory A, Kra RS (eds) Calibration 1993. *Radiocarbon* 35(1):1-23.

Van Zeist W, Bottema S (1977) Palynological investigations in western Iran. *Palaeohistoria* 19:19-95.

Van Zeist W, Bottema S (1991) Late Quaternary Vegetation of the Near East. Beihefte zum Tübinger Atlas des Vorderen Orients, Reihe A, Nr. 18. Ludwig Reichert Verlag Wiesbaden.

Van Zeist W, Woldring H (1978) A Postglacial pollen diagram from lake Van in East Anatolia. Review of Palaeobotany and Palynology 26:249-276.

Van Zeist W, Woldring H, Stapert D (1975) Late Quaternary vegetation and climate of southwestern Turkey. *Palaeohistoria* 17:53-143

Weiss H (1994) Causality and Chance: The Origins and Collapse of Third Millennium Mesopotamian Civilization. Kemer contribution

Wirth E (1971) Syrien, eine geographische Landeskunde, Wissenschaftliche Länderkunden, Band 4/5. Darmstadt

Woldring H, Bottema S, Aytuğ B (1993) Late Quaternary vegetation history of Abant. In: Demınz, H . Özhatay N (eds) OPTIMA, Proceedings of the fifth meeting at Istanbul, 8-15 September 1986. Istanbul, pp. 467-471

Woodward FI (1987) Climate and Plant Distribution. Cambridge Studies in Ecology. Cambridge University Press Cambridge.

Yeni Türkiye Atlası (1977) MSB Harita Genel Müdürlüğü, Ankara.

Middle and Late Holocene Vegetation and Climate Changes in Peninsular Italy

Donatella MAGRI

Dipartimento di Biologia Vegetale
Università "La Sapienza"
P.le Aldo Moro, 5
00185 Roma
Italy

Abstract

Pollen-stratigraphic data and radiocarbon dates are examined for seven sites in the Italian peninsula (Lago Grande di Monticchio, Valle di Castiglione, Lago di Vico, Lago Lungo, Lagaccione, Lago Padule and Agoraie di Fondo), in order to outline the history of vegetation during the second half of the Holocene. Palaeoclimatic interpretations from the pollen evidence are supported by archaeological, palaeohydrological and lithostratigraphical data from lacustrine deposits. Three main vegetation changes have been identified: a) around 5000 BP diffusion of Fagus in central and northern Italy and of Abies in southern Italy are interpreted as an effect of increased wetness; b) around 3700 BP there is a temporary reduction of tree abundance at all sites in central Italy that, on the basis also of a marked change in the hydrologic regime of lakes, is interpreted as a dry event; c) around 2500 BP there is clear evidence of cultivations, however without any dramatic change in the density and distribution of woodland.

Introduction

The aim of this paper is to identify the most significant vegetational changes in the Italian peninsula during the second half of the Holocene, as a result of climate variations and human activity.

The climate of the region is clearly conditioned by the thermostatic action of the Mediterranean Sea, and at the same time is influenced by the Apennine chain that runs across the peninsula and bars the way to westerly air streams, causing high rainfall at some sites. In relation also to the heterogeneity of the geography and climate, the flora of the region is particularly rich and the vegetation diversified.

NATO ASI Series, Vol. I 49
Third Millennium BC Climate Change
and Old World Collapse
Edited by H. Nüzhet Dalfes, George Kukla,
and Harvey Weiss

Besides, in the Italian peninsula a continuous succession of human settlements has certainly modified the natural environment, through agricultural, sylvicultural and pastoral practices, and development of urbanization. However the scale of impact of human activity on the vegetation of this region during prehistorical times is still little known.

The pollen data from seven sites (Lago Grande di Monticchio, Valle di Castiglione, Lago di Vico, Lago Lungo, Lagaccione, Lago Padule and Agoraie di Fondo) are synthetically presented and discussed. These sites are located between 41° and 45° N, at elevations between 40 m and 1350 m a.s.l. and less than 90 km from the sea (Fig.1).

For each site (Figs 2-8) the percentages of trees and herbs, in a summary diagram, are shown together with the percentages of Abies, Fagus and Quercus. The starting times of cultivation of Castanea and Juglans are indicated by arrows.

All the ages in the text are given as uncalibrated radiocarbon years before the present.

Fig. 1. Location of pollen sites mentioned in text.

Pollen data

Lago Grande di Monticchio is a crater lake belonging to the Monte Vulture volcanic complex, located about 110 km ENE of Naples (Fig. 1). The published pollen diagram from Lago Grande di Monticchio (Watts, 1985) is a preliminary work. Multidisciplinary investigations on new cores have been carried out (Newton and Dugmore, 1993; Robinson, Shimmield and Creer, 1993; Zolitschka and Negendank, 1993; Turton, 1993; Giraudi and Narcisi, 1994), or are still in progress (e.g. pollen analysis by W.A. Watts, tephrochronology by B. Narcisi).

In Fig. 2 a synthetic pollen diagram redrawn and simplified from Watts (1985) is presented. The published radiocarbon dates (Watts, 1985) are affected by hard-water effect, so in Fig. 2 tentative ages are given (in brackets). The pollen diagram from Lago Grande di Monticchio indicates that the landscape has been densely forested throughout the Holocene. Only a slight opening of the woodland is recorded in the topmost layers, in correspondence with clear signs of cultivation (cf. start of Castanea and Juglans). During the middle Holocene the forest composition shows an appreciable variation, as populations of Abies, Alnus and Carpinus expanded while oaks underwent a clear reduction. This change is most likely to be related to moister climatic conditions. As regards the indicators of human activity,

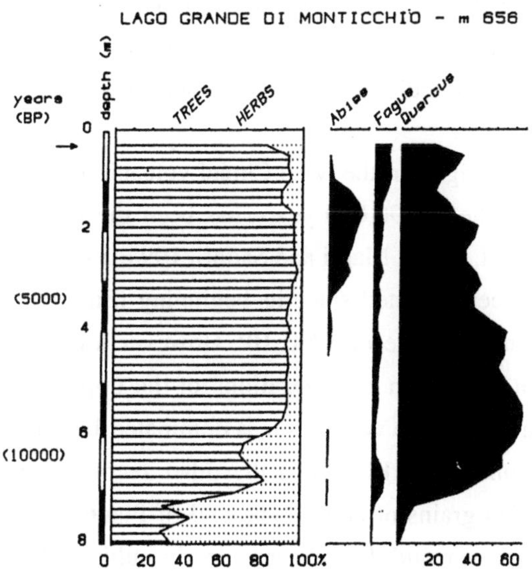

Fig. 2. Summary pollen diagram from Lago Grande di Monticchio (redrawn and simplified from Watts, 1985). The arrow indicates the start of Castanea.

pollen of cereal type, as well as Plantago, occur scattered throughout the whole period and back to the Pleistocene. Thus, these pollen types appear not to have the indicator value of human activity that they would have in central and northern Europe.

Valle di Castiglione is a crater lake belonging to the Colli Albani volcanic district; it is located 20 km east of Roma (Fig. 1). Multidisciplinary investigations have been carried out on a 250,000-year core from the lacustrine sediments of Valle di Castiglione, concerning primarily pollen analysis and lithostratigraphy (cf. Follieri, Magri and Sadori, 1988, 1989; Follieri, Magri and Narcisi, 1990; Narcisi et al., 1992). A synthetic Holocene pollen diagram is represented in Fig. 3. The percentages of arboreal pollen are generally much lower than at the other sites examined, certainly in relation to the drier climatic conditions at Valle di Castiglione due to its low elevation. It clearly appears that starting from the beginning of the Holocene, there is a general trend towards more forested landscapes. The highest percentages of trees are reached in the second half of the Holocene, from about 5000 BP onwards. An appreciable recession of forest cover is found around 3500 BP, after which episode tree percentages again display high values. The forest composition shows the dominance of oaks, both deciduous and evergreen, throughout the Postglacial, and a clear increase of Fagus from about 5000 BP, this also indicating moister climatic conditions. Single grains of cereal type pollen and of Plantago are found as from the middle Pleistocene; Castanea shows signs of cultivation in the upper part of the diagram.

Lago di Vico is another lake of volcanic origin, in the Vico-Cimini volcanic district, 55 km NNW of Roma. The core, spanning the last 90,000 years, has been the subject of palynological investigations (Follieri et al., 1995; Magri and Sadori, 1994, 1995). At Lago di Vico the Holocene has always been forested since its beginning (Fig. 4). A moderate deforestation is recorded around the date of 3700 BP. However after this episode forest recovers very well, so that the pollen concentrations of the uppermost layers are among the highest of the sequence. The vegetation is dominated by oaks, with some Fagus increasing already around 9000 BP. As in the other records from central Italy, Abies is almost absent. Pollen grains of cereal type and of Plantago are found throughout the record, while Castanea and Juglans are clearly cultivated as from the radiocarbon date 2630+95 BP.

Lago Lungo is located in the Rieti Plain, an intermontane depression of the Apennine chain barred by a travertine threshold, whose growth gave origin to the lake. Frank Holocene lacustrine sediments, deposited in a period when the travertine barrier was high, have ages between about 9000 and 3700 BP, after which time the lithostratigraphical characteristics, and the mollusc and diatom records, point to a change towards a fluviolacustrine environment,

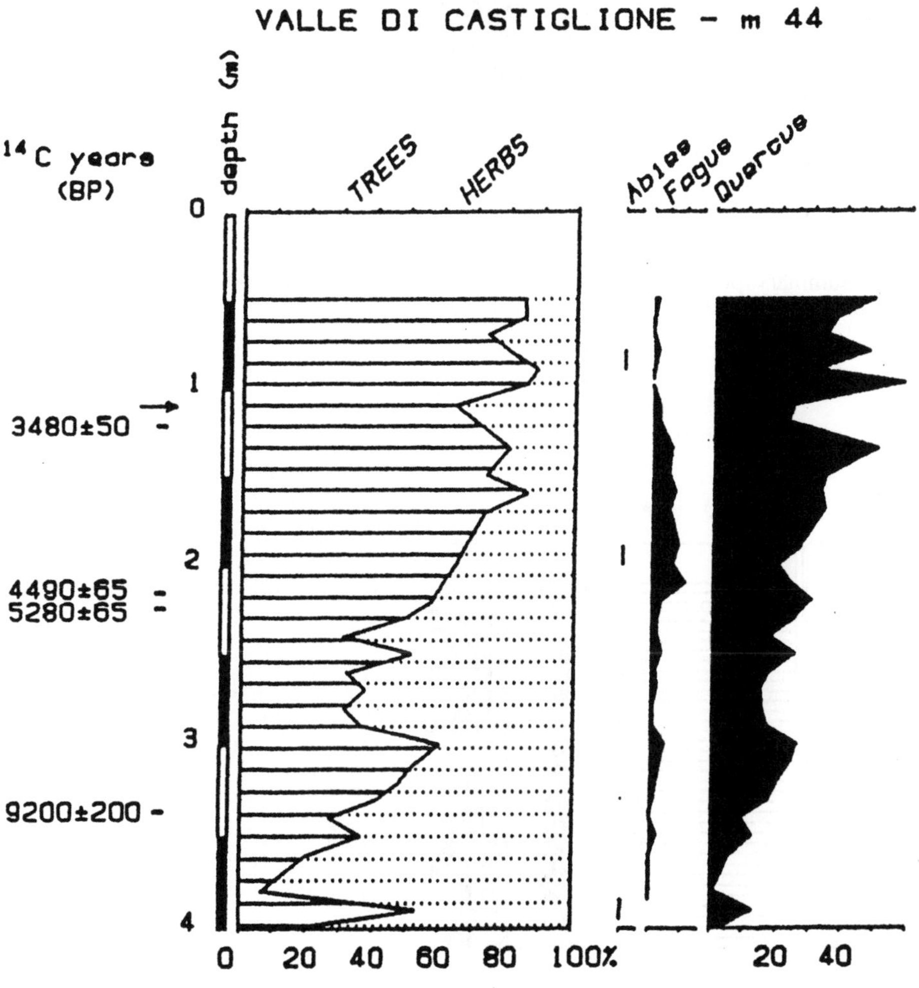

Fig. 3. Summary pollen diagram from Valle di Castiglione (Alessio et al. 1986). The arrow indicates the start of cultivation of Castanea and Juglans.

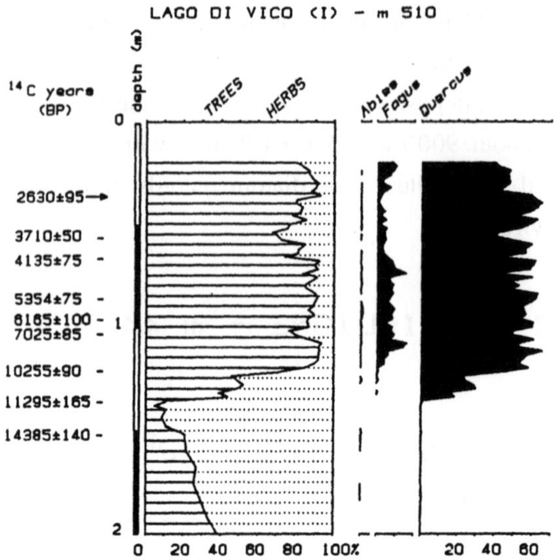

Fig. 4. Summary pollen diagram from Lago di Vico (Magri and Sadori, 1995). The arrow indicates the start of Castanea and Juglans.

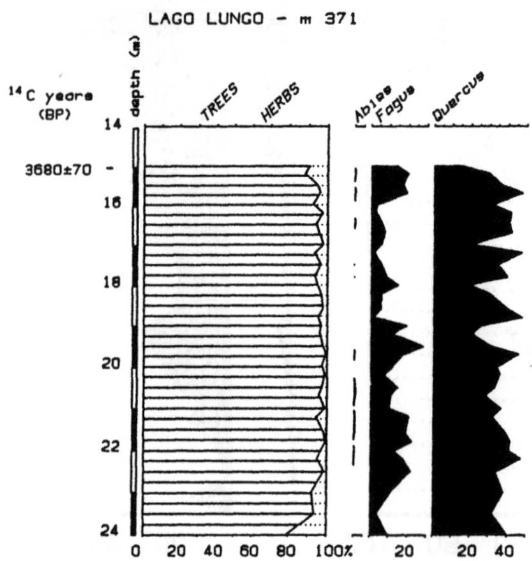

Fig. 5. Summary pollen diagram from Lago Lungo (Sadori, 1994).

indicating a lowering of the travertine barrier (Calderoni et al., 1994). The results of pollen analysis (Sadori, 1994) are shown in the simplified diagram of Fig. 5. They indicate that during the mid-Holocene, and up to the radiocarbon date 3680+70 BP, the area around the lake was covered by a dense oak-dominated forest. No appreciable variations are recorded in the vegetation cover and composition through this time-period. Nor can any clear indicators of human activity be traced.

Lagaccione is a small crater lake located 2 km west of Lago di Bolsena, belonging to the Monti Vulsini volcanic complex. The whole record spans about 100,000 years (Follieri, Magri and Narcisi, 1993; Follieri et al., 1995), but in Fig. 6 only the Holocene portion of the record from a preliminary core (Magri, 1989) is represented. The radiocarbon dates belong to a second parallel core which is being studied in more detail, and which can be well correlated with the first one by means of palynology. Also at Lagaccione the Holocene is characterized by dense forest vegetation, dominted by oak, and with abundant Corylus and Fagus. The arboreal pollen percentages, always high, undergo only one clear reduction starting around 3750 BP. Towards the top of the core, when also pollen of cultivated Castanea and Olea is found, forest recovers and trees again show very high percentage values.

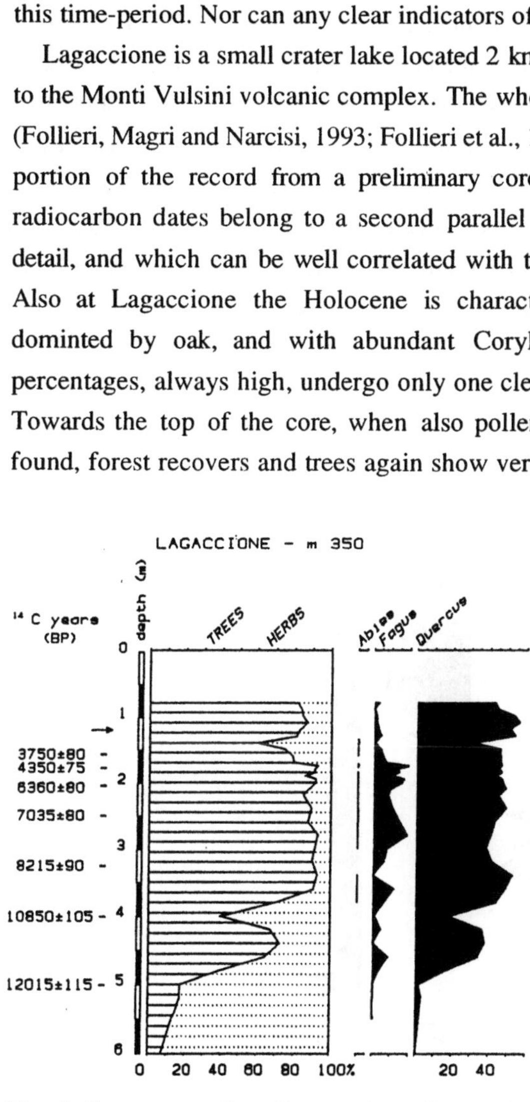

Fig. 6. Summary pollen diagram from Lagaccione (modified from Magri, 1989). The arrow indicates the start of Castanea.

Lago Padule is located at a high elevation on the northern Apennines, about 35 km NE of La Spezia. In Fig. 7 a synthetic pollen diagram redrawn and simplified from Lowe and Watson (1993) is represented. As in the other cases examined, the vegetation of the Postglacial is characterized by a well-developed forest, showing no appreciable sign of decline until about 2800 years ago. There were however some changes in the composition of the dominant trees: around 5300 BP Fagus populations immigrated and progressively spread to the expenses of Abies. A marked rise in Fagus at the nearby site of Prato Spilla (Lowe, 1992), dated to about 5000 BP, is interpreted by Lowe et al. (1994) as an important anthropogenic effect. However the pollen diagram from Padule (Lowe and Watson, 1993) does not report any evidence for cultivated plants. Among the so-called anthropogenic indicators, the pollen diagrams from the nearby site of Prato Spilla (Lowe,1992) record only a moderate increase of Plantago and Rumex, which are at any rate present from the Lateglacial.

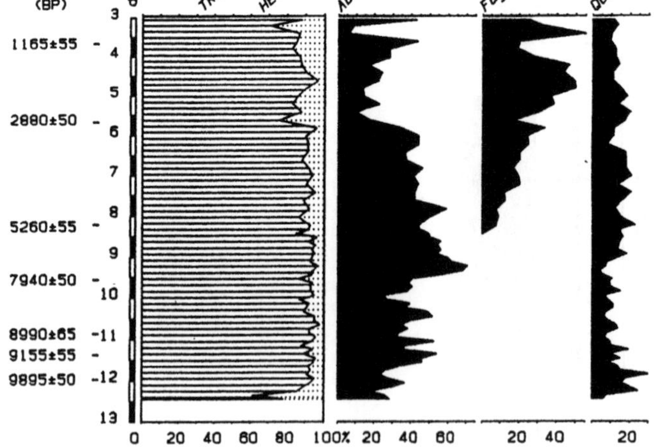

Fig. 7. Summary pollen diagram from Lago Padule (redrawn and modified from Lowe and Watson, 1993).

The Agoraie lakes are kettle-holes located at a high elevation on the northern Apennines, approx. 50 km NE of Genoa. In Fig. 8 a synthetic pollen diagram from Agoraie di Fondo is redrawn and simplified from Cruise (1990a). The peat inception is somewhat older than 4200 BP (Cruise, 1990b). Erratic variations of Filipendula (in black in the summary

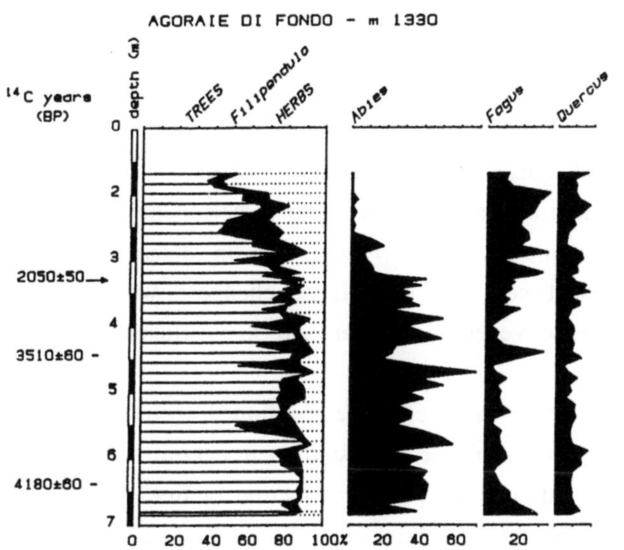

Fig. 8. Summary pollen diagram from Agoraie di Fondo (redrawn and modified from Cruise, 1990a). The arrow indicates the start of cultivation of Castanea and Juglans.

diagram of Fig. 8) and Cyperaceae have been interpreted as stream movements across the site which resulted in major fluctuations in the representation of mire-edge communities with concomitant changes in the curves of woody taxa. The pollen data indicate that from an age older than 4000 years to about 2000 BP a tall forest community of Abies, Fagus and Quercus was established, and most likely had no appreciable variations. After about 2000 BP the pollen spectra are indicative of the continuation of woodland conditions in which Abies played a diminishing role and Fagus expanded. It is after 2000 BP that the first finds of Castanea and Juglans are recorded. The appreciable increase of non arboreal plants at the top of the sequence is again due to changes in stream and mire edge communities.

Discussion

The Italian peninsula appears to have been densely forested during the whole Postglacial.

The mid-Holocene is characterized by only slight vegetation changes, not always recorded at all of the sites studied. In no case are clear trends towards deforestation recorded; on the contrary in some cases there is evidence for denser vegetation cover (cf. Valle di Castiglione, Fig. 3). Around 5000 BP, that is around the transition from Neolithic to Eneolithic, changes of forest composition are found at three sites: at Lago Grande di Monticchio there is an appreciable increase of Abies, at Valle di Castiglione a clear expansion of Fagus, and at Lago Padule immigration of Fagus at the expense of Abies. The behaviour of these trees, in apparent discrepancy, rules out that human activity may have played an important role in determining the geographical distribution of fir and beech. Instead, a climate change towards moister conditions might explain the dense vegetation cover and the diffusion of trees requiring humidity.

Another vegetational change which deserves to be mentioned occurred around 3700-3500 BP. This is clearly visible at all the sites studied in central Italy, and is characterized by an appreciable, although temporary, reduction of tree abundance. The question whether this recession of woodland was natural or human-induced should be interpreted also in the light of archaeological, palaeohydrological and sedimentological data. In fact there is a wealth of data in central Italy indicating a marked change in the hydrologic regime of lakes, which were lowered by several metres, so that many middle bronze age settlements, abundant around lakes and rivers, are nowadays submerged, in particular at Lago di Albano (Chiarucci, 1985), Lago di Bracciano (Fugazzola Delpino, 1982; Mantero, 1993), Lago di Bolsena (Fioravanti, 1991) and Lago di Mezzano (Petitti and Mitchell, 1993). At Lago Lungo (Fig. 5) lithostratigraphy and diatom analysis indicate a clear change from lacustrine to fluvio-lacustrine environment around 3700 BP, most likely in relation to a lowering of the travertine threshold of Le Marmore, responsible for the upstream deposition of lacustrine sediments (Calderoni et al., 1994). The deposition and erosion of travertine, caused by chemical-physical, morphologic and biological factors are generally connected with climatic conditions. At Lago Lungo it has been shown (Calderoni et al., 1994) that over the last interglacial-glacial cycle low heights of the travertine barrier always occurred during periods when climatic conditions were unfavourable to forest vegetation, that is in dry or relatively dry periods.

On the basis of palynological, archaeological, palaeohydrological and litho-bio-statigraphical observations it appears that a period of marked dryness occurred in central Italy around 3700-3500 BP. However this event is not recorded at the other sites examined in southern and northern Italy (Lago Grande di Monticchio, Lago Padule and Agoraie di Fondo), possibly because these lakes are located at higher elevations than the sites in central Italy.

A third event of interest in the development of the late Holocene history of vegetation in Italy occurred around 2500 BP. At this time at most sites there is clear evidence of cultivations. However the impact of human activity at that time does not seem to have been devastating for the natural environment, and in most cases there is no complete destruction of woodland.

At the time of prehistoric cultures Italy was densely inhabited; the oldest evidence for agriculture through plant macroremains is older than 7000 BP (Follieri, 1977-1982). However unequivocal signs of human activity are very rare in the pollen diagrams from the Italian peninsula. This result is unexpected and surprising (Magri, 1995), above all because in northwest Europe (Birks, 1986) it has long been recognized that during the second half of the Holocene the natural patterns of vegetational changes were disrupted and changed by the ever-increasing importance of human activity, and in particular by the appearance of agriculture.

Acknowledgement

Research was supported by the EC Environment Program 1991-1994, contract EV5V-0027.

References

Alessio M, Allegri L, Calderoni G, Cortesi C, Dai Pra G, De Rita D, Follieri M, Improta S, Magri D, Narcisi B, Petrone V, Sadori L (1986) 14C dating, geochemical features, faunistic and pollen analyses of the uppermost 10 m core from Valle di Castiglione (Rome, Italy). Geologica Romana 25: 287-308

Belluomini G, Manfra L, Proposito A (1993) Lago di Bolsena: datazione con il Carbonio-14 di reperti sommersi e fluttuazioni oloceniche del livello delle acque. Il Quaternario 6: 265-270

Birks HJB (1986) Late-Quaternary biotic changes in terrestrial and lacustrine environments, with particular reference to north-west Europe. In BE Berglund (ed) Handbook of Holocene Palaeoecology and Palaeohydrology, 3-65. Wiley Chichester

Calderoni G, Carrara C, Ferreli L, Follieri M, Gliozzi E, Magri D, Narcisi B, Parotto M, Sadori L, Serva L (1994) Palaeoenvironmental, palaeoclimatic and chronological interpretations of a late-Quaternary sediment core from Piana di Rieti (central Apennine, Italy). Giornale di Geologia 56/2 in press

Chiarucci G (1985) Materiali dell'etÖ del Bronzo nelle acque del Lago di Albano. Archeologia Laziale VII 11: 34-39

Cruise GM (1990a) Pollen stratigraphy of two Holocene peat sites in the Ligurian Apennines, northern Italy. *Review of Palaeobotany and Palynology* 63: 299-313

Cruise GM (1990b) Holocene peat initiation in the Ligurian Apennines, northern Italy. Review of Palaeobotany and Palynology 63: 173-182

Fioravanti A (1991) Trentatre anni di ricerche subacquee di geoarcheologia nel Lago di Bolsena. Bollettino di Studi e Ricerche (Bibliot. Comun. Bolsena) 6: 13-30

Follieri M (1977-1982) Le piÓ antiche testimonianze dell'agricoltura neolitica in Italia meridionale. Origini 11: 337-344

Follieri M, Giardini M, Magri D, Sadori L (1995) Palynostratigraphy of the last glacial period in the volcanic region of central Italy. Quaternary International in press

Follieri M, Magri D, Narcisi B (1990) A comparison between lithostratigraphy and palynology from the lacustrine sediments of Valle di Castiglione (Roma) over the last 0.25 MA. Memorie della SocietÖ Geologica Italiana 45: 889-891

Follieri M, Magri D, Narcisi B (1993) Palaeoenvironmental investigations on maar lakes sediment cores from Lazio (central Italy). In: JFW Negendank, B Zolitschka (eds) Palaeolimnology of European Maar Lakes. Lecture Notes in Earth Sciences 49: 95-107. Springer Heidelberg

Follieri M, Magri D, Sadori L (1988) 250,000-year pollen record from Valle di Castiglione (Roma). Pollen et Spores 30: 329-356

Follieri M, Magri D, Sadori L (1989) Pollen stratigraphical synthesis from Valle di Castiglione (Roma). Quaternary International 3/4: 81-84 (issued 1990)

Fugazzola Delpino MA (1982) Rapporto preliminare sulle ricerche condotte dalla Soprintendenza Archeologica dell'Etruria Meridionale nei bacini lacustri dell'apparato vulcanico sabatino. Bollettino d'Arte 4 (suppl. Archeologia Subacquea): 123-149

Giraudi C, Narcisi B (1994) Ricerche paleoclimatiche in ambiente lacustre. ENEA RT/AMB/94/31

Lowe JJ (1992) Lateglacial and early Holocene lake sediments from the northern Apennines, Italy-pollen stratigraphy and radiocarbon dating. *Boreas* 21:193-208

Lowe JJ, Davite C, Moreno D, Maggi R (1994) Holocene pollen stratigraphy and human interference in the woodlands of the northern Apennines, Italy. The Holocene 4(2): 153-164

Lowe JJ, Watson C (1993) Lateglacial and early Holocene pollen stratigraphy of the northern Apennines, Italy. Quaternary Science Reviews 12: 727-738

Magri D (1989) Palinologia di sedimenti lacustri olocenici a Lagaccione, presso il Lago di Bolsena. Giornale Botanico Italiano 123: 297-306

Magri D (1995) Some questions on the late-Holocene vegetation of Europe. The Holocene 5(3), in press

Magri D, Sadori L (1994) Late-Quaternary history of vegetation at Lago di Vico (central Italy). Giornale Botanico Italiano 128(1): 434

Magri D, Sadori L (1995) Late-Pleistocene and Holocene pollen stratigraphy at Lago di Vico (central Italy). Vegetation History and Archaeobotany, submitted

Mantero D (1993) Rinvenimenti preistorici in localitÖ Arrone-La Torre, Anguillara Sabazia (RM): considerazioni su una stratigrafia lacustre del Lago di Bracciano. Origini 17: 107-124

Narcisi B, Anselmi B, Catalano F, Dai Pra G, Magri G (1992) Lithostratigraphy of the 250,000 year record of lacustrine sediments from the Valle di Castiglione crater, Roma. Quaternary Science Reviews 11: 353-362

Newton AJ, Dugmore AJ (1993) Tephrochronology of core C from Lago Grande di Monticchio. In: JFW Negendank, B Zolitschka (eds) Palaeolimnology of European Maar Lakes. Lecture Notes in Earth Sciences 49: 333-348. Springer Heidelberg

Petitti P, Mitchell E (1993) Dati preliminari sulla topografia dell'abitato sommerso del Lago di Mezzano. In: Vulcano a Mezzano - Insediamento e produzioni artigianali nella media Valle del Fiora nell'etÖ del Bronzo. 17-31. Comune di Valentano Museo Civico

Robinson C, Shimmield GB, Creer KM (1993) Geochemistry of Lago Grande di Monticchio, S Italy. In: JFW Negendank, B Zolitschka (eds) Palaeolimnology of European Maar Lakes. Lecture Notes in Earth Sciences 49: 317-332. Springer Heidelberg

Sadori L. (1994) Pollen analysis. In: G Calderoni et al., Palaeoenvironmental, palaeoclimatic and chronological interpretations of a late-Quaternary sediment core from Piana di Rieti (central Apennine, Italy). Giornale di Geologia, in press

Turton I (1993) Palaeomagnetic investigations of Lago Grande di Monticchio, southern Italy. In: JFW Negendank, B Zolitschka (eds) Palaeolimnology of European Maar Lakes. Lecture Notes in Earth Sciences 49: 377-392. Springer Heidelberg

Watts WA (1985) A long pollen record from Laghi di Monticchio, Southern Italy: a preliminary account. J geol Soc London 142: 491-499

Zolitschka B, Negendank JFW (1993) Lago Grande di Monticchio (southern Italy)-a high resolution sedimentary record of the last 70,000 years. In: JFW Negendank, B Zolitschka (eds) Palaeolimnology of European Maar Lakes. Lecture Notes in Earth Sciences 49: 277-288. Springer Heidelberg

Environmental Changes in Central Europe 5000-3000 BP

Leszek STARKEL

Department of Geomorphology and Hydrology
Institute of Geography, Polish Academy of Sciences
ul. sw. Jana 22
31-018 Kraków
Poland

Abstract

In different systems such as lakes, glaciers, rivers, forest communities the climatic fluctuations are reflected with various rate and delay. In Central Europe were distinguished phases of strengthening of westerlies, more humid and with higher frequency of extreme events (5.4-4.9, 4.6-4.2, 3.3-3.0 ka BP). These phases alternate wirh more stable and drier ones. Those changes in the atmospheric circulation coincide with phases of high volcanic activity. A characteristic feature of the Subboreal in global scale is the increase of climatic stability, more frequent repetition of such cooler and wetter phases alternating with drier ones superimposed on general trend to the cooling. It coincides with the decline of the monsoonal activity and of penetration of the westerlies in the arid zone of SW Asia causing the expansion of the deserts during much drier phases.

Keywords: Subboreal periods, Central Europe, environmental changes, synchronous climatic fluctuations, causes of rapid changes, cooling or aridisation

Introduction

General tendency towards aridification in the lower latitudes between 3 - 5000 yr BP reflected in lake recessions and desiccations of rivers (Servant 1983, Street-Perrott and Grove 1979), was simultaneous with a cooling trend in the higher latitudes reflected in the Neoglacial advances of glaciers (Goldthwait 1966, Denton and Karlen 1973), rebuilding of the permafrost (Baulin et al. 1984) and retreat of the forest margins (Khotinsky 1984).

In the temperate latitudes (45-55°N) of Central Europe, characterized by oceanic air masses, this change was not so sharp due to its modification by the superimposed

NATO ASI Series, Vol. I 49
Third Millennium BC Climate Change
and Old World Collapse
Edited by H. Nüzhet Dalfes, George Kukla,
and Harvey Weiss

second order climatic fluctuations (Magny 1993) as well as by the gradually increasing human activity (Starkel 1987).

In this paper a variety of information will be reviewed on environmental changes in Central Europe, in the area between Rhine and Dnieper and from the Baltic and the North Sea up to the Alps, Carpathians and the Pannonian Basin inclusively. Records of the climatic and hydrological changes are registered in pollen diagrams, and by the rate of peat growth, fluctuations of lake levels, advances and retreats of mountain glaciers and vertical zones accompanying them (the upper tree line, snow line and solifluction) in activity of mass movements, precipitation of the calcareous tufa, fluvial accumulation and erosion, in soil profiles, mollusca and rodents and finally in continuous, climatically controlled records of ^{18}O and tree rings. The compiled data make possible to distinguish several phases of various climatic parameters during the time span of 5-3 ka BP. These phases correlate more or less with the sequence of changes in other parts of Europe (Fig. 1).

Evolution of plant communities

The Atlantic-Subboreal boundary put just at 5 ka BP coincides with the retreat of the *Quercetum mixtum* with the characteristic *Ulmus* decline (Firbas 1950, Frenzel 1966). The Subboreal is marked by the gradual transformation of forest communities. In the belt of the Central European mountains and uplands new mixed communities with the dominance of *Fagus* and *Abies*, and at higher elevations with *Picea* developed. The statistical evaluation of about 20 palynological diagrams in the Polish territory has helped in the detection of shorter periods of relatively rapid changes in the forest communities at 5.1-4.9 ka BP, 4.4-4.1 ka BP, 3.7-3.4 ka BP and 3.3-2.8 ka BP. While the third period seems to be connected with the warming favouring the expansion of *Carpinus* or *Fagus* the other ones indicate the retreat of several species which signify the cooling and the rise in humidity. Especially distinct there was the expansion of *Picea* with the characteristic culmination at the lower elevation of the Carpathians (Szczepanek - in: Gil et al. 1974) and the Sudetes (Madeyska - in Ralska-Jasiewiczowa 1989) between 4.5 and 4.0 ka BP. In the Alps three phases (5.3-5.1 ka BP, 4.6-4.4 ka BP and after 3.3 ka BP) with the distinct lowering of the upper tree line (Bortenschlager 1982, Burga 1987) were demonstrated (Fig. 2).

In the lowlands of Poland and Western Ukraine the turn to moister conditions is manifested by the increase of an accumulation rate of peat and by the beginning of

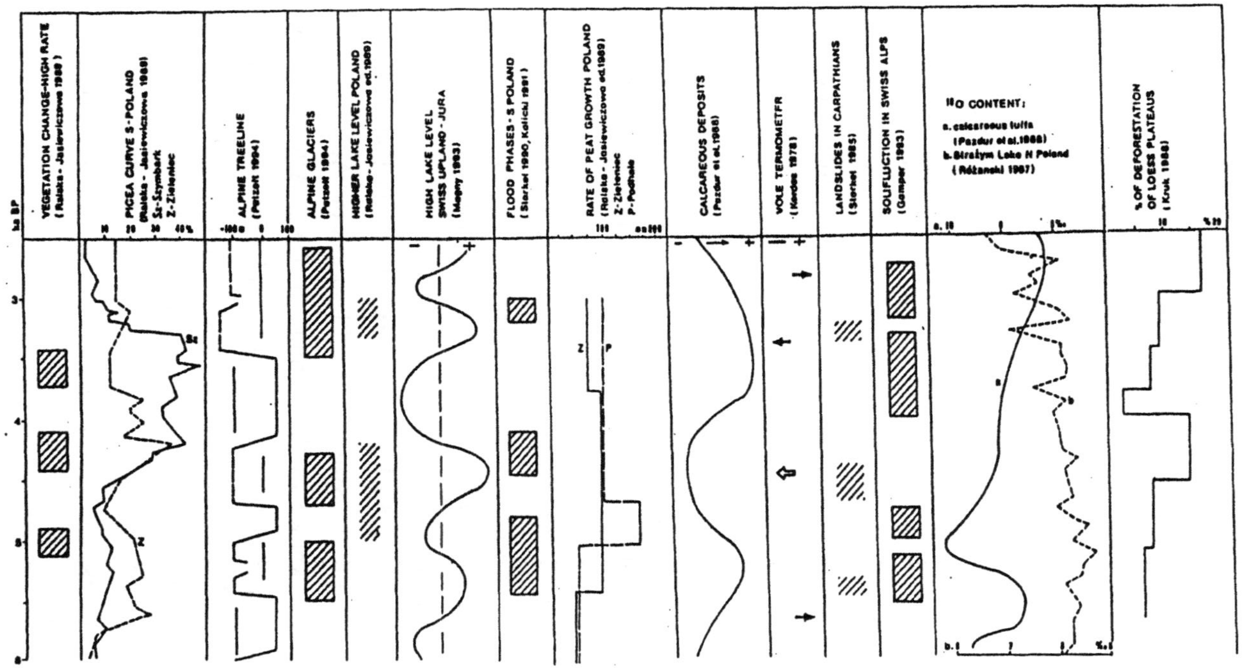

Fig.1. Coincidence of various environmental changes in Central Europe between 6.0 and 2.5 ka BP (after various authors).

A - phases of main changes in the vegetation cover in Poland B - pollen curves of Picea at 2 localities in the Carpathians (Sz) and Sudetes (Z), C - fluctuations of the alpine treeline in the Austrian Alps, D - phases of advances of glaciers in the Austrian Alps, E - phases of lake transgression in Poland, F - fluctuations of lake level in Swiss alpine foreland, G - phases of higher flood frequency, H - peat growth rate in Polish Carpathians (P) and Sudetes (Z), I - rate of calcareous precipitation, J - "vole termometer"- small mammals indicating changes in temperature, K - phases of frequent landslides overflysch, L - phases of frequent solifluctional lobes over fossil soils, M - O curves for calcareous tuffa in limestone plateau and N-Polish lake, N - estimated percent of deforestation on loess plateaus in S-Poland.

formation of the raised bogs (¯urek 1986, Ralska-Jasiewiczowa 1989, Bezusko et al. 1988). In Byelorussia the highest rates were registered earlier, before the end of the Atlantic phase (Elovichova et al. 1988). In the peatbogs of the Netherlands Casparie (1972) discovered a rise of ground water table after 5 ka BP.

The peatbogs of the Subboreal time register many recurrence layers (aridity and oxidation), difficult to correlate (Firbas et al. 1958), which were very seldom observed during the earlier stages. In the Danish bogs Aaby (1976) discovered the rhythmicity of hydrological changes with duration of cycles of 260 years each, starting from 4600 a BP.

Variations in lake levels

Recent tests for correlation of the changes in lake levels across Europe omit the samples from Central Europe (Harrison et al. 1993). A lot of information is contained in the lakes from the Alpine foreland and the Polish Lowland. In the Swiss Plateau in Lake Biel and others the rapid rises were registered after the Atlantic low level, at the beginning and at the end of the Subboreal (Amman 1975). The collections of records from the Alps foreland and Jura Mts show very distinct rises separated by regressions which are dated for 5.0-4.7 ka and 4.0-3.5 ka BP (Joos 1987, Magny 1993) (Fig. 1, 2).

At the beginning of the Subboreal the lakes of the Polish Lowland register a rise in a deposition rate accompanied by a water level rise (Ralska-Jasiewiczowa and Starkel 1988, Ralska-Jasiewiczowa ed. 1989, Niewiarowski 1990). In several lakes this rise is delayed till 4.5-4.0 ka BP. The next rise started ca. 3.0 ka BP. The studies of diatoms indicate also the phases of the water level rise at ca. 5.0, 3.8 and 2.8 ka BP (Bogacewicz-Adamczak 1987). In Southern Poland several mires were flooded (by groundwaters) in the early Subboreal (Ralska-Jasiewiczowa ed. 1989).

Advances of the Alpine glaciers

The close of the Atlantic period (5.3-5.1 ka BP) is marked by distinct advances of the glaciers which are well dated in the Swiss (Zoller 1977) and the Austrian Alps (Patzelt 1977). This advance is known as the Rotmoos I phase. After the brief recession the next advance (Rotmoos II), dated at ca. 4.7-4.4 ka BP followed. After the break which was one thousand year long the advance of the Löbben phase (3.4-3.2 ka BP) followed. The oscillations of glaciers were accompanied by downwind shifting of the upper tree line by 100-150 m (Burga 1987) as well as by expansion of permafrost and solifluction processes (Furrer et al. 1975, Gamper 1993). Also the examination of the widths of tree rings documents variations in the upper forest limit

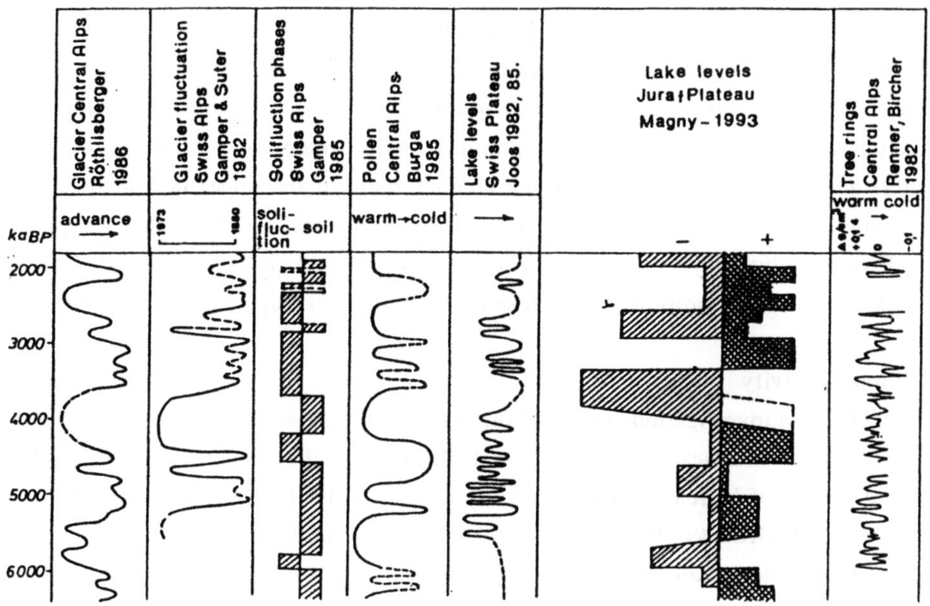

Fig.2. Changes of various parameters in the Alps (compiled after various authors).

(Bircher 1982, Renner 1982) and in the cooling of 5.3-5.0 ka BP, about 4.9 ka PB and 3.3 ka BP (Fig. 2).

Landslides

In the mountains, especially those built of the flysch rocks or the unconsolidated Neogene, there are many active and fossil landslides. In the Polish Carpathians several of them were formed or rejuvenated at the transition from the Atlantic to the

Subboreal, indicating a more humid phase with continuous rains or rainy seasons (Gil et al. 1974, Starkel 1985). The other common feature are the rockfalls, expressed in a from of layers of coarse debris beneath the limestone cliffs in the Schwabian Jura, and in Czech and Slovak Karst (Brunnacker 1975, Loûek 1991). These may be connected with reactivation of frost weathering in deep karstic canyons during cool phases.

Precipitation of the calcareous tufa

The deposition of tufas, well known from various limestone plateaus (Brunnacker 1975, Pazdur 1988, Loûek 1991), stopped at the beginning of the Subboreal; this may be explained by the activation of fluvial erosion in the valley floors or of slope processes (mentioned above). The return to precipitation of tufa indicates the increase in temperature combined with higher humidity, documented also by findings of foam sinters at the low elevations which are characteristic of the upper belt with annual rainfall exceeding 900mm (Loûek 1991). The rendzina soils or chernozems with the late Bronze cultural layers found from Thuringia to Slovakia in loess plateaus indicate the change to drier conditions at the close of the Subboreal (Jäger and Loûek 1983).

Fluvial activity

The detailed investigations on facial differences, avulsions and cut-off of meanders and clusterings of black oaks in the Vistula drainage basin enable one to distinguish several phases of increased fluvial activity about 5.5-4.8 ka BP, 4.5-4.2 ka BP and 3.2-3.0 ka BP (Kalicki 1991, Starkel 1994, Fig. 3). The well dated avulsion in the Grobla Forest older than 5090±110 a BP (Starkel et al 1991) is especially spectacular. The clusterings of the buried black oaks are of similar age: 4.5-4.4 ka BP (Kr1piec 1992). In the younger phase many trees show the tears of the bark which are characteristic of snowmelt floods with ice breaks. In the karstic niches of the Dunajec gorge Loûek (1977) found the marks of high floods, divided by the cultural late Bronze horizon which, in turn, indicates a low flood intensity during the younger part of the Subboreal. In the Oder catchment the paleochannels or channel deposits were dated at 5.0-4.4 ka BP (Florek 1978) in the lower Bóbr valley, at 4.1-4.0 ka BP (Kozarski 1991) in the middle Warta valley as well as they were dated in the Prosna valley where the bankfull river discharges were reconstructed for various paleochannels (Rotnicki 1991). The Prosna paleomeander dated at 4920±70 a BP had the discharge of 18.6 m^3s^{-1}, that dated at 4140±220 a BP as much as 45.7 m^3s^{-1}, and that from the close of the Subboreal (2860±90 a BP) only 9.85 m^3s^{-1}. It means that the paleodischarges were extremely high during the second half of the fifth millennium BP.

In the small catchment of the Ner and Moszczenica rivers (at the interfluve between the Vistula and Oder) the phase of increased fluvial activity was recently discovered (Turkowska 1988, Kamiñski 1992). In the Neris valley, tributary of the Nieman several abandoned channels on the higher terrace were dated between 6 and 4 ka BP and on the lower floodplain the oldest paleomeander was dated at about 2.9 ka BP (Gaigalas and Dvoreckas 1987). The higher terrace was probably dissected during the early Subboreal floods. In the upper Main valley, on the base of the black oaks clusters Schirmer (1983) distinguished different ages of the main terrace fills: one of them between 7.0 and 5.4 ka BP, and the second 4.5-3.2 ka BP. In the Austrian sector of the Danube valley Fink (1977) described the fill dated at 5.1-4.5 ka PB. In the Danube tributary valleys draining the Alps Schreiber (1985) distinguished two episodes 5.0-4.9 and ca 4.6 ka BP (the Lech river) and earlier Brunnacker (1978) discovered the signs of rapid erosion dated to the mid of the fifth millennium BP (the Salzach valley). Downstream in the Iron Gate across the Carpathians a late Bronze cultural layer shows the phase of rare and lower floods (Brunnacker 1971). Also in the corner of the Hungarian Plain the systems of abandoned paleomeanders of the Tisza river were dated at the transition of the Atlantic/Subboreal and at the close of the Subboreal (Borsy and Felegyhazi 1983).

Many dendrochronological records collected by Becker (1982) in the Danube and Rhine river basins indicate a somewhat similar clustering of black oaks between 5.0-4.3 ka BP, 3.9-3.1 ka and 3.0-2.7 ka BP. In the Aar river basin at the Alpine foreland a substantial transformation of the channel pattern took place at the Atlantic/Subboreal transition (Wohlfarth and Amman 1991). In the valley floors of the French Alps the increased fluvial activity was dated between 5.0 and 4.7 ka BP (Jorda 1982).

Animal remains

The investigations undertaken by Loûek (1982) and later by Alexandrowicz (1983) have shown the beginning of transformation of malacological assemblages at the Atlantic/Subboreal transition, yet a distinct change is visible in the upper Subboreal and at the transition to the Subatlantic, visible in the restriction of forest and other habitats and their replacement by species adapted to the cultivated areas (synanthropic). Also the rodents found in the caves of Hungary and Slovakia indicate the cooling and a probable rise of precipitation (Kordos 1983).

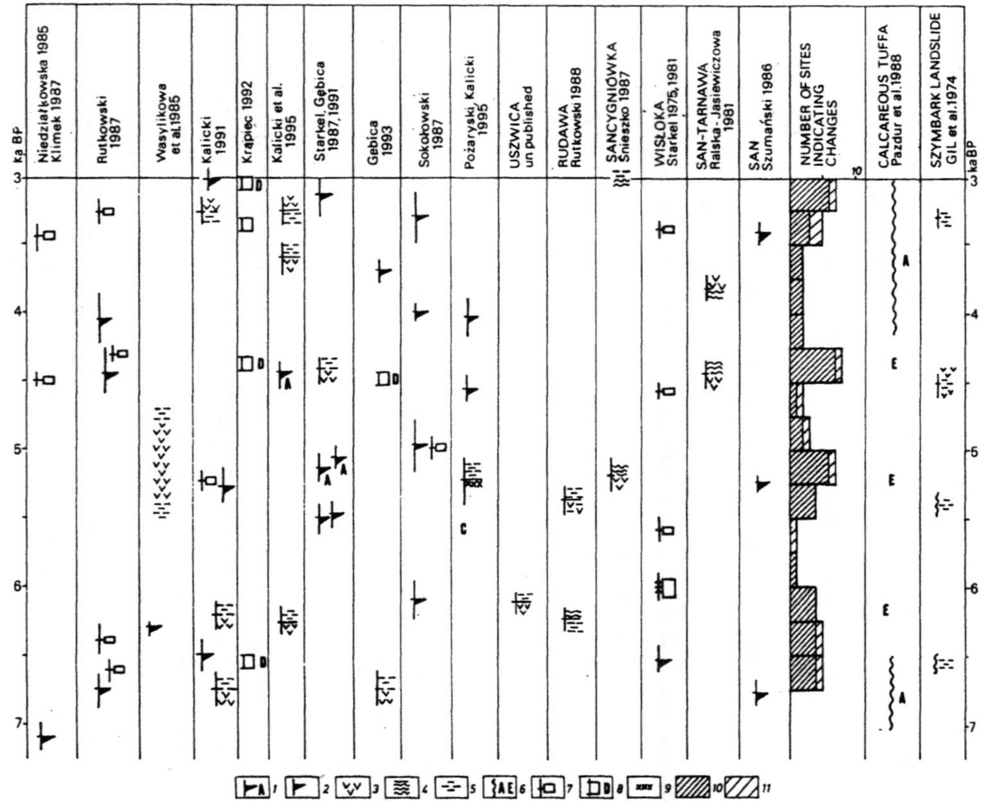

Fig.3. Records on flood phases between 7 and 5 ka BP in the upper Vistula basin.

Signs: 1. radiocarbon age of the bottom of the paleochannel fill with avulsion, 2. the same - not clear avulsion, 3. peat, 4. gyttya, 5. overbank facies, 6. calcareous tuffa deposition or erosion, 7. dated black oaks, 8. series of fallen oaks dated by dendrochronological method, 9. fossil soil, 10. number of sites indicating changes in fluvial activity, 11. other dated sites.

Oxygen Isotope Records

Significant falls in temperature are recorded in the ^{18}O profiles of the Greenland ice cores ca. 5.5 and 2.9 ka BP in calendar years (Dansgaard et al. 1984). To be more specific, the entire ^{18}O profile from Götland registers a rapid decline in temperature about 5 ka BP (Mörner and Wallin 1977). A similar decline, although not so well dated, was found in the lake cores in Northern Poland (Ró¿añski 1987). The cooling of about 5.1-5.0 ka BP was distinguished in several sections of the calcareous tufa in Southern Poland (Pazdur et al. 1988).

Human activity

In Central Europe the discussed period of 5-3 ka BP is the time of the flourishing of the Neolithic and Bronze Age cultures. Forest clearance and cultivation caused gradual transformation of ecosystems, acceleration of water runoff, soil erosion and the rise of the groundwater level in the plains. These changes undoubtedly coincided with the climatic fluctuations. In the upper and middle course of the Vistula river lower horizons of the alluvial loams are dated at 5-4 ka BP (Falkowski 1975, Biernacki 1975). Also in the tributary valleys draining the loess plateaus the change from calcareous to clay muds took place before 5 ka BP (Œnieszko 1985). In the Morava river basin the beginning of loamy aggradation was dated between 3.7 and 3.2 ka BP, indicating the progress of agriculture during that time (Havlicek 1991). In the valley of the German Mittelgebirge a different time of alluviation was combined with a different time of occupation by human (Jäger 1982).

But there are also archeological findings indicating changes in terrain occupation which may be explained by climatic changes. Simultaneously with the acceleration of soil erosion the settlements of the Funnel Beaker Culture, dated at 4600-4100 a BP (3200-2500 BC), expanded from the valley floors on slopes and interfluves (Kruk 1980). This coincides with the cooler and wetter phases. Similar new findings at the Carpathian Foothills (Machnik 1988) are explained as leaving of cold, inversional and flooded valley floors and occupation of fertile, dry and warm habitats at higher elevations.

Climatic fluctuations 5-3 ka BP in Central Europe

Different systems react to climatic changes with a different rate and a delay. While the equilibrium of the fluvial system may be disturbed by the higher frequency of extreme floods (Starkel 1983), forest communities may preserve relict features for a long time; lakes react to multiannual trends in a water balance, then mountain glaciers differently and individually reflect variations in snow falls and in summer melting (Patzelt 1985). Taking under consideration the precision of the radiocarbon datings up to 100 years, it may be concluded that the various processes described above show relatively well their synchronousness indicating similar pattern of fluctuations of temperature, rainfall, runoff etc. over the whole Central Europe. This fact allows us to distinguish several climatic phases between 5 and 3 ka BP.

The decline of the Atlantic period (5.4-4.9 ka BP) is characterized by the increase of oceanicity and drop of temperature reflected in the advances of glacier, rise of lake water level, drastic changes in a river channel pattern and simultaneous rapid drop of ^{18}O curve, reduction of precipitation of the calcareous tufa and the beginning of transformation of rich forest communities of the Atlantic period.

Later, there was a short, 200-300 years long (4.9-4.7 or 4.6 ka BP) return to a warmer climate with rare floods, which is visible in the retreat of the Alpine glaciers and lowering of lake levels at their foreland.

Next, cooler and wetter phase was registered first in the Alps and at their foreland (4.7-4.3 ka BP) and later in the Polish territory (4.6-4.2 ka BP). It is well reflected in the lowering of the vegetation vertical zones (expansion of *Picea* in the Carpathians and Sudetes).

After that there was a relatively long period (4.2-3.3 ka BP) of more stable warm climate, favourable for expansion of *Carpinus, Fagus* and *Abies*. The increasing degradation of natural geoecosystems facilitated soil erosion and deposition of the alluvial loams.

The end of the period discussed, i.e. 3.3-3.0 ka BP, characterizes the readvance of a wetter and probably cooler(?) climate, reflected in the fluvial activity and a next advance of glaciers. These data, showing the tendency to aridisation, are in contradiction to the records from the limestone plateaus from Bohemia and Slovakia (Jäger and Loûek 1983). It is probably the effect of the coincidence of transition to pannonian-type climate and deforestation.

The climatic phases described have their main source in the atmospheric circulation changes of various duration, connected with the strengthening or weakening of the westerlies and reflected in alternation of drier and warmer phases with wetter and relatively cooler phases (cf. Magny 1993).

Correlation with other parts of Europe

When studying climatic variations one refers most frequently to the records of glaciated high latitudes and Scandinavia. The maximum acid fallout registered in ice cores (Hammer et al. 1980) indicates the existence of three peaks of volcanic activity ca. 5, 1, 4.6 and 3.1 ka BP, which, after Bryson (1989), caused the rise of totals and intensity of precipitation. These peaks coincide with the advances of the Scandinavian mountain glaciers (Nessje and Johannessen 1992). Although the glaciers in the Alps and Scandinavia in the mid-Holocene are not synchronous, they correlate extremely well beginning from the Subboreal. However, it is not the case of lake level fluctuations which in Southern Sweden show different culminations ca. 5.5-5.3 and about 4.0 ka BP (Digerfeldt 1988, Harrison et al. 1993). The rivers of Finland registered much lower discharges during the Atlantic than at present, yet very rare abandoned channels (in this regions) are dated just at the Atlantic/Subboreal transition (Koutaniemi 1991, Mansikkaniemi 1991).

In the central part of Russia (between Moscow and St. Petersburg) there were also registered the rises of lake levels (Dolukhanov 1977) and three wetter phases reflected in the pollen diagrams of peatbogs ca. 4.9, 4.6 and 3.2 ka BP (Khotinsky 1984, Klimanov 1990). It is likely that the latitudinal belt across Europe had a similar rhythmicity of a precipitation regime, with a slight delay towards the east. Simultaneously, far to the north in the Petchora basin the thermal profile across the permafrost shows the rise of temperature correlated with the Subboreal (Kondratieva et al. 1993).

Western Europe does not exhibit so distinct hydrological changes (Lamb 1977) which might have been caused by a more humid climate in general. The studies in the southern part of the Netherlands have shown a more humid phase as late as 4.3-4.0 ka BP (Bohncke and Vandenberghe 1991).

When compared with the Atlantic period the characteristic feature of the Subboreal over the whole Europe is the increase of climatic instability. More frequent and shorter are the returns of cooler and wetter phases with glacial advances as well as warmer ones superimposed on the slow tendency towards global cooling after the Atlantic optimum.

Discussion

Trying to put the presented picture on the wide Euroasiatic background it may be concluded that this pulsating character of climatic changes coincides with fluctuations of monsoonal precipitations in the Southern Asia (Bryson 1992, this volume) and in the Eastern Africa (Bonnefille et al. 1990). These variations are visible in the passing of threshold values in various systems like lakes, rivers or plant communities (Fig. 4). Especially spectacular is the shift of ecotonal boundaries as well as of vertical belts in the mountains. The crossing of thresholds may be local or regional in scale, it may happen once without any return (expansion of desert) or repeat several times (such as advances of glaciers in Middle-European mountains).

Between these two humid zones in Eurasia is located the dry region of the Near East and Middle East with general tendency to aridisation during the last 5 ka BP. It is the zone (being in the reach of two humid air masses in summer from the Indian ocean, in winter from the Atlantic (COHMAP 1988). From various evidence and modelling (Sinclair and Bryson - this volume) we see that these invasions of the humid air from the far distance since ca 5 ka BP were more and more rare and restricted. This phenomenon seems to be the main cause of the gradual expansion of desert and semidesert in time (Fig. 5). One of such fluctuations with retreat of extension of humid air masses caused in the Near East probably so rapid and distinct drop in precipitations (Fig. 4) that the desert belt expanded, trees retreated to the mountain refugies and contributed to the collapse of the Old World cultures.

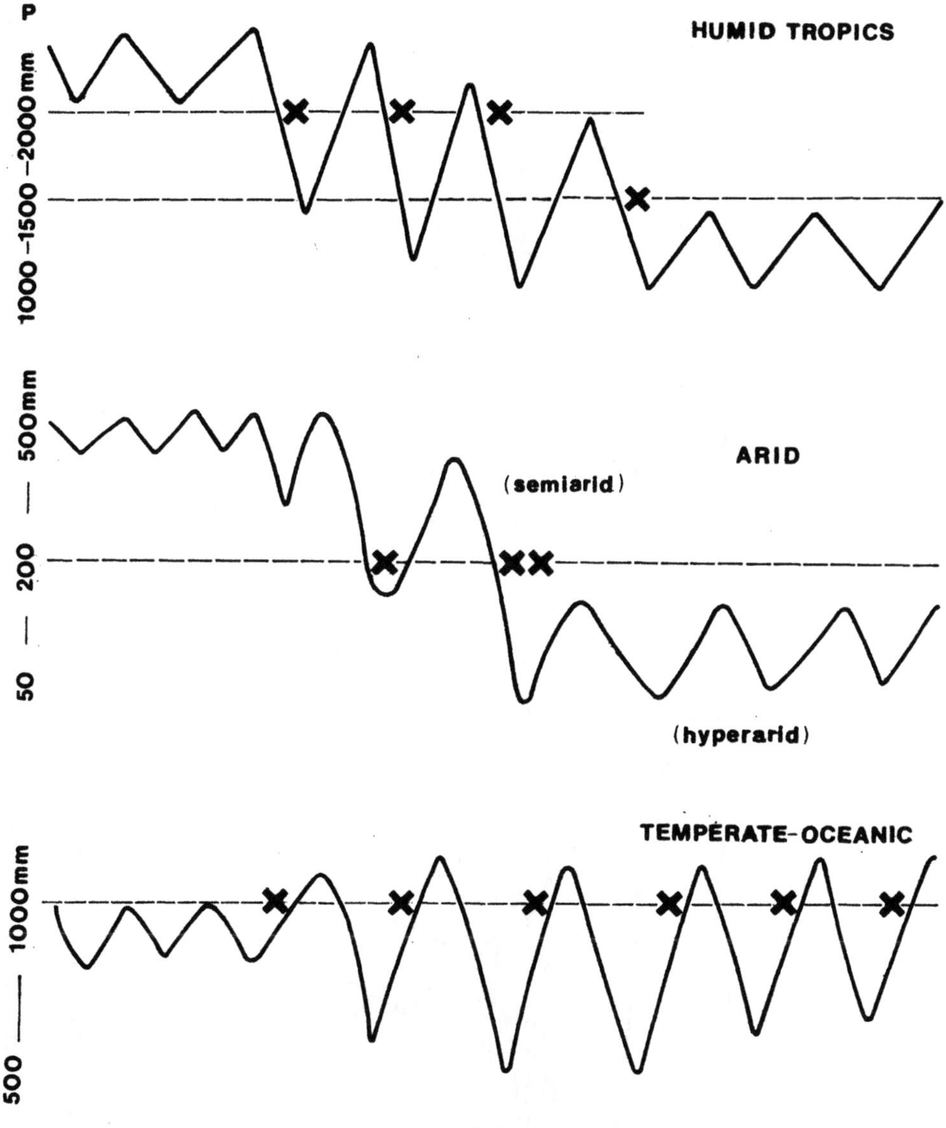

Fig.4. Scheme of general trend and long-term fluctuations of precipitation in various climatic zones between 8-0 ka BP. Broken lines indicate the levels of substancial changes in hydrology and ecosystems. Stars indicate the crossings of treshold values causing substancial changes of ecosystems.

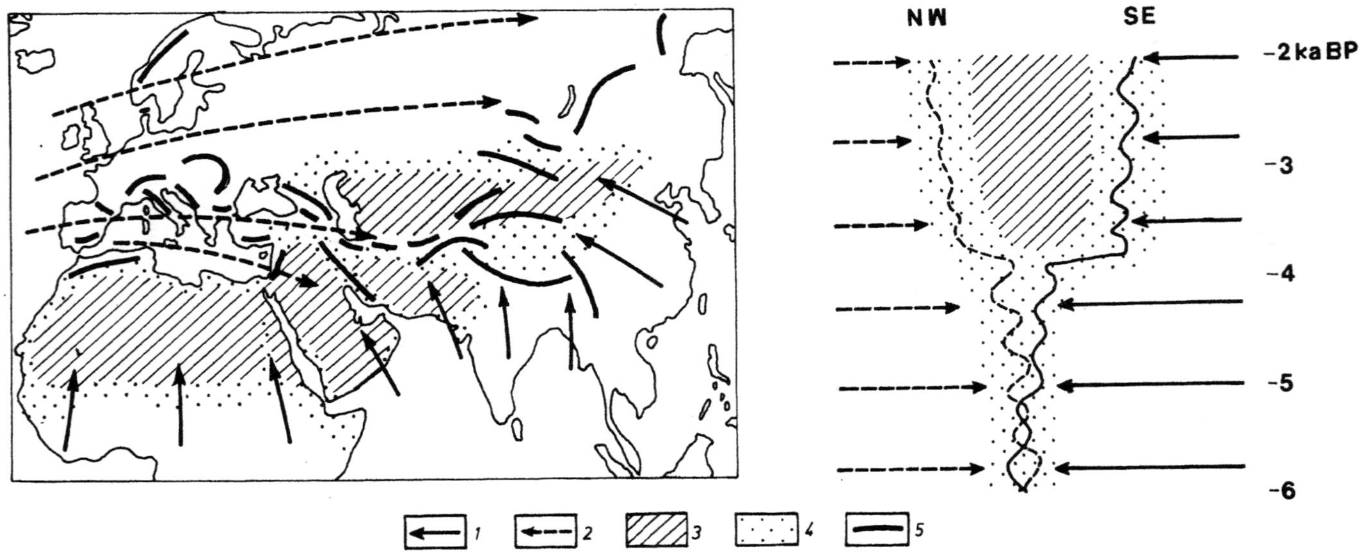

Fig.5. Simplified map of present rainfall tracks and deserts and NW-SW Eurasian transect showing the expansion of the arid zone in relation to the weakening of monsoonal and westerlies air mass circulation.

Signs: 1. monsoonal summer rains, 2. western winter rains, 3. desert zone, 4. semidesert zone, 5. mountain barriers.

References

Aaby B. (1976) Cyclic climatic variations in climate over the past 5500 years reflected in raised bogs. Nature 263: 281-284

Alexandrowicz S.S. (1983) Malacofauna of Holocene calcareous sediments of the Cracow Upland. Acta Geologica Polonica 33, 1-4: 117-158

Ammann-Moser B. (1975) Vegetationskundliche und pollenanalytische Untersuchungen aur den Heidenweg im Bielersee. Beiträge zur Geobotanischen Landesaufnahme der Schweiz 56: 1-76

Baulin V.V., Belopukhova Ye.B. and Danilova N.S. (1984) Holocene permafrost in the USSR, in: Late Quaternary Environments of the Soviet Union, A.A.Velichko (ed.). Univ. of Minnesota Press 87-94

Becker B. (1982) Dendrochronologie und Palaoekologie subfossilen Baumstamme aus Flussablagerungen; ein Beitrag zur nacheiszeitlichen Auenentwicklung im südlichen Mitteleuropa, in: Mitteilungen der Kommission fur Quartarforschung der Österreichischen Akademie der Wissenschaften 5: 1-120

Biernacki Z. (1975) Holocene and late pleistocene alluvial sediments of the Vistula river near Warsaw. Biuletyn Geologiczny 19 Warsaw University 199-219

Bircher W. (1982) Klima und Gletschergeschichte des Saastales: Glazial morphologische und dendroklimatologische Untersuchungen. Physische Geographie, 9.

Bogaczewicz-Adamczak B. (1987) Development of the lakes in the Bory Tucholskie area (NW Poland) since the Atlantic Mesolithic Period, in: Methods for the investigation of lake deposits: paleoecological and palaeoclimatological aspects, (ed.) M.Kabaliene, Vilnius 19-29

Bohncke S. and Vanderberghe J. (1990) Small river basins of S-Netherlands, in: L.Starkel, K.J.Gregory and J.B.Thornes: Fluvial processes in the temperate zone during the last 15 000 years. J.Wiley and Sons 253-281

Bonnefille R., Roeland J.C. and Guiot J. (1990) Temperate and rainfall estimates for the past 40 000 years in equatorial Africa. Nature 346: 347-349

Borsy Z. and Felegyhazi E. (1983) Evolution of the network of water courses in the North-Eastern part of the Great Hungarian Plain from the end of the Pleistocene to our days. Quaternary Studies in Poland 4, Poznağ, 115-124

Bortenschlager S. (1982) Glacial fluctuations and changes in forest limit in the Alps, in: Gaillard M.J.ed. IGCP 158, Lundqua Report 27: 43-45

Brunnacker K. (1971) Geologisch - pedologische Untersuchungen. In: Lepenski Vir am Eisernen Tor. Fundamenta, Reihe A 3: 20-32

Brunnacker K. (1975) Aktivierungen des fluviatilen Geschehens im Holozän nördlich der Alpen. Biul. Geologiczny U.W., 19, Warszawa, 149-156

Brunnacker K. (1978) Der Niederrhein im Holozän, Fortschritte der Geologie In: Rheinland und Westfalen 28: 399-440

Bryson R.A. (1989) Late Quaternary volcanic modulation of Milankovitch climate forcing. Theoretical and Applied Climatology 39: 115-125

Bryson R.A. (1992) A macrophysical model of the Holocene intertropical convergence and jetstream positions and rainfall for the Saharan Region. Meteorology and Atmospheric Physics 47: 247-258

Bryson R.A. (1994) Proxy indicators of Holocene winter rains in Southwest Asia compared with simulated rainfall. Abstracts of Papers, NATO-ARW on Third Millenium BC Abrupt Climate Change and Old World Social Collapse, Kemer, Turkey, 7 p.

Burga C.A. (1988) Swiss vegetation history during the last 18 000 years. New Phytologist 110: 581-602

Casparie W.A. (1972) Bog development in Southeastern Drenthe (The Netherlands). Rijks Univ.Groningen 271 p.

COHMAP members (1988), Climatic changes of the last 18 000 years: observations and model simulations. Science 241: 1043-1052

Dansgaard W., Johnsen S.J., Clausen H.B., Dahl-Jensen D., Gundestrup N., Hammer C.U. and Oeschger H. (1984) North Atlantic climatic oscillations revealed by deep Greenland ice cores, in: J.E.Hansen and T.Takahaski eds. Geophysical Monograph 29, Climate Processes and Climate Sensitivity, A.G.U., Washington D.C. 288-298

Denton G.H. and Karlen W. (1973) Holocene climatic variations - their pattern and possible cause. Quaternary Research 3, 2: 155-205

Digerfeldt G. (1988) Reconstruction and regional correlation of the Holocene lake-level fluctuations in Lake Bysjon, South Sweden. Boreas 17, 2: 165-182.

Dolukhanov P.M. (1977) The Holocene history of the Baltic sea and ecology of prehistoric settlement. Baltica 6: 227-244, Vilnius

Elovicheva Y. and Bogdel J. (1987) Reconstruction of palaeoclimatic and vegetation of the Byelorussian Holocene using bog and lake deposit data, in: Palaeohydrology of the temperate zone. Vol.2, IGCP-158, Soviet Working Group 152-166

Falkowski E. (1975) Variability of channel processes of lowland rivers in Poland and changes of the valley floors during the Holocene. Biuletyn Geologiczny U.W., 19, Warszawa, 45-78

Fink J. (1977) Jüngste Schotterakkumulationen im Österreichischen Donauabschnitt, in: Dendrochronologie und postglaziale Klimaschwankungen in Europa, Erdwiss. Forschung 13, Wiesbaden, 190-211

Firbas F. (1949-1952) Spät- und nacheiszeitliche Waldgeschichte Mitteleuropas nordlich der Alpen, 480 and 256 pp

Firbas F., Münnich K.O., Wittke W. (1958) 14C Datierungen zur Gliederung der nacheiszeitlichen Waldentwicklung und zum Alter von Rekurrenzflächen im Fichtelgebirge. Flora 146: 512-520

Florek W.(1978) Próba analizy zmian cech geometrycznych meandrów wspólczesnych i kopalnych na przykladzie dolnego Bobru (Sum: An attempt to analyse changes in geometrical characteristics of present-day and fossil meanders: a case study of the Lower Bóbr). Przegl. Geogr. 50, 4: 643-660

Frenzel B. (1966) Climatic change in the Atlantic/sub-Boreal transition on the Northern Hemisphere: botanical evidence, in: Proc Symposium on World Climate from 8 000 to O.B.C. Royal Meteorological Society London, 99-123

Furrer G., Leuzinger H., and Amman K.. (1975) Klimaschwankungen wöhrend des alpinen Postglazials im Spiegel fossiler Böden. Vierteljahrschr. Naturf. Ges., Zürich 120, 1: 15-31

Gaigalas A. and Dvareckas V. (1987) Geomorphological structure and development of river during Last Glaciation and Holocene in the South peribaltic area, palaeohydrology of temperate zone. Rivers and Lakes, Tallin 99-110

Gamper M., (1993) Holocene solifluction in the Swiss Alps: dating and climatic implications. Special Issue ESF Project, European Paleoclimate and Man, ed. B.Frenzel. Akad Wiss. Mainz 1-9

Gil E., Gilot E., Kotarba., Starkel L., Szczepanek K. (1974) An early Holocene landslide in the Niski Beskid and its significance for paleogeographical reconstructions. Studia Geomorph. Carpatho-Balcanica 8: 69-83

Goldthwait R.P. (1966) Evidence from Alascan glaciers of major climatic changes, Proc.Symp.World Climate fromn 8 000 to O.B.C. London 40-53

Harrison S.P., Prentice I.C. and Guiot J. (1993) Climatic controls of Holocene lake-level changes in Europe. Climate Dynamics 8: 189-200

Hammer C.U., Clausen H.B. and Dansgaard W. (1980) Greenland ice sheet evidence of post-glacial volcanism and its climatic impact. Nature 288: 230-235

Havlicek P. (1991) The Morava River Basin during the last 15 000 years

Jäger K.D. (1982) Stratigraphische Belege fur Klimawandlungen im mitteleuropäischen Holozän. Zeitschrift für geologische Wissenschaften 10, 6, Berlin 799-809

Jäger K.D., Loûek V. (1983) Paleohydrological implications on the Holocene development of climate in Central Europe based on depositional sequences of calcareous fresh-water sediments. Quaternary Studies in Poland 4: 81-89

Jorda M. (1993) Histoire des paleoenvironments tardi- et postglaciares Sud-Alpins de moyenne altitude. Essai de reconstitution cinematique, in: Geomorphologie et Amenagement de la Montagne Hommage a P.Gabert. Bulletin 42, Centre de Geomorphologie, CNRS Caen 99-111

Joos M. (1982) Swiss Midland-lakes and climatic changes, in: Harding A. (ed.), Climatic change in later Prehistory, Edinburgh Univ.Press 44-51

Kalicki T. (1991) The evolution of the Vistula river valley between Cracow and Niepolomice in Late Vistulian and Holocene times, in: Evolution of the Vistula river valley during the last 15 000 years, part.IV. Geogr. Stud. Spec. Issue 6: 11-37

Kamiñski M. (1993) Póznoplejstoceñska i holoceñska transformacja doliny Moszczenicy jako rezultat zmian œrodowiska naturalnego oraz dzialalnozœci czlowieka (Late Pleistocene and Holocene evolution of Moszczenica valley as the results of natural changes and human activity), doctor Thesis, Dept. of Geography, Lódz University

Khotinsky N.A. (1984) Holocene vegetation history, in: Late Quaternary Environments of the Soviet Union, A.A.Velichko (ed.). Univ of Minnesota Press 179-200

Klimanov W.A., (1990) Kolichestvennye characteristiki Klimata severnoy Evrazji v pozdnelednikovye (Quantitative characteristics of the Northern Eurasia in the Lateglacial - in Russian). Izvestya Akad.Nauk SSSR, seria geograf. 4: 116-126

Kondratieva K.A., Khrutzky S.F. and Romanovsky N.N. (1993) Changes in the extent of permafrost during the Late Quaternary period in the territory of the former Soviet Union. Permafrost and Periglacial Processes 4: 113-119

Kordos L. (1987) Climatic and ecological changes in Hungary during last 15,000 years, in: Holocene environment in Hungary, M. Pecsi and L. Kordos eds. Geograph. Institute, Budapest 11-24

Koutaniemi L. (1991) Glacio-isostatically adjusted palaeohydrology the river Ivalojoki and Oulanka - joki, Northern Finland, in: Temperate Palaeohydrology, eds. L.Starkel, K.J.Gregory and J.B.Thornes. J.Wiley 65-78,

Kozarski S., (1991) Warta - a case study of lowland river, in: Temperate Palaeohydrology, eds. L.Starkel, K.J.Gregory, J.B. Thornes. J.Wiley 189-215

Krlpiec M. (1992) Skale dendrochronologiczne póznego holocenu poludniowej i centralnej Polski (sum. Late Holocene tree-ring chronologies of South and Central Poland). Geologia 18, 3 Wyd. AGH, Kraków 37-119

Kruk J. (1980) Gospodarka w Polsce poludniowo-wschodniej w V-III tysilcleciu p.n.e. (Land use in the South-eastern Poland during V-III millenia B.C. - in Polish). IHKM, Ossolineum

Lamb H.H. (1977) Climate. Present, past and future vol.2. Climatic history and the future. Methuen 835 pp.

Lozek V., (1977) Stratigrafie a mekkysi jeskynní vyplne v prulomu Dunajce v Pieninach (sum. Stratigraphy and molluscan fauna of a cave fill in the Dunajec Gorge, Pieniny Mts). Ceskoslovensky Kras, Praha 28: 75-83

Lozek V. (1982) Faunengeschichliche Grundlinien zur spat - und nacheiszeitlichen Entwicklung der Molluskenbestände in Mitteleuropa. Rozpr.CSAV, Rada Matern. a prirod.ved. 92, 4: 1-106

Lozek V. (1991) Palaeogeography of limestone areas, in: Temperate Palaeohydrology eds.L.Starkel, K.J.Gregory and J.B.Thornes. Wiley 413-429

Machnik J. (1988) Wyniki najnowszych badañ archeologicznych w Karpatach polskich oraz ich znaczenie dla innych dyscyplin naukowych (New results of archeological studies in Polish Carpathians in Polish). Rocznik Oddzialu PAN w Krakowie 83-104

Magny M. (1993) Holocene fluctuations of lake levels in the French Jura and sub-Alpine ranges and their implications for past general circulation pattern. The Holocene 3, 4 306-313

Mansikkaniemi H. (1991) Development of river valleys originated by emergence S-Finland. Case study of the river Kyronjoki, in: Fluvial processes in the temperate zone during the last 15 000 years eds. L.Starkel, K.J.Gregory and J.B.Thornes.J.Wiley, 79-104

Mörner N.A., Wallin B. (1977) A 10 000 year temperate record from Gotland, Sweden. Palaeogeography, Palaeoclimatology, Palaeoecology 21: 113-138.

Niewiarowski W. (1990) Hydrological changes in the light of palaeolake studies, in: Evolution of the Vistula river valley during the last 15,000 years, part.3, ed.L.Starkel. Geographical Studies, Spec.Issue 5: 170-178

Patzelt G. (1977) Der zeitliche Ablauf und das Ausmass postglazialer Klimaschwankungen in den Alpen. In: Dendrochronologie und postglaziale Klimaschwankungen in Europa. Erdwiss. Forschung 13, Wiesbaden, 249-259

Patzelt G. (1985) The period of glacier advances in the Alps, 1965 to 1980. Zeitschrift für Gletscherkunde und Glazialgeologie, 21: 403-407

Pazdur A., Pazdur M.F., Starkel L. and Szulc J. (1988) Stable isotopes of the Holocene calcareous tufa in S-Poland as paleoclimatic indicators. Quaternary Research 30, 2: 177-189

Ralska-Jasiewiczowa M. (ed.) (1989) Environmental changes recorded in lakes and mires of Poland during the last 13 000 years, part III. Acta Palaeobotanica 26, Kraków, 1-120

Ralska-Jasiewiczowa M. and Starkel L. (1988) Record of the hydrological changes during the Holocene in the lake, mire and fluvial deposits of Poland. Folia Quaternaria 57, Kraków 91-127

Renner F. (1982) Beitrage zur Gletschergeschichte der Gotthardgebietes und dendroclimatologische Analysen an fossilen Hölzern. Physische Geographie 8, 210 p.

Rotnicki K. (1991) Retrodiction of palaeodischarges of meandering and sinuous alluvial rivers and its palaeohydroclimatic implications, in: Temperate palaeohydrology, eds. L.Starkel, K.J.Gregory, J.B.Thornes. J.Wiley 431-471

Rózański K. (1987) The ^{18}O and ^{13}C isotope investigations of carbonate sediments from the Lake Strazym (Brodnica Lake District). Acta Palaeobotanica 27, 1: 277-282

Röthisberger F. (1986) 10 000 Jahre Gletschergeschichte der Erde, Aarau, Verlag Sauerlander

Schirmer W. (1983) Criteria for the differentiation of late Quaternary river terraces. Quaternary Studies in Poland 4: 199-205

Schreiber U. (1985) Das Lechtal zwischen Schongau und Rain im Hoch- Spat - und Postglazial, Geolog. Institut Univ. Koln, Sonderverofentlichungen 58: 1-192

Servant M. (1983) Sequences continentales et variations climatiques Evolution du Bassin du Tchad au Cenozoique Superieur. Travaux et Documents de l'ORSTOM 159: 573 p.

Sinclair L. and Bryson R. (1994) Preliminary results of joint historical and climatological high resolution modelling study of E.Mediterranean Area, Part T: Egypt, Mesopotamia and Palestine. Abstracts of Papers, NATO-ARW on Third Millenium BC Abrupt Climate Change and Old World Social Collapse, Kemer, Turkey, 5p.

Starkel L. (1983) The reflection of hydrologic changes in the fluvial environment of the temperate zone during the last 15 000 years. In: Background to palaeohydrology: A Perspective ed. K.J.Gregory. J.Wiley, 213-235

Starkel L. (1985) The reflection of the Holocene climatic variations in the slope and fluvial deposits and forms in the European mountains. Ecologia Mediterranea 11: 91-97.

Starkel L. (1987) Anthropogenic sedimentological changes in Central Europe. Striae 26: 21-29 Uppsala

Starkel L. (ed.) (1990) Evolution of the Vistula river valley during the last 15 000 years vol.III, Synthesis. Prace Geogr. IGiPZ PAN, Special Issue 5, Warsaw

Starkel L. (1994) Frequency of floods during the Holocene in the Upper Vistula Basin. Studia Geomorphologica Carpatho-Balcanica 27-28: 3-13

Starkel L., Gêbica P., Niedzialkowska E., Podgórska-Tkacz A. (1991) Evolution of both the Vistula floodplain and lateglacial - early Holocene palaeochannel systems in the Grobla forest (Sandomierz Basin). In: Evolution of the Vistula river valley during the last 15 000 years, part.IV. Geogr.Stud., Spec.Issue 6: 87-99

Street-Perrott F.A. and Grove A.T. (1979) Global maps of lake-level fluctuations since 30 000 BP. Quaternary Research, 12: 83-118

Snieszko Z. (1985) Paleogeographie de l'Holocene dans le bassin de la Sancygniówka, in Polish, res.). Acta Geogr.Univ. Lodz. 51: 1-106

Snieszko Z. (1987) The Late Vistulian and Holocene fluvial deposits of the middle Nidzica river in area of D. In: Evolution of the Vistula river valley during the last 15 000 years, vol.II. Geographical Studies, Special Issue 4: 87-94

Turkowska K. (1988) Rozwoj dolin rzecznych na Wyzynie Lódzkiej w póznym czwartorzêdzie. (Evolution des vallees fluviatiles sur le Plateau de Lódz au cours du Quaternaire tardif). Acta Geographica Lodziensia 57

Wohlfarth B. and Ammann B. (1991) The history of the Aare river and the forealpine lakes in Western Switzerland, in: Temperate Palaeohydrology eds. L. Starkel, K. J.Gregory and J.B.Thornes. J.Wiley 301-318

Zoller H. (1977) Alter und Ausmass postglazialer Klimaschwankungen in den Schweizer Alpen. Erdwissensch.Forschung 13: 271-281

Zurek S. (1986) Szybkoœæ akumulacji torfu i gytii w profilach torfowisk i jezior Polski na podstawie danych 14C. (Accumulation rate of peats and gyttjas in the profile of peatlands and lakes of Poland (as based on the radiocarbon dating)). Przegl. Geogr. 58, 3: 459-477

Development of Sediments, Soils, Erosional Events, Molluscan and Vertebrate Assemblages in Connection with Human Impact in Central Europe During the Time Span 3,000 - 5,000 BP.

Vojen LOZEK

Geological Institute of the Czech Academy of Sciences
Rozvojova 135
16500 Praha 6
Czech Republic

Abstract

In central Europe there are numerous sites situated in karstlands and other calcareous areas that have provided evidence of depositional, pedogenic and erosional processes with which are associated molluscan and vertebrate faunas as well as human settlements. The evidence presented here demonstrates that the beginning of the "climatic optimum" (Latest Boreal-Early Atlantic) represents the most humid phase of the Holocene, immediately prior to the Neolithic occupation. Its decline was very abrupt as is documented by the sharp boundary between the Early Atlantic travertines and clastic sediments with early Neolithic pottery in many mid-European caves. The foam sinter horizons in caves and rock-shelters correspond to soil horizons with decalcified fine earth as well as to $CaCO_3$ precipitates of calcareous springs, i.e. to a standstill phase in slope sedimentation (colluvia). In contrast, the decline of the Bronze Age (ca 1250-700 BC) is characterized by fossil soil horizons and scree intercalations in travertine sequences as well as by coarse debris and boulder scree accumulations in caves and talus deposits which indicate a dry but unbalanced (seasonally cold) climate. The timespan in the focus of the conference (3000 - 2000 BC) shows moist and dry oscillations represented by soils and clastic intercalations in travertine sequences and by humic screes in caves. Archaeologically this phase corresponds to the "Eneolithic," a period characterized by numerous migrations and cultural changes. In Bohemia, the Late Eneolithic settlements are rather closely confined to the chernozem area and thus quite restricted in comparison with the areas cultivated during the Neolithic. Such phenomena as well as the decline of several animal species highly sensitive to

NATO ASI Series, Vol. I 49
Third Millennium BC Climate Change
and Old World Collapse
Edited by H. Nüzhet Dalfes, George Kukla,
and Harvey Weiss

environmental changes herald the start of significant climatic changes. which culminated later during the final Bronze Age.

Introduction

The purpose of this paper is to review some of the progress that has been made in recent years in our knowledge of the environmental development of Central Europe in the Postglacial. It is based on numerous sites in karstlands and other calcareous areas which have provided evidence of depositional, pedogenic and erosional processes that are found in association with molluscan and vertebrate faunas and with human settlements (Lozek, 1982, Lozek and Horácek, 1992). Mollusca reliably reflect climatic and vegetational conditions, whereas the solution and secondary precipitation of $CaCO_3$ is indicative of humid phases with reduced clastic slope sedimentation (colluvium). The sites investigated are situated in several altitudinal zones at elevations of 200-1800 m. Best records of climatic oscillations come from xerothermic karst uplands at 200-600 m and from the borders of chernozem areas. These have been settled and permanently cultivated since early Neolithic times, i.e. since the 6th millennium BC. The evidence presented in this review covers the whole time-span Late Glacial-Present which enables the phase in question to be interpreted in a broad context.

Development of sediments

The terrestrial facies is represented by infillings of karst cavities and rock shelters as well as by accumulations under rock cliffs and at the foot of slopes. The main sedimentation factor is the gravitational downslope movement of clastic material or the breakdown from rock and cave walls. These processes are controlled by both, the climate factors (frost, precipitation), and vegetation, the latter tending to considerably control the slope transport.

All of these accumulations show a relatively standard depositional sequence:

(1) The pre-Alleröd phase of the Late Glacial is characterized by the decline of loess formation during which the slope sedimentation was strongly reduced, This is documented by layers of rather pure loess covered by Postglacial coarse screes (Lozek, 1988).

(2) During the Late Glacial the loess accumulation was generally replaced by the deposition of fine debris with loess-like matrix (Lozek, 1985).

(3) With Postglacial warming, the clasts became progressively coarser and the matrix more humic. Locally dark humic rendzina soil sediments occur.

(4) During the Boreal a rapid increase can be observed in travertine precipitation at cave entrances, rock-shelters and even at the foot of

vertical cliffs. The $CaCO_3$ crusts form on mosses and lichens covering the moistened rock walls and fall down to the bottom or infiltrate directly the floor deposits.

(5a) This process culminated in the early Atlantic when a striking whitish horizons of loose carbonate sinter- the so called "foam sinter" were deposited. In most cases they are archaeologically and osteologically sterile and document a temporary moistening of the caves and rock-shelters. An analogous process occurs up till today in highly humid mountain areas of Central Europe (Lozek, 1984).

(5b) At the same time, there is much less of scree formation which is associated with decalcification of the matrix and can be observed on open slopes, which indicates that the pedogenesis exceeds the slope deposition or transport (Lozek, 1988).

(6) In the following phase the accumulation of coarser scree increases, characterized generally by a brownish gray humic matrix with increasing $CaCO_3$-content. In caves the change from the calcareous precipitation to clastic deposition is very abrupt, as documented by humic clastic layers with early Neolithic pottery separated by a sharp boundary from the underlying foam sinter. This kind of sedimentation is also characteristic of the 5th millennium BP.

(7) With the close of the Bronze Age, i.e. at the turn of the 4th and 3rd millennium BP, the coarse screes with free interstices were formed, which indicate an intensive mechanical destruction of rock outcrops and breakage from rock walls and cave ceilings The matrix is high in humus and very loose; complete molluscan shells are common (Lozek, 1988).

(8) The youngest phase is characterized by less coarse scree with loose humouse calcareous matrix.

The above summary thus suggests that the 1st and the 5th sedimentation phases are characterized by a considerable reduction of slope transport representing standstill phases within the clastic slope deposition. The semiterrestrial to aqueous facies is represented by floodplain sediments and spring or marsh travertines.

In general, the fluvial sediments show a structure similar to that of the overland deposits which occur in the corresponding catchment basin. This is because their material is mostly derived from adjacent terrestrial environments. This is particularly true of valley floor deposits of small streams. The greater the distance from the source, the greater is the difference between alluvial and slope deposits. This is particularly true of large lowland rivers.

Floodplain deposits are characterized by intercalations of various limnic and organic sediments, in certain cases even of the $CaCO_3$-precipitations (tufa, alm, bog lime, lake marl). These form infillings of former oxbow lakes, or are concentrated in marshy areas within the floodplain.

Here attention is drawn to valley floor sediments of smaller streams which reflect the character of adjacent terrestrial environments and in a number of cases provide very valuable paleoenvironmental evidence. For instance, we may mention the floodplain sequence at the site Srbsko-Waterfalls in the Bohemian Karst (Lozek, 1974) which consists of four units, each corresponding to altogether different environments:

(a) - The basal complex consists of loess with intercalations of intermittently sedimented stream gravels. The corresponding environment was an ephemeral brook in the loess steppe. This complex is capped by the Late Glacial loess-like loams with scree.

(b)- On this substrate a steppe soil of chernozem character developed under a parkland as documented by the incorporated snail fauna.

(c) - The soil is covered by loose travertine deposited in a calcareous swamp which formed in the earlier phase of the Holocene climatic optimum. This corresponds to the humidity maximum in the early Atlantic. The soft travertines grade into screes with loamy matrix coming from the adjacent slopes. The upper part of this unit formed during the 5th millennium BP in Epiatlantic.

(d) - The youngest sediments consist of mixed alluvial and colluvial humic loams with medium coarse scree derived particularly from eroded cultivated soils deposited by occasional floods which are characteristic of this region.

The standstill phase in the clastic deposition is here represented both by the chernozem and the overlying calcareous tufa. It corresponds to the dry Boreal as well as to the wet Early Atlantic. The phase of rock outcrops destruction is represented by higher content of coarse scree in the basal horizon of the uppermost unit. However, the floodplain deposits present a rather wide variety of lithofacies. For instance, at the nearby site of Karlstejn-Krabina, the standstill phase is represented by thick brown decalcified clayey soil.

In general, the floodplain deposits of Central Europe are strongly affected by plowing and grazing in the adjacent landscape. This impact was responsible for the increasing accumulation of colluvial loams from fields and pastures whose deposits form the upper part of floodplains and are called "Auelehm" or "Klock," i.e. the floodplain loam (Jäger, 1962). The underlying strata are of different character due to a high proportion of sediments reflecting quiet environments, such as fens, soft travertines, lake marls, floodplain soils or water-laid clays.

Freshwater limestones are represented by precipitations of $CaCO_3$ from springs and small brooks forming the so-called tufas. The soft travertine sequences provide evidence of the intensity of slope sedimentation and moisture oscillations documented by scree and soil intercalations. In various areas of Central Europe deposits occur, whose lower part consists of thick pure hard travertine, whereas their upper part is characterized by a number of soil or scree intercalations in a predominantly loose travertine sediment. The sequences reflect climatic oscillations during the Late Atlantic and Early Subboreal of the classical stratigraphic scale. This complex corresponds to the interval which Jäger (1969) proposed to call the "Epiatlantic" phase. These oscillations culminated in the formation of a distinct humic soil, which mostly includes scree, and contains at many sites pottery from the Late Bronze Age. It documents a dry phase at the turn of the 4th and 3rd millennium BP (second and first BC), which represents the driest spell of the Holocene as well as the sharpest drop from the climatic optimum (=Subboreal sensu Jäger, 1969; Jäger and Lozek, 1968). The 5th millennium BP coincides with the late Epiatlantic dry and moist oscillations.

Stratigraphical position of buried soils

As already noted, the depositional sequences listed above include horizons of buried soils which formed during the standstill phase in clastic sedimentation. The most

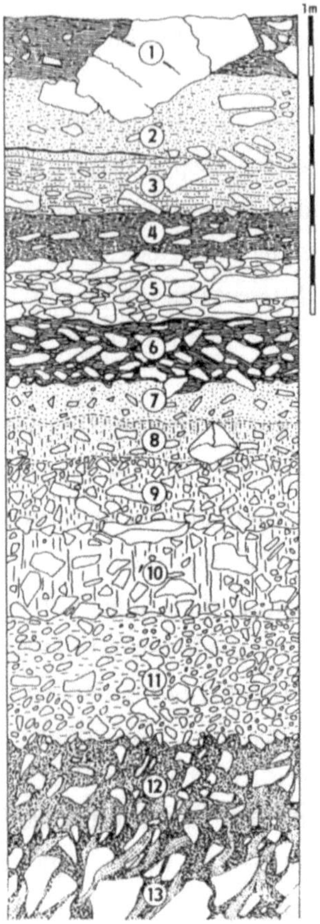

Fig. 1. Profile of the entrance deposits in the Great Ruzin Cave near Kosice (Slovakia) 1 - humus-rich, black loose loam with scree; 2 - brownish gray loamy "foam sinter" with coarse flat stones (X- intercalation of white fine foam sinter with black humous band on top); 3- gray loamy foam sinter with increasing humus content in its lower part; 4 - grayish black compact humic loam, sinter infiltrations, charcoal, prehistoric pottery; 5 - horizon of coarse flat blocks poor in finer matrix; 6 - black humic loam with medium to coarse scree (at the 5/6 boundary concentrations of Neolithic pottery - Bükk-culture); 7 - grayish white, rather pure foam sinter with small angular fragments. Indicative of the extremely humic phase immediately preceding the first Neolithic occupation); 8 - grayish ochreous clayey loam with foam sinter admixture; 9 - ochreous clayey loam rich in medium to coarse angular fragments; 10 - light ochreous homogeneous loam of loess-like character with less numerous rather coarse fragments (equivalent to the youngest loess at higher elevations); 11 - predominantly medium to coarse scree with numerous rounded fragments and light grayish ochreous matrix; 12 - dark reddish brown loam with medium coarse fragments; 13 - brownish black phosphate loam with coarse limestone blocks (zoogenic sediment of cave bears).
Age of the layers:1 - Subrecent; 2-3 - Subatlantic; 3/4 - Subboreal: 4, 5 and partly 6 - Epiatlantic; 6 (base) - Late Atlantic; 7 - Early Atlantic (humidity maximum); 8-9 - Late Glacial, Preboreal, Boreal; 10 - Late Weichselian Pleniglacial (Last Glacial loess optimum); 11-12 - Middle and Early Weichselian Pleniglacial; 13 - Early Weichselian,

important phase of pedogenesis corresponds to the humidity culmination in final Boreal and early Atlantic time which abruptly declined immediately prior to the Neolithic occupation. At that time, brown soils with decalcified fine grain fraction developed even on slopes, where otherwise an intensive transport of calcareous detritus occurred. In caves and rock-shelters these soils grade into the "foam sinter" horizons which document a highly humid phase with maximum movement of $CaCO_3$ in solutions. This process resulted on the one hand in intensive decalcification, on the other hand in rapid precipitation of $CaCO_3$. It was comparatively short-lived and did not last more than some 1500 years.

Besides this strongly developed soil there also occur humic rendzina or chernozem soils which formed at the beginning of the Holocene at places with reduced slope sedimentation (e.g. the above-mentioned site of Srbsko-Waterfalls). Also the above-mentioned soil horizons in travertine sequences, particularly the Late Bronze Age soil, show rendzina character. Humic soils also occur in floodplain sequences, particularly a strong chernozemic soil formed in the early Holocene (e.g. Pavlov, Vasátko and Lozek, 1973.) The critical time-span of the 5th millennium BP thus falls stratigraphically between the decalcified soils of Atlantic age and humic Subboreal soils formed when periodical slope sedimentation took place.

Erosional events

The depositional sequences discussed above, particularly the fan and the alluvial deposits, in which clastic horizons alternate with carbonate precipitates , show two depositional patterns of different structure: the first one is characterized by thick deposits of Late Glacial to Early Atlantic age, with the younger layers strongly reduced or lacking; the second one begins with the Early Atlantic horizon capped by a thick series of younger strata. In such cases the older strata are mostly lacking, i.e. the Atlantic horizon immediately overlies the bedrock. This fact suggests that in the phase of maximum humidity during the final Boreal and Early Atlantic, a short, but very intensive phase of downcutting took place, with the sheet erosion not manifested (Jäger and Lozek, 1983). In contrast to with younger erosional events, which mostly resulted from the activities of prehistoric man, this phenomenon had purely natural causes. Anthropogenic erosion can be traced in settled and cultivated areas, where it is reflected by an increasing volume of

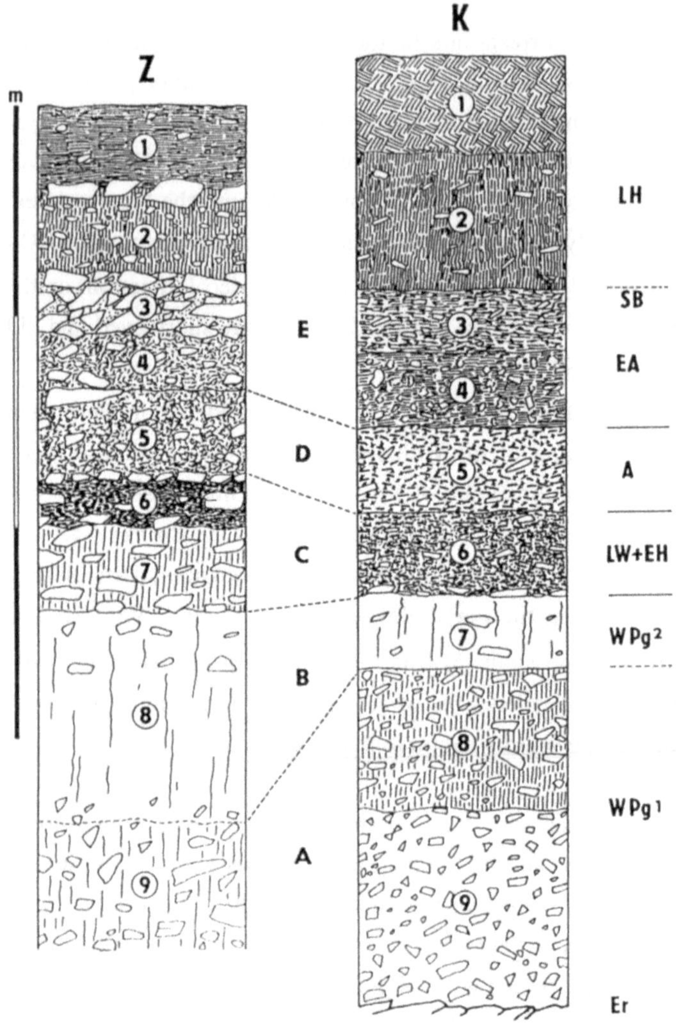

Fig. 2. Slope deposits at Zadni Kopanina (Z) and Krabina near Karlstejn (K) in the Bohemian Karst.
Chronology: LH - Late Holocene, SB - Subboreal, EA - Epiatlantic, A- Atlantic, EH - Early Holocene, LW- Late Glacial (Weichselian), WPg^2 - late Weichselian Pleniglacial, WPg^1 - middle (to early) Weichselian Pleniglacial, Er - erosional surface of the bedrock.
Sedimentology: Common at both sites:1-2 - humic loams with debris, 3-4 - brownish grey loams rich in debris, partly coarse, 5 - weathered clayey brown loams, rather poor in debris, with decalcified fine matrix; Separately: Z6, Z7 and K6 - gray humic loams with debris; Z8 and K7 - typical light ochreous loess with sparse stones; Z9- brown loess rather rich in stones; K8 and K9 - angular scree with loess-like matrix. A to E: lithologic and faunal zones. Scale in meters,

alluvial and colluvial loams derived from cultivated fields. This process was naturally supported by climatic imbalance at the decline of the Bronze Age.

Of particular interest is the erosion and subrosion of travertine deposits which are mostly associated with the decline of their accumulation and took place during the Late Holocene. At present, the great majority of large such deposits are no longer active.. This phenomenon has obviously natural causes and may be due to a secular climate shift following the Subboreal dry oscillation (sensu Jäger, 1969). The time-span of the 5th millennium BP shows no traceable downcutting phase apart from repeated minor episodes of sheet erosion and formation of scree fans.

Development of fauna

All of the sediments discussed above include a rich malacofauna (i.e. molluska), and the fills of karst cavities are also rich in vertebrates. Faunal changes show the following developmental pattern (Lozek, 1982; Horácek and Lozek, 1988):

The Late Glacial is characterized by predominantly tolerant open-ground communities with local admixture of some resistant elements of the boreal taiga, such as *Discus ruderatus* (Fér.) *Vertigo substriata* (Jffr.) or *Perpolita petronella* (L. Pfr.) A number of Pleistocene small mammals survive: *Microtus gregalis* and *nivalis, Ochotona,* and at the transition to the Holocene, also *Sicista betulina*.

During the Early Holocene (Preboreal-Boreal) these zoocoenoses are increasingly enriched in thermophilous species living in semi-open warm woodlands (*Helix pomatia* L., *Aegopinella minor* (Stab.)), cherozem (*Chondrual tridens* (Müll.)) as well as rocky steppes (*Granaria frumentum* (Drap.)), and particularly in mantle formations, i.e. in woodland-grassland ecotones (*Bradybena fruticum* (Müll.), *Euomphalia strigella* (Drap.)). There also appear more demanding mammal species, such as *Clethrionomys* or *Gliridae*, and in the final phase also thermophilous reptiles, for instance *Emys* or *Lacerta viridis*. Steppes and parklands were gradually replaced by warm woodland. The culmination of these ecotone elements corresponds to the hazel maximum.

The Atlantic stage represents an important turning point of floral and faunal development. Steppe assemblages rapidly decline, while the retreat of the taiga species is coupled with the invasion of a number of demanding elements, particularly inhabitants of mesic to damp closed forests.

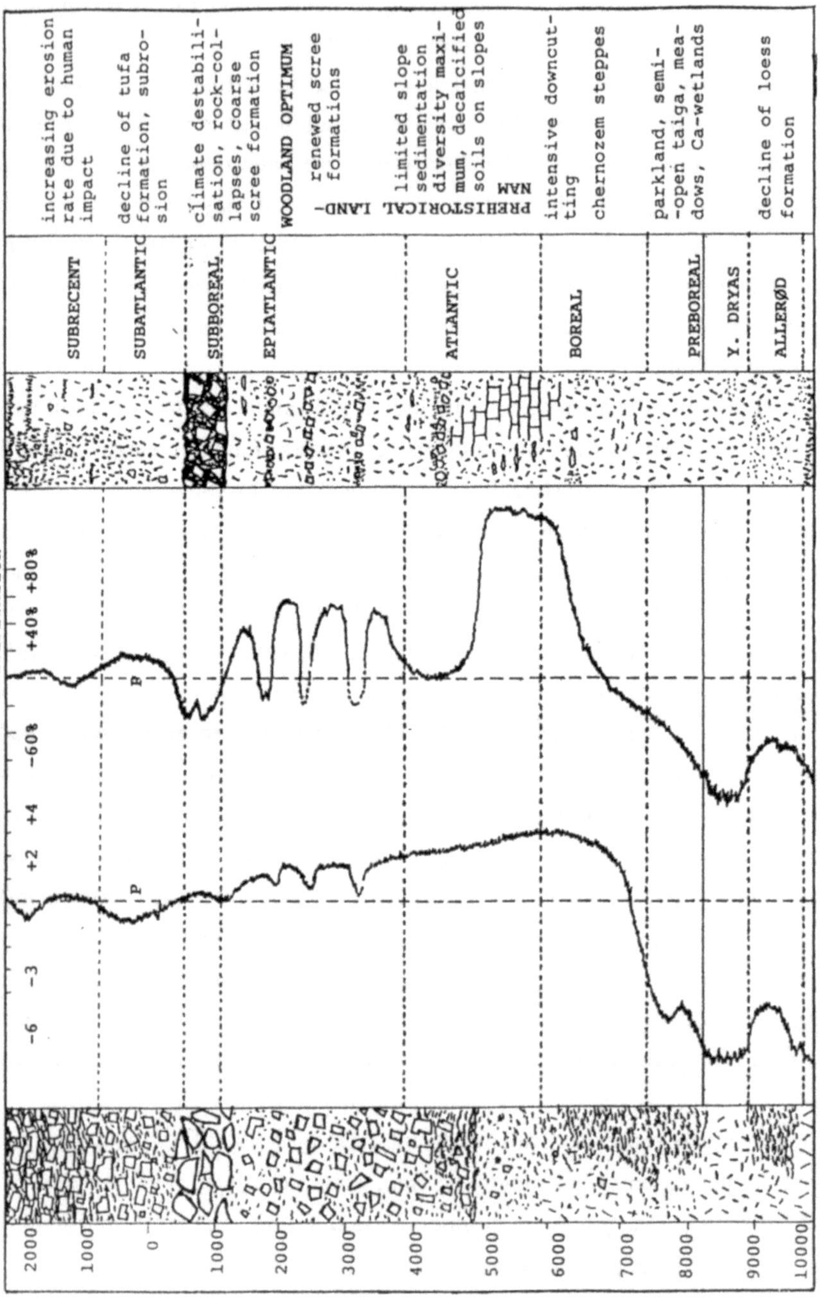

Fig. 3. - Reconstructed precipitation, temperature and sediment development of the Bohemian Holocene. The time scale is in AD and BC calendar years.

The Epiatlantic is characterized by the disappearance of steppe and taiga species and by culmination of woodland assemblages (woodland optimum). There appear also some highly thermophilous species living in extremely warm habitats, for instance *Truncatellina claustralis* (Grd.) on protected limestone rocks.

With the Subboreal in sense of Jäger (1969) commenced the depauperization of natural woodland communities. It is coupled with gradual immigrations of the so-called modern elements, i.e. of predominantly thermophilous species which never invaded Central Europe in the pre-Holocene times and which found favorable living conditions in the cultivated landscape in a warm and damp climate.

In non-cultivated areas the 5th millennium BP is characterized by the maximum development of woodland biocoenoses, while in cultivated regions it is marked by the retreat of woodland elements coupled with the re-expansion of open ground species. These include several former steppe faunal elements and new modern immigrants confined to specific environments of deforested landscape.

Impact of human activities

The Holocene differs from all comparable phases of the Pleistocene (i.e. interglacials) by its environmental development affected by human impact. In Central Europe the Atlantic stage represents the crucial phase in which the first "landnam" took place. The Neolithic farmers occupied the chernozem area including its periphery, i.e. the territory in which steppe patches and open warm woodland persisted the longest. Neolithic people cleared much of the marginal woods and changed the settled areas into predominantly open landscape prior to the later expansion into fully developed ecosystems of closed woodland. A number of former steppe species could thus survive in the chernozem belt; among snails for instance *Chondrula tridens* (Müll.) and *Helicopsis striata* (Müll.), among mammals the hamster. The regions occupied in the Neolithics have preserved their steppe character until the present and owing to this fact, their flora and fauna have been enriched by a number of modern immigrants from the south. The cultivated area of the Neolithic landnam was considerably extended during the younger Bronze Age by deforestation and cultivation of additional large forested areas. At that time plowing and grazing gave rise to extensive sheet erosion which destroyed the original forest soils which at this stage were already well developed and deeply decalcified. These soils were in dismantled partly eroded areas replaced by poorly developed calcareous soils (pararendzinas). This process resulted in an extensive secondary expansion of steppe elements as well as in the retreat of many sensitive forest species and retrograde soil development. Further expansion of cultivated areas followed in the Medieval periods when even some mountain areas were deforested and cultivated.

For these reasons, during the Postglacial only a few areas were not settled and remained permanently covered by forest after the great woodland expansion during the Early Atlantic when the last undisturbed natural development took place.

It is of interest that the 5th millennium BP (third BC) corresponds to the time when no striking environmental changes are traceable by multiple earth science methods, whereas from the archaeological point of view this time span represents a key phase, which in Central Europe is characterized by extensive migrations and complicated alternations of a numerous cultures. It should be stressed that some important Eneolithic cultures, for instance in Bohemia, the Corded Ware and Bell Beaker cultures, occupied a considerably smaller area than that of the first Neolithic farmers. Apart from a few minor exceptions their cultivated territory coincides with chernozem areas, which is true also of the Rivnác culture. It may be mentioned that at that time the people of the Baden culture penetrated a number of mountain caves in the West Carpathians (e.g. Mazarná) at times when the formation of foam sinter was briefly interrupted, suggesting a dry phase. Only in the beginning of the Bronze Age, at the time of the Unetice culture in the early half of the 4th millennium BP a re-expansion of the cultivated landscape took place and several so far unsettled areas came under use.. This process culminated during the decline of the Bronze Age. The development discussed above suggests that in the time span in question, some environmental changes probably took place, although unrecorded by flora, fauna or depositional processes. The reduction of settled areas can be interpreted in terms of a temporary local climatic deterioration.

Summary

The purpose of this study was to examine the setting of the 5th millennium BP (third millennium BC) in the broader context of Postglacial environmental development in central Europe in the light of paleontological, sedimentological and pedological results.

The starting point was the sudden climatic change immediately prior to the first Neolithic occupation. The very humid and warm phase of the Early Atlantic characterized by a standstill phase in clastic sediments and by the formation of the soft travertine locally called foam sinter in caves was suddenly replaced by purely clastic sedimentation coupled with the appearance of the first Neolithic farmers. In Bohemia this event can be dated to about 5000 BC when the deposition of screes including a loamy matrix with increasing humus content replaced the sinter precipitation. The evidence from travertine sequences in footslope positions shows that the scree formation took place during oscillations characterized by alternating clastic and carbonate sedimentation. This phase falls into the timespan in question, the third millennium BC (fifth BP). It shows no striking depositional events, but is characterized by the beginning of the decline of some climate-sensitive snails.

This development culminated in a dry period spanning about 500 years at the boundary of the 4th and 3rd millennium BP, marked by the formation of a distinct buried soil which includes Late and Terminal Bronze Age pottery. At that time the intensive accumulation of travertines ceased, suggesting the start of a trend leading to the present-day regime characterized by climate cooler and considerably more continental than in the Holocene climatic optimum. First signs of the process may be observed during the 5th millennium BP Certain climatic reversal at that time is also suggested by the complicated social situation during the Eneolithic, documented by the alternation of various cultures, extensive migrations and even by the reduction of the settled area to the chernozem belt, i.e. its decrease in comparison with the Neolithic period. Thus, it may be believed that at that time an environmental change commenced which later culminated at the decline of the Bronze Age. Alternatively, the environmental and climate changes at could have taken place elsewhere in Europe or Asia, with Bohemia under stress of migrating tribes driven from their original lands by recurrent droughts.

References

Horácek, I., Lozek, V. (1988) Palaeozoology of the Mid-European Quaternary past: scope of the approach and selected results. Rozpravy Ceskoslovenske Akademie Ved MPV 98, 4: 102pp. Praha.

Jäger, K-D, Lozek, V. (1962) Über Alter und Ursachen der Auelehmablagerung thüringischer Flüsse. Preahistorische Zeitschrift 40: 1-59, Berlin.

Jäger, K-D (1969). Climatic Character and Oscillations of the Subboreal Period in the Dry Regions of the Central European Highlands, Proceedings of the VII Congress INQUA 16: 38-42. Washington.

Jäger, K-D, and Lozek V. (1968) Beobachtungen zur Geschichte der Karbonatdynamik in der holozänen Warmzeit . Ceskoslovensky Kras. 19: 5-20, Praha.

Jäger, K-D, and Lozek V. (1983) Paleohydrological implications of the Holocene development of climate in Central Europe based on depositional sequences of calcareous fresh-water sediments. Quaternary Studies in Poland, 4: 81-89, Warszawa-Poznan.

Lozek, V. (1965) The Formation of Rock Shelters and Foam Sinter in the High Limestone Carpathians. Problems of the Speleological Research: 73-84, Praha.

Lozek, V. (1974) Príroda Ceského krasu v nejmladsi geologické minulosti (Landschaftsgeschichte des Böhmischen Karstes in der jüngsten geologischen Vergangenheit). Bohemia Centralis, 3: 175-194, Praha.

Lozek, V. (1982) Faunengeschichtliche Grundlinien zur spät- und nacheiszeitlichen Entwicklung der Molluskenbestände in Mitteleuropa. Rozpravy Ceskoslovenské Akademie Ved MPV, 4: 106 pp., Praha.

Lozek, V. (1984) The Foam Sinter as Palaeoclimatic Indicator. Ceskoslovensky Kras 34: 7-14, Praha.

Lozek, V. (1985) The Site of Souteska and its significance for Holocene climatic development. Ceskoslovenky kras 36: 7-22, Praha.

Lozek, V. (1988) Slope deposition in karst environments of Central Europe. Ceskoslovensky Kras, 39: 15-33, Praha.

Lozek, V.(1992) Sit' opernych profilu kvyvoji krajiny Ceského krasu. (Netz von Stützprofilen zur Landschaftsgeschichte des Böhmischen Karstes). Bohemia Centralis, 21: 47-67, Praha.

Lozek, V. Horácek, I (1992) Slovensky kras ve svetle kvartérni geologie (Slovak Karst in the light of Quaternary geology). Slovensky Kras 30: 29-56, Martin.

Vasátsko, J., Lozek, V. (1973) Der holozäne Bodenkomplex von Pavlov und seine Bedeutung für die Lanschaftsgeschichte des südmährischen Tschernosemgebietes. Zpravy Geografického Ustavu CSAV 10, 7: 1-10, Brno.

High Resolution Simulations of Regional Holocene Climate: North Africa and the Near East

Reid A. BRYSON * and Robert U. BRYSON **

*Center for Climatic Research,
University of Wisconsin,
Madison, WI 53706 USA.
**Archaeoclimatology Consultants
P.O Box 2084
Davis, CA 95617 USA.

Abstract

In the present work we have used synoptic climatology to extend previous studies for site-specific, two-century specific paleoclimate modeling. We call this approach Archaeoclimatology, for it is intended primarily to serve Quaternary field scientists and archaeologists.

Standard Milankovitch-type calculations were used to estimate the past irradiance. A global volcanicity index based on radiocarbon dated eruptions was then constructed and compared with the measured annual mean aerosol optical for the last century to estimate the past aerosol optical depth for two or five century averages. This was used to estimate the attenuation of the incoming sunlight, and the glacial area was used to estimate the albedo of the hemisphere. An estimate of the mean seasonal hemispheric temperature was then possible.

Using the Lettau protocol monthly hemispheric mean temperatures, evapo-transpiration, and net radiation values could then be estimated.

Assuming that the equatorial temperatures are nearly constant, the meridional temperature gradient becomes a function of the mean hemispheric temperature. Estimating the vertical lapse rate as a function of season, the latitude of dynamic instability of the westerlies could then be estimated and in turn the latitude of the subtropical anticyclones and the jetstream latitude for the sector.

Most local climates can be estimated from such "centers of action". Tests suggest that the estimated past positions combined with synoptic climatology give modeled paleoclimates that fit the field evidence quite well.

NATO ASI Series, Vol. I 49
Third Millennium BC Climate Change
and Old World Collapse
Edited by H. Nüzhet Dalfes, George Kukla,
and Harvey Weiss

Climate Models

There are two fundamentally different ways to model the climate of the earth. The first, the basis of the widely used "General Circulation Models" (GCMs), is microphysical in nature. This means that such relationships as the equations of motion are used which refer primarily to forces and influences on individual parcels of air. In general, this approach starts with these very small segments of the atmosphere and works up to global weather patterns through iteration. They usually require iteration with a rather short time step, and thus for climatic simulations may require millions of iterations. These models are quite sophisticated and very complicated, requiring great computational power.

The second method is simpler and may not require iteration. This is the macrophysical approach. Relationships of large scale are used, such as the Rossby long wave equations, the thermal wind relationships, or the Z–criterion derived from the work of Smagorinsky (1963). The perspective is reversed relative to the GCMs in a model that uses global phenomena to estimate climatic patterns on human scales. Models of this sort are also complicated but require only seconds to run on a microcomputer.

The Macrophysical Model

The hierarchical model which is discussed in the following paragraph required first a model of global glacial volume, from which the glaciated area could be calculated and the ice albedo effect estimated (Bryson and Goodman, 1986). An hemispheric temperature model was then constructed using solar irradiance as modulated by volcanic aerosols (Bryson, 1988). Going from hemispheric or global models to regional models requires that some broadscale relationships be used. For example, one knows that the westerlies must become stronger with height, up to about 10 km, because the earth is colder at the poles than at the equator, and the upward rate of increase is proportional to the magnitude of the temperature gradient. Since equatorial temperatures change little even from glacial to nonglacial times, the average hemispheric temperature and the temperature gradient must both depend primarily on highlatitude temperatures. (For the aerosol record, see the Appendix).

Using the temperature gradient, the simulation may be carried a step farther to simulate the past monthly latitudes of the subtropical anticyclones, the monthly latitudes of the Inter-Tropical Convergence (ITC) in the Saharan region, the monthly latitudes of the jetstream, and the monthly rainfall in portions of the Saharan region, and the like.

Calculating the Latitude of the Subtropical Anticyclones and Jet Stream

The northern part of North Africa has winter rains and summer drought, the rains diminishing from west to east. The winter rains are associated with the southward winter expansion of the circumpolar westerlies and the passage of the southern edges of cyclonic storms in the westerlies. In summer, as the circumpolar vortex shrinks, the subtropical anticyclone in the North Atlantic expands and extends eastward, bringing the Mediterranean coast under divergent flow from the northeast quadrant of the anticyclone (LaFontaine et al., 1990). The winter rains diminish inland towards the heart of the Sahara.

South of the central Sahara, summer rains penetrate inland from the Gulf of Guinea, associated with the invasion of moist maritime air behind the Inter-Tropical Convergence (ITC). The rains increase southward for some distance from the ITC position and then decrease until close to the coast where the monsoonal rains cannot be distinguished from those associated with coastal effects (Ilesanmi, 1971). How far the monsoon rains penetrate into the interior clearly is related to the latitude attained by the subtropical anticyclone in its annual march. Thus both the winter rains of the north and the monsoon of the south may be related to the position of the subtropical anticyclone. It has been shown by Smagorinsky (1963) that the outer edge of the circumpolar westerly vortex should, according to his formulation, be determined by the vertical rate of variation of the temperature, and by the northsouth or meridional temperature gradient.

Since the outer edge of the circumpolar vortex is the latitude of the subtropical anticyclones at the same height, Smagorinsky's formulation gives an estimate of the latitude of the anticyclones at the surface. The outer edge of the westerlies aloft is closely related to the latitude of the "jet stream", so by nearly the same formulation we can estimate the jet stream position. Both of these latitudes are crucial to estimates of mid-latitude climates.

Calculating the Latitude of the Intertropical Convergence and Monsoon Rain

Assuming that the present is the key to the past, the relationships discussed above may be applied to model the history of the Subtropical Anticyclone and, from empirical relationships, to obtain estimates of the latitude of the Inter-Tropical Convergence in North Africa and elsewhere. The simulated history of the August latitude of the ITC is given in figure 1. The range of latitudes calculated agrees well with commonly accepted shifts in the northern limit of significant vegetation.

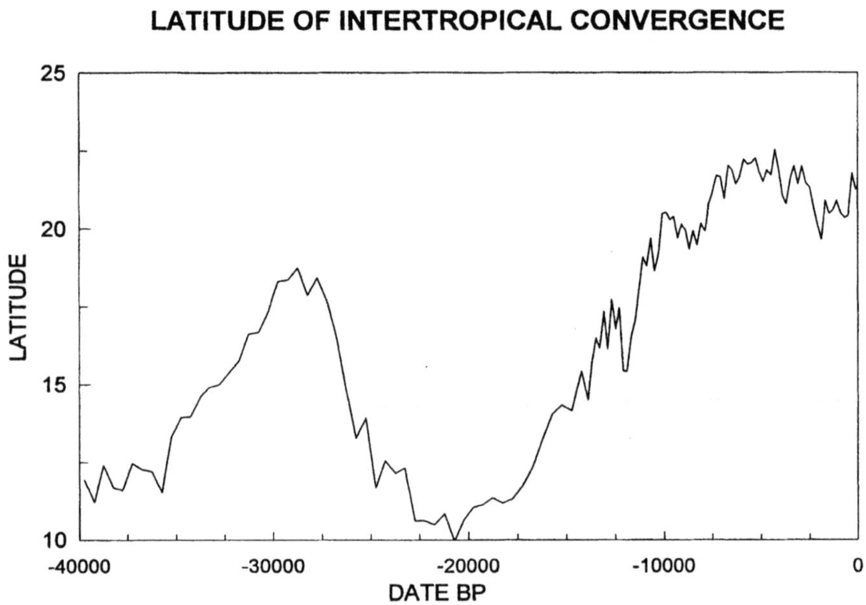

Fig. 1 The simulated history of the August latitude of the ITC calculated using the "Z-Criterion".

Up to this point, the latitude of the ITC has been taken as the mean between 10°W and 30°E. However, the ITC is not a straight line across North Africa. Its latitude varies with longitude and with month. Assuming that the variation with longitude is due to the shape of the continent, which has not changed dramatically within the postglacial period, we can use the present intraannual eastwest variation of the latitude as a function of the mean to estimate past latitudes of the ITC at various longitudes. These are shown for selected past times in figure 2. The figure suggests that while the climatic history of the Ahaggar and Tibesti might be expected to differ, the Air Massif (Adrar Bous), Chad, and Darfur regions might have had similar rainfall histories.

Modeling the Monsoon Rainfall

Ilesanmi (1971) has shown that for some distance south of the Intertropical Convergence the rainfall increases linearly in the NigeriaTogo sector. Thus if we have modeled the latitude of the ITC, we can use that relationship to model the summer monsoon rainfall at those latitudes that are within the linear regime distance south of the ITC.

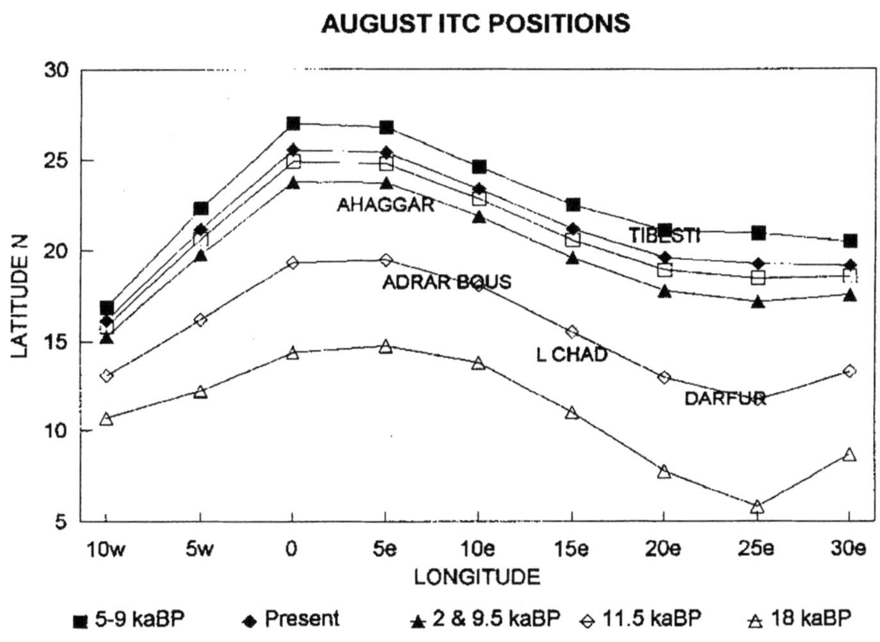

Fig. 2 Simulated shape of the ITC across north Africa at selected past times.

Using the concept of the Ilesanmi work, the history of summer monsoon rainfall at a fixed latitude may be calculated (Figure 3).

Let us compare the modeled subSaharan rainfall history with field data There are several sorts of field information with which to compare the model results, but no very good sets of proxy data. We shall compare the results with data on occupation of the area, on the realistic assumption that without rainfall the region is essentially uninhabitable.

All radiocarbon dates for occupation sites within the western North Africa region were extracted from a global data base of dates referring to named cultures within the area. These were then sorted into latitude band groups, and the number of dated occupations per century were counted. As might be expected, the number of occupation dates increased towards the present, in part because older sites are more likely to have been lost by burial or erosion. However the dates diminish from a

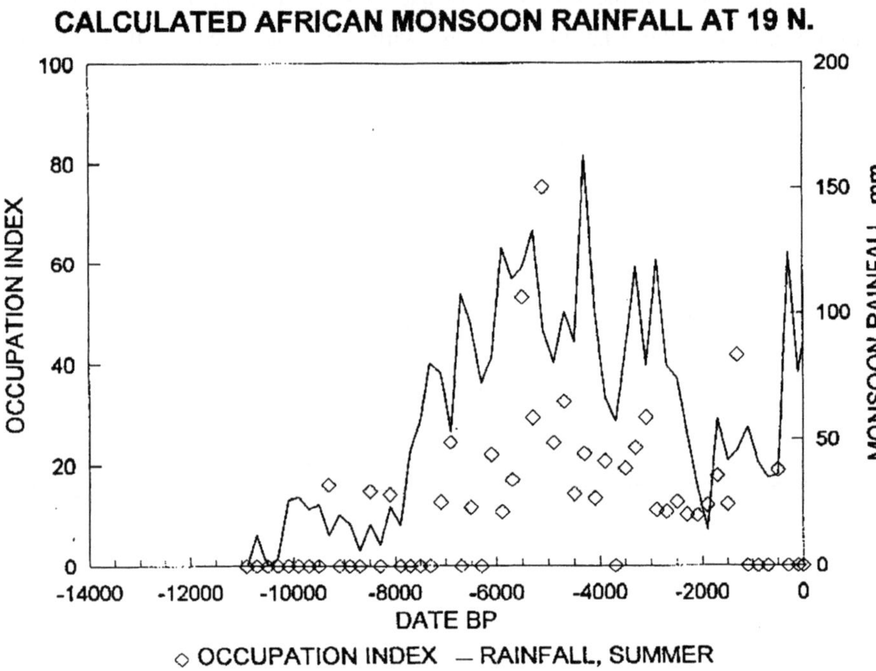

Fig. 3 Simulated history of the mean monsoon rainfall at 19° N in North Africa, compared with the Occupation Index.

maximum towards the present within the historical period, for an investigator is far less likely to incur the expense of running a radiocarbon date if the date is already known because of its historical context. An analysis of this type was done by Geyh and Jakel (1974), but with too little latitudinal specificity for the present purposes.

This bias of date numbers towards the early historic period can be compensated for by fitting a suitable curve to the entire assembly of dates and then expressing the date numbers per century as an index defined as the departure of the observed date numbers from the "expected" number (based on the fitted curve), divided by the expected number, plus one to make all indices positive. A curve suitable for expressing the "expected" number is the "serpentine curve" (the Occupation Index).

The Occupation Index for the Sahelian region, scaled for convenience, is included in figure 3. There are essentially no sites more than about 9000 years old. For this zone, when the model simulation indicates essentially no rainfall, there are no people, all of the occupation of the region being restricted to the period of indicated rainfall adequate to support some vegetation.

Rognon (1976) has summarized a large amount of field evidence for the Saharan region, as well as attempting a dynamic interpretation of the atmospheric circulation that might have influenced the climates of the last 40,000 years in the area. His summary, as well as the summaries of others (e.g. McIntosh and McIntosh, 1981) indicates maximum rainfall in subSaharan Africa in the period from about 10,500 BP (10,450 BCE) to about 6000 BP (3400 BCE) . Most authors cited suggest a decline from 6000 to about 2000 BP (100 BCE) or so rather than a decrease as rapid as the increase at the onset of the Holocene. The modeled results shown in figure 3 are in accord with these summaries. Several of the authors cited in these summaries also found the absolute minimum of Holocene rainfall to be about 2000 BP with a small maximum a couple of centuries ago. This also accords with figure 3.

Probably the most complete examination of the field evidence for the Sahel, and indeed for the entire Sahara, is that of Petit Maire et al. (1993). Their map shows that all indicators of moisture such as large animals, swamps, trees, intensive human use, etcetera, have dates nearly in the 9500-2000 BP (8580-100 BCE) range in the Sahelian band. This in turn indicates that the present model is more in accord with the field evidence than the results obtained by the National Center for Atmospheric Research CCM1 mainframe model. Those results suggest roughly twice the present rainfall amount for the entire past 18,000 years, including the present (Kutzbach, 1994).

The Problem of Mountainous Terrain and Mesoclimates

The modeling of climates in mountainous terrain is relatively undeveloped. Most models of the climate treat mountainous areas as broad uplands with no local relief. However, the pattern of almost any map of climatic parameters in montane regions is dominated by local rather than broad features. In order to treat climatic reconstruction or simulation in a sitespecific sense, therefore, we will use a method which simulates regional climates and then relates this simulation to local or sitespecific climates.

The elements of this method are contained in the work of Mitchell (1969) and Bryson and Mitchell (1986). In essence this work showed, that as long as one stayed within a physiographic region, there was a close relationship between the climatic variables and elevation, latitude and longitude.

As pointed out by Blodget (1857), the western Cordilleran region of the United States can be thought of as a plateau at about 5000 feet (roughly 1500 m) elevation with mountains rising from it. We therefore reduced all the local climatic data in each climatic division to the equivalent 2000 m value, using multiple regression against latitude, longitude, elevation, and a dummy variable simulating exposure. This was done for each month. The dummy variable, empirically assigned (1 or 0), was quite stable through the year and changed only slightly during the course of the year. The mean monthly values of the temperature and precipitation were then simulated for the district equivalent mean at 2000 m. At this point the sitespecific climatic history at the ground level could be calculated from the simulated 2000 m mean by the equations derived from the current field observations.

An inherent assumption contained in this approach is that the pattern of climate as a function of location within a small region is invariant. How well the method worked can only be judged by the results of comparison with field data.

As an example of the application of the method, we will consider the simulation of past climates in the mountainous Pacific Northwest of the United States. Simulation of the past climate of this region is much more complex than simulating the African monsoon, because of its west coast position athwart the westerlies and its mountainous terrain. There are two anticyclones to deal with, and two branches of the jet stream. The climate is also sensitive to the longitude of the major ridges and troughs in the westerlies. In principle, however, it is a similar problem. Enough progress has been made to indicate that at least a tentative climatic history can be constructed for testing by the archaeologists expert in the prehistory of the region (Bryson and Bryson, 1994).

Figures 4a and b display the results of the output of the local model driven by the macrophysical hemispheric climate model for two archaeological sites east of the Cascade Range in Oregon. The most dramatic event suggested by the results for the southernmost site, 35DS557, is the onset of summer rainfall at the beginning of the

Holocene, rising to about triple the present value during the Atlantic climatic episode. The second to most recent point on the graph suggests somewhat wetter summers during the subepisode that has been named the NeoAtlantic and is popularly known as the "Little Ice Age". This of course extends into the historic period and can be easily verified by reference to the local weather records.

The northern site, 35WS231, on the other hand, shows a quite different simulated rainfall history, despite the fact that the sites are less than 100 miles apart. The two sites are on opposite sides of the climatic divide that runs eastward from the Cascades to the Blue Mountains of eastern Oregon. The northern region, which contains 35-WS231, is the Columbia Basin and appears to have a climatic history like the coast. Site 35DS557, south of the climatic divide, appears to act more like the entire northwestern Great Basin.

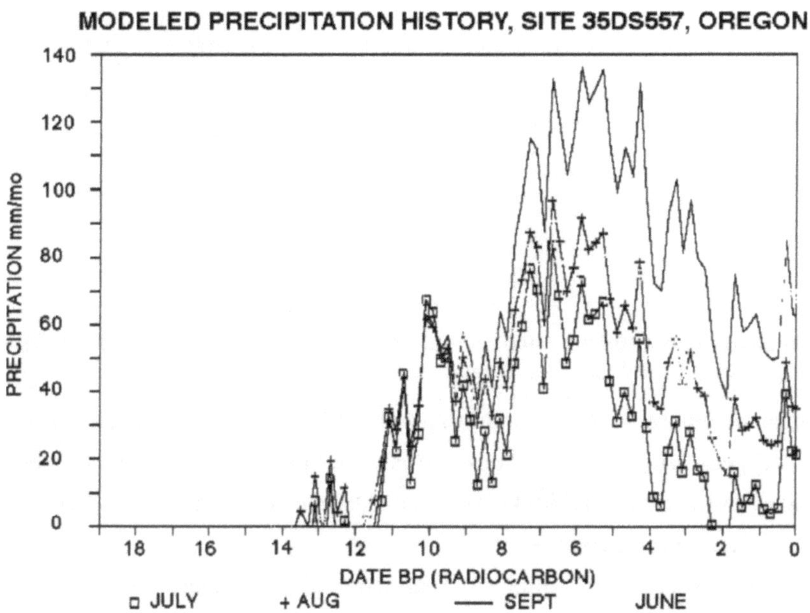

Fig. 4a Simulated rainfall history for the summer months at site 35DS557, in south-central Oregon east of the Cascade Range.

The Near East

Analysis of the modern climate of Syria shows that the rainfall monthly amounts are closely related to the latitude of the Mediterranean branch of the jetstream relative to the station in question. The relationship is quite sharp: monthly precipitation is relatively uniform in quantity north of the jet and absent south of the jet. Since the jetstream positions were modeled in connection with the African study cited above, the estimation of monthly precipitation at specific sites is quite straightforward. The modeled annual sum of the monthly values is shown in figure 5 for Kameshli, Syria, a town close to the location of the famous Tell Leilan site. (Weiss *et al.*, 1993)

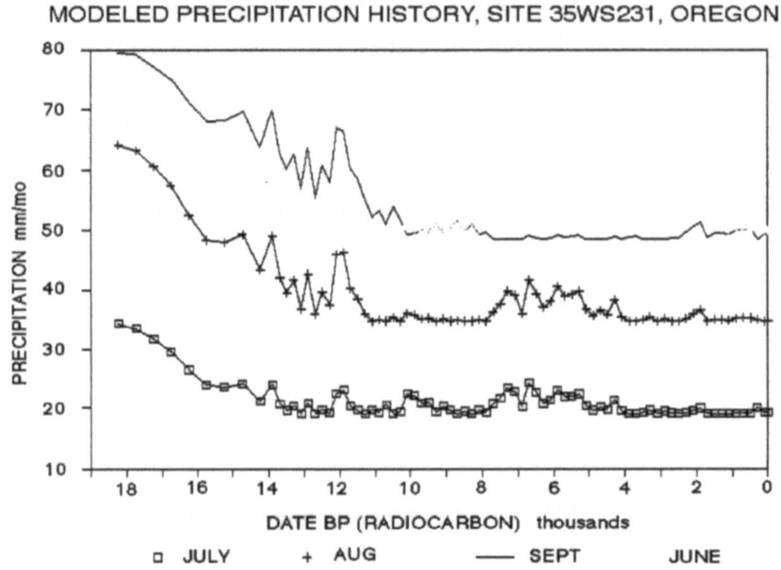

Fig. 4b Simulated rainfall history for the summer months at site 35WS231, in north-central Oregon east of the Cascade Range.

For comparison, the peak Nile flood at Aswan was simulated. The bulk of the Nile discharge at peak flood (95%) comes from the Blue Nile which carries the precipitation from northern Ethiopia. Most of the remainder comes from the White Nile, i.e. Lake Victoria (Waterbury, 1979).

To model the peak flood, which is a measure of how much flooding of the agricultural lands would have occurred, the monthly rainfall on the headwaters of the Blue and White Niles was simulated individually using the same approach as was described above for the Sahel region. This was possible because the processes are similar. The monthly rainfall is demonstrably related to the position of the ITC, though not as simply as in West Africa nor in the same way for northern Ethiopia as for the Lake Victoria region. This graph may be compared with field data contained in Hassan, 1985.

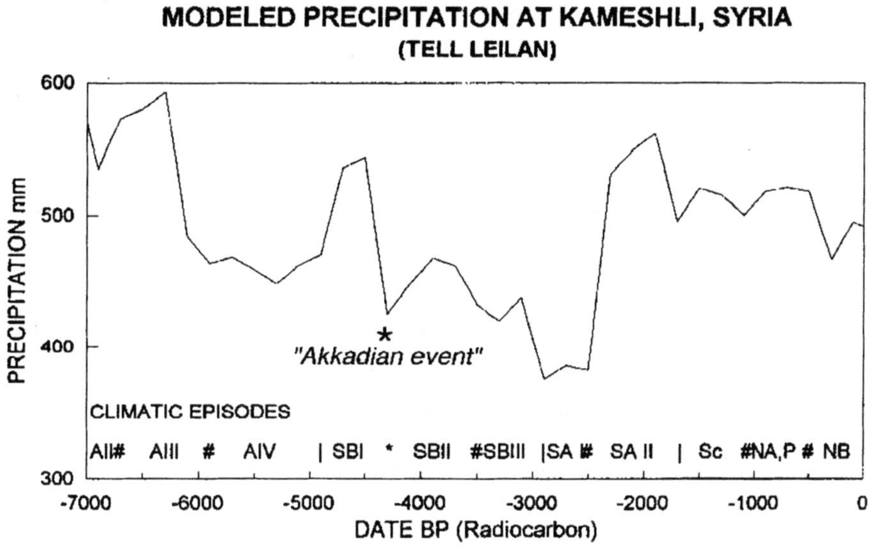

Fig. 5 Simulated mean annual rainfall at Kameshli, Syria, near the site of Tell Leilan. Global climatic episodes and subepisodes are indicated at bottom (See Table 1).

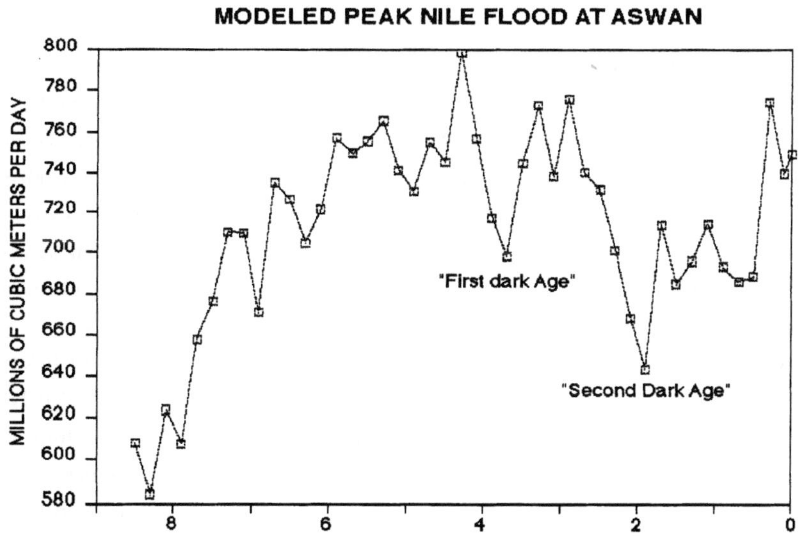

Fig. 6 Simulated peak flood of the Nile at Aswan. The modeled precipitation at the headwaters of the White and Blue Niles were modeled, combined 5%/95%, and scaled to the value of the past century.

The other main tributary, the Atbara, was assumed to be parallel in behavior to the Blue Nile, which it is near in terms of source.

Figures 5 and 6 illustrate a fundamental climatic principle, namely that the climate is a function mostly of the radiation from the sun. In those centuries when the attenuation of the sunshine was great, due to volcanic aerosols, the earth was relatively cold, i.e. winterlike. In winterlike centuries the circumpolar vortex is expanded and the jet stream is equatorward of normal, as are the subtropical anticyclones. This enhances the winter precipitation on the headwaters of the Tigris and Euphrates Rivers, but suppresses the monsoon rains that feed the Nile. The agricultural water supplies of Mesopotamia and of Egypt are therefore out of phase, changing the relative agricultural economic bases of the two ancient centers of power.

Table 1. Holocene Climatic Episodes

CLIMATIC EPISODE	SUBEPISODE	PROVISIONAL TERMINI	POSSIBLE SUBDIVISION TERMINI	CHARACTER
			Present	
	Modern			Maximum warmth ca. 1945 (NH)
	**************19151920 AD		1915 AD	Cool n. hemisphere
	NeoBoreal		1885	Mildn. hemisphere
	("Little Ice Age")		1820	Generally cold n. hemisphere
			1765	Mild n hemisphere
Post				
SubAtlantic			1720	Coldest 16001630, 16701720 (NH)
			1600	rapid drop of temperature, esp. N Atlantic area
	**************1550 AD			Warm N Atlantic
	Pacific		1400	Cooler N America N Atlantic
	**************11501200 AD (900 BP)			

NeoAtlantic 1000 "Medieval warm period"

**************700750 AD (1250 BP)

Scandic Character largely unknown

****************300400 AD (1700 BP)

SubAtlantic SA III

**************2050 BP (50100 BC) Beginning of 2000 yr general decline of N. Amer. summer temperature.

SA II

************** 2430 BP (ca. 500 BC) Highly unstable, weak transition.

SA I monsoon NW India, freq. drought in China

****************2850 BP (ca. 950 BC)

SubBoreal SB II Monsoon failure NW India, expanded tundra, glacial ad vances

**************3700 BP (2100 BC)

SB I

****************4400 BP (29003000 BC)

Atlantic ("Climatic Optimum") A IV probably warmest postglacial summers

*********** 6000 BP (ca. 4900 BC)

A III Cooler than A I or A IV

**************6700 BP (5595BC)

A II

**************7050 (5900 BC)

A I Warm summers, cold winters (NH).

****************7900 BP (ca. 6700 BC)

Boreal B II Formation of

Cochrane glacial moraines

***************8500 BP (7500 BC)

B I

*****************9150 BP (8100 BC)

PreBoreal

Monsoons expand dramatically, rapid warming

*****************10800 BP (10780 BC)

Note: The transition times between episodes were calibrated from radiocarbon data, using the CALIB program of Stuiver and Reimer (1993) except for those in the past few centuries, which were derived mostly from instrumental data. The accuracy of the division points decreases with increasing age. (Modified from Bryson and Padoch, 1981).

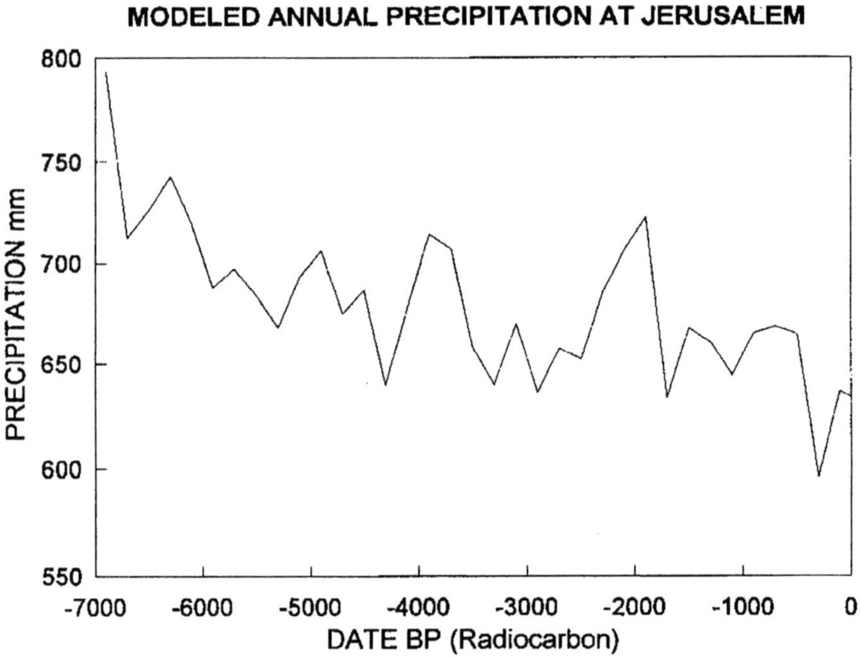

Fig. 7 Simulated annual rainfall history for Jerusalem. Note especially the differences between this simulation and that for Kameshli, Syria (Fig. 5).

Between Egypt and Syria lies the former Palestine and its longtime capital city of Jerusalem. The simulated precipitation history of Jerusalem (Fig. 7) shows a similar pattern to that for Kameshli, as it should, for both receive precipitation associated with storms in the westerlies, but with significant differences. For example, the Israelites appear to have suffered a drought after the northern Mesopotamians. However, their recovery appears to have been more dramatic, and they may have had a more erratic rainfall history in the past two and a half millennia.

Conclusion

In order to adequately address human response to climate change, it is first necessary to consider that change on human temporal and geographic scales. It appears that it is possible to produce useful sitespecific, centuryspecific simulations of past climates, and to do it on a microcomputer. How useful can only be answered in light of field data and further improvements. So far, comparisons between archaeological data and these climate simulations have been very encouraging

There are many shortcomings of the model as presented here. Much work needs to be done on the improvement of the aerosol optical depth record developed in the earliest stages of the model, for example. Another known shortcoming is the too slow rate of Northern Hemisphere temperature rise an the 18 ka-8ka BP (18 ka-7 ka BCE) period, which probably produces some spurious early Holocene trends in other derived parameters. Many of the relationships between hierarchical levels of the model have been linearized to produce a first test of the method. These linear approximations must be improved. In no case, however, was a statistical relation used without some understanding of the physical relation it represented. In essence, regression was used to calibrate the output to reality. It is also likely that the variance is suppressed, and that the actual historical fluctuations of climate were greater than shown here.

In the meantime, the results should be taken only as a first guess at what might have happened, a test of the consistency of hypotheses, a source of hypotheses to be tested against field data, and an encouragement to pursue the approach farther.

Appendix

Appended below are the volcanic and aerosol data used in the construction of the estimated aerosol optical depth history refered to on page 1. The first table (App 1) is a list of radiocarbon dated eruptions from which a subset of all unique events was extracted by eliminating duplicate dates from a common location.

The second table (App 2) is the observed annual hemispheric mean aerosol optical depth (AOD), taken from Goodman (1984). This was used throughout the studies summarized here in the conversion of the *volcanicity index* into a long history of aerosol optical depth (Bryson, 1988). This 40000 year long estimate of AOD is also given below in table App 3 .

Table App 1

RADIOCARBON DATED VOLCANIC ERUPTIONS Compiled by R.A. Bryson

LAB ID	RC V/NR *	LAB NAME	LAB #	DATE	LAT	LONG
1	23 2	A	1525	>37700	15	91
1	23 2	A	1526	>37700	15	92
1	23 2	A	1527	>37700	15	91
1	23 2	A	808	1670	39	119
121	27 2B	AC	280	7500	-38	71
4	20 3	ANU	247	2200	-4	-152
4	20 3	ANU	646	18100	-38	-141
4	20 3	ANU	86	1865	-16	-168
4	20 3	ANU	87	1980	-16	-168
4	20 3	ANU	88	1725	-16	-168
4	20 3	ANU	89	2030	-16	-168
118	18 1	AU	112	19900	66	165
118	18 1	AU	113	7070	66	164
139	QR 37-2	BETA	2316	10520	59	138
139		BETA	26604-6	11425	57	154
139	QR 37-2	BETA	26728	12290	58	136
139	QR 37-2	BETA	4987	10720	59	137
139	QR 37-2	BETA	4988	9510	59	137
139	QR 40-1	BETA	5233-4-6	33677	61	157
7	13 2	BIRM	181	663	28	17
7	10 2	BIRM	35	4672	38	25
7	16 3	BIRM	399	3520	13	-24
7	16 3	BIRM	417	28500	28	17
7	10 2	BIRM	51	3658	17	63
7	10 2	BIRM	52	23566	17	62
19	21 3	F	108	31200	41	-15
23	11 2	GAK	1176	4820	9	-148
23	11 2	GAK	1177	9150	9	-148
23	11 2	GAK	1179	12800	9	-148
23	11 2	GAK	1180	17450	9	-148
23	11 2	GAK	1181	26100	9	-148
23	11 2	GAK	1183	4210	9	-148
23	11 2	GAK	1240	3620	31	-131
23	15 1	GAK	1455	6720	35	-136
23	4	GAK	163	3380	35	-135
23	5	GAK	203	3800	35	-139
23	5	GAK	204	27800	36	-135
23	5	GAK	205	12700	41	-141
23	5	GAK	207	24950	35	-139
23	5	GAK	211	22000	31	-131
23	5	GAK	224	24100	35	-139
23	5	GAK	225	30200	35	-134
23	5	GAK	226-227	>30000	40	-141
23	5	GAK	230	8650	33	-132
23	5	GAK	231	>30000	31	-131
23	5	GAK	240	26600	36	-135
23	5	GAK	247	6460	43	-144
23	5	GAK	248	7190	44	-145
23	5	GAK	252	>30000	35	-139
23	5	GAK	253	2830	35	-139
23	5	GAK	255	21200	35	-139
23	7	GAK	352	2420	35	-139
23	7	GAK	391	3800	35	-139
23	7	GAK	409	19600	37	-138
23	7	GAK	411	4790	37	-138
23	7	GAK	442	690	35	-139
23	7	GAK	457	17900	37	-138
23	8	GAK	472	24500	31	-130
23	7	GAK	476	21230	35	-139
23	9	GAK	519	32200	43	-143
23	8	GAK	542	6800	-9	-148
23	8	GAK	543	15600	-9	-148
23	8	GAK	544	20100	-9	-148
23	8	GAK	545	7930	-9	-148
23	8	GAK	633	890	36	-139
23	8	GAK	635	470	36	-139
23	8	GAK	636	630	36	-139
23	9	GAK	712	29200	44	-142
23	9	GAK	713	31900	43	-142
23	9	GAK	714	32200	43	-142
25	14 2	GIF	1492	8900	46	-3
25	14 2	GIF	1497	8100	46	-3
25	14 2	GIF	1498	8400	46	-3
25	14 2	GIF	1499	8200	46	-3
25	14 2	GIF	1500	7800	46	-3
25	14 2	GIF	1501	8200	46	-3
25	14 2	GIF	1529	2050	46	-3

25	14 2	GIF	1553	8000	46	3
25	16 1	GIF	1904	7400	33	-25
25	16 1	GIF	2117	13200	46	-3
25	16 1	GIF	2349	3890	46	-3
25	16 1	GIF	2355	2030	46	-3
25	28 1	GIF	2448	1840	38	-15
25	28 1	GIF	25	3200	16	62
25	24 3	GIF	2554	7400	12	43
25	28 1	GIF	2640	11770	45	-4
25	28 1	GIF	2642	>35000	46	-3
25	28 1	GIF	2643	>35000	45	-4
25	28 1	GIF	2775	1320	38	-15
25	28 1	GIF	2776	3230	38	-15
25	28 1	GIF	2777	6100	38	-15
25	28 1	GIF	2778	4280	38	-15
25	28 1	GIF	2965	1650	15	61
25	28 1	GIF	3014	>35000	16	62
25	28 1	GIF	3015	14950	16	62
25	28 1	GIF	3016	25500	16	62
25	28 1	GIF	3031	>35000	16	62
25	28 1	GIF	3032	21340	16	62
25	28 1	GIF	3033	200	16	62
25	28 1	GIF	3034	2800	16	62
25	28 1	GIF	3035	3450	16	62
25	28 1	GIF	3069	18100	38	-15
25	28 1	GIF	3070	1810	38	-15
25	28 1	GIF	3071	1760	38	-15
25	28 1	GIF	3257	2820	15	61
25	28 1	GIF	3427	5460	38	-15
25	28 1	GIF	3428	2840	38	-3
25	28 1	GIF	3429	8140	38	-3
25	28 1	GIF	3589	6900	46	-3
25	28 1	GIF	3638	8760	46	-3
25	28 1	GIF	4007	>36000	46	-3
25	28 1	GIF	4203	1030	41	-14
25	28 1	GIF	4205	1750	41	-14
25	28 1	GIF	4206	1600	41	-14
25	28 1	GIF	4250	1590	41	-14
25	28 1	GIF	4251	8740	41	-14
25	28 1	GIF	4342	3200	16	62
25	28 1	GIF	4343	>37000	16	62
25	28 1	GIF	4344	>37000	16	62
25	28 1	GIF	4345	12700	16	62
25	28 1	GIF	4346	29800	16	62
25	28 1	GIF	4347	17800	16	62
25	28 1	GIF	4357	>35000	-21	-56
25	28 1	GIF	4375	17050	41	-14
25	28 1	GIF	4376	9760	41	-14
25	28 1	GIF	4377	1810	41	-14
25	28 1	GIF	4378	7910	41	-14
25	28 1	GIF	4379	8470	41	-14
25	28 1	GIF	4483	1280	41	-14
25	28 1	GIF	4484	1800	41	-14
25	28 1	GIF	4485	5530	41	-14
25	28 1	GIF	4486	3600	41	-14
25	28 1	GIF	4487	14120	41	-14
25	28 1	GIF	4488	11400	41	-14
25	28 1	GIF	4517	3760	41	-14
25	28 1	GIF	4618	14180	38	-15
25	28 1	GIF	4700	3500	16	62
25	28 1	GIF	4702	1700	16	62
25	28 1	GIF	4704	460	16	62
25	28 1	GIF	4705	2550	16	62
25	28 1	GIF	4706	13640	16	62
25	28 1	GIF	4707	>30000	16	62
25	28 1	GIF	4708	24300	16	62
25	28 1	GIF	4709	>38500	16	62
25	28 1	GIF	4712	4400	16	62
25	28 1	GIF	4713	4600	16	62
25	28 1	GIF	4714	24200	16	62
25	28 1	GIF	4715	23800	16	62
25	28 1	GIF	4716	2850	16	62
25	28 1	GIF	4717	21730	16	62
25	28 1	GIF	4718	23450	16	62
25	28 1	GIF	4719	13800	16	62
25	28 1	GIF	4720	14500	16	62
25	28 1	GIF	4721	23200	16	62
25	28 1	GIF	4723	7700	16	62
25	28 1	GIF	4724	8500	16	62
25	28 1	GIF	4725	>38500	16	62
25	28 1	GIF	4729	>38500	16	62
25	28 1	GIF	4730	3600	16	62
25	28 1	GIF	4731	2700	16	62
25	28 1	GIF	4732	>38500	16	62
25	28 1	GIF	4733	200	17	62
25	28 1	GIF	4735	680	16	62
25	28 1	GIF	4738	450	16	62
25	28 1	GIF	4751	4780	15	61
25	28 1	GIF	4752	>40000	14	61
25	28 1	GIF	4753	22100	14	61
25	28 1	GIF	4754	2250	15	61
25	28 1	GIF	4755	1700	15	61
25	28 1	GIF	4756	4040	15	61
25	28 1	GIF	4757	3960	15	61
25	28 1	GIF	4758	79.9	15	61
25	28 1	GIF	4759	11500	15	61
25	28 1	GIF	4760	>40000	14	61
25	28 1	GIF	4761	26500	15	61
25	28 1	GIF	4762	2160	15	61
25	28 1	GIF	4763	4710	15	61
25	28 1	GIF	4764	12400	15	61
25	28 1	GIF	4765	21100	15	61
25	28 1	GIF	4766	2300	15	61
25	28 1	GIF	4767	10280	15	61
25	28 1	GIF	4768	5070	15	61
25	28 1	GIF	4769	3660	15	61
25	28 1	GIF	4773	1670	15	61
25	28 1	GIF	4774	2020	15	61
25	28 1	GIF	4775	4530	15	61
25	28 1	GIF	4779	640	15	61
25	28 1	GIF	4780	4620	15	61
25	28 1	GIF	4781	1720	15	61
25	28 1	GIF	4782	6630	14	61

25	28 1	GIF	4865	24500	-21	-55
25	29 1	GIF	4866	26700	-21	-56
25	28 1	GIF	4867	30500	-21	-56
25	28 1	GIF	4868	>40000	-21	-56
25	28 1	GIF	4869	30700	-21	-56
25	28 1	GIF	4870	21900	-21	-56
25	28 1	GIF	5096	3170	41	-14
25	28 1	GIF	5097	1900	41	-14
25	28 1	GIF	5098	1890	41	-14
25	28 1	GIF	5099	8110	41	-15
25	28 1	GIF	5132	16280	45	-4
25	28 1	GIF	5133	15000	45	-4
25	28 1	GIF	5262	6870	41	-14
25	28 1	GIF	5264	3530	41	-14
25	28 1	GIF	5265	1550	41	-14
25	28 1	GIF	5266	470	41	-14
25	28 1	GIF	5267	2750	41	-14
25	29 1	GIF	5268	330	41	-14
25	28 1	GIF	5326	16800	38	-15
25	28 1	GIF	5327	>35000	38	-15
25	28 1	GIF	5328	22600	38	-15
25	28 1	GIF	5371	20500	38	-15
25	28 1	GIF	5375	23500	38	-15
25	28 1	GIF	5432	>37000	-21	-56
25	28 1	GIF	5433	31700	-21	-56
25	28 1	GIF	5559	>42000	-21	-55
25	28 1	GIF	5587	22480	38	-15
25	28 1	GIF	5591	20300	38	-15
25	28 1	GIF	6152	7750	45	-3
25	28 1	GIF	6229	3490	46	-3
25	28 1	GIF	6230	8150	46	-3
25	28 1	GIF	6231	8110	46	-3
116	10 2	GIN	119	5555	53	-159
116	10 2	GIN	120	3300	53	-159
116	10 2	GIN	160	4220	44	-146
116	10 2	GIN	184	7600	56	-160
116	10 2	GIN	89	23600	51	-39
27	12 2	GRN	4892	5830	-47	-38
27	12 2	GRN	4893	2530	-47	-38
28	13 2	GSC	1072	5250	76	90
28		GSC	141	9560	50	114
28		GSC	1564	1750	62	140
28		GSC	1568	1280	62	140
28		GSC	161	10620	50	113
28	7	GSC	206	7510	50	119
28	7	GSC	214	6270	50	119
28	8	GSC	298	3390	50	121
28		GSC	321	7340	49	123
28		GSC	332	9330	52	117
28		GSC	343	1240	64	139
28	8	GSC	345	3410	51	120
28		GSC	369	5430	50	113
28		GSC	400	1990	64	140
28		GSC	408	1200	61	130
28		GSC	479	>22200	50	119
28		GSC	524	>42900	64	137
28		GSC	529	2440	51	122
28		GSC	531	2670	52	117
28		GSC	531	2670	52	117
28	9	GSC	566	1340	58	131
28	12 1	GSC	832	1220	50	117
28	13 2	GSC	863	4740	76	86
30		GX	645	>35000	19	99
34		HV	4241	38895	19	98
34		HV	4242	515	19	98
34		HV	4243	7645	19	98
34		HV	4244	12060	19	98
34		HV	4245	20735	19	98
34		HV	4757	7405	19	98
34		HV	4758	7690	19	98
34		HV	4883	4805	19	99
34		HV	4884	7715	19	98
34		HV	5813	3075	19	98
34		HV	6746	3095	19	98
34		HV	6746	16765	19	98
34		HV	6748	17800	19	98
34		HV	6959	35600	19	100
34		HV	7173	5075	19	99
34		HV	7174	25750	19	99
34		HV	7175	1365	19	98
34		HV	7176	2670	19	98
35	23 3	I	10183	40000	3	102
35	23 3	I	10369	3950	70	146
35	23 3	I	10371	4890	71	146
35	10 2	I	1505	6150	44	121
35	11 1	I	3158	6190	52	119
35	11 1	I	3159	7670	52	119
35	12 1	I	3674	6670	48	123
35	14 1	I	4101	9030	-38	-143
35		I	529	5570	59	156
35	18 2	I	6192	1145	3	-102
35	18 2	I	6193	33250	3	-102
35	18 2	I	6472	36500	3	-102
37	15 2	IRPA	122	1600	-41	72
37	15 2	IRPA	123	25320	-41	73
38	15 1	ISGS	105	8620	-23	66
38	23 3	ISGS	389	1690	37	104
120	26 2	JGS	100	39000	35	-139
120	26 2	JGS	16	13600	37	-139
120	26 2	JGS	37	13600	36	-138
120	26 2	JGS	46	2905	34	-139
120	26 2	JGS	89	34000	33	-140
120	26 2	JGS	99	>41600	35	-139
40	10 2	K	1166	6220	64	21
40	2	K	139	2310	66	17
40	2	K	139	2310	66	17
40	2	K	140	4030	66	18
40	2	K	141	6610	66	18
40	29 2	K	1542	9140	62	7
122	28 3	KSU	21	2560	35	-139
122	28 3	KSU	22	2950	35	-139
122	28 3	KSU	23	19190	36	-139
122	28 3	KSU	26	1130	34	-139

122	28 3	KSU	370	25320	43	-142
122	28 3	KSU	372	14410	43	-142
122	28 3	KSU	390	24610	35	-136
122	28 3	KSU	42	1460	35	-139
122	28 3	KSU	43	1550	35	-139
122	28 3	KSU	44	1290	35	-139
122	28 3	KSU	49	2350	35	-139
122	28 3	KSU	50	1760	35	-139
122	28 3	KSU	57	1650	35	-139
122	28 3	KSU	60	41900	36	-139
122	28 3	KSU	62	1790	35	-139
122	28 3	KSU	64	1050	35	-139
122	28 3	KSU	671	22200	35	-138
122	28 3	KSU	73	24330	35	-139
122	28 3	KSU	74	2570	35	-139
122	28 3	KSU	75	1430	35	-139
122	28 3	KSU	76	2520	35	-139
122	28 3	KSU	77	310	35	-139
122	28 3	KSU	78	1650	35	139
122	28 3	KSU	79	2700	35	-139
122	28 3	KSU	81	24520	35	-139
122	28 3	KSU	83	1110	35	-139
122	28 3	KSU	84	1150	35	-139
122	28 3	KSU	85	1350	35	-139
122	28 3	KSU	86	1120	35	-139
122	28 3	KSU	88	1480	35	-139
122	28 3	KSU	91	1130	35	-139
49	19 3	LU	1166	1100	64	21
49	19 3	LU	1167	1190	64	21
49	19 3	LU	1168	1180	64	21
49	19 3	LU	1169	1150	64	21
49	19 3	LU	1170	1290	64	21
49	29 3	LU	2599	11530	64	22
49	29 3	LU	2601	7800	64	20
51	25 1	LY	2196	1630	11	61
51	25 1	LY	2197	2200	11	61
51	27 2B	LY	2369	5325	-18	-55
51	17 1	LY	691	3430	46	-3
52		M	21	6500	43	122
52		M	292	24000	15	91
52	1	M	851	3890	13	61
58	10 2	N	360	1950	34	-139
62	21 3	NY	565	1280	46	-3
62	21 3	NY	566	1510	46	-3
63	5	NZ	1	1820	-39	-176
63	5	NZ	10	930	-38	-177
63	5	NZ	11	9270	-37	-175
63	5	NZ	113	2030	-39	-177
63	5	NZ	12	20670	-39	-177
63	5	NZ	128	1450	-40	-174
63	5	NZ	156	250	-38	-176
63	5	NZ	158	1760	-40	-177
63	5	NZ	159	1750	-40	-177
63	5	NZ	16	13870	-9	-148
63	5	NZ	160	1300	-38	-177
63	5	NZ	161	1780	-38	-176
63	5	NZ	162	1830	-38	-174
63	5	NZ	163	1840	-38	-176
63	5	NZ	164	1890	-38	-176
63	5	NZ	165	1900	-38	-176
63	5	NZ	166	840	-38	-176
63	5	NZ	168	1900	-39	-176
63	5	NZ	169	900	-38	-176
63	5	NZ	170	1800	-39	-175
63	5	NZ	172	1800	-38	-177
63	5	NZ	173	1750	-38	-176
63	5	NZ	174	1800	-38	-176
63	5	NZ	175	1850	-39	-176
63	5	NZ	176	1900	-39	-176
63	5	NZ	177	2500	-39	-176
63	5	NZ	179	3420	-39	-176
63	5	NZ	180	3150	-39	-176
63	5	NZ	181	<200	-39	-176
63	5	NZ	182	2800	-39	-176
63	5	NZ	183	1800	-39	-177
63	5	NZ	184	2400	-39	-177
63	5	NZ	185	8850	-39	-177
63	5	NZ	186	2500	-39	-176
63	5	NZ	190	1190	-4	-152
63	5	NZ	204	<200	-5	-147
63	5	NZ	215	29000	-37	-175
63	5	NZ	216	28000	-37	-175
63	5	NZ	217	28000	-37	-175
63	5	NZ	218	>42000	-37	-175
63	5	NZ	219	>40000	-37	-175
63	5	NZ	221-222	525	-37	-175
63	5	NZ	223	>30000	-37	-175
63	5	NZ	224	>42000	-37	-175
63	5	NZ	225	9000	-37	-175
63	5	NZ	227	>36000	-37	-175
63	5	NZ	28	5920	-39	-173
63	5	NZ	287	2800	-40	-176
63	5	NZ	288	2800	-40	-176
63	5	NZ	289	3400	-40	-176
63	5	NZ	3	1640	-39	-176
63	5	NZ	33	4830	-38	-141
63	5	NZ	330	20000	-38	-176
63	5	NZ	331	34400	-39	-174
63	5	NZ	333	5370	-37	-176
63	5	NZ	37	1780	-38	-175
63	5	NZ	370	8390	-37	-176
63	5	NZ	371	385	-38	-175
63	5	NZ	372	8020	-41	-175
63	5	NZ	373	39600	-46	176
63	5	NZ	375	715	-14	172
63	5	NZ	376	1850	-14	172
63	5	NZ	377	1915	-14	171
63	5	NZ	378	>40000	16	61
63	5	NZ	379	34000.1	15	61
63	5	NZ	49	8200	-39	-174
63	5	NZ	64	360	-39	-174
63	5	NZ	7	2400	-40	-175
63	5	NZ	8	2420	-40	-175

63	5	NZ	82	2000	-39	-176
68	23 2	P	2792	3670	36	-25
68	23 2	P	2793	3300	36	-25
68	26 2	P	3128	16090	36	-24
68	26 2	P	3131	39100.9	36	-24
72	28 3	PTA	463	48500.1	-27	-32
72	28 3	PTA	488	48500.1	-27	-32
72	28 3	PTA	719	42000	-27	-32
73	2	Q	418	2345	-40	10
75	19 1	R	1003	560	41	-14
75	19 1	R	1004	450	41	-14
75	18 3	R	1050	26700	41	-14
75	18 3	R	1051	29000	41	-14
75	18 3	R	1052	6220	41	-14
75	18 3	R	1053	14100	41	-14
75	18 3	R	1054	19620	41	-14
75	18 3	R	1055	28850	41	-14
75	19 1	R	1056	16520	41	-15
75	34 1	R	149	7680	38	-16
75	13 2	R	577	33000	41	-15
75	13 2	R	578	11820	41	-14
75	13 2	R	580	12280	41	-14
75	13 2	R	581	12280	41	-14
75	13 2	R	582	12680	41	-14
75	13 2	R	583	11770	41	-15
75	13 2	R	584	11910	41	-15
75	13 2	R	585	11830	41	-15
75	13 2	R	591	3700	41	-14
75	13 2	R	592	4520	41	-14
75	13 2	R	593	5270	41	-14
75	13 2	R	594	8400	41	-14
75	13 2	R	595	5880	41	-14
75	13 2	R	596	4720	41	-14
75	13 2	R	597	3820	41	-14
75	13 2	R	682	4000	41	-14
75	13 2	R	703	4360	41	-14
75	13 2	R	704	10740	41	-14
75	13 2	R	705	11650	41	-14
75	13 2	R	706	11800	41	-14
75	15 1	R	707	3650	41	-14
75	13 2	R	708	3520	41	-14
75	15 1	R	708	3680	41	-14
75	15 1	R	709	3670	41	-14
75	15 1	R	709	3830	41	-14
75	15 1	R	710	3710	41	-14
75	15 1	R	711	3700	41	-14
75	15 1	R	711	3710	41	-14
75	15 1	R	712	3780	41	-14
75	15 1	R	712	3790	41	-14
75	15 1	R	713	3640	41	-14
75	15 1	R	713	3760	41	-14
75	15 1	R	715	1800	41	-15
75	13 2	R	716	27800	41	-15
75	13 2	R	716	29300	41	-15
75	15 1	R	785	4070	41	-14
75	15 1	R	826	12280	41	-14
75	15 1	R	827	11240	41	-14
75	15 1	R	827	11360	41	-14
75	15 1	R	828	8510	41	-14
75	15 1	R	828	8620	41	-14
75	15 1	R	829	4800	41	-14
75	15 1	R	830	3510	41	-14
75	15 1	R	830	3610	41	-14
75	16 3	R	935	25100	41	-15
75	16 3	R	937	7870	41	-15
75	16 3	R	938	3870	41	-15
75	16 3	R	939	1630	41	-15
75	16 3	R	940	4340	41	-15
75	18 3	R	941	25200	41	-14
75	18 3	R	967	40000	41	-14
76		RL	1243	17100	46	122
76		RL	1244	20500	46	122
76		RL	1245	25850	46	122
76		RL	1246	22450	46	122
76		RL	1247	32250	46	122
76		RL	1248	30650	46	122
76		RL	1249	27200	46	122
76		RL	1250	25650	46	122
79	26 2	S	1046	630	55	129
79	17 3	S	582	2775	51	122
117	6	SA	30	8100	15	61
117	6	SA	81	5650	15	61
117	6	SA	82	19500	15	61
117	6	SA	83	1670	15	61
117	6	SA	90	7650	46	-3
117	6	SA	91	1210	-39	-176
117	6	SA	94	8580	46	-3
83	16 3	SMU	86	10330	8	-39
86	9	ST	1605	3710	61	-16
86	9	ST	1799	525	66	18
86	9	ST	1996	575	63	-11
86	9	ST	1997	1100	63	-11
86	9	ST	1998	505	63	-8
86	9	ST	1999	2775	63	-8
86	9	ST	2059	2730	63	-8
86	9	ST	2063	1100	63	-11
86	9	ST	2064	490	63	-11
86	9	ST	2065	330	63	-8
86	9	ST	2066	3050	63	-8
86	9	ST	2078	8455	62	7
86	9	ST	2084	3600	63	-8
86	9	ST	2087	6875	63	-8
86	9	ST	2106	3700	62	-12
86	9	ST	2107	575	62	-12
86	9	ST	2108	1090	62	-12
86	9	ST	2119	1535	62	-12
86	9	ST	2132	1585	62	7
86	9	ST	2133	3075	62	7
86	9	ST	2134	3350	62	7
86	9	ST	2135	900	62	7
86	9	ST	2136	3520	62	7
86	9	ST	2137	6065	66	18
86	9	ST	2139	3650	62	7

86	9	ST	2140	2585	62	7
86	9	ST	2151	2650	62	7
86	9	ST	2155	3450	62	7
86	9	ST	2156	3800	62	7
86	5	ST	813	2820	64	20
86	5	ST	814	2660	64	20
88		SUA	697	780	-6	-143
89	6	T	394	770	65	16
95	16 3	TK	104	26500	31	-131
95	10 1	TK	21	870	30	-138
95	16 3	TK	88	25800	32	-130
95	16 3	TK	89	25200	32	-130
95	16 3	TK	98	23500	32	-131
98	17 1	TX	1128	36750	0	8
98	17 1	TX	1130	39560	0	8
98	17 1	TX	1131	39100	0	8
98	14 2	TX	1164	2090	43	113
98	14 2	TX	1165	2360	43	113
98	17 1	TX	1268	22360	37	105
98	17 1	TX	1269	13860	37	105
98	20 2	TX	1302	14260	44	113
98	20 2	TX	1303	3840	44	114
98	20 2	TX	1304	11460	39	113
98	20 2	TX	1305	3600	39	112
98	20 2	TX	1418	44000.9	43	113
98		TX	1518	24410	19	100
98		TX	1519	25620	19	100
98		TX	1520	24930	19	100
98		TX	1521	25020	19	100
98		TX	1522	25250	19	100
98		TX	1523	23940	19	100
98		TX	1524	24260	19	100
98	19 2	TX	1555	2580	0	78
98		TX	1594	24400	19	100
98		TX	1596	24590	19	100
98		TX	1597	24440	19	100
98		TX	1598	21170	19	100
98		TX	1600	11950	19	100
98		TX	1601	11850	19	100
98		TX	1602	11470	19	100
98		TX	1603	11050	19	100
98	30 2	TX	1679	1710	19	99
98	30 2	TX	1680	5270	19	99
98	30 2	TX	1681-1682	3365	19	99
98	30 2	TX	1683	1690	19	99
98	20 2	TX	1736	2170	43	113
98	30 2	TX	1783	14580	19	99
98	19 2	TX	2323	2500	14	89
98	19 2	TX	2324	1970	14	90
99	1	U	3	1940	64	18
99	1	U	33	6910	80	-19
99	1	U	34	4150	80	-19
99	1	U	36	6620	80	-18
99	1	U	38	7960	80	-19
101	5	UCLA	131	7940	47	119
101	5	UCLA	228	1536	19	99
101	8	UCLA	908	920	38	119
103	17 1	UGA	449	3375	38	119
103	17 1	UGA	450	1210	37	119
103	17 1	UGA	451	1175	37	119
103	17 1	UGA	604	3320	37	119
103	17 1	UGA	605	3770	37	119
104	20 2	UM	1027	169.9	15	61
104	20 2	UM	1028	1845	15	61
104	20 2	UM	1029	765	15	61
104	20 2	UM	1030	4500	15	61
104	20 2	UM	1031	7745	15	61
104	20 2	UM	1032	7365	15	61
104	20 2	UM	1033	1900	15	61
104	20 2	UM	1034	3150	15	61
104	20 2	UM	1035	535	15	61
104	20 2	UM	1036	4375	15	61
104	20 2	UM	1037	545	15	61
104	20 2	UM	1038	5250	15	61
104	20 2	UM	1039	1625	15	61
104	20 2	UM	1040	1625	15	61
104	20 2	UM	1085	1280	15	61
104	20 2	UM	1086	2550	15	61
104	20 2	UM	1087	1800	15	61
104	20 2	UM	1088	4990	15	61
104	20 2	UM	1089	2470	15	61
104	20 2	UM	1090	2260	15	61
104	18 2	UM	376	2670	15	61
104	18 2	UM	377	5190	15	61
104	18 2	UM	378	5310	15	61
104	18 2	UM	379	8400	15	61
104	18 2	UM	380	440	15	61
104	18 2	UM	381	1040	15	61
104	18 2	UM	382	540	15	61
104	18 2	UM	383	25120	15	61
104	18 2	UM	384	2470	15	61
104	18 2	UM	385	2430	15	61
104	18 2	UM	386	36100	15	61
104	18 2	UM	387	7410	15	61
104	18 1	UM	394	1230	15	61
104	18 1	UM	395	3110	15	61
104	18 1	UM	396	2020	15	61
104	18 1	UM	397	4940	15	61
104	18 2	UM	410	4020	15	61
104	18 2	UM	411	3990	15	61
104	18 2	UM	426	2150	15	61
104	18 2	UM	427	4230	15	61
104	18 2	UM	428	20240	15	61
104	18 2	UM	429	24550	15	61
104	18 2	UM	430	310	15	61
104	18 2	UM	431	2560	15	61
104	18 2	UM	432	3940	15	61
104	18 2	UM	433	1140	15	61
104	19 3	UM	856	21185	15	61
104	19 3	UM	873	4335	13	61
104	19 3	UM	874	4325	13	61
104	19 3	UM	875	2700	13	61
104	19 3	UM	876	2480	13	61

104	19 3	UM	877	635	13	61
104	19 3	UM	878	555	13	61
104	19 3	UM	879	615	13	61
104	19 3	UM	880	1045	13	61
104	19 3	UM	881	4165	13	61
104	19 3	UM	885	9175	15	61
104	19 3	UM	935	615	15	61
104	19 3	UM	936	575	15	61
119	23 2	USGS	105	6090	44	121
119	23 2	USGS	106	5870	44	121
119	23 2	USGS	107	6160	44	121
119	25 1	USGS	316	6040	60	150
119	25 1	USGS	317	10380	60	150
119	25 1	USGS	318	2270	43	113
119	25 1	USGS	344	1550	44	121
119	25 1	USGS	382	13070	39	123
119	25 1	USGS	431	2730	60	150
119	25 1	USGS	755	1340	44	121
119	19 3	USGS	80	6150	44	121
119	31 1	USGS	858	1110	42	122
119	31 1	USGS	865	8420	48	122
119	31 1	USGS	866	1790	48	122
119	31 1	USGS	967	53000	47	123
105	18 2	UW	165	4470	20	155
105	18 2	UW	166	4790	20	155
105	18 2	UW	188	4500	20	155
105	18 2	UW	189	570	20	155
105	18 2	UW	213	29700	20	156
105	18 2	UW	214	80	20	156
105	18 2	UW	215	2190	20	156
105	18 2	UW	219	37200	20	156
105	18 2	UW	220	31900	20	156
105	18 2	UW	253	445	47	122
105	18 2	UW	254	690	47	122
105	18 2	UW	270	22150	20	156
105	18 2	UW	283	4980	47	122
105	18 2	UW	312	6490	46	123
105	18 2	UW	314	4800	46	123
105	18 2	UW	315	5420	46	123
108	6	W	1116	4000	47	122
108	20 2	W	1431	650	38	119
108	9	W	1483	400	59	156
108	9	W	1587	2350	47	122
108	9	W	1644	12750	48	113
108	9	W	1739	8570	57	135
108	9	W	1751	4680	46	122
108	9	W	1752	3510	46	122
108	20 2	W	1908	23940	19	98
108	20 2	W	1909	8240	19	98
108	20 2	W	1911	25920	19	98
108	20 2	W	1912	5750	19	98
108	20 2	W	1923	7450	19	98
108	20 2	W	1925	17650	19	98
108	20 2	W	1927	8110	19	98
108	11 1	W	1944	8310	45	110
108	20 2	W	1980	4350	19	98
108	20 2	W	1995	>40000	19	98
108	19 1	W	2007	25390	67	164
108	19 1	W	2168	1720	44	121
108	20 2	W	2213	7980	47	118
108	19 2	W	2228	1000	41	122
108	19 2	W	2231	5400	47	121
108	19 2	W	2232	4600	47	121
108	19 2	W	2235	1000	41	122
108	19 2	W	2257	1200	41	122
108	19 2	W	2261	1010	47	121
108	19 2	W	2277	1860	46	122
108	19 2	W	2403	610	46	122
108	19 2	W	2413	18560	46	122
108	19 2	W	2422	6730	47	122
108	19 2	W	2423	6440	47	122
108	19 2	W	2424	6380	47	122
108	19 2	W	2436	2200	46	123
108	19 2	W	2437	5770	47	122
108	19 2	W	2439	2780	47	122
108	19 2	W	2440	2780	47	122
108	19 2	W	2441	11880	46	122
108	20 2	W	2486	8000	45	110
108	19 2	W	2527	1740	46	122
108	19 2	W	2529	2580	46	122
108	19 2	W	2540	20350	46	122
108	19 2	W	2541	2670	46	122
108	19 2	W	2542	2850	46	122
108	19 2	W	2549	3350	46	122
108	19 2	W	2552	3180	47	122
108	19 2	W	2587	8300	46	122
108	19 2	W	2653	36000	46	122
108	13 2	W	2655	11800	46	122
108	19 2	W	2661	37600	46	122
108	19 2	W	2675	2960	46	122
108	19 2	W	2677	3900	46	122
108	19 2	W	2702	8430	46	122
108	20 2	W	2777	1390	44	121
108	20 2	W	2780	14360	45	110
108	19 1	W	2800	9350	66	165
108	19 1	W	2801	11610	66	165
108	19 1	W	2802	9410	66	165
108	19 1	W	2803	10370	66	165
108	19 1	W	2806	14490	67	164
108	19 2	W	2814	9230	41	122
108	19 2	W	2829	2930	46	122
108	19 2	W	2832	11700	46	122
108	19 2	W	2863	1890	46	122
108	19 2	W	2866	11900	46	122
108	19 2	W	2867	5860	46	122
108	19 2	W	2868	12110	46	122
108	19 2	W	2870	11550	46	122
108	19 2	W	2871	2130	46	122
108	19 2	W	2872	2220	46	122
108	19 2	W	2873	2840	46	122
108	19 2	W	2874	460	46	122
108	19 2	W	2875	2900	46	123
108	20 2	W	2894	13650	45	110

108	19 2	W	2923	2200	46	122
108	19 2	W	2924	1780	46	122
108	19 2	W	2925	1850	46	122
108	19 2	W	2933	530	49	122
108	19 2	W	2934	110	49	122
108	19 2	W	2944	5980	49	122
108	21 2	W	2970	<200	19	155
108	19 2	W	2971	6650	49	122
108	19 2	W	2972	10350	49	122
108	19 2	W	2976	36000	46	122
108	19 2	W	2977	2060	46	122
108	19 2	W	2978	2100	46	122
108	19 2	W	2980	2910	46	122
108	20 2	W	2981	1040	38	119
108	19 2	W	2983	13130	46	122
108	19 2	W	2989	510	46	122
108	19 2	W	2990	1620	46	122
108	19 2	W	2991	8900	46	122
108	19 2	W	2993	1150	46	122
108	21 2	W	3000	1330	19	155
108	21 2	W	3018	450	19	155
108		W	3133	12120	47	120
108		W	3136	13650	47	120
108	21 2	W	3467	<200	19	155
108	21 2	W	3468	<200	19	155
108	21 2	W	3480	<200	19	155
108	21 2	W	3483	<200	19	155
108	21 2	W	3485	<200	19	155
108	21 2	W	3487	11780	19	156
108	21 2	W	3790	420	20	155
108	21 2	W	3793	450	20	155
108	21 2	W	3798	4820	19	155
108	21 2	W	3800	350	19	155
108	21 2	W	3801	22500	19	155
108	21 2	W	3803	4050	20	155
108	21 2	W	3809	10680	19	155
108	21 2	W	3811	330	19	155
108	21 2	W	3813	7950	19	156
108	21 2	W	3814	22800	19	155
108	21 2	W	3827	1130	19	155
108	21 2	W	3831	3480	19	155
108	21 2	W	3836	2830	19	155
108	21 2	W	3840	9500	19	155
108	21 2	W	3841	2950	19	155
108	21 2	W	3842	450	19	155
108	21 2	W	3843	<200	19	155
108	21 2	W	3844	4340	19	155
108	21 2	W	3845	2880	19	155
108	21 2	W	3849	340	19	155
108	21 2	W	3850	1810	19	155
108	21 2	W	3853	8550	19	155
108	21 2	W	3855	4210	19	155
108	21 2	W	3856	1040	19	155
108	21 2	W	3857	1470	19	155
108	21 2	W	3858	1400	19	155
108	21 2	W	3859	2080	19	155
108	21 2	W	3860	670	19	155
108	21 2	W	3862	5650	20	155
108	21 2	W	3870	350	19	155
108	21 2	W	3871	230	19	155
108	21 2	W	3873	6620	19	155
108	21 2	W	3876	2190	19	155
108	21 2	W	3879	830	19	155
108	21 2	W	3880	530	19	155
108	21 2	W	3881	260	20	155
108	21 2	W	3884	3610	19	155
108	21 2	W	3885	2160	19	155
108	21 2	W	3886	<200	19	155
108	21 2	W	3898	590	19	156
108	21 2	W	3910	1270	19	156
108	21 2	W	3930	6160	19	156
108	21 2	W	3935	31100	19	156
108	21 2	W	3937	<200	19	155
108	21 2	W	3938	210	19	155
108	21 2	W	3941	450	20	155
108	21 2	W	3999	730	19	155
108	21 2	W	4006	290	19	155
108	21 2	W	4008	2300	19	156
108	21 2	W	4009	3900	19	156
108	21 2	W	4012	740	19	156
108	21 2	W	4014	10820	19	156
108	21 2	W	4015	2180	19	156
108	21 2	W	4017	2010	19	156
108	21 2	W	4019	26410	19	156
108	21 2	W	4022	1860	19	156
108	21 2	W	4025	640	19	156
108	21 2	W	4047	910	19	155
108	21 2	W	4049	420	19	155
108	21 2	W	4116	1840	19	155
108	21 2	W	4117	7300	19	156
108	21 2	W	4118	580	20	155
108	21 2	W	4119	1800	19	155
108	21 2	W	4121	13210	19	156
108	21 2	W	4128	31400	19	156
108	21 2	W	4132	3900	19	156
108	21 2	W	4135	5160	19	156
108	21 2	W	4137	890	19	156
108	21 2	W	4142	2440	19	156
108	21 2	W	4152	3800	19	156
108	21 2	W	4156	740	19	156
108	21 2	W	4160	10290	19	156
108	21 2	W	4161	2000	19	156
108	21 2	W	4162	310	19	155
108	21 2	W	4163	1980	19	156
108	21 2	W	4164	3750	19	156
108	21 2	W	4174	2890	20	155
108	21 2	W	4175	300	20	155
108	21 2	W	4177	3690	20	155
108	21 2	W	4183	300	19	155
108	21 2	W	4184	<200	19	155
108	21 2	W	4186	220	20	156
108	21 2	W	4198	<200	19	156
108	21 2	W	4210	9170	19	156

108	31 2	W	4343	1280	20	155
108	31 2	W	4536	5250	20	155
108	31 2	W	4620	14500	20	155
108	31 2	W	4621	3140	20	155
108	31 2	W	4623	10610	20	155
108	31 2	W	4624	3380	20	155
108	31 2	W	4627	13350	20	155
108	31 2	W	4629	3360	20	155
108	31 2	W	4631	1110	20	155
108		W	477	2000	40	155
108		W	478	2070	40	155
108		W	482	8065	64	20
108		W	486	6120	47	118
108	31 2	W	4884	9780	20	155
108	31 2	W	4890	23840	20	155
108	31 2	W	4971	14080	20	155
108	31 2	W	4973	14370	20	155
108	31 2	W	4975	9020	20	155
108	31 2	W	4977	14530	20	155
108	31 2	W	4981	1470	20	155
108	31 2	W	5072	10320	20	155
108	31 2	W	5075	24240	20	155
108	31 2	W	5077	4410	20	155
108	31 2	W	5098	570	20	155
103	33 1	W	5158	840	38	26
103	33 1	W	5160	11230	38	25
103	33 1	W	5161	1020	38	26
103	33 1	W	5165	17160	38	26
103	33 1	W	5170	21160	38	26
103	33 1	W	5172	15180	38	25
103	33 1	W	5175	>29000	38	26
108	33 1	W	5186	260	38	25
103	33 1	W	5188	22060	38	25
103	33 1	W	5189	20890	38	26
103	33 1	W	5190	6520	38	25
103	33 1	W	5205	1100	38	25
103	33 1	W	5247	10620	38	25
103	33 1	W	5251	4990	38	26
103	33 1	W	5257	5150	38	25
103	33 1	W	5259	23000	38	25
103	33 1	W	5262	980	38	26
103	33 1	W	5264	295	38	25
103	33 1	W	5265	2700	38	26
108	31 2	W	5278	1740	20	155
103	33 1	W	5406	1860	38	26
103	33 1	W	5409	18600	38	25
103	33 1	W	5411	5140	38	25
103	33 1	W	5414	3800	38	25
103	33 1	W	5416	15190	38	25
103	33 1	W	5420	8700	38	26
103	33 1	W	5424	4480	38	26
103	33 1	W	5426	4960	38	25
103	33 1	W	5427	26500	38	25
103	33 1	W	5430	500	38	26
103	33 1	W	5431	5380	38	25
103	33 1	W	5433	1790	38	26
103	33 1	W	5434	4890	38	25
103	33 1	W	5435	1120	38	25
103	33 1	W	5436	4700	38	25
103	33 1	W	5437	5020	38	25
103	33 1	W	5438	4700	38	25
108	31 2	W	5597	1140	20	155
108	31 2	W	5599	7230	20	155
108	31 2	W	5604	5081	3685	20
155	2	W	564	4700	47	122
108	31 2	W	5670	1500	20	155
108	2	W	573	3620	62	148
108	33 1	W	6131	1010	38	26
103	33 1	W	6132	1044	38	26
103	33 1	W	6137	1350	38	26
103	33 1	W	6140	1100	38	26
103	33 1	W	6143	2800	38	26
103	33 1	W	6145	2460	38	26
103	33 1	W	6149	1280	38	26
103	33 1	W	6151	>30000	38	26
103	33 1	W	6154	1250	38	26
103	33 1	W	6174	4160	38	25
103	33 1	W	6175	780	38	25
103	3	W	905	17360	19	155
108	3	W	907	10140	19	156
108	3	W	912	9130	64	21
108	3	W	913	8170	64	21
108	3	W	922	1640	47	122
108	20 2	W	923	20000	39	112
108	20 2	W	931	12560	40	112
108	28 1	WIS	0	11615	58	136
109	QR 37-2	WIS	1950	12280	58	136
109	29 2	WK	106	241	6175	-37
129	29 2	WK	214	6210	-38	-175
-176	29 2	WK	215	1730	-38	-175
129	29 2	WK	227	7375	-38	-175
129	29 2	WK	229	7650	-38	-175
129	29 2	WK	231	9910	-38	-175
129	29 2	WK	232	10000	-38	-175
129	29 2	WK	233	12200	-38	-175
129	29 2	WK	235-236	12850	-38	-175
129	29 2	WK	236	12600	-38	-175
129	29 2	WK	237-238	14700	-38	-175
129	29 2	WK	242-244	6317	-37	-176
129	29 2	WK	263	10600	-34	-177
129	29 2	WK	425	1920	-39	-174
129	29 2	WK	491-351	9910	-34	-177
129	29 2	WK	492	8710	-34	-177
129	29 2	WK	493	5440	-34	-177
129	29 2	WK	494-495	7470	-34	-177
129	29 2	WK	496	4490	-39	-177
129	29 2	WK	497-501	4695	-34	-177
129	29 2	WK	498-500	3067	-39	-177
129	29 2	WK	505	5800	-38	-175
129	29 2	WK	507-506	2520	-38	-175
129	29 2	WK	511	13450	-38	-175
129	29 2	WK	514	12700	-38	-175
129	29 2	WK	515	12450	-38	-175

129	29 2	WK	516	12300	-38	-175
129	29 2	WK	517	11700	-38	-175
129	29 2	WK	518	11050	-38	-175
129	29 2	WK	519	10100	-38	-175
129	29 2	WK	520	8930	-38	-175
129	29 2	WK	521	9020	-38	-175
129	29 2	WK	523	8360	-38	-175
129	29 2	WK	529-530	13175	-38	-175
129	29 2	WK	531	12800	-38	-175
129	29 2	WK	535-536	2455	-38	-175
129	29 2	WK	537-538	2710	-38	-175
129	29 2	WK	539-540	3680	-38	-175
129	29 2	WK	541-542	4480	-38	-175
129	29 2	WK	543-544	5245	-38	-175
129	29 2	WK	545-546	5850	-38	-175
129	29 2	WK	547-548	8005	-38	-175
129	29 2	WK	549	14750	-38	-175
129	29 2	WK	550	14650	-38	-175
129	29 2	WK	561	10650	-38	-175
129	29 2	WK	562	5850	-38	-175
129	29 2	WK	567	13800	-38	-175
129	29 2	WK	570-571	7170	-38	-175
129	29 2	WK	572-573	12500	-38	-175
129	29 2	WK	574-575	11750	-38	-175
129	29 2	WK	586	>35000	-35	-174
129	29 2	WK	587	>30000	-35	-174
129	29 2	WK	590	>35000	-35	-174
129	29 2	WK	604	5410	-35	-174
129	29 2	WK	610	3660	-39	-177
129	29 2	WK	611	5510	-39	-177
129	29 2	WK	612	8560	-38	-176
129	29 2	WK	626	8030	-35	-174
129	29 2	WK	659	15850	-38	-176
129	29 2	WK	660	4850	-38	-176
129	29 2	WK	661	3950	-38	-176
129	29 2	WK	664	3810	-38	-176
129	29 2	WK	705	7520	-38	-176
129	29 2	WK	706-707	8310	-38	-176
129	29 2	WK	708-709	10895	-38	-176
129	29 2	WK	711	8000	-38	-176
129	29 2	WK	713	8990	-38	-176
129	29 2	WK	714-715	12015	-38	-176
129	18 1	WSU	1445	390	47	118
129	18 1	WSU	1445	400	47	118
111	18 1	WSU	1446	1800	47	118
111	18 1	WSU	1448	900	47	118
111	18 1	WSU	1449	1140	47	118
111	18 1	WSU	1450	2300	47	118
111	18 1	WSU	1451	3690	47	118
111	18 1	WSU	1451	4180	47	118
111	18 1	WSU	1452	5380	47	118
111		WSU	1548	11200	46	114
111		WSU	155	12000	48	124
111		WSU	1554	11300	46	114
111		WSU	1747	3780	65	149
111		WSU	468	22335	19	99
111	11 2	Y	1410	4670	0	-30
111	1	Y	182	6690	-54	67
113	1	Y	188	8905	-53	70
113	1	Y	189	9380	-53	70
113	11 2	Y	2303	1990	62	142
113	11 2	Y	2304	1850	62	142
113	1	Y	573	4090	13	61
113		Y	85	2720	66	18
113		Y	86	4100	66	18
113		Y	87	1940	66	17
113	13 2	?	0	10600	64	18
113	34 1	Rome	148	>41000	39	-15
0	34 1	Rome	149	7680	38	-15
	34 1	NZA	293	18990	43	-172

*The second column indicates the volume and issue number of **Radiocarbon** which contains the data given. QR indicates **Quaternary Research**. Negative values for latitude and longitude mean southern and western hemispheres respectively.

Table App 2

Mean Annual Northern Hemisphere Aerosol Optical Depth (Goodman, 1984)

YEAR	**AOD**	**YEAR**	**AOD**	**YEAR**	**AOD**	**YEAR**	**AOD**
		1910	0.005	1940	0.007	1970	0.030
		1911	0.010	1941	0.006	1971	0.027
		1912	0.071	1942	0.018	1972	0.032
1883	0.015	1913	0.071	1943	0.012	1973	0.030
1884	0.048	1914	0.030	1944	0.013	1974	0.031
1885	0.066	1915	0.007	1945	0.009	1975	0.040
1886	0.044	1916	0.015	1946	0.019	1976	0.022
1887	0.013	1917	0.030	1947	0.019	1977	0.013
1888	0.024	1918	0.010	1948	0.022	1978	0.012
1889	0.028	1919	0.020	1949	0.025	1979	0.025
1890	0.035	1920	0.011	1950	0.027	1980	0.033
1891	0.042	1921	0.006	1951	0.024		
1892	0.032	1922	0.011	1952	0.025		
1893	0.005	1923	0.021	1953	0.030		
1894	0.027	1924	0.012	1954	0.031		
1895	0.010	1925	0.024	1955	0.032		
1896	0.018	1926	0.002	1956	0.028		
1897	0.013	1927	0.012	1957	0.027		
1898	0.015	1928	0.018	1958	0.021		
1899	0.015	1929	0.023	1959	0.018		
1900	0.019	1930	0.016	1960	0.029		
1901	0.010	1931	0.017	1961	0.032		
1902	0.016	1932	0.002	1962	0.029		
1903	0.068	1933	0.006	1963	0.046		
1904	0.038	1934	0.008	1964	0.052		
1905	0.021	1935	0.004	1965	0.047		
1906	0.026	1936	0.013	1966	0.051		
1907	0.039	1937	0.009	1967	0.046		
1908	0.012	1938	0.010	1968	0.049		
1909	0.010	1939	0.008	1969	0.031		

Table App 3

Reconstructed Aerosol Optical Depth Derived from Above Two Tables

DATE BP	OPT. DEPTH	DATE BP	OPT. DEPTH	DATE BP	OPT. DEPTH
0	0.019544	-9700	0.010861	-27750	0.023635
-100	0.012021	-9900	0.007887	-28250	0.011828
-300	0.015819	-10100	0.011056	-28750	0.015776
-500	0.023021	-10300	0.024240	-29250	0.015780
-700	0.020269	-10500	0.007887	-29750	0.015784
-900	0.014623	-10700	0.018328	-30250	0.023687
-1100	0.024598	-10900	0.011476	-30750	0.015789
-1300	0.020844	-11100	0.015285	-31250	0.019742
-1500	0.016073	-11300	0.023133	-31750	0.019744
-1700	0.029433	-11500	0.023592	-32250	0.019745
-1900	0.026188	-11700	0.040229	-32750	0.011840
-2100	0.022699	-11900	0.041177	-33250	0.011840
-2300	0.018099	-12100	0.020731	-33750	0.011840
-2500	0.021090	-12300	0.025500	-34250	0.007887
-2700	0.018812	-12500	0.007887	-34750	0.007887
-2900	0.021072	-12700	0.026494	-35250	0.035560
-3100	0.016655	-12900	0.007887	-35750	0.019747
-3300	0.016958	-13100	0.022603	-36250	0.019747
-3500	0.024572	-13300	0.012921	-36750	0.011840
-3700	0.029466	-13500	0.018213	-37250	0.023700
-3900	0.020168	-13700	0.029059	-37750	0.023700
-4100	0.015974	-13900	0.013310	-38250	0.007887
-4300	0.021037	-14250	0.025975	-38750	0.031606
-4500	0.015309	-14750	0.012670	-39250	0.015793
-4700	0.024529	-15250	0.007887	-39750	0.035560
-4900	0.017161	-15750	0.010536		
-5100	0.013371	-16250	0.013431		
-5300	0.016400	-16750	0.016561		
-5500	0.016697	-17250	0.019907		
-5700	0.013966	-17750	0.017224		
-5900	0.015750	-18250	0.014312		
-6100	0.019279	-18750	0.014497		
-6300	0.014624	-19250	0.011277		
-6500	0.013115	-19750	0.018290		
-6700	0.025921	-20250	0.029111		
-6900	0.015351	-20750	0.015086		
-7100	0.011748	-21250	0.026159		
-7300	0.015876	-21750	0.022698		
-7500	0.018218	-22250	0.034099		
-7700	0.024984	-22750	0.015448		
-7900	0.023360	-23250	0.023134		
-8100	0.030745	-23750	0.030915		
-8300	0.024432	-24250	0.050349		
-8500	0.027436	-24750	0.023402		
-8700	0.020517	-25250	0.046827		
-8900	0.018330	-25750	0.031326		
-9100	0.021377	-26250	0.023554		
-9300	0.019037	-26750	0.019662		
-9500	0.016525	-27250	0.011819		

References

Blodget, Lorin (1857) **Climatology Of The United States**, Philadelphia, Lippincott, 536 pp.

Bryson, R.A. and V. Mitchell (1986) "Fundamental Large Scale Spatial Structure of Mean Precipitation in the Montane United States: Technique," In **Proc. of International Symposium on the QinghaiXizang Plateau and Mountain Meteorology, Beijing, China, May 2024, 1984**. pp.561579.

Bryson, R.A., and B.M. Goodman (1986) "Milankovitch and Global Ice Volume Simulation". Theoretical and Applied Climatology, 37, 2228.

Bryson, R. A. and Padoch, C. (1981) 'On the Climates of History', in Climate and History, R.I.Rotberg and T.K. Rabb eds., Princeton, N.J., Princeton Univ. Press, pp.318.

Bryson, R.A. (1988) "Late Quaternary Volcanic Modulation of Milankovitch Climate Forcing" Theoretical and Applied Climatology 39, 115125.

Bryson, R.A. (1992) A Macrophysical Model of the Holocene Intertropical Convergence and Jetstream Positions and Rainfall for the Saharan Region" Meteor. and Atmos. Phys. 47, 247258

Bryson, Robert U. and R.A. Bryson (1994) "A Comparison of Cultural Evidence and Simulated Holocene Climates of the Pacific Northwest" Abstracts of the 59th Annual Meeting. Soc for Amer. Archaeology, Anaheim, CA, Apr.2024, 1994, p.29.

Goodman, B.M. (1984) **The Climatic Impact Of Volcanic Activity**. Ph.D.Thesis,Department of Meteorology, University of Wisconsin Madison, 245pp.

Geyh, M.A. and D. Jakel (1974) "Late glacial and Holocene Climatic History of the Sahara Desert derived from a statistical assay of 14C dates". Palaeogeogr., Palaeoclimatol., Palaeoecol. **15(3),** 205208.

Hassan, F, (1985) "Holocene lakes and prehistoric settlements of the Western Faiyum, Egypt". Jour. of Archaeological Research **13,** 483501.

Ilesanmi, Oluwafemi O. (1971) "An empirical formulation of an ITD rainfall model for the tropics A case study of Nigeria", Journal of Applied Meteorology, **10,** 882-891.

Kutzbach, J.E. (1994) Personal communication.

LaFontaine, C.V., R.A. Bryson and W.M. Wendland (1990) "Airstream Regions of North Africa andthe Mediterranean", Jour. Climate **3(3)**, 366372.

McIntosh, S.K. and R.J. McIntosh (1981) "West African Prehistory". Amer. Scientist **69(6)**, 602 613.

Mitchell, V.L. (1969) **The Regionalization of Climate in Montane Areas.** Ph.D. Dissertation, Dept. of Meteorology, The University of Wisconsin Madison, 147pp.

PetitMaire, N., N. Page, and J. Marchand (1993) **The Sahara in the Holocene** (Map).Laboratoire de Geologie du Quaternaire, CNRS, Luminy Case 907, 13288, Marseille

Rognon, P. (1976) "Essai d'Interpretation des variations climatiques au Sahara depuis 40.000 ans" Revue de Geographie physique et de Geol. dynamique **XVIII (23),** 251-282.

Smagorinsky, J. (1963) "General circulation experiments with the primitive equations, I: The basic experiment". Monthly Weather Rev. **91**, 99-164.

Waterbury, John (1979) **Hydropolitics Of The Nile Valley.** Syracuse, NY, Syracuse University Press 301pp.

Weiss, H., M.A. Courty, W. Wetterstrom, F. Guichard, L. Senior, R. Meadow, and A. Curnow (1993) "The Genesis and Collapse of Third Millennium North Mesopotamian Civilization". Science **261,** 995-1004.

Background to Mid-Holocene Climatic Change in Anatolia and Adjacent Regions

Rhodes FAIRBRIDGE[1], Oğuz EROL[2], Mehmet KARACA[3] and Yücel YILMAZ[3]

[1] Columbia University and NASA, Goddard Institute for Space Sciences, 2880 Broadway New York, NY 10025, USA

[2] Institute of Marine Sciences, University of İstanbul, Müşküle Sok. No: 1 34470 Vefa, Istanbul, Turkey.

[3] Istanbul Technical University, Maden Fak. Genel Jeoloji ABD., 80626 Maslak Istanbul, Turkey.

Abstract

The Holocene climate of Anatolia is strongly seasonal. It has been modulated by the astronomic effects of global origin. Because of progressive shift of the orbital precession, the summer seasons have grown slightly cooler, with reduced convectional precipitation and thunderstorms while winter seasons are, at low elevations, milder and wetter. However, climatic cycles typically of about 208.5 and 514 years periodicity (of the latter, about 20 in 10,700 yr) appear to be influenced by the trajectories of the southern Jet Stream, which may swing north or south, either at times to the Black Sea latitudes, at others to the eastern Mediterranean. During its northern swings (warmer global cycles) there are often drought conditions in the south and in northern Syria and Mesopotamia. The Third Millennium BC cultural disasters appear to reflect droughts engendered by the extreme enhancement of the first effect by the shorter-term cycles of the second type. The special position and elevation of Anatolia goes back to its geologic history and plate-tectonic evolution, briefly reviewed here. Differential uplift of the interior with marine basins to the north, west and southwest create a unique climatological setting which has established the physical basis for the Pleistocene, Holocene and Subboreal events.

Introduction

Plate-tectonic antecedents began shaping the Anatolian Block during the mid-Tertiary with the collision of Africa-Arabia with the Eurasiatic continent, creating orogenic belts on either side of an ancient platform. As new seaways begun to open

NATO ASI Series, Vol. I 49
Third Millennium BC Climate Change
and Old World Collapse
Edited by H. Nüzhet Dalfes, George Kukla,
and Harvey Weiss

to the north (Paratethys) and south (eastern Mediterranean) in the early Oligocene, the subcontinent of Asia Minor began to emerge, with tropical soils reflecting a former Inter-Tropical Convergence Zone (ITCZ) at about 40°N. In the early Miocene the region became completely isolated from the former Tethys, the ITCZ shifted to about 30°N and the region became semi-arid. Tropical monsoonal climates returned in the middle to late Miocene, as shown by red latosols on an etchplain surface that is still preserved in places (Erol's D-1) at high elevations. Renewed aridity in the latest Miocene was caused by the Messinian tectonic isolation of the Mediterranean; uplift in the Anatolian interior initiated a "basin and range" Turkish "ova", environment with extensive pedimentation (D-II) in a semi-arid climate. With the Pliocene came renewed uplift (D-III) in a strong seasonal climate, reflected by the savanna-type fauna (camels, etc.) of the Villafranchian and younger pediment-type generation of surfaces (D-IV) continued to develop.

Early Quaternary in Anatolia shows a continuation of Villafranchian conditions and glacis-type terraces developed. The basin-range structure of the interior evolved, with deepening lake basins and uplifted plateaus. The late Quaternary saw the global development of the strong Milankovitch-type climatic fluctuations. The interior lake basins became rimmed by terraces, which were stepped down as the graben structures subsided and climates fluctuated between the glacial and "pluavial" (rainy, wet) phases and the semi-arid interglacials. Only the highest ranges (Tauride and Pontide belts) were affected by mountain glaciers, but periglacial conditions (day-night/thaw-freeze) created widespread colluvial soils. Coastal regions became deeply incised during glacio-eustatic lowered base levels, the gradients being increased repeatedly by neotectonics with differential uplift and subsidence. In the glacial stages extremely cold and semi-arid conditions extended even to the (now-submerged) coasts. In the warm cycles, strongly seasonal climates, as today, led to deep dissection of the landscapes and heavy sedimentation near the coast and in the interior basins.

Physical Environmental Change from Eocene to Pleistocene

A. Plate-Tectonic Setting

The Anatolian Plate is principally a block faulted Platform, which is mowing westwards between the North Anatolian Fault in the north and the East Anatolian Fault in the east-southeast. To the north and south of these principal strike-slip faults there are respectively the North Anatolian Mountains and the Taurus Mountains, which are mainly built up of nappes and thrusts, and their uplift is still continuing. The Black Sea in the North and Mediterranean Sea in the south of this Plate are back-arc basins .

B. Tectono-Geomorphic Evolution

1. Eocene

The latest oceanic environment that bridges the eastern Mediterranean with the Indian Ocean via the Southeast Anatolia-Iran existed during the middle Eocene time. The gradual demise followed by complete elimination of the southern branch of the Neo-Tethyan Ocean floor occurred during the latest middle Eocene time about 40 million years ago. This led to the collision between the Arabian plate and Laurasian plate. After the initial collision, the following collision between the plates eliminated the remnant seas left after the ocean(s), which lasted to the end of the early Miocene time (about 22 million years ago). The continent/continent collision eventually formed the Bitlis-Zagros suture and associated orogenic system during the Miocene (Şengör and Yılmaz, 1981; Görür et. al., 1991; Lüttig and Steffens, 1976; Okay et al., 1994).

The complete elimination of the Neo-Tethyan ocean, about 40 m. years ago abruptly terminated the circum-equatorial warm oceanic currents in this part of the world. As a result of this no more warm waters of the Indian Ocean passed through the Anatolia and the adjacent regions before it reached the Mediterranean system. Cessation of this circulation appears to have influenced greatly the climatic conditions . As a result of this cold front moved southwards into the previously warmer regions. This was the beginning of a gradual change of the climate from the warm-humid tropical to subtropical conditions .

During this time, the geomorphological character of the land stretches surrounding the remnant seas were denudational lowlands. The relics of these surfaces, which were developed during this palaeotectonic phase, are only to be observed as "fossil unconformity surfaces", which was buried under younger sediments. Therefore, there is no relic of them on the recent surface of the land. Only some exhumed stretches of it may reappear as very limited exposures (Erol and Tüfekçi, 1993).

2. Late Eocene-Oligocene

Late Eocene-Oligocene is the time of progressive uplift and consequent erosion in Anatolia. As a result of this the Anatolia became a Himalayan type plateau. Some relatively narrow and low basins of this age filled by partly marine, partly continental detritic or even gypsiferous sediments, which reflect some local dry inland climate. The relics of denudational surfaces are observed also as buried (fossil) surfaces. But some planation surfaces, which are observed today at the summit are of some older mountains, may be attributed to this period, and therefore they are indicated with the symbol "D O" by Erol (1981; 1983; 1987; 1991). But they are in very limited dimensions.

3. Early Miocene

Oligocene seaway of Tethys now completely closed and gave way to the beginning of a new and complicated phase. During this period new rifting system began locally as exemplified in the south and central Anatolia where sea incursions occurred, whiles the north - south compression causing gradual uplift and subsequent elimination of the remnant seas in the rest of the region. Therefore, the Miocene marine faunas begin to evolve independently of Indian Ocean and Mesopotamia Province. Upper Mesopotamia (in SE of Turkey) now a nearly closed inland sea marked by gypsum and salt deposits ("Lower Fars Group"). Climate was arid to subtropical arid and ITCZ boundary shifted from c. 40°N to about 30°N and summer monsoonal rains ended in the north.

During the earlier phases of the Early Miocene, in the southern part of the Anatolian land, that is at along the Inner Tauride belt of today, some remnants of an ancient denudational surface, which identified by Erol (1981; 1991) as DIa surface, had developed on Early Miocene (mainly on Burdigalien) marine sediments, but they are not easily to observe in the north, that is in the area of Central Anatolia. Because marine sediments of Lower Miocene age are not found there. This may be the result of two reasons: a tectonic uplift in the north and shift of ITCZ towards the south, and therefore formation of that surface might had been ceased.

4. Middle Miocene .

The Anatolian block begun to emerge as a single "micro continent" during the Middle Miocene. The climate became more humid again as the ITCZ shifted north once more. On the emerged land, a deep and red weathering profile ("latosol") developed, remnants of which are preserved in places on a planation surface ("etchplane"), identified by Erol (1963; 1981; 1991) as "D-I surface", which is found today in the interior of the country, uplifted to 1400-1500 m. Especially in western Anatolia the palinological study of lignite beds in the lake basins indicate a humid-warm climate, or some stretches steppe woodland (Gemici and Yılmazer, 1992; Akgün, 1992; Benda, 1971; Atalay 1992) and even tropical to savanna fauna (Aslan, 1992; Kaya, 1987).

This extensive surface is the latest product of the Palaeotectonic regime of Anatolia and its distortion during the following period was the first indication of a new tectonic phase, that is the Neotectonic regime. Therefore, it is numbered by Erol as "D I" surface. D means "Dönem" in Turkish and used to indicate the earliest or "first landform generation". Towards the margins of the Anatolian block these Middle Miocene surfaces are found distorted, i.e. downwarped, faulted and cut off by deep valleys and the sea. The main cause of these distortions is the continent/continent collision at the end of Serravalien. Thus the transition from "denudational surface formation" (=etchplane) regime to the formation of faulted intramountain basins or "Ova regime" of Şengör and Yılmaz (1980; 1981) and this transition of Palaeotectonic/neotectonic regimes also caused a major climatic change and transition from denudational surface formation to erosional-fluvial valley formation. It is well known in geomorphology that the denudational surface formation is a characteristic process for tropical climates and linear erosional valley formation is a characteristic process for temperate climates. However, this major change in tectonic, climatic and geomorphic processes were happened gradually in Anatolia during the upper Miocene.

5. Upper (Tortonian) and Terminal Miocene (Messinian).

Following the collision at the end of Serravalian, the major morphological features of Anatolia began to take sape from the Upper Miocene onwards. This is closely associated with continuing north-south convergence between the Arabia and the Anatolia, and this initially formed the North Anatolian and East Anatolian faults and then westward escape of the Anatolia from the point of emergence; the Karlıova basin. The westward escape caused the transpressional and transtentional basins in the central Anatolia and the north-south extensional basins; the graben system in the West Anatolia.

During the earlier phases, that is during the Tortonian, of this tectonically transitional period of Upper Miocene the humid-warm climate of tropical character, has started to change gradually as the new down faulted basins developed. In these gradually deepening basins lakes were developed and lignite formations, and even minerals such as borate, trona etc., which were originating from volcanic centers surrounding the basins, were also deposited there. Due to the tectonically uprising basin margins, and of course deepened basin bottoms, the denudational surfaces surrounding those lakes gradually became the form of inclined pediment surfaces, and the climate too became gradually drier. Thus the tropical Mid-Miocene climate had changed to subtropical conditions. The presence of flourished Hipparion fauna also indicates that the vegetation cover was still considerable (Rögl and Steininger, 1984; Martyn 1992).

Messinian tectonic activity in the western Betic orogenic belt in the vicinity of Gibraltar temporarily closed the Mediterranean. Sea level dropped to about - 2000 m and along tectonic valleys some of the coastal rivers became deeply incised. Climate at the time became hyperarid as Anatolia became part of a massive continent, with Africa and Arabia to the south and Eurasia to the north . This was favorable to an extremely dry continental regime. Anhydrite and rocksalt accumulated offshore, but in a few spots they are exposed by uplift near the shore. A partial dry land denudation surface (D-II systems acc. to Erol 1981; 1991) with playa lakes, bajada sediments, pediments and conical hills was developed during this time. These surfaces are found today at an elevation of 1200-1300 m in Central Anatolia, but they became lower and lower towards the coastal area and even submerge under the waters of the Mediterranean Sea (Ryan and Cita, 1978; Erol 1991).

6. Pliocene

This period was marked by a return a high world eustatic level of the sea (maximum 100-200 m), and this ameliorating climatic element resulted increasing river erosion in Anatolia . But this erosion opposed by further tectonism resulting a new phase of uplift and a fluvial erosional - depositional surface (D-III systems acc. to Erol, 1981; 1991) on older bedrock formations and younger red fluvial deposits of Pliocene basins. The parts of this surface are found today at about 150-200 m relative elevation above the recent coastline and inland valley bottoms and they are to be observed in Central Anatolia at about 1100-1150 m. This suggests an important change from a perennially warm humid Miocene climate to a strongly seasonal Pliocene climate, which is dry in the summer . Palaeontological evidence supports this climatological conclusion, in the form of very distinctive Villafranchian fauna (incl.

camel and savanna species), appears all across Southern Europe (its type area is in northern Italy) and generally persists, with little diagnostic change, into the early Quaternary. One might identify this evidence of progressive desiccation with the global onset of glacial activity in different regions, already during the Pliocene. The development of the giant mid-latitude ice sheets such as the Laurentian and Fennoscandian only followed in mid-Quaternary.

7. Early Quaternary

In this stage, the Villafranchian facies, with red fluvial deposits and the associated fauna and flora, continued and expanded . A final Anatolia-wide planation surface of fluvial origin developed:

D IV, Late Villafranchian, with its maximum surface at 1050 m in Central Anatolia, grading down in places near the coast. In the interior of the Anatolian Plateau a mosaic fault blocks appeared during the progressive uplift . In the down-dropped blocks numerous lakes were formed, notably Lake Van in the east, Tuz Gölü (Salt Lake) in southwest, Lake Burdur and a few others in the northwest (Erol, 1978). These provide excellent evaporation/precipitation (E/P) meters. They were first described as Pleistocene (eg. "Diluvial" of Pfannenstiel, 1944; Louis, 1938) and as "Pluvial" by Morrison (in Fairbridge, 1968, p. 873 - 883). Multiple shore terraces give evidence of the Quaternary cycles of pluviosity versus aridification. Some of the oldest and highest were regarded as Pliocene (Lake Van: Klear, 1965). Some were complicated by karst development and some spectacular travertine deposits. Some of the lake waters were at times fresh (or mildly brackish) to support populations of the clam Dreissencia, which was widespread in the Paratethys late Tertiary basins from the Vienna Basin to the Black Sea. On Burdur Lake, nine terraces are counted; only the last two lower are dated as late Pleistocene and Holocene (Erol, 1978; 1984; Roberts, 1982).

In the Mediterranean region at 18,000 BP, the high lake levels in the Near East coinciding with pollen evidence for steppe vegetation may be explained by a southern shift of the westerlies in Summer. Such a shift would have brought cloudiness, occasional rains, and low evaporation rates into the region and would have reduced the strength of the summer subtropical high that prevails there today. Winters would have been cold and precipitation reduced because of strong northeasterly airflow (COHMAP, 1988; Petit-Maire et al., 1994).

Sand dunes developed during the lake recession stages, pointing not only the availability of sand, but also to the extreme aridity, which made it possible to mobilize the sand.

Subsidence of the graben-type basins in this region continued through the Quaternary. It is tectonically analogous to the North American basin-range complex and interior plateaus (with Great Salt Lake and the pluvial lakes Bonneville and Lahontan). In places over 1500 m of sediment accumulated. The terraces show progressive (step-like) drops from over 100 m above the present lake level. Pre-Holocene terraces are marked in places by Palaeolithic tools.

8. Late Quaternary.

In the coastal belt successive erosion surfaces mark the glacial/interglacial (Milankovitch) cycles. In the warm cycles, sea level was high and caused coastal rivers to back up, reducing their gradients and creating thick wedges of sediments in the estuaries ("thalassostatic" deposits). Each erosion surface (in calcareous terrain) developed a massive calcrate crust, the surface of which is often plastered by beach rock or breccia of broken shells. On this surface, also cemented by calcrate, are veneers of periglacial (cryoclastic) angular debris as a lime-cemented breccia. This phenomenon demonstrates that the intensely cold periglacial climate began before the eustatic fall of the sea level that reflected the massive build-up of continental ice sheets.

According to Erol (1978) due to the climatic change and the successive phases of tectonic uplift of Anatolian Block, the Late Quaternary marine and fluvial terraces, and also continental soils (Erol, 1993b) and sediments are clearly different from the earlier Quaternary glacis - type terraces and other landform generations.

Part 2. Climate of Holocene

The term of Holocene is an epoch in geological chrono-stratigraphy corresponding to the last 10,000 of the planet Earth's 4.5 billion yr history. It is the last and briefest time unit in the 2 billion-yr Quaternary period (Fairbridge, 1968, p. 525). The critical warming and sea-level rise in the world ocean (the Holocene boundary, by international convection) began with the pollen-defined termination of the Younger Dryas (or 'tundra stage") and beginning of the Preboreal biozone; in deep-sea deposits it appears to be approximately -9770 or 11,720 BP (about 10,900 BP in radiocarbon yr).

The mid-Holocene is not formally defined but corresponds broadly to the Subboreal Chronozone of the Blytt-Sernander system. Its boundaries are not very specific; the lower one marks the transition from the peak of the warm-wet postgacial "climatic optimum", the Atlantic Chronozone, when the mean summer temperature in Scandinavia was about 2.5 °C higher than today. The Subboreal was there drier and cooler; in turn, it is terminated by the Sub Atlantic Chronozone, which in general is like today, cool and wet.

Meriç et al. (1991) found that Ayamama closed Basin (located nearby to western Istanbul) was belong to Late Quaternary period. Analyzing data at that region they showed that two wide-spread diagenetically grown gypsum crytal zones detected in the bore hole indicate that the region was twice subjected to intense evaporation during Holocene under the dry and hot climatic conditions.

Middle and Late Holocene fluctuations in sea-level are vigorously denied by many workers, but in favorable areas the evidence seems incontrovertible. The amplitude of the oscillations is strongly amplified by tidal mechanisms, e.g. in France, which are not applicable in the Mediterranean; in a low-tidal range area like Sweden, where the Earth's crust is rising due to post-glacial isostatic response.

Chronology of Middle East Climatic trends and Sea-Level during Subboreal:

Cool: -3385 to -3250 (135 yr *Sumerian*): Winter rains in Anatolia and Northern Sahara; Bahama or Treompan Emergence (low sea-level). Gerzean in lower Egypt (firs irrigation: forced by low Nile floods);

Warm: -3250 to -2900 (350 yr); Dry in Anatolia, Northern Mesopotamia and Northern Sahara; strong monsoonal rains in southern Sahara (Herding Neolithic Peoples); high Nile `floods (Early Dynastic in Egypt). In Northern Europe there was glacial retreat.

Cool: -2900 to -2710 (190 yr; proposed name *Memphis*): Winter rains in Northern Sahara, lower Egypt (Old Kingdom at Memphis; Nile delta widely exposed for cultivation by low sea level; likewise other deltas; Interior Anatolia wet and streams fed by snow-melt. Rotmoos II or Piora neoglacial in Alps. In Sweden, a glacial readvance occurred about -1850.

Warm: -2710 to -2345 (365 yr): Dry in Anatolia and northern Crescent; Nile floods high and Fayum irrigated. Minor sea-level rise (no major transgressions). There was a corresponding glacier retreat in Scandinavia and about 2300 BC (varve years) there was desiccation in the peat areas.

Cool: -2345 to -2205 (140 yr; proposed name *Akkadian*): Winter rains in the Northern Crescent; Nile floods weak and Fayum reduced.

Warm: -2205 to -1650 (555 yr): Widespread desiccation in Anatolia and northern fringe of "Fertile Crescent", as well as northern Africa, with widespread agricultural abandonment, the Mesopotamian Collapse. Strong monsoonal rains returned eventually in East Africa, Sahel and (at times) to the lower Mesopotamian plains. In northern latitudes there was widespread glacial melting (peaks -2250 and -1925), and sea level reached 2-3m above present in many areas, the Younger Peron Stage (Fairbridge, 1961). The sea level rise brought severe agricultural constraints in the irrigated delta lands of lower Mesopotamia as well as to the Nile delta and coastal Anatolia. In contrast, the irrigated alluvial plains of the Euphrates and Tigris (e.g. Ur III, -2150 to -2050) prospered. Likewise the Nile and Fayum (sometimes) was favored; During some (less than 50 yr) intervals very heavy monsoonal rains fell in the Nile headwaters. Not far upstream, at Abka, a 14C-dated oyster shows the Nile at 8260 BP to have 12 m above present maximum. From this and many other dates, it is clear that the monsoon rains have been progressively dropping (in a fluctuating sequence) ever since the early Holocene.

Cool: -1670 to -1555 (115 yr; proposed name *Mycenaean* or *Theban*): winter rains returned to the fertile Crescent, e. g. including the dry-farming lands of Upper Mesopotamia and modern Syria. There was a short- term but major neoglacial advance in Sweden and a corresponding fall in sea level which exposed large deltaic areas in coastal Anatolia, Egypt and southern Mesopotamia to intense cultivation.

Warm: -1650 to -1410 (210 yr): Another cycle of desiccation in the northern rim of the Fertile Crescent as well as Anatolia, and Cyprus. It was marked there by a weakening of the Hititite empire, and the end of the Middle Bronze Age. In Egypt it was associated with the Second Intermediate period. The same warming trend was not unfavorable in the Aegean and Greece. About the year 1600 saw the *Mycenaean* culture at its zenith.

In 1623 BC occurred the caaclysmic explosion of Santorini (*Thera*) which showered the Middle East with ash, leaving e. g. a 4 cm thick layer at Gölhisar (A lake basin in southwestern Anatolia (Roberts et al. 1994). It appears to have produced darkened skies and red rains in Egypt. Climatically, one would expect a chilling dust cloud and for several years Anatolian tree rings show severe effects and even frost rings. An acidity horizon is shown in Greenland ice cores. It is easy enough for certain cultural archaeologists and historical specialists to attribute a nearly simultaneous breakdown of several civilized societies in the region to this single catastrophic event. An eruptive event such as Tambora, Krakatau or Pinatoba (during the last two centuries) had global climatic repercussions, but they only lasted 2-3 yr, and their politico-economic consequences were almost zero, except in the immediate vicinity. There seems little doubt that cultural and economic centers in the Aegean and Anatolia were undergoing a serious decline for several decades before Santorini eruption.

Cool: -1410 to -1205 (205 yr): Plentiful winter rains returned to Anatolia and the northern belt of the Fertile Crescent. In Egypt, the New Kingdom flourished. Egyptian power spread out, but was constrained after a defeat at the battle of *Kadesh* (in western Syria) in 1288 BC. In Eastern Mediterranean in general it was the late Bronze Age.

Warm: -1205 to -815 (390 yr): yet another period occurred in the Fertile Crescent. From the 14C flux data, it appears that there may been a double peak in the aridity, with a moister and cooler interval, -1100 to -950. This was a dark age in Greece. Perhaps the most dramatic cultural feature was the Egyptian Collapse beginning about 1200. Good monsoonal rains in the south favored the growth and expansion of African cultures from Upper and Middle Nile. In Mesopotamia the center of gravity moved south, as summer rains and irrigation technology supported the Neo-*Babylonian* culture in the region between the modern Baghdad and Basra. A higher global sea level (about 1.0-1.5 higher than today) brought the coastline somewhat upstream of modern Basra, so that *Babylon* on the lower Euphrates became a partly maritime city. The improved summer rains also brought prosperity to the southern of Iran, the *Fars* province (formerly *Assyrian*) where the ancestral *Persians* appear to have driven by droughts, first from north of the Caucasus, and then from the Iranian Plateau. Droughts beginning around 1200 BC and the plateau put pressure on the Persians also to push westward into Anatolia and thus added military destruction to the drought hardships suffered by the widely scattered centers of *Hittite* culture (notably Beyşehir). The Persian invasion spelled the final coup-de-grace for the once powerful *Hittites*. In a domino effect, the *Dorians* from western Anatolia invaded Greece soon after 1200 BC.

Cold: -815 to -685 (130 yr). A short, but extremely cold interval (to judge from ^{14}C flux rate). Cold winters and wet conditions appear to have been widespread from central Iran to the Anatolian Plateau and Macedonia. The Greeks (of Athens and other city states), however, prospered with the good rainfall and developed their maritime trade ranging from the Black sea and the Bosphorus to Cirenaica, Libya, Tunisia and Sicily. Sedimentary progradation of the western and southern deltaic plains in Anatolia (which means the "land of the sunrise" in Greek, "*Anadolu*" in Turkish) was aided by a small drop of sea level, but commerce was interrupted by the isolation of ancient port cities (Erol, 1976; Kraft et al., 1980). Some of these harbors may have been isolated by neotectonic motions, but, for most, widespread synchronicity points to a eusatic-climatic cause. This low sea level was named Pelham Bay Emergence from a site near New York City (a tree stump dated 2830 BP: Fairbridge, 1961).

Warm: -685 to -406 (280 yr). This was the last phase of Subboreal (SB-2 or SB-3 in the Scandinavian-Baltic region), characterized by a warm and mild climate. In Anatolia and many parts of the "Fertile Crescent" it was marked by numerous droughts that may have been a forcing factor that spurred mass migrations and military operations from afflicted areas.

The end of the Subboreal marks the beginning of the SubAtlantic Chronozone, which starts with a short but sharp cool phase (-405 to -315), and continues with many smaller oscillations up to the Little Ice Age of the present millennium and the problematic prospects of the 21st century AD.

Part 3. Present Climate of Anatolia

Mainly Anatolian continent is located in the transition between the Ferrel Cell, where the baroclinic disturbances are dominant activity, and the descending branch of Hadley Cell of the General Circulation of the Atmosphere. The typical land feature of these descending areas is that these lands are arid or semi-arid zones of the Earth. Inland of the Anatolia, particularly the southern part, also has similar features like the other places of the world, inland of southern part of Unites States, the North Africa. Most of the Anatolia climate is seasonal and is under the variations of the Mediterranean climate. Because the inland of Anatolia is isolated from maritime influences due to high mountains parallel to shorelines.

The Köppen definition of a Mediterranean climate is, in simple terms, one in which winter rainfall is more than three times summer rainfall, but for much of the region summer rainfall is practically zero. A seasonal cycle in which contrast in almost all climate variables is well pronounced, can be defined. July, August and September are characterized by warm, dry conditions associated with a strong high pressure ridge which pushes eastwards from the Azores subtropical high over the Mediterranean. The axis of this ridge is displaced southward over Egypt by a through which extends from the Persian Gulf area northwestwards towards Greece and which is associated with Indian summer monsoon depressions. In October the rainy season begins associated with a change on the area wave pattern of the upper westerlies from a four to a three wave pattern on the 5-day time scale and on upper air flow which is characterized by cyclonic disturbances and low area pressure in the Mediterranean with higher pressure to the east associated with Siberian high. In spring, as the main features of the upper flow begin to move northward from their southern most winter positions, the rainy season continues. By May, polar front and associated strong upper-air westerly flow is sufficiently for north that its influence is removed, and the subtropical highs and associated ridges once more exert their influence.

Precipitation, although mainly associated with cyclonic disturbances that originate in the Mediterranean Basin, is strongly influenced by local orographic effects. Very few surface cyclones can be traced back to Atlantic Origins. In fact, most depressions originate in four specific regions is effective on Turkey. A region lies in the western Mediterranean, producing "the Gulf of Genoa" depressions which only occasionally move eastward for enough to affect Turkey. Central and Eastern Mediterranean Basin depressions, that are winter and spring phenomena, are most important for Turkey. Occasionally, Central Basin depressions may be traced back to the Gulf of Genoa, but this is not the general rule. Central Basin depressions tend to move to the Northwestern Turkey (around 10 occasions per year) or eastwards (eight per year) where further intensification may occur in the eastern Basin area. Eastern Basin depressions that move to northeast can only affect Turkey. The fourth trajectory of disturbances that is especially effective in winter, in the origin of the middle Europe and Balkans (Deniz and Karaca, 1995). Its influence is sensible in Turkey, when it moves towards southeast associated with North European cold air after its cold front passage. That is the kind of depressions that brings much snow to Turkey. The winter mean surface pressure pattern shows features which result from these cyclogenetic aspects. On very rare occasions, monsoon air masses may bring summer rain to the southeastern Turkey.

The formation of Mediterranean depressions is partly determined by history excursions of the polar front jet and the European through, modified land-sea temperature differences which favors cyclogenesis over the area (La Fontaine and Bryson, 1990). Eastern Basin depressions are often associated with cold northerly airflow and lee cyclogenesis. The movement of depressions is not well understood. In the western Mediterranean, depressions are frequently steered along the Mediterranean front, the temperature contrast which results from colder continental air moving over the warmer sea. This front is particularly strong in spring. Roughly half of the Central Basin depressions are steered over the Black Sea, and there is some evidence of steering by the upper flow along the axis of the subtropical jet. The movement of the eastern Basin depressions may be determined by the zonality of the upper low and by the strength of Siberian high.

References

Akgün, F. (1992) Palynological age revision of the Neogene Soma coal basin. 6th Congress of the Geological Society of Greece on emphasis on the Geology of the Aegean, May 25-27, 1992 Athens.

Arslan, F. (1992) Felis cf. catus (Carnivora-Mammalia) from the Late Miocene of Şerefköy (Muğla - Yatağan). *Geological Bulletin of Turkey*, 35: 47 - 50 . Ankara.

Atalay, I. (1992) The Palaeogeography of the Near East (from Late Pleistocene to Early Holocene) and Human Impact. 38 pp. Izmir.

Benda, L. (1971) Principles of the palynologic subdivision of the Turkish Neogene. *Newsletters Stratigraphy* 1.3: 23-26. Leiden.

COHMAP Members (1988) Major climatic changes of the last 18,000 years: observations and model simulations. *Science,* 241: 1043-52.

Deniz, A. and Karaca, M., (1995) Analysis of cyclone tracks over Turkey, *İ. T.Ü. Dergisi*, Vol. 1-2, (in Turkish, in print).

Erol, O. (1978) The Quaternary history of the lake basins of Central and Southern Anatolia. In Brice W.C. (Ed.) The environmental history of the Near and Middle East since the last Ice Age: 111-139. Academic Press.

Erol, O. (1981) Neotectonic and Geomorphologic evolution of Turkey In Fairbridge R.W. (Ed.) Neotectonics. Zeitschr. für Geom. Suppl. Bd. 40: 193 - 211.

Erol, O. (1983) Die naturaeumliche Gliederung der Türkei. Beiheftezum Tübinger Atlas des Vorderen Orients . Herausgegeben in Auftrag des Sonderforschungsbereichs 19 von H. Blume und W. Frey . Reihe A (Naturwissenschaften) Nr 13 . 245 s. Wiesbaden L. Reichert Verlag (TAVO-Karte A VII 2).

Erol, O. (1991) Geomorphological evolution of the Taurus Mountains, Turkey. Zeitschr. für Geom. N.F. Supp. Bd 82: 99 - 109.

Erol, O. and Tüfekçi, K. (1993a) Ilgün gölü dolayının jeomorfolojisi. The geomorphology of the surroundings of the Ilgın Lake (Abstract). *Jeomorfoloji Dergisi*, 20: 1 -12 . Ankara.

Erol, O. (1993b) Dating of Pleistocene Caliche formations and their significance in the geomorphology of Turkish red soils. 2nd International Meeting on "Red Mediterranean Soils" 3-9 May 1993, Adana, Turkey. Abstracts: 90. (Text is in print in Catena).

Erol, O., Akkan, E., Elibüyük, M. and Doğu, A.F. (1987) The Present and Quaternary natural environmental conditions in the Lower Euphrates region. Lower Euphrates Project 1978 - 1979 activities. METU Lower Euphrates Project Publications. Series 1. No 3: 15 - 23. Levha-Plate 1. Ankara .

Fairbridge, R. W. (1968) Encyclopedia of Geomorphology. Reinhold Publ. Co. New York.

Gemici, Y. and Yılmazer, Ü. (1992) Macro and Micro fossil flora of the Akşehir (Tire - Izmir) Neogene basin (Abstract) . *Doğa, Turkish Journal of Botany* 16: 383 . Ankara

Görür, N., Şengör, A. M. C., Sakınç, M., Tüysüz, O., Yiğitbaş, E., Oktay, F. Y., Engin, S., Okuroğulları, A. H. and Özgül, K. (1991) Late Triassic-late Miocene palaeogeographic evolution of Turkey and its surroundings. Ozan Sungurlu Symp. Proceedings: 174 -189 . Ankara.

Kaya, T. (1987) Middle Miocene Anchiterium and Aceratherium found in Tire, İzmir. *Journal of Faculty of Science* Ege University . Series B . 9/1: 11 - 15 . İzmir.

Klear, W. (1965) Geomorphologische Untersuchungen in den Randgebirgen des Van-Sees (Ostanatolien). Zeitschr. f. Geomorph. N.F.9: 346-355

LaFontaine, C. V. and Bryson, R. A. (1990) Airstream Regions of North Africa and the Mediterranean, *Jour. Climate*, Vol. 3, 366, 372.

Louis, H. (1938) Eiszeitliche Seen in Anatolien. Z. Ges. Erdkunde, Berlin 7-8: 267-285.

Lüttig, G. and Steffens, P. (1976) Palaeogeographical Atlas of Turkey from the Oligocene to the Pleistocene. 64 s. 1: 1,5 million. Hannover.

Martyn, D. (1992) Climates of the World, 370 pp. Elsevier Publishing Company, Warsaw.

Meriç, E., F. Y. Oktay, M. Sakınç, D. Gülen and A. İnal, (1991) Sedimentary geology and palaeoecology of Ayamama (Bakırköy) (in Turkish), *Bulletin of the Faculty of Engineering, Cum. Üniv.*, Serie A-Earth Sciences, V. 8, No 1.

Okay, A. I., A. M. C. Şengör and N. Görür, (1994) Kinematic history of the opening of the Black Sea and its effect on the surrounding regions. *Geology*, 22: 267-270.

Petit-Maire, N., Sanlaville, P. and Yan, Z. W. (1994) Changements globaux et palaeomoussons: la zone de transition afro-asiatique au course des derniers 140 000 ans. *Geochronique* 50: 16 - 19.

Roberts, N. (1982) Lake levels as an indicator of Near Eastern palaeo-climates: a preliminary appraisal . In Bintliff J L, Zeist W. v (Eds.) Palaeoclimates,

Palaeoenvironments and Human Communities in the Eastern Mediterranean Region in Later Prehistory: 234 - 265

Rögl, F. and Steininger, F. F. (1984) Neogene Paratethys, Mediterranean and Indo - Pacific Seaways . In Brenchley, P. (ed.) Fossils and Climate: 171 - 200 . Wiley.

Ryan, W.B.F. and Cita, M.B. (1978) The nature and distribution of Messinian erosional surfaces. Indicators of a several-kilometer-deep Mediterranean in the Miocene. *Marine Geology,* 27: 193-230.

Şengör, A. M. C. (1980) Outlines of the neotectonics of Turkey. Turkish Geological Society Conferences No 2. 40 pp. Ankara.

Şengör, A. M. C. and Yılmaz, Y. (1981) Tethyan evolution of Turkey: A plate tectonic approach. *Tectonophysics,* 75: 181-241.

Middle to Late Holocene Changes in Tropical Africa and Other Continents: Paleomonsoon and Sea Surface Temperature Variations.

Jean MALEY

Paléoenvironnements & Palynologie (CNRS/ISEM & ORSTOM)
Université de Montpellier 2, MONTPELLIER-34095 France

Abstract

In tropical Africa the transition from the Middle Holocene (ca. 7000 to 3800 years BP) climate to the Late Holocene (ca. 3800 BP to present) was abrupt and marked by major paleoenvironmental shifts. The sediment and lake level changes in forested equatorial belt of tropical Africa, in the Sahel and Sahara and in the tropical areas of other continents (India, Australia and South America) are interpreted in terms of paleoclimate based on comparison made between the African paleodata and some present-day rainfall anomalies. These anomalies are associated with sea surface temperature (SST) variations and are closely linked to the intensity of upwelling and trade winds. We conclude that the abrupt changes in the tropical belt at the beginning of the Late Holocene could be related to the variation of the Zonal Walker Circulation, mainly the El Niño-Southern Oscillation (ENSO. The SST opposition between northern and southern Atlantic Ocean has a major influence on climatic changes in tropical Africa. Moreover, the predicted global warming during the next century could lead in tropical Africa to climate induced large rain forest destructions.

Introduction

In Tropical Africa the changes from the Middle Holocene (ca. 7000 to 3800 years BP) to the Late Holocene (ca. 3800 BP to present day) were abrupt and with major paleoenvironmental modifications. First we present the different changes in tropical Africa: the equatorial regions of the forest sector and of East Africa, next the dry north tropical sector, essentially the Sahara. Second we point out the changes observed in tropical areas of the other continents (India, Australia and South America). In order to interpret the paleoclimatic changes, a comparison is

NATO ASI Series, Vol. I 49
Third Millennium BC Climate Change
and Old World Collapse
Edited by H. Nüzhet Dalfes, George Kukla,
and Harvey Weiss

made for Africa between these paleodata and some present day rainfall anomalies. Such anomalies are associated with Sea Surface Temperature (SST) variations which are linked mainly to changes in the intensity of the upwelling and trade winds. Two main conclusions are made: 1, The abrupt changes observed in the tropical belt at the beginning of the late Holocene could be related to variations of the Zonal Walker Circulation, mainly the El Niño-Southern Oscillation (ENSO). 2, The SST opposition in the Atlantic Ocean dipole has a larger influence today and during the late Holocene on the precipitations of western and central tropical Africa than that of the ENSO mode.

Moreover, the predicted global warming during the next century could lead in Africa to large rain forest destructions.

1) Changes in African equatorial forest.

In the African equatorial forest the main lowland sites are two crater lakes, the Lake Bosumtwi in Ghana (6°30'N-1°25'W) and the Lake Barombi Mbo in west Cameroon (4°40'N-9°24'E). The data were obtained in studying sediment cores, mainly sedimentology, palynology and isotopes geochimistry, particularly δ13C. The chronology is based on classical radiocarbon datates. The sediment cores of these two lakes go back to about 25/28, 000 BP. The Lake Bosumtwi present-day level is at the altitude of ca. 100 m but its overflow level is at the altitude of ca. 205 m. The lacustrine deposits inside the crater between these two altitudes were studied by Talbot et al. (1977, 1984), permitting us to reconstruct a curve of the main lacustrine fluctuations. The Lake Barombi Mbo level, which is at the altitude of ca. 300 m, is controlled by an overflow functioning from the base of the core studied, excepted during a few minor regressions of some meters between 20 and 10, 000 BP (Giresse et al. 1994).

In the regions around these two lakes the last maximum forest extension began at ca. 9500 BP. In Bosumtwi the sediment shows quasiannual laminations, except ca. 9000 to ca. 3700 yr BP when it is unlaminated and of sapropel type (Talbot et al., 1984). In that period the lake reached very high levels, particularly after 7000 BP when it overflowed. The reappearance of the laminations at ca. 3740 BP coincided with an abrupt fall of the lake level by about 130 m (Talbot et al., 1984; Maley, 1991)(Fig.1). Between ca. 3000 and 2000 BP a new transgression observed at a lower level than during middle Holocene. The large regression phase between ca. 3700 and 3000 BP observed without forest retreat ession around the lake (Maley, 1991). However, at this time there was probably a first retreat of the northern edge of the forest (today the closest edge is situated ca. 70 km to the north) when a new opening of the "Dahomey Gap" where the savannas interrupt the forest block and reach the seacoast (in the present-day Togo and Bénin) (Maley, 1991).

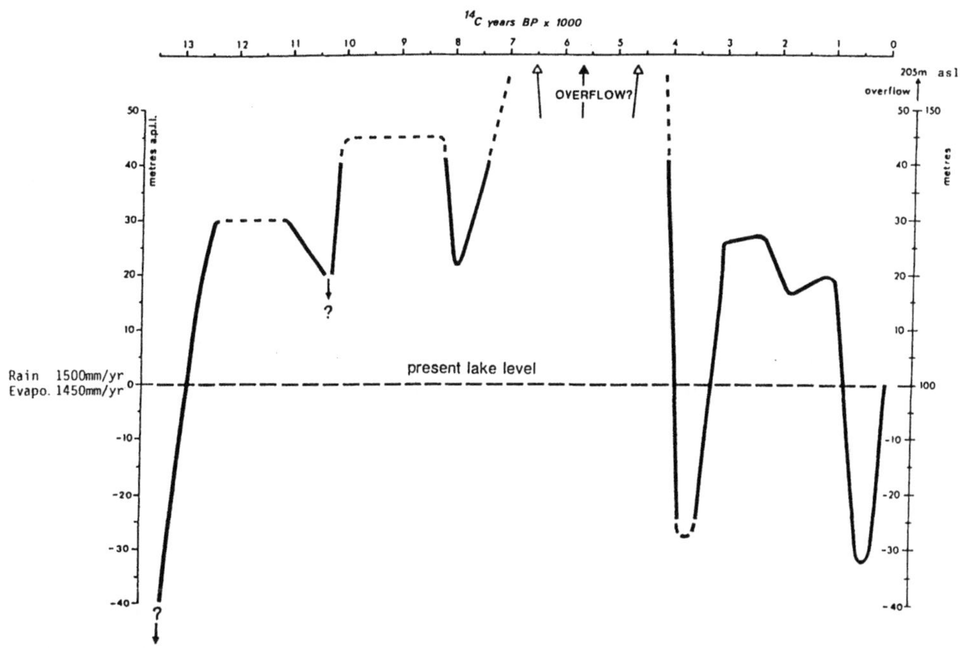

Figure 1 - Lake Bosumtwi (Ghana) level variations during late Pleistocene and Holocene (Talbot & al., 1984, fig.6); at left, the current values of rainfall and evaporation; meters a.p.l.l. = above present lake level; a.s.l. = above sea level (Maley, 1991).

For the Barombi Mbo the Holocene forest extent changed abruptly around 2800 BP and the maximum forest retreat occurred between 2500 and 2000 BP (Maley, 1992, 1996; Maley & Brenac, 1996). The same phenomenon was observed in several sites of central Atlantic Africa, in west and south Cameroon for the lakes Mboandong (Richards, 1986), Ossa (Reynaud-Farrera et al., 1996; Nguetsop and

Servant-Vildary, 1996) and Njupi (Zogning, to be published), but also in south west Congo for the lakes Kitina, Sinnda and Songolo (Vincens et al., 1996). During the same period (2800 to 2000 BP) extensive erosion was frequently observed, particularly near the rivers, with coarse sediments (sands, gravels, pebbles) constituting the base of a low terrace (Maley, 1981, 1992). After ca. 2000 BP the deposits of this terrace were fine, well stratified, and extending until the recent centuries (Maley, 1992). A new forest extension began in west Cameroon after ca. 2000 BP and possibly later in other sites.

For the transition period from middle to late Holocene the evolution of two arboreal pollen taxa may be noticed. First, the Caesalpiniaceae, typical family of the evergreen and wetter forests, reached maximum values between ca. 4500 and 3000 BP in the sites of west and south Cameroon (Barombi Mbo, Mboandong, Ossa) (Fig.2). Secondly, *Podocarpus*, a typical element of the montane cloud forests, presented in these sites a synchronous extension phase. In Barombi Mbo a strong maximum is bracketed between 3800 and 3400 BP (dated at 3690±100 BP)(Fig.2). In Congo, north of Brazzaville, the upper part of a pollen spectrum beginning around the end of the Pleistocene and terminating at the present time (Elenga et al., 1991), present a positive fluctuation of the *Podocarpus* pollen curve which could be tentatively linked to this event.

2) Changes in East Africa.

In East and South Africa the montane formations dominated by tropical gymnosperms are relatively frequent and characterized by *Podocarpus* and *Juniperus*. These formations are favoured by the regular presence of stratiform cloud covers and fog, corresponding to cloud forests (Kerfoot, 1968; Moll, 1972; Maley & Elenga, 1993; Maley, 1996). In East Africa, Perrott (1982) pointed out that around the beginning of late Holocene *Podocarpus* exhibited a well characterized and apparently synchronous maximum between different mountains (Sacred Lake in Mount Kenya, Mount Elgon, Ruwenzori, Aberdare, etc). Perrott (1982) was able to date the abrupt beginning of this *Podocarpus* spike at ca. 3720 BP. This spike was found also in other sites, as in the Lake Bogoria, Kenya (Tiercelin,

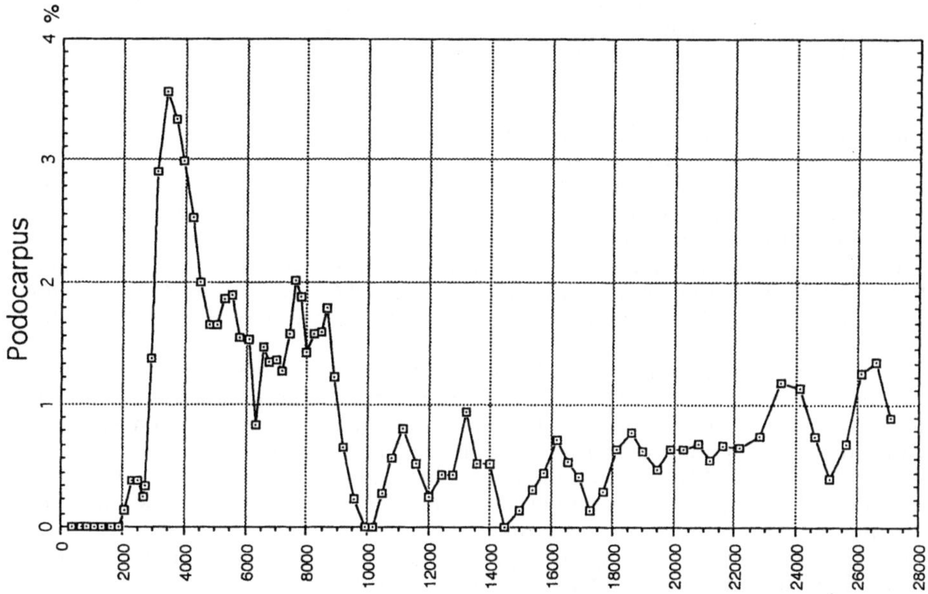
Podocarpus
4 %
3
2
1
0
0
2000
4000
6000
8000
10000
12000
14000
16000
18000
20000
22000
24000
26000
28000

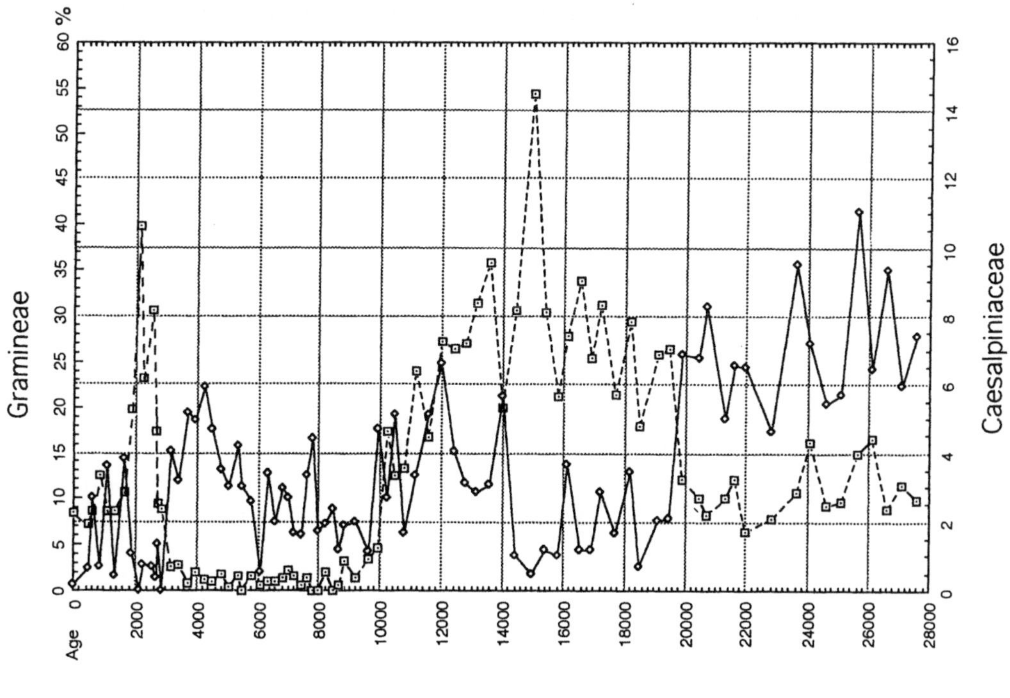
Gramineae
Caesalpiniaceae
Age
60 %
55
50
45
40
35
30
25
20
15
10
5
0
16
14
12
10
8
6
4
2
0
0
2000
4000
6000
8000
10000
12000
14000
16000
18000
20000
22000
24000
26000
28000

Figure 2 - Pollen curves (relative percentages) of 3 taxa from the Lake Barombi Mbo (West Cameroon) quaternary deposits (core BM-6)(Maley, 1996). Interpolated ages from 12 radiocarbon dates. For the geological and pollen data of this site see Brenac (1988); Maley (1989, 1991, 1992); Giresse & al. (1991, 1994); Maley & Brenac (1996 and in preparation).

- Gramineae (dotted line) - This taxon composes almost the total of non aquatic herbaceous pollen and it characterizes the open environments of savanna type. After its maximum extension during the last Arid Maximum (ca. 20, 000 to 10, 000 BP), the very low percentages between 9500 to 3000 BP point out the regional almost total disappearance of open environments and, in consequence, a maximum extension of the rain forest. A new and relatively short aridification observed abruptly between 3000 and 2000 BP, culminating between 2500 and 2000 BP. After 2000 BP and until the present day, there was a new forest extension but not so dense as that before.
- Caesalpiniaceae (total of the Caesalpiniaceae pollen)(solid line) - This Family is typical for the evergreen and wetter primary rain forests. The production of pollen grains is very low in this Family and for this reason relatively low percentages are indicative of some importance. During the Holocene, its maximum extension observed between 4500 and 3000 BP. During the late Holocene Arid phase (3000 to 2000 BP) this Family and the related sempervirent rain forest regressed strongly. Between 2000 and the present day its recovery was weak.
- *Podocarpus* (3 samples running averages). This tree is a typical element of the montane cloud forests. This formation was particularly well developped during the Holocene time with a strong maximum bracketed between 3800 and 3400 BP. Synchronous maxima observed as well in other sites of Cameroon and in East Africa .

Vincens et al., 1987). This event is apparently synchronous with that evidenced in south Cameroon and central Congo (§ 1).

Other paleoenvironmental data obtained by the study of fossil stromatolites from the Lake Tanganyika, indicate that its level lowered abruptly by 10 m around 3800 BP (Casanova & Hillaire-Marcel, 1992). Then from around 2000 BP the lake level rose progressively In Lake Bogoria (Kenya) a 4 m regression began also around 3800 BP (Tiercelin, Vincens et al., 1987), just as in Burundi (Jolly, 1993). This regression must be linked to the large one which observed in the same time in Bosumtwi (§ 1).

So, it appears that the date of ca. 3800/3700 BP (uncalibrated ^{14}C) is a key-date in tropical Africa for abrupt changes going to drier or wetter climates. A change to wetter climates observed also at this time in the Sahara.

3) Changes in the Sahara.

In the southern and central regions of the Sahara most authors concluded that there was a major aridification associated with important lacustrine regressions observed near the end of the middle Holocene, between ca. 4500 and 4000 BP:

in north Mali at Taoudenni (22/23°N) and Ine Sakane (20°40'N) (Hillaire-Marcel et al., 1983; Fabre & Petit-Maire, 1988),

in the Ténéré and the Chadian Sahara (Servant, 1973; Maley, 1981, 1982),

in the eastern Sahara (Pachur & Kröpelin, 1989; Haynes et al., 1989; Pachur & Hoelzmann, 1991).

Many data show, however, that this aridification was interrupted between ca. 3800/3700 and 3000 BP with the spreading of a new wet phase. Several precise data were obtained in the Chad basin. In the Ténéré at Agadem Servant (1973) described an 1.45 m thick lacustrine formation, with diatomitic clays, deposited in a lake of almost 20 m depth. One date on charcoal at 3550 ±110 BP was obtained 45 cm above the base. Between Fachi and Bilma other diatom deposits were dated 3350±200 and 3180±200 BP (Faure, 1966). At Nguigmi, west of Lake Chad, Gasse (1988) and Gasse & Van Campo (1994) showed that a lacustrine phase began shortly after 3810±100 BP. Considering these dates and the stratigraphic data, the beginning of this lacustrine phase must be situated around 3700/3800 BP. In the centre of Chad, 5 dates were obtained from this formation at several points in the

Palaeochad depression which range from ca. 3540 and 3000 BP (Servant, 1973; Ergenzinger, 1978; Maley, 1981). During the maximum of this transgression, resulting partly from large influxes of the palaeo-Chari, the lake levels were situated between the altitudes of 285 to 290 m (Maley, 1981; Servant & Servant, 1983), meaning that the depth of this lake was greater than 100 m in the northern part of the Paleochad (Djourab-Borkou region).

In the eastern Sahara, the Nile reached very high flood levels (+ 6 to +8 m above present level in Nubia, cf. Maley, 1981, p.206), confirmed by the very high levels of Lake Moeris (Fayum depression, northern Egypt), fed mainly by the Nile and dated between 3800 and 3000 BP (Schild, 1987). The rainfall on the Ethiopian Plateau which fed the Blue Nile floods and also Lake Asal, downstream the Awash River, resulting in a transgressive phase (3300±130 BP: Gasse & Fontes, 1989) and similarly for the Lake Turkana, downstream the Omo river (3250±150 BP: Owen et al.1982).

Similar lake extensions have been found also in other points of the Sahara, such as in the south of Mauritania (3170 ±105 BP: Faure, 1969), in central Mali, north of Tombouctou, related to the palaeo-Niger floods (3530 and 3460±200 BP: Petit-Maire & Riser, 1987), in the north of the Grand Erg Occidental, Algeria, constituting the last part of a thick lacustrine series, with 4 dates going from 3380±150 to 3080±255 BP (Gasse et al., 1987), in central Libya (3775±200 BP: Pachur et al.1987; 3420±230 BP: Pachur, 1975) and at Kufra where a lacustrine phase ended around 3000±50 BP (Edmunds et al., 1979). The archaeological data are numerous for this period in Libya, Niger and eastern Sahara. The sites of Uan Muhuggiag, southwestern Libya, is important because a thick and well characterized layer exhibit 4 dates going from 3810±80 to 3720±90 BP (Barich, 1987; Pazdur, 1993). In the western Ténéré, Niger, Roset (1987) pointed out that the Tenerean Neolithic tradition ended around 3800 BP and around 3600 BP began the new Iwelen tradition, related to a palaeo-Berber migration, developing an original culture that lasted until ca. 3000 BP. In eastern Sahara, there were no true lacustrine deposits between ca. 3800 and 3000 BP, but during this period the civilisations of Kerma and Merowe expanded, that were characterized particularly by cattle keeping. This points clearly the regular development of grazing grounds and therefore the occurrence of some regular rainfall (Pachur & Kröpelin, 1989; Pachur & Hoelzmann, 1991). Based on the determination of numerous charcoal samples, Neumann (1993) had reconstructed the Holocene vegetation of the eastern Sahara and described an arboreal phase between ca. 4000 to 3300 BP (in uncalibrated radiocarbon years).

After 3000 BP and until ca. 2600 BP a general aridification took place in the Sahara regions. Ballouche and Neumann (1995) traced out this event in the western Sahel. Next, between ca.2500 and 2000 BP, during the development of the First Iron Age, indications of strong erosion are noted in many places, such as in the Chad basin (Servant, 1973; Maley, 1981), in the southern Red Sea Hills (Mawson & Williams, 1984) and in the wet savannas and the forest regions (Maley, 1992)(§ 1). This widespread erosion phase indicates an important climate change characterized by a strong seasonality. (Maley, 1992; Maley et al., 1996).

4) Climatic changes at 3800/3700 BP in the tropical zone of other continents.

In South America, specifically the paleoclimatic events in Amazonia and in the contiguous Andes, the data presented by Van Der Hammen (1991) show a flood increase of the Amazon between 3500 and 2700 BP, while for the Rio Magdalena this increase began as early as 3900 and peaked around 3700/3600 BP. A similar trend is observed in Lake Titicaca: its level was very low from about 8000 BP and rose abruptly by ca. 20 m around 3800 BP (Martin et al., 1993). In the eastern lowland Amazonian rain forest at Carajas the pollen and charcoal data have shown large forest clearings from ca. 8000 to 4000 BP; a new forest extension began around 3700 BP (Absy et al., 1991; Servant et al., 1993; Siffedine et al., 1994). These data are strongly in contrast to what happened apparently simultaneously in wet tropical Africa, for instance in Lake Bosumtwi which was very high in middle Holocene and dropped abruptly about 130 m around 3700 BP (§ 1)(Fig.1) or in Lake Tanganyika which fall also abruptly by 10 m around 3800 BP (§ 2).

In India, pollen, micropaleontological and isotopic (δ13C) data obtained from the study of a marine core taken near the mouth of the Kalinadi River (Western Ghats), disclose the fluctuations of the vegetation and indirectly of the monsoon for almost 5000 years (Caratini et al., 1994). The regression of the forest environments began between ca. 4000 and 3500 BP and the domination of savannas in the catchment of this river was established by ca. 2200 BP. In Rajasthan, western India, Bryson & Swain (1981) have reconstructed the rainfall fluctuations by using pollen data from two lake deposits. They have shown that during the Harapan-Indus civilisation (ca. 5000 to 3700 BP), a part of the annual rains fell during the winter time. The abrupt end of this civilisation at ca. 3700 BP coincided with the drying up of some lakes (the phenomenon was opposite in the Sahara (§ 3) by reduction or disappearence of the summer monsoon rains, but possibly with the persistence of winter rains (Bryson & Swain, 1981).

In tropical north Australia, the paleoenvironmental data are numerous. In the rain forests of north-east Queensland, Lake Euramoo fell abruptly around 3700 BP (Kershaw, 1970); in the west of the Gulf of Carpentaria (Groote Eylandt), the monsoon rainfall decreased also abruptly around 3700 BP (Shulmeister & Lees, 1995). After ca. 2000 BP a new increase of the precipitation followed in many sites.

5) Paleoclimatic interpretations. Role of Sea Surface Temperature (SST) variations.

In broad outline the climatic variations in the tropical zone are linked to two main types of circulation: one is mainly meridional, the Hadley circulation, which controls partly the monsoon and the rainfall systems by the annual shifting of the ITCZ (Inter-Tropical Convergence Zone), the other is zonal, the East-West Walker circulation, which is related to the ENSO, or El Niño-Southern Oscillation, which is particularly powerful in the equatorial belt of the Pacific Ocean regions. These two circulation systems transfer the produced energy and compensate for the earth's rotation (Flohn, 1971; Cadet & Garnier, 1988; Philander, 1989).

In the tropical zone a large part of the climate dynamic is related to the air-sea interaction processes and particularly the moisture exchange at their interface. Another particularity in some sectors of the tropics is the important seasonal variation of Sea Surface Temperature (SST). This oceanic phenomenon affects the upwelling system which carry relatively deep (50 to 100 m below surface) oceanic water to the surface. Because these deep waters are colder than those at the surface, the upwelling cool the SST below the latitudinal average (Flohn, 1971, 1983). The main forcing of the upwelling is the seasonal stress of the northern or southern trade winds which blow from the subtropical anticyclones toward the ITCZ (Servain et al., 1985). In this way the upwelling of cold waters are infact remotely governed by the activity of the subtropical anticyclonic cells. The SST variations could be large, for example in the Gulf of Guinea (Fig.3), and have a major impact on the climate, particularly on rainfall variations in the adjacent African continent (Hirst & Hastenrath, 1983; Nicholson & Entekhabi, 1987; Mahé & Citeau, 1993; Fontaine & Bigot, 1993); the same phenomenon is effective also around the Pacific Ocean (Gordon & Hunt, 1991). The SST variations in the North and South Atlantic and also in the global ocean influence the subsaharan rainfall (Folland et al., 1986; Palmer, 1986; Druyan, 1991; Gunn, 1991).

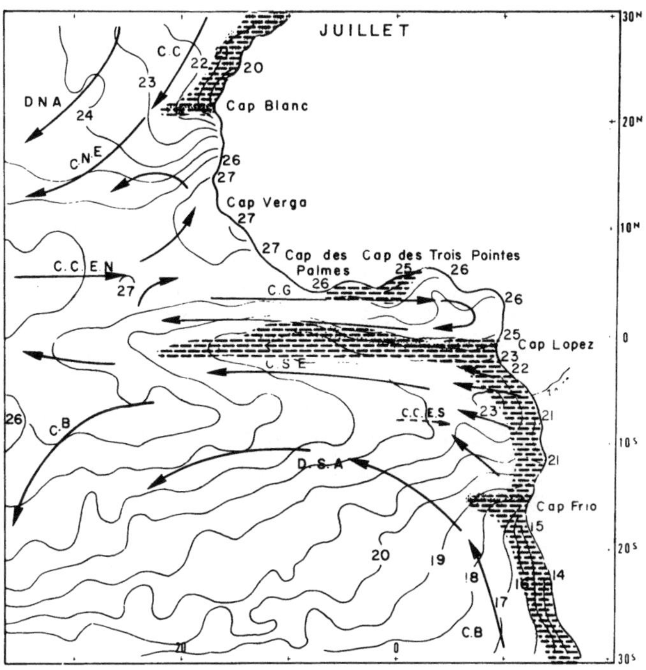

Figure 3 - The upwelling (shaded areas) and the related Sea Surface Temperature distribution (isolines in degrees C) in the Guinea Gulf and in the tropical Atlantic Ocean for July. The arrows indicate the main currents (from Wauthy, 1983, fig.2A).

An example of this relation between SST variations and climate is found in the appearance of the "little dry season", generally two months from mid-July to mid-September, in the rain forest belt north of the Gulf of Guinea (Drochon, 1976; Bakun, 1978; Hisard, 1980; Maley, 1989) (Fig.4).

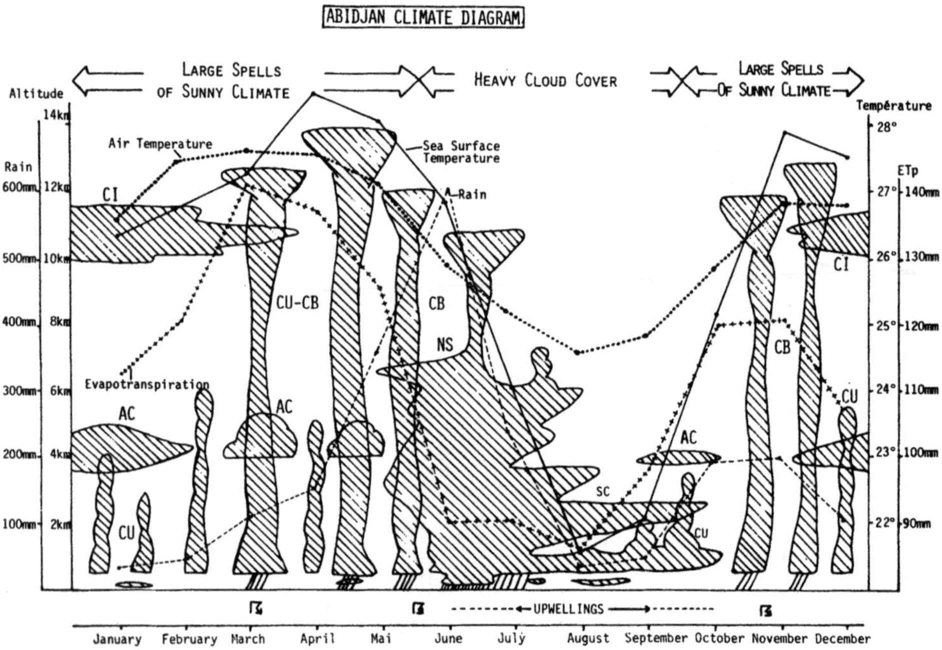

Figure 4 - Diagram of seasonal climatic changes in Abidjan (Ivory Coast); this station is representative of the western Guinean sector of the rain forest block. The diagram shows the succession of the main types of clouds and the principal elements of the climate: rain, evapotranspiration, air temperature and offshore sea surface temperature (Maley, 1989; adapted from Drochon, 1976).

This dry season, which is also the coolest period of the year, depends directly on the presence in the lower troposphere of non-precipitating stratiform clouds, generated by the upwelling of cold waters at this time of the year (Drochon, 1976). The SST is below the temperature during the monsoon, the vertical humidity gradient is reversed and evaporation is replaced by condensation; at the same time convection is replaced by subsidence induces the formation of stratiform clouds instead of cumuliform (Flohn, 1983). On the adjacent continent, the subsidence maintained aloft by the anticyclonic conditions spreading northwards and eastwards during the southern winter explains the persistance of the stratiform clouds far inland above the rain forest region. East of the Gulf of Guinea, closer to the equator in the regions of Gabon and Congo, the impact of the upwelling is larger because the same chain of oceanic and atmospheric anomalies prevails during four months; the spread of stratiform clouds can reach the Congo-Zaire basin (Maley, 1989; Maley & Elenga, 1993). Following the trade wind stress, the upwelling, and in consequence the SST, vary annually (Fig.5). In some particular years (for example 1968 or 1984) "warm" waters persisted because the upwelling were weak or absent (Fig.5) - a phenomenon similar to the "El Nino" of the eastern Pacific Ocean (Hisard, 1980; Merle, 1980). During these "warm" years, this season was not dry whatsoever inland, but instead it was very wet with heavy rainfall, mainly of convective type (Hisard, 1980; Shannon et al. 1986; Maley & Elenga, 1993).

The East-West Walker circulation is strongly influenced by such SST anomalies which during maxima events, like between 1982 to 84, are able to reverse the direction of the zonal circulation and by this way to have a major influence on the rainfall processes in the equatorial belt (Flohn, 1987b; Whetton et al., 1990; Gordon & Hunt, 1991; Beltrando & Camberlin, 1995). The warm SST anomalies of the "El Niño" type appear currently with a relatively irregular periodicity between 2 to about 5 years.

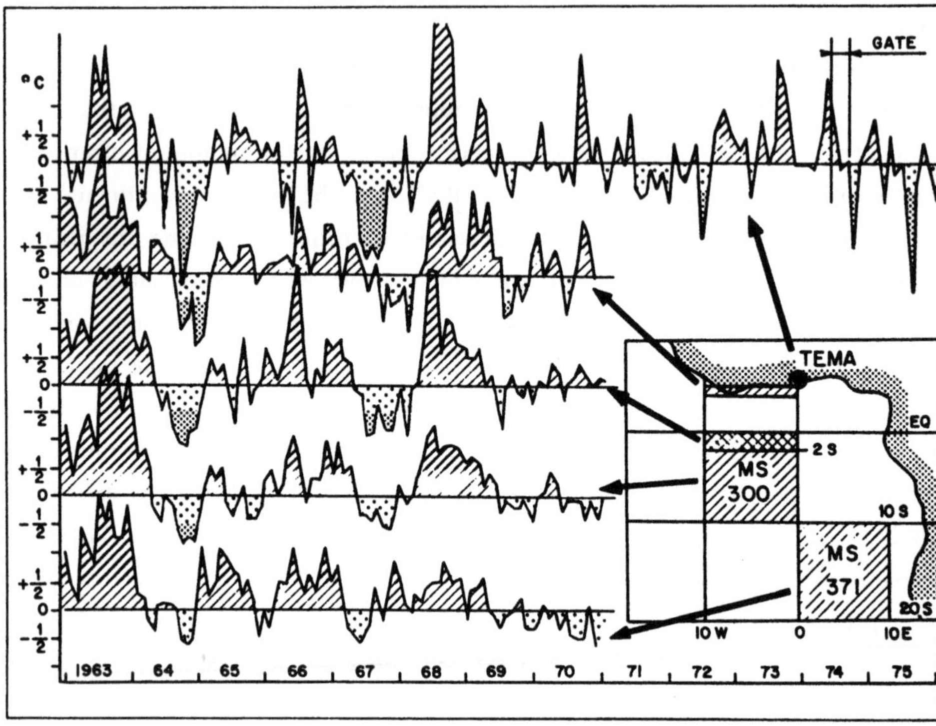

Figure 5 - Interannual Sea Surface Temperature (SST) anomalies for some successive years of the recent past in the Guinea Gulf . Temperature anomalies: oblique shading, positive anomalies ("warm" waters related to weak upwelling); large dots: negative anomalies from 0 to 0.5°C ("cold" waters related to normal upwelling); small dots: large negative anomalies of more than 0.5°C ("cold" waters related to strong upwelling). The recorded anomalies concern, from north to south: Tema, a coastal station south of Ghana, and four marine areas (homogeneized Nansen data)(Figure adapted from Merle, 1980, fig.9). This author notes that the choice of the marine areas has been mainly determined by the density and quality of the data. The coherence of anomalies throughout the Guinea Gulf is the principal feature of this figure.

In conclusion, the upwelling phenomenon can be considered as a key for explaining abrupt climatic changes (Flohn, 1983, 1986, 1987a). Maley (1987, 1989) has shown already that at 18, 000 BP there was a major SST cooling in the Gulf of Guinea (5 to 8° C below current values: Prell et al., 1976; Mix et al., 1986), that was related to very strong upwelling. This was responsible for important lessening of rainfall which induced forest destruction and fragmentation. At high northern latitudes, large SST anomalies were also considered responsible for abrupt climatic changes, such as during the Younger Dryas event (Rind et al., 1986).

5.2) Variations of Sea Surface Temperature as key for abrupt climatic change at 3800/3700 BP in the tropics.

5.2.1) Changes in lowland African rain forest north of the Gulf of Guinea.

To try to understand what happened at 3800/3700 BP in Lake Bosumtwi and its surroundings (§ 1), one must first examine the present-day phenomena. Today this lake is stratified with anoxic waters below approximately 10m, with a maximum water depth of ca. 80 m. Almost each year an overturn of the deep anoxic waters occurs (Beadle, 1974). Because the laminations of the sediment are an annual phenomenon (Talbot et al., 1984; Maley, 1991), we can deduce that the laminations must be related to the yearly overturn of deep waters. Studies done on this phenomenon by Beadle (1974) and Whyte (1975) have shown that the overturn is mainly related to the lowering of air temperature from mid-July to mid-September, affecting as well the water column which becomes more or less homothermal. At the same time, the wind stress of the southwesterlies is at its maximum, and accrodingly the unstable waters are overturned (Beadle, 1974; Whyte, 1975). The decrease of air temperature during this season is directly related to the non-precipitating stratiform cloud cover generated by the upwelling of cold sea-water offshore in the Gulf of Guinea and by the associated atmospheric phenomena related to the southern winter, particularly the maximum wind stress of the southwesterlies. We can conclude therefore that the unlaminated interval of the core, which also corresponds to the high lake levels, can be related to an absence or reduction of the "little dry season" and in consequence to the prolongation of the rainy season (Maley, 1987, 1989, 1991). Offshore in the Gulf of Guinea one can deduce also that this period was characterized by the absence or reduction of upwelling and so by relatively warm SST ("El Nino" type phenomenon, see Hisard, 1980; Merle, 1980). The recent "warm" years 1968, 1984 and 1987, could be analogs for such periods (Fontaine & Bigot, 1993; Maley & Elenga, 1993)(cf. Fig.5).

In consequence the abrupt change at 3800/3700 BP could be due probably to a sudden return of the upwelling with all the associated phenomenons. One can deduce also that the early and middle Holocene high lake level with the associated forest extension evident around Bosumtwi and Barombi Mbo (§ 2) and also during middle Holocene in western Congo (Schwartz et al., 1991; Vincens et al., 1996), could be related to relatively warm waters in the Gulf of Guinea (Maley, 1987, 1989, 1991). To check this interpretation there are few well-dated Holocene marine data. However, because the SST fluctuations are coherent throughout the Gulf of Guinea from its northern sector to the Benguela region (Fig.5), we can use for this purpose a core studied by the CLIMAP Group and taken in the latter region, RC13-228 (offshore Swakopmund in Namibia; 22°20'S - 11°12'E; 3204 m water depth)(Morley & Dworetzky, 1993). In this core the SST estimates were obtained by transfer functions based on the abundance of several radiolarian species (Morley & Dworetzky, 1993). In the 1.5 m core top three radiocarbon dates were obtained (13, 375±435; 8740±185; 4725±170 BP)(Fig.6). The last glacial maximum at 18, 000 BP is very obvious and at the top the last 2 millennia are probably missing. The resolution is rather poor in the Holocene, but by interpolation of these dates and because the SST variations are large, the curves show clearly that these were warm to very warm waters during the southern winter (July to September), with 4°C above the present day value, from ca. 9000 to 7000/6500 BP and ca. 1°C above the present value from ca. 6000 to 4500 BP (the cooling trend between ca. 7000/6500 to 6000 BP remains above the present-day values)(Fig.6).

After 4500 BP, we observes an abrupt cooling peaking around 3800/3500 BP, with winter and summer temperatures below the present-day values until ca. 3000 BP. This cooling trend resulted from an increase of the trade winds which could be linked to a temperature drop in the higher latitudes of the Southern Hemisphere, particularly in Antarctica (Ciais et al., 1992). After 3000 BP the SST exhibited a warming trend up to the top of the recovered sediment, being estimated around 2000 BP. Moreover, isotopic studies ($\delta^{18}O$) and radiocarbon dates on marine shells (two species of *Patella*) from the Benguela region (Cohen et al., 1992) confirm these trends between 4500 and 2000 BP. Thus, these first marine data support the interpretation given above. However, more marine cores with more dates and a better resolution are needed to go further into the paleoclimatic interpretation of the Holocene climatic variations in tropical Africa.

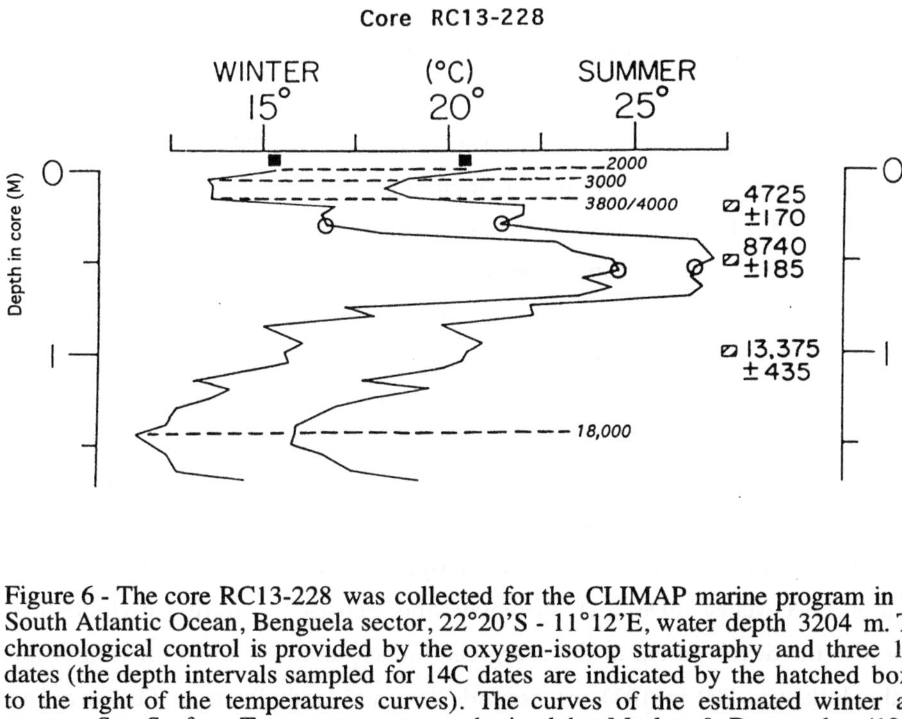

Figure 6 - The core RC13-228 was collected for the CLIMAP marine program in the South Atlantic Ocean, Benguela sector, 22°20'S - 11°12'E, water depth 3204 m. The chronological control is provided by the oxygen-isotop stratigraphy and three 14C dates (the depth intervals sampled for 14C dates are indicated by the hatched boxes to the right of the temperatures curves). The curves of the estimated winter and summer Sea Surface Temperatures were obtained by Morley & Dworetzky (1993) with transfer functions based on the abundance of Radiolarian species (Morley, 1979). The open circles designate the 6000 and 9000 BP interpolated dates and estimated temperatures. Modern SSTs (averages for February/southern Summer and August/ southern Winter) are indicated by the solid squares above each curve. The dashed lines represent some interpolated dates (italics): 18, 000 BP (also based on the oxygen-isotop stratigraphy), 3800/4000 BP, 3000 BP and 2000 BP. Figure adapted from Morley & Dworetzky (1993, Fig.6.2).

In the Barombi Mbo region and in the Congo sector of the rain forest block the climatic aridification between ca. 3000 and 2000 BP is out of phase with that in the Bosumtwi region (3800/3700 to 3000 BP). The erosive feature of the period ca. 3000 to 2000 BP in the first sector (Maley, 1992; § 1) points the mainly convective character of the rains, which is consistent with the warming trend of the SST after 3000 BP (Fig.6). A major increase of convective rains in this sector could lead to a decrease in the total rainfall and can so explain this climatic aridification. The recent year 1984 exhibited such features in West Cameroon and could be an analog (Suchel, 1988; Maley & Elenga, 1993). The extensive retreat of the rain forest at this time (Fig.2) could have been accelerated by penetration of fires spreading out from the savannas; samples of charcoal dated from this period were collected in the eastern part of the forest block (Hart et al., to be published; J. & F.Runge, to be published) and also in Gabon (Schwartz, 1992; Oslisly, 1995).

5.2.2) Changes in uplands of rain forest and in East Africa.

At 3800/3700 BP one pointed out that synchronous to the decrease in lake levels and rainfall in the western part of the rain forest block (Bosumtwi sector) and in East Africa, also in the rain forest sector of south Cameroon (and possibly central Congo north of Brazzaville, § 1) there has been a strong increase in the stratiform cloud forests characterized by the development of *Podocarpus* (§ 1 & 2). The climatic link between these two phenomena seems to be the stratiform clouds which are responsible on one hand for the lowering of the Bosumtwi Lake levels (§ 5.2.1), and on the other, when these clouds hang regularly on the tops of hills and mountain slopes, for the development of *Podocarpus* cloud forests (Kerfoot, 1968; Maley, 1996)(§ 2). Moreover, the increase in stratiform clouds can favor also the development of the evergreen forests, as today in south Cameroon on the hills (ca. 700 to 1000 m) around Yaoundé and down to Ebolowa and Sangmelima (Letouzey, 1985; Achoundong, 1985; Brenac, 1988) and probably on the hills spreading in the western central part of the rain forest block (Monts de Cristal, Chaillu, Mayombe, etc) (Maley et al., 1990; Maley, 1996). The maximum reached by the Caesalpiniaceae pollen between 4500 and 3000 BP (§ 2) is probably related also to this last phenomenon.

The strong increase of the *Podocarpus* Montane formation in East Africa synchronous with that in countries adjacent to the Gulf of Guinea point to the same phenomenon and particularly to a link with a SST lowering not only in the Gulf of Guinea but also probably in the Indian Ocean (cf. Flohn, 1987b; Beltrando & Camberlin, 1995)(§ 5.2.3).

5.2.3) General paleoclimatic interpretations.

Synchronous with the events observed in equatorial Africa between 3800/3700 and 3000 BP (§ 1, 2), there has also been the development of a wet phase in the Sahara region (§ 3). This wet phase seems to have been linked to the increase of monsoon rainfall, but also partly to the increase of cyclonic rainfall (i.e. tropical depressions)(§ 4) during the cool season, particularly in the spring. These particular rainfall in India as well in the Sahara are linked to the activity of the Subtropical Jet Stream (Flohn, 1971; Maley, 1977, 1981; Bryson & Swain, 1981). During this period the opposite climatic trend between the forest belt and the Sahara can be compared to some anomalous patterns on a yearly scale ("anomalies") during the wettest recent period, the1950s. During this decade the yearly monsoon rainfall was above normal in the Sahel and the south Sahara, and below normal in the rain forest and the wet savanna belt south of ca. 10°N (Nicholson, 1980; Janicot, 1991; Fontaine & Bigot, 1993). These last authors have shown that this rainfall pattern, particularly in August, is associated with anomalous SST not only in the Gulf of Guinea but also in the global ocean. These SST patterns were characterized by warm SST in the North Atlantic (but not in the polar sector) and cold SST on the equatorial and southern parts of the global ocean (Atlantic, Indian and Pacific) (Fontaine & Bigot, 1993; Trzaska et al., 1994). During the period between 3800/3700 and 3000 BP we can deduce the following SST patterns:

first in the Gulf of Guinea a general cooling, which is supported by the SST estimates obtained in the Benguela sector on the core RC13-228 (Morley & Dworetzky, 1993)(§ 5.2.1; Fig.6)

second in the North Atlantic a slight warming appears on core SU 81-18 (off Portugal) and core NA 87-22 (Rockall Plateau)(Duplessy et al., 1992), third the abrupt climatic deterioration during this period in East Africa (§ 2), India and tropical Australia (§ 4) could regionaly be related to some SST cooling in the adjacent seas,

fourth in tropical South America the strong rainfall increases during this period (§ 4) could be related to SST warming in the eastern Pacific.

These SST patterns are consistent in the equatorial belt with a Zonal Circulation of the El Nino mode (sensu Newell & Hsiung, 1987: a period dominated by this circulation type) (Fig.7a). For the middle Holocene the data presented above for equatorial Africa (wet phase until 3800/3700 BP) and equatorial South America (a relatively dry phase until 3800/3700 BP) point to a reverse Zonal Circulation of the La Niña mode (Fig.7b). However the abrupt changes at 3800/3700 BP seem due to an increase of the Hadley Circulation and particularly of the trade winds which induced strong upwelling in the equatorial zone. These SST coolings seem to be the main forcing leading at 3800/3700 BP to the reverse of the Zonal Circulation.

McGlone et al.(1992) have already estimated that only after 7000 BP (or 8000 BP, see § 4) the ENSO/Zonal Circulation of the El Niño mode could have had a major influence around the Pacific ocean and only in the late Holocene (probably after 3800/3700 BP) this phenomenon played currently its important role. This scenario is very schematic and concerns mainly the tropical regions, but because the changes at 3800/3700 BP seem to have been global, the temperate and polar regions should be included (see other chapters of this book).

Recent climatological studies (diagnostic analyses) have shown that the SST variations in the Atlantic Ocean and particularly the SST opposition between the northern and southern Atlantic, forming a dipole, have a larger influence than the ENSO mode on the precipitations of the western and central parts (as far as the Rift) of tropical Africa (Fontaine & Bigot, 1993; Trzaska et al., 1994; Wotling et al, 1995). The dominant influence of the SST in the Atlantic dipole appears clearly on present day precipitation anomalies in tropical Africa at an annual scale for the period 1951-1990, particularly the two precipitation anomalies associated with opposite SST in the Atlantic dipole. The geographic pattern of these two present day precipitation anomalies (not shown; see the References cited above) have been compared to two schematic maps illustrating the opposite climatic trends which occured for two successive periods in the late Holocene discussed in this chapter:

- between ca. 3700 and 3000 BP (Fig. 8a)
- between ca. 2800 and 2000 BP (Fig. 8b)

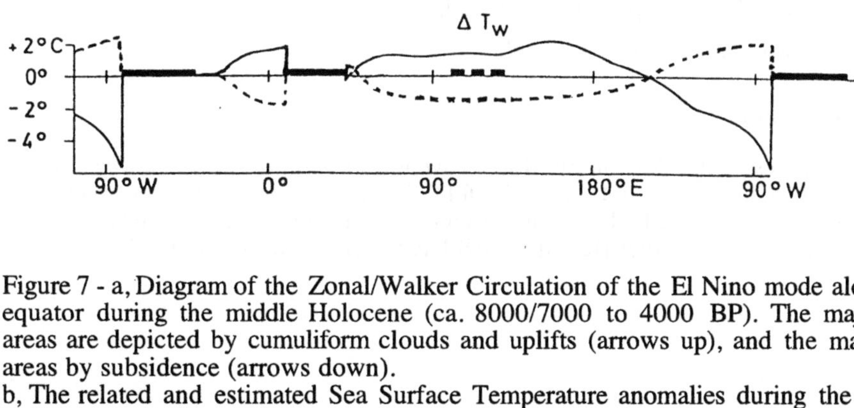

Figure 7 - a, Diagram of the Zonal/Walker Circulation of the El Nino mode along the equator during the middle Holocene (ca. 8000/7000 to 4000 BP). The major wet areas are depicted by cumuliform clouds and uplifts (arrows up), and the major dry areas by subsidence (arrows down).
b, The related and estimated Sea Surface Temperature anomalies during the middle Holocene (solid line).
- The estimated Sea Surface Temperature anomalies during the first period (3800/3700 to 3000 BP) of the late Holocene (dashed line), corresponding to a Zonal - Walker Circulation of the La Nina mode (not shown). Figure modified and adapted from a diagram of Flohn (1971, fig.7) for the present day time.

The pattern of these two sets of maps is very similar (Maley et al., 1996). This fact reinforces the interpretations presented above (§ 5.2.1 & 5.2.2), concerning the evolution of SST in the Gulf of Guinea and their impact on the palaeoclimates of tropical Africa. This fact shows also that the pattern and the mechanism of major climatic anomalies seems to be apparently similar at an annual or centuries to millenium scale.

6) Future of the African rain forest.

In the framework of the predicted global warming during the next century, climatological models show that a 4°C rise in the global air temperature would lead inland to a 30% increase in evaporation with only 12% more rain (Rind, 1995). The low and subtropical latitudes would be particularly affected by such phenomenons leading to more frequent dry events (Rind, 1995). The catastrophic retreat and destruction of the African rain forest which took place between ca. 3000 and 2000 BP, particularly after ca. 2800 BP, with an estimated large increase of convective rainfall (§ 5.2.1; 5.2.3), could be an analog of what could happen during the predicted global warming.

To assess the reliability of the reconstructed late Holocene events, more radiocarbon dates and also more precise data from other sites are indeed necessary. To improve the paleoclimatic interpretations presented above and also to modelize the Holocene paleoclimates, relatively precise maps of SST estimates at a millenia scale are urgently needed.

Acknowledgements

The researches in the African rain forests were done within the framework of the Programs GEOCIT (ORSTOM) and ECOFIT (ORSTOM & CNRS). S.Bigot and B.Fontaine kindly provided advice and present day data. This is Institut des Sciences de l'Evolution de Montpellier (ISEM/CNRS URA-327) contribution 95-070.

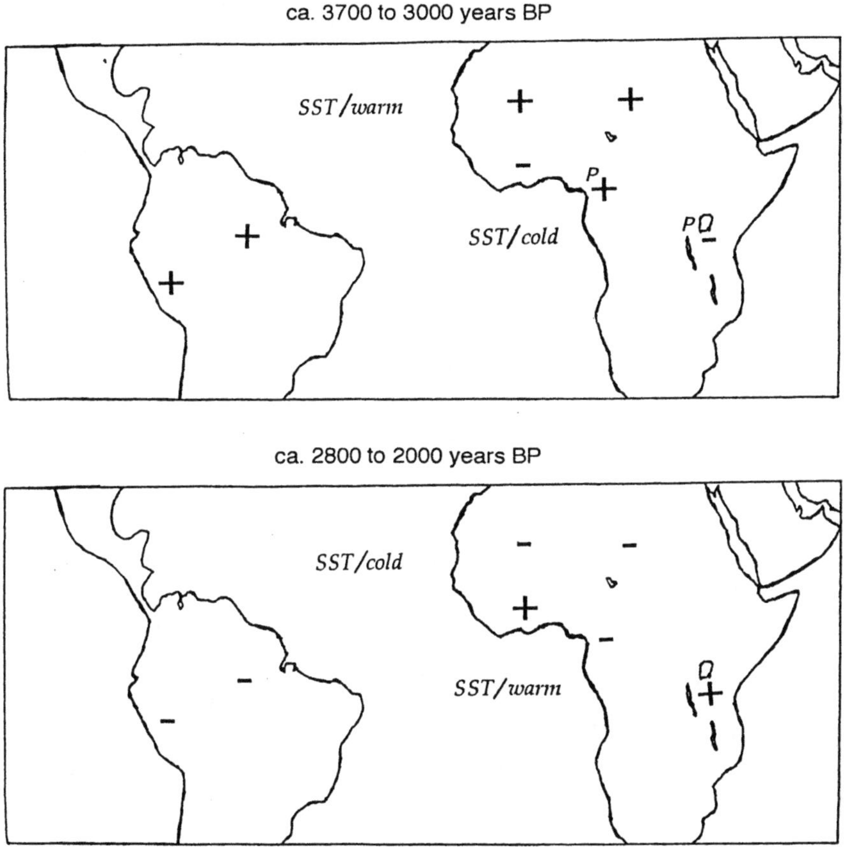

Figure 8 - Sketch map of the estimated trend of precipitation/evaporation (P/E) (negative or positive) for tropical Africa and South America. a - period 3700 to 3000 years BP; *P* indicates sectors where *Podocarpus* pollen grain maxima have been registered. b - period 2800 to 2000 years BP.

References.

Absy, M.L., Cleef, A.L., Fournier, M., Martin, L., Servant, M., Sifeddine, A., Da Silva, M.F., SoubiesS, F, Suguio, K., Turcq, B. & Van Der Hammen, Th. 1991 - Mise en évidence de quatre phases d'ouverture de la forêt dense dans le sud-est de l'Amazonie au cours des 60.000 dernières années. Comptes Rend. Acad. Sc., Paris, II, 312: 673-678.

Achoundong, G. 1985 - Etude écologique et floristique de la végétation des collines de Yaoundé au-dessus de 1000m. Thèse Ecologie Végétale, Univ. Yaoundé, 301 pp.

Bakun, A. 1978 - Guinea current upwelling. Nature, 271: 147-150.

Ballouche, A. & Neumann, K. 1995 - A new contribution to the Holocene vegetation history of the West African Sahel: pollen from Oursi, Burkina Faso and charchoal from three sites in northeast Nigeria. Veget. Hist. Archaeobota., 4: 31-39.

Barich, B.E. 1987 - Archeology and environment in the Libyan Sahara. The excavations in the Tadrart Acacus 1978-1983. British Archaeological Rep. Intern. Series, 368. Oxford.

Beadle, L.C. 1974 - The inland waters of Africa. An introduction to tropical limnology. Longman, London, 365 p.

Beltrando, G. & Camberlin, P. 1995 - Les sécheresses en Ethiopie et dans l'Est de l'Afrique. Sécheresse, Paris, 6: 87-94.

Brenac, P. 1988 - Evolution de la végétation et du climat dans l'Ouest Cameroun entre 25.000 et 11.000 ans BP. *Actes Xème Symposium Ass. Palynologues Langue Française*, Trav.Sect.Sci. & Tech.Inst. Français, Pondichéry, 25: 91-103.

Bryson, R.A. & Swain, A.M. 1981 - Holocene variations of monsoon rainfall in Rajasthan. Quat.Res., 16: 135-145.

Cadet, D. & Garnier, R. 1988 - L'oscillation australe et ses relations avec les anomalies climatiques globales. La Météorologie, (7)21: 4-18.

Caratini, C., Bentaleb, I., Fontugne, M., Morzadec-Kerfourne, M.T., Pascal, J.P. & Tissot, C. 1994 - A less humid climate since ca. 3500 yr B.P. from marine cores off Karwar, western India. Palaeogeogr., Palaeoclimatol., Palaeoecol., 109: 371-384.

Casanova, J. & Hillaire-Marcel, C. 1992 - Late Holocene hydrological history of Lake Tanganyika, East Africa, from isotopic data on fossil stromatolites. Palaeogeogr., Palaeoclimatol., Palaeoecol., 91: 35-48.

Ciais, P., Petit, J.R., Jouzel, J., Lorius, C., Barkov, N.I., Lipenkov, V. & Nicolaiev, V. 1992 - Evidence for an early Holocene climatic optimum in the Antarctic deep ice-core record. Climate Dynamics, 6: 169-177.

Cohen, A.L., Perkington, J.E., Brundrirt, G.B. & Van Der Merwe, N.J. 1992 - A Holocene marine climate record in Mollusc shells from the southwest African coast. Quat.Res., 38: 379-385.

Drochon, A. 1976 - Données climatologiques au sol et en altitude pour la station d'Abidjan. *Note 55 de l'ASECNA* (Agence Sécurité Navigation Aérienne).

Druyan, L.M. 1991 - The sensitivity of sub-Saharan precipitation to Atlantic SST. Climatic Change, 18: 17-36.

Duplessy, J.C., Labeyrie, L., Arnold, M., Paterne, M., Duprat, J. & Van Weering, T.C. 1992 - Changes in surface salinity of the North Atlantic Ocean during the last deglaciation. Nature, 358: 485-488.

Edmunds, W.M. & Wright, E.P. 1979 - Groundwater recharge and palaeoclimate in the Sirte and Kufra basins, Libya. J.Hydrology, 40: 215-241.

Elenga, H., Vincens, A. & Schwartz, D. 1991. Présence d'éléments forestiers montagnards sur les Plateaux Batéké (Congo) au Pléistocène supérieur: nouvelles données palynologiques. Palaeoecology of Africa, 22: 239-252.

Ergenzinger, P.J. 1978 - Das Gebiet des Enneri Misky im Tibesti Gebirge, Tchad. Erläuterungen zu einer geomorphologischen Karte. Berliner Geogr.Abhandl. 23, 57 p.

Fabre, J. & Petit-Maire, N. 1988 - Holocene climatic evolution at 22-23°N from two palaeolakes in the Taoudenni area (Northern Mali). Palaeogeogr., Palaeoclimatol., Palaeoecol., 65: 133-148.

Faure, H. 1966 - Evolution des grands lacs sahariens à l'Holocène. Quaternaria, 8: 167-176.

Faure, H. 1969 - Les lacs quaternaires du Sahara. Intern.Symp.Paleolimnology, 17: 131-146.

Flohn, H. 1971 - Tropical circulation pattern. Bonner Meteo Abhandl., 15, 55 p.

Flohn, H. 1983 - A climate feedback mechanism involving oceanic upwelling, atmospheric CO2 and water vapour. in *Variations in the global water budget.* A.Street-Perrott & al. eds., D.Reidel Pub., 403-418.

Flohn, H. 1986 - Singular events and catastrophes now and in climatic history. Naturwissenschaften, 73: 136-149.

Flohn, H. 1987a - Air-Sea interaction processes as models for abrupt climatic changes. in *Abrupt climatic change.* W.H.Berger & L.D.Labeyrie eds., D.Reidel Pub., 23-30.

Flohn, H. 1987b - East African rains of 1961/62 and the abrupt change of the White Nile discharge. Palaeoecology of Africa, 18: 3-18.

Folland, C.K., Palmer, T.N. & Parker, D.E. 1986 - Sahel rainfall and worldwide sea temperatures, 1901-85. Nature, 320: 602-607.

Fontaine, B. & Bigot, S. 1993 - West African rainfall deficits and sea surface temperatures. Intern.J.Climato ., 13: 271-285.

Gasse, F. 1988 - Diatoms, palaeoenvironments and palaeohydrology in the western Sahara and the Sahel. Würzb.Geogr.Arb ., 69: 233-254.

Gasse, F., Fontes, J.C., Plaziat, J.C., Carbonel, P., Kaczmarska, I., De Deccker, P., Soulié-Marsche, I., Callot, Y. & Dupeuble, P.A. 1987 - Biological remains, geochemistry and stable isotopes for the reconstruction of environmental and hydrological changes in the Holocene lakes from north Sahara. Palaeogeogr., Palaeoclimatol., Palaeoecol., 60: 1-46.

Gasse, F. & Fontes, J.C. 1989 - Palaeoenvironments and Palaeohydrology of a tropical closed lake (Lake Asal, Djibouti) since 10, 000 yr B.P. Palaeogeogr., Palaeoclimatol. Palaeoecol., 69: 67-102.

Gasse, F. & Van Campo, E. 1994 - Abrupt post-glacial climate events in West Asia and North Africa monsoon domains. Earth & Planet.Sc.Letters, 126: 435-456.

Giresse, P., Maley, J. & Kelts, K. 1991. Sedimentation and palaeoenvironment in crater Lake Barombi Mbo, Cameroon, during the last 25, 000 years. Sedimentary Geol., 71, 151-175.

Giresse, P., Maley, J. & BrenacC, P. 1994. Late Quaternary palaeoenvironments in the Lake Barombi Mbo (Cameroon) deduced from pollen and carbon isotopes of organic matter. Palaeogeogr., Palaeoclimatol., Palaeoecol., 107: 65-78.

Gordon, H.B. & Hunt, B.G. 1991 - Droughts, floods, and sea-surface temperature anomalies: a modelling approach. Intern.J.Climato., 11: 347-365.

Gunn, J. 1991 - Influence of various forcing variables on global energy balance during the period of intensive instrumental observation (1958-1987) and their implications for paleoclimate. Climatic Change, 19: 393-420.

Haynes, C.V., Eyles, C.H., Pavlish, L.A., Ritchie, J.C. & Rybak, M. 1989 - Holocene palaeoecology of the Eastern Sahara: Selima oasis. Quat.Sc.Rev., 8: 109-136.

Hillaire-Marcel, C., Riser, J., Rognon, P., Petit-Maire, N., Rosso, J.C. & Soulié-Marsche, I. 1983 - Radiocarbon chronology of Holocene hydrologic changes in Northeastern Mali. Quat. Res., 20: 145-164.

Hirst, A.C. & Hastenrath, S. 1983 - Atmosphere-Ocean mechanisms of climate anomalies in the Angola-tropical Atlantic sector. J.Phys.Oceano ., 13: 1146-1157.

Hisard, P. 1980 - Observation de réponse de type 'El Nino' dans l'Atlantique tropical oriental, Golfe de Guinée. Oceanologica Acta, 3: 69-68;

Janicot, S. 1991 - Variabilité des précipitations en Afrique de l'Ouest et circulations quasi stationnaires durant une phase de transition climatique. La Météorologie, 7, 36: 41-45.

Jolly, D. 1993 - Evolution et dynamique des ecosystèmes du Burundi. Pollen et statistique. *Thèse Sc*. Univ. Aix-Marseille, 137 p.

Kerfoot, O. 1968 - Mist precipitation on vegetation. Forestry Abstracts, Oxford, 29: 8-20.

Kershaw, A.P. 1970 - A pollen diagram from Lake Euramoo, north-east Queensland, Australia. New Phytol., 69: 785-805.

Letouzey, R. 1985. Notice de la carte phytogéographique du Cameroun au 1/500.000. Inst. *Carte Intern. Végétation*, Toulouse & *Inst. Rech. Agron.*, Yaoundé.

McGlone, M.S., Kershaw, A.P. & Markgraf, V. 1992 - El Nino/Southern Oscillation climatic variability in Australasian and South American paleoenvironmental records. in El Nino, Historical and paleoclimatic aspects of the Southern Oscillation, H.F.Diaz & V.Markgraf eds., Cambridge Univ. Press, 435-462.

Mahé, G. & Citeau, J. 1993 - Interactions between the ocean, Atmosphere and continent in Africa, related to the Atlantic monsoon flow. General pattern and the 1984 case study. Veille Climatique Satellitaire, 44: 34-54.

Maley, J. 1977 - Palaeoclimates of Central Sahara during the early Holocene. Nature, 269, 573-577.

Maley, J. 1981. Etudes palynologiques dans le bassin du Tchad et paléoclimatologie de l'Afrique nord-tropicale de 30.000 ans à l'époque actuelle. Travaux & Documents ORSTOM, 129, 586 p.

Maley, J. 1982. Dust, clouds, rain types and climatic variations in tropical North Africa. Quat. Res., 18: 1-16.

Maley, J. 1987. Fragmentation de la forêt dense humide africaine et extension des biotopes montagnards au Quaternaire récent: nouvelles données polliniques et chronologiques. Implications paléoclimatiques et biogéographiques. Palaeoecology of Africa, 18: 307-334.

Maley, J. 1989. Late Quaternary climatic changes in the African rain forest: the question of forest refuges and the major role of sea surface temperature variations. In *Paleoclimatology and Paleometeorology: modern and past patterns of global atmospheric transport*. Leinen, M. & Sarnthein, M. eds. NATO Adv.Sc.Inst., Ser.C, Math. & Phys.Sc., 282, 585-616, Kluwer Acad. Publ., Dordrecht.

Maley, J. 1991. The African rain forest vegetation and palaeoenvironments during late Quaternary. Climatic Change, 19: 79-98.

Maley, J. 1992. Mise en évidence d'une péjoration climatique entre ca. 2500 et 2000 ans BP en Afrique tropicale humide. Bull.Soc.Géol.France, 163: 363-365.

Maley, J. 1996 - The African rain forest: main characteristics of changes in vegetation and climate from the upper Cretaceous to the Quaternary. in *Essays on the Ecology of the Guinea-Congo rain forest*. I.J. Alexander, M.D. Swaine & R. Watling eds, The R. Soc. Edinburg, Biol.Sc., 104B, 31-73.

Maley, J., Livingstone, D.A., Giresse, P., Thouveny, N., Brenac, P., Kelts, K., Kling, G., Stager, C., Haag, M., Fournier, M., Bandet, Y. & Zogning, A. 1990a. Lithostratigraphy, Volcanism, Paleomagnetism and Palynology of Quaternary lacustrine deposits from Barombi Mbo (West Cameroon): preliminary results. J.Volcan. & Geoth.Res., 42: 319-335.

Maley, J., Caballé, G. & Sita, P. 1990b. Etude d'un peuplement résiduel à basse altitude de *Podocarpus latifolius* sur le flanc Congolais du Massif du Chaillu. Implications paléoclimatiques et biogéographiques. Etude de la pluie pollinique actuelle. In *Paysages Quaternaires de l'Afrique centrale Atlantique*, Lanfranchi, R. & Schwartz, D. eds., 336-352, Didactiques, ORSTOM, Paris.

Maley, J. & Elenga, H. 1993. Le rôle des nuages dans l'évolution des paléoenvironnements montagnards de l'Afrique tropicale. Veille Climatique Satellitaire, Lannion, 46: 51-63.

Maley, J., Brenac, P. Bigot, S. & Moron, V. 1996 - Variations de la végétation et des paléoenvironnements en forêt dense africaine au cours de l'Holocène. Impact de la variation des températures marines de surface. in *Dynamique à long terme des Ecosystèmes forestiers intertropicaux*. Symposium ECOFIT, Mars 1996, Bondy: 321-324.

Maley, J. & Brenac, P. 1996 - Vegetation and climatic changes in the forests of West Cameroon during the last 25, 000 years. *9th Intern.Palynological Congress*, Houston, Abstract: 94-95.

Martin, L., Fournier, M., Mourguiart, P., Sifeddine, A., Turcq, B., Absy, M.L. & Flexor, J.M. 1993. Southern Oscillation signal in South American palaeoclimatic data of the last 7000 years. Quat. Res., 39: 338-346.

Mawson, R. & Williams, M.A.J. 1984 - A wetter climate in eastern Sudan 2, 000 years ago. Nature, 309: 49-51.

Merle, J. 1980 - Variabilité thermique annuelle et interannuelle de l'Océan Atlantique équatorial est. L'hypothèse d'un 'El Nino' Atlantique. Oceanologica Acta, 3: 209-220.

Mix, A.C., Ruddiman, W.F. McIntyre, A. 1986 - Late Quaternary paleoceanography of the tropical Atlantic, 2: The seasonal cycle of Sea Surface Temperatures, 0-20, 000 years B.P. Paleoceanography, 1: 339-353.

Moll, E.J. 1972 - The current status of the mistbelt mixed *Podocarpus* forest in Natal. Bothalia, 10: 595-598.

Morley, J.J. 1979 - A transfer function for estimating paleoceanographic conditions based on deep-sea surface sediment distribution of Radiolarian assemblages in the South Atlantic. Quat.Res., 12: 381-395.

Morley, J.J. & Dworetzky, B.A. 1993 - Holocene temperature patterns in the South Atlantic, southern, and Pacific Oceans. in *Global climates since the Last Glacial Maximum*. Wright, H.E. & al. eds., Univ. Minnesota Press, 125-135.

Neumann, K. 1993 - Holocene vegetation of the Eastern Sahara: charcoal from prehistoric sites. in *Environmental change and human culture in the Nile basin and northern Africa until the second Millennium B.C.* Krzyzaniak L., Kobusiewicz M. & Alexander J. eds., Poznan Archaeological Museum, 153-169.

Newell, R.E. & Hsiung, J. 1987 - Factors controlling free air and ocean temperature of the last 30 years and extrapolation to the past. in *Abrupt climatic change*. W.H. Berger & L.D.Labeyrie eds., D.Reidel Pub., 67-87.

Nguetsop, F. & Servant-Vildary, S. 1996 - Variations hydrologiques du lac Ossa (Ouest Cameroun) au cours des 5000 dernières années à partir de l'étude des diatomées. in *Dynamique à long terme des Ecosystèmes forestiers intertropicaux*. Symposium ECOFIT, Mars 1996, Bondy: 243-244.

Nicholson, S.E. & Entekhabi, I.D. 1987 - Rainfall variability in equatorial and southern Africa: relationships with sea surface temperatures along the southwestern coast of Africa. J. Climate & Appl.Meteo., 26: 561-578.

Nicholson, S.E. 1980 - The nature of rainfall fluctuations in subtropical West Africa. Monthly Weather Rev., 108: 473-487.

Oslisly, R. 1995 - The middle Ogooué valley: cultural changes and palaeoclimatic implications of the last four millenia. Azania, 24/25: 324-331.

Owen, R.B., Barthelme, J.W., Renaut, R.W. & Vincens, A. 1982 - Palaeolimnology and archaeology of Holocene deposits north-east of Lake Turkana, Kenya. Nature, 298: 523-529.

Pachur, H.J. 1975 - Zur spätpleistozänen und holozänen Formung auf der Nordabdachung des Tibestigebirges. Die Erde, 106: 21-46.

Pachur, H.J., Roper, H.P., Kröpelin, S. & Goschin, M. 1987 - Late Quaternary hydrography of the eastern Sahara. Berliner Geowiss.Abhandl., 75: 331-384.

Pachur, H.J. & Kröpelin, S. 1989 - L'aridification du Sahara oriental à l'Holocène moyen et supérieur. Bull.Soc.Géol.France (8)5: 99-107.

Pachur, H.J. & Hoelzmann, P. 1991 - Paleoclimatic implications of Late Quaternary lacustrine sediments in Western Nubia, Sudan. Quat.Res ., 36: 257-276.

Palmer, T.N. 1986 - Influence of the Atlantic, Pacific and Indian Oceans on Sahel rainfall. Nature, 322: 251-253.

Pazdur, M.F. 1993 - Evaluation of radiocarbon dates of organic samples from Uan Muhuggiag and Ti-n-Torha, Southwestern Libya. in *Environmental change and human culture in the Nile basin and northern Africa until the second Millennium B.C.* Krzyzaniak L., Kobusiewicz M. & Alexander J. eds., Poznan Archaeological Museum, 43-47.

Petit-Maire, N. & Riser, J. 1987 - Holocene palaeohydrography of the Niger (river). Palaeoecology of Africa, 18: 135-141.

Perrott, R.A. 1982 - A high altitude pollen diagram from Mount Kenya: its implications for the history of glaciation. Palaeoecology of Africa, 14: 77-83.

Philander, G. 1989 - El Niño and La Niña. American Scientist, 77: 451-459.

Prell, W.L., Gardner, J.V., Be, A.W. & Hays, J.D. 1976 - Equatorial Atlantic and Caribbean foraminiferal assemblages, temperatures and circulation: interglacial and glacial comparisons. Geological Soc.America Memoir, 145: 247-266.

Reynaud-Farrera, I., Maley, J. & Wirrmann, D. 1996. Végétation et climat dans les forêts du Sud-Ouest Cameroun depuis 4770 ans B P: analyse pollinique des sédiments du lac Ossa. C.R.Acad.Sc., Paris, sér.2a, 322: 749-755.

Richards, K. 1986. Preliminary results of pollen analysis of a 6, 000 year core from Mboandong, a crater lake in Cameroun. Hull Univ.Geography Dept. Misc. Ser., 32: 14-28.

Rind, D., Peteet, D., Broecker, W., McIntyre, A. & Ruddiman, W. 1986 - The impact of North Atlantic sea surface temperatures on climate: implications for the Younger Dryas cooling (11-10 ky). Climate Dynamics, 1: 3-33.

Rind, D. 1995 - Drying out the Tropics. New Scientist, 5: 36-40.

Roset, J.P. 1987 - Néolithisation, Néolithique et post-Néolithique au Niger nord-oriental. Bull.Ass.Fr.Et.Quaternaire, 1987: 203-214.

Schild, R. 1987 - Unchanging contrast ? The late Pleistocene Nile and Eastern Sahara. In *Prehistory of arid North Africa. Essays in honor of Fred Wendorf*. Close A.E. ed., 13-27. Southern Methodist Univ. Press, Dallas.

Schwartz, D., Guillet, B. & Dechamps, R. 1991 - Etude de deux flores forestières mi-Holocène (6000-3000 BP) et subactuelle (500 BP) conservées in situ sur le littoral Pontenegrin (Congo). In *Paysages Quaternaires de l'Afrique centrale Atlantique*, Lanfranchi, R. & Schwartz, D. eds., 283-297, Didactiques, ORSTOM, Paris.

Schwartz, D. 1992 - Assèchement climatique vers 3000 B.P. et expansion Bantu en Afrique centrale atlantique: quelques réflexions. Bull.Soc.géol.France, 163: 353-361.

Servain, J., Picaut, J. & Busalacchi, A.J. 1985 - Interannual and seasonal variability of the tropical Atlantic Ocean depicted by sixteen years of sea-surface temperature and wind stress. in *Coupled Ocean-Atmosphere models*. Elsevier Publ., 211-237.

Servant, M. 1973 - Séquences continentales et variations climatiques: Evolution du bassin du Tchad au Cénozoïque supérieur. *Thèse Sc*. Univ. Paris. Trav. & Docu.ORSTOM, 159 (1983): 573 p.

Servant, M. & Servant, S. 1983 - The lacustrine environment and its evolution. Paleolimnology of an upper Quaternary endorheic lake in Chad basin. in *Lake Chad.*, Carmouze, J.P. & al. eds., W.Junk Publ., The Hague, 11-26.

Servant, M., Maley, J., Turcq, B., Absy, M.L., Brenac, P., Fournier, M. & Ledru, M.P. 1993. Tropical forest changes during the late Quaternary in African and South American lowlands. Global Planet.Change, 7: 25-40.

Shannon, L.V., Boyd, A.J., Brundrit, G.B. & Taunton-Clark, J. 1986 - On the existence of an El Nino-type phenomenon in the Benguela system. J.Marine Res ., 44: 495-520.

Shulmeister, J. & Lees, B.G. 1995 - Pollen evidence from tropical Australia for the onset of an ENSO-dominated climate at c. 4000 BP. The Holocene, 5: 10-18.

Sifeddine, A., Fröhlich, F., Fournier, M., Martin, L., Servant, M., Soubiès, F., Turcq, B., Suguio, K. & Volkmer-Ribeiro, C. 1994. La sédimentation lacustre indicateur de changements des paléoenvironnements au cours des 30.000 dernières années (Carajas, Amazonie, Brésil). C.R.Acad.Sc., 2, 318: 1645-1652.

Suchel, J.B. 1988 - Les anomalies climatiques des années 1983 et 1984 au Cameroun. Publ.Ass.Intern.Climatologie, Univ. Aix en Provence, 1: 131-142.

Talbot, M.R. & Delibrias, G. 1977 - Holocene variations in the level of Lake Bosumtwi, Ghana. Nature, 268: 722-724.

Talbot, M.R., Livingstone, D.A., Palmer, P.G., Maley, J., Melack, J.M., Delibrias, G. & Gulliksen, S. 1984, Preliminary results from sediment cores from Lake Bosumtwi, Ghana. Palaeoecology of Africa, 16: 173-192.

Tiercelin, J.J., Vincens, A. & al. 1987 - Le demi-graben de Baringo-Bogoria, Rift Gregory, Kenya. 30.000 ans d'hystoire hydrologique et sédimentaire. Bull. Centres Rech. Explor. Prod. Elf-Aquitaine, 11: 249-540.

Trzaska S., Roucou P., Bigot S., et Fontaine B., 1994. Variabilité du cycle de l'eau en août en Afrique de l'Ouest en réponse aux modes propres de variabilité des SST dans le modèle ARPEGE-Climat. in *Les ateliers de modélisation de l'atmosphère*, Météo-France, Toulouse, 337-344.

Van der Hammen, Th. 1991 - Palaeoecological background: Neotropics. Climatic Change, 19: 37-47.

Vincens, A., Elenga, H., Schwartz, D., De Namur, C., Bertaux, J., Fournier, M. & Dechamps, R. 1996 - Histoire des écosystèmes forestiers du sud-Congo depuis 6000 ans. in *Dynamique à long terme des Ecosystèmes forestiers intertropicaux*. Symposium ECOFIT, Mars 1996, Bondy: 291-294.

Wauthy, B. 1983 - Introduction à la climatologie du Golfe de Guinée. Océanographie tropicale, ORSTOM, 19: 103-138.

Whetton, P., Adamson, D. & Williams, M. 1990 - Rainfall and river flow variability in Africa, Australia and East Asia linked to El Nino - Southern Oscillation events. Geol.Soc.Australia Symp.Proc. 1: 71-82.

Whyte, S.A. 1975 - Distribution, trophic relationships and breeding habits of the fish populations in a tropical lake basin, Lake Bosumtwi, Ghana. J.Zool., London, 177: 25-56.

Wotling G., Mahé G., Lebarbe L., et L'Hote Y., 1995. Analyse par les vecteurs régionaux de la variabilité spatio-temporelle des précipitations annuelles liées à la mousson africaine.Veille Climatique Satelllitaire, 52: 58-73.

Late Holocene Vegetational Changes in South-West Cameroon

Isabelle REYNAUD-FARRERA

Laboratoire Paléoenvironnements et Palynologie
ISEM - URA CNRS 327
Université Montpellier II
Place E. Bataillon
34095 Montpellier cedex 5
France

Abstract

The detailed palynological analysis of a core recovered in Lake Ossa, localized in the wet rain forests of Southwest Cameroon, provides new information on the vegetation and climate history of this region during the last 4800 years B.P.. Pollen counts, showing a rich and well-diversified microflora, indicate the persistance of forest environment during all this period. Between around 4800 to 2800 yr BP, the Biafrean and mountain forests were well represented in pollen spectra. Then, at around 2800 yr B.P. (estimated age), a major change in landscape occured indicated by the substitution of the primary forest by a colonizing one rich in pioneer species (*Alchornea, Macaranga*). Later there was appearance and extension of the oil palm (*Elaeis guineensis*) which culminated around 1600 yr BP.

Introduction

Over the last 10 years, the understanding of the evolution of west African vegetation during the late Holocene has improved tremendously. However, compared to investigations carried out in east Africa, data of the vegetation history of west-central Africa are still very poor. Therefore, a field campaign, founded by the French program ECOFIT focused on the evolution of humid African and American tropical forests during the last 10 000 years BP, was conducted in south-west Cameroon. Several cores were drilled in Lake Ossa (3°45'-3°52'N - 9°58'-10°4'E). Results of the pollen analyses of sequence OW4 are presented in this paper. They document, for the first time, the uninterrupted botanical and climatic history of this region from 4770 yr BP (in uncalibrated radiocarbon years) until today.

NATO ASI Series, Vol. I 49
Third Millennium BC Climate Change
and Old World Collapse
Edited by H. Nüzhet Dalfes, George Kukla,
and Harvey Weiss

Modern environmental setting

Lake Ossa (3°45'-3°52'N - 9°58'-10°4'E) (fig1) is located in south-west Cameroon, at 8m above the sea level. With a mean depth of 3m and a surface of around 4500 hectares, this lake is mainly feeded by temporary streams (Wirrmann, 1992). It is situated within the Atlantic coastal rainforest characterized by two typical arboreal species, *Lophira alata* and *Sacoglottis gabonensis* (Letouzey, 1968, 1985). Today, the human impact is very strong in this area. Over the last century, a large plantation of *Hevea brasiliensis* and *Elaeis guineensis* (Oil Palm) was established on the western side of the lake.

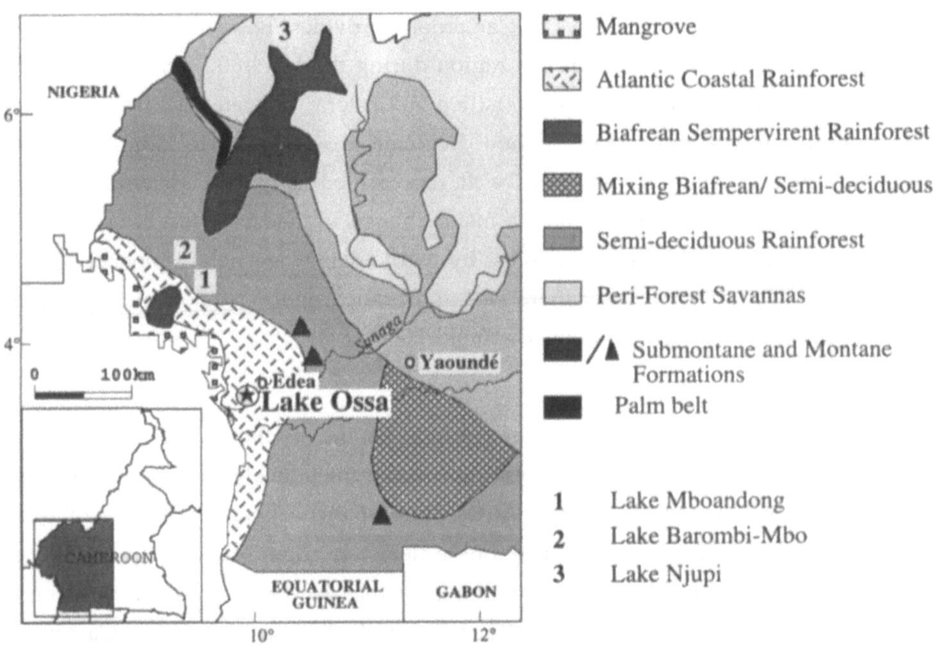

Fig. 1. Location of Lake Ossa, South-West Cameroon. Major vegetation types of the study area (from Letouzey, 1968, 1985, simplified)

The modern climatic conditions prevailing in this region are a variant of Guinean equatorial type. They are characterized by mean annual rainfall of 3000 mm and by a long rainfall season from March to November interrupted by a short lowering in June, and a 3 months dry season from December to February.

Material and methods

The studied core, OW4, consists of 5.56m of sediment, drilled in the western part of the lake by vibrocoring (Martin and Flexor, 1987). The sediment core is quite homogeneous and correspond to, from the bottom to the top, a compact silty clay, without laminations. Seven radiocarbon dates made on total organic matter content were performed on this core (Beta Analytic Inc., Miami, USA (Beta); Utrecht University, Netherlands (Utc)). The dates obtained, corrected with reference to $\delta^{13}C$, are presented in table 1. These seem consistent with the stratigraphic position of the samples.

Tab. 1. Radiocarbon ages from core OW4, lake Ossa

Dated Material	Depth (cm)	Laboratory Number	Conventional ^{14}C age (BP)
Organic matter	6,8 - 9	BETA73082	90±60
Organic matter	122,9 - 125,1	BETA73083	1890±60
Organic matter	174	Utc-3911	2442±43
Organic matter	243 - 245,2	BETA73084	3330±60
Organic matter	358,8 - 361	BETA73085	3880±60
Organic matter	523,6 - 525,8	BETA73086	4580±60
Organic matter	548,3- 550,5	BETA73087	4770±60

Twenty six distinct levels have been sampled and analysed, the mean time resolution between two samples being of around 200 years. The sediments were processed using the standard method described by Faegri and Iversen (1975). In all the analysed levels, the microflora was rich and diverse. A total of 190 pollen taxa have been identified and classified in relation to the physiognomy (AP: Arboreal Pollen; NAP: Non Arboreal Pollen).

Palynological results

The synthetic pollen diagram (figure 2) includes only the main abundant taxa and therefore which are considered as the most important for the palaeoenvironmental interpretation. According to the regularity of the sedimentation, the age of the different levels is estimated by interpolation between the radiocarbon datings. As they were present in considerable numbers, pollen of Cyperaceae and ferns spores were excluded from the pollen sums before the relative percentages of each taxa have been calculated.

The curve of AP/NAP, calculated for all samples, show a dominance of arboreal taxa from 4770 until today.

Two main pollen zones have been recognized, each of them being subdivided in two sub-zones.

- Zone II (ca.4770 yr to ca. 2730 yr BP): During this period, the most abundant arboreal taxa are Caesalpiniaceae (10-18%), Sapotaceae (5-7%) and *Martretia* (10%), typical from the Biafrean evergreen rainforest. Afromontane elements, with essentially *Podocarpus latifolius* and *Olea capensis* (on average 4% and 2,5%), and Atlantic coastal rainforest components, such as *Lophira alata* and *Sacoglottis gabonensis*, are also well represented in this section.

* sub-zone IIb (ca. 4770 to ca. 4150 BP): In addition to pollen taxa mentioned previously, pioneer taxa such as *Alchornea* (5-12%), indicator of disturbed environments, and *Holoptelea, Piptadeniastrum* and *Triplochiton,* indicators of semi-deciduous forest, are present. During the same time, Caesalpiniaceae develops (7 to 15%). For herbaceous taxa, the only aquatic component, Cyperaceae (15 to 40%) increases.

* sub-zone IIa (ca. 4150 yr to 2730 yr BP): Caesalpiniaceae, Sapotaceae and *Martretia* pollen grains reach their maximum while the lowest percentages of *Alchornea* are recorded. *Holoptelea* and *Piptadeniastrum* strongly decrease as well as the herbaceous taxa Amaranthceae/ Chenopodiaceae.

-Zone I (ca. 2730 yr BP to Present): Pollen zone I marks the significant change of the vegetation.

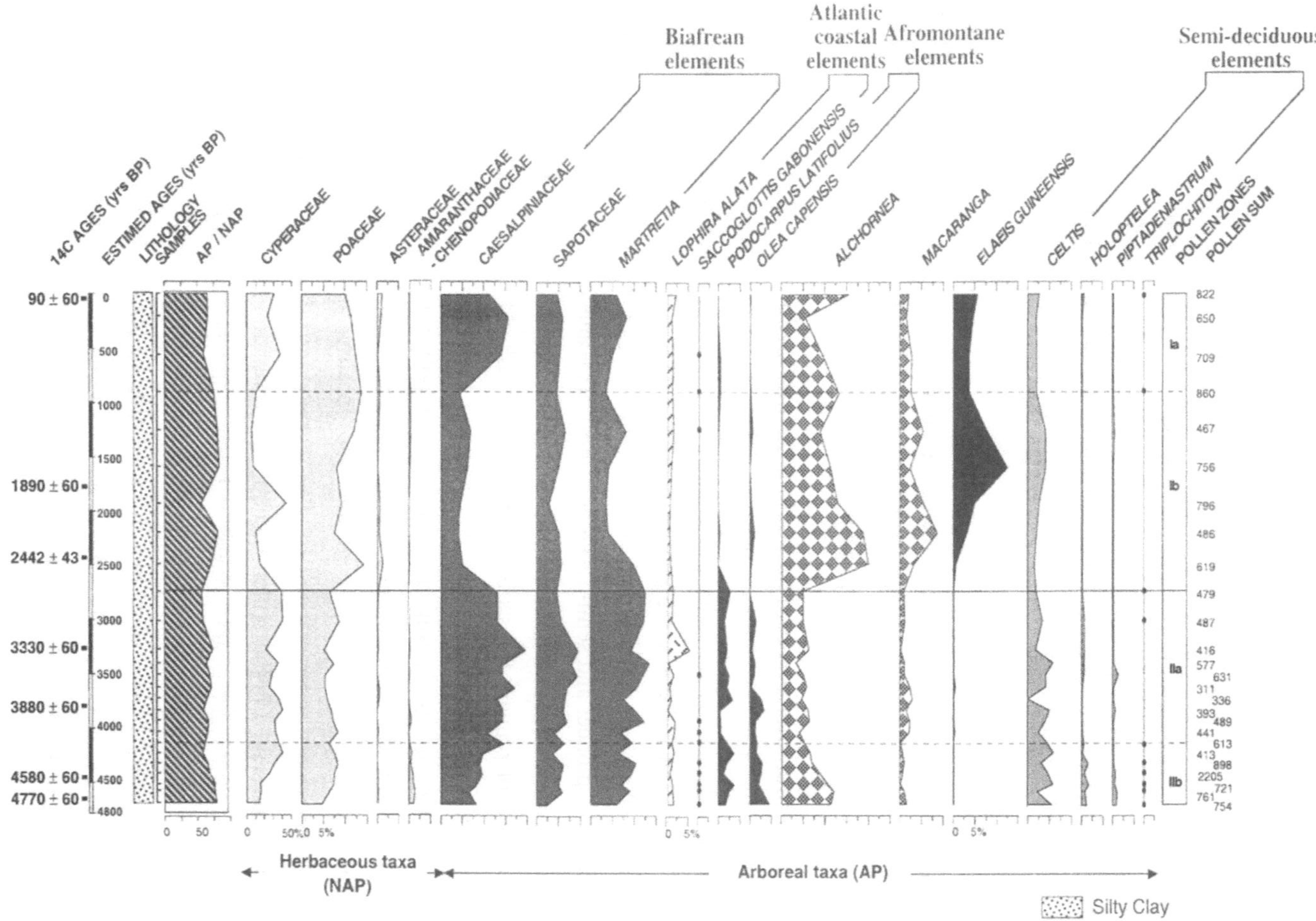

Fig. 2. Synthetic pollen diagram from core OW4, lake Ossa (South-west Cameroon)

* sub-zone Ib (ca.2730 to 700 BP): Afromontane and Biafrean arboreal taxa decrease considerably. *Podocarpus latifolius* and *Olea capensis* become very scarce (<1%). At the same time, a strong increase of pioneer taxa, mainly *Alchornea* and *Macaranga*, with a maximum between 2500 and 2000 yr B.P., is recorded. Around 2000 yr B.P., *Elaeis guineensis* (Oil palm), whose few pollen grains was present in some samples of sub-zone I, greatly increases. Its percentages reach a maximum value around 1600 yr BP. Among the herbaceous taxa, percentages of Poaceae increase at the bottom of the sub-zone, and then decrease. However, they are higher those from the zone II.

-sub-zone Ia (ca.700 yr BP to Present): In this sub-zone, the percentages of typical tree taxa from Biafrean evergreen (perennial) rainforest, especially Caesalpiniaceae, and herbaceous aquatic taxa increase. *Lophira alata* and *Sacoglottis gabonensis* are well represented again. *Podocarpus latifolius* and *Olea capensis* disappear.

Environmental changes and paleoclimatic implications

Pollen analysis of the deposits recovered from Lake Ossa clearly shows several environmental changes in this locality since 4770 yr BP. However, during this period, the persistance of forest environment is recorded.

In spite of a phase of minor degradation before ca.4150 yr BP shown by the occurrence of pioneer and semi-deciduous forest taxa, and the lowest percentages of Caesalpiniaceae, the microflora indicates a relative stable environment with a great development of primary forests between ca. 4770 BP and ca.2730 yr BP. The high representation of Biafrean elements, particulary of Caesalpiniaceae which have a low pollen production as demonstrated in modern pollen rain studies in Cameroon (Brenac, 1988; Reynaud-Farrera, 1995), suggests that the Biafrean evergreen rainforest was more developed around the locality than now. Moreover, the Afromontane forest, characterized by *Podocarpus latifolius* and *Olea capensis,* is also well represented in the pollen spectra. This formation is favoured by the regular presence of stratiform cloud covers and fog, corresponding to the cloud forest (Kerfoot, 1968; Maley and Elenga, 1993; Reynaud-Farrera, 1995). Presently, small spots of montane forests, spreading at an altitude of 900-1200 meters, are present 100km eastward from Lake Ossa. However no pollen of *Podocarpus* and *Olea* are identified in the modern pollen spectra in the surrounding of the lake (Reynaud and

Maley, 1994; Reynaud-Farrera, 1995). Then, as for the Biafrean forest, this vegetation may have been more developed than today. Nevertheless, by comparison with dispersal pollen curves of *Podocarpus* and *Olea* (Vincens, 1982), it is assumed that this vegetation did not reach Lake Ossa.

Contemporanously to the area of Ossa, a large extension of the humid rain forests is recorded during this period in pollen sequences in West Central Africa. This development was recently described from Lake Barombi-Mbo, Cameroon (Maley, 1995) where Caesalpiniaceae and Podocarpus reached a maximum between ca.4500 and ca.3000 BP. In Congo, the forest seemed to be denser than today in Batéké region (Elenga et al., 1991), and the coastal plain, presently occupied by savannas, was covered with a temporary flooded rainforest (Dechamps et al., 1988; Schwartz et al., 1990; Elenga et al, 1992). A similar event was also recognized in East Africa between ca.9000 yr and ca. 4000 yr BP (Kendall, 1969; Bonnefille et al., 1991; Jolly and Bonnefille, 1991; Vincens, 1989, 1993; Sowumni, 1991; Ssemmanda and Vincens, 1993). This period is now well documented. The great development of the humid rain forests recorded in tropical Africa must be related to humid climatic conditions with an increase in rainfall as high levels of lakes attest (Butzer et al., 1972; Street and Grove, 1979; Gasse et al., 1980; Servant and Servant-Vildary, 1980; Talbot et al., 1984).

From 2730 yr BP onwards, an abrupt change in the landscape of Lake Ossa occurred. All the forest types were affected synchronously. A great perturbation is recorded with the extension of *Alchornea* and *Macaranga* pioneer taxa, an increase of Poaceae and a notable decrease of pollen taxa belonging to the Biafrean and montane forests. Thus, these data suggest a substitution of the primary forest by a colonizing one, rich in pioneer species, at low and high altitudes. This change is related to the introduction of the arid phase previously recognized in the Congo and the Cameroon at around ca. 3000 BP (Elenga and Vincens, 1990; Elenga et al., 1992, 1994; Maley, 1992; Schwartz, 1992; Giresse et al., 1994).This climatic event was also detected in East Africa between ca. 4000 BP and ca. 2500 BP (Vincens, 1986, 1989; Hamilton, 1987; Bonnefille and Riollet, 1988; Roche et al., 1988; Ssemmanda and Vincens, 1993; Jolly et al., 1994). This dry phase is responsible for the extension of disturbed and/or open types of vegetation. It led to local fragmentation of the African forest, as has been observed in some forested areas in Congo (De Foresta, 1990;

Schwartz et al., 1990), and to a complete drying up of some lakes such as Lake Sinnda in the Niary valley (Congo) (Vincens et al., 1994).

A great development of the oil palm coincides with this dry phase. Since that time, the number of oil palm (*Elaeis guineensis*) pollen is rising, up to a peak around 1600 BP (13%) and then decreased until recently. An extension of *Elaeis guineensis* from ca. 2800 yr BP onwards has been already reported from other sites in West Central Africa and interpretated as an indicator of human activities (Elenga et al., 1992; Schwartz, 1992). However, reliable data from Western and South-Western Cameroon lead us consider this interpretation with serious cautions. At present time, the percentages of oil palm from surface samples of Lake Ossa reach only 7% (Reynaud et Maley, 1994) in spite of the presence of a large plantation of *Elaeis guineensis* on the western side of the lake, and of prevailing westerly winds. Between ca.2500 yr and ca.1500 yr BP, high percentages of *Elaeis guineensis* have been recorded from lakes Barombi-Mbo (Maley, same volume), Mboandong (Richards, 1986) and Njupi (Zogning et al., in press) (Western Cameroon) similar to those obtained in the fossil pollen spectra of Lake Ossa. Moreover, Letouzey (1978) described a forest belt "Palm belt" in the northern part of West Cameroon forest. This Palm belt, 150 km long and approximately 15 km wide, expands at elevations of 500 - 800 m. Letouzey considers this vegetal formation as natural because of criteria such as the absences of trees classically linked to humans (Letouzey, 1978, 1985). *Elaeis guineensis*, being an heliophilous species, develops naturally in gaps. It is assumed that, favoured by the abrupt climatic change at ca.2730 BP, the expansion of this species could be related to a pioneer behaviour. Later, with the migration of the Bantu populations within the rainforest (Schwartz, 1992), its cultivation was possible.

Since ca.700 BP, there has been a return of the Biafrean and Atlantic coastal rainforests to the south Cameroon which characterizes more favourable conditions, probably similar to the current ones.

Conclusion

The pollen sequence OW4 from Lake Ossa, south-west Cameroon, provides new information about the environmental history of west central Africa during the period ca. 4770 yr BP - Present. These data clearly show that the vegetation of this region was strongly affected by climatic changes during this time interval, as well as other regions of the intertropical Africa. The main changes occurred at around 2730 yr BP.

The Biafrean evergreen rainforest located at low and mid elevations was strongly disturbed, the montane forests disappeared while pioneer taxa spread widely. This seems to be due to arid conditions, which were previously recognized in other studies.

Acknowledgements

The research is part of ECOFIT, a programme funded by CNRS and ORSTOM. The author is undebted to M. Servant, chairman of Ecofit; J. Maley for help during preparation of the manuscript; D.D. Rousseau for comments on earlier versions; S. Servant and F. Nguestop for discussions and encouragement; D. Wirrmann for providing samples from core OW4; D. Duzer for technical assistance. The author is grateful to Dr George Kukla for giving the opportunity to present this work here and to Prof. Rhodes Fairbridge who helped with the editing. This is contribution 96-115 of Institut des Sciences de l'Evolution (UMR CNRS 5554) de Montpellier.

References

Bonnefille R and Riollet G (1988) The Kashiru pollen sequence (Burundi). Paleoclimatic implications for the last 40 000 yr BP in tropical Africa. *Quaternary Research* 30: 19-35.

Bonnefille R, Riollet G and Buchet G (1991) Nouvelle séquence pollinique d'une tourbière de la crête Zaïre-Nil (Burundi). *Rev Paleobot Palynol* 67: 315 - 330.

Brenac P (1988) Evolution de la végétation et du climat dans l'ouest Cameroun entre 25 000 et 11 000 ans BP. Act 10th Symp APLF Bordeaux, *Trav Sec Sci Tech Inst Fr Pondichery* 25: 91 - 103.

Butzer KW, Isaac GL, Richardson JL and Washbourn-Kamau CK (1972) Radiocarbon Dating of East African Lake Levels. *Science* 175: 1069 - 1076.

Dechamps R, Guillet B and Schwartz D (1988) Découverte d'une flore forestière mi-Holocène (5800-3100 BP) conservée in situ sur le littoral ponténégrin (RP du Congo). *CR Acad Sci Paris* 306 II: 615 - 618.

De Foresta H (1990) Origine et évolution des savanes intramayombiennes (RP du Congo) II. Apports de la botanique forestière. In: R Lanfranchi and D Schwartz (Editors), *Paysages Quaternaires de l'Afrique Centrale Atlantique*. ORSTOM Paris: 236 - 355.

Elenga H, Schwartz D and Vincens A (1994) Pollen evidence of late quaternary vegetation and inferred climate changes in Congo. *Palaeogeog. Palaeoclimatol. Palaeoecol.* 107: 345 - 356.

Elenga H and Vincens A (1990) Paléoenvironnements quaternaires récents des Plateaux Batéké (Congo): étude palynologique des dépôts de la dépression de Bilanko. *In:* R Lanfranchi and D Schwartz (Editors), *Paysages Quaternaires de l'Afrique Centrale Atlantique*. ORSTOM Paris: 271 - 282.

Elenga H, Vincens A and Schwartz D (1991) Présence d'éléments forestiers montagnards sur les Plateaux Batéké (Congo) au Pléistocène supérieur: nouvelles données palynologiques. *Palaeoecology of Africa* 22: 239 - 252.

Elenga H, Schwartz D and Vincens A (1992) Changements climatiques et action anthropique sur le littoral congolais au cours de l'Holocène. *Bull. Soc. Géol. France* 163 1: 83 - 90.

Faegri K and Iversen J (1975) Textbook of Pollen Analysis. Blackwell Oxford, pp. 295.

Gasse F, Rognon R and Street FA (1980) Quaternary History of the Afar and Ethiopian Rift Lakes. *In:* Williams MAJ and Faure H (Editors), *The Sahara and the Nile*. Balkema Rotterdam: 361 - 400.

Giresse P, Maley J and Brenac P (1994) Late quaternary palaeoenvironnements in the Lake Barombi-Mbo (West cameroon) deduced from pollen and carbon isotopes of organic matter. *Palaeogeog. Palaeoclimat. Palaeoecol.* 107: 65 - 78.

Hamilton AC (1987) Vegetation and climate of Mt Elgon during the late Pleistocene and Holocene. *Palaeoecology of Africa* 18: 283 - 304.

Jolly D and Bonnefille R (1991) Diagramme pollinique d'un sondage holocène de la Kuruyange, Burundi (Afrique Centrale). *Palaeoecology of Africa* 22: 265-274.

Kendall RL (1969) An Ecological History of the Lake Victoria Basin. *Ecol Monogr* 39: 121 - 176.

Jolly D, Bonnefille R and Roux M (1994) Numerical interpretation of a high resolution Holocene pollen record from Burundi. *Palaeogeog. Palaeoclimatol. Palaeoecol.* 109: 357-370.

Kerfoot O (1968) Mist precipitation on vegetation. *Oxford Forestry Abstracts* 29: 8-20.

Letouzey R (1968) Etude phytogéographique du Cameroun. (Paris) Lechevalier: pp. 508.

Letouzey R (1978) Notes phytogéographiques sur les palmiers du Cameroun. *Adansonia* 18: 293 - 325.

Letouzey R (1985) Notices de la carte phytogéographique du Cameroun au 1:500.000. IRA Yaoundé et Inst Cart Intern Végétation Toulouse.

Maley J (1992) Commentaires sur la note de D Schwartz. Mise en évidence d'une péjoration climatique entre ca 2500 et 2000 ans BP en Afrique tropicale humide. *Bull. Soc. Géol. France* 163 3: 363 - 365.

Maley J (1995) Holocene changes in the African rain forest: Paleomonsoon and sea surface temperature variations. Congrès INQUA Berlin Paleomonsoon Symp 10, Abstract: 1p.

Maley J and Elenga H (1993) Le rôle des nuages dans l'évolution des paléoenvironnements montagnards de l'Afrique tropicale. *Veille Climatique Satellitaire* 46: 51 - 63.

Martin L and Flexor JM (1987) Vibro-testemunhador leve: construçao e possibilidades. *2° Congresso Ass Brasileira Estudos do Quaternario* Publ sp 1: 1-2.

Reynaud I and Maley J (1994) Histoire récente d'une formation forestière du sud-ouest Cameroun à partir de l'analyse pollinique. *C. R. Academy Sc. Paris* 317: 575 - 580.

Reynaud - Farrera I (1995) Histoire des paléoenvironnements forestiers du sud-Cameroun à partir d'analyses palynologiques et statistiques de dépôts Holocènes et actuels. Thèse Univ Montpellier II: 239 p.

Richards K (1986) Preliminary results of pollen analysis of a 6 000 years core from Mboandong, a crater lake in Cameroon. *Hull Univ. Geogr. Dept. Misc.* Ser 32: 14 - 28.

Roche E, Bikwemu G and Ntaganda Ch (1988) Evolution du paléoenvironnement quaternaire au Rwanda et au Burundi. Analyse des phénomènes morphotectoniques et des données sédimentologiques et palynologiques. Inst Fr Pondichéry, trav sec sci tech t XXV: 105 - 123.

Schwartz D (1992) Assèchement climatique vers 3000 BP et expansion Bantu en Afrique centrale atlantique: quelques réflexions. *Bull Soc Géol France* 163 3: 353 - 361.

Schwartz D, Guillet B and Dechamps R (1990) Etude de deux flores forestières mi-holocène (6000 - 3000 BP) et subactuelle (500 BP) conservées in situ sur le littoral pontenegrin (Congo). *In:* R Lanfranchi and D Schwartz (Editors), Paysages Quaternaires de l'Afrique Centrale Atlantique. ORSTOM Paris: 283 - 297.

Servant M and Servant-Vildary S (1980) L'environnement quaternaire du bassin du Tchad. *In:* Williams MAJ et Faure H (Editors), *The Sahara and the Nile.* Balkema Rotterdam: 133 - 162.

Servant M, Charles-Dominique P, and Guillaumet JL (1993) ECOFIT (ECO systèmes et Paléoécosystèmes des Forêts Inter-Tropicales), Lettre du Programme Environnement CNRS 10: 31 - 33.

Sowunmi A (1991) Late quaternary environments in equatorial Africa: palynological evidence. *Palaeoecology of Africa* 22: 213 - 238.

Ssemmanda I and Vincens A (1993) Végétation et climat dans le bassin du lac Albert (Ouganda, Zaïre) depuis 13 000 ans BP: apport de la palynologie. *CR Acad Sci Paris* 316 II: 561 - 567.

Street FA and Grove A (1979) Global maps of lake-level fluctuations since 30 000 yr BP. *Quaternary Research* 12: 83 - 118.

Talbot MR, Livingstone DA, Palmer PG, Maley J, Melack JM, Delibrias G and Gulliksen S (1984) Preliminary results from sediment cores from Lake Bosumtwi, Ghana. *Palaeoecology of Africa* 16: 173 - 192.

Vincens A (1982) Palynologie, environnements actuels et plio-pleistocènes à l'est du lac Turkana (Kenya). Thèse Doct. Etat Sc. Univ. Aix-Marseille II: 244 p.

Vincens A (1986) Diagramme pollinique d'un sondage Pléistocène supérieur - Holocène du lac Bogoria (Kenya). *Rev. Palaeobot. Palynol.* 47: 169 - 192.

Vincens A (1989) Les forêts claires zambéziennes du bassin Sud-Tanganyika. Evolution entre 25 000 et 6 000 ans BP. *CR Academy Sci. Paris* 308 Sér II: 809 - 814.

Vincens A (1993) Nouvelle séquence pollinique du lac Tanganyika: 30 000 ans d'histoire botanique et climatique du Bassin Nord. *Rev. Palaeobot. Palynol.* 78: 381 - 394.

Wirrmann D (1992) Le lac Ossa: une monographie préliminaire. *Rev. Géogr. du Cameroun* XI 1: 27 - 38.

Zogning A, Giresse P, Maley J and Gadel F (in press) The Late Holocene paleoenvironment in the lake Njupi area, West Cameroon; implications regarding the Nyos history. *J. African Earth Sciences:* 18p.

$\delta^{18}O$ and Trace Element Measurements as Proxy for the Reconstruction of Climate Changes at Lake Van (Turkey): Preliminary Results -

Gerry LEMCKE and Michael STURM

EAWAG, Swiss Federal Institute for Environmental Science and Technology, Dübendorf, Switzerland

Abstract

The sediments of Lake Van, the fourth largest terminal lake on earth, located at the eastern end of the Taurus Mountain Range show an undisturbed continuous record of chemically precipitated carbonate varves, which provide:

a, a yearly carrier for changes in environmental proxy like $\delta^{18}O$, Sr/Ca and Mg/Ca of authochthonous precipitated carbonates which allow to detect short term climatic variations,

b, a continuous, non floating varve chronology back to 13,700±356 yrs B.P.

Calculations based on an isotopic-hydrologic balance model show that temperature fluctuations are less important for the isotope enrichment in Lake Van than changes of the relative humidity. Thus a paleohumidity curve of Lake Van can be estimated and compared to a Mg/Ca record which is found to be a proxy for lake salinity.

Lake Van sediments indicate a first period of cold and arid climate between 12,600-10,460 yrs B.P. During this time humidity decreased by 0.17 or 30% (no temperature shift) or 20%, if temperature is reduced by 5 °C.

A second phase of climatic change also occurred: in a first period (4,190-3,040yrs B.P.), slightly reduction in lake level and humidity mark a regional climate shift toward more continental climate with reduced precipitation. In a second period (3,040-2,000 yrs B.P.), precipitation increased again, but humidity remained at a level reduced 5% below that encountered today.

NATO ASI Series, Vol. I 49
Third Millennium BC Climate Change
and Old World Collapse
Edited by H. Nüzhet Dalfes, George Kukla,
and Harvey Weiss

Introduction

The water content and the chemical composition of terminal lakes are controlled mainly by the relation of water input by rivers and precipitation and the water loss by evaporation. Changes of temperature, humidity and input/evaporation ratio result in lake level changes, which in turn directly influence the salinity of the water. Thus terminal lakes represent a sensitive tool to reconstruct climate and the environmental changes of the past (Street-Perrott & Roberts, 1983; Chivas et al., 1993; Fontes et al., 1993; Barker et al., 1994).

The regional climatic situation at Lake Van, the fourth largest terminal lake on earth (Degens & Kurtmann, 1978; Landmann et al., 1994), is strongly influenced by changes of the position of the northern branch of the jet stream and by the extent of the subtropical high pressure belt. Both of them control the boundary between the Mediterranean climate and continental steppe climate. Salinity of Lake Van records each shift of these atmospheric circulation (precipitation) patterns. Today the lake receives 1000 mm annual precipitation on its S and SW shores and less than 400mm on its N and E part (Zohary, 1973; Frey & Kürschner, 1989; van Zeist & Bottema, 1991).

Lake Van covers an area of 3522 km^2. It has a maximum water depth of 460 m and a water volume of 576km^3. It is situated in the East Anatolian Highlands of Turkey, where the Afro/Arabian Plate from the South meets the Eurasian Plate from the North and East. This tectonic situation is reasonable for recent volcanism and hydrothermal activity in the whole region (Degens & Kurtmann, 1978; Kipfer et al., 1994). There are two semi-active volcanoes in the vicinity of the lake: Nemrut Dağ (3050 m) and Süphan Daghi (2800m). It is assumed that the last major eruption (Nemrut Dağı) occurred 1441 A.D. (Kempe, 1977). High evaporation rates and the influx of dissolved ions from hydrothermal sources explain the high alkalinity of the lake water (155 $meql^{-1}$), a pH of 9.8 and a salinity of 21.4‰ (Kempe et al., 1991; Wüest et al., 1994). The water is highly supersaturated with regard to all alkaline earth elements. Mass balance calculations done by Reimer et al. (1994) show that the Ca content of the sediments can be quantitatively explained with autochthonous precipitated carbonates. Most of the carbonates are precipitated during summer and early fall when a substantial amount of surface water evaporates, after its dilution by Ca-rich water originated from spring snow melt. This has two consequences for the lake system:

- The annual precipitation of carbonates during summer forms a distinct light layer, whereas the deposition of detrital particles and organic material forms a dark layer during the subsequent winter and early spring. Such couplets of light and dark layers are believed to represent one year and are thus true varves as described by Lotter (1989) and others. First discovered in Lake Van by Kempe (1977) during the first international Lake Van expedition, these regular varve lamination is clearly visible back to the end of the last glaciation (Landmann et al., 1994). That provides a world-wide unique year by year datable archive which covers the Late Glacial/Holocene transition.
- Yearly precipitating carbonates provide the tracer to detect environmental or climate changes on a high resoluted scale. Thus the amount of trace metals (Sr, Mg) incorporated in the carbonate lattice of autochthonous precipitated inorganic carbonates is a proxy for the salinity of a lake (Morse & Mackenzie, 1990; Barker et al., 1994). Stable oxygen isotope records describe the hydrological situation and, additional, the atmospheric circulation patterns influencing the lake system from which the carbonates precipitate (McKenzie, 1993; Casanova & Hillaire-Marcel, 1993; Fontes et al., 1993).

In this paper a simple hydrologic/isotopic balance model ($\delta^{18}O$) which is calibrated with instrumentally measured data of the last 60 years is connected with results of trace metal measurements (Sr, Mg) of authochthonous carbonates of Lake Van. The aim of the present study is to evaluate the main driving forces such as humidity, responsible for the shifts found in the sediment record, and their possible effect on lake level changes on a high resoluted time scale.

Methods

Coring, geochemistry and mineralogy in general

The long piston core (Kelts et al., 1989) Van 90-10 (length 845.2 cm) used for this study was taken during the third international Lake Van expedition in 1990 from the south western part (Tatvan basin, 38°32.4'N, 42°48.0'E) of Lake Van at a water depth of 420 m.

The mineral composition of 279 samples was determined by X-ray diffraction (Cuk$_\alpha$ radiation, Scintag). Quartz was used to normalise shifts in peak positions caused by an appliance drift. The amount of quartz was quantified after Behbehani

(1987). Particular inorganic carbon (PIC) represented by calcite, aragonite and dolomite in the X-ray chart was used as an internal standard. PIC values were determined by the reaction of 20 mg (mean value) of sediment in 2n-HCl to form CO_2 and subsequent potentiometric titration (Couloumeter 5011, Laborlux S.A.). Different mixtures of pure calcite and aragonite were prepared to create a calibration curve for the aragonite/calcite ratio.

To determine the major elements a *t*otal *d*isintegration method (TD) was adjusted as suggested by Heinrichs & Herrmann (1990). After adding 2 ml HF, 3 ml HNO_3 and 1 ml $HClO_4$, 50mg of sediment were disintegrated under pressure with stepwise increasing temperature (50-180°C) for 8 hours. Subsequently the acids were evaporated and the remaining salts were dissolved with 2 n HNO_3 (1 ml) and diluted with 50 ml double distilled water. The metal content was determined by ICP-AES (Spectro Analytical Instruments).

Varve counting

To establish a continuos varve chronology similar to other sites (see, e.g., Lotter et al., 1991; Hajdas, 1993) varves of Lake Van sediments were counted on thin sections of the entire core. Core samples of 30cm length were dehydrated with acetone and subsequently embedded in a 4-component epoxy resin (Lamoureux, 1994). After hardening, the 30cm blocks were cut to single chips, fixed on 5x8 cm microscope slides and polished down to 30 μm thick overlapping thin sections. Overlapping of thin sections was used to correlate consecutive slides to each other by comparing a distinct set of laminae of the two slides. Beginning with the top of a section the correlated slides were marked each 0.5 cm and the varves within were counted.

Cross-checks of the total length of all thin sections with the length of the core section measured before preparation showed a max. error of ±0.3 cm. Missing material between core sections was determined by comparison of the distance between two marker-beds in the core, cut by a core cap, and the distance between these markers in a reference core. The resulting loss was multiplied by the mean sedimentation rate of the equivalent core section and added. Correlation and counting was repeated after a year. The mean deviation of the two results lies within ±2.6%. No varves can be distinguished below a sediment depth of 790 cm (13,699±356 yrs B.P.). The sediment age of the lowest part of the sediment was estimated by multiplying the length of the remaining core with the mean sedimentation rate of the last section (8.4 varves/5 mm).

The bottom layer of core Van 90-10 has thus been calculated to have an age of 14,720±380 yrs B.P. All ages given in the following text refer to calendar years before 1950 [cal yrs B.P.].

Sr/Ca and Mg/Ca

Samples for trace element analyses were taken every 5 cm corresponding to a mean time interval of 83.4 years. To dissolute only the carbonate fraction of the bulk sediment sample an *a*cetic *a*cid *d*isintegration (AAD) was applied. 5ml 1.0mol acetic acid were added to an average of 19.2mg (187 samples) of water-soluble salt-free sediment. The samples were disintegrated in 25 ml pre cleaned (1 mol HNO_3) centrifuge tubes under permanent shaking for 60min. The suspension of dissolved carbonates and remaining silica minerals were cleared by ultra centrifugation. 1ml of the overlying pure solution was mixed with 0.5 ml $HNO_{3conc.}$, 1 ml Rh (0.1 ppm) as internal standard and 15ml double distilled water. Ca, Mg an Sr were determined simultaneously on a Perkin Elmer ELAN 5000 ICP-MS. X-ray analyses were used to check if the remaining fraction was carbonate free. Although AAD is just strong enough to dissolve the carbonate minerals, it must be kept nevertheless in mind that even such a weak disintegration will remove some ions like e.g. Mg, that are weakly adsorbed on surface active minerals like clay. To estimate a possible error standard sediment samples were 19 times independently disintegrated and measured: The error was ±4.5% (Mg), ± 3.3% (Ca) and ± 3.1% (Sr).

Stable oxygen isotopes

The stable oxygen isotope record presented here is based on 294 measurements of the CO_2 isotope ratio of autochthonous carbonates. Each of the sample integrated 16.7 years (mean value). The distance between two samples varied between 1 to 5 cm (representing 16.7 and 83.4 a, respectively). After extracting the water soluble salt, CO_2 of 1-3 mg sediment was generated adding 100% phosphoric acid at 50°C. Additional removing of organic carbon was not necessary as cross-checked pairs of samples show. In each case one of these samples was roasted under vacuum for 120 min. at 400°C. The $^{18/16}O$-isotope ratio was determined with a VG Micromass 903 triple collector mass spectrometer of the Geological Institute, ETH Zürich. The results are plotted in the standard δ-notation expressed in ‰, as deviation of the isotope ratio of the sample from the international PDB (*Pee Dee Belemnite*) standard. All measurements were corrected after Craig (1957). The mean analytical error of $\delta^{18}O$ was ±0.29‰.

Carbonate and isotope chemistry in Lake Van, theoretical concept

Sr/Ca and Mg/Ca

The trace element concentration of an autochthonously precipitated carbonate mineral depends on its availability in the solution and on the kind of carbonate which is precipitating.

Equation 1 gives the relation between the ratio of the amount of trace metals (m_t) and the Ca ions (m_{Ca}) in the carbonate mineral ($m^{carb.}$) and the liquid phase ($m^{sol.}$). $k_t^{carb.}$ gives the distribution coefficient.

$$\frac{m_t^{carb.}}{m_{Ca}^{carb.}} = k_t^{carb.} \cdot \frac{m_t^{sol.}}{m_{Ca}^{sol.}} \qquad (1)$$

Concentrations of Sr and Mg in lake water are a function of the salinity values if the ratio of Sr/Ca and Mg/Ca of the inflowing water is kept constant and if the Ca concentration of the lake water is hold on a constant value. In Lake Van Ca concentration is fixed on a very low level (3.52 mg/l; Reimer et al., 1994) because of the annual carbonate precipitation. The element ratio of the inflowing water is given by the mean composition of the surrounding rocks which is assumed to have not been changed during post glacial times. Lattice absorption of precipitating carbonates or, to a lesser degree, adsorption on active surfaces of clay minerals, are the only mechanisms to transport Sr and Mg ions form the water to the sediment.

If the distribution coefficient is known, ratios of Sr/Ca and Mg/Ca of a precipitated carbonate are proxies for changes of the Sr and Mg concentrations of the water. In the case of a terminal lake the enrichment of an element in the water is either a simple accumulation process or caused by salinity changes, e.g. due to lake level changes. Salinity changes are only recorded in the carbonates if in eq. 2 $\frac{dC_e}{dt}$ is much greater than $\frac{dC_a}{dt}$.

$$\frac{dC_{eff.}}{dt} = \frac{dC_e}{dt} - \frac{dC_a}{dt} \qquad (2)$$

Here $C_{eff.}$ is the *eff*ective change of trace metal concentration in the water body where C_e and C_a describe shifts in concentration caused by changes in *e*vaporation and enrichment by *a*ccumulation (residence time), respectively.

To correct for the fact that all carbonates which precipitate in Lake Van are predominately a mixture of calcite and aragonite equation 1 becomes (3) in case of Sr incorporation and (4) in case of Mg incorporation.

$$\frac{m_{Sr}^{sol.}}{m_{Ca}^{sol.}} = \frac{\frac{m_{Sr}^{mix.}}{m_{Ca}^{mix.}}(\omega+1)}{k_{Sr}^{cc}+k_{Sr}^{ara}\omega} \qquad (3)$$

$$\frac{m_{Sr}^{sol.}}{m_{Ca}^{sol.}} = \frac{m_{Mg}^{mix.}}{m_{Ca_{tot}-Ca_{ara}}^{mix.}} \cdot \frac{1}{k_{Mg}^{cc.}} \qquad (4)$$

The element concentration and the distribution coefficient are given by m and k, respectively. The sub indices indicate the element, the super indices, *sol.*, *mix.*, *ara* and *cc* give the matrix: *sol*ution, carbonate *mix*ture, *ara*gonite and *c*al*c*ite. Ca_{tot} is the *tot*al amount of Ca dissoluted by acidic acid disintegration, Ca_{ara} is the amount of Ca which is fixed by aragonite precipitation. The aragonite/calcite-ratio determined by X-ray diffraction is given in ω. The distribution coefficient used for k_{Sr}^{cc} and k_{Sr}^{ara} are 0.06 and 1.17 (Pingitore & Eastman, 1986; Kinsman & Holland, 1969). Mucci & Morse (1983) show that $Mg/Ca_{carb.}$ increases linear with $Mg/Ca_{sol.}$, if $Mg/Ca_{sol.}$ is greater than 7.5 and Oomori et al. (1987) give a value of 0.012 (T=10°C) for k_{Mg}^{cc}. If the Mg-calcite content exceeds 35 mol.% a there is a clear trend towards a non linearity, which results in an excess Mg incorporation in carbonates (Lahann & Siebert, 1982). Therefore is has to be kept in mind that the Mg/Ca ratios of autochthonous carbonates are sensitive to reconstruct salinity changes but in terms of a numerical calibration these ratios are at the best a semi quantitative proxy.

Stable oxygen isotopes

The isotope ratio of a carbonate sample, in case of Lake Van a mixture of aragonite and calcite ($^hI/^lI_{mix}$), is related to the δ-notation [‰] by equation 5. $^hI/^lI$ expresses the ratio of the *h*eavy isotope to the *l*ight one. $^hI/^lI_{PDB}$ is the isotope ratio (Craig, 1957) of the international PDB standard.

$$\frac{{}^{h}I}{{}^{l}I}_{mix} = \frac{{}^{h}I}{{}^{l}I}_{PDB} \cdot \left(\frac{\delta}{1000} - 1\right) \qquad (5)$$

Small but significant differences occur between calcite and aragonite during the water-carbonate fractionation (Rubinson & Clayton, 1969; Tarutani et al. 1969). The isotope ratio of a theoretically pure calcite (${}^{h}I/{}^{l}I_{cc}$) can be calculated using equation 6.

$$\frac{{}^{h}I}{{}^{l}I}_{cc} = \frac{\frac{{}^{h}I}{{}^{l}I}_{mix}(\omega + 1)}{1 + \omega\gamma} \qquad (6)$$

ω describes the aragonite/calcite-ratio. For γ the value of Tartuani et al. (1969) is used. Equation 6 put again in eq. 5, allow to calculate the $\delta^{18}O$ value of such a theoretical calcite. All $\delta^{18}O$ values referred in the text are normalised to calcite.

The relation between the isotope ratio of the lake water and the precipitating calcite follows eq. 7,

$$\alpha_{cc-water} = \frac{\frac{{}^{h}I}{{}^{l}I}_{cc}}{\frac{{}^{h}I}{{}^{l}I}_{water}} \qquad (7)$$

whereby the temperature dependent fractionation factor α is calculated after O´Neil (1967).

The hydrological and isotopic balance of Lake Van is described by (8):

$$\frac{dV_l}{dt}\delta_l + \frac{d\delta_l}{dt}V_l = F_{in}\delta_{in} - F_{out}\delta_{out} - F_e\delta_e \qquad (8)$$

V_l is the water *v*olume of the *l*ake, δ represents the $\delta^{18}O$ value and F the water flux. The indices (*in*, *out*, *e*) indicate *i*nput, *o*utput and *e*vaporation. Although Lake Van is a terminal lake the term $F_{out}\delta_{out}$ is kept in the equation to allow the simulation of lake level high stands with possible outflow (Lake Van has a natural outflow 94 m above current lake level; Schweizer, 1975).

Modified by Gat & Bowser (1991)δ_e follows the Craig-Gordon (1965) linear resistance model, equation 9.

$$\delta_e = \frac{\psi^*\delta_l - h\delta_a - \varepsilon}{(1-h) + \frac{\Delta\varepsilon}{1000}} \qquad (9)$$

Where h stands for the relative humidity. The equilibrium fractionation factor ψ^* is estimated after Gonfiantini (1980). The kinetic enrichment factor $\Delta\varepsilon$ is related to humidity by $\Delta\varepsilon = \xi(1 - h)$, where, after Gonfiantini (1986), ξ is a constant factor. The value for ε can be calculated with $\varepsilon = (1 - \psi^*)1000 + \Delta\varepsilon$ (Gat & Bowser, 1991). δ_a is the isotopic composition of the *a*tmospheric moisture, which is close to the equilibrium $\delta_a = -\varepsilon^* + \delta_p$ for continental locations (Gat, 1995). ε^* is given by $\varepsilon^* = \varepsilon - \Delta\varepsilon$. Temperature is the most important climatic parameter next to humidity who influences the isotope enrichment and it is hidden behind ψ^* and h, both are directly controlled by temperature.

Using (9) equation (8) can be rewritten as (10)

$$\frac{dV_l}{dt}\delta_l + \frac{d\delta_l}{dt}V_l = -\delta_l A^* + B^* \quad (10)$$

where $A^* = -F_{out} - F_e\frac{\psi^*}{x}$, $B^* = F_{in}\delta_{in} + F_e\frac{h\delta_a+\varepsilon}{x}$ and $x = 1 - h + \frac{\Delta\varepsilon}{1000}$.

In a terminal lake any change in the hydrological input/evaporation-ratio, χ, leads to changes of the lake level resulting in changes of the lake volume. There is no hydrological steady state. If χ is held constant between two given ages, represented by two single measurements of $\delta^{18}O$ within the sediment profile, equation (10) can be simplified and rewritten (11) as

$$\frac{d\delta_l}{dt} = -\delta_l A + B \quad (11)$$

where $A = \frac{A^*-m_{(i)}}{V_{(i)}}$, $B = \frac{B^*}{V_{(i)}}$ and $m_{(i)} = \frac{F_{in}-F_{e(i)}}{V_{(i)}}$. h and F_{in} are given input values, thus $V_{(i)}$ follows using (12) to estimate $F_{e(i)}$. i marks a discrete time step [a] of the model with i=0 for the today situation .

$$F_{e(i)} = C_s(1-h)v_tA_{l(i)} \quad (12)$$

C_s is the saturation concentration of the water vapour at the liquid-air boundary, $A_{l(i)}$ the lake surface *a*rea and v_t is the exchange *v*elocity for the water vapour transition (Schwarzenbach et al., 1993). Any volume change $V_{(i+1)}$ within the next time step is calculated using F_{in} and $F_{e(i)}$. $A_{l(i+1)}$ is calculated by interpolation if $V_{(i+1)} < V_{(0)}$, using a bathymetric map of Lake Van (Landmann et al., 1994) and if $V_{(i+1)} > V_{(0)}$ by extrapolation using the data given for the shelf area (0-40 m water depth).

After integrating eq. 11 for the *s*teady *s*tate value of the lake water at a given temperature, humidity follows $\delta_{l\,ss} = -\frac{B}{A}$.

Results

Figure 1 summarises the results of the measurements of Ca, Mg and Sr using TD technique and the AAD method. Nearly identical values are found for Ca_{TD}-Ca_{AAD}, Sr_{TD}-Sr_{AAD} and Ca_{AAD}-Sr_{AAD}. It can thus be assumed that Ca and Sr is quantitatively fixed in easy soluble minerals like carbonates. Contrary to Mg, of which most of the time is found to be strongly fixed by silica minerals (montmorillonit-15 Å and -18 Å). An exception is the time between 10800-10460 yrs B.P. Similar values for Mg_{TD} and Mg_{AAD} indicate that almost all Mg is kept by weak bindings.

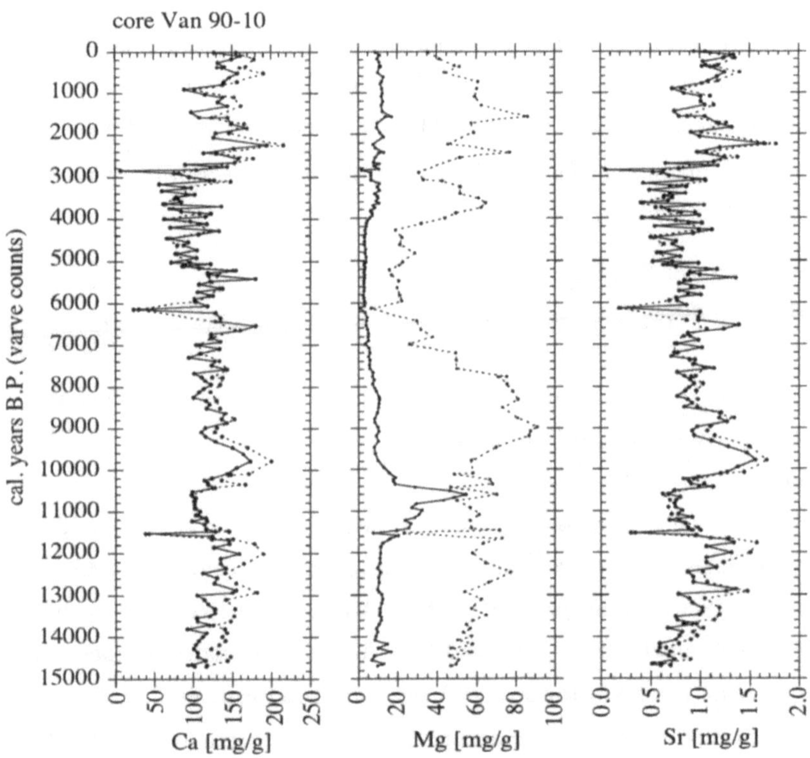

Fig. 1 Comparison of the concentrations of Ca, Mg and Sr used as chemical proxy for salinity changes within the Lake Van record. Solid line: element content found after the *a*cetic *a*cid *d*isintegration (AAD); dotted line: element content determined using HF-disintegration (TD).

The data measured by AAD were used to calculate (eq. 3, 4) the theoretical Sr/Ca and Mg/Ca ratios of Lake Van palaeowater. All values of Mg/Ca and Sr/Ca in the text refer to such a calculation.

An outstanding covariation is observed between the $\delta^{18}O$ record and the Mg/Ca ratio. No correlation is visible between $\delta^{18}O$ and Sr/Ca (Fig. 2).

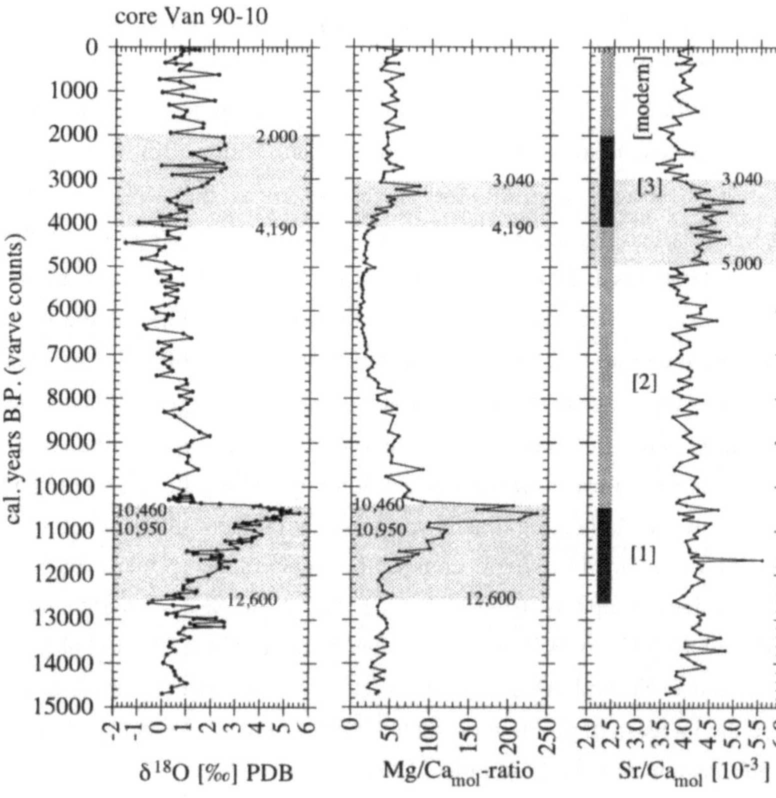

Fig. 2 Oxygen isotope record of autochthonous precipitated carbonates related to the theoretical lake water ratio of Mg/Ca and Sr/Ca. Phases with increased element ratios or an enrichment of $\delta^{18}O$ are shaded grey. The bar in the Sr/Ca graph marks the time period referred to stages in the text.

The $\delta^{18}O$ record (Fig. 2) with a mean value of 1.32‰ show two main shifts from the modern isotope ratio of 0.78‰. A maximum isotope enrichment with values of 5.6‰ is observed between 10,950-10,460 yrs B.P. This extraordinary enrichment occurs at the end of a stepwise increase, starting about 12,600 yrs B.P. [1]. Such steps are observed, in each case after a short phase of $\delta^{18}O$ depletion, at 11,450 yrs B.P. and 10,950 yrs B.P. The high values at 10,460 yrs B.P. are followed by an abrupt decrease within the time span of max. 180 years. Continuously decreasing values are measured subsequently between 10,460-4,190 yrs B.P. [2]. Mean values of $\delta^{18}O$ of 0.37‰ are little lighter than values of 0.78‰ characterising the modern situation of the last 2,000 years. A second distinct maximum of isotope enrichment can be observed between 4,190-2,000 yrs B.P. $\delta^{18}O$ reaches 2.48‰ [3]. Similar to the older event a slow continuous increase is stopped by a sudden step back within a few decades.

The Mg/Ca shows generally the same pattern as the $\delta^{18}O$ record. The only disagreement is that the second enrichment period [3] ends already at 3,090 yrs B.P. and not at 2,000 yrs B.P. as it is indicated by the isotope curve. No correlation is visible between the Sr/Ca and the oxygen isotope values. Whereas the former shows an increasing factor up to 20, the Sr/Ca varies within a band of only 25% (Fig. 2).

Al_{TD}, Na_{TD} reflect changes of allochthonous detrital material brought into the lake by tributaries. These elements represent typically silica rocks like, e.g., clay and pumice that are not formed in the water column or within the sediment. There are no substantial changes indicating drastic variations in the detrital sediment input (Fig. 3). The peaks are explained by single, conspicuous tephra layers. Quartz also can be taken as tracer for detrital input by tributaries. The sudden increase of quartz between 12,000 yrs B.P. and 10,000 yrs B.P., however, cannot be attributed to an enhanced detrital influx, because Na and Al do not correlate with the quartz curve. It is therefore assumed the increase of quartz represents times of expanded wind activity and aeolian sediment transport to the lake (Courty, 1994).

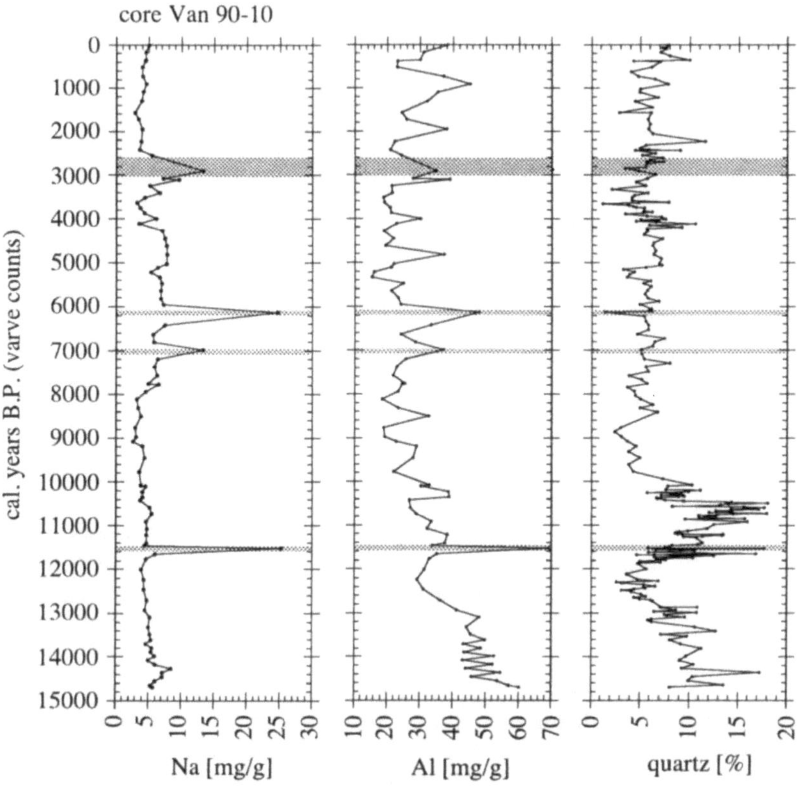

Fig. 3 Amounts of Na, Al and quartz found in the sediment record of Lake Van. Na and Al have been analysed using a HF total disintegration technique and quarts was measured by X-ray diffraction. Peaks in the Na, Al record are the effect of distinct single layers of volcanic tephra (shadowed horizons).

Discussion

Oxygen isotopes

A simple one box model is used to evaluate the different driving forces for an enrichment of isotopes in the Lake Van system.

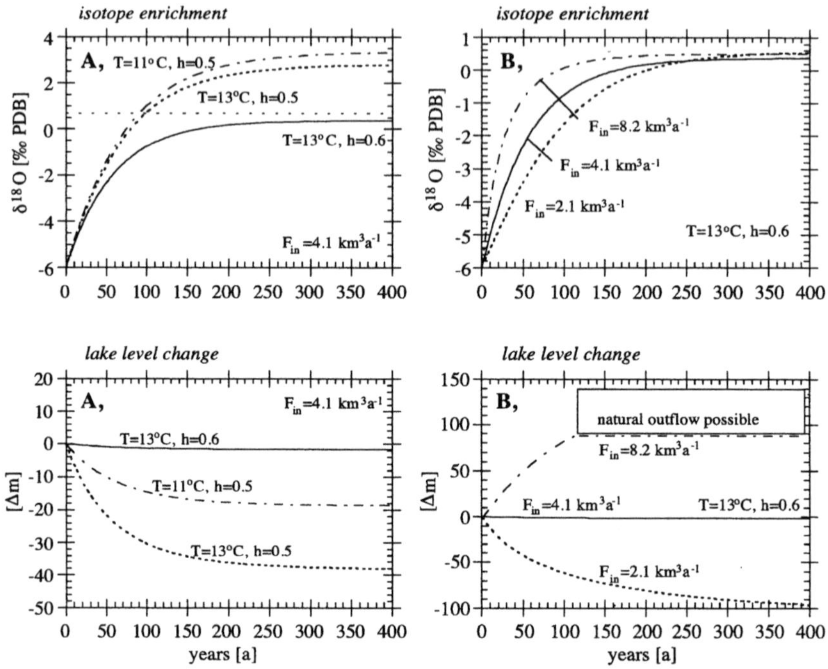

Fig. 4 Simulation of different climatic settings and their influence on the isotope enrichment of autochthonous carbonates precipitating of Lake Van and the resulting lake-level changes. $\delta^{18}O$ gives the value of calcite in equilibrium with lake water under given conditions for T, h, and F_{in}. Δm is the relative lake-level change.

A. variation of humidity and temperature with constant influx.

B. variation of the influx volume.

δ_a varies between -21.41‰ (11° C) and -21.04‰ (150 C) SMOW. δ_{in} is -10.91‰ SMOW (Schöll & Faber, 1978). Shadowed area: Level of natural outflow, if the lake level would be 94 m higher as todays level (Schweizer, 1964).

The base of the model is formed by the well defined hydrology of this closed basin and by the mixing regime of Lake Van water of at least one complete turn over cycle within two years (Kipfer et al., 1994). To check the paleohydrological relevance of the model, hydrological and isotope data of ´today´ (last 60 years) have been used as input data. Calculated steady state values for precipitating carbonates in equilibrium with the lake water ($\delta^{18}O_{model}$, 0.93‰) are in good agreement with values, measured in the uppermost sediment layer ($\delta^{18}O_{sed.}$, 0.78‰). It is therefore assumed that the model is appropriate to evaluate the effects of changes of temperature, humidity and run-off on the hydrological and isotopic system of Lake Van. Figure 4 summarises these results related to isotope enrichment and/or lake level changes, respectively.

table 1 Changes of relative humidity and lake level for different temperatures and a constant water input ($F_{in} = 4.1 km^3a^{-1}$, after Kempe, 1977) for the main stages of Lake Van record. Humidity (h) is adjusted to fit $\delta^{18}O_{model}$ to $\delta^{18}O_{sed.}$ depending on temperature (T). The oxygen isotope value of the atmospheric moisture, d_a, varies between -21.41‰ (11°C) and -21.04‰ (15°C) SMOW. δ_{in} is -10.91‰ SMOW (Schöll & Faber, 1978). For T (1929-1965) and h (1971) of the modern stage the values from March to November are used (data given by Kempe, 1977).

stage	time [yrs B.P.]	T [°C]	h	d18O [‰ PDB] sed.	d18O [‰ PDB] model	lake level change [D m]
modern	0-2,000	13.8	0.58	0.78	0.93	0
3	4,190-2,000	15	0.49	2.48	2.47	- 58
		13	0.51		2.49	- 34
		11	0.53		2.49	- 7
2	10,460-4,190	15	0.58	0.37	0.37	- 31
		13	0.60		0.37	+ 2
		11	0.63		0.37	+ 32
1	12,600-10,460	15	0.36	5.6	5.62	- 89
		13	0.38		5.63	- 66
		11	0.41		5.63	- 44

A shift of temperature in the order of -5°C as it is assumed, e.g., for the Younger Dryas period by Dansgaard (1993) results in an isotopic shift of just +2.3 ‰ in Lake Van. To generate heavy $\delta^{18}O$ values of up to +5.6‰ as they have been measured between 10,950-10,460yrs B.P. (Fig. 2) would require an additional drastic decrease of the relative humidity. Other minor variation of the isotope record could be explained with temperature changes alone. That leads to uncertainties in the reconstruction of lake-level changes that are only based on the oxygen isotope variation. Temperature influences to a great level humidity because it directly controls the maximum water vapour saturation state of the water. In contrast the atmospheric vapour concentration just above the water-atmosphere interface is variable. Low humidity can occur in arid warm climate as well as in cold dry conditions. Low temperature will increase the isotope fractionation processes but decrease the evaporative water flux that controls the lake level for a given influx. The different effects of temperature on the water level of Lake Van for different environmental conditions and also the temperature effect on the relative humidity needed to explain the isotope enrichment for the main stages (Fig. 2) are summarised in table 1.

Within some limits, decreasing humidity reflects increasing aridity, representing less precipitation and less river discharge. Therefore, the values given in table 1 for lake level changes are minimum estimations, where only the effects of changing humidity on the evaporative flux are taken into account. Decreasing oxygen isotope values of the *a*tmospheric moisture (δ_a) or of the *p*recipitation (δ_p) are found with increasing aridity, continentallity or during times of general climate cooling like the Younger Dryas event (11,000-10,000 yrs B.P.). Low values of δ_a and δ_p support a reduction of $\delta^{18}O$ of the lake water. In Lake Van, however, such a depletion of $\delta^{18}O$ is not found during this time (Fig. 2). Considering these effects the isotope stratigraphy can be interpreted as a record of relative humidity changes over the lake area during the last 14,720 years (Fig. 5).

Sr/Ca and Mg/Ca

There is a perfect correlation between Mg/Ca and $\delta^{18}O$, but no covariance is found between Sr/Ca and $\delta^{18}O$.

The Sr flux (0.13 $gm^{-2}a^{-1}$) calculated based on the Sr concentration of the authochthonous precipitated carbonates of the topmost sediments in Lake Van is about the same as the Sr flux (0.15 $gm^{-2}a^{-1}$) given by Reimer et al. (1994) using the mean concentrations of the river discharge. That mean Sr is quantitatively removed by precipitating carbonates (Stabel et al., 1986), most of it build into aragonite lattices. The Sr concentration of the lake water (C_r, in eq. 2) is hold on a very low level. There is not enough Sr remaining in the lake water to monitor salinity changes. If all Sr precipitates together with the Ca, both of them brought by tributaries, than the Sr/Ca of the carbonates is equivalent to the ratio of these elements in the incoming water.Thus the very stable relation between Ca and Sr shown in Fig. 3 is a result of an unchanged element ratio of the influx over the last 14720 years. Only between 3,000-5,000 yrs

B.P. light increasing Sr/Ca indicate a shift toward higher relative Sr concentrations of the water.

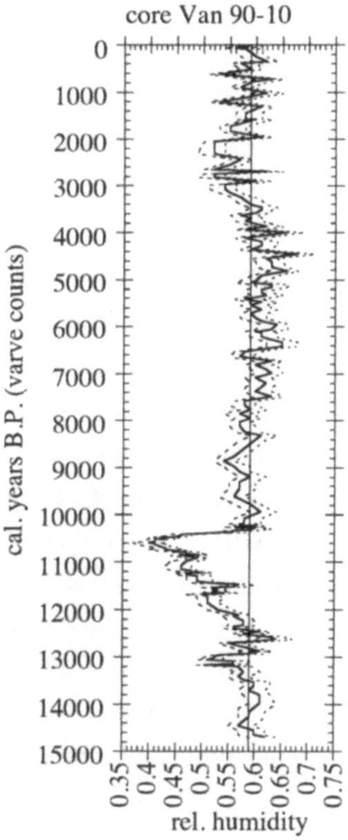

Fig. 5 Reconstruction of the relative humidity of Lake Van area over the last 14720 years. The calculation is made for different temperatures by fitting the steady state isotope value estimated for different values of humidity to the isotope value measured for the authochthonous precipitated carbonates of the related age. It has to be taken into account that a single run does not take into consideration any shifts in temperature, changes in the influx and its isotope ratio as well as any variation of the isotopic constitution of the atmospheric moisture.

solid line: calculation for 13°C; narrow dotted line: calculation for 11°C; wide dotted line: calculation for 15°C.

Input variablesare: δ_a, which varies between -21.41‰ and -21.04‰ SMOW depending on temperature (11°C to 15°C) . δ_{in} is -10.91‰, SMOW (Schöll and Faber, 1978). F_{in}=4.1km ^{3}a^{-1}; The vertical line marks the humidity of today (0.58; Kempe, 1977)

In the case of Mg the mass balance calculation shows that today only 16% of the diluted Mg (6.59 gm^{-2}a^{-1}, Reimer et al., 1994) supply precipitates with autochthonous carbonate sedimentation (1.02 gm^{-2}a^{-1}). 84% of the yearly Mg input by rivers are added to the lake water reservoir of current 6.3 10^{13} g Mg. That is a summing up of 0.03 % every year. This is about 200 times less than due to a salinity rise of e.g. 6 % corresponding to a lake level reduction of 10m. In general Mg/Ca ratios are a sensitive proxy for long term salinity changes of Lake Van.

For stage 1 and 3 (Fig. 3) the values for Mg/Ca are 7 times and 3 times higher than today. In terms of a lake level change this corresponds to a lake level lowering of 100 m and 20 m, respectively. Carbonates of the form $Ca_{0.29}Mg_{0.71}[CO_3]_2$ should precipitate out of a solution like proposed for 10,420 yrs B.P.

Besides aragonite however, high magnesium calcite with an average of 12 mol.% Mg is found precipitating out of Lake Van water. This is not enough to balance the Mg content. Moreover, analyses of X-ray diffraction reveals wide blurred peaks that show considerable amounts of high magnesium carbonates of a dolomite, huntite, magnesite like composition But there is, however no continuous mixing line between calcite and these minerals.

If Mg and Ca are quantitatively fixed by carbonates, the amount of these elements can be expressed in terms of magnesite ($MgCO_3$) and calcite ($CaCO_3$), as end members of a mixture, and the content of inorganic carbon can be calculated. Such a procedure may lead to an overestimation of inorganic carbon, if excess Mg is adsorbed e.g. on clay minerals. Otherwise there is a strong evidence that all Mg is fixed by carbonates (Fig. 6). Because there is no indication of an increasing Mg input into the lake for stages 1 and 3 (constant values for Al, Na, Sr/Ca), it is proposed that the additional amount of Mg within the sediments which can not be explained by a linear Mg incorporation in calcite reflect a change in the Mg scavenging and fixation process. Therefore the following model is suggested: during phases of decreasing humidity and precipitation or increasing evaporation the lake level was lowered. That resulted in an increase of areas where former lake bottom sediments with high porosity and highly supersaturated pore water came in contact with the atmosphere. This leads to the formation of high Mg containing carbonates of dolomite like composition, which are not found directly precipitating out of a water column (Schlanger, 1965, Usdowski, 1967).

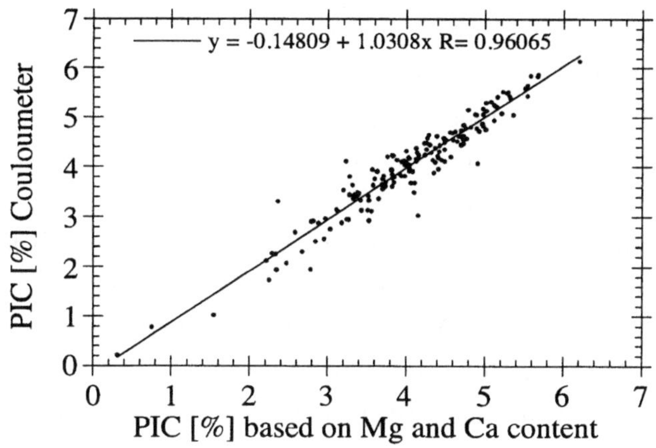

Fig. 6 Correlation between the content of particular inorganic carbon (PIC) measured directly (Coulometer) and the PIC content which results if one assume that all Ca and Mg is fixed by *carbonate minerals with $MgCO_3$ and $CaCO_3$ as end members*.

Synthesis and regional climatic context

Shifts in the stable isotope and carbonate geochemistry records allow a reconstruction of humidity and lake level changes (if the influx volume is kept constant) dependent on temperature.

Between 12,600 yrs B.P. and 10,460 yrs B.P. the relative humidity decreased stepwise. Starting with values similar or slightly higher as today (0.58) humidity decreased by 0.17 or 30% till 10,460yrs B.P., if temperature is assumed to have been unchanged (11°C). If temperature retreated at the same time by e.g. 5°C humidity at Lake Van would have been reduced by 20% to explain $\delta^{18}O$ values of 5.6% in Lake Van sediments. A substantial increase of Mg/Ca during the same time supports the assumption of a considerable lower lake level due to the remarkable reduction of humidity which has forced evaporation and less precipitation.

Cold and dry climatic conditions assumed for Lake Van between 12,600 to 10,460 yrs B.P. are in good agreement with palynological results for the Middle East. In the south Jordan Valley a gradual retreat of oak pollen indicates a drop in precipitation and a shift to dryer and colder conditions peaking at about 10,600 yrs B.P. (Baruch &

Bottema, 1991; Wright, 1993). A temperature reduction of about 2-3°C and significantly less humid conditions are suggested by van Zeist & Bottema (1991) on the basis of pollen stratigraphies of Lake Urmia and Lake Zeribar. The long term climate at Lake Van is controlled by reversed circulation mechanisms: strongly seasonal Mediterranean climate with moist westerly winds in the winter and arid conditions in the summer and on the other hand north-easterly anticyclonic winds. The latter are driven by increasing importance of high pressure belts with decreasing temperatures over north-eastern Europe and Siberia. These north-easterly winds inject cold and dry air to the south. These conditions, causing the return of cold, glacial conditions in Central Europe (Younger Dryas), have likewise induced dry climate in the Near East (Moore & Hillman, 1992). Furthermore they are also considered to be the main factor for an important arid phase in western Tibet (Gasse et al., 1991; Fontes et al., 1993). The equivalent dry period at Lake Van seems to be connected to the same circulation pattern indicating a strengthening of cold, dry air from the North to Lake Van area, beginning 12,600 yrs BP, lasting at 10460 yrs BP. This period roughly matches the Younger Dryas cold period in the northern hemisphere.

After a sudden retreat back to values similar like today humidity continuously rose between 10,460 yrs B.P. and 4,190 yrs B.P. A ´moist period´ with humidity higher than today is observed starting around 8,200 yrs BP.

By compiling Holocene pollen diagrams of the Western Iran and south-eastern Turkey van Zeist & Bottema (1991) found a strong correlation between the onset of forest vegetation and significant increase in precipitation and humidity. Forest vegetation has developed between 8,000-5,050 yrs BP at Lake Van and Lake Sögütly. That is synchronous with the onset of moist climatic conditions as indicated by the humidity reconstruction at Lake Van.

A second phase of $\delta^{18}O$ enrichment began at about 4,190 yrs B.P. reaching max. values between 3,040 yrs B.P. and 2,000 yrs B.P. The increase of 2.1‰ marks a humidity reduction of 0.1 or 15% (3,040 yrs B.P.) if it is assumed that temperature was constant during this period. On the other hand the increase of heavy isotopes could be explained by a temperature reduction of about 2.8°C. Additionally there is a difference between the Mg/Ca record and the $\delta^{18}O$ curve. Beginning at 4,190 yrs B.P. Mg/Ca is slowly rising for 1,050 years and jumps back (3,040 yrs B.P.) to modern values (0.78‰) within about 100 years just when $\delta^{18}O$ reach maximum values. These

$\delta^{18}O$ exist until 2,000 yrs B.P. Thus the period between 4,190 yrs B.P. and 3,040 yrs B.P. seems to be controlled by slowly decreasing humidity (temperature) and a significant shift of precipitation. The first causes an increases of the $\delta^{18}O$, the latter leads to a drop of lake level as indicated by Mg/Ca. Between 3,040 yrs B.P. and 2,000 yrs B.P. humidity (temperature) is stabilised on a lowered level compared to today. To explain the reduction of Mg/Ca, precipitation must be increased again. The evaporative loss of water caused by lower humidity necessary to explain the $\delta^{18}O$ is neglectable compared to the effect, which a reduced influx would have on the water volume of the lake. Evaporation could be easily compensated by increased precipitation (table 1, Fig. 4).

Archaeological and mineralogical evidence for drastic climatic change between 4,200 yrs B.P. and 3,900 yrs B.P. in northern Mesopotamia is given by Weiss et al. (1993) and Weiss & Courty (1993). It has also been discussed for eastern Europe (Gerasimenko, 1994), Tibet (Gasse et al., 1991) as well as for the Caribbean (Curtis & Hoddel, 1993). Nothing comparable is found for sites in central and northern Europe (Dansgaard, 1993; Zolitschka, 1994).

Conclusion

Lake Van provides a unique, annually datable, varved sediment record to reconstruct climate change in Eastern Turkey and the Near East. In combination with high resoluted sampling it is possible to detect and date short term climatic fluctuation within the Holocene.

A hydrologic/isotopic balance model is used to show that the $\delta^{18}O$ enrichment of Lake Van carbonates is mainly affected by changes of the relative humidity over the lake area. Because effects of temperature variability on the isotope enrichment of Lake Van are relatively small, it became possible to calculate a record of local changes of paleohumidity.

To get information about lake level changes of Lake Van, trace metal incorporation in autochtonous carbonates are estimated. They are assumed to be a proxy for lake salinity. Periods of high Mg/Ca values correspond with phases of $\delta^{18}O$ enrichment. A non linear incorporation of Mg in calcite restrict the possibility to use Mg/Ca as a direct measure of paleosalinity. High Mg/Ca are assumed to indicate very low lake level because only at this stages extended shelf areas provide a reactive zone for the formation of authighene, high Mg containing carbonates (dolomite).

Between 11,800-10,460 yrs B.P Lake Van sediments indicate cold and arid climate. Minimum values for humidity of ≈0.41 and lake level reduction of at least ≈-45 m are calculated for a temperature of 11°C. The period is suggested to be connected to the Younger Dryas cold period in the northern hemisphere.

There is a second phase of climatic change in the Late Holocene, starting at 4,190 yrs B.P. Slightly reduced humidity and lake levels are indicated by $\delta^{18}O$ and Mg/Ca until 3,040yrs B.P. That has been interpreted by a regional climate shift toward more continental climate with reduced precipitation between 4,190-3,040 yrs B.P. For the time between 3,040 yrs B.P. and 2,000 yrs B.P., $\delta^{18}O$ remains on a high level (low humidity) whereas Mg/Ca drops to modern values That indicates a restoration of the former precipitation pattern with slightly increased water supply to Lake Van after 3,040 yrs B.P

Acknowledgements

We thank S. Kempe (Darmstadt), head of the third international Lake Van Expedition (1990) and G. Landman and A. Reimer for making unpublished data available. A. Stahel, J.A. McKenzie, S. Bernasconi and A. Fischer (all ETH Zürich) for providing their laboratories for X-ray analyses and stable isotope measurements, including a lot of helpful discussions. C. Stengel and A. Zwyssig are acknowledged for laboratory help, D.M. Imboden, D.M. Livingstone, A.F. Lotter and C. Ohlendorf for advice and critical discussion.

This work has been supported by an EAWAG L+F grant Nr. 30-1.

References

Barker, P.A., N. Roberts, H.F. Lamb, S. van der Kaars and A. Benkaddour, 1994. Interpretation of Holocene lake-level change from diatom assemblages in Lake Sidi, Middle Atlas, Marocco. *Journal of Paleolimnology*. 12: 223-234.

Baruch, U. and S. Bottema, 1991. Palynological evedence for climatic changes in the Levant ca. 17,000 - 9,000 BP. In: O. Bar-Yosef and F.R. Valla, (ed.), The Nutufian culture in the Levant. *Ann Arbir, Michigan, Prehistory Press*. pp. 11-20.

Behbehani, A.R., 1987. Sedimentations- und Klimageschichte des Spät- und Postglazials im Bereich der nördlichen Kalkalpen (Salzkammergutseen, Österreich). *Göttinger Arbeiten zur Geologie und Paläontologie*. 34:

Casanova, J. and C. Hillaire-Marcel, 1993. Carbon and Oxygen Isotopes in African Lacustrine Stromatolites: Paleohydrological Interpretation. *Geophysical Monograph*. 78: 123-133.

Chivas, A.R., P. De Decker, J.A. Cali, A. Chapman, E. Kiss and J.M.G. Shelley, 1993. Coupled stable-isotope and trace-element measurements of lacustrine carbonates as paleoclimatic indicators. In: P.K. Swart, K.C. Lohmann, J.A. McKenzie and S.

Savin, (ed.), Climate change in continental isotopic records. *Washington, USA, American Geophysical Union*. pp. 113-122.

Courty, M.-A., 1994. The scenario of enviromental degradation in Tell Leilan region (NE Syria) during the late third millenium abrupt climate change. NATO ARW on Third Millenium BC abrupt climate change and old world social collapse. *Kemer, Turkey: abstract volume*.

Craig, H., 1957. Isotopic standards for carbon and oxygen and correction factors for mass-spectrometric analysis of carbon dioxide. *Geochim. Cosmochim. Acta*. 12: 133-149.

Craig, H. and L.I. Gordon, 1965. Deuterium and oxygen 18 variations in the ocean and the marine atmosphere. *Conference on stable isotopes in oceanographic studies and paleotemperatures, Spoletto*. 9-130.

Curtis, J.H. and D.A. Hodell, 1993. An Isotopic and trace element study of ostracods from Lake Miragoane, Haiti: A 10.500 year record of paleosalinity and paleotemperature changes in the Caribbean. In: P.K. Swart, K.C. Lohmann, J.A. McKenzie and S. Savin, (ed.), Climate change in continental isotopic records. Washington, USA, *American Geophysical Union*. pp. 135-152.

Dansgaard, W., S.J. Johnsen, H.B. Clausen, D. Dahl-Jensen, N. Gundestrup, C.U. Hammer, C.S. Hvidberg, J.P. Steffensen, A.E. Sveinbjörndottir, J. Jouzel and G. Bond, 1993. Evidence for general instability of past climate from a 250-kyr ice-core record. *Nature*. 364: 218-220.

Degens, E.T. and F. Kurtman, 1978. *The Geology of Lake Van*. Ankara, MTA Press.

Epstein, S., R. Buchsbaum, H.A. Lowenstam and H.C. Urey, 1951. Carbonate-water isotopic temperature scale. *Bull. Geol. Soc. America*. 62: 417-426.

Fontes, J.C., F. Melieres, E. Gibert, L. Qing and F. Gasse, 1993. Stable isotope and radiocarbon balances of two Tibetian lakes (Sumxi Co, Longmu Co) from 13.000 B.P. *Quaternary Science Reviews*. 12: 875-887.

Frey, W. and H. Kürschner, 1989. Vorderer Orient. 1:8000000, Karte A VI 1. Tübinger Atlas des Vorderen Orients. *Wiesbaden, Dr. Ludwig Reichert Verlag*.

Gasse, F., M. Arnold, J.C. Fontes, M. Fort, E. Gilbert, A. Huc, Li Bingyan, Li Yuanfang, Li Qing, F. Méliéres, E. Van Campo, Wang Fubao and Z. Qingsong, 1991. A 13,000-year climat record from western Tibet. *Nature*. 353: 742-745.

Gat, J.R., 1995. Stable isotopes and the water balance of fresh and saline lakes. In: A. Lerman, D.M. Imboden and J.R. Gat, (ed.), Physics and chemistry of lakes. *Heidelberg, Berlin, New York, Springer Verlag*. pp. 139-166.

Gat, J.R. and C. Bowser, 1991. The heavy isotope enrichment of water in coupled evaporative systems. *Stable Isotope Geochemistry*. 3: 159-168.

Gerasimenko, N., 1994. Enviromental and climatic changes between 3 and 5 kyr PB in eastern Ukraine. NATO ARW on Third Millenium BC abrupt climate change and old world social collapse. *Kemer, Turkey: abstract volume*.

Gonfiantini, R., 1986. Enviromental isotopes in lake studies. In: P. Fritz and J.C. Fontes, (ed.), Handbook of enviromental isotope geochemistry. *Amsterdam, Oxford, New York, Tokyo, Elsevier*. pp. 113-163.

Hajdas, I., 1993. Extension of the Radiocarbon Calibration Curve by AMS Dating of Laminated Sediments of Lake Soppen and Holzmaar. *PhD Thesis, ETH Zürich* Nr. 10157. pp. 147.

Heinrichs, H. and A.G. Herrmann, 1990. Praktikum der analytischen Geochemie. *Berlin, Heidelberg, New York, Springer Verlag*. pp. 699.

Kelts, K. and U. Briegel, 1986. The limnogeology-ETH coring system. *Schweizerische Zeitschrift für Hydrologie*. 48: 104-115.

Kempe, S., 1977. Hydrography, Warven-Chronologie und organische Geochemie des Van See, Ost Türkei. *Mitt. Geol. Paläontol. Inst. Univ. Hamburg*. 47: 125-228.

Kempe, S., J. Kazmierczak, G. Landmann, T. Konuk, A. Reimer and A. Lipp, 1991. Largest known mikrobialites discovered in Lake Van, Turkey. *Nature*. 349: 605-608.

Kinsman, D.J.J. and H.D. Holland, 1969. The co-prcipitation of cations with $CaCO_3$-IV. The co-precipitation of Sr with aragonite between 16° and 96°C. *Geochimica et Cosmochimica Acta*. 33: 1-17.

Kipfer, R., A.-H. W., H. Baur, M. Hofer, D.M. Imboden and P. Signer, 1994. Injection of mantle type helium into Lake Van (Turkey): the clue for quantifying deep water renewal. *Earth and Planetary Science Letter*. 125: 357-370.

Lahann, R.W. and R.M. Siebert, 1982. A kinetic model for distribution coefficients and application to Mg-calcites. *Geochimica et Cosmochimica Acta*. 46: 2229-2237.

Lamoureux, S.F., 1994. Embedding unfrozen sediments for thin section preparation. *Journal of Paleolimnology*. 10: 141-146.

Landmann, G., A. Reimer and S. Kempe, 1994. Die Sedimente des Van See. *DFG Projekt Wo 395, 2-1 bis 2-4, final report*.

Lotter, A.F., 1989. Evidence of annual layering in Holocene sediments of Soppensee, Switzerland. *Aquatic Sciences*. 52: 19-30.

Lotter, A.F., B. Ammann, J. Beer, I. Hajdas and M. Sturm, 1991. A step towards an absolute time-scale for the Late-Glacial: annually laminated sediments from Soppensee (Switzerland). In: E. Bard and W.S. Broecker, (ed.), The Last Deglaciation: Absolute and Radiocarbon Chronologies. *Heidelberg, Springer-Verlag*. pp. 45-68.

McKenzie, 1993. Pluvial conditions in the eastern Sahara following the penultimate deglaciation: imlications for changes in atmospheric circulation patterns with global warming. *Palaeo*. 103: 95-105.

Moore, M.M.T. and G.C. Hillman, 1992. The Pleistocene to Holocene transition and human economy in Southwest Asia: the impact of Younger Dryas. *American Antiquity*. 57: 482-494.

Morse, J.W. and F.T. Mackenzie, 1990. Geochemistry of sedimentary carbonates. *New York, Elsevier Science Publishing Co*.

Mucci, A. and J.W. Morse, 1983. The incorporation of Mg and Sr into calcite overgrowths: influences of growth rate and solution composition. *Geochimica et Cosmochimica Acta*. 47: 217-233.

O'Neil, J.R., R.N. Clayton and T.K. Mayeda, 1969. Oxygen isotope fractionation in divalent metal carbonates. *J. Chem. Phys.*. 51: 5547-5558.

Oomori, T., H. Kaneshima and Y. Maezato, 1987. Distribution coefficient of Mgions between calcite and solution at 10-50°C. *Marine Chemistry*. 20: 327-336.

Pingitore Jr., N.E. and M.P. Eastman, 1986. The coprecipitation of Sr with calcite at 25° and 1 atm. *Geochimica et Cosmochimica Acta*. 50: 2195-2203.

Reimer, A., G. Landmann and S. Kempe, 1994. Wasserchemie des Van Sees, seiner Zuflüsse und der Porenwässer. *DFG Projekt Wo 395/2-1 bis2-4, final report.*

Rubinson, M. and R.N. Clayton, 1969. Carbon-13 fractionation between aragonite and calcite. *Geochim. Cosmochim. Acta*. 33:

Schlanger, S.O., 1965. Dolomite-evaporite relations on Pacific islands. *Sci. Rep. Tohoku Univ*. 37: 15-29.

Schöll, M. and E. Faber, 1978. Oxygen isotope study of Lake Van sediments and waters. In: E.T. Degens and F. Kurtman, (ed.), *The geology of Lake Van*. Min. Res. Explor. Inst. Türkey.

Schwarzenbach, R.P., P.M. Gschwend and D.M. Imboden, 1993. Enviromental organic chemistry. *New York, John Wiley &Sons, Inc*. pp. 681

Schweizer, G., 1975. Untersuchungen zur Physiogeographie von Ostanatolien und Nordwestiran. Geomorphologische, klima- und hydrogeographische Studien im Van See- und Rezaiyehsee-Gebiet. *Tübinger Geographische Studien*. 60/9:

Stabel, H.-H., J. Küchler-Krischun, J. Kleiner and P. Merkel, 1986. Removal of strontium by coprecipitation in Lake Constance with calcite. *Naturwissenschaften*. 73: 551.

Street-Perrot, F.A. and N. Roberts, 1983. Fluctuation in closed basin lakes as an indicator of past atmospheric circulation patterns. In: F.A. Street Perrot, M. Beran and R. Ratcliffe, (ed.), Variations in the global water budged. *D. Reidel Pblishing Company*. pp. 331-345.

Tarutani, T., R.N. Clayton and T.K. Mayeda, 1969. The effect of polymorphism and magnesium substitution on oxygen isotope fractionation between calcium carbonate and water. *Geochimica et Cosmochimica Acta*. 33: 987-996.

Usdowski, H.E., 1967. Die Genese von Dolomit in Sedimenten. Mineralogie und Petrographie in Einzeldarstellungen. *Berlin, Heidelberg, New York,, Springer Verlag*.

Weiss, H., M.-A. Courty, W. Wetterstrom, F. Guichard, L. Senior, R. Meadow and A. Curnow, 1993. The Genesis and Collapse of Third Millennium North mesopotamian Civilisation. *Science*. 261: 995-1004.

Weiss, H. and M.A. Courty, 1993. The genesis and collaps of the Akkadian Empire: the accidental refraction of historical law. In: M. Liverani, (ed.), Akkad: The first world empire. Padua: Sargon.

Wright, H.E.J., 1993. Enviromental determinism in Near Eastern prehistoric archaeology. *Current Antropology*.

Zeist van, W. and S. Bottema, 1991. Late Quaternary vegetation of the Near East. Beihefte zum Tübinger Atlas des Vorderen Orients. *Wiesbaden, Dr. Ludwig Reichert Verlag*.

Zohary, M., 1973. Geobotanical Foundations in the Middle East. Stuttgart, Amsterdam,

Zolitschka, B., 1994. Paleoclimatic and anthropogenic influences on the formation of annually laminated sediments from Lake Holzmaar, Germany. NATO ARW on Third Millenium BC abrupt climate change and old world social collapse. *Kemer, Turkey: abstract volume*.

Climate Change at the End of the Third Millennium BC - Evidence from Varved Lacustrine Sediments

Bernd Zolitschka[1,2] & J.F.W. Negendank[2]

[1]Fachbereich VI/Geologie, Universität Trier, D-54286 Trier, Germany

[2]GeoForschungsZentrum Potsdam, Telegrafenberg A26, D-14473 Potsdam, Germany

Abstract

The sedimentary record of Lake Holzmaar provides an absolute varve chronology confirmed by AMS-14C data. Before 2500 BC and after 1800 BC depositional processes were controlled by both human and palaeoclimatic influences. Between the end of the Neolithic and the onset of the Bronze Age no anthropogenic interference with the environment is suggested by the archaeological data. An increase in dry density and magnetic susceptibility during the period 2500 to 1800 BC (in sideral years) is therefore, attributed to palaeoclimatic causes. These sediment changes are interpreted to be the result of increased runoff related to an increase of precipitation and/or colder temperatures which is in agreement with other proxy records from northern and central Europe, indicating cold and wet conditions at this time.

Introduction

The understanding of short-term century- to millennium-scale climatic changes is essential to further the knowledge of natural climate variability as a baseline upon which present and future anthropogenic influences are superimposed. Intervals like the "Little Ice Age" (Bradley & Jones, 1993) and the Younger Dryas (Bard & Broecker, 1991) are the best-studied climatic changes of this kind in the late Quaternary. But still there are serious problems of how such climatic fluctuations have to be correlated on a regional or a global scale. Such information is necessary to understand the forcing of these intermediate frequencies in climate variability. The lack of an accurate dating is one of the major obstacles for comparing different palaeoclimatic proxy records from all over the world. Resolution of most records is restricted to several centuries or possibly millennia, making the deciphering of century to millennium-scale climatic fluctuations very difficult or impossible. Therefore, high resolution records are needed from lacustrine and marine sediments,

NATO ASI Series, Vol. I 49
Third Millennium BC Climate Change
and Old World Collapse
Edited by H. Nüzhet Dalfes, George Kukla,
and Harvey Weiss

from ice cores, corals, and tree rings to improve our knowledge about these climatic variations.

Recently, archaeological investigations in combination with soil stratigraphic and soil dynamic analyses in Mesopotamia suggested an increase in aridity and dust storms for the time 2200 BC to 1900 BC (Weiss et al., 1993). A relatively precise chronological framework of this study is provided by archaeological data. This climatic deterioration had severe consequences for the rain-fed agricultural societies in the Near East which basically collapsed due to the lack of precipitation. At the end of the third millennium BC not only agricultural civilizations in Mesopotamia collapsed but also societies in the Nile and the Indus valleys (Kukla, in this volume) pointing to a more regional rather than a local signal. If this climatic event is indeed of regional or global importance, it should be recorded in other high resolution palaeoclimatic proxy archives. The scope of this study is to find out, whether annually laminated lacustrine sediments from Lake Holzmaar (Germany) give any evidence for a climatic event around 2000 BC in central Europe. The information obtained will be compared with other European and African palaeoclimatic proxy-data to validate its regional significance.

Results

Lake Holzmaar (425 m a.s.l.; 6°53'E, 50°7'N) is located in the West Eifel Volcanic Field of the Rhenish Massif, 45 km northeast of the city of Trier, Germany (Fig. 1). The lake is circular in shape and was formed by phreatomagmatic maar eruptions ca. 40,000 to 70,000 years ago (Büchel, 1993). The dimictic, holomictic, and mesotrophic to eutrophic lake is 20 m deep and has a catchment area of 2.06 km≤.

Several sediment cores were recovered from the center of this lake (Fig. 1) to a maximum depth of 31.6 m. The uppermost 11.4 m consist of 13,800 organic varves (Zolitschka 1990), whereas the basal part is composed of clastic periglacial varves and extends back to 22,500 years (Brauer 1994, Brauer et al. 1994). The annual nature of the organic varves was confirmed by their internal composition (Zolitschka 1991). More than 40 AMS-14C-dates were made to check the varve chronology (Hajdas et al. 1995). After dendro-calibration of these radiocarbon dates, a comparison with varve ages revealed an offset of +878 years between both absolute chronologies. Significant discrepancies occur only between 3500 and 5000 cal. BP, such that the lower part of the Lake Holzmaar varve chronology must be considered

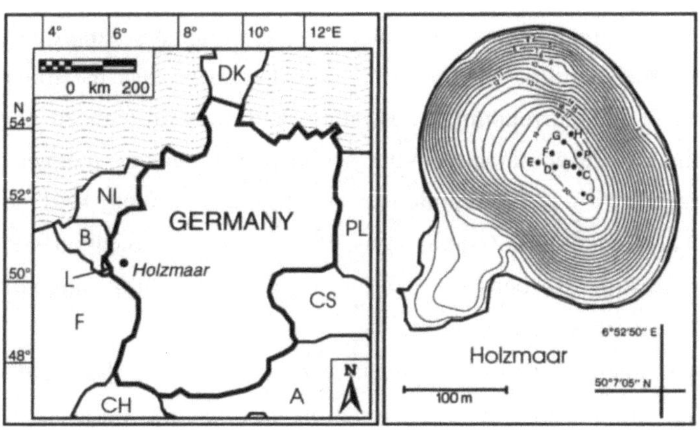

Fig. 1: Location and bathymetric map of Lake Holzmaar with indicated coring sites (isobaths in meters).

a floating chronology (Hajdas et al., 1995). These difficulties are not due to any hiatus but to extremely thin annual laminations that are poorly developed and preserved. Recent attempts to recover new sediment cores with better varve preservation from other locations of the same lake were not completely successful. Although it was possible to narrow down the zone with indistinct varves to from +878 to +346 years (Zolitschka, 1996), but the varve chronology beyond 5000 cal. BP remains floating. Nevertheless, this floating chronology is in good agreement with other radiocarbon-dated varve records until 12.000 cal. BP, like Lake Gosciaz in Poland (Goslar et al., 1995) or Lake Soppensee in Switzerland (Hajdas et al., 1993). Applying the corrected time scale to palaeomagnetic and pollen data, a good agreement with other regional records is obtained (Haverkamp & Beuker 1993, Stockhausen 1996, Usinger & Wolf 1991).

Varved lacustrine sediments provide two different and independent sources of information. First, an internal absolute chronology in calendar (sidereal) years based on varve counts. Second, it provides a time series of varve thickness measurements. Varve thickness data were smoothed by calculating mean values for each centimeter. These sedimentation rate data (Fig. 2) reflect processes within the lacustrine environment, like fluctuations of algal productivity (autochthonous sedimentation) and processes within the catchment, such as soil erosion (allochthonous sedimention). Primarily, these processes are controlled by palaeoclimatic conditions but since the beginning of the Neolithic (ca. 5300 BC) man started to clear the forest in the catchment, modified the surrounding vegetation and thus influenced the composition of the sediments. In combination with sediment dry density, magnetic susceptibility (Stockhausen, 1996), selected pollen spectra (Usinger & Wolf, 1991) as well as historic and archaeological data (Anonymous, 1977; Lüning & Stehli, 1989) an attempt was made to distinguish between natural (climatic) and anthropogenic causes for the fluctuations in sedimentation rates.

Relatively stable environmental conditions prevailed during most of the middle Holocene. Between BC 6000 and 750 sedimentation rates show only slight variations related to human activities in the catchment (Fig. 2). Anthropogenic influences modify varve thickness through cultural eutrophication causing increased algal blooms (autochthonous sedimention) and, at that time to a much lesser extent, through enhanced soil erosion increasing the second important allochthonous sediment component. Conditions changed considerably with the beginning of the Iron Age (750 BC), when man started to change the environment severely. Production of charcoal for iron melting and working reduced tree populations rapidly. From now onwards, except for the "Migration Period" and the early Middle Ages (400 to 950 AD), vegetation and soils were no longer in equilibrium with the prevailing climatic conditions. The resulting destabilization of the catchment caused erosion of top soils which were washed into the lake. A nearly threefold increase in sedimentation rates was the consequence (Fig. 2).

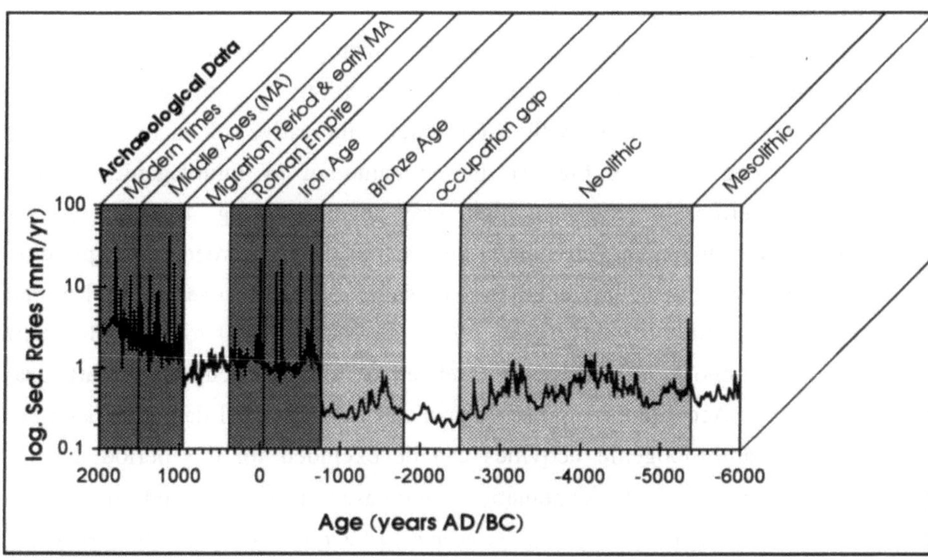

Fig. 2: Sedimentation rates in comparison with periods of settlement (shaded) according to the archaeological record. Dark shading indicates severe human influences.

The severe human perturbation in the catchment of Lake Holzmaar since 750 BC is not only reflected in the sedimentation rates. The pollen records (Fig. 3) of non-arboreal pollen contents (NAP) and sorrel (*Rumex*) increased as well, giving further evidence for very intense woodland clearance during the Iron Age, as well as during the Roman Empire, the Middle Ages, and modern times. During the "Migration Period" forest regained its territory as indicated by a peak of birch (*Betula*) pollen which indicates the beginning of a succession towards woodland. In general, the peak of birch pollen is an approximation for the intensity of the preceeding forest clearing, indicating a steady intensification of land use since Neolithic times.

Prior to 750 BC there are only minor increases in the non-arboreal pollen contents usually succeeded by birch peaks. In combination with the absence of *Rumex* pollen grains this points to less intense methods of cultivation and probably an incomplete removal of the trees. Especially the period 2500 to 1800 BC does not show any sign of human activities in the pollen data, thus corroborating the archaeological record of the immediate surroundings of the site.

Additional information about allochthonous (minerogenic) deposition is obtained from dry density and magnetic susceptibility data (Fig. 4). Both records show a certain degree of covariation with high values indicative of increased clastic deposition (shaded areas in Fig. 4). If sedimentation of minerogenic deposits coincides with periods of settlement (light gray shading in Fig. 4), enhanced soil erosion, resulting from human activities in the catchment, is regarded as the principal cause. Nevertheless, climatic influences may be superimposed or even control human settlement activities to a certain degree. Only one period (2500 to 1800 BC) with increased allochthonous influence is not related to human occupation (dark gray shading in Fig. 4). According to archaeological and palynological data enhanced soil erosion due to anthropogenic activities can be excluded for this period. Natural causes remain as a viable explanation. Increased runoff caused by higher precipitation and/or colder temperatures seems to be responsible for the increased amount of allochthonous deposition. A marked difference between high minerogenic influx related to enhanced soil erosion and increased clastic deposition related to higher runoff was recognized in the relationship of dry density with sedimentaton rates. Periods with settlement activities show high dry densities which coincide with high sedimentation rates (light shading in Fig. 4). This is a combined result of cultural eutrophication (increased autochthonous deposition) and soil erosion (increased allochthonous deposition). However, the period around 2000 BC with increased runoff but without human occupation is different: high dry density is related to low sedimentation rates (dark shading in Fig. 4). This can be explained by a decline of autochthonous productivity due to a lack in nutrient supply in combination with unfavourable (colder/wetter) climatic conditions and by lower sediment transfer rates of natural erosion as compared to human accellerated soil erosion.

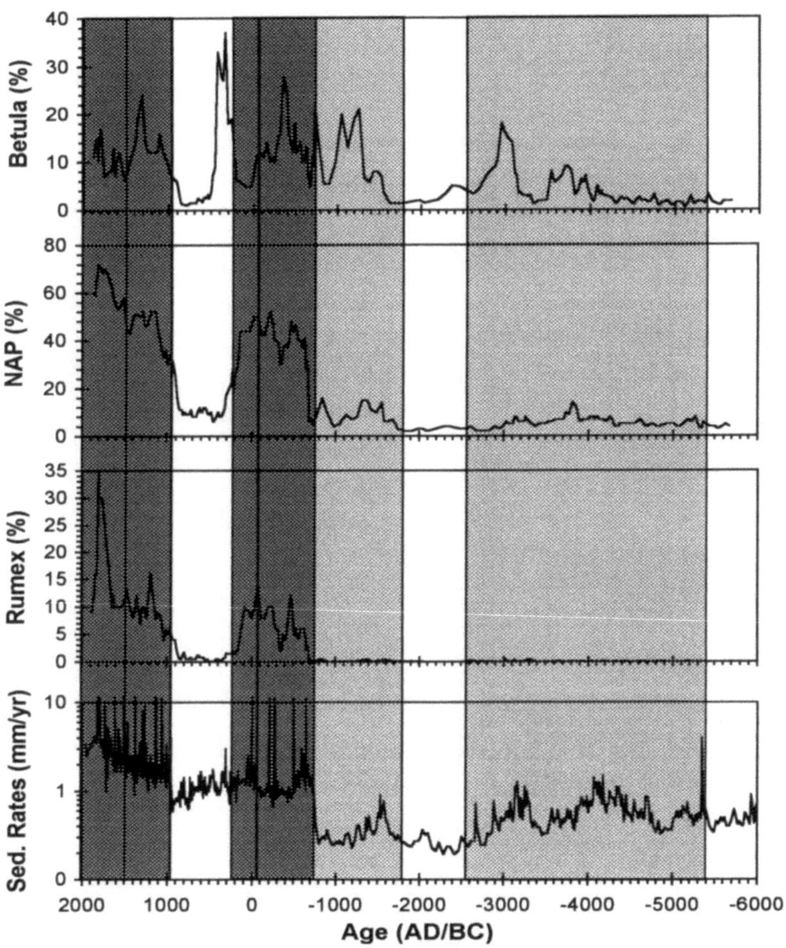

Fig. 3: Sedimentation rates in comparison with sorrel (*Rumex*), birch (*Betula*), and non-arboreal (NAP) pollen contents (Usinger & Wolf, 1991). Shading indicates periods of settlements (cf. Fig. 2).

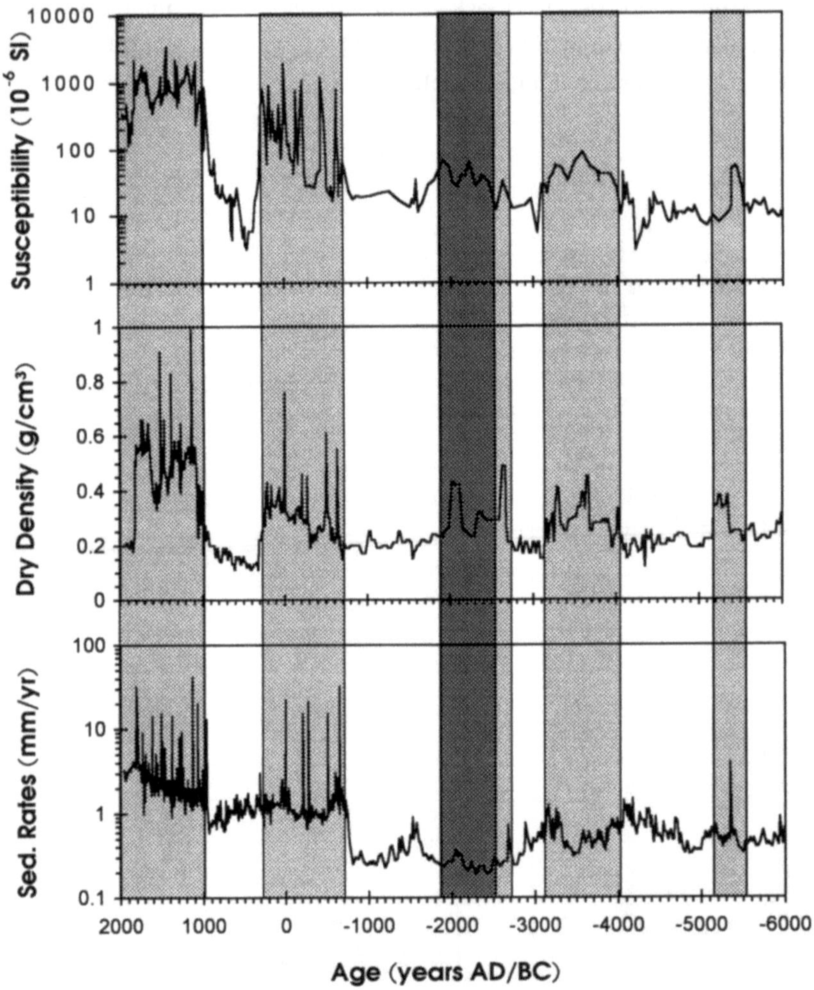

Fig. 4: Sedimentation rates in comparison with dry density and magnetic susceptibility (Stockhausen, 1996). Light shading indicates increased allochthonous (minerogenic) deposition during periods of settlement whereas dark shading marks increased allochthonous deposition occurring without human influence.

Discussion

Increased dry density and magnetic susceptibility of the Lake Holzmaar record between 2500 and 1800 BC are interpreted in terms of higher runoff related to increased rainfall and/or colder temperatures peaking during the last century of the third millennium BC. Questions may arise about the climatic interpretation of proxy parameters, like dry density and magnetic susceptibility. Many processes leading to a higher flux of minerogenic particles into the lake, like eolian or volcanic activities, slumps or turbidites, exist, but they can be ruled out because the detailed microstratigraphic (microscopic) sediment inspection gives no evidence for such processes. Still there is the possibility of rare events, like landslides in the catchment, that increase the sediment availability and thus average climatic conditions can transfer far more suspended sediment than usual. On the other hand, such an event should also create turbidites or slumps which are absent. Another uncertainty remains: regional validity. As long as the record is unreplicated, doubts in the confidence of the interpretation should remain. There is no other well studied varved record available yet which might prove or disprove the above interpretation. Therefore, this problem has to be overcome by looking at other palaeoclimatic proxy archives and comparing their information with those from Lake Holzmaar.

In northern Europe glaciers of Jostedalsbreen (Norway) reached their first maximum extension after the Holocene climatic optimum at 3500 yr BP (1800 BC) (Nesje et al., 1991). At 3600 yr BP (1900 BC) glaciers in the Austrian Alps (Patzelt, 1977) and in the Swiss Alps (Zoller, 1977) advanced to the "Löbben" stage. There is also coincidence with lake level fluctuations in the French Jura and Pre-Alps (Magny, 1993) which document the most severe lake level rise of the the last 6000 yr for the "Pluvis" transgression around 3600 yr BP (1900 BC). All these evidences point to a distinct climatic deterioration: cooling in combination with increased precipitation which favour glacier advances and rising lake levels at the beginning of the second millennium BC. From varved sediments of Lake Saki (Crimea, Ukraine) another palaeoclimatic proxy exists. Thickness variations of these annual laminations were interpreted in terms of precipitation indicating high rainfall for 2300 BC (Schostakowitsch, 1936). This interpretation is still in discussion. Varve thickness at the site of Lake Sacki may also be regarded as the result of increased dustiness (Kukla 1994, pers. comm.) and therefore indicate increased aridity.

Clear evidence for an increase in aridity is reported from Spain (Julia et al., 1994). At the site of Ebre evergreen oak pollen have a maximum at 3900 yr BP (BC 2500) and at the site of Elx an increase in polyhaline biota occurs at 3980 yr BP (BC 2600). Both records are interpreted as the result of increases in aridity. Climatic deterioration and not human influence is also the reason for a dramatic reduction in arboreal pollen contents in the sediments of three Italian lakes from the Rome area around 3500 yr BP (1800 BC) (Magri & Follieri, 1989; Magri in this volume).

From Egypt low Nile floods are reported for 2200 BC which decreased agricultural productivity and destabilized the Egyptian state (Hassan, in this volume). This is related to a decrease of precipitation in equatorial Africa resulting in less runoff within the Nile catchment basin: this decrease is also shown by marked lake-level drops around 3500 yr BP (1800 BC) in four tropical African lakes (Street-Perrott & Harrison, 1984). A correlation was also found between low Nile discharge and cold climates in Europe (Hassan, 1981); and marked by global fall of sea-level of advance of glaciers ("Neoglacial II" of Fairbridge, 1976, p. 542).

Combining all the varied information it becomes evident that northern and central Europe experienced predominantly cold and wet climatic conditions around 2000 BC confirming that Lake Holzmaar sediments carry a regional palaeoclimatic signal. On the other hand the Mediterranean region, the Near East, and northern Africa experienced more arid conditions during that time. Neglecting the limited accuracy of dating for some of the records the difference between the northern and southern data sets might be explained by a response time lag of the glaciers, reaching the maximum extension at the end of a cold period. This is corroborated by the Lake Holzmaar record which shows first indications of rainfall intensification already around 2500 BC and thus agrees with the data from the Mediterranean and from Africa.

In conclusion, during the second half of the third millennium BC cooling and increased precipitation occurred in high latitudes, while mid- and low latitudes can be characterized by drier conditions. How this relates to the general circulation of the atmosphere remains to be discussed.

Acknowledgements

We greatly appreciate the comments made by Hartwig Löhr and Rhodes Fairbridge on an earlier draft of this manuscript. The study was supported by a grant from the German National Science Foundation (DFG Ne 154/28-1).

References

Anonymous, (1977) Führer zu vor- und frühgeschichtlichen Denkmälern, Vol 33: Südwestliche Eifel, Mainz; pp 341.

Bard E, Broecker WS (eds) (1991) The last deglaciation: Absolute and radiocarbon chronologies. NATO ASI Ser. I2: 1-344.

Bradley, R. S., Jones, P. D. (1993) "Little Ice Age" summer temperature variations: their nature and relevance to recent global warming trends. *The Holocene* 3: 367-376.

Brauer A (1994) Weichselzeitliche Seesedimente des Holzmaares - Warvenchronologie des Hochglazials und Nachweis von Klimaschwankungen. Documenta naturae 85: 1-210

Brauer A, Hajdas I, Negendank JFW, Rein B, Vos H, Zolitschka B (1994) Warvenchronologie - Eine Methode zur absoluten Datierung und Rekonstruktion kurzer und mittlerer solarer Periodizitäten. Geowiss 12: 325-332

Büchel G (1993) Maars of the Westeifel. In: Negendank JFW, Zolitschka B (eds), Paleolimnology of European maar lakes, Lecture Notes in Earth Sci 49: 1-13

Fairbridge, R. W. (1976) Effects of Holocene climate change on some tropical geomorphic processes. *Quat. Res.*, 6: 529-556.

Goslar, T., Arnold, M., Bard, E., Kuc, T., Pazdur, M. F., Ralska-Jasiewiczowa, M., Rozanski, K., Tisnerat, N., Walanus, A., Wicik, B. & Wieckowski, K. (1995): High concentration of atmospheric 14C during the Younger Dryas cold episode. Nature 377, 414-417

Hajdas, I., Ivy, S.D., Beer, J., Bonani, G., Imboden, D., Lotter, A., Sturm, M. & Suter, M. (1993): AMS radiocarbon dating and varve chronology of Lake Soppensee: 6000 to 12,000 14C years BP. Climate Dynamics 9, 107-116

Hajdas I, Zolitschka B, Ivy-Ochs SD, Beer J, Bonani G, Leroy SAG, Negendank JFW, Ramrath M, Suter M (1995) AMS radiocarbon dating of annually laminated sediments from Lake Holzmaar, Germany. Quat Sci Rev 14: 137-143

Hassan FA (1981) Historical Nile floods and their implications for climatic change. Science 212: 1142-1145

Hassan FA (this vol) Mid-to-late Holocene climatic change between 3 and 5 ka BP in NE Africa and its impact on culture change. NATO-ASI Ser I

Haverkamp B, Beuker T (1993) A palaeomagnetic study of maar-lake sediments from the Westeifel. In: Negendank JFW, Zolitschka B (eds), Paleolimnology of European maar lakes, Lecture Notes in Earth Sci 49: 349-365

Julia R, Negendank JFW, Seret G, Brauer A, Burjachs F, Endres C, Giralt S, ParÈs JM, Roca JR (1994) Holocene climatic change and desertification in the western Mediterranean region. Terra Nostra (Schr Alfred-Wegener-Stiftung) 1/94: 81-84

Kukla G (this vol) Paleoclimatic data from the second and third millennium BC: overview and the problems. NATO-ASI Ser I.

Lüning J, Stehli P (1989) Die Bandkeramik in Mitteleuropa: von der Natur- zur Kulturland-schaft. Spektrum Wiss 1989: 78-88.

Magny M (1993) Holocene fluctuations of lake levels in the French Jura and sub-Alpine ranges, and their implications for past circulation patterns. *The Holocene* 3: 306-313.

Magri D (in this Vol.) Mid-Holocene vegetation changes in Italy. NATO-ASI Ser I.

Magri D, Follieri M (1989) Caratteri della biostratigrafia pollinica dell 'Olocene in Italia centrale. *Mem. Soc. Geol. Italia* 42: 147-153.

Nesje A, Kvamme M, Rye M, Lovli R (1991) Holocene glacial and climate history of the Jostedalsbreen region, western Norway: evidence from lake sediments and terrestrial deposits. *Quat. Sci. Rev.* 10: 87-114.

Patzelt G (1977) Der zeitliche Ablauf und das Ausmaß postglazialer Klimaschwankungen in den Alpen. In: Frenzel B (ed), Dendrochronologie und Klimaschwankungen in Europa, Steiner (Wiesbaden); 248-259.

Schostakowitsch WB (1936) Geschichtete Bodenablagerungen der Seen als Klima-Annalen. Meteorol Zeitschr 5/1936: 176-182.

Stockhausen H (1994) New palaeomagnetic secular variation data of three maar lakes from the Westeifel (Germany) for the last 12,000 years. Terra Nostra (Schr Alfred-Wegener-Stiftung) 1/94: 90-98.

Stockhausen H (1996) Die paläomagnetische Säkularvariation holozäner Maarsedimente der Westeifel. PhD Thesis Univ Münster, 216pp.

Street-Perrott FA, Harrison SP (1984) Temporal variations in lake levels since 30,000 years BP - an index of the global hydrological cycle. In: Hansen JE, Takahshi T (eds), Climate processes and climate sensitivity, AGU (Wash., DC); 118-129.

Usinger H, Wolf A (1991) Biostratigraphie mit Hilfe der Pollenanalyse und einige Aussagen zur klimatischen Entwicklung und Besiedlung durch den Menschen. In: Negendank JFW, DFG-Abschlußbericht zum Projekt "Westeifel-Maare und Quartärstratigraphie", 12-15.

Weiss H, Courty M-A, Wetterstrom W, Guichard F, Senior L, Meadow R, Curnow A (1993) The Genesis and Collapse of third millennium north Mesopotamian civilization. Science 261: 995-1004.

Zolitschka B (1990) Spätquartäre jahreszeitlich geschichtete Seesedimente ausgewählter Eifelmaare. Documenta naturae 60: 1-226

Zolitschka B (1991) Absolute dating of late-Quaternary lacustrine sediments by high resolution varve chronology. *Hydrobiology* 214: 59-61

Zolitschka, B. (1996): Paläoklimatische Bedeutung laminierter Sedimente. Habilitation Thesis, Universität Potsdam, 196 p.

Zoller H (1977) Alter und Ausmaß postglazialer Klimaschwankungen in den Schweizer Alpen. In: Frenzel B (ed), Dendrochronologie und Klimaschwankungen in Europa, Steiner (Wiesbaden); 271-281.

Environmental Vulnerability of Early Societies: Some Reflections on Modeling Issues

H. Nüzhet DALFES

İstanbul Technical University
Department of Geology
80626 Maslak Istanbul
Turkey

Abstract

Understanding the vulnerability of early societies to climatic fluctuations and local environmental degradation requires development of a hierarchy of system dynamical models. It is through the development and verification process of such models that we will arrive at explanations of abrupt demographic and/or cultural changes documented by archaeology. Any modeling process will have to ask questions and make decisions about the definition of the system of interest, the appropriate spatial and demographic aggregation level to be adopted and details of the physical, ecological and cultural processes to be included.

Introduction

Humans and climate change has always been a subject of active research and scientific debate. Whether the question pertains to the evolution of African primates leading to hominids, or to drastic social changes in already complex socioeconomic human systems like in the subject matter of this book, links have never been easy to establish.

Furthermore, although the interaction of humans with their physical/biological environment can be thought, at first, as one-way interaction, rising consciousness about environmental issues is contesting this basic premise. It became nowadays an accepted fact that with the advent industrial revolution, the interaction of climate and humans became bi-directional; humans have been modifying the components of the earth system at *global* and *regional* scales. Any careful analyses of the pre-industrial era will provide

NATO ASI Series, Vol. I 49
Third Millennium BC Climate Change
and Old World Collapse
Edited by H. Nüzhet Dalfes, George Kukla,
and Harvey Weiss

us with many examples were human activities did have serious impacts on *local* environmental systems which in turn might have *synergetically* brought additional vulnerabilities to global/local scale changes.

The causal route that goes from climate to biological evolution or to social response intersects the grounds of many disciplines. These disciplines are usually at very different levels of intellectual 'maturity' and certainly do not speak the same language. Therefore any attempt to build a solid causal chain need an integration platform. Our claim is that the methodology of *system dynamics* (Forrester, 1969) can help to bridge above-mentioned disciplinary gaps. As it will discussed in this paper, many bits of information can be brought together and through model sensitivity analyses key parameters and/or structures can be discerned through formal model building and simulation.

Systems thinking is not new to environmental issues nor to archaeology. In the 1970's system dynamical models constituted the scientific basis for then influential reports of the *Club of Rome* (see Meadows *et al.,* 1972, Meadows *et al.,* 1992 for detailed accounts). There is a current revival of the global multidisciplinary modeling approach under the theme of *integrated assessment* (CIESIN, 1995).

The application of systems approach came into archaeology in the late 1960's (Flannery, 1968) and played a central role in the development of *New Archaeology.* Later, there has been several examples of the use of simulation techniques within the systems framework as *explanatory tools* for drastic social/cultural changes (Hosler *et al.*, 1977; Lowe, 1985). Johnson and Gould (1984) using this methodology attacked the very problem of this volume: can climate fluctuations compounded with other local environmental degradational processes (such as soil salinization and silting associated with irrigation) explain the breakup of Mesopotamian civilizations?

This paper will attempt to formulate a series basic questions to be asked for system dynamical modeling of the response of an early agrarian society to climate fluctuations. The modeling questions will be mostly limited to demographic and ecological/agricultural aspects; no attempt to venture into the dynamics of cultural change *per se* will be made. We will conclude by summarizing the key issues the proposed modeling program has to address.

Why model?

It is certain that to establish causal links between climate change and history of human societies one has to go beyond mere correlations. We have to *explain* how at given time a climate event might have led to a given social/cultural consequence. Toward this aim, we have to first put together a set of *elements* and *interactions* with *appropriate characteristic time and space scales* that will constitute our physical/cultural *system*. A particular abstract *representation* of this system will be our *model*.

A model can have a representation in a variety of 'languages': natural languages, graphics, or mathematical equations, *etc*. Our main task is to deduce the implications of our particular model for the time/space behavior of the system of interest. The implications have to be concrete enough to be comparable to (paleoenvironmental and archaeological) observations; for our model has to be *refutable*.

When the number of elements and interactions increases, and the variables become quantitative, or at least, semi-quantitative, the natural language becomes ineffective and the need arises for mathematical/computational frameworks to handle the complexity. For our specific problem this is the case. This ability to handle quantitative relationships opens many alleys:

a. One can "borrow" present-day process models for a more accurate description (parameterization) of the interactions among the elements of the studied system, *i.e.*, today might guide us to understand the past.

b. One can calibrate (and this is necessity, see below) our model or submodels against quantitatively reconstructed paleoenvironmental parameters.

c. One can encounter many surprises of the world of complex and/or nonlinear systems (for a popular account of this new realm that numerical computation opened for us, see Gleick, 1987) and discover the potential inherent unpredictability in such systems.

What to model and how to model: methodological issues

The starting point is the definition of the system, i.e., of its elements and interactions. As stated above, the particular problem of climatic/environmental vulnerability dictates that both the elements of the 'physical environment' and appropriate parts of the socioeconomic system has to be included (Fig. 1). The system has to encompass appropriate local/regional geospheric and biospheric elements as factors in the food production and as interfaces to global/regional-scale climate change.

It is essential for any systematic modeling program should adopt a hierarchical approach. Instead of building a single model, there are many demonstrated advantages (as it has been advocated very early in the field of climate modeling, see Schneider and Dickinson, 1974) in building a series of models going from the simple to complex. Each additional level of disaggregation or each new dimension will bring its own set of parameterization problems, but will also enable the modeler to estimate the sensitivity of the results to model complexity. For example, the impact of food availability on the demographics can be first modeled with the vague carrying capacity concept; the next level of complexity can incorporate an explicit relation between nutritional level and vital statistics such as fertility, mortality, etc.

The socioeconomic system has to be accounted for with some detail, at least to include the demographics and agricultural decision making. Ultimately, the 'social learning' processes and through them cultural change should be simulated.

It is crucial for the model to simulate among others some specific variables that we will call *observables* that are reconstructable from field observations and laboratory analyses. Time series of various estimators of population size or density, land use extent, crop mixtures, *etc*., should be comparable to model simulated system variables.

The boundary of the system in the geospheric-biospheric realm is the interface through which external 'inputs' such as fluctuating climate will be applied. Model's mathematical/computational structure should permit a range of climate parameters to be defined in terms of kind (temperature, precipitation, wind, *etc*.) and in time granularity.

The boundary on the socio-cultural-economic side can be much more fuzzy. Decision making, and risk perception and 'learning' (Rosen, this volume) in the context of cognitive systems is certainly interesting, but challenging modeling task.

For a complete assessment of the impact of regional-scale climatic change one must deal with a distributed system (*e.g*., a network) of communities spread on the 'climate and resources landscape' and take into account the movement of goods and humans, *e.g*., *trade* and *migration* processes. In such a case, the system can be composed

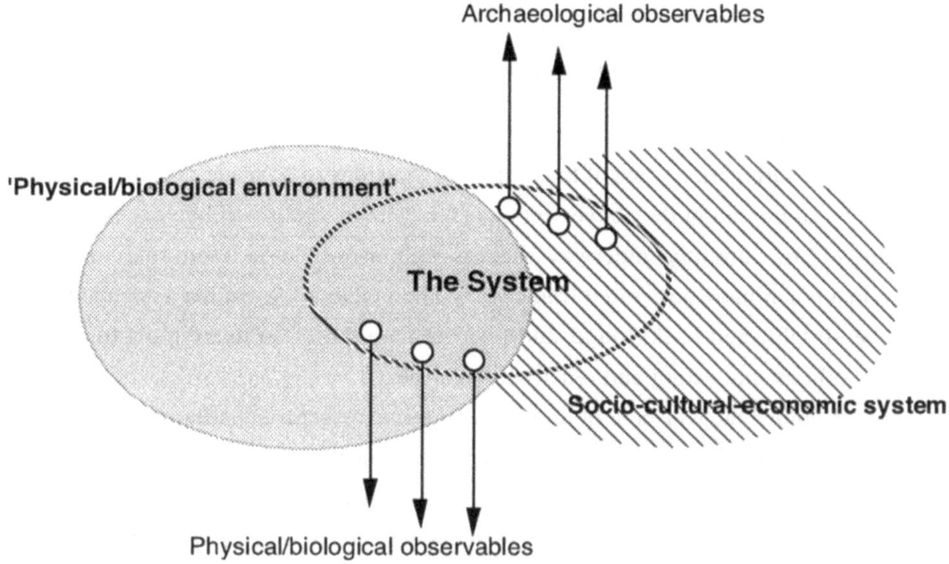

Fig. 1 Defining the system to be modeled with respect to the geopheric-biospheric environment and socio-cultural-economic system. The state of the system can be 'monitored' through a set of reconstructed observables.

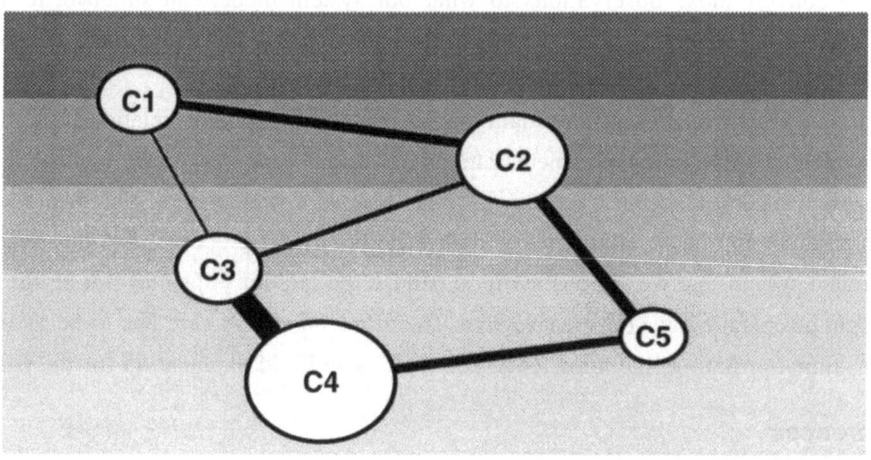

Fig. 2. A network of trading communities spread on a climate and/or other resource gradient.

replicate of the unit model built and tested for a single community. Population and food 'stocks' of each community will be interconnected.

Concluding comments

We tried to summarize some of the modeling decisions to be made and methodological issues to be addressed in the above sections. We believe that it will be very helpful if the modeling efforts currently underway such as Wilkinsons' (Wilkinson, this volume) can be cast in a much more structured framework. The methodology of the systems dynamics has a well-established background and there is a wealth of software tools to help with model development, testing and sensitivity analyses.

Another point that will benefit from quantitative or semi-quantitative frameworks is related to the paleoenvironmental and archaeological reconstructions. It should be always explicit that there is no such thing as a model-free reconstruction. In all cases, many assumptions are made about the system (whether a vegetational community or a human settlement) to convert the 'raw' observations to physically, biologically or anthropologically meaningful variables. Therefore, between the system model that we are trying to built and the 'raw' observations, there are always implicit model layers that are used to convert those observations to what our system model can simulate. It should nevertheless be preferable whenever possible to avoid this layer and have the system model produce directly observable quantities and/or features.

As stated before, present-day data on geospheric-biospheric relationships should guide us in the developing parameterizations of various processes to be incorporated in the system model. One must be warned that there is a fundamental pitfall in such an uniformitarian approach: although basic physical/biological processes are the same, the integrated system that we are observing to deduce the relationships may not be the same or might be operating at a different regime. Therefore, an utmost care has to be exercised when 'time translating' the knowledge about today to build explanation for the past.

References

Consortium for International Earth Science Information Network (CIESIN). (1995) *Thematic Guide to Integrated Assessment Modeling of Climate Change* [online]. University Center, Michigan.
CIESIN URL: http://sedac.ciesin.org/mva/iamcc.tg/TGHP.html

Flannery, K. V. (1968) Archaeological Systems Theory and the Early Mesoamerica, in *Anthropological Archaeology inthe Americas* (B. J. Meggers, ed.), 67-87. Anthropological Society of Washington.

Forrester, J. W. (1969) *Urban Dynamics*. The MIT Press: Massachusetts.

Gleick, J. (1987) *Chaos: Making a New Science*. Viking Penguin: New York.

Johnson, D.L. and Gould H. (1984) The effect of climate fluctuations on human populations: a case study of Mesopotamian Society, in *Climate and Development* (A. Biswas, ed.), Tycooly International Pub.: Dublin.

Hosler, D.H., Sabloff, J.A. and Runge, D. (1977) Simulation model development: a case study of the Classic Maya collapse, in *Social processes in Maya prehistory* (N.Hammond, ed.), 553-90. Academic Press: New York & London.

Meadows, D. and Meadows, D. and Randers, J. and Behrens, W. W. (1972) *The Limits to Growth*. Universe Books: New York, NY.

Meadows, D. H., Meadows D., Randers J. (1992) *Beyond the Limits: Confronting Global Collapse, Envisioning a Sustainable Future*. Post Mills, VT: Chelsea Green Publishing Co.

Lowe, J.W.G. (1985) *The dynamics of apocalypse: a systems simulation of the Classic Maya collapse*. University of New Mexico Press: Albuquerque.

Rosen, A. M. (this volume) Environmental change and human adaptational failure at the end of the Early Bronze Age in the Southern Levant: A view from the Bronze Age Upper Mesopotamia, in *Third Millenium BC Climate Change and Old World Social Collapse* (H. N. Dalfes, G. Kukla and H. Weiss, eds.), Springer-Verlag: Heidelberg.

Schneider, S. H. and Dickinson, R. (1974) Climate modeling. *Rev. Geophys. Space Phys.*, 12, 447-493.

Wilkinson, T. J. (this volume) Environmental fluctuations, agricultural production and collapse: A view from the Bronze Age Upper Mesopotamia, in *Third Millenium BC Climate Change and Old World Social Collapse* (H. N. Dalfes, G. Kukla and H. Weiss, eds.), Springer-Verlag: Heidelberg.

Was the Holocene Climate Uniquely Benign or Is the Eemian Catching Cold from the Reinforcement Syndrome Virus?

George KUKLA

Lamont Doherty Earth Observatory of Columbia University,
Palisades, NY 10964, USA

Abstract

It has been recently proposed, based on the comparison between the upper and lower part of the GRIP Greenland ice core that the Holocene climate is uniquely stable and benign in contrast to the supposedly highly variable Eemian climates. This conclusion is questionable. The evidence presented here shows that with very high probability the Eemian lasted considerably longer than the Holocene and that the climates of the elapsed part of the current interglacial in many parts of the world were far from uniform and certainly no more benign than in the corresponding segment of the last interglacial. If the future natural climate development were to follow the Eemian blueprint, the start of the next glacial would be still many millennia ahead.

Introduction

In his masterpiece on the magnetic polarity time scale, Watkins (1971) described his "reinforcement syndrome", which states that "the evidence for short events will be forthcoming, whether or not they have actually occurred". According to Watkins, it is more difficult to prove that a given behavior has not taken place, than to show that it did. Therefore the publication of data confirming an original mistaken finding is considerably more likely than publication of the data which would disprove it. Thus, the initial false conclusion is reinforced.

It now appears that the last interglacial is under the attack of the same phenomenon. Although seriously at variance with the nearby Summit GISP core, the oxygen isotope record of the GRIP core led researchers (Dansgaard, et al.,1993) to conclude that the Eemian climate was punctuated throughout by short cold spells which made it very different from the supposedly stable and uniformly benign climate of the last 10 millennia. This conclusion is at variance with the conventional concept of the last interglacial, originally defined by Jessen and Milthers (1928) as an interval with a

NATO ASI Series, Vol. I 49
Third Millennium BC Climate Change
and Old World Collapse
Edited by H. Nüzhet Dalfes, George Kukla,
and Harvey Weiss

vegetational and hence climatic evolution similar to that of the Postglacial, including a clear expansion of thermophilous trees.

Watkins in his 1971 paper called for caution in interpreting data unconfirmed by parallel sampling. While it would appear prudent to apply such a rule to the ice cores as well, in Greenland it may be already too late. The trap has been set and the hunt for the Eemian cold spikes is already on. Surprises such as the one on Siberian winters which supposedly punctuated the Eemian *Carpinus* zone (Field et al,1994) keep popping up in the Earth Science journals, strengthening by circular reasoning the ice core based conclusion (e.g. Cortijo et al, 1994; Thouveny et al.,1994; Lauritzen,1995).

Problem

Was the Eemian climate so different from the Holocene and so much more unstable? Or are we witnessing an attack of the reinforcement syndrome virus (RSV) on the Eemian record? In order to agree with the conclusion on the Holocene-Eemian climate contrast, we have to accept four points, namely:

(1) The global climates during the Holocene were uniformly benign.

(2)) The basal section of the GRIP ice core correlates with the Eemian and is the undisturbed complete representation of this interval.

(3) The GRIP ice core oxygen isotope data accurately represent global climate, rather than reflecting the variations of regional oceanic and atmospheric water transport.

(4) And finally, the elapsed section of the Holocene is of about the same duration as the Eemian, so that the two periods can be justly compared.

None of these assumptions is certain. Although the isotopic composition of the Holocene ice in some cores in Greenland and Antarctica varies relatively little, this is not the case for all the cores (Johnsen et al.,1995) Records of temperature and precipitation variations of the last ten millennia in other parts of the world are even less uniform than the ice records. Large fluctuations of lake levels and mountain glaciers are described from all continents. A cool, dry spell in the Early Holocene is indicated in proxy climate data from various environments (Blunier et al,1995 and references within). Similar fluctuations during the Late Holocene have been reported earlier ((Petit-Maire et al.,1990; Maley, 1977; Denton and Karlen,1973) and are discussed in detail in the present volume. There is little justification to consider the Holocene climates uniformly benign.

The record of the so-called Eemian interval in the GRIP ice core differs considerably from the corresponding section in the nearby GISP core. It was proposed that one if not both of the records are mechanically disturbed by the ice tectonics (Alley et al.,1995). In this situation it is far fetched to claim that one of the records represents the climate of the last interglacial whereas the other does not.

The oxygen isotope ratio in the ice cores responds principally to the changing temperature of the locally deposited snow but is also affected by the source area of the precipitation and the distance and level of the moisture transport. The mean monthly values of the ratio in the contemporary precipitation in Thule and in recent snow in Camp Century is some 20 to 25 per mil, whereas the difference between the peak glacial and peak interglacial values in the Camp Century core is only about 15 per mil (Johnsen, et al.,1972) . Thus, if the season during which the bulk of precipitation arrived at the site changed, the isotopic composition of the corresponding ice layer may be quite different, although the average annual temperature at the site would have remained the same. It has been argued that the source of the Greenland precipitation may fluctuate among different parts of the world ocean including low latitudes (Newell,1994) , a fact which would considerably affect the isotopic record. It has been also revealed that while the Summit cores show the relatively uniform oxygen isotope record through the Holocene, this is not true for the cores elsewhere in Greenland, such as eg. in Renland, which indicates considerable cooling in the Late Holocene (Johnsen and Dansgaard, 1992).

Duration of Eemian

A critical question however, is the length of the Eemian. Is the Holocene warm interval equally long as the Eemian? This issue has an important bearing on the forecast of future climates. If the two warm intervals were approximately of the same duration and if the natural climate developments in the interglacials followed the same basic periodic pattern, then the advent of the next glacial would be near. The increase of the carbon dioxide in the atmosphere might even be beneficial, since it could retard the natural cooling trend.

However, the length of the last interglacial is not yet accurately known. It is mostly considered to be about 11, 000 years, based on the count of pollen bearing varved diatomites in Bispingen, Germany and on the estimated duration of the Marine Isotope

Stage (MIS) 5e, the supposed equivalent of the Eemian interglacial in the deep sea sediments. Both these assumptions hvae uncertain validity

In Bispingen in Luhetal in northwest Germany, within the type region of the last interglacial (Jessen and Milthers,1928; Fairbridge,1972), Muller(1974) was able to count 1,900 annual laminations through the early part of the last interglacial diatomite. Using the ratio of silica to organic matter in the remainder of the deposit, he then estimated the total duration of the Eemian to approximately 11 thousand years. His estimate differs considerably from the 18 thousand years obtained by Dabrowski (1971) in the Glowczyn deposit in Poland. Dabrowski's figure is based on the per volume pollen concentration, calibrated by measurements on recent sediments in similar depositional and floristic environments. An even longer duration of 20 thousand years has been proposed for the last interglacial in La Grande Pile (Guiot, et al.,1989) based on comparison of the pollen spectra with the pollen record of the deep sea core SU 8132 off Portugal (Turon,1984). The estimate of the great length of the Grande Pile interglacial record seems reasonable, given the relatively uniform lithology of the deposit, with no indication of any significant variation of sedimentation rates in the early glacial and interglacial segment (Woillard and Mook,1982) and the limited thickness of the Melisey steppe episode (Beaulieu and Reille, 1992). These latter estimates seriously conflict with the conventional one-to-one correlation of the Eemian with the deep sea substage MIS 5e which is only about 14 thousand years long (Shackleton,1969; Mangerud, et al.,1979, Martinson, et al.,1987)

La Grande Pile and the deep sea core V21-191 There are additional observations which indicate that the Eemian in France lasted about twice as long as the elapsed part of the Holocene. The fluctuations of the tree pollen ratio in La Grande Pile appears to closely mimic the variation of the cold water foraminifer *Neogloboquadrina pachyderma sinistra* in the deep sea core V29-191 situated in eastern North Atlantic at 54° 16' N and 16° 47' W.

In figure 1 we compared the paleoceanographic information on the marine isotope stage MIS 5 obtained in this core with the corresponding interval of the lacustrine pollen- rich sequence at La Grande Pile (GP) in France. Both records are essentially continuous and lithologically relatively uniform, so that substantial differences of sedimentation rates within MIS 5 are unlikely (McManus,1994 for the ocean and

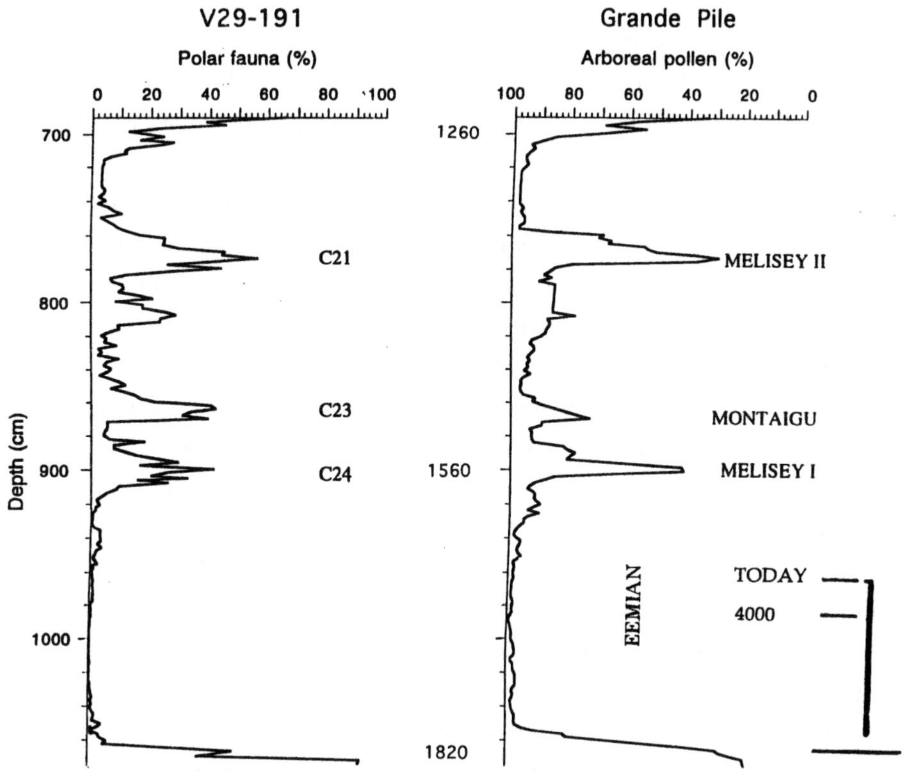

Figure 1: Percentage of left coiling *Neogloboquadrina pachyderma sinistral* in the N. Atlantic core V29-191 .[C 21-24, cold water intrusion events from McManus et al., 1994] Compared with the percent of arboreal pollen in La Grande Pile pollen record (Woillard, 1978). Plotted on the original depth scale and aligned at the MIS 4/5 and 5/6 boundaries assuming linear depth to time ratio. The equivalent position of the elapsed part of the Holocene and of the 4000 BP dessication event in the glacial/interglacial cycle is shown on the right.

Beaulieu,1992. for the GP record). The data were therefore plotted on their original depth scales.

The boundaries of the mild zone correlative with MIS 5 were defined by Molfino et al.(1984) in the GP composite core X and XIV at the depth of 1830 and 1195 cm.

The Grande Pile sequence includes in its lower part the sediments of the last interglacial, locally labeled Lure (Woillard,1975), but shown convincingly to represent the Eemian. The onset of the interglacial vegetation is recorded in the thinly laminated gyttjas of the core GP X overlying glacial varves. The replacement of the closed forest, represented in the pollen spectrum by over 90% tree pollen, by an open steppe vegetation, the conventional upper boundary of the last interglacial (Menke and Tynni, 1984), is at the depth of 1562 cm, at the base of the 10 cm thick Melisey I band of silty gyttja with eolian admixture (Seret et al,1992) and steppe pollen assemblage (Woillard,1978) .

The ocean core has a fairly well differentiated oxygen isotope record, obtained from the benthic foraminifera *Cibicidoides*. which enables a reliable recognition of Marine Isotope Stage (MIS) boundaries 4/5 and 5/6. Following convention (Emiliani, 1955), the upper boundary of MIS 5 has been placed at the core depth of 650 cm and the MIS 5/6 boundary at 1070 cm (McManus et al 1994).

Sea surface conditions are represented in the core by varying amounts of ice rafted detritus and cold water planktonic foraminifera *N. pachyderma sinistral*. Six polar water invasions are indicated by the increased proportion of the cold water foraminifera. The oldest, labeled C24, is at 900 cm in the upper part of the MIS 5d. It is considerably shallower than the MIS 5d heavy oxygen isotope peak at the depth of 930 cm (McManus,1994) . An almost complete lack of cold water indicators in the deep sea core below the C24 peak confirms that the sea surface in the central part of North Atlantic at that time was relatively warm. This includes the time when the oxygen isotope ratio in the benthic foraminifera corresponds to MIS 5d, a substage usually interpreted as marking the first episode of the land -ice build-up.

The comparison of the GP vegetational succession with the surface ocean indicators in V29-191, aligned at the MIS 5 boundaries and made on the original depth scales reveals the close correspondence of the arboreal pollen count with the variation of the left coiling *N. pachyderma.* The number of the anomalies, their depth spacing and the relative amplitude of the signals correlate reasonably well. The tree pollen minima have their counterparts in the *N. pachyderma* spikes. There could be little doubt that the Melisey I steppe episode corresponds to the C24 polar water intrusion, the Montaigu event to C23 and the Melisey II to C21. The thinness of the steppe interlayers Melisey I and Melisey II corresponds well to the equally thin marine equivalents.

It follows from our data that the chronostratigraphic position of the Eemian, as recognized in the French pollen record, is in disagreement with the conventional climatostratigraphic model of MIS 5e and MIS 5d. It has been commonly assumed that the glacial onset is coeval with the MIS 5e/5d boundary (Mangerud et al., 1979).

In GP this cannot be the case. The demise of the interglacial woodland, if indeed correlative with the polar intrusion C24, is clearly younger than the heavy isotope minimum of substage MIS 5d dated by (Martinson et al.,1987) to 110.7 kyr BP.

The close correlation of the two records from vastly different sedimentary environments implies that the sedimentation rates within the MIS 5 zone at both sites were sufficiently uniform to justify tentative chronology based on relative thickness . This is not unusual in the case of the deep sea core but may be surprising in a lake. However, Woillard and Mook (1982) demonstrated with a set of radiocarbon dates that the GP sedimentation rates in pleniglacial and early glacial differed very little. If that is the case in such strikingly different climatic regimes, it is highly unlikely that the rates would have significantly varied within the much more uniform interglacial and early glacial environments.

Through linear interpolation of the V29-191 depth record within the MIS 4/5 and MIS 5/6 boundaries, dated by Martinson et al. (1987) respectively to 73.9 and 129.8 kyr BP, the age estimate of 107 kyr BP was obtained for the C24 polar outbreak and 111 kyr BP for the heavy isotope peak of MIS 5d. The latter figure is very close to the Martinson's 110.7 kyr value for that level.

The base of the eolian silt Melisey I in La Grande Pile sequence,which defines the end of the GP Eemian, and is according to our figure apparently coeval with the *N. pachyderma s*. peak C24, is therefore with high probability approximately 107 kyr old. The Eemian interglacial, as recognized in its Lure equivalent in France, lasted from about 130 kyr BP to 107 kyr BP and was some 23 millennia long. This is twice the amount of time given for Bispingen diatomites.

Discussion

Thiie above conclusion is in obvious disagreement with the conventional model of the Eemian on two counts. First, how could the length of an interglacial established in a varved deposit be underestimated by several millennia? Second, how could a warm environment in western Europe coexist with a build-up of continental ice reflected by the MIS 5d substage?

The question arises whether the conventional chronostratigraphic correlations of the pollen bearing Eemian deposits across Europe are correct. While the Eemian was originally named after the marine clays with temperate Lusitanian fauna in the Netherlands by (Harting, 1875), its palynological zonation was defined in detailed studies at the localities in Denmark and northwestern Germany by Jessen and Milthers (1928), Selle, (1962) and others. The type area of the palynostratigraphically defined Eemian is in Jutland, not at La Grande Pile.

The Eemian vegetational sequence follows a fairly similar pattern all over Europe. It ends with the so-called katathermal zone of coniferous taiga, whose length is estimated to be at least some 2 millennia (Frenzel and Bludau,1987). Because of the similarity, most palynologists consider the onset of the open glacial environments in western Europe to be,by and large, synchronous (Menke and Tynni,1984). Judging from the diachroneous oscillations of vegetational belts in Holocene (Zagwijn,1994), this is by no means guaranteed. It is possible that the coniferous woodlands in the type area of the Eemian in Jutland were replaced by open vegetation earlier than in the Vosges. However it is difficult to visualize an interval at least some six thousand years long, during which the lush deciduous forests with evergreen undergrowth at La Grande Pile would coexist with the harsh steppes and tundras in Northern Germany, only 300 km away.

It is much more likely that the Bispingen based figure is seriously underestimated. This indeed could have happened. Muller (1974) was able to count only 1,900 years of laminae through the early part of the Eemian diatomites which contain up to five times more carbonate than in the rest of the deposit. The sedimentation rates through the bulk of the interglacial was than estimated based on the ratio of silica to organic matter, apparently without correction for the carbonate.

Climate in the second half of the interglacial

The last interglacial climate, as reflected in GP, was relatively benign, but by no means uniform. The increase of Taxus in the early part of the interglacial at a core depth of 1770 cm has been interpreted as a sign of a significant cold spell (Guiot, et al.,1993). The interval between 1605 and 1670 cm has been considered to represent the interglacial optimum due to the high percentage of evergreen shrubs, which require mild winters (Iversen,1944; Aaby and Tauber,1995). This was however already the time in the the second half of the interglacial in which the deciduous elements were declining, gradually replaced by conifers. An abrupt transition from the mixed deciduous to a coniferous forest recorded at a depth of 1600 cm was studied in detail by Woillard (1979). It has been linked to a severe climate. The shift occurred a few thousand years prior of the end of the interglacial proper. Woillard estimated that the event took 150 + 75 years assuming that Lure lasted 11 thousand years. However, since the Eemian is GP is about twice as long, Woillard's estimate must be revised to approximately 300 + 150 years. At a later occasion, Woillard and Frenzel (1991) and Frenzel and Bludau (1987) analyzed an apparently identical feature in the annually banded sediments at Krumbach, Germany. There the "Woillard event" is so sharp that Frenzel and Bludau called it a hiatus. It is preceded by at least 1500 years of the gradually deteriorating mixed forest and followed by at least 1650 years of katathermal coniferous woodland.

Could additional abrupt short episodes of adverse climate at other times during the interglacial remained undetected by palynologists? Field et al (1994) reported such an interval about half way through the *Carpinus* zone in Bispingen, in which the January mean temperature should have dropped to - 20°C. Their conclusion is based on the Muller's (1974) original pollen count. While the reality of such estimate, based on transfer functions of current pollen rain is highly questionable (Aaby and Tauber,1995; Zagwijn,1994) the potential occurrence of abrupt cold or dry spells in the second half of the interglacial is not a priori impossible.

Comparison of Holocene with the Eemian.

The described decline of the deciduous forests in GP started some 13 thousand years after the onset of the last interglacial aligned with Termination II. This would roughly correspond to the current chronological position of the Holocene with respect to Termination I (fig 1). Within the periodic glacial/interglacial cycle the Holocene interglacial may therefore be only half way through.. The interval of the Late Holocene climate instability at the turn of the third and fourth millennium BC would then correspond to the time of the Eemian climatic optimum, still well before the glacier build-up start of MIS 5d.

Obviously then, any notion of unstable Eemian climates contrasting with supposedly stable Holocene has to be carefully reconsidered. Only the first half of the Eemian may be justly compared with the present interglacial. As already observed by Zagwijn (1994), there is no reason to conclude that the climate of the elapsed part of the Holocene was any more stable than the corresponding part of the last interglacial.

The developments of the second half of the Eemian may represent a potential analog of the next several millennia of natural climate development heading into next glacial. It would be particularly useful to achieve more insight into the climate change mechanism of those times. It appears that the ocean changes in the northernmost North Atlantic may have affected the Greenland ice and the deep water chemistry without leaving obvious traces in the vegetation of western and central Europe. Indications are, that the apparently rapid climatic deteriorations were followed by much more gradual environmental relaxation in a period 1 to 3 thousand years apart (Beaulieu and Reille, 1992). A scenario which would best fit our data from the last few millennia of the Eemian interglacial involves a gradual ice build-up in the high latitudes punctuated by periodic sudden ice discharges (Bond and Lotti, 1995; Broecker 1994). Ocean changes apparently preceded the environmental shifts on land in the middle latitudes by several millennia. It was the ocean that was apparently instrumental apparently instrumental in bringing the interglacial/glacial shift to land..

In summary, there is little reason for accepting the model of a uniquely favorable Holocene climate contrasting with the highly variable Eemian. Furthermore, if the Eemian were to serve as the analog of the current warm period, then the natural onset of glacial environments on land would lay far in the future.

References

Aaby B and. H Tauber (1995) Eemian climate & pollen. *Nature* 376: 27.

Beaulieu J-L and M Reille (1992) *Quater. Sci. Rev.* 11: 431.

Blunier T et al (1995) Variations in atmospheric methane concentration during the Holocene epoch. *Nature* 374: 46-49.

Bond GC.and R Lotti (1995) Iceberg discharges into North Atlantic on millennial time scales during last glaciation. *Science* 267: 1005.

Broecker WS (1994) Massive iceberg discharges as triggers for global climate change. *Nature*: 372: 421-424.

Cortijo E. et al (1994) *Nature* 372: 446-449.

Dansgaard W et al (1993) Evidence for general instability of past climate from a 250-kyr ice-core record. *Nature* 364: 218-220.

Beaulieu J-L de and M.Reille (1992) Last climatic cycle at La Grande Pile (Vosges, France): new pollen profile. *Quaternary Science Reviews* 11: 431.

Fairbridge RW (1972) Climatology of a glacial cycle. *Quat. Res.* 2: 283-302.

Field M H et al (1994) *Nature* 371: 779-783.

Frenzel B. and Bludau W. (1987) On the duration of the interglacial transition at the end of the Eemian interglacial (deep sea stage 5 e): Botanical and sedimentological evidence. in: Abrupt Climatic Change- Evidence and Implications . Berger WH, Labeyrie LD eds NATO ASI Series C 216: 151-162 Reidel Dodrecht.

Guiot J. et al. (1993) *Paleogeogr., Paleoclim., Paleoecol.* 114: 1.

Guiot J. et al. (1989) A 140,000 year continental climate reconstruction from two European pollen records. *Nature* 338: 309-313.

Harting P.(1874) Versl en Meded. Afd. Naatuurk. 2nd Reeks Deel VIII: 282.

Iversen J (1944) Viscum, Hedera and Ilex as climate indicators. Geol. For. Stockh. Forh. 66: 463-483.

Jessen K. and V Milthers (1928) Stratigraphical and paleontological studies of interglacial freshwater deposits in Jutland and North-West Germany. Danmarks Geologiske Undersogelse IIrke 48.

Johnsen SJ.et al. (1972) Oxygen isotope profiles through the Antarctic and Greenland ice sheets. *Nature* 235: 429-434.

Lauritzen S-E. (1995) High-resolution paleotemperature proxy record for the last interglaciation based on Norwegian speleothems. *Quat. Res.* 43: 133-146.

Mangerud J et al. (1979) *Nature* 277: 189.

Martinson DG et al. (1987) Age dating and the orbital theory of the ice ages: Development of a high-resolution 0 to 300,000 year chronostratigraphy. *Quat. Res.* 27: 1-29.

McManus JF.et. al. (1994) High-resolution climate records from North Atlantic during last interglacial. *Nature* 371: 326.

Menke B Tynni R (1984) *Geol. Jb.* A76: 3.

Molfino BL. Heusser and G, Woillard (1984) Frequency components of a Grande Pile pollen record: Evidence of precessional orbital forcing. In: Milankovitch and Climate Part 1 Brger AL et al eds NATO ASI, Reidel, Dordrecht, 391-404.

Muller H (1974) Pollenanalytische Untersuchungen und Jahresschichtenzahlungen an der eem-zeitlichen Kieselgur von Bispingen Luhe. Geol. Jahrb A21: 149-169.

Newell R E and Zhu Y (1994) Tropospheric Rivers: A one-year record and a possible application to ice core data. *Geophys. Res. Letters* 21: 113-116.

Petit-Maire.N et al. (1990) . *Paleogeogr. Paleoclim.Paleoecol.*79: 333.

Selle W (1962) Geologische und vegetationskundliche Unersuchungen and einigen wichtigen Vorkommen les letzten interglazials in Nordwestduetschland. Geol. Jb 79: 295-352.

Seret G Guiot J Wansard G Beaulie JL de Reille M (1992) Tentative palaeoclimatic reconstruction linking pollen & sedimentology in La Grande Pile (Vosges, France). *Quat. Sci. Rev.* 11: 425-430.

Shackleton NJ (1969) The last inerglacial in the marine and terrestrial records. Proc.of Royal Soc. London B174: 135-154.

Thouveny N.et al. (1994) *Nature* 371: 503.

Turon J-L.(1984) Direct land/sea correlations in the last interglacial complex. *Nature* 309: 673-676

Watkins N (1971) Geomagnetic polarity events and the problem of "The Reinforcement Syndrome". Comments on Earth Sciences: *Geophysics* 2: 36-42, London.

Woillard GM(1975) Recherches Palynologiques sur le Pleistocene dans l'Est de la Belgique et dans les Vosges Lorraines. Acta Geogr. Lovan. 14: 1.

Woillard GM (1979) Abrupt end of the last interglacial s.s. in north-east France. *Nature* 281: 558-562.

Woillard GM and FrenzelB (1991) Forest Changes at the end of the Last Interglacial. In: B. Frenzel editor: Klimageschichtliche Probleme der letzten 130,000 Jahre. Palaoklimaforschung: Bd.1: Fischer Verlag. Stuttgart: 37- 50.

Woillard GM (1978) Grande Pile peat bog: A continuous pollen record for the last 140,000 years. *Quat. Res.* 9: 1-21.

Woillard GM. and WG Mook (1982) Carbon-14 dates at Grande Pile: correlation of land and sea chronologies. *Science* 215: 159-161.

Zagwijn W. H. (1994) Reconstruction of climate change during Holocene in western & central Europe based on pollen records of indicator species. *Veget.Hist. Archeobot* 3: 65.

Late Third Millennium Abrupt Climate Change and Social Collapse in West Asia and Egypt

Harvey WEISS

Yale University
New Haven CT 06520-8236

Abstract.

The palaeoenvironmental record for the 2200 BC abrupt climate change is synthesized. Alternative explanations for synchronous and extended Old World social collapse are examined and rejected. Quantification of the abrupt climate change is necessary if we are to understand its social consequences.

Paleoenvironmental Data for Abrupt Climate Change

The 2200 BC aridification event has been observed in lake level records that extend from west Africa to west Tibet (Gasse and van Campo 1994). In western Asia, in both the regions dependent upon the Mediterranean westerlies for precipitation and those dependent upon the monsoon, the climate record for the event has considerable radiometric and archaeological-historical chronology precision. The range of records for Palestine (Dead Sea), the Nile and East Africa (e.g., Lake Turkana), alongside the range of other Old World data available in this volume, indicate that the general circulation model extends the impact of this event across W Asia and Europe, and probably into E Asia as well. A dissenting opinion is expressed by Butzer (this volume) who believes that the aridity implied by the Lake Van and Lake Zeribar records would have been circumscribed by the Tigris-Euphrates drainage basin.

1. Mesopotamia. There are now four independent observations of an unusual dust event across Mesopotamia at ca. 2200 BC. These observations fall across a north-south transect from Lake Van to the Gulf of Oman.

a. Lemcke and Sturm (this volume, Fig. 3) document an abrupt doubling of the quartz content of Lake Van core 90-10, from ca. 6% to ca. 12%, or a tripling in other Van sediment records (Lemcke 1994), from 4200 to 4000 BP. This spike is synchronous with initiation of the k^{18}O enrichment phase at 4190 cal yr BP (Lemcke and Sturm, this volume, Fig. 2). As Butzer (this volume) notes, the Lemcke and Sturm

NATO ASI Series, Vol. I 49
Third Millennium BC Climate Change
and Old World Collapse
Edited by H. Nüzhet Dalfes, George Kukla,
and Harvey Weiss

oxygen isotope record from the Van varves indicates a pronounced dry spell ca. 2350-2075 BC. Together these suggest, with Courty (1994), that the 2200 BC phase of decreased precipitation was synchronous with increased wind turbulence and aeolian dust transport to Lake Van. Sampling at 84 year intervals may have precluded observation of Na and Al peaks that are considered to be effects of volcanic tephra (Fig. 3).

b. In northeast Syria, or northern Mesopotamia, the Tell Leilan dust event, and it relationship to glass shard (possibly tephra) fallout, was identified by Courty and dated by Weiss (Weiss et al. 1993); new data for this are now available (Courty and Weiss, this volume). Independent confirmation for the dust event may also be provided by Wendy Matthews soil micromorphology analyses at Tell Brak, 50 kilometers SW of Tell Leilan. The event at Tell Brak is synchronous with site abandonment, on a scale still to be determined, in the late Akkadian period, i.e., at the same time as the Leilan event (Matthews, Matthews, McDonald 1994; Matthews 1994).

c. From his surface reconnaissance in southern Iraq, Adams reports that "in at least some formerly settled and cultivated areas that are now desert...there appears to have been a substantial net lowering through erosion of the land surface that obtained at the end of the third millennium BC (Adams 1981: 10); "a major cycle of wind erosion may have occurred before the Third Dynasty of Ur" (Adams 1981: 31).

d. deMenocal and Cullen have documented a five-fold increase in windblown, dolomite Mesopotamian dust, at ca. 2300 BC, within a sediment core from the Gulf of Oman. Tephra associated with the dust peak is now undergoing major and trace element analysis (Weiss 1996). A quartz spike, similar to that at Van, also occurs (H. Cullen, personal communication). Cullen is now examining other Gulf cores for complementary information.

Dust and dust storms increase in frequency after drought years. Therefore, the essentially synchronous and unique dust spikes, as period of dust storms, observed along this transect reflect a major aridification event.

2. Palestine In the eastern Mediterranean, the exceptionally arid climate stage 4 of the Dead Sea Holocene record, beginning abruptly at ca. 2200 BC, is represented by a ca. 100 meter drop in Dead Sea level (Frumkin et al. 1991: figs. 7, 8). "Abruptness" in this case is defined by six radiocarbon dates, with interpolation for the date of the

nadir. This is confirmed by the speleothem isotope record which documents a drying period ca. 2450-1900 BC (Issar, et al. 1992).

3. Nile flood behavior The monsoon, responsible for the precipitation that sustains lake levels in east Africa and Nile River flow, also provides the precipitation that controls Indus flow. Lake Turkana, the best proxy for Nile flow, underwent "an abrupt change in lake conditions at 3900 ybp (14C time scale). At this time the oxygen isotope values gradually shift to heavier values....Mathematical modeling suggests that while lake conditions changed abruptly from open- to closed-basin status, the atmospheric circulation over the drainage basin shifted gradually over a period of 2000 years after the 4 ky BP "event" (Johnson 1995: 011B-8). These data, as well as the similar data provided by Halfman, Jones and Finney (1994), amplify those from the earlier studies cited by Butzer, e.g., Halfman et al. (1992), that also indicate minimal lake levels at 2250-2200 BC.

4. Indus Valley Although Possehl (this volume) points to the absence of climate change data for the Indus Valley in the third millennium, Bentaleb et al. (this volume) provide new dated data for monsoon variation. Mughal (1990) and others, including now Louis Flam, have repeatedly pointed to late third millennium changes in the course of the Indus as responsible for the changes in settlement pattern, beginning ca. 2200 BC, known as the Indus collapse. Such alterations are the likely product of decreases in stream flow.

There are two explanations for abrupt changes in the monsoonal climate, that could have affected both the Nile and the Indus. One relates these abrupt changes to ENSO teleconnections (Quinn 1992). But the monsoon is controlled by same forces as the westerlies that control SW Asian climate. Abrupt changes in the monsoonal climate are closely coupled with rapid changes in the north Atlantic, where ice sheet decay or shelf ice surges create winter surface air pressure anomalies (Sirocko 1995: ii; Sirocko, et al. 1996; Bond 1995).

5. Paleobotany The quality of 2200 BC abrupt climate change records varies considerably, but none so much as the paleobotanical one (Bottema, this volume). Why this is the case remains to be explained. Many paleobotanists, prior to the realization that their record often misses the 2200 BC event, had already concluded that the paleobotanical record for the Holocene in West Asia was extremely unreliable (Bottema, Entjes-Nieborg, van Zeist 1990). Nevertheless, the Lake Huleh core (van

Zeist and Bottema 1991: fig. 37) and the Lake Zeribar core (van Zeist and Bottema 1991: fig. 14) both indicate an abrupt decline in arboreal pollen ca. 2200-1950 BC.

Settlement Pattern Data:

1. The Habur Plains A post-Akkadian, pre-Habur ware ceramic assemblage has yet to be identified upon the Habur Plains in either surface survey or excavated contexts (Weiss and Courty 1993, 1994). The only evidence for occupation in this region during the period ca. 2200-1900 BC is epigraphic (i.e., the "kingdom of Urkish and Nagar," and possibly the still undated houses that in some areas lie above the Akkadian abandoned public buildings at Tell Brak (Eidem and Warburton in press).

2. Northern Iraq in the Tell al-Hawa survey area of the northern Iraq, approximately 150 km from the northern of Tell Brak, Wilkinson and Tucker (1995) have not distinguished terminal Ninevite 5 Leilan IIId), from pre-Akkadian (Leilan IIa), from Akkadian (Leilan IIb), from post-Akkadian collapse (hiatus) occupations. It is, therefore, difficult to identify settlement changes in this region during this interval. The significant occupational hiatus at one of the region's major cities, Tell Taya level 5 (Reade 1982), is synchronous with the abandonment at Tell Leilan, although Butzer (this volume) believes that this region's settlement pattern is "not consistent with that around Leilan.".

3. South and west of the Habur Plains the pattern of adaptive response was clearly migration to riverine environments with sustainable agriculture. Hence the tripling of Ur III village-level settlement in southern Mesopotamia (Adams 1981: 142), and the expansion of settlement along the middle and upper Euphrates at Ur III period cities such as Tell Sweyhat and Carchemish. Unusual fluvial incursions during this period along the upper Euphrates may have been responsible for regional abandonment at Jerablus Tahtani and the population concentrations at nearby cities such as Carchemish (Peltenburg 1995). Synchronously, dry farming regions like those around Titrish (Algaze et al 1992) and Arslantepe (Frangipane 1993) suffered serious abandonment. Titrish also provides geomorphological evidence for synchronous aridification (Rosen, this volume). Further west, erosion layers at Ilıpınar have an EB III terminus post quem, implying EB IV abandonment (Gerard 1995: 72-75). Significantly, for the cultural and environmental history of western Syria, the extensive excavations at Tell Mardikh now document partial, if not major, site abandonment during this period (Matthiae 1995: 88). With rare exceptions, such as

Khirbet Iskander situated along perennial water courses, and Khirbet al-Umbashi, designed to trap water flows (Braemer and Echallier 1995), Palestine was devoid of towns during the EB IV period.

4. To the north and east, similar responses seem evident in the synchronous population transfers in Turkmenistan, from the Kopet dağ region to Margiana (Kohl 1992, P'yankova 1994), and Iranian plateau sites like Shahr-i Sokhta where settlement in this period fell from 80 to five hectares (for documentation, Weiss 1996b), and the sudden Harappan urban population density increases, a first response to the abrupt climate change, followed by "collapse" (R. Meadow, personal communication).

Alternative Explanations

The robust geoclimatic data for abrupt climate change ca. 2200 BC appear in East Africa, Palestine, Anatolia, northern Mesopotamia, southern Mesopotamia, and the Gulf of Oman. Ignoring these data requires alternate explanations of the archaeological data for synchronous collapse from the Aegean to the Indus.

Butzer had previously examined the paleoclimatic record in detail and concluded that "only three first order [climatic] anomalies have been identified that apparently affected the entire Near East, namely the 'dry shifts' of around 3000, 2200, and 1300 BCE.... the three most incisive negative changes recorded in Lakes Van and Zeribar (around 3000, 2200, and 1300 BCE)--evidently controlled by extratropical trends--parallel the most dramatic reductions of Nile flood waters--as controlled by tropical circulation. Synchronisms on such a continental scale, across 40 degrees of latitude, imply major anomalies of the general atmospheric circulation, potentially of a global scale. They underscore that these were incisive episodes of major ecological significance, perceptible to some degree or other throughout the Near East." (Butzer 1995: 135)

Butzer (this volume) believes that there is no data whatsoever for an abrupt climate change at 2200 BC. Apart from the strengths of the abrupt climate change data, are there strong alternate explanations for the collapse phenomena of 2200 BC?

The well known economic decline late in the reign of Pepi II was followed by the political and economic decline of the First Intermediate Period that featured [Nile] flood failures and droughts. Negus (1986) has observed Old Kingdom gradual drought followed by radical declines during the two hundred years of First Intermediate Period: four low-Nile sequences, each of two to three decades duration, each sequence containing one or more years of pronounced flood failure and severe famine. The Egyptian data suggest at least two periods of flood failure, and correlation with the southern Mesopotamian Akkadian and Ur III collapses. Although Butzer (this volume) suggests "excessively high Nile floods during a 600 year period," there was, as both he and Hassan (this volume) note, an abrupt Nile flood minimum at 2200 BC that is documented in the geoclimatic data from both Qasr el-Sagha and Birket Qarun at ca. 2200 BC.

The historical synchronism between the First Intermediate Period and the Akkadian-Ur III collapses is precise, in spite of the ca. 120 year margin of error within Mesopotamian historical chronology for the 3rd millennium and ca. 100 year "margin of historical dating error" for Old Kingdom and [First Intermediate Period] Egypt (Kitchen 1991: 202). Kitchen places Dynasties VII through X at 2190-2010 BC Using the middle chronology, and reducing the Gutian period through the synchronism Ur-namma/Puzur-in u inak, the period ar-kali- arri through Ibbi-Sin is 2190-2004 BC

Butzer (this volume) believes that Egyptian "prosperity, and indirectly royal power" was linked to the Byblos trade that "was probably applied to redistribution and even credit at the pharoah's disposal." The destruction of Byblos, possibly in late ED III or early in the reign of Sargon, antedates the First Intermediate Period. The trade was restricted to luxury and elite items such as timber, oil, wine, and did not contribute to Egyptian "prosperity." This was a surplus-consuming trade, not a surplus-producing trade. State surplus was expended, not accumulated. Profit, such as there may have been, was collected on the Byblian end, while the trade expenses accumulated on the Egyptian end. There was no relationship here similar to that which produced Assyrian profits in Anatolia and Assur for the textile and tin trade (Adams 1974). The subsequent replacement of Egyptian with Mesopotamian or Anatolian influences at Byblos reflects the collapse of the sources of Egyptian state wealth: Nile irrigation agriculture.

There is a broad consensus among Palestinian archaeologists that towns, with but a few rare exceptions, were abandoned during the EB IV period (Esse 1991; Palumbo 1991, Finkelstein 1991; Dever 1992). The rate of this process is not certain; it may have begun as many as 100 years prior to its completion at the beginning of EB IV (Seger 1989).

Elsewhere Butzer agrees that the hallmark of EB IV is "a changing settlement archaeology: disappearance of EB II-III towns, with total abandonment of some areas, and new small settlements in others...although EB IV sites remained concentrated near springs or running water." There is an excellent synchronism with the 2200 BC abrupt climate change, recorded very finely in the Dead Sea lake levels (Frumkin et al 1991). The synchronism of EB IV (beginning ca. 2250 or 2200 BC) with the First Intermediate Period (beginning ca. 2190 BC according to Kitchen 1991) is presented throughout Ehrich (1992). There is little evidence for the Egyptian destruction of the towns of EB II-III Palestine.

Linkages between unexplained "militarism" and the collapses of Egypt, Palestine and Mesopotamia, north and south, are problematic. For Butzer, "the destruction of Byblos probably proved fatal, cutting Egypt off from the Near Eastern exchange network." The destruction of Byblos was due to Sargon's strategy "to extend Mesopotamian hegemony to Syria and the Mediterranean coast," and was based on windfall profit taking" (destruction and looting) ...that would have left the Akkadian economy incapable of growth and therefore fragile." But the stages of Mesopotamian expansion were (1) late ED III - Sargon conquest and looting, followed by 2. Rimush, Manishtushu, Naram-Sin and successors, who established and maintained, briefly, the extended network of massive fortresses that controlled Akkadian agro- imperialism across the dry-farming belt surrounding southern Mesopotamia (Weiss and Courty 1993). There are no Mesopotamian - Palestinian - Egyptian economic linkages, nor "banking," here.

Another alternative explanation presented by Wilkinson (1994, this volume) takes the well-known periodicity of drought in contemporary northern Mesopotamia to explain the Leilan aridification event of 4000 years ago. This may have been a normal event, a few years of drought, but in the ancient case coincident with the strains of overpopulation relative to available agro-technologies. Intensive cultivation of the limited areas surrounding towns and cities would have generated soil deterioration, in Wilkinson's analysis (1994, this volume). The north Mesopotamian collapse was therefore the product not of a catastrophic event, but a social systems failure relative to environmental stress.

A significant crux is the calculation of cultivated lands of third millennium BC settlements. Wilkinson interprets the map of Habur region "routes rayonnantes" (Van liere and Lauffray 1954/55) derived from early 1950s aerial photography to define "the enclosed cultivation" or "cultivated zones associated with bronze age settlement" (Wilkinson 1994: 561). These "routes rayonnantes," however, do not link third millennium BC settlements; they are the roads used by shepherds and their flocks in the 1940s and 50s (Weiss 1996c). The limits of cultivation for the Akkadian imperialization prior to sudden collapse are the distributions of large Akkadian dominated cities that controlled the countryside and its agro- production. These suggest that the entire eastern region of Habur Plains, all the regions encompassed by the hinterlands of Brak, Mozan and Leilan, were brought under Akkadian agro-production (Weiss 1992).

Challenges

Within the past five years new tools and new data for archaeologists, climatologists, and historians have brought us to the edge of a new era in the study of global and hemispheric climate change and its cultural impacts. The climate of the Holocene, previously assumed static, now displays a surprising dynamism, which has affected the agricultural bases of pre-industrial societies. The list of Holocene climate alterations and their socio-economic effects has rapidly become too complex for brief summary. Any list would need to include, however, the tenth to fourteenth century Medieval Warm Epoch (Hughes and Diaz 1994) that promoted viticulture in England, cereal agriculture in Iceland (Street-Perrot 1994), the collapse of Norse Greenland settlement (McGovern 1990) and the demise of Anasazi agriculture in the Southwest (Fish and Fish 1994).

Still earlier, in the Andes, the late 6th century Moche civilization suffered a 30 year drought followed by severe El-Niño flooding; the Moche capital was destroyed, field and irrigation systems swept away, and widespread famine ensued (Shimada et al. 1991). Between 1000 and 1100 AD the Tiwanaku empire collapsed with the deterioration of its regional agricultural systems. Chronic drought, now documented in the Quelccaya ice cap, was too severe and long-lasting for the Tiwanaku agro-engineeers (Kolata 1994: 297; Ortloff and Kolata 1993). And now the first hard data for the role of climate change in the 750- 1000 AD Maya collapse: prolonged drought beginning at ca. 800 AD, a two hundred year period identified as the driest episode of the past 8000 years (Hoddell, Curtis, Brenner 1995).

The subject of this volume, 2200 BC abrupt climate change, brings the study of Holocene climate dynamics to the Old World and forces reconsideration of Old World climate - culture dynamics within ancient civilizations. One month prior to our publication (Weiss et al. 1993) of the Tell Leilan abrupt climate change study, Sirocko et al. (1993) had identified within the marine sediment record a century scale dust flux event and aridification, ca. 3500 BC, over North Africa and Arabia. The ramifications of this event for Mesopotamian and Egyptian state formation remain to be determined with archaeological studies that can document the event and its social effects "on the ground." The Tell Leilan abrupt climate change study differed from the Sirocko study in that it was "on the ground" both climatically and archaeologically, and had almost half a century of background data, hypotheses, and argumentation within the literatures of Mesopotamian, Palestinian and Egyptian archaeology, as well as the archaeology of the Aegean, central and eastern Europe, the Ukraine and Russia. Hence the challenge of that study and the challenge which subsequent data collection and synthesis presents for traditional epigraphers historiography of this period, and especially the collapse of Akkad. Traditional third millennium historiography (e.g., Yoffee, 1995; Michalowski, 1993) remains isolated from synchronous developments in adjacent regions and the realities of dynamic soils and landscapes, the economics of imperialized agro-production, and abrupt climate change.

The challenges ahead, however, for archaeologists, geoarchaeologists, and climatologists are impressive. As the contributions to this volume make clear, the [quantification] of that which has been identified in various archaeological and natural records is a necessary first step towards the resolution of still conflicting data sets, the full description of the 2200 BC abrupt climate change, and its eventual explanation. Outstanding quantitative issues are the chronology of the beginning of the abrupt climate change as well as for its problematic terminus, geoclimatic processes of wind turbulence, dust deposition, glass shard/tephra deposition, aridification, and river flow, across the range of Old World environments during this time frame.

Hundreds of years after the event, a cuneiform collection of "prodigies," omen predictions of the collapse of Akkad, preserved the record that "many stars were falling from the sky" (Bjorkman 1973: 106). Closer to the event, perhaps as early as 2100 BC, the author of the Curse of Akkad alluded to 'flaming potsherds raining from the sky" (Attinger 1984). Davis (1996) has reminded us of Clube and Napier's impact theory, and asked "Where is the archaeological and geological evidence for the role of their 'Taurid Demons' in human history?" The abrupt climate change at 2200 BC, regardless of an improbable impact explanation, situates hemispheric social collapse in a global, but ultimately cosmic, context.

Acknowledgments.

I thank Peter deMenocal and Heidi Cullen for permission to cite their preliminary research results, and M.-A. Courty, Peter deMenocal, Gerard Bond, Ray Bradley, Heidi Cullen, Nüzhet Dalfes and George Kukla for many informative discussions. I remain responsible for errors of fact or emphasis.

References

Adams, Robert Mc C. (1974) Anthropological Perspectives on Ancient Trade, *Current Anthropology* 15.3: 239-258.

Adams, Robert Mc C. (1981) Heartland of Cities. Chicago: University of Chicago Press.

Algaze, G., A. Mısır, and T. Wilkinson (1992) Şanlıurfa Museum/University of California Excavations and Surveys at Titris Höyük, 1991: a preliminary report Anatolica 18: 33-60.

Attinger, P. (1984) Remarques à propos de la "Malédiction d'Accad" *Revue d'Assyriologie* 78: 99-121.

Bjorkman, Judith (1973) Meteors and Meteorites in the Ancient Near East, *Meteoritics* 8: 91-132.

Bond, Gerard (1995) Millennial Pacing of Holocene Ice-Rafting Events Between Iceland and Greenland, *EOS* supplement November 7, 1995: 011B-1.

Bottema, S., G. Entjes-Nieborg, W. van Zeist (1990) Man's Role in the Shaping of the Eastern Mediterranean Landscape. Rotterdam: Brookfield.

Braemer, F. and J.-C. Echallier (1995) La marge désertique en Syrie du Sud du IIIe millànaire in S. van der Leeuw, ed., L'homme et la degradation de l'environment. Sophia Antipolis: Editions APDCA. pp. 345-356.

Butzer, Karl (1995) Environmental change, climatic history, and human modification in Jack M. Sasson et al., eds., Civilizations of the Ancient Near East. New York: Scribner's. pp. 123-151.

Colodny, R.G. (1977) Logic, Laws and Life. Pittsburgh: Univ. Pittsburgh Press.

Courty, M., -A. (1994) Le cadre peleogeographique des occupations humaines dans le bassin du Haut Khabur, Syrie du Nordest. Paleorient 20: 21-60.

Davis, Mike (1996) Cosmic Dancers on History's Stage? The Permanent Revolution in the Earth Sciences New Left Review 217: 48-84.

Dever, Wiiliam G. (1992) Pastoralism and the End of the Urban Early Bronze Age in Palestine in O. Bar-Yosef and A. Khazanov, eds., Pastoralism in the Levant. Maison: Prehistory Press. pp. 83-92.

Ehrich, Robert W., ed (1992) Chronologies in Old World Archaeology. 3rd ed. Chicago: University of Chicago Press.

Eidem, J. and D. Warburton in press In the land of Nagar: a survey around Tell Brak Iraq in press.

Esse, Douglas (1991) Subsistence, Trade and Social Change in Early Bronze Age Palestine. Chicago: Univeristy of Chicago Press.

Finkelstein, I. (1991) The Central Hill country in the Intermdeiate Bronze Age Israel *Exploration Journal* 41: 14-45.

Frangipane, M. (1993) Melid Reallexikon der Assyriologie 8: 42-52.

Fish, S. and P. Fish (1994) Prehistoric desert farmers of the *Southwest Annual Review of Anthropology* 23: 86-109.

Frumkin, A., M. Magaritz, I. Carmi, I. Zak (1991) The Holocene climatic record of the salt caves of Mount Sedom, Israel *The Holocene* 1.3: 191- 200.

Gasse, F. and E. Van Campo (1994) Abrupt post glacial climate events in West Asia and North Africa monsoon domains *Earth Planetary Science Letters* 126: 435-456.

Gérard, Fréd (1995) La secteur sud-ouest de Ilıpınar Orient-Express 1995.3: 72-75.

Halfman, J. D., D.F. Jacobson, C.M. Canella, K.A. Haberyan, B.P. Finney (1992) Fossil diatoms and the mid to late Holocene paleolimnology of Lake Turkana, Kenya: a reconnaissance study *J. of Paleolimnology* 7: 23-35.

Halfman, John D., Thomas C. Johnson and Bruce P. Finney (1994) New AMS dates, stratigraphic correlations and decadal climatic cycles for tyhe past 4 ka at Lake Turkana, Kenya *Palaeogeography, Palaeoclimatology, Palaeoecology* 11: 83-98.

Hoddell, David A., J.H. Curtis, M. Brenner (1995) Possible role of climate in the collapse of the Classic Maya civilization *Nature* 375: 391-394.

Hughes, Malcolm K. and H.F. Diaz, eds., (1994) The Medieval Warm Period, *Climate Change* 26.2-3: 109-342.

Issar, A.S., Y. Govrin, M.A. Geyh, E. Wakshal, M. Wolf (1992) Climate changes during the Upper Holocene in Israel Israel *Jour. Earth Sciences* 40: 219-223

Johnson, T.C. (1995) Abrupt Climate change in Holocene Records from African Lakes, *EOS* supplement November 7, 1995: 011B-8.

Kohl, Philip L. (1992) Central Asia (Western Turkestan): Neolithic to early Iron Age in R. Ehrich, ed. 1992: 179-195.

Kolata, Alan L. (1993) The Tiwanaku: portrait of an Andean civilization. Cambridge, Ma.: Blackwell.

Lemcke, G. (1994) Geochemistry of varved sediments of Van Gölü, Turkey. Abstract. NATO-ARW Third Millennium Abnrupt Climate Change and Old World Social Collapse.

Matthews, Roger (1994) Imperial Catastrophe or Local Incident? An Akkadian Hoard from Tell Brak, Syria, *Cambridge Archaeological Journal* 4.2: 290-316.

Matthews. R.J., W. Matthews, H. McDonald (1994) Excavations at Tell Brak, 1994 Iraq 55: 177-194.

Matthiae, Paolo (1995) Tell Mardikh-Ebla (Siria), campagna di scavi 1994 *Orient-Express* 1995.3: 86-88.

McGovern, Thomas H. (1990) The archaeology of the Norse Atlantic Annual Review of Anthropology 19: 331-351.

Michalowski, Piotr (1993) Memory and Deed: the historiography of the political expansion of the Akkad state in M. Liverani, ed., Akkad: the first empire. Padua: Sargon. pp. 69-90.

Mughal, Rafique (1990) The Decline of the Indus Civilization and the Late Harappan Period in the Indus Valley, *Lahore Museum Bulletin* 3.2: 1-17.

Negus, Anne Lynne (1986) The Fall of the Old Kingdom: a great African drought? Ph.D. diss. University of California at Los Angeles.

Ortloff, C.R. and A.L. Kolata (1993) Climate and Collapse: agro- ecological perspectives on the decline of the Tiwanaku state Journal of Archaeological *Science* 20: 195-221.

Palumbo, Gaetano (1990) The Early Bronze Age IV in the Southern Levant. Rome: CMAO 3.

Peltenburg (1995) Rescue Excavations at Jerablus Tahtani, Syria, 1995 *Orient-Express* 1995.3: 70-72.

P'yankova, L. (1994) Central Asia in the Bronze Age: sedentary and nomadic cultures *Antiquity* 68: 355-372.

Quinn, W.H. (1992) A study of Southern Oscillation Related Climatic Activity for AD 622 - 1990 in H.F. Diaz and V. Markgraf, eds., El- Niño: historical and paleoclimatic aspects of the Southern Oscillation. Cambridge: Cambridge Univ. Press. pp. 119-149.

Reade, Julian (1982) Tell Taya in J. Curtis, ed., Fifty Years of Mesopotamian discovery. London: British School of Archaeology in Iraq. pp. 72-78.

Shimada I., C.B. Schaaf, L.G. Thompson, E. Moseley-Thompson (1991) Cultural impacts of severe droughts in the prehistoric Andes World, *Archaeology* 22: 247-270.

Sirocko, Frank (1995) Abrupt change in monsoonal climate: evidence from the geochemical composition of Arabian Sea sediments. Habilitation thesis, University of Kiel.

Sirocko, F., M. Sarnthein, H. Erlenkeuser, H. Lange, M. Arnold, J.C. Duplessy (1993) Century scale vents in climate over the past 24,000 years, *Nature* 364: 322-324.

Sirocko, F., D. Garbe-Schànberg, A. McIntyre, B. Molfino (1996) Teleconnections between the subtropical monsoons and high-latitude climates during the last deglaciation, *Science* 272: 526-529.

Street-Perrot, F.A. (1994) Drowned trees record dry spells, *Nature* 369: 518.

Van Liere, W.J. and J. Lauffray (1954/55) Nouvelles prospection archéologique dans la Haute Jezireh syrienne, *Annales achéologiques de Syrie* 4: 129-148.

Van Zeist, W. and S. Bottema (1991) Late Quaternary Vegetation of the Near East. Wiesbaden: Reichert Verlag.

Weiss, Harvey (1992) Habur Triangles Third Millennium Urban Settlement in Subir NABU (Paris) 1992.4: 91-94

Weiss, Harvey (1996a) Desert Storm The Sciences May-June: 30-36.

Weiss, Harvey (1996b) Causality and Chance Late Third Millennium Collapse in Southwest Asia: in: O. Rouault, ed., 1996 La Djezireh et l'Euphrate Syriens. Paris: ERC.

Weiss, Harvey (1996c) Tell Leilan Region Settlement in the Third Millennium BC. International Symposium on the Syrian Djezireh. Deir ez-Zor, Syria. April 20, 1996.

Weiss, Harvey and M.-A. Courty (1993) The Genesis and Collapse of the Akkadian Empire: the accidental refraction of historical law in M. Liverani, ed., Akkad: the first world empire. Padua: Sargon. pp. 131- 156.

Weiss, H., M.-A. Courty, W. Wettersrom, F. Guichard, L. Senior, R. Meadow, A. Curnow (1993) Science 261: 995-1003.

Weiss, Harvey and M.-A. Courty (1994) Entre Droite àpigraphique et Gauche archéologique, y-a-t-il une place pour la science? Les nouvelles de l'archéologie 57: 33-41.

Wilkinson, T.J. (1994) The Structure and Dynamics of Dry-Farming States in Upper Mesopotamia Current Anthropology 35.5: 483-505.

Wilkinson, T.J. and D.J. Tucker (1995) Settlement Development in the North Jezira, Iraq. Warminster: British School of Archaeology in Iraq.

Yoffee, Norman (1995) Political economy in early Mesopotamian states Annual Review of Anthropology 24: 281-311.

Subject Index

F

G

H

I

J

K

L

Y

Z

The ASI Series Books Published as a Result of Activities of the Special Programme on Global Environmental Change

This book contains the proceedings of a NATO Advanced Research Workshop held within the activities of the NATO Special Programme on Global Environmental Change, which started in 1991 under the auspices of the NATO Science Committee.

The volumes published as a result of the activities of the Special Programme are:

Vol. 1: **Global Environmental Change.** Edited by R. W. Corell and P. A. Anderson. 1991.
Vol. 2: **The Last Deglaciation: Absolute and Radiocarbon Chronologies.** Edited by E. Bard and W. S. Broecker. 1992.
Vol. 3: **Start of a Glacial.** Edited by G. J. Kukla and E. Went. 1992.
Vol. 4: **Interactions of C, N, P and S Biogeochemical Cycles and Global Change.** Edited by R. Wollast, F. T. Mackenzie and L. Chou. 1993.
Vol. 5: **Energy and Water Cycles in the Climate System.** Edited by E. Raschke and D. Jacob. 1993.
Vol. 6: **Prediction of Interannual Climate Variations.** Edited by J. Shukla. 1993.
Vol. 7: **The Tropospheric Chemistry of Ozone in the Polar Regions.** Edited by H. Niki and K. H. Becker. 1993.
Vol. 8: **The Role of the Stratosphere in Global Change.** Edited by M.-L. Chanin. 1993.
Vol. 9: **High Spectral Resolution Infrared Remote Sensing for Earth's Weather and Climate Studies.** Edited by A. Chedin, M.T. Chahine and N.A. Scott. 1993.
Vol. 10: **Towards a Model of Ocean Biogeochemical Processes.** Edited by G. T. Evans and M.J.R. Fasham. 1993.
Vol. 11: **Modelling Oceanic Climate Interactions.** Edited by J. Willebrand and D.L.T. Anderson. 1993.
Vol. 12: **Ice in the Climate System.** Edited by W. Richard Peltier. 1993.
Vol. 13: **Atmospheric Methane: Sources, Sinks, and Role in Global Change.** Edited by M. A. K. Khalil. 1993.
Vol. 14: **The Role of Regional Organizations in the Context of Climate Change.** Edited by M. H. Glantz. 1993.
Vol. 15: **The Global Carbon Cycle.** Edited by M. Heimann. 1993.
Vol. 16: **Interacting Stresses on Plants in a Changing Climate.** Edited by M. B. Jackson and C. R. Black. 1993.
Vol. 17: **Carbon Cycling in the Glacial Ocean: Constraints on the Ocean's Role in Global Change.** Edited by R. Zahn, T. F. Pedersen, M. A. Kaminski and L. Labeyrie. 1994.
Vol. 18: **Stratospheric Ozone Depletion/UV-B Radiation in the Biosphere.** Edited by R. H. Biggs and M. E. B. Joyner. 1994.
Vol. 19: **Data Assimilation: Tools for Modelling the Ocean in a Global Change Perspective.** Edited by P. O. Brasseur and J. Nihoul. 1994.

Vol. 44: **Decadal Climate Variability: Dynamics and Predictability.** Edited by D. L.T. Anderson and J. Willebrand. 1996.

Vol. 45: **Radiation and Water in the Climate System: Remote Measurements.** Edited by E. Raschke. 1996.

Vol. 46: **Land Surface Processes in Hydrology: Trials and Tribulations of Modeling and Measuring.** Edited by S. Sorooshian, H. V. Gupta, and J. C. Rodda. 1997.

Vol. 47: **Past and Future Rapid Environmental Changes: The Spatial and Evolutionary Responses of Terrestrial Biota.** Edited by B. Huntley, W. Cramer, A. V. Morgan, H. C. Prentice, and J. R. M. Allen. 1997.

Vol. 48: **The Mathematics of Models for Climatology and Environment.** Edited by J. I. Díaz. 1997.

Vol. 49: **Third Millennium BC Climate Change and Old World Collapse.** Edited by H. N. Dalfes, G. Kukla, and R. Weiss. 1997.

Springer-Verlag and the Environment

We at Springer-Verlag firmly believe that an international science publisher has a special obligation to the environment, and our corporate policies consistently reflect this conviction.

We also expect our business partners – paper mills, printers, packaging manufacturers, etc. – to commit themselves to using environmentally friendly materials and production processes.

The paper in this book is made from low- or no-chlorine pulp and is acid free, in conformance with international standards for paper permanency.

MIX
Papier aus verantwortungsvollen Quellen
Paper from responsible sources
FSC® C105338

If you have any concerns about our products,
you can contact us on
ProductSafety@springernature.com

In case Publisher is established outside the EU,
the EU authorized representative is:
Springer Nature Customer Service Center GmbH
Europaplatz 3, 69115 Heidelberg, Germany

Printed by Libri Plureos GmbH
in Hamburg, Germany

[H.A.S.C. No. 108–10]

DESTRUCTION OF THE U.S. CHEMICAL WEAPONS STOCKPILE—PROGRAM STATUS AND ISSUES

HEARING

BEFORE THE

TERRORISM, UNCONVENTIONAL THREATS AND CAPABILITIES SUBCOMMITTEE

OF THE

COMMITTEE ON ARMED SERVICES
HOUSE OF REPRESENTATIVES

ONE HUNDRED EIGHTH CONGRESS

FIRST SESSION

HEARINGS HELD
OCTOBER 30, 2003

U.S. GOVERNMENT PRINTING OFFICE

98–849 WASHINGTON : 2005

For sale by the Superintendent of Documents, U.S. Government Printing Office
Internet: bookstore.gpo.gov Phone: toll free (866) 512–1800; DC area (202) 512–1800
Fax: (202) 512–2250 Mail: Stop SSOP, Washington, DC 20402–0001

TERRORISM, UNCONVENTIONAL THREATS AND CAPABILITIES SUBCOMMITTEE

JIM SAXTON, New Jersey, *Chairman*

JOE WILSON, South Carolina
FRANK A. LoBIONDO, New Jersey
JOHN KLINE, Minnesota
JEFF MILLER, Florida
ROSCOE G. BARTLETT, Maryland
MAC THORNBERRY, Texas
JIM GIBBONS, Nevada
ROBIN HAYES, North Carolina
JO ANN DAVIS, Virginia
W. TODD AKIN, Missouri
JOEL HEFLEY, Colorado

MARTY MEEHAN, Massachusetts
JIM TURNER, Texas
ADAM SMITH, Washington
MIKE McINTYRE, North Carolina
CIRO D. RODRIGUEZ, Texas
BARON P. HILL, Indiana
SUSAN A. DAVIS, California
JAMES R. LANGEVIN, Rhode Island
RICK LARSEN, Washington
JIM COOPER, Tennessee

THOMAS HAWLEY, *Professional Staff Member*
JEAN REED, *Professional Staff Member*
UYEN DINH, *Professional Staff Member*
WILLIAM NATTER, *Professional Staff Member*
CURTIS FLOOD, *Staff Assistant*